P9-CEP-563

COMMON GRAPHS

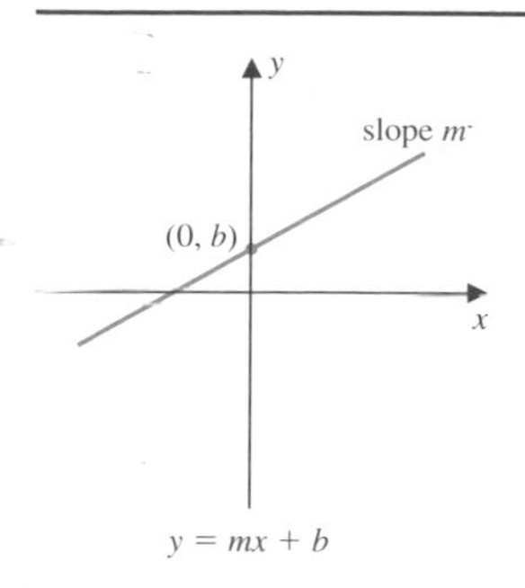

$y = mx + b$

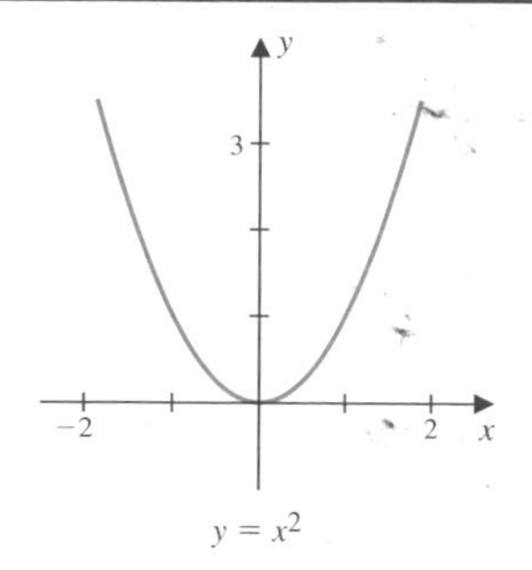
$y = x^2$

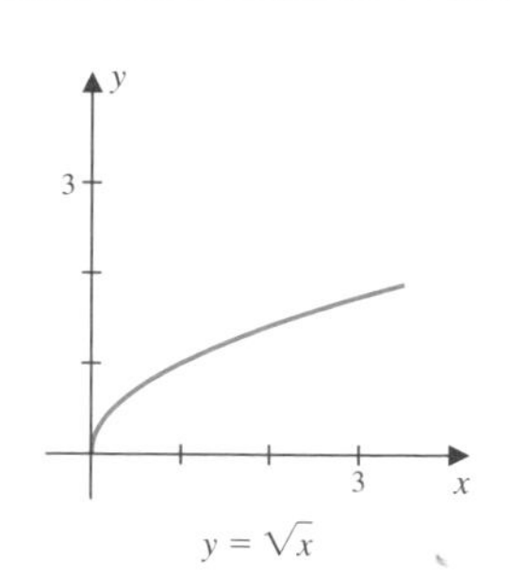
$y = \sqrt{x}$

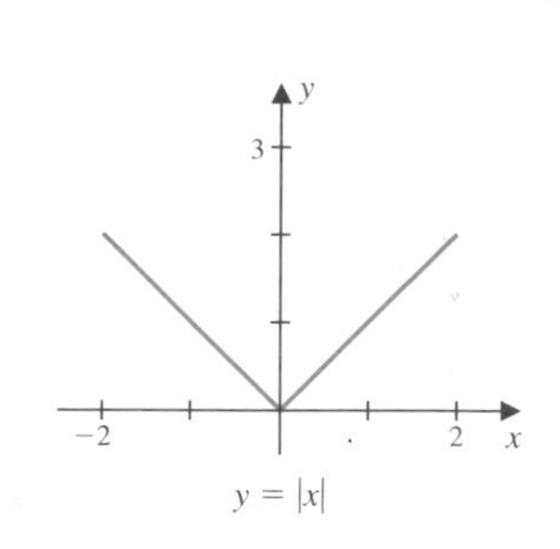
$y = |x|$

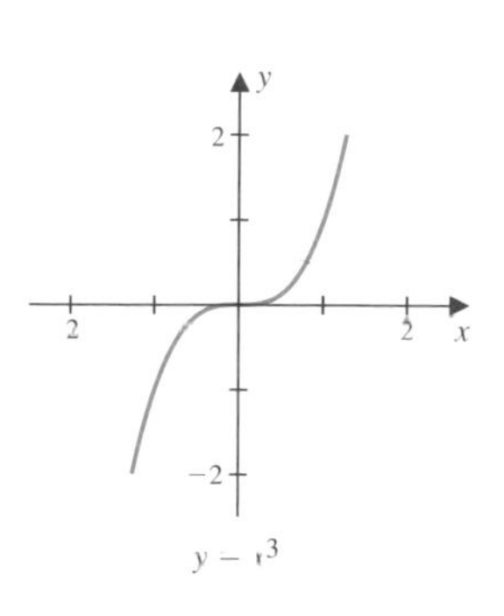
$y = x^3$

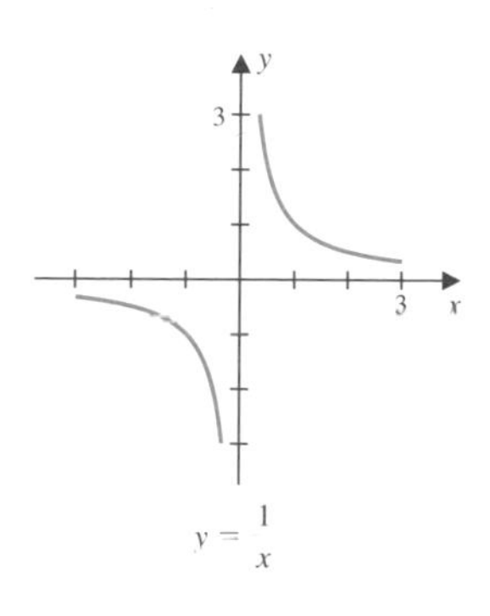
$y = \dfrac{1}{x}$

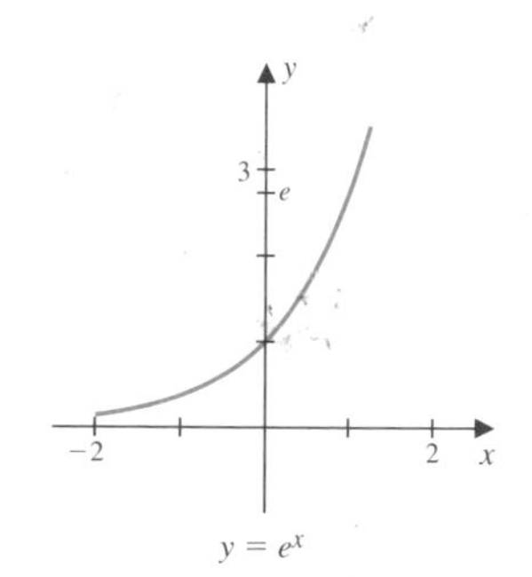
$y = e^x$

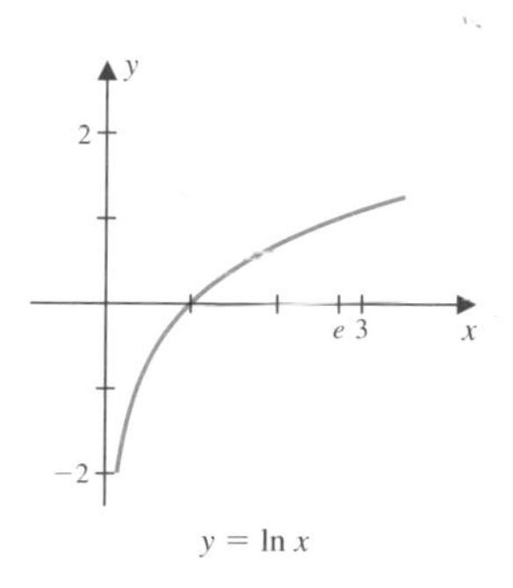
$y = \ln x$

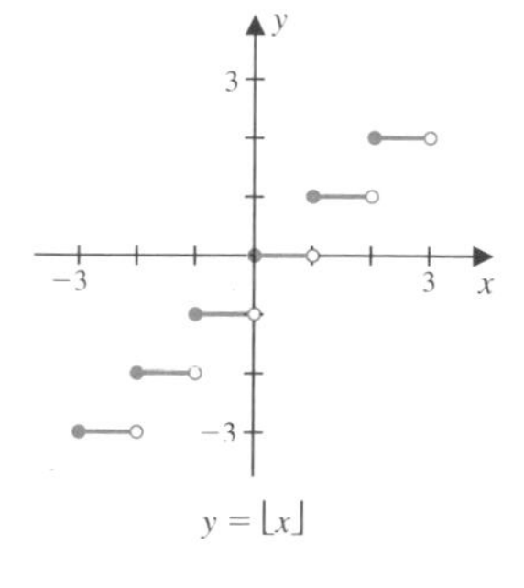
$y = \lfloor x \rfloor$

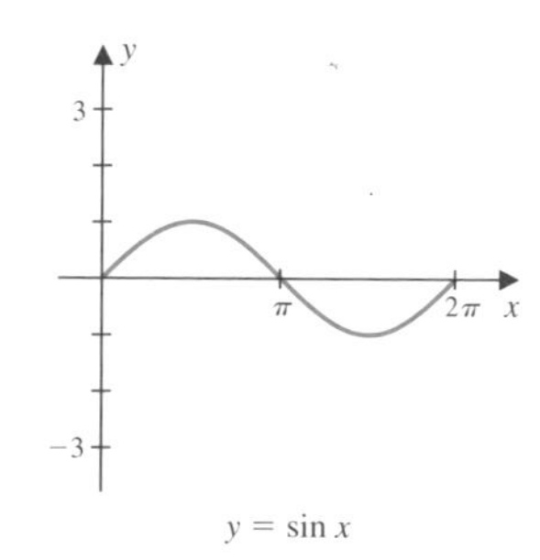
$y = \sin x$

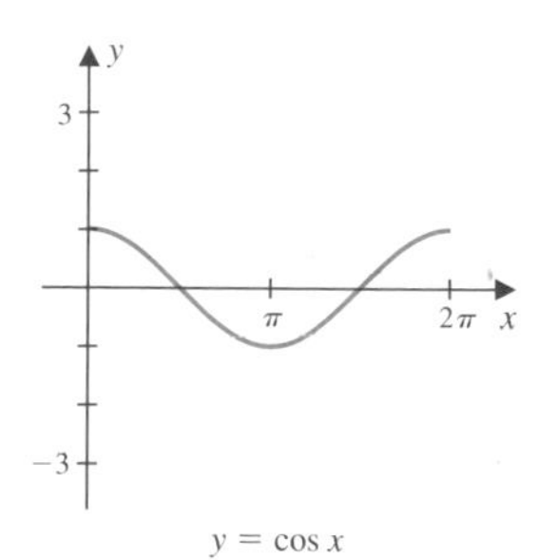
$y = \cos x$

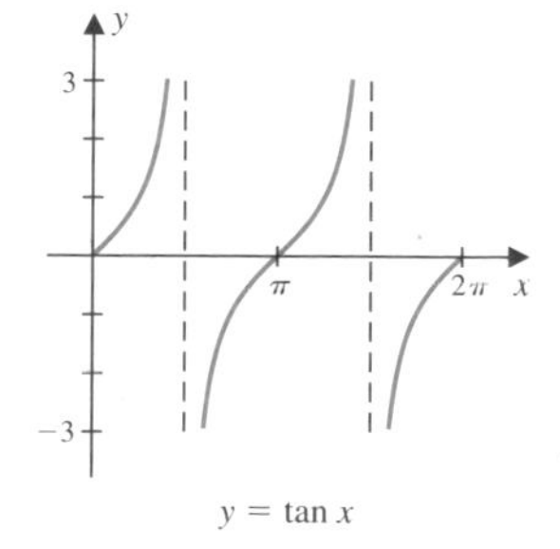
$y = \tan x$

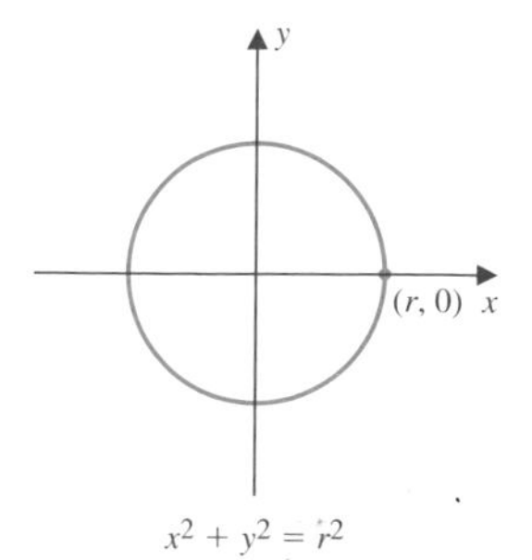

$x^2 + y^2 = r^2$

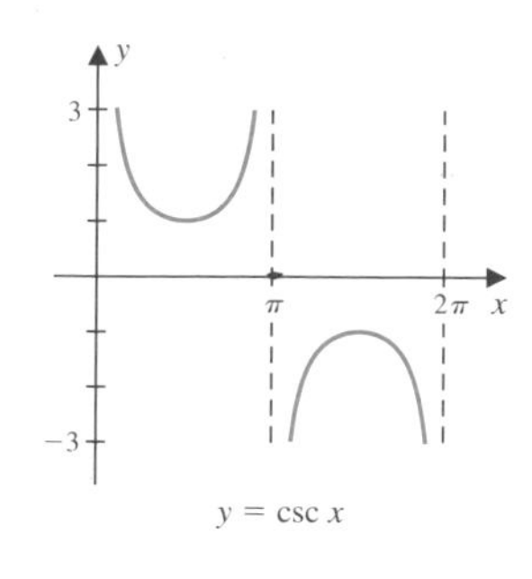
$y = \csc x$

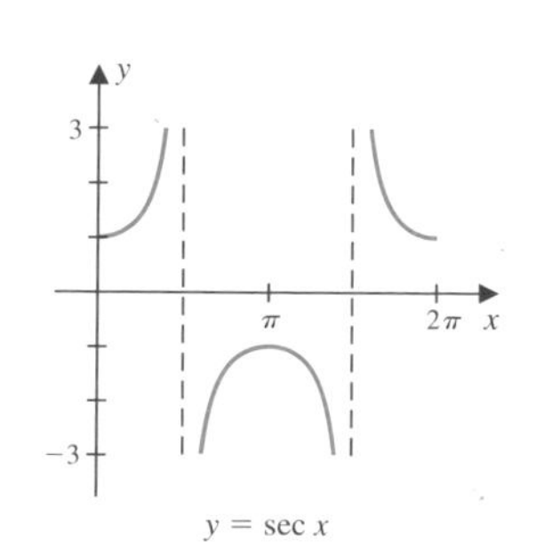
$y = \sec x$

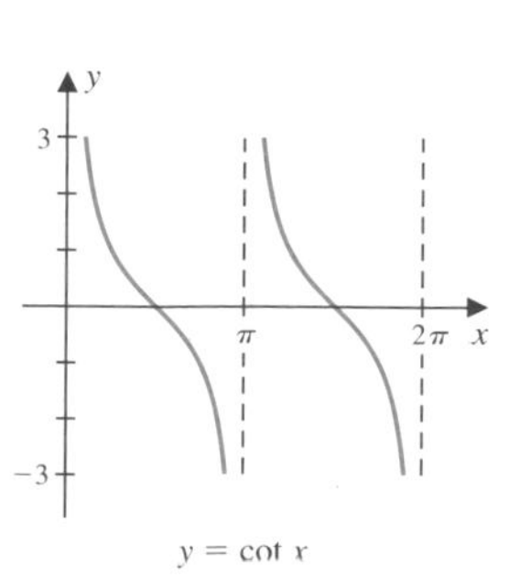
$y = \cot x$

PreCalculus

PreCalculus

Fifth Edition

J. Douglas Faires
Youngstown State University

James DeFranza
St. Lawrence University

BROOKS/COLE
CENGAGE Learning™

Australia • Brazil • Japan • Korea • Mexico • Singapore • Spain • United Kingdom • United States

PreCalculus, **Fifth Edition**

J. Douglas Faires, James DeFranza

Acquisitions Editor: Gary Whalen

Developmental Editor: Stacy Green

Assistant Editor: Cynthia Ashton

Editorial Assistant: Naomi Dreyer

Media Editor: Lynh Pham

Marketing Manager: Myriah FitzGibbon

Marketing Assistant: Angela Kim

Marketing Communications Manager: Darlene Macanan

Content Project Manager: Jennifer Risden

Design Director: Rob Hugel

Art Director: Vernon Boes

Print Buyer: Linda Hsu

Rights Acquisitions Specialist: Roberta Broyer

Production Service: MPS Limited, a Macmillan Company

Text Designer: Kim Rokusek

Photo Researcher: Bill Smith Group

Copy Editor: Laurene Sorenson

Illustrator: Scientific Illustrators

Cover Designer: Denise Davidson

Cover Image: Martin Child/ Robert Harding World Imagery/ Getty Images

Compositor: MPS Limited, a Macmillan Company

© 2012, 2007 Brooks/Cole, Cengage Learning

ALL RIGHTS RESERVED. No part of this work covered by the copyright herein may be reproduced, transmitted, stored, or used in any form or by any means graphic, electronic, or mechanical, including but not limited to photocopying, recording, scanning, digitizing, taping, Web distribution, information networks, or information storage and retrieval systems, except as permitted under Section 107 or 108 of the 1976 United States Copyright Act, without the prior written permission of the publisher.

For product information and technology assistance, contact us at **Cengage Learning Customer & Sales Support, 1-800-354-9706.**

For permission to use material from this text or product, submit all requests online at **www.cengage.com/permissions.**

Further permissions questions can be e-mailed to **permissionrequest@cengage.com.**

Library of Congress Control Number: 2010929056

ISBN-13: 978-0-8400-6862-0

ISBN-10: 0-8400-6862-X

Brooks/Cole
20 Davis Drive
Belmont, CA 94002-3098
USA

Cengage Learning is a leading provider of customized learning solutions with office locations around the globe, including Singapore, the United Kingdom, Australia, Mexico, Brazil, and Japan. Locate your local office at: **www.cengage.com/global.**

Cengage Learning products are represented in Canada by Nelson Education, Ltd.

To learn more about Brooks/Cole, visit **www.cengage.com/brookscole**

Purchase any of our products at your local college store or at our preferred online store **www.cengagebrain.com**

Printed in China by China Translation & Printing Services Limited.
2 3 4 5 14 13 12 11

Contents

Preface

PreCalculus, Fifth Edition, is designed for a one-term course to prepare students for the standard university calculus sequence. This may be a strange opening sentence to include in this Preface, because every precalculus book should have this as its goal. However, a quick review of the university precalculus books on the market reveals that nearly all are over 800 pages long, which is far more material than can reasonably be covered in one term. There is so much unnecessary algebra and trigonometry review material in most precalculus books that the precalculus course is often dominated by these topics at the expense of the graphing and function analysis that is needed for calculus.

In *PreCalculus,* Fifth Edition, we have concentrated on the concepts that will be specifically needed in calculus. Algebra and trigonometry review topics are covered in the context in which they are seen in calculus. Instructors can adjust the extent of this review material based on the level of their class, and, when appropriate, can refer to the more extensive review material presented in the Student Study Guide and on the book's website,

http://www.math.ysu.edu/~faires/PreCalculus/

The mathematical preparation of the students entering colleges and universities is much broader than in the past. More students are arriving with a university-level calculus background. At the other end of the spectrum, there is an increasing number of students who, although they might have taken college preparatory courses in high school, are not prepared to do the type of analysis that is required to successfully complete a university calculus sequence. These students often have some knowledge of the elementary computational techniques of calculus, and many are quite proficient in the use of graphing calculators, but they have not had a comprehensive analysis and elementary functions course while in high school, and they are often weak in graphing techniques. They need the calculus-context review that a good precalculus course provides.

Universities, particularly the predominantly commuter campuses, have seen a substantial increase in serious students with a nontraditional background—students who have been away from mathematics for a while but are seriously interested in science and engineering. Many of these students had not considered science-oriented careers when in high school and need to take preparatory mathematics courses before they can enter their intended major subject. Most universities now offer a wide range of remedial courses to serve these students' needs, but precalculus is not a remedial course. Rather, precalculus presents this review material from a perspective that will enable students to succeed in calculus.

TRANSITION TO CALCULUS

To make the transition from precalculus to calculus as smooth as possible, the terminology presented and level of exposition should closely match that commonly used in calculus. There are a sufficient number of new and important terms, concepts, and applications for the student to master in calculus without the added difficulty of learning new ways to express old ideas. You will find that this precalculus book leaves no gap between the end of precalculus and the beginning of calculus, and is of a length that will reasonably permit it to be covered in one term.

This book includes material from algebra, geometry, and trigonometry that is sufficient to fill holes in a student's background, but this review material is interwoven into the book rather than presented as a remedial block. This permits the student to review these topics in the manner they will be used in calculus. In addition, this approach does not mislead those who need more review than a true precalculus course can provide.

The terminology used in the book parallels that used in calculus books. Both authors have experience in this area and have spent many years working with students who need the subject material presented in this book. The examples and exercises are presented in the way the student will see them in calculus, and the concepts are presented in a student-friendly manner using an intuitive approach. Examples are used to reinforce the concepts, which are further reinforced by the exercise sets. In addition, there are exercise sets at the end of each chapter that preview calculus, so that students can see the type of material they will encounter in their next mathematics course.

TECHNOLOGY IN PRECALCULUS AND CALCULUS

The book includes a significant amount of material that is appropriate for use with graphing calculators and computer algebra systems. Although these devices are helpful for developing mathematical intuition and visualizing important concepts, we have not assumed that they are essential for understanding the basic concepts of precalculus. We treat graphing devices as an important tool, not as a substitute for the analysis that is necessary for a complete understanding of calculus and its applications. To successfully complete the calculus sequence, students need to feel comfortable with the graphs and behavior of a wide variety of basic functions and equations. Our primary use of graphing devices is to take a problem farther than would be reasonable without this tool.

Exercises that require technology are placed near the end of exercise sets, where they will not disrupt the sequencing of exercises that do not use technology. Section 1.5, entitled Using Technology to Graph Equations, is the only section in the book that specifically deals with technology. This is an optional section that is not referenced in the remainder of the text.

EXERCISE SETS

Before starting to write this edition, we once again went through every exercise set to be sure that the techniques covered in the examples are sufficient for students to work all the routine exercises (generally the first 70%). Where necessary, we then modified the exercise sets, rewrote text examples, and added text examples. We also had all the exercise sets reviewed by undergraduate student tutors in precalculus, to be certain that the topics that cause difficulty for students are thoroughly reviewed.

Our general philosophy with regard to the exercise sets is that

- An average student should be able to work the first 20–25% of the exercises without referring to the text material so that confidence in knowledge is gained.
- The next 25–30% of the exercises in a section will likely require the student to review the examples in the section to pick up any subtleties of the material. It is expected that some of these problems will be discussed in a subsequent class or recitation session.
- The next 25% of the exercises will require most students to look at the text material to solve the problems.
- The final exercises extend the theory in the section and give applications. None of these will be required for an understanding of subsequent material in the book.

The Review Exercises are designed to ensure that students have a firm grasp of the material presented in the chapter. For example, exercises that extend the theory and applications of the material discussed in a chapter are always placed in the exercises at the ends of the sections, not in the place where the student should look when doing a chapter review.

At the end of each chapter there are two additional exercise sets. The first of these, called Exercises for Calculus, provides students with a preview of the types of problems they will see in a calculus course. These are used to emphasize that the setup of many calculus problems involves concepts from precalculus. The second exercise set is a true/false Chapter Test. The questions in these sets are designed to help students connect concepts and reinforce the facts presented in the chapter.

WEB MATERIALS

We have made a concerted effort to keep the review material in the book to a minimum to ensure that the topics that are essential in calculus can be covered thoroughly. We also have resisted the temptation to include somewhat extraneous topics that we particularly like but that do not quite fit into the philosophy of the book. However, we recognize that not everyone will agree on those topics that are needed in precalculus, just as we don't always agree ourselves.

To meet the needs of as many instructors and students as possible, we have added PDF files on our website that contain additional material written in the same format as the text. Currently the website features the following topics:

- Descartes' Rule of Signs
- Systems of Equations
- Rotations of Axes
- Sequences and Geometric Series
- Vectors

We will continue to add material to this website when we feel it might be useful, even if it is not covered in a majority of precalculus courses. If a school is particularly interested in including one of these chapters in the text, this can be arranged through the publisher.

MODIFICATIONS FOR THE FIFTH EDITION

Trigonometry

Calculus and other scientific applications require trigonometry to be viewed in terms of functions of real numbers. However, many high school graduates have studied trigonometry primarily in terms of right triangles. For this edition we decided to review right-angle trigonometry earlier in Chapter 4 so that students will feel more comfortable with the subject before they encounter the unit circle definition that is needed for calculus. This required a number of changes from previous editions, but we are pleased with the result. We hope and expect that this approach will make for an easier transition for students but will still emphasize a functional approach to trigonometry that is consistent with the other topics in the text.

Additional Examples and Applications

In *PreCalculus,* Fifth Edition we have added, both to the examples and exercise sets, many applications that emphasize the importance and relevance of precalculus and calculus. These applications have been highlighted and placed at the end of each section and can be included or not at the discretion of the instructor and/or depending on the interest of the student. The applications cover a wide range of areas, including biology, ecology, medicine, physics, economics, geometry, engineering, archeology, optimization, social science, finance, space science, and mathematics. Applications at early stages of a student's mathematical training need to be designed carefully to

- Illustrate the utility of the mathematical material.
- Demonstrate the connections of mathematics with other areas of study.
- Be sufficiently clear and uncomplicated that the student does not become confused or discouraged.

This is a difficult set of criteria to satisfy, but we are confident that the time and effort we have spent preparing these new applications meets our goals.

Exercise Set Changes

All of the exercise sets have been carefully reviewed, as in every previous revision, to ensure that the material covered in the section and the techniques presented in the examples are sufficient to permit students to work the exercises. In addition, we have changed approximately 30% of the routine exercises, those in the first half of each exercise set. Although the problems are similar to those in previous editions, this change gives an instructor who has used previous editions a new source of exercises to assign. Exercises from previous editions can be used for quizzes or additional assignments.

Principles of Problem Solving

Teaching problem-solving skills, in the context of precalculus, has been a priority in every edition of this text. To better highlight some of the many important techniques of problem solving, we have added in the margin of the text a feature called Principles of Problem Solving. These principles are explicitly used when needed later in the text. Our hope is that making these principles explicit will help students develop systematic procedures for setting up and solving routine as well as application problems.

ACCURACY

Every problem and example in the book has been checked by the authors as well as two independent accuracy checkers. One of the accuracy checkers was a student at Youngstown State University whom we asked to be particularly aware of situations where statements, even though correct, might be misinterpreted by students. Faculty often use a higher degree of formal rigor and logic than do students, and the mathematical vocabulary that faculty use is not always familiar to students. We wanted to be sure that this book is truly clear to the intended audience but uses the mathematically precise statements that students will see in calculus.

FOUR-COLOR FORMAT

In the third edition we introduced a modest four-color format. This worked well in the third edition and has been enhanced in this edition. There is an important pedagogical reason for adopting the four-color format used in this book. Our major theme is to emphasize the problem-solving strategy of breaking new and difficult problems into a series of smaller, less difficult, and more familiar problems. This is particularly the case in our approach to graphing. We consistently construct a small collection of functions whose graphs it is essential to know. Then we illustrate how techniques such as scaling, translation, and reflection can be used to construct graphs of many other functions. Having a variety of colors permits us to make the individual steps much clearer. We have consistently shown the base graphs in cyan, the first translation in red, and the next in yellow. Our students have found this to be a natural color transition. It permits them to better see precisely how the final graph is constructed and shows that each individual step is not complex. The artist and the production editor have spent many hours creating what they feel is the best choice of colors to distinguish the various graphs, taking into consideration even seemingly minor details that we, as authors, would not have considered.

Aside from the graphs, you will see very little other use of color in the book. It is used only where it adds to the understanding of the material or helps in the highlighting of important topics.

RELEVANCE

Beginning with the first edition of *PreCalculus,* our primary goal was to present the essential concepts and techniques needed for a head start in calculus in a package that permitted instructors and students to cover the essential material in one term. The material had to be presented in a manner that was user friendly and easily read by students who did not need a comprehensive course in algebra and trigonometry. We have remained dedicated to and consistent with these goals. Although we have added new applications and exercises where we and our advisors felt they were needed, we have not substantially changed the length of the text. The essential concepts and techniques have remained basically unchanged. The new material has been added to enhance and often clarify this material and to further emphasize the relevance of mathematics.

STUDENT STUDY GUIDE

The Student Study Guide provides more detail on algebra, trigonometry, and other background material, as well as numerous supplemental examples and worked-out exercises. We have found that this material is particularly useful for students with a nontraditional background and those who have been away from mathematics for a while. It is also helpful for students who don't have ready access to an instructor, such as those taking the course in a distance-learning environment.

The format of the material in the Guide is similar to that of the book. We have written all of this material ourselves so there will be no transition problems from the text to the Guide, which can be the case when the supplemental material is not written by the authors of the textbook.

We realize that the cost for text material can be a hardship for students, so we have included a PDF file of the first chapter of the Student Study Guide on our website. This permits students to determine if the Guide is sufficiently useful for them. If students are interested, they can obtain the entire Guide from the publisher's website.

Also included in the Student Study Guide are two copies of an examination that students can take to test their readiness for precalculus and for calculus. We expect that students will be successful in a precalculus course based on this book if they score 16 or higher on the 40-question test before taking the precalculus course. Students will be well prepared for a university calculus sequence if they score 30 or higher after they take the precalculus course.

We suggest that students try one of the examinations before taking their precalculus course and the other after completing it, and we predict that they will see a significant improvement.

For those who do not have access to the Student Study Guide, the examinations are also available on the book's website.

SUMMARY

In *PreCalculus,* Fifth Edition:

- The focus is on the essential prerequisites needed for calculus, with the assumption that students taking the course will take a calculus course in their next term.
- Graphing and the analysis of functions are used to smooth the transition to calculus.
- Algebra and trigonometry review is woven through the text to help students see where the gaps in their background must be filled before they take calculus.
- Explanations in the margin alert students to material that applies directly to calculus.
- New concepts are presented using relevant applications, reinforced first with examples and then with exercises.

- Technology is integrated throughout as a tool to extend knowledge, never as a substitute for the mastery of basic concepts that will be needed in calculus.
- Important applications of technology are highlighted throughout the text, even though the use of technological devices is not required.
- The amount of essential material has been kept to a length that can reasonably be covered in a single term.
- Numerous applications have been added to reinforce the relevance of the mathematics.
- Principles of Problem Solving comments have been added to help students approach this subject with more confidence.
- A modest four-color format is used to improve explanations of complicated subjects.
- Additional course material will be added to the book's website when it is helpful to instructors and students. This allows flexibility while keeping the core material at a reasonable length.

We hope you have a productive and pleasant experience with *PreCalculus,* Fifth Edition. If you have any suggestions that will help us improve your experience, please do not hesitate to contact us at the e-mail addresses listed at the end of this Preface.

ACKNOWLEDGMENTS

We have been fortunate to have received many constructive comments regarding the previous editions of this book, and we have tried to incorporate all suggestions that we felt would enhance the learning experience for students in precalculus. We would particularly like to thank the following individuals for their helpful advice on this and past editions. Those reviewing the book for this edition are distinguished by an asterisk symbol *.

Sofia Agrest, College of Charleston
Mark Ashbrook, Arizona State University
June Bjercke, San Jacinto College
David Collingwood, University of Washington
Jeff Dintz, University of Vermont
Gerri M. Dunnigan, University of North Dakota
Thomas Eynden, Thomas More College
David Ferrone, University of Connecticut*
John Gosselin, University of Georgia
Michael B. Gregory, University of North Dakota
Glenda Haynie, North Carolina State University
John Herron, University of Montevallo
Matthew Isom, Arizona State University
Andrew D. Jones, Florida A&M University
Karla Karstens, University of Vermont
Annela Kelly, University of Louisiana at Monroe
Harvey Keynes, University of Minnesota
David W. Kinsey, University of Southern Indiana
William A. Kirk, University of Iowa

Elena Kravchuk, The University of Alabama at Birmingham
Estela S. Llinas, University of Pittsburgh at Greensburg
Nancy Matthews, University of Oklahoma
Lee McCormick, Pasadena City College
Sanford Miller, State University of New York at Brockport
Doris J. Mohr, University of Southern Indiana
Carl Mueller, Georgia Southwestern State University
Dongwen Qi, Georgia Southwestern State University*
Nancy Ressler, Oakton Community College
Phillip Schmidt, Northern Kentucky University
Charlotte Skinner, University of Cincinnati—Raymond Walters College*
Yorck Sommerhäuser, University of South Alabama*
Louis Talman, Metropolitan State College of Denver*
Douglas Thomasey, Lynchburg College*
Michael P. Trapuzzano, Arizona State University
Benton Tyler, University of Montevallo
Dennis Zill, Loyola Marymount University

We would also like express our appreciation to Jena Baun, a student at Youngstown State University, who did much of the editing of the Instructor's Manual, Student Study Guide, and Test Bank, as well as to Krista Foster who checked the accuracy of the answers at the final stage. We are grateful for the time and effort that they took to ensure the accuracy of this material.

Doug Faires
(faires@math.ysu.edu)

Jim DeFranza
(jdefranza@stlawu.edu)

To the Student

In a short time you will be taking a course in calculus, an exciting subject that has applications in every quantitative discipline. Calculus provides a systematic way to describe how a change in one quantity affects another.

Engineers use calculus to determine how the forces of wind and water affect the stability of a structure, how heat and fluids flow, how current affects various electronic devices, and so many other applications that they are too numerous to list. Scientists use calculus to study issues such as population problems, the effect of toxic substances on the environment, and the rates of chemical reactions.

In business, the growth of the economy is determined with methods based on calculus, and future trends are predicted. In fact, most statistics that are a common part of our everyday life have their basis in calculus techniques. Calculus is truly the keystone of quantitative study. Calculus is not difficult to master, but it requires a solid background in, and conceptual feel for, the precalculus topics of algebra, geometry, trigonometry, and, most important, graphing. Many students studying calculus for the first time at the university level have difficulty because they are not prepared for the way precalculus topics are used in calculus.

The material in this book will give you a solid foundation for the study of calculus. You will see many topics that are familiar but that use new notation and a different perspective than you have seen in the past. The topics presented in this book are directly related to the study of calculus, and the perspective we give is the same as the one that will be used in your calculus courses. Read the material carefully and work the exercises. By doing so, you will master these concepts and gain a great head start in your study of calculus.

Functions

1

Calculus Connections

Contrary to some rumors you may have heard, calculus is not a hard subject. Most of the difficulties that students have in calculus can be traced directly to new applications of topics from precalculus. To illustrate, let us consider a typical applied problem from calculus.

Suppose that you are working for a company that makes containers, and are asked to design an open plastic box with a square base that holds a specified volume. Let's see how we can approach the solution to this problem.

The figure shows some possible boxes that might meet the requirements. Should we choose one of these, and if so, which one? Although it was not stated explicitly that the box chosen should use the minimum amount of plastic, this is likely what your boss had in mind. We could produce a few boxes and choose the one using the least amount of material, but this would be time-consuming and a waste of material. Also, it might be difficult to explain why the box you chose is the best possible one for the job.

© Image copyright Rafael Ramirez Lee, 2009. Used under license from Shutterstock.com

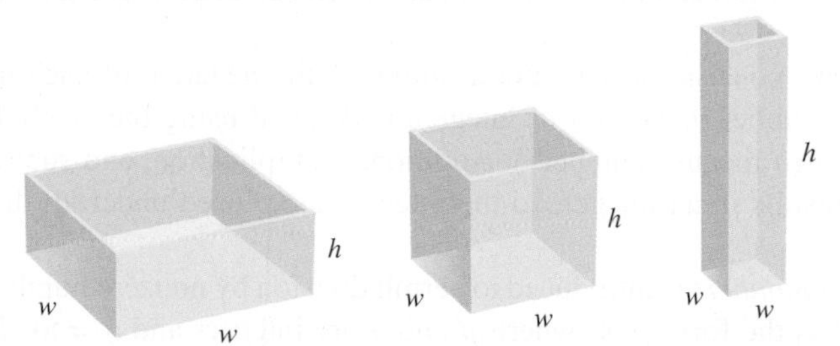

The calculus solution to this problem requires that we first determine equations for the volume, $V = w^2h$, and for the surface area, $S = w^2 + 4wh$. Solving the volume equation for h gives

$$h = \frac{V}{w^2}.$$

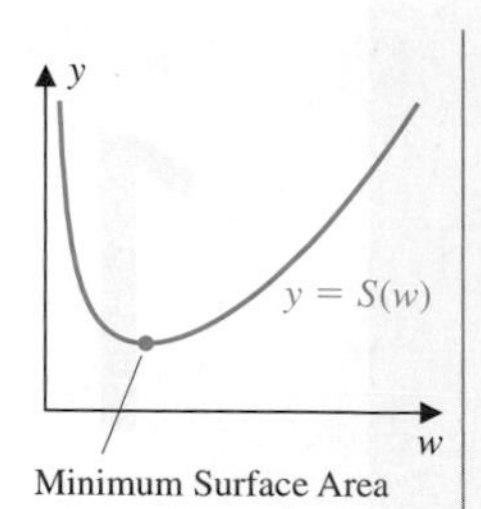

Minimum Surface Area

This can be substituted into the equation for surface area to produce

$$S(w) = w^2 + 4w\left(\frac{V}{w^2}\right) = w^2 + \frac{4V}{w}.$$

So the surface area S of the box is a *function* of the width w; each positive value of w gives a specific amount of required material $S(w)$. We need to choose w so that $S(w)$ is a minimum, which is easy to do using calculus. When we have this value of w, the equation $V = w^2h$ gives us the appropriate value of h. The figure shows a typical graph of $y = S(w)$, which illustrates that a number w does exist with $S(w)$ a minimum.

We can approximate the solution using the graph, and the exact solution can easily be found using calculus. The difficult part of the problem is not the calculus solution, it is the construction of the function $S(w)$. This is a precalculus problem of the type we will consider throughout the book. In this chapter we see what functions are and what they look like, and begin to examine in detail the functions that are commonly used in calculus and beyond.

1.1 INTRODUCTION

The functions we will study in precalculus—because they are needed in calculus—are transformations of the set of real numbers. This section contains a short development of the *real numbers*. We begin with the most basic set of numbers, the *natural numbers*.

The set of **natural numbers** consists of the counting numbers, $1, 2, 3, \ldots$, and is denoted by the symbol $\mathbb{N}$. Any pair of natural numbers can be added or multiplied and the result is another natural number. This is sometimes expressed by saying that set $\mathbb{N}$ is *closed* under the operations of addition and multiplication.

The operation of subtraction is needed for many purposes, but the set of natural numbers is *not closed* under subtraction. For example, if we subtract the natural number 3 from the natural number 7, we get the natural number 4, but subtracting 7 from 3 is impossible if we must stay within the set of natural numbers. To permit subtraction, we expand the set of natural numbers to the set of *integers*, a set that remains closed under addition and multiplication, and is closed under subtraction as well.

The set of **integers** consists of the natural numbers, the negative of each natural number, and the number 0. The set of integers is denoted using the symbol $\mathbb{Z}$ (the German word *zählen* means "number"). Addition, multiplication, and subtraction of two integers results in an integer, so the integers are closed under all these operations.

The set of *rational numbers* is introduced to permit division by nonzero numbers. A **rational number** has the form p/q, where p and q are integers and $q \neq 0$. The set of rational numbers is denoted $\mathbb{Q}$ (for quotient) and is closed under addition, multiplication, subtraction, and, except when the denominator is 0, under division.

The rational numbers satisfy all the common arithmetic properties but fail to have a property called *completeness*; that is, there are some essential numbers missing from the set. At least as early as 400 B.C.E. Greek mathematicians of the Pythagorean school recognized that $\sqrt{2}$, the length of a diagonal of a square with sides of length 1, is not a rational number. (See Figure 1.)

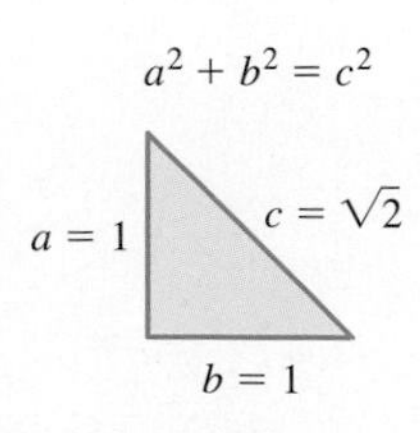

FIGURE 1

The nonrational numbers, or **irrational numbers**, were originally said to be *incommensurable* because they could not be directly compared to the familiar rational numbers. There are many irrational numbers, including $\sqrt{5}$, π, and $-\sqrt[3]{3}+1$. The discovery of irrational numbers resulted in a profound change in ancient mathematical thinking. Before that time it was assumed that all quantities could be expressed as proportions of integers, and integers in some cultures had a mystical property.

A precise definition of completeness needs the concept of limits, an important topic in calculus.

The set $\mathbb{R}$ of **real numbers** consists of the rational numbers together with the irrational numbers. This set is described most easily by considering the set of all numbers that are expressed as infinite decimals. The rational numbers are those with expansions that terminate or that eventually repeat in sequence, such as

$$\tfrac{1}{2} = 0.5, \qquad \tfrac{1}{3} = 0.\bar{3}, \qquad \tfrac{16}{11} = 1.\overline{45}, \quad \text{or} \quad -\tfrac{123}{130} = -0.9\overline{461538},$$

where the bar indicates that the digit or block of digits is repeated indefinitely. Numbers with decimal expansions that have no repeating blocks are irrational.

1.2 THE REAL LINE

The material in the next few sections will likely be familiar from your previous mathematics courses. But in these sections we establish notation and a number of definitions that will be used throughout the text.

The fact that each real number can be written uniquely as a decimal gives us a way to associate each real number with a distinct point on a **coordinate line**. We first choose a point on a horizontal line as the origin and associate the real number 0 with the origin. We designate some point to the right of 0 as the real number 1. The positive integers are then marked with equal spacing consecutively to the right of 0. The negative integers are marked, with this same spacing, to the left of 0. Nonintegral real numbers are placed on the line according to their decimal expansions.

Figure 1 shows an x-coordinate line and the points associated with a few of the real numbers.

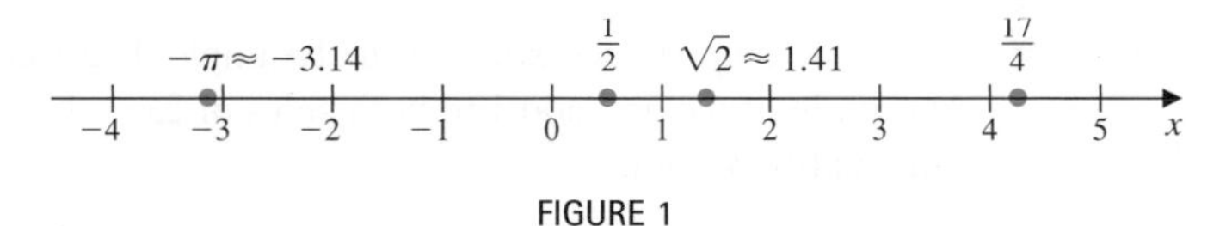

FIGURE 1

The coordinate-line representation of the real numbers is so convenient that we frequently do not explicitly distinguish between the points on the line and the real numbers that these points represent. Both are called the set of real numbers and are denoted $\mathbb{R}$.

Inequalities

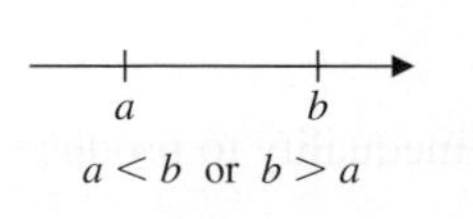

FIGURE 2

The relative position of two points on a coordinate line is used to define an inequality relationship on the set of real numbers. We say that a is less than b, written $a < b$, when the real number a lies to the left of the real number b on the coordinate line. This is also expressed by stating that b is greater than a, written $b > a$, as shown in Figure 2.

The notation $a \leq b$, or $b \geq a$, is used to express that a is either less than or equal to b. The following properties of inequalities can be verified by referring to the coordinate-line representation of the real numbers a, b, and c.

Inequality Properties

- Precisely one of $a < b$, $b < a$, or $a = b$ holds.
- If $a > b$, then $a + c > b + c$.
- If $a > b$ and $c > 0$, then $ac > bc$.
- If $a > b$ and $c < 0$, then $ac < bc$.

Note that the fourth Inequality Property states that the inequality sign must be reversed when both sides of the inequality are multiplied by a negative number. For example,

$$4 > 3, \quad \text{but} \quad -8 = (-2)4 < (-2)3 = -6.$$

These rules are used to solve problems involving inequality relations.

EXAMPLE 1 Find all real numbers satisfying $2x - 1 < 4x + 3$.

Solution First add -3 to both sides of

$$2x - 1 < 4x + 3 \quad \text{to produce} \quad 2x - 4 < 4x.$$

Now add $-2x$ to each side to isolate x on one side, giving

$$-4 < 2x.$$

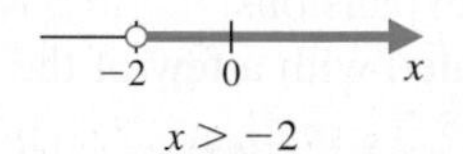

FIGURE 3

Finally, multiply both sides by $\frac{1}{2}$, which gives the solution, $-2 < x$. This can also be written as $x > -2$, as shown in Figure 3. ■

The steps in the solution to Example 1 are not the only way to proceed. For example, if we first add 1 to both sides of $2x - 1 < 4x + 3$ and then subtract $4x$, the inequality becomes

$$-2x < 4.$$

Now divide both sides by -2. The division is by a negative number, so we reverse the inequality to obtain $x > -2$.

EXAMPLE 2 Find all real numbers x satisfying $-1 < 2x + 3 \leq 5$.

Solution This inequality relation is a compact way of expressing that x must satisfy both of the inequalities

$$-1 < 2x + 3 \quad \text{and} \quad 2x + 3 \leq 5.$$

Proceeding as in Example 1, we add -3 to both sides of each inequality to produce

$$-4 < 2x \quad \text{and} \quad 2x \leq 2.$$

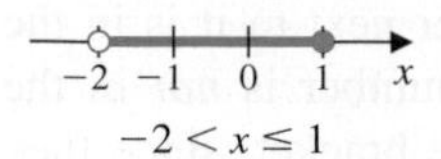

FIGURE 4

Multiplying these inequalities by $\frac{1}{2}$ gives

$$-2 < x \quad \text{and} \quad x \le 1.$$

This last set of inequalities can be expressed compactly as $-2 < x \le 1$, which is shown in Figure 4. ■

Intervals

Interval notation is preferred in calculus to describe sets of real numbers lying between two given numbers.

Interval notation is a convenient way to represent certain important sets of real numbers. For real numbers a and b with $a < b$, the **open interval** (a, b) is defined as

$$(a, b) = \{x \mid a < x < b\},$$

and read "the set of real numbers x such that x is greater than a and less than b."

When the *endpoints*, a and b, of the interval are included in the set, it is called a **closed interval** and denoted

$$[a, b] = \{x \mid a \le x \le b\}.$$

An interval that contains one endpoint but not the other is said to be a *half-open* interval (although it could just as well be called half-closed). So the intervals

$$(a, b] = \{x \mid a < x \le b\}$$

and

$$[a, b) = \{x \mid a \le x < b\}$$

are both half-open.

The *interior* of an interval consists of all the numbers in the interval that are not endpoints. The intervals (a, b), $[a, b]$, $(a, b]$, and $[a, b)$ all have the same interior, which is the open interval (a, b).

In addition to the intervals with finite endpoints, the **infinity symbol**, ∞, is used to indicate that an interval extends indefinitely. Such an interval is said to be **unbounded**. The intervals

$$[a, \infty) = \{x \mid x \ge a\}$$

and

$$(a, \infty) = \{x \mid x > a\}$$

are *unbounded above* since they contain no largest real number. The intervals

$$(-\infty, a] = \{x \mid x \le a\}$$

and

$$(-\infty, a) = \{x \mid x < a\}$$

are *unbounded below*. The interval $(-\infty, \infty)$, which represents the set $\mathbb{R}$ of all real numbers, is unbounded both above and below.

In general, a square bracket indicates that the real number next to it is in the interval. A real number next to a parenthesis indicates the number is *not* in the interval. The symbols $-\infty$ and ∞ are *never* next to a square bracket, since they are only symbols and do not represent real numbers. Notice also that

- In any use of interval notation, the symbol on the left must be less than the symbol on the right.

Table 1 summarizes the interval notation.

TABLE 1

Interval Notation	Set Notation	Graphic Representation
(a, b)	$\{x \mid a < x < b\}$	a b x
$[a, b]$	$\{x \mid a \le x \le b\}$	a b x
$[a, b)$	$\{x \mid a \le x < b\}$	a b x
$(a, b]$	$\{x \mid a < x \le b\}$	a b x
(a, ∞)	$\{x \mid a < x\}$	a x
$[a, \infty)$	$\{x \mid a \le x\}$	a x
$(-\infty, b)$	$\{x \mid x < b\}$	b x
$(-\infty, b]$	$\{x \mid x \le b\}$	b x
$(-\infty, \infty)$	$\mathbb{R} = \{x \mid -\infty < x < \infty\}$	x

EXAMPLE 3 **a.** Express the interval $[-3, 2)$ using inequalities.

b. Express the inequality $-\infty < x \le 2$ using interval notation.

Solution **a.** The interval $[-3, 2)$ represents all real numbers between -3 and 2, including -3, indicated by the square bracket, and excluding 2, indicated by the parenthesis. So $-3 \le x$ and $x < 2$. Both conditions hold so we can write $-3 \le x < 2$. To graph the inequality $-3 \le x < 2$ we shade all the points on the real line between -3 and 2, use a solid dot at -3 to indicate that it is included, and use an open circle at 2 to indicate that it is not included, as shown in Figure 5.

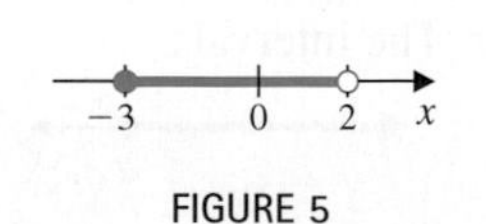

FIGURE 5

b. The inequality $-\infty < x \le 2$ represents all real numbers less than or equal to 2. This is equivalent to the interval notation $(-\infty, 2]$, whose graph is shown in Figure 6. ■

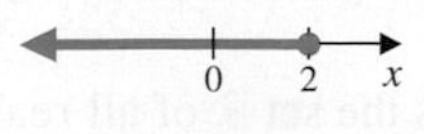

FIGURE 6

Interval notation can be used to give an alternative expression for the answers to Examples 1 and 2.

- In Example 1, the inequality was satisfied when $-2 < x$, that is, for x in $(-2, \infty)$.
- In Example 2, the inequality was satisfied when $-2 < x \leq 1$, that is, for x in $(-2, 1]$.

The next example involves a quadratic inequality that can be solved algebraically in a manner similar to the method used in Examples 1 and 2. It is easier, however, to use a graphical technique involving the coordinate line. This is the method of choice for solving the numerous inequalities that arise in a calculus problem.

EXAMPLE 4 Find all values of x that satisfy the inequality $(x - 1)(x - 3) > 0$.

Solution Figure 7 is a **sign chart** for the inequality $(x - 1)(x - 3) > 0$. It is used to determine where the inequality is positive and where it is negative.

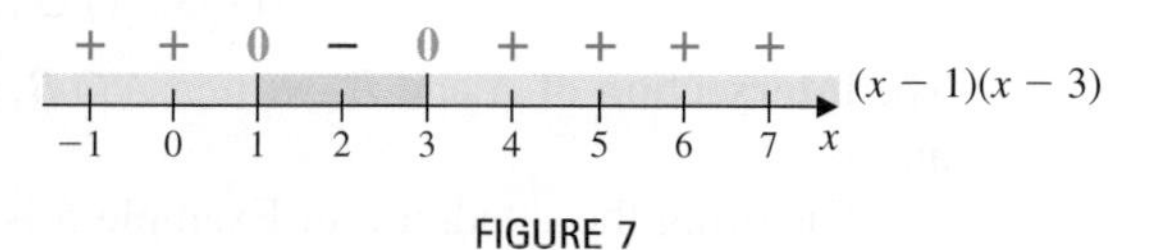

FIGURE 7

Solving inequalities is often needed in calculus, when determining where functions are defined or where graphs are increasing and decreasing.

To construct the sign chart, first determine where each of the factors is 0, namely, at $x = 1$ and $x = 3$. The inequality will be always positive or always negative on the intervals separated by the values where the factors are 0.

Select any number inside one of the intervals. For example, choose $x = 0$, in $(-\infty, 1)$, and evaluate the left side, giving

$$(0 - 1)(0 - 3) = (-1)(-3) = 3 > 0.$$

The value of the inequality is positive when $x = 0$, so it is positive on the entire interval $(-\infty, 1)$. The value of the inequality is negative when $x = 2$ because

$$(2 - 1)(2 - 3) = (1)(-1) = -1 < 0,$$

so the inequality is negative on the entire interval $(1, 3)$. Similarly, the value is positive when $x = 4$ because

$$(4 - 1)(4 - 3) = (3)(1) > 0.$$

So the inequality is positive on $(3, \infty)$. ■

EXAMPLE 5 Find all values of x that satisfy the inequality $x^2 - 4x + 5 > 2$.

Solution This problem is solved by first changing the inequality into one that has 0 on the right side, and then factoring the term on the left side. It will then be in a form similar to the inequality in Example 4.

Subtracting 2 from both sides of the inequality, we see that

$$x^2 - 4x + 5 > 2 \quad \text{implies that} \quad x^2 - 4x + 3 > 0.$$

We now factor the quadratic $x^2 - 4x + 3$ by determining constants a and b so that $x^2 - 4x + 3 = (x + a)(x + b)$. Because

$$x^2 - 4x + 3 = (x + a)(x + b) = x^2 + (a + b)x + a \cdot b,$$

we must have $a + b = -4$ and $a \cdot b = 3$.

This implies that $x^2 - 4x + 3 = (x-1)(x-3)$. So to solve the original inequality, we must have $(x-1)(x-3) > 0$. Since this factored form is the same as the inequality in Example 4, the solution is as shown in Figure 7. ■

The sign chart in Figure 7 also tells us that

$$x^2 - 4x + 3 = (x-1)(x-3) < 0 \text{ when } 1 < x < 3.$$

Unions and Intersections

The answer to Examples 4 and 5 can also be expressed using interval notation, but it requires the introduction of the union symbol. The **union** of two sets A and B, written $A \cup B$, is the set of all elements that are in either A or in B (or in both). So in interval notation the answer in Example 4 is

$$(-\infty, 1) \cup (3, \infty).$$

The **intersection** of A and B, written $A \cap B$, is the set of elements that are in both A and B.

Factoring the quadratic in Example 5 is relatively easy because there are only two integer possibilities for a product of the constant term $3 = 1 \cdot 3 = (-1) \cdot (-3)$. Once we determine this, it is clear which one to use because the coefficient of x in the quadratic $x^2 - 4x + 3$ must be -4.

When there are more factoring possibilities for the constant term, experimentation is needed. For example, suppose that we want to factor the quadratic $x^2 + 7x + 12$. Since

$$12 = 1 \cdot 12 = 2 \cdot 6 = 3 \cdot 4 = (-1) \cdot (-12) = (-2) \cdot (-6) = (-3) \cdot (-4)$$

we have many more possibilities. However, the coefficient of x is positive, so we can disregard all the negative factors. In addition, $7 = 3 + 4$, so the only possibility is $(x+3)$ and $(x+4)$. Notice that we do, in fact, have $(x+3)(x+4) = x^2 + 7x + 12$.

There are other techniques that can be used to simplify the factoring process, but we will postpone these until Chapter 3.

EXAMPLE 6 Find all values of x for which $(x-1)(x-3)(5-x) \le 0$.

Solution The three factors separate the real line into four separate regions,

$$x < 1, \quad 1 < x < 3, \quad 3 < x < 5, \quad \text{and} \quad x > 5.$$

To check the region where $x < 1$, substitute $x = 0$ in the inequality to obtain

$$(0-1)(0-3)(5-0) = (-1)(-3)(5) = 15 > 0.$$

Since the inequality is positive at $x = 0$, the inequality is positive on the entire region $x < 1$. Checking the values when x is 2, 4, and 6 shows that the inequality alternates on the other regions as shown in the sign chart in Figure 8.

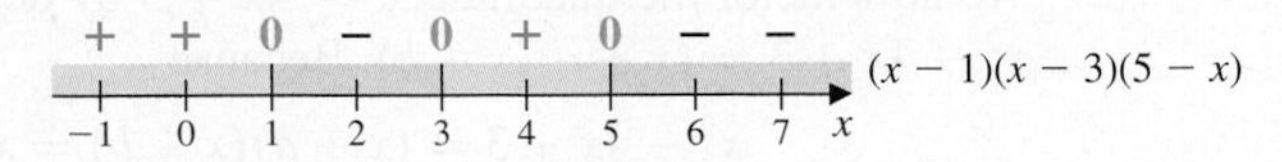

FIGURE 8

The solution to this inequality is written in interval notation as

$$[1, 3] \cup [5, \infty).$$ ■

The sign chart technique can also be applied to factored quotients.

EXAMPLE 7 Find all values of x for which $\dfrac{x^2 - 4}{x(4 - x)} \geq 0.$

Solution The numerator of the quotient can be factored to give

$$\frac{x^2 - 4}{x(4 - x)} = \frac{(x + 2)(x - 2)}{x(4 - x)}.$$

This quotient is zero when the numerator $x^2 - 4 = 0$, which occurs at $x = 2$ and $x = -2$. The quotient is undefined when the denominator is 0, which occurs at $x = 0$ and $x = 4$.

The four factors in the numerator and the denominator of the quotient separate the real line into five separate regions, $x < -2$, $-2 < x < 0$, $0 < x < 2$, $2 < x < 4$, and $x > 4$.

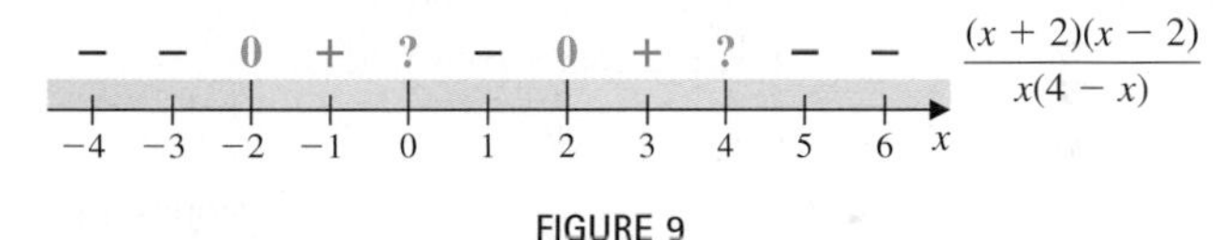

FIGURE 9

The sign chart in Figure 9 shows the possibilities. We have used the symbol ? as a shortcut to indicate that the quotient is undefined at $x = 0$ and $x = 4$. The chart gives the solution as $[-2, 0) \cup [2, 4)$. ■

Absolute Values

The **absolute value** of a real number x, denoted $|x|$, describes the distance on the coordinate line from the number x to the number 0. The definition follows.

Absolute Value

$$|x| = \begin{cases} x, & \text{if } x \geq 0, \\ -x, & \text{if } x < 0. \end{cases}$$

A number and its negative have the same absolute value. For example, $|2| = 2$ and $|-2| = -(-2) = 2$. This means when solving an equation of the form

$$|\square| = a$$

we should determine the values that satisfy either

$$\square = a \quad \text{or} \quad \square = -a.$$

Notice, that since the absolute value of a quantity is always greater than or equal to zero, if $a < 0$, then the equation $|\square| = a$ has no solutions.

EXAMPLE 8 Determine all values of x for which $\left|\dfrac{x - 2}{2x + 1}\right| = 3.$

Solution We need to find all values of x that satisfy either

$$\frac{x-2}{2x+1} = 3 \quad \text{or} \quad \frac{x-2}{2x+1} = -3.$$

The first equation implies that when $x \neq -\frac{1}{2}$, we have

$$x - 2 = 6x + 3.$$

So $-5 = 5x$ and $x = -1$ is one solution.

The second equation implies that when $x \neq -\frac{1}{2}$, we have

$$x - 2 = -6x - 3.$$

So $7x = -1$ and $x = -\frac{1}{7}$ is the only other solution. ■

The absolute value of x is $|x| = \sqrt{x^2}$, so $d(x_1, x_2) = \sqrt{(x_1 - x_2)^2}$. A similar formula for the distance between points in the plane is given in the next section.

The absolute value of a number gives a measure of its distance from 0, as shown in Figure 10(a). Using this as a guide, we can define the **distance** $d(x_1, x_2)$ between the real number x_1 and the real number x_2, as shown in Figure 10(b).

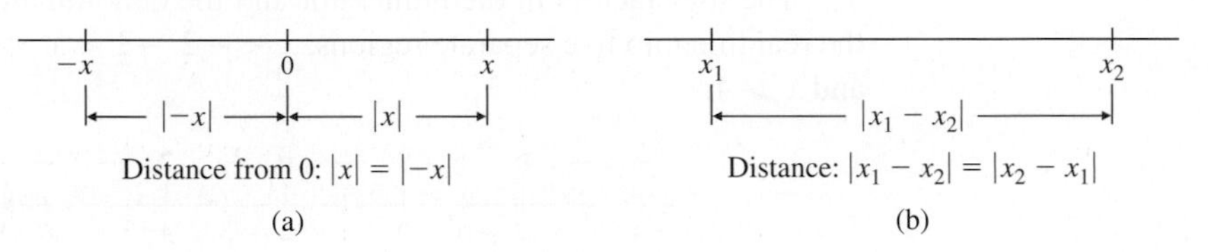

FIGURE 10

The Distance from x_1 to x_2

The distance between real numbers x_1 and x_2 is $d(x_1, x_2) = |x_1 - x_2|$.

For example,

$$d(5, 1) = d(1, 5) = 4 \quad \text{and} \quad d(-3.2, 4.5) = d(4.5, -3.2) = 7.7.$$

The distance formula is also used to determine a formula for the *midpoint* of two numbers, as illustrated in Figure 11.

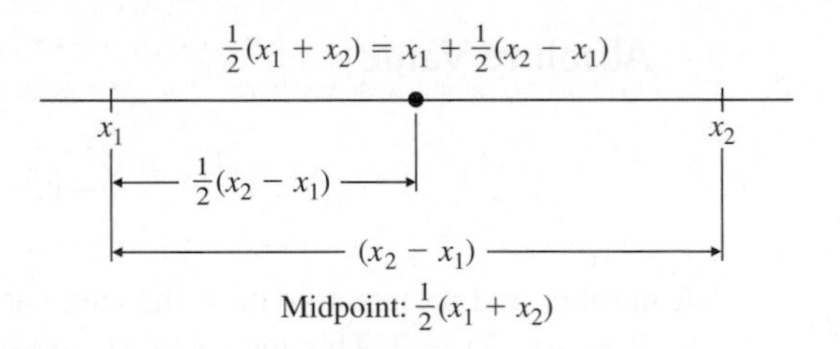

FIGURE 11

The midpoint of two numbers is just their average.

The Midpoint Formula

The midpoint of the line segment joining x_1 and x_2 is $\frac{1}{2}(x_1 + x_2)$.

For example, the midpoint of the interval $[-1.2, 5.6]$ is

$$\tfrac{1}{2}(5.6 + (-1.2)) = \tfrac{1}{2}(4.4) = 2.2.$$

Absolute Value Properties

The third condition will be used frequently in calculus. It implies that $|x| < a$ precisely when the distance from x to the origin is less than a.

- $|ab| = |a|\,|b|$ and, if $b \neq 0$, $\left|\dfrac{a}{b}\right| = \dfrac{|a|}{|b|}$.
- $||a| - |b|| \leq |a + b| \leq |a| + |b|$.
- $|x| < a$ if and only if $-a < x < a$.
- $|x| > a \geq 0$ if and only if $x < -a$ or $x > a$.

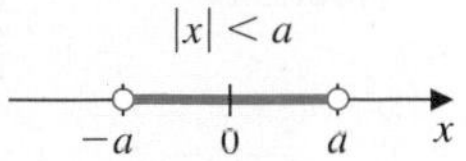

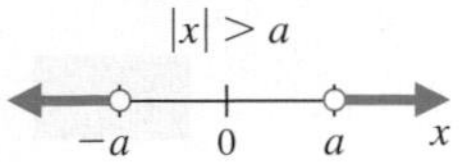

EXAMPLE 9 Find all values of x that satisfy $|2x - 1| < 3$.

Solution Suppose we replace x with $2x - 1$ and a with 3 in the third Absolute Value Property. Then we have

$$|2x - 1| < 3 \quad \text{if and only if} \quad -3 < 2x - 1 < 3.$$

This implies that we need to solve both the inequalities

$$-3 < 2x - 1 \quad \text{and} \quad 2x - 1 < 3.$$

Adding 1 to both sides of each of the inequalities gives

$$-2 < 2x \quad \text{and} \quad 2x < 4.$$

Multiplying the inequalities by $\frac{1}{2}$ gives the solution

$$-1 < x \quad \text{and} \quad x < 2,$$

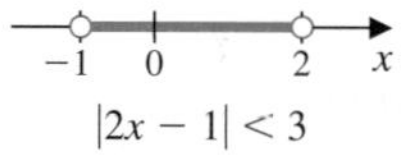

FIGURE 12

as shown in Figure 12.

Note that the numbers that are outside the interval shown in Figure 12 satisfy the opposite inequality $|2x - 1| \geq 3$. ■

The problem in Example 9 can also be solved using the coordinate line and the distance property of the absolute value of the difference of two real numbers. Since

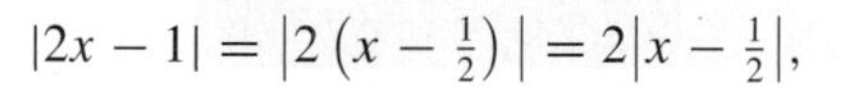

$$|2x - 1| = \left|2\left(x - \tfrac{1}{2}\right)\right| = 2\left|x - \tfrac{1}{2}\right|,$$

the inequality $|2x - 1| < 3$ is equivalent to the inequality

$$\left|x - \tfrac{1}{2}\right| < \tfrac{3}{2}.$$

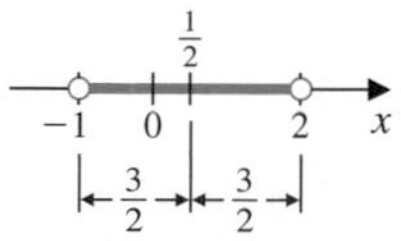

FIGURE 13

So, as shown in Figure 13, x satisfies the condition $|2x - 1| < 3$ precisely when

$$d\left(x, \tfrac{1}{2}\right) < \tfrac{3}{2}.$$

EXAMPLE 10 Find all values of x that satisfy $|-3x + 2| > 8$.

Solution The Absolute Value Properties imply that

$$|-3x + 2| > 8 \quad \text{if and only if} \quad -3x + 2 > 8 \text{ or } -3x + 2 < -8.$$

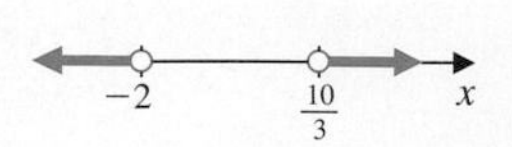

FIGURE 14

Subtracting 2 from both sides of each of the latter inequalities gives

$$-3x > 6 \quad \text{or} \quad -3x < -10.$$

Multiplying these inequalities by $-\frac{1}{3}$ (remembering to reverse the inequality) gives the solution, shown in Figure 14,

$$x < -2 \quad \text{or} \quad x > \tfrac{10}{3}. \qquad \blacksquare$$

Applications

EXAMPLE 11 When a charity conducts a fund-raising campaign, it is common to describe the progress with respect to the midpoint value (half the goal). Find an expression for the amount raised relative to the midpoint value.

Solution Let R be the amount the charity wants to raise. Then the midpoint value is $M = R/2$. We need an expression describing A, the amount raised, relative to M.

Before considering the general question, let's look at a specific example. If the amount the charity wants to raise is $R = \$200{,}000$, then the midpoint amount is $M = \$100{,}000$. When the current amount raised is \$78,000, the charity reports that it is \$22,000 from half the goal. When the current amount raised is \$150,000, they report that they are \$50,000 beyond half the goal.

Principles of Problem Solving:

- Analyze some examples.
- Consider different cases.

In the general situation, let C be the amount currently raised. There are two cases that can occur.

Case 1: If $C \le M$, they report that they are below half the goal by

$$A = M - C \text{ dollars.}$$

Case 2: If $C \ge M$, they report that they are above half their goal by

$$A = C - M \text{ dollars.}$$

Because $C - M = -(M - C)$, in either case this can be expressed as

$$A = |M - C| = \left|\frac{R}{2} - C\right|. \qquad \blacksquare$$

EXERCISE SET 1.2

In Exercises 1–8, express the interval using inequalities, and sketch the numbers in the interval.

1. $[-1, 5]$ **2.** $[-3, -1)$

3. $(-\sqrt{3}, \sqrt{2}]$ **4.** $[-\sqrt{5}, -\sqrt{2})$

5. $(-\infty, 3)$ **6.** $(-\infty, 0]$

7. $[\sqrt{2}, \infty)$ **8.** $(-2, \infty)$

In Exercises 9–16, express the inequalities using interval notation, and sketch the numbers in the interval.

9. $-2 \le x \le 3$ **10.** $2 < x \le 6$

11. $2 \le x < 5$ **12.** $-2 \le x < 4$

13. $x < 3$ **14.** $x \le -2$

15. $x \ge 3$ **16.** $x > -2$

*In Exercises 17–20, find **(a)** the distance between the points and **(b)** the midpoint of the line segment connecting them.*

17. 3 and 7 **18.** 4 and 7

19. −3 and 5 **20.** −4 and −1

In Exercises 21–26, factor the quadratic equation.

21. $x^2 + 3x + 2$ **22.** $x^2 + 7x + 6$

23. $x^2 + 5x + 6$ **24.** $x^2 - 9x + 8$

25. $x^2 + 4x - 12$

26. $x^2 - 4x - 12$

*In Exercises 27–30, find (**a**) the intersection, and (**b**) the union of the intervals.*

27. $[-1, 3]$ and $(0, 4)$

28. $[-3, -1]$ and $[-2, 2)$

29. $(-\infty, 0)$ and $(-2, 3]$

30. $[1, \infty)$ and $[-3, 3)$

In Exercises 31–56, use interval notation to list the values of x that satisfy the inequality.

31. $x + 3 < 5$

32. $x - 4 < 9$

33. $2x - 2 \geq 8$

34. $3x + 2 \geq 8$

35. $-3x + 4 < 5$

36. $-2x - 4 \geq 10$

37. $2x + 9 \leq 5 + x$

38. $-3x - 2 < 3 - x$

39. $-1 < 3x - 3 < 6$

40. $-3 < 2x + 1 \leq 2$

41. $(x + 1)(x - 2) \geq 0$

42. $(x - 1)(x + 3) < 0$

43. $x^2 - 4x + 3 \leq 0$

44. $x^2 - 2x - 3 > 0$

45. $(x - 1)(x - 2)(x + 1) \leq 0$

46. $(x - 1)(x + 2)(x - 3) \geq 0$

47. $x^3 - 3x^2 + 2x \geq 0$

48. $x^3 - 3x^2 - 4x < 0$

49. $x^3 - 2x^2 < 0$

50. $x^3 - 2x^2 + x > 0$

51. $\dfrac{x + 3}{x - 1} \geq 0$

52. $\dfrac{x - 2}{x + 1} \leq 0$

53. $\dfrac{x(x + 2)}{x - 2} \leq 0$

54. $\dfrac{x + 2}{x(x - 2)} > 0$

55. $\dfrac{(1 - x)(x + 2)}{x(x + 1)} > 0$

56. $\dfrac{(1 - x)(x + 3)}{(x + 1)(2 - x)} \leq 0$

In Exercises 57–60, solve the inequality. (Hint: First rewrite the inequality by setting one side to 0.)

57. $\dfrac{1}{x} \leq 5$

58. $-2 \leq \dfrac{1}{x}$

59. $\dfrac{2}{x - 1} \geq \dfrac{3}{x + 2}$

60. $\dfrac{2}{x - 1} - \dfrac{x}{x + 1} \leq -1$

In Exercises 61–64, find all values of x that solve the equation.

61. $|5x - 3| = 2$

62. $|2x + 3| = 1$

63. $\left|\dfrac{x - 1}{2x + 3}\right| = 2$

64. $\left|\dfrac{2x + 1}{x - 3}\right| = 4$

In Exercises 65–72, solve the inequality and write the solution using interval notation.

65. $|x - 4| \leq 1$

66. $|4x - 1| < 0.01$

67. $|3 - x| \geq 2$

68. $|2x - 1| > 5$

69. $\dfrac{1}{|x + 5|} > 2$

70. $\left|\dfrac{3}{2x + 1}\right| < 1$

71. $|x^2 - 4| > 0$

72. $|x^2 - 4| \leq 1$

73. Show that if $0 < a < b$, then $a^2 < b^2$. (*Hint*: Show first that $a^2 < ab$ and $ab < b^2$.)

74. Solve the equation $|x - 1| = |2x + 2|$.

75. Degrees Celsius (C) and degrees Fahrenheit (F) are related by the formula $C = \frac{5}{9}(F - 32)$.

 a. What is the temperature range in the Celsius scale corresponding to a temperature range in Fahrenheit of $20 \leq F \leq 50$?

 b. What is the temperature range in the Fahrenheit scale corresponding to a temperature range in Celsius of $20 \leq C \leq 50$?

76. Calculus can be used to show that a ball thrown straight upward (neglecting air resistance) from the top of a building 128 feet high, with an initial velocity of 48 feet/second, has a height, in feet, of

$$h(t) = -16t^2 + 48t + 128$$

t seconds later. For what time interval will the ball be at least 64 feet above the ground?

77. You are contemplating investing \$50,000 between two different investments. One of the investments, a stable one, returns 5% annually. A more risky one returns 9% annually. You need a return of at least 7.5%. What is the maximum amount of the money that you can invest at the 5% rate?

1.3 THE COORDINATE PLANE

Calculus is the study of change between variables, and graphs help us visualize this change. To sketch graphs we first need a coordinate system on which to place the graphs.

In Section 1.2 we saw how the set of real numbers relates to points on a coordinate line and how this relationship permits us to solve problems more easily. In this section we consider ordered pairs of real numbers and their relation to the coordinate plane.

Each point in the plane is associated with an ordered pair of real numbers. First, an arbitrary point in the plane is associated with (0, 0) and designated the *origin*. Then horizontal and vertical lines are drawn, intersecting at the origin. The horizontal line is called the first-coordinate axis, or, commonly, the *x-axis*. The vertical line is called the second-coordinate axis, or the *y-axis*. (See Figure 1.)

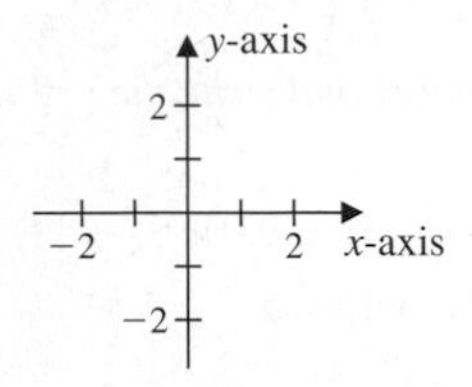

FIGURE 1

A scale is placed on both axes. Since the x-axis is the same as the coordinate line introduced in the previous section, it is again labeled with positive numbers to the right of the origin and negative numbers to the left. Labeling on the y-axis is similar. The numbers above the origin are labeled as positive and the ones below the origin as negative.

The ordered pair (a, b) is associated with the point of intersection of the vertical line drawn through the point a on the x-axis and the horizontal line drawn through b on the y-axis. (See Figure 2.) We will not generally make a distinction between an ordered pair and the point it represents in a coordinate plane.

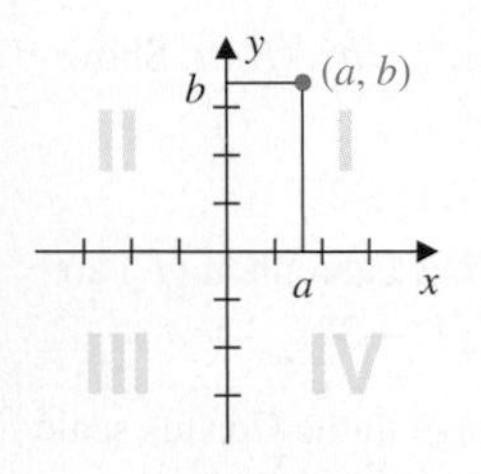

FIGURE 2

The x- and y-axes divide the plane into four regions, or *quadrants*, labeled as shown in Figure 2. The set of all ordered pairs of real numbers is denoted $\mathbb{R} \times \mathbb{R}$, or $\mathbb{R}^2$, and the plane determined by the x- and y-axes is called the *xy-plane*.

The coordinate plane is also called the Cartesian plane, and a rectangular coordinate system is called a Cartesian coordinate system. These names honor the versatile mathematician, philosopher, and physicist René Descartes (1596–1650), whose name in Latin was Renatus Cartesius. He introduced analytic geometry to the mathematical world in *La géométrie*, an appendix to his treatise on universal science.

EXAMPLE 1 Sketch the points in the coordinate plane associated with the ordered pairs $(1, 2)$, $(-1, 3)$, $(-2, -\pi)$, and $(\sqrt{2}, -\sqrt{3})$.

Solution To plot a point in the plane we locate the intersection of the vertical line through the first coordinate on the x-axis and the horizontal line through the second coordinate on the y-axis. These points are shown in Figure 3. ■

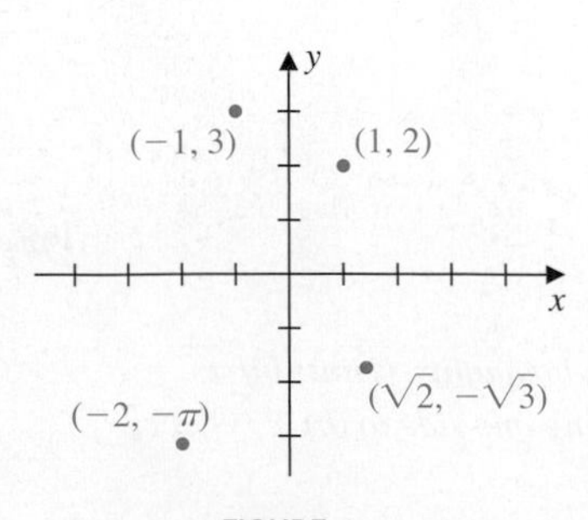

FIGURE 3

EXAMPLE 2 Sketch the points in the xy-plane that satisfy both $x > 2$ and $y \le 1$.

Solution The points satisfying the single inequality $x > 2$ are shown in Figure 4(a). The dashed line at $x = 2$ indicates that these points are not in the set. The points satisfying $y \le 1$ are shown in Figure 4(b). The points satisfying both conditions are in the shaded region shown in Figure 4(c). ■

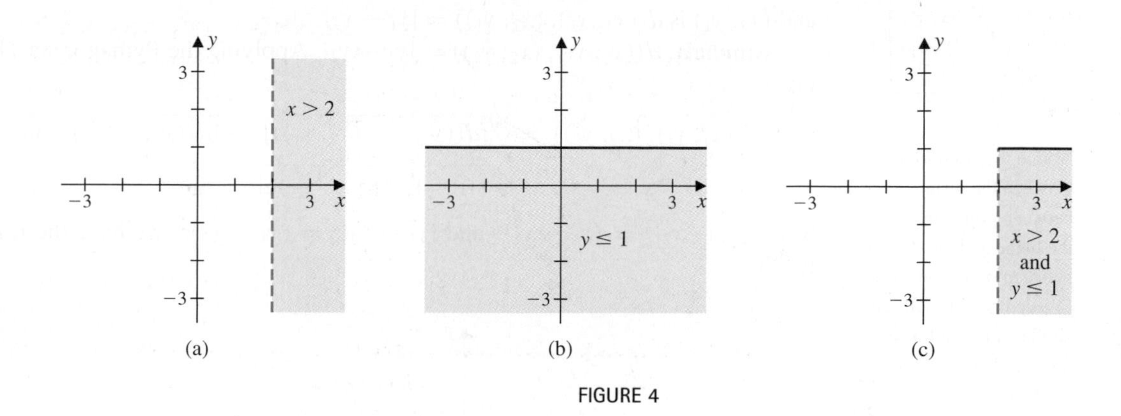

FIGURE 4

EXAMPLE 3 Sketch the points in the xy-plane that satisfy both the inequalities

$$1 \le |x| \le 2 \quad \text{and} \quad -1 < y < 3.$$

Solution The inequality $1 \le |x| \le 2$ implies that the distance from the x-coordinate of the point to the y-axis is between 1 and 2 units. These are the points that lie on or between the vertical lines $x = -2$ and $x = -1$, together with those that lie on or between the vertical lines $x = 1$ and $x = 2$. This region is shown shaded in Figure 5(a).

The points whose y-coordinates satisfy $-1 < y < 3$ lie strictly between the dashed horizontal lines $y = -1$ and $y = 3$, shown in Figure 5(b). The points satisfying both conditions are in the shaded regions shown in Figure 5(c). ■

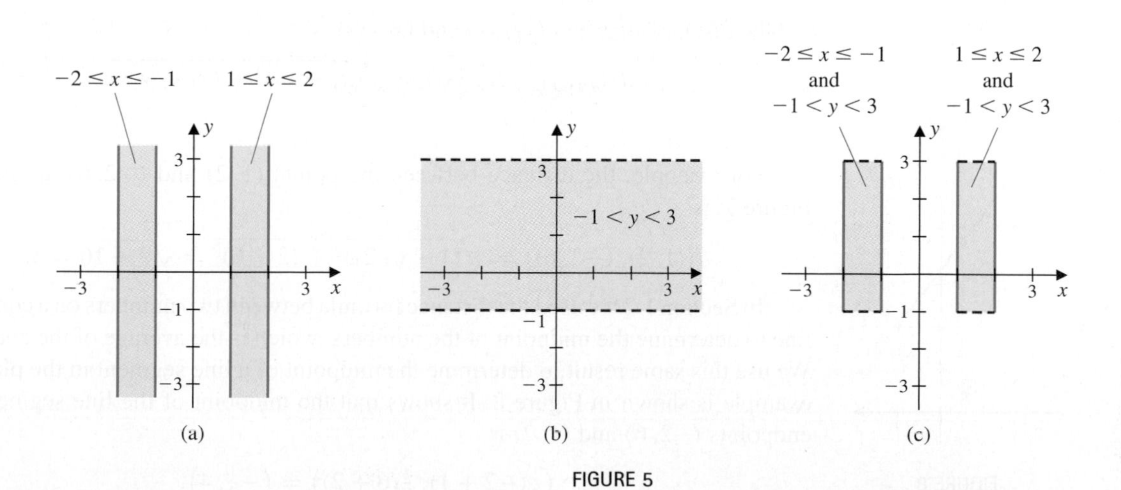

FIGURE 5

Distance between Points in the Plane

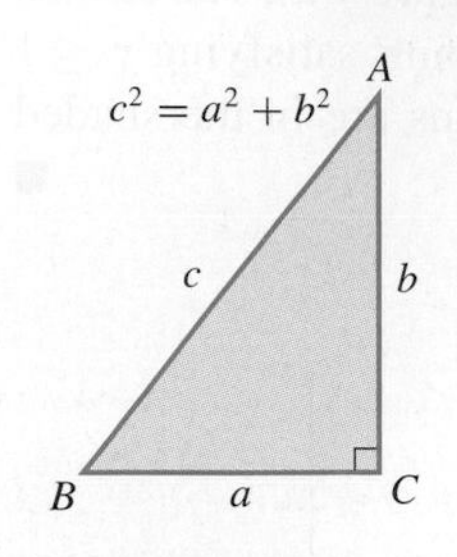

Notice the similarity between the distance between points in the plane and the distance between points on a line. In calculus you will see a similar formula for the distance between points in space.

The distance between two ordered pairs of real numbers (x_1, y_1) and (x_2, y_2) is found by introducing the third point (x_2, y_1) and applying the Pythagorean Theorem, as shown in Figure 6.

On the x-axis we have $d(x_1, x_2) = |x_1 - x_2|$, so the distance between (x_1, y_1) and (x_2, y_1) is $d((x_1, y_1), (x_2, y_1)) = |x_1 - x_2|$.

Similarly, $d((x_2, y_1), (x_2, y_2)) = |y_1 - y_2|$. Applying the Pythagorean Theorem gives

$$\begin{aligned} d((x_1, y_1), (x_2, y_2)) &= \sqrt{[d((x_1, y_1), (x_2, y_1))]^2 + [d((x_2, y_1), (x_2, y_2))]^2} \\ &= \sqrt{|x_1 - x_2|^2 + |y_1 - y_2|^2}. \end{aligned}$$

Since $|x_1 - x_2|^2 = (x_1 - x_2)^2$ and $|y_1 - y_2|^2 = (y_1 - y_2)^2$, we have the following result.

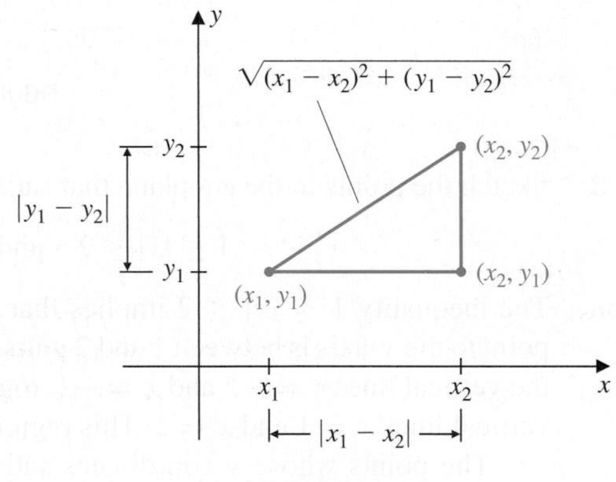

FIGURE 6

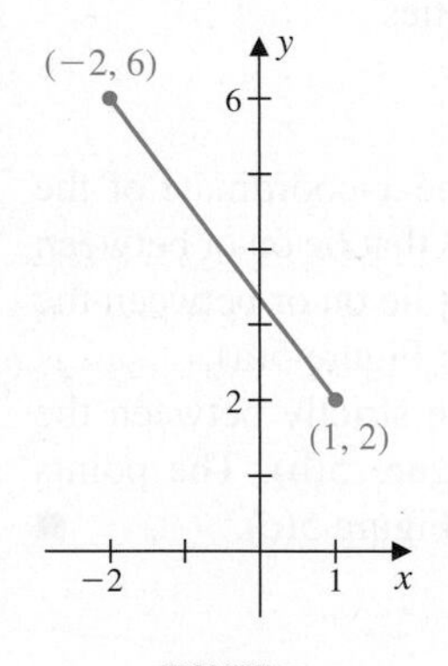

FIGURE 7

Distance between Points in the Plane

The distance between (x_1, y_1) and (x_2, y_2) is

$$d((x_1, y_1), (x_2, y_2)) = \sqrt{(x_1 - x_2)^2 + (y_1 - y_2)^2}.$$

For example, the distance between the points $(1, 2)$ and $(-2, 6)$, as shown in Figure 7, is

$$d((1, 2), (-2, 6)) = \sqrt{(1 - (-2))^2 + (2 - 6)^2} = \sqrt{9 + 16} = 5.$$

In Section 1.2 we used the distance formula between two numbers on a coordinate line to determine the midpoint of the numbers, which is the average of the endpoints. We use this same result to determine the midpoint of a line segment in the plane. An example is shown in Figure 8. It shows that the midpoint of the line segment with endpoints $(-2, 6)$ and $(1, 2)$ is

$$\left(\tfrac{1}{2}(-2 + 1), \tfrac{1}{2}(6 + 2)\right) = \left(-\tfrac{1}{2}, 4\right).$$

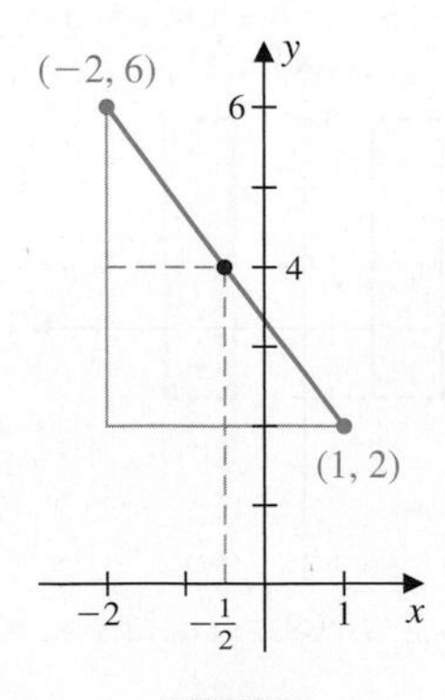

FIGURE 8

The Midpoint Formula

The midpoint of the line segment with endpoints (x_1, y_1) and (x_2, y_2) is

$$\left(\tfrac{1}{2}(x_1 + x_2), \tfrac{1}{2}(y_1 + y_2)\right).$$

Circles in the Plane

The distance formula for points in the plane can also be used to obtain the equation of a circle. A **circle** is the set of all points whose distance from a given point, the *center*, is a fixed distance, the *radius*. Figure 9 shows that a point (x, y) will be on the circle with center (h, k) having radius r precisely when

$$r = d((x, y), (h, k)) = \sqrt{(x - h)^2 + (y - k)^2}.$$

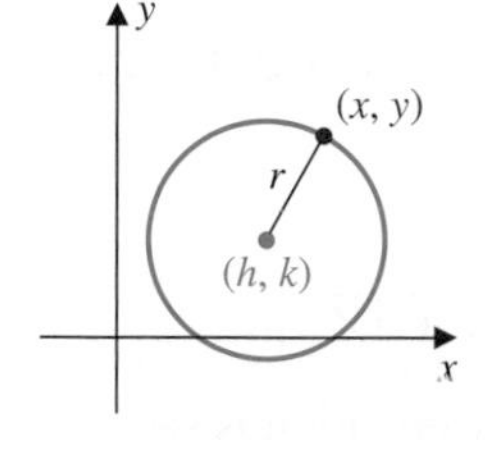

FIGURE 9

Squaring both sides produces the formula in its most familiar, or *standard*, form.

Circles in the Plane

The point (x, y) lies on the circle with center (h, k) and radius r precisely when

$$(x - h)^2 + (y - k)^2 = r^2.$$

This is the **standard form** of the equation of a circle.

The circle with center (h, k) is simply the unit circle scaled by the factor r and then moved h units horizontally and k units vertically.

We call a circle whose radius is 1 a **unit circle**. The applications involving the unit circle whose center is at $(0, 0)$ are so extensive that this circle is frequently known as *the* unit circle. The points (x, y) on this unit circle are those that satisfy the equation $x^2 + y^2 = 1$.

EXAMPLE 4 Determine an equation for the circle of radius 3 centered at $(-1, 2)$.

Solution The equation has the form

$$(x - (-1))^2 + (y - 2)^2 = 9 \quad \text{or} \quad (x + 1)^2 + (y - 2)^2 = 9.$$

Expanding the factors on the left side of the equation gives

$$x^2 + 2x + 1 + y^2 - 4y + 4 = 9.$$

This simplifies to $x^2 + y^2 + 2x - 4y = 4$. ■

EXAMPLE 5 Sketch the set of points that satisfies the inequality

$$1 < (x - 2)^2 + (y - 1)^2 \le 4.$$

Solution The equation

$$(x - 2)^2 + (y - 1)^2 = 1$$

describes the circle with center $(2, 1)$ and radius 1. The equation

$$(x - 2)^2 + (y - 1)^2 = 4$$

describes the circle with center (2, 1) and radius 2. A point satisfies

$$(x-2)^2+(y-1)^2>1$$

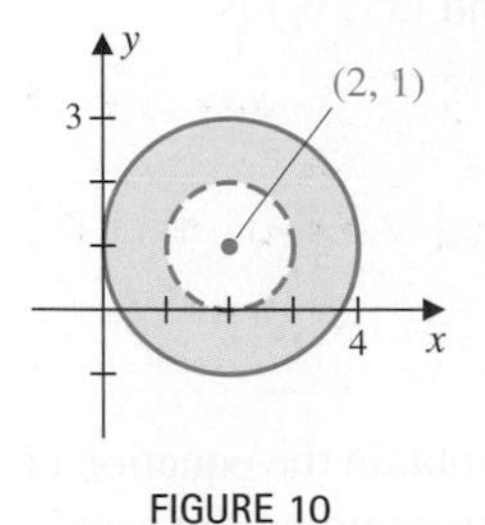

FIGURE 10

if its distance from the point (2, 1) is greater than 1. Similarly, a point satisfies

$$(x-2)^2+(y-1)^2\leq 4$$

if it lies inside or on the circle. So the points satisfying

$$1<(x-2)^2+(y-1)^2\leq 4$$

lie inside the circle of radius 4, including the points on the circle, and outside the circle of radius 1, excluding the points on the inner circle, as shown in Figure 10. ■

EXAMPLE 6 Find the equation of the circle with center $(-2,-1)$ that passes through the point $(-3, 1)$.

Solution The circle has center $(-2,-1)$, so its equation has the form

$$r^2=(x-(-2))^2+(y-(-1))^2=(x+2)^2+(y+1)^2.$$

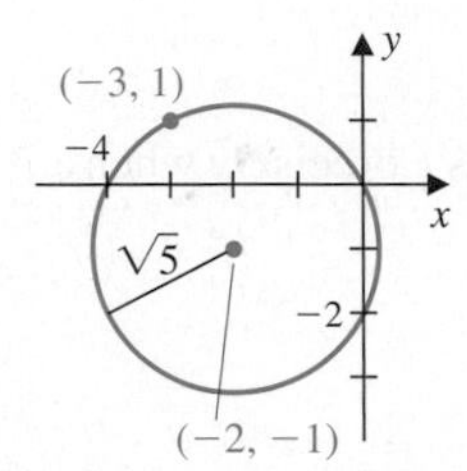

FIGURE 11

It remains to find the radius r. Since $(-3, 1)$ is on the circle, the equation is satisfied when $x=-3$ and $y=1$. So

$$r=\sqrt{(-3+2)^2+(1+1)^2}=\sqrt{1+4}=\sqrt{5}.$$

The circle shown in Figure 11 has equation

$$(x+2)^2+(y+1)^2=5.$$ ■

Completing the Square

Completing the square follows from the fact that

$$(x+a/2)^2 = x^2+ax+(a/2)^2.$$

All circles in the plane have equations of the form

$$x^2+y^2+ax+by=c.$$

We use a technique called **completing the square** to determine the center and radius of the circle. First group the x-terms and the y-terms separately and rewrite the equation as

$$x^2+ax+\quad y^2+by\quad =c.$$

A space is left after the terms ax and by because we will be adding constants that will make the x- and y-terms perfect squares.

The constants we need are $(a/2)^2$ and $(b/2)^2$, respectively. These terms must also be subtracted to ensure the equation has not changed. So the equation becomes

$$x^2+ax+\left(\frac{a}{2}\right)^2-\left(\frac{a}{2}\right)^2+y^2+by+\left(\frac{b}{2}\right)^2-\left(\frac{b}{2}\right)^2=c.$$

This simplifies to

$$\left(x+\frac{a}{2}\right)^2-\left(\frac{a}{2}\right)^2+\left(y+\frac{b}{2}\right)^2-\left(\frac{b}{2}\right)^2=c$$

or to

$$\left(x+\frac{a}{2}\right)^2+\left(y+\frac{b}{2}\right)^2=c+\frac{a^2}{4}+\frac{b^2}{4}=\frac{4c+a^2+b^2}{4}.$$

Hence, the circle has center $(-a/2, -b/2)$ and radius $\frac{1}{2}\sqrt{4c+a^2+b^2}$, provided, of course, that $4c+a^2+b^2>0$.

There is no need to memorize the form of the resulting equation; just follow the procedure illustrated in Example 7.

EXAMPLE 7 Sketch the graph of the circle with equation

$$x^2+y^2+4x-6y=3.$$

Solution The equation can be regrouped as

$$x^2+4x \quad +y^2-6y \quad =3.$$

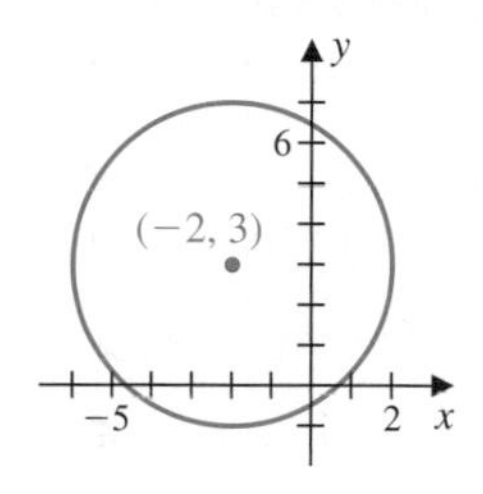

FIGURE 12

Principles of Problem Solving:

- Draw a picture to show the unknowns.
- Use equations to describe the problem.

To complete the square on the x-terms and y-terms, we add and subtract, respectively,

$$\left(\tfrac{4}{2}\right)^2=2^2=4 \quad \text{and} \quad \left(-\tfrac{6}{2}\right)^2=(-3)^2=9.$$

This produces the equation

$$(x^2+4x+4)-4+(y^2-6y+9)-9=3$$

or

$$(x+2)^2+(y-3)^2=3+4+9=16.$$

The equation of the circle in standard form is

$$(x-(-2))^2+(y-3)^2=4^2,$$

which describes the circle with center $(-2, 3)$ and radius 4 that is shown in Figure 12. ■

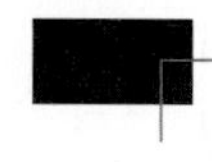

Applications

EXAMPLE 8 A gas pipeline is to be constructed between points A and B on opposite sides of a river, as shown in Figure 13. A surveyor first determines a third point C on the same side of the river as point A so that the lines between A and C and between B and C are perpendicular. The distance from A to C is 300 feet and the distance from B to C is 750 feet. How much pipe is required?

Solution The problem asks for the distance between the points A and B. Let z denote this distance, and $x=300$ and $y=750$ be the measured distances.

Because ΔACB is a right triangle we can use the Pythagorean Theorem to find z. This gives

$$\begin{aligned} z^2 &= x^2+y^2 \\ &= (300)^2+(750)^2 \\ &= 652500, \quad \text{so} \\ z &= \sqrt{652500}\approx 808 \text{ feet,} \end{aligned}$$

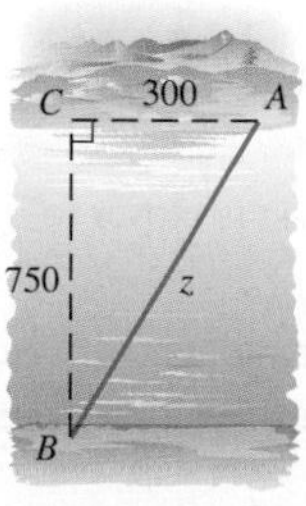

FIGURE 13

and the project requires approximately 808 feet of pipe. ■

EXERCISE SET 1.3

In Exercises 1–4, sketch the listed points in the same coordinate plane.

1. $(1, 0), (0, 1), (-1, 0), (0, -1)$
2. $(0, 3), (-1, 3), (3, -2), (-3, -1)$
3. $(2, 3), (-2, -3), (2, -3), (-2, 3)$
4. $(5, -10), (10, 20), (-20, 10), (-20, -10)$

*In Exercises 5–8, find **(a)** the distance between the points and **(b)** the midpoints of the line segments joining the points.*

5. $(2, 4), (-1, 3)$
6. $(-3, 8), (5, 4)$
7. $(\pi, 0), (-1, 2)$
8. $(\sqrt{3}, \sqrt{2}), (\sqrt{2}, \sqrt{3})$

In Exercises 9–22, indicate on an xy-plane those points (x, y) for which the statement holds.

9. $x = 5$
10. $y = -2$
11. $x > 1$
12. $x < -2$
13. $x \geq 1$ and $y \geq 2$
14. $x < -3$ and $y < -4$
15. $-3 < y \leq 1$
16. $-1 \leq x \leq 2$
17. $2 \leq |x|$
18. $|y + 1| > 2$
19. $-1 \leq x \leq 2$ and $2 < y < 3$
20. $-3 < x \leq 1$ and $-1 \leq y \leq 2$
21. $|x - 1| < 3$ and $|y + 1| < 2$
22. $|x - 2| \leq 4$ and $|y + 3| < 7$

In Exercises 23–28, find the standard form of the equation of the circle, and sketch the graph.

23. center $(2, 0)$; radius 3
24. center $(0, 2)$; radius 3
25. center $(-2, 3)$; radius 2
26. center $(-1, 4)$; radius 4
27. center $(-1, -2)$; radius 2
28. center $(-2, -1)$; radius 3

In Exercises 29 and 30, find the equation of the circle shown in the figure.

29.

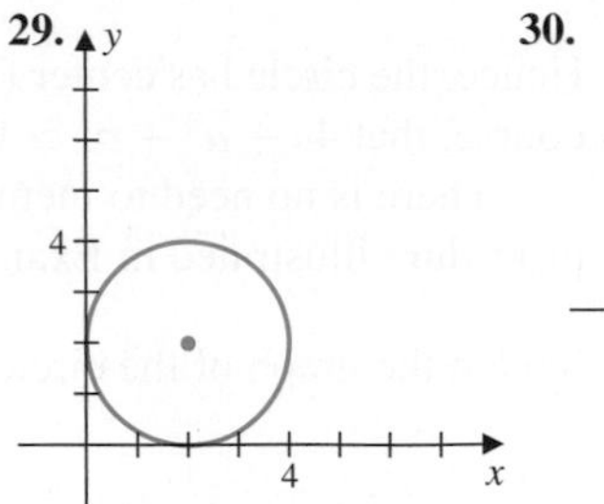

30.

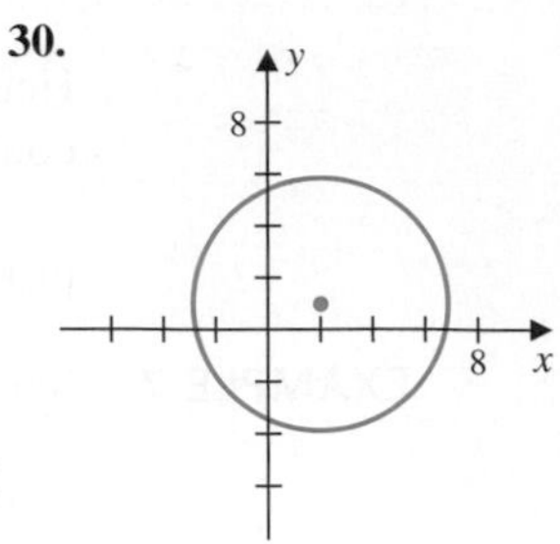

*In Exercises 31–36, **(a)** find the center and radius of each circle, and **(b)** sketch its graph.*

31. $x^2 + y^2 = 9$
32. $x^2 + y^2 = 2$
33. $x^2 + (y - 1)^2 = 1$
34. $(x + 1)^2 + y^2 = 9$
35. $(x - 2)^2 + (y + 1)^2 = 9$
36. $(x - 1)^2 + (y + 2)^2 = 16$

In Exercises 37–42, complete the square on the x and y terms to find the center and radius of the circle.

37. $x^2 - 2x + y^2 = 3$
38. $x^2 + y^2 + 4y = -3$
39. $x^2 + 2x + y^2 - 4y = -4$
40. $x^2 - 2x + y^2 + 4y = 4$
41. $x^2 - 4x + y^2 - 2y - 4 = 0$
42. $x^2 + 4x + y^2 + 6y + 9 = 0$

In Exercises 43–48, sketch the region in the xy-plane.

43. $\{(x, y) \mid x^2 + y^2 \leq 1\}$
44. $\{(x, y) \mid (x - 1)^2 + y^2 > 2\}$
45. $\{(x, y) \mid 1 < x^2 + y^2 < 4\}$
46. $\{(x, y) \mid 4 \leq (x - 1)^2 + (y - 1)^2 \leq 9\}$
47. $\{(x, y) \mid x^2 + y^2 \leq 4 \text{ and } y \geq x\}$
48. $\{(x, y) \mid x^2 + y^2 \geq 4 \text{ and } y \leq x\}$
49. Which of the points $(-7, 2)$ and $(6, 3)$ is closer to the origin?
50. Which of the points $(1, 6)$ and $(7, 4)$ is closer to $(3, 2)$?
51. **a.** Find the distances between the points $(-1, 4)$, $(-3, -4)$, and $(2, -1)$, and show that they are vertices of a right triangle.

    b. At which vertex is the right angle?

52. Show that the points $(2, 1)$, $(-1, 2)$, and $(2, 6)$ are vertices of an isosceles triangle.

53. Find a fourth point that will form the vertices of a rectangle when added to the points in Exercise 51. Is the point unique?

54. Find a fourth point that will form the vertices of a parallelogram when added to the points in Exercise 52. Is the point unique?

55. Find an equation of the circle with center $(0, 0)$ that passes through $(2, 3)$.

56. Find an equation of the circle with center $(1, 3)$ that passes through $(-2, 4)$.

57. Find an equation of the circle with center $(3, 7)$ that is tangent to the y-axis.

58. Find a point on the y-axis that is equidistant from the points $(2, 1)$ and $(4, -3)$.

59. Find an equation of the circle whose center lies in the second quadrant, that has radius 3, and that is tangent to both the x-axis and the y-axis.

60. Find an equation for the circle whose center lies in the first quadrant, that has radius 2, and that is tangent to the lines $x = 1$ and $y = 2$.

61. Find the area of the region that lies outside the circle $x^2 + y^2 = 1$ and inside the circle $x^2 + y^2 = 9$. (*Note*: Area formulas are on the back inside cover.)

62. Find the area of the region that lies outside the circle $(x - 1)^2 + y^2 = 1$ and inside the circle $(x - 2)^2 + y^2 = 4$.

63. Indicate on an xy-plane those points for which $|x| + |y| \leq 4$.

64. Indicate on an xy-plane those points for which $|x - 1| + |y + 2| \leq 2$.

65. Find the area of the region containing the points satisfying $|x| + |y| \geq 1$ and $x^2 + y^2 \leq 1$.

66. Find the area of the region containing the points satisfying $|x - 1| + |y + 2| \geq 2$ and $(x - 1)^2 + (y + 2)^2 \leq 4$.

67. Towns A and B are connected by two roads that intersect at a right angle, as shown in the figure. The highway department needs to repair the roads and has decided instead to construct a new road that connects the two towns directly. The cost of eliminating the two existing sections of road is estimated at \$50,000 and the cost of constructing the new road is \$200,000 per mile. Estimate the cost of the project.

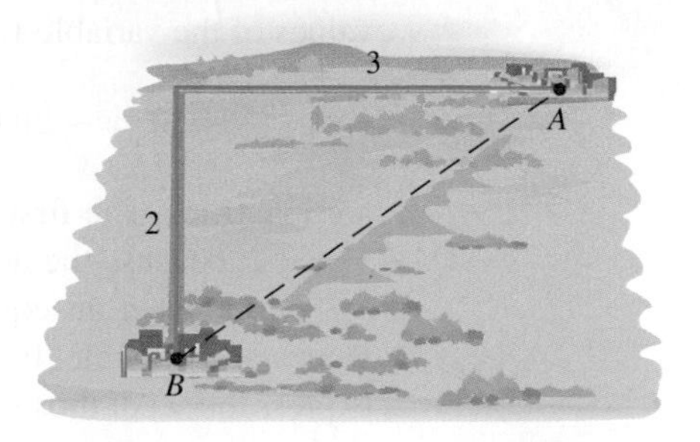

68. Two ships leave port 2 hours apart. Ship A leaves first and travels due east at 15 miles per hour and ship B leaves 2 hours later due south at 8 miles per hour. When ship A is 5 hours out of port, estimate the distance between the two ships.

69. A gas pipeline has to be constructed between points A and B, as shown in the figure. Roads connect points A and C and points C and B and the line can be buried alongside the road at \$200,000 per mile. The pipe can also be buried directly between A and B at a cost of \$150,000 per mile, but the construction must avoid the swamp shown in the figure. The decision has been made to bury the pipe along the road for 3 miles and then run it directly to point B. Estimate the savings using this alternate route. Estimate the cost per mile to bury the line off-road that would make the route along the entire existing road more economical.

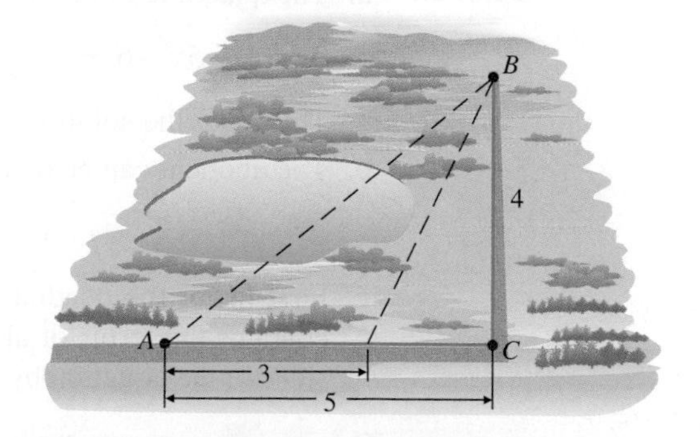

1.4 EQUATIONS AND GRAPHS

We use an **equation** to indicate that two mathematical expressions are equivalent. Sometimes the equation is an **identity**, which means that the equation is true for all values of the variable for which the equation is defined. For example, the equations

$$|x^2 - 4| = |x - 2||x + 2|, \quad |x| = \sqrt{x^2}, \quad \text{and} \quad \frac{1}{x^2 - 4} = \frac{1}{(x - 2)(x + 2)}$$

are identities. The first two hold for all real numbers x. The third holds only when $x \neq \pm 2$, because the denominator must be nonzero for the expression to be defined.

More often an equation is **conditional**, which means that it is true for some values of a variable, but not all. For conditional equations we need to determine and express the set of those values of the variable for which the equation is true. Equations such as

$$3x + 2 = 5, \quad x^2 - 3x + 2 = 0, \quad \text{and} \quad x^2 = -1$$

are conditional. The first has the single solution $x = 1$. The second is solved by factoring

$$0 = x^2 - 3x + 2 = (x - 1)(x - 2)$$

to give the solutions $x = 1$ and $x = 2$. The third conditional equation, $x^2 = -1$, has no real number solutions, because the right side is negative and the left side cannot be negative.

EXAMPLE 1 Determine all values of x that satisfy the following conditional equations.

a. $x^2 - 5x - 6 = 0$

b. $\dfrac{x - 3}{x^2 + x + 1} = 0$

c. $\dfrac{1}{x + 1} + 1 = \dfrac{1}{x^2 - x - 2}$

d. $\sqrt{x - 6} + \sqrt{x - 1} = 5$

Solution **a.** The quadratic in the equation $x^2 - 5x - 6 = 0$ is first factored to give

$$0 = x^2 - 5x - 6 = (x + 1)(x - 6), \quad \text{which implies} \quad x + 1 = 0 \quad \text{or} \quad x - 6 = 0.$$

This gives the solutions $x = -1$ and $x = 6$.

b. The quotient can be 0 only when the numerator $x - 3$ is 0. In this case

$$x - 3 = 0 \quad \text{and the solution is} \quad x = 3.$$

c. First factor the quadratic as $x^2 - x - 2 = (x - 2)(x + 1)$. The factored form contains the terms of all the denominators in the equation, so we multiply both sides of the equation by $(x - 2)(x + 1)$ to produce

$$(x - 2)(x + 1)\left(\frac{1}{x + 1} + 1\right) = 1.$$

This simplifies to

$$x - 2 + x^2 - x - 2 = 1, \quad \text{so} \quad x^2 = 5 \quad \text{and} \quad x = \pm\sqrt{5}.$$

d. Rewriting the equation with one radical on each side gives

$$\sqrt{x-6} = 5 - \sqrt{x-1}.$$

Squaring both sides produces

$$x - 6 = 25 - 10\sqrt{x-1} + x - 1, \quad \text{so} \quad 10\sqrt{x-1} = 30,$$

and $\sqrt{x-1} = 3$. Squaring both sides once more gives

$$x - 1 = 9, \quad \text{so} \quad x = 10.$$

Notice that $x = 10$ is indeed a solution since $\sqrt{10-6} = \sqrt{4} = 2$, and that $5 - \sqrt{10-1} = 5 - 3 = 2$. ■

We needed to verify that the value of x obtained in part (d) of Example 1 was truly a solution, because the squaring operations could have produced solutions to the final equation that were not solutions to the original equation. These are called *extraneous* solutions. For example, suppose that the equation in part (d) was instead

$$\sqrt{6-x} - \sqrt{1-x} = -5,$$

and that the operations similar to those in the example were performed. Square both sides of

$$\sqrt{6-x} = \sqrt{1-x} - 5 \quad \text{to obtain} \quad 6 - x = 1 - x - 10\sqrt{1-x} + 25,$$

which simplifies to $\sqrt{1-x} = 2$. Squaring again produces

$$1 - x = 4, \quad \text{so} \quad x = -3.$$

However, when we substitute this value back into the original equation we find that it is not actually a solution because

$$\sqrt{6-(-3)} - \sqrt{1-(-3)} = \sqrt{9} - \sqrt{4} = 3 - 2 \neq 5.$$

As a consequence, there are no real numbers that satisfy the original equation $\sqrt{6-x} - \sqrt{1-x} = -5$.

Conditional equations might involve more than one variable, but the objective is the same. In the case of two variables, the objective is to determine which collection of ordered pairs satisfies the equation and gives a representation of the solution. For example, the conditional equation in the two variables x and y given as $y = 2x + 1$ has as its solutions those pairs of real numbers of the form $(x, 2x + 1)$, where x can be any real number. A few of the solutions to this equation, then, are $(0, 1)$, $(2, 5)$, and $(-3, -5)$.

Graphs of Equations

Graphing can add to your understanding of a problem. Draw a picture as the first step to solving a problem.

The **graph of an equation** in the variables x and y consists of the set of points (x, y) in the xy-plane whose coordinates satisfy the equation.

EXAMPLE 2 Sketch the graph of the equation $2x + 3y = 6$.

Solution One way to obtain a rough sketch of the graph of an equation is to first find the coordinates of several points, as we have done in Table 1. Then plot the points in the

TABLE 1

x	y
0	2
1	$\frac{4}{3}$
2	$\frac{2}{3}$
3	0
−1	$\frac{8}{3}$

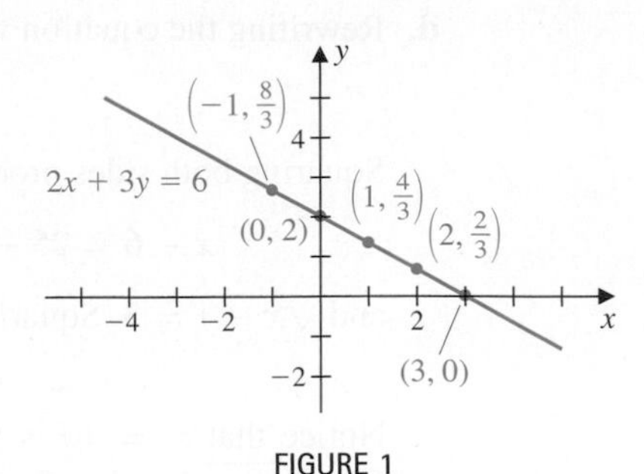

FIGURE 1

coordinate plane and connect those points, as best you can, with a curve. Figure 1 indicates that this process produces a straight line for the graph of the equation $2x + 3y = 6$. In Section 1.7 we will see that all equations of the form $Ax + By = C$ have graphs that are straight lines. ■

The points (0, 2) and (3, 0) shown on the graph of the line in Example 2 are the **axis intercepts** of the graph. They are found by setting one of the variables to zero and solving for the other. The ***x*-intercept** of the graph of $2x + 3y = 6$ occurs when $y = 0$. Hence

$$x\text{-intercept:} \qquad 2x = 6 \quad \text{so} \quad x = 3.$$

The ***y*-intercept** of the graph of $2x + 3y = 6$ occurs when $x = 0$. Hence

$$y\text{-intercept:} \qquad 3y = 6 \quad \text{so} \quad y = 2.$$

In general, all points of the form $(0, y)$ that satisfy an equation are called the y-intercepts of its graph, and the points of the form $(x, 0)$ that satisfy an equation are called the x-intercepts. When these points can be easily determined, they should be plotted on the graph.

EXAMPLE 3 Sketch the graph of the equation

$$y = \frac{x^2 + x - 2}{x - 1}.$$

Solution First notice that the numerator of the fraction can be factored to give

$$y = \frac{x^2 + x - 2}{x - 1} = \frac{(x - 1)(x + 2)}{x - 1} = x + 2, \quad \text{provided that } x \neq 1.$$

This means that the graph of

$$y = \frac{x^2 + x - 2}{x - 1}$$

is the same as the graph of $y = x + 2$, except it is not defined when $x = 1$.

As in Example 2, we plot several points on the graph of the equation and connect the points with a curve. The x-intercept occurs when $x = -2$, and the y-intercept occurs when $y = 2$. The result is the straight line shown in Figure 2.

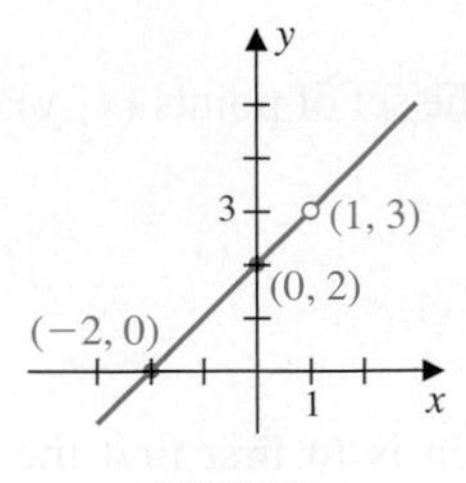

FIGURE 2

The open circle at (1, 3) indicates that this point is missing from the graph. ■

EXAMPLE 4 Consider the graphs of the following equations.

a. $y = x^2$ **b.** $x = y^2$ **c.** $x^2 + y^2 = 1$

Solution **a.** The graph of $y = x^2$ is the parabola shown in Figure 3(a). This graph was obtained by plotting representative points that satisfy the equation and then connecting the points with a smooth curve.

b. This equation is the same as the equation in part (a) except that the roles of the two variables are interchanged. So the graph of $x = y^2$ is the parabola shown in Figure 3(b).

c. We know from Section 1.3 that the graph of the equation $x^2 + y^2 = 1$ is the unit circle shown in Figure 3(c). ■

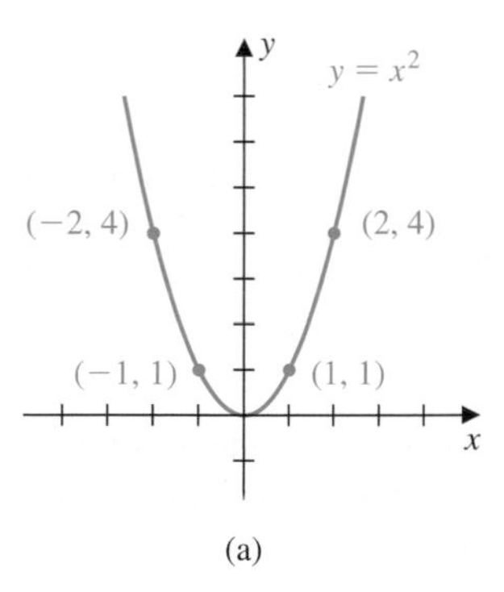

(a)

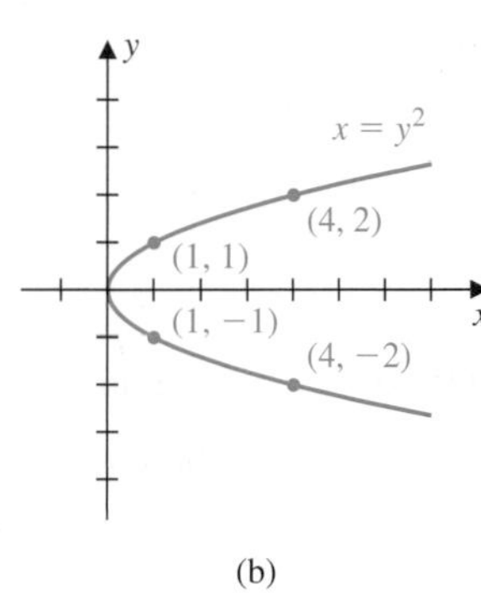

(b)

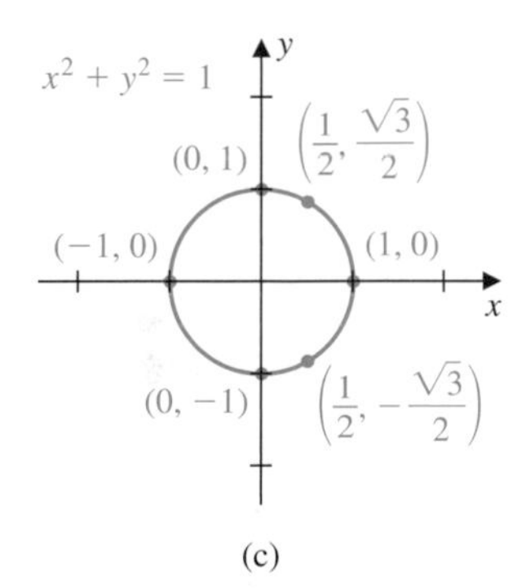

(c)

FIGURE 3

EXAMPLE 5 Use the graph of $x = y^2$ shown in Figure 3(b) to sketch the graph of $y = \sqrt{x}$.

Solution Solving the equation $x = y^2$ for y gives two solutions

$$y = \sqrt{x} \quad \text{and} \quad y = -\sqrt{x}.$$

The graph of $y = \sqrt{x}$ is the part of the graph of $x = y^2$ shown in Figure 3(b) that lies above the x-axis. Recall that the square root is defined only for positive values of x, and that when $x > 0$ the value $\sqrt{x}$ is the positive square root of x. The graph is shown in Figure 4. ■

y
3
$y = \sqrt{x}$
6 x

FIGURE 4

Symmetry of a Graph

The graphs of the equations in Example 4 illustrate a feature known as *symmetry to a line*. A graph is **symmetric to a line** when the portion of the graph on one side of the line is the mirror image of the portion on the other side. Symmetry of a graph to a line is easy to determine when the line is one of the coordinate axes.

Coordinate Axis Symmetry of a Graph (see Figure 5)

Axis symmetry is also called *symmetry with respect to the axis.*

- The graph of an equation has **y-axis symmetry** if $(-x, y)$ is on the graph whenever (x, y) is on the graph.
- The graph of an equation has **x-axis symmetry** if $(x, -y)$ is on the graph whenever (x, y) is on the graph.

The graph of $y = x^2$ in Figure 3(a) has y-axis symmetry because $(-x)^2 = x^2$. The graph in Figure 3(b) of $x = y^2$ has x-axis symmetry because $(-y)^2 = y^2$. The graph of the circle with equation $x^2 + y^2 = 1$ in Figure 3(c) has both x-axis and y-axis symmetry.

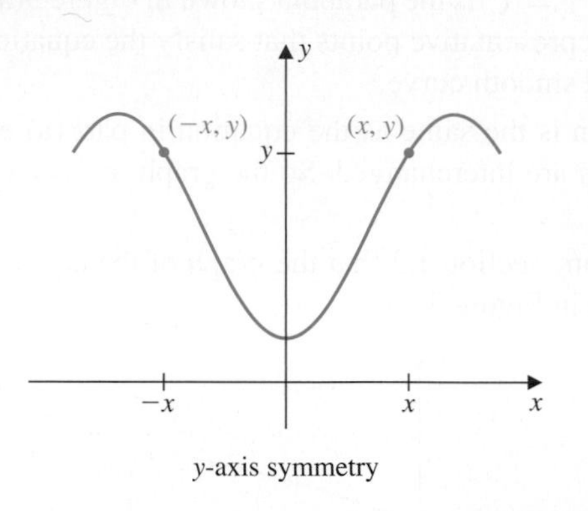

y-axis symmetry

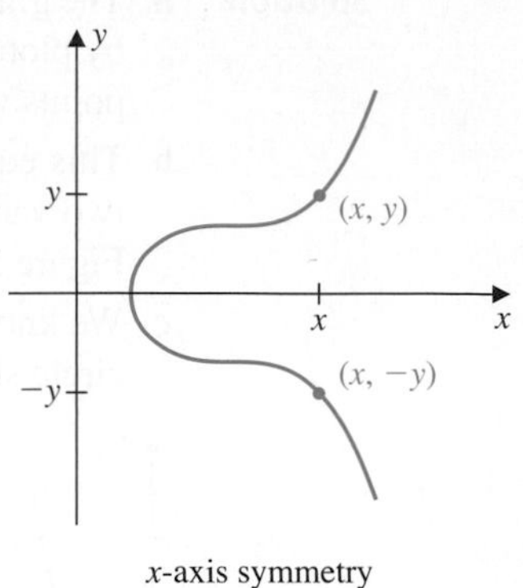

x-axis symmetry

FIGURE 5

EXAMPLE 6 Use symmetry to sketch the graph of $y = |x|$.

Solution When $x > 0$, the graph is the same as $y = x$. In addition, the absolute value satisfies $|-x| = |x|$. This means that $(-x, y)$ is on the graph whenever (x, y) is on the graph, and the graph has the y-axis symmetry shown in Figure 6. ■

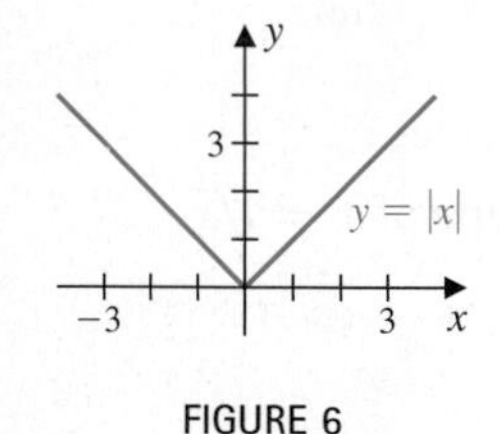

FIGURE 6

Symmetry is also defined *with respect to a point* in the plane. In this case, the graph has a mirror reflection property with respect to the point. This feature might be difficult to detect for arbitrary points in the plane, but it is easy when the point is the origin. A graph having this symmetry is shown in Figure 7. The unit circle shown in Figure 3(c) also has this property.

Origin Symmetry of a Graph

The graph of an equation has **origin symmetry** if $(-x, -y)$ is on the graph whenever (x, y) is on the graph.

Origin symmetry is also called *symmetry with respect to the origin.*

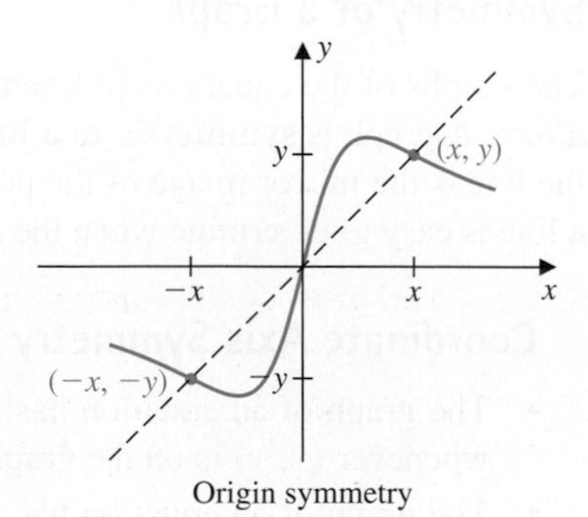

Origin symmetry

FIGURE 7

EXAMPLE 7 Determine any symmetry properties of the *cubing* function $y = x^3$.

Solution Suppose (x, y) satisfies the equation $y = x^3$. Since $(-x)^3 = -x^3 = -y$, the point $(-x, -y)$ also satisfies the equation, and the graph has origin symmetry. For example, the point $(2, 8)$ is on the graph of $y = x^3$, as is $(-2, -8)$ since $(-2)^3 = -8$.

Figure 8 shows a graph of $y = x^3$ along with several points on the graph. Later in this chapter we will introduce more aids to graphing. You will learn other valuable techniques when studying calculus. ■

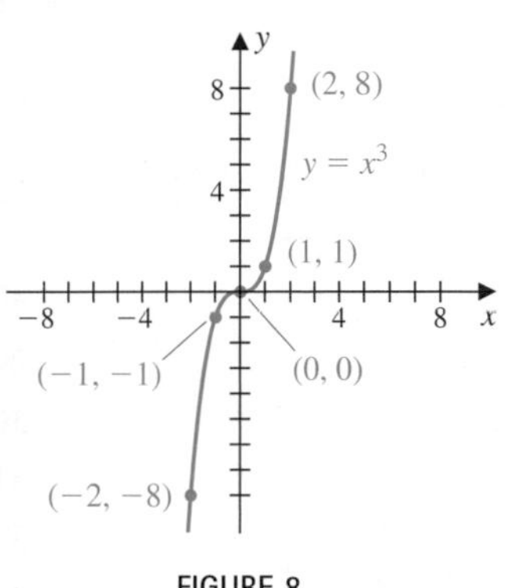

FIGURE 8

Applications

EXAMPLE 8 A homeowner wants to create a garden surrounded by a brick border contained in a plot with an area of 384 square feet. The garden is to be rectangular with a length twice its width. The brick border is 4 feet wide on all sides. **(a)** Find the dimensions of the garden, and **(b)** estimate the number of 10-inch by 4-inch bricks required for the border.

Solution

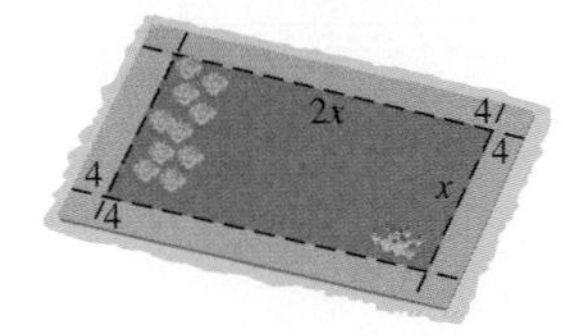

FIGURE 9

a. The situation is shown in Figure 9, where x denotes the width of the garden. The length of the garden is $2x$ and the border is 4 feet, so the area, in square feet, of the garden with its brick border is

$$(x + 8)(2x + 8) = 2x^2 + 24x + 64.$$

This is specified to be 384, so

$$2x^2 + 24x + 64 = 384, \quad \text{which simplifies to} \quad x^2 + 12x - 160 = 0.$$

Factoring the quadratic equation $x^2 + 12x - 160$ gives

$$0 = x^2 + 12x - 160 = (x - 8)(x + 20), \quad \text{so} \quad x = 8 \quad \text{or} \quad x = -20.$$

Since x must be positive, the garden has width 8 feet and length $2 \cdot 8 = 16$ feet.

b. The area of the garden is $8 \cdot 16 = 128$ square feet, and the garden together with its border has area 384 square feet, so the area of the walkway is $384 - 128 = 256$ square feet. Each 10-inch by 4-inch brick has a face area of

$$10 \text{ inches} \left(\frac{1 \text{ foot}}{12 \text{ inches}}\right) \cdot 4 \text{ inches} \left(\frac{1 \text{ foot}}{12 \text{ inches}}\right) = \frac{5}{18} \text{ square feet.}$$

So the required number of bricks is

$$\frac{256}{5/18} = \frac{256 \cdot 18}{5} \approx 922. \quad \blacksquare$$

EXERCISE SET 1.4

In Exercises 1–8, specify any axis or origin symmetry of the graphs.

1.

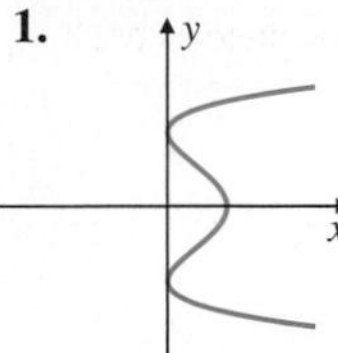

2.

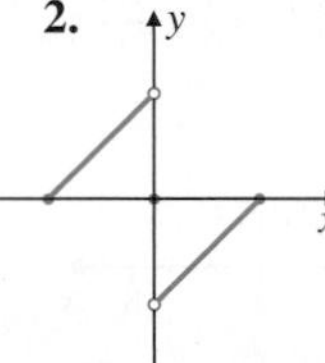

3.

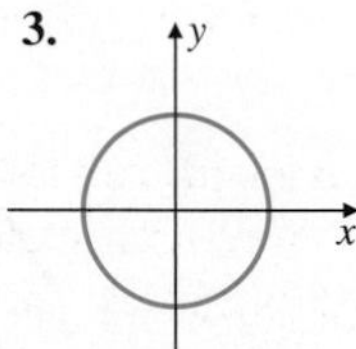

4.

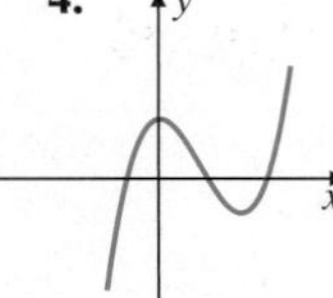

5.

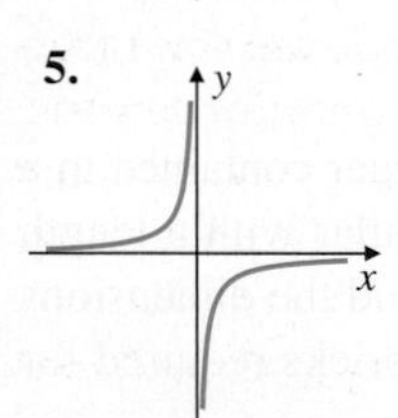

6.

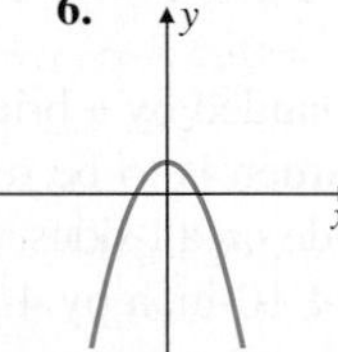

7.

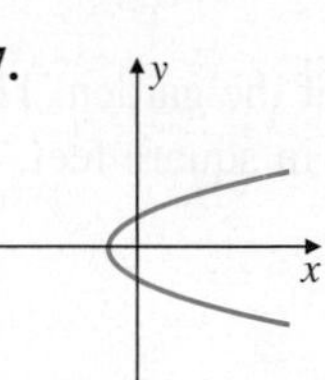

8.

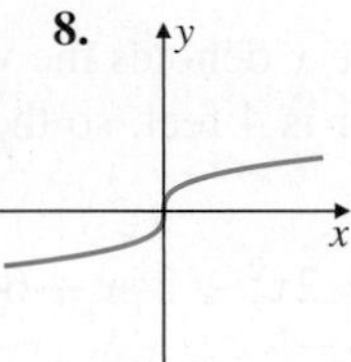

In Exercises 9–38, determine any axis intercepts and describe any axis or origin symmetry.

9. $y = x + 3$

10. $y = 2x - 3$

11. $x + y = 1$

12. $-2x - y = 2$

13. $y = x^2 - 3$

14. $y = x^2 + 2$

15. $y = 1 - x^2$

16. $2y = x^2$

17. $y = -2x^2$

18. $y = 3 - 3x^2$

19. $x = y^2 - 1$

20. $x = y^2 - 4$

21. $y = x^3 + 1$

22. $3y = x^3$

23. $y = -x^3$

24. $y = -x^3 + 1$

25. $y = \dfrac{(x+3)(x-3)}{x-3}$

26. $y = \dfrac{(x+3)(x+1)}{x+1}$

27. $y = \dfrac{x^2 - x - 6}{x+2}$

28. $y = \dfrac{x^2 + 2x - 3}{x+3}$

29. $y = \sqrt{x} + 2$

30. $y = \sqrt{x-1}$

31. $x^2 + y^2 = 4$

32. $(x-1)^2 + y^2 = 1$

33. $y = \sqrt{9 - x^2}$

34. $y = -\sqrt{9 - x^2}$

35. $y = |x|$

36. $y = |x - 1|$

37. $y = |x| - 1$

38. $y = 2 - |x|$

In Exercises 39–42, complete the graph in the figure so the curve has the specified symmetry.

39. x-axis symmetry

40. y-axis symmetry

41. origin symmetry

42. x-axis and y-axis symmetry

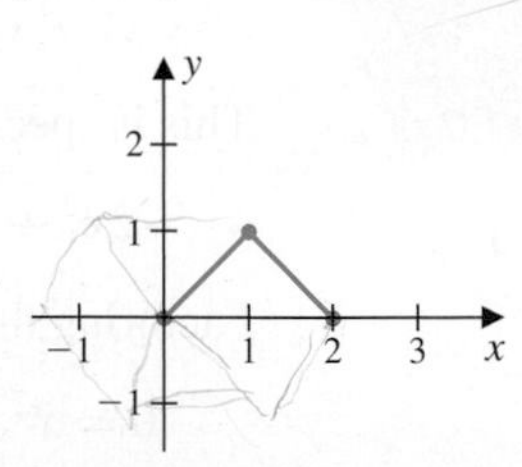

43. Find the distance between the points of intersection of the graphs $y = x^2 + 1$ and $y = 2$.

44. Find the distance between the points of intersection of the graphs $y = x^2 - 3$ and $y = x + 3$.

45. Determine three consecutive positive integers whose sum is 156.

46. Determine two consecutive positive integers whose squares sum to 925.

47. Find the dimensions of the rectangle whose area is 12 and whose perimeter is 14.

48. Find the area of a square whose side length is one fourth the value of its area.

49. A automobile radiator has 10 quarts of water with a 10% concentration of antifreeze. How much of the liquid in the radiator should be drained and replaced with pure antifreeze to ensure that the radiator contains water that has a 30% concentration of antifreeze?

50. Metal alloy A contains 20% copper, and metal alloy B contains 45% copper. How many pounds of each alloy should be combined to produce 100 pounds of a new alloy that contains 35% copper?

51. Symmetry was discussed in the text with respect to the x-axis, the y-axis, and the origin. Show that if a graph has x-axis and y-axis symmetry, it must also have origin symmetry.

52. You are contemplating investing $10,000 between two bond funds. One fund is less risky, and expected to return 5% annually. The riskier fund is expected to return 8% annually. If you would like an overall return of 6%, how much should you place in each fund?

1.5 USING TECHNOLOGY TO GRAPH EQUATIONS

Calculators with extensive graphing capabilities cost no more than a standard textbook and are easily worth their price. In addition, powerful computer algebra systems such as Maple, Mathematica, Mathcad, and Derive are available on most campuses, and all these contain sophisticated graphing techniques. Technology is becoming more advanced and cheaper, and portions of these computer algebra systems are now incorporated into calculators and even into phones. In this book we take a generic approach, indicating where technology can be useful without giving instructions specific to any particular device. Technology is not necessary for an understanding of the material in this book, but it will be helpful if you use it intelligently.

Graphing calculators and computer algebra systems sketch graphs of equations quickly by plotting as many points on the graph as the resolution of the screen will permit. When using a graphing device to plot an equation, you must be careful to ensure that all the interesting aspects of the graph have been displayed. Using technology without an understanding of the underlying concepts can result in accepting misleading information. This is particularly true in the case of plotting curves.

In calculus you will need a familiarity with the graphs of a library of functions (see the inside front cover). Technology can help in visualizing curves but an understanding of the concepts underlying how graphs are constructed is essential.

The plots in this section and throughout this book were generated using a computer algebra system. A rectangular portion of the plane is called a **viewing rectangle** for a plot and is defined by specifying the range of values for x and the range of values for y. We denote a viewing rectangle specified by the inequalities

$$a \le x \le b \quad \text{and} \quad c \le y \le d \qquad \text{as } [a, b] \times [c, d].$$

Choosing an appropriate viewing rectangle is essential for obtaining a representative plot of an equation, and understanding the concepts of graphing is essential for recognizing when you have a good representation. The following examples show how to use technology to plot curves and how to choose an appropriate viewing rectangle to maximize the information.

EXAMPLE 1 Sketch the graph of the equation $y = x^2 + 5$ in the following viewing rectangles.

a. $[-3, 3] \times [-3, 3]$ **b.** $[-6, 6] \times [-6, 6]$

c. $[-10, 10] \times [-3, 30]$ **d.** $[-50, 50] \times [0, 1000]$

Solution The graphs for these viewing rectangles are shown in Figure 1. In part (a) the viewing rectangle does not contain any portion of the graph, because $x^2 \geq 0$ implies that $x^2 + 5 \geq 5$, and the range of y-values lies outside the viewing rectangle. We see better representations of the graph in parts (b), (c), and (d).

A good viewing rectangle is needed to give an accurate graph. Modify your initial viewing window to refine the view.

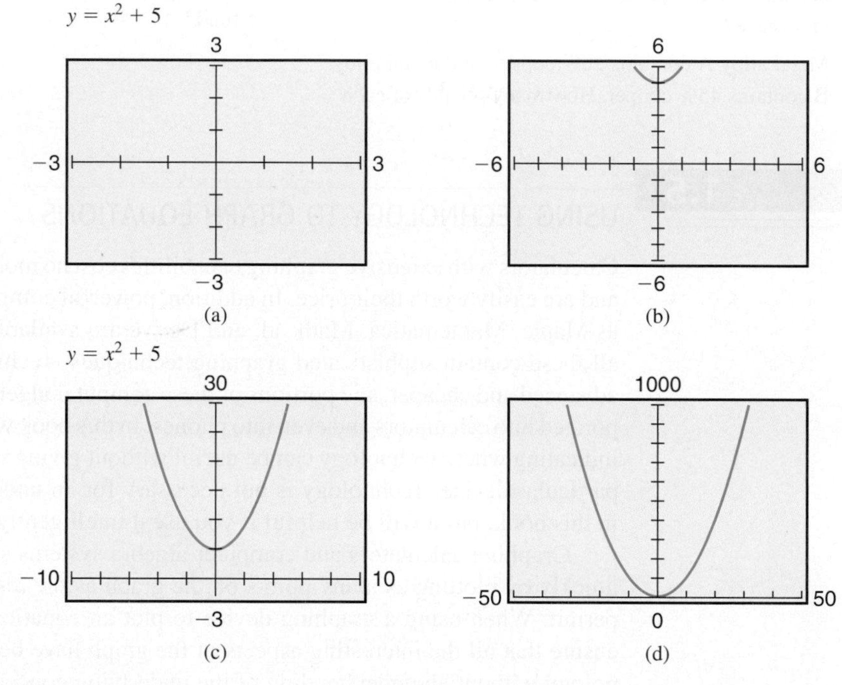

FIGURE 1

Pictures that are more complete appear in (c) and (d), although in part (d) the y-scale is so large that it appears the graph passes through the origin instead of through $(0, 5)$. ■

EXAMPLE 2 Sketch the graph of the equation $y = x^3 - 25x$.

Solution We begin by experimenting with viewing rectangles. In Figure 2(a), we chose a viewing rectangle $[-5, 5] \times [-5, 5]$, but we strongly suspect that the graph does not consist of the several straight lines that appear in this plot. To obtain a more representative plot, we need to extend the range of points to reveal the graph outside this viewing rectangle. In parts (b) and (c), we have used viewing rectangles $[-10, 10] \times [-25, 25]$, and $[-10, 10] \times [-50, 50]$. With some confidence we accept the plot in (c) as representative of the true curve. It appears from this figure that the graph has a local low point near $x = 3$ and a local high point near $x = -3$.

Finding where a graph is locally high or locally low is one of the first applications in calculus.

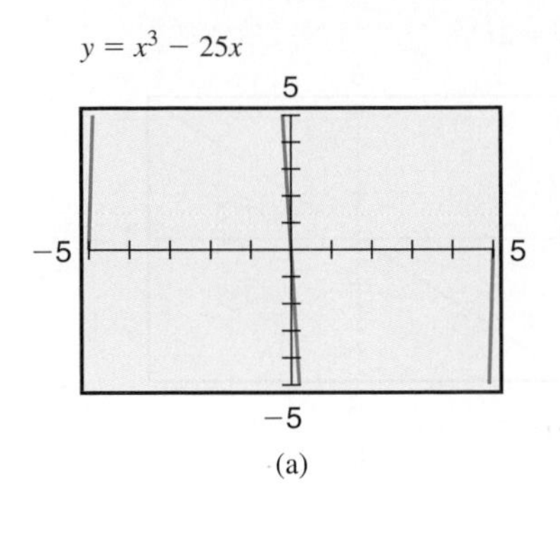

(a)

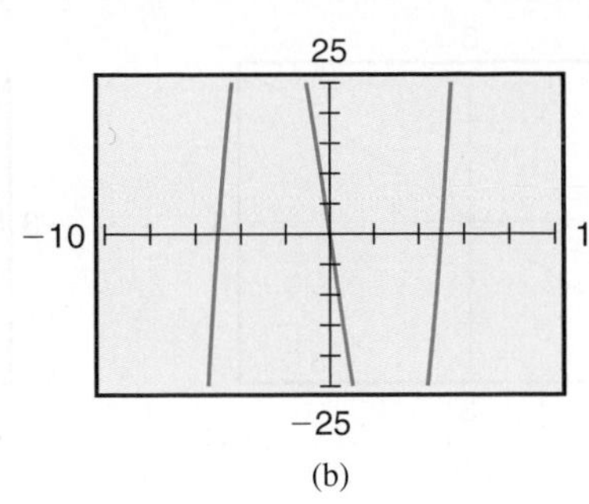

(b)

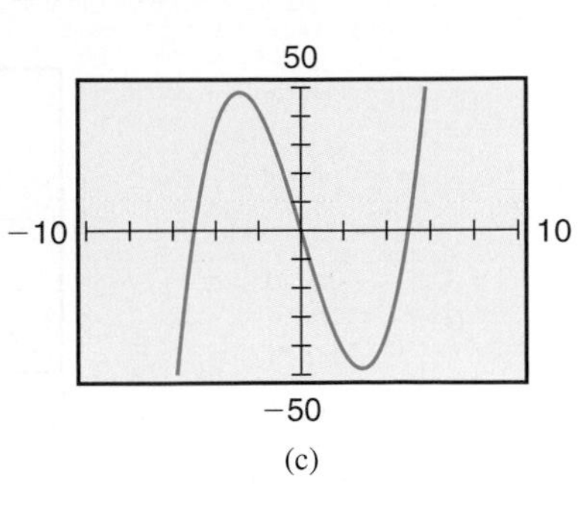

(c)

FIGURE 2

In Chapter 3 we will study these specific types of equations and will see that the plot given here does reveal the important features of the curve. You will completely analyze curves of this type when you study calculus. ■

EXAMPLE 3 Sketch the graph of $y = kx$ for $k = \pm 1, \pm 2, \pm 3, \pm\frac{1}{2}, \pm\frac{1}{3}$, and discuss the effect of the constant k on the graph.

Solution Figure 3(a) shows the plots for some positive values of k, and Figure 3(b) shows the plots for negative values of k. The figures indicate that the graph of the equation $y = kx$ is always a straight line that passes through the origin. For positive k, the graphs rise, or increase, from left to right, and for negative k, they fall, or decrease, from left to right. The larger the magnitude of k, the steeper the incline or decline. ■

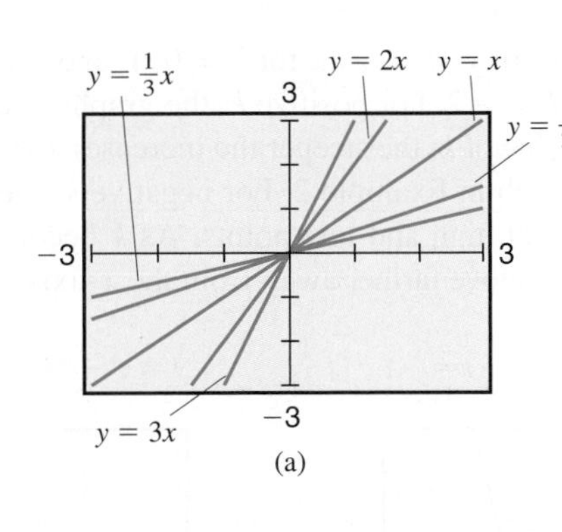

(a)

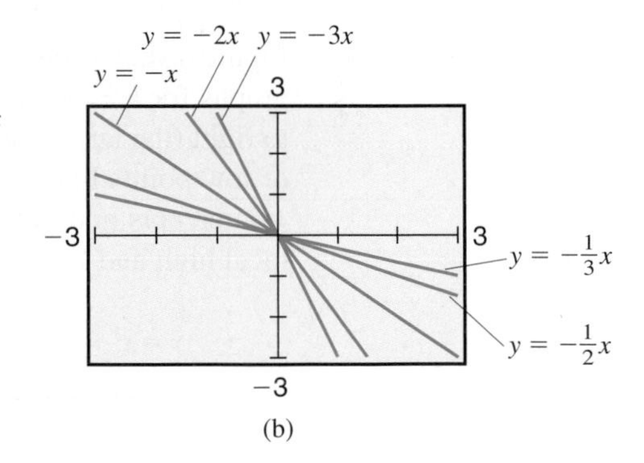

(b)

FIGURE 3

EXAMPLE 4 Sketch the graph of $y = x$ in the viewing rectangle $[-3, 3] \times [-3, 3]$, and then sketch the graph of $y = 2x$ in the viewing rectangle $[-3, 3] \times [-6, 6]$.

Solution The graphs are shown in Figures 4(a) and 4(b). It appears that the two graphs are the same, but we know this is not the case since the graph of $y = 2x$ increases twice as fast as the graph of $y = x$. The visual inconsistency occurs because the scale on the y-axis is twice as large for the plot of $y = 2x$ as it is for the plot of $y = x$. This makes the plots look the same, even though they are not. ■

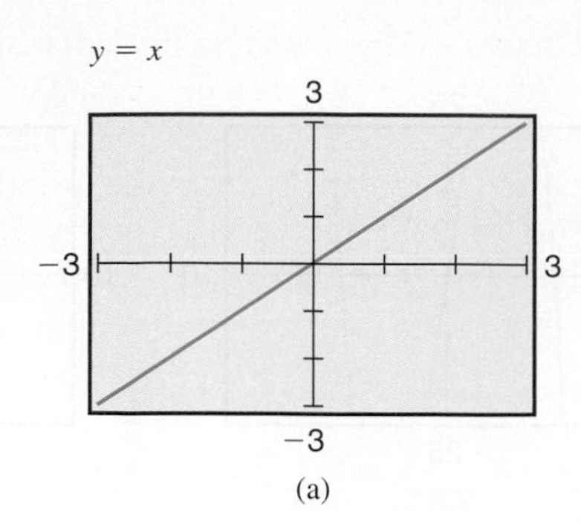

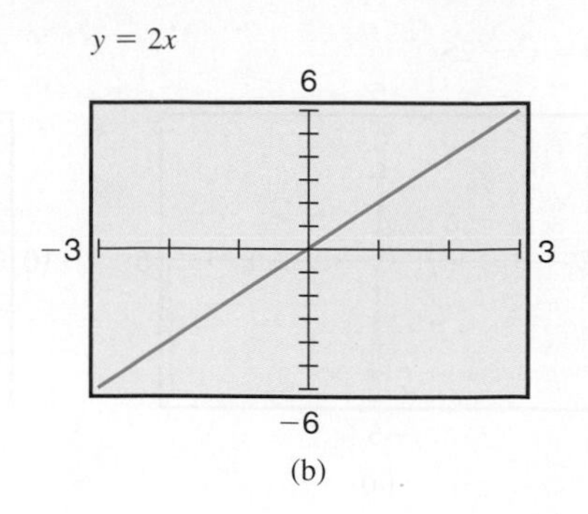

FIGURE 4

EXAMPLE 5 Sketch the graph of $y = x^3 + kx$ for several values of k and describe the effect of k on the curve.

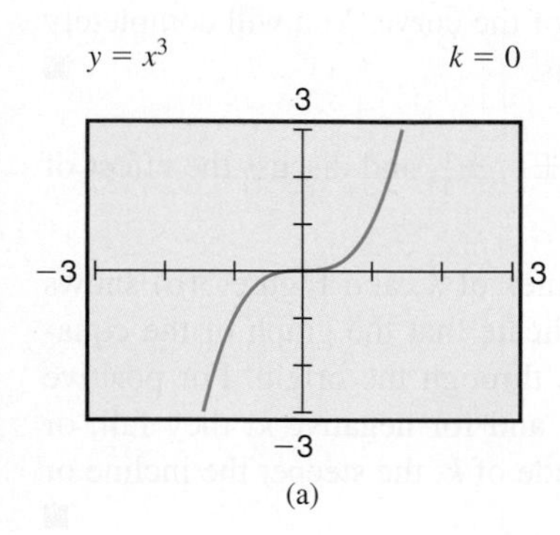

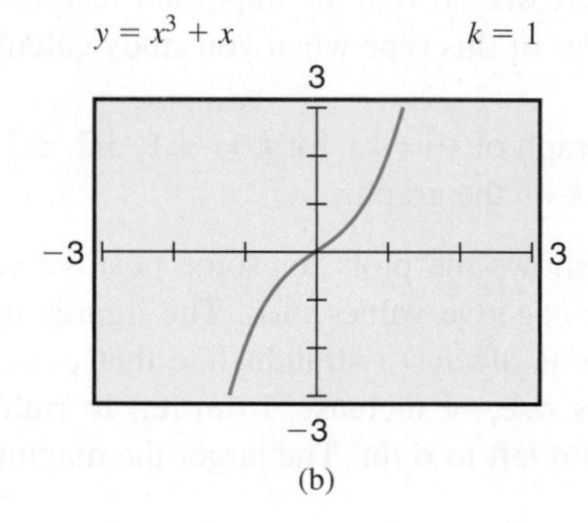

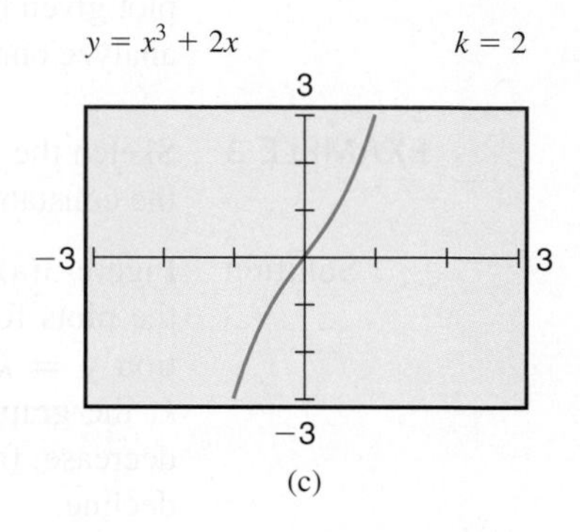

FIGURE 5

Solution Figure 5 shows the graphs of $y = x^3 + kx$ for $k = 0$, 1, and 2, and Figure 6 shows the graphs for $k = -1$ and $k = -2$. For positive k, the graphs rise continually from left to right (the larger the value of k, the steeper the increase) and do not have local high or low points like the graph in Example 2. For negative k, the graphs have nonzero x-intercepts and both local high and low points. As k becomes more negative, the local high and low points move farther away from the x-axis. ■

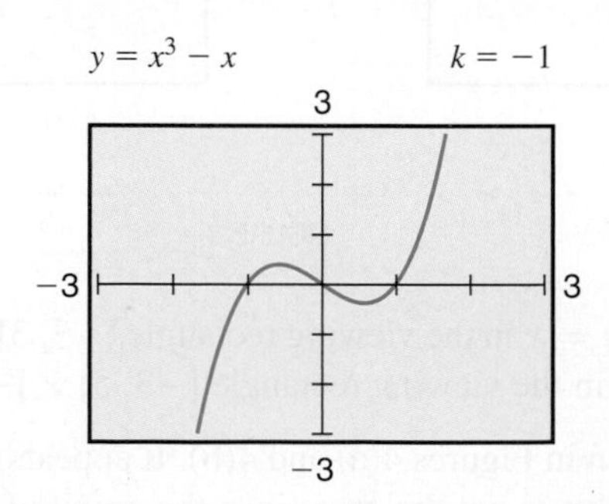

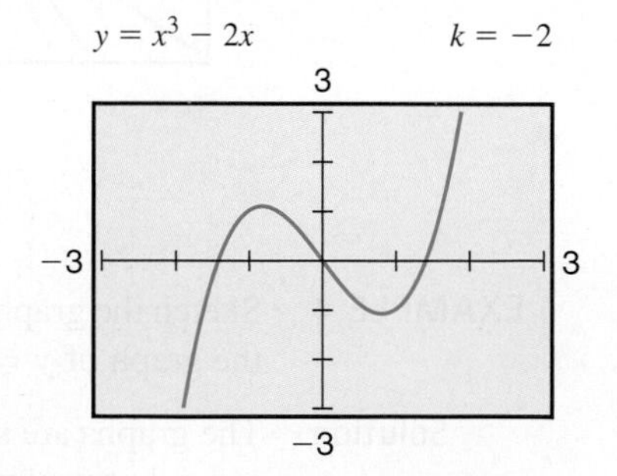

FIGURE 6

EXAMPLE 6 Sketch the graph of

$$y = \frac{x+2}{x-1}.$$

Solution Figure 7 shows three views of the graph of the equation using different viewing rectangles. In Figure 7(a), it is clear the viewing rectangle is not sufficient to show all the interesting parts of the graph. Parts (b) and (c) give successively better views. The graph appears to be broken near $x = 1$, where the original equation is not defined. However, the vertical lines in (b) and (c) are not part of the graph.

Always be skeptical when using a graphing device. Simply plotting points and connecting them may lead to incorrect graphs. In this case, Figure 7(c) appears to be the best at representing the graph, even though it shows a vertical line through $x = 1$ that is not a part of the graph. ■

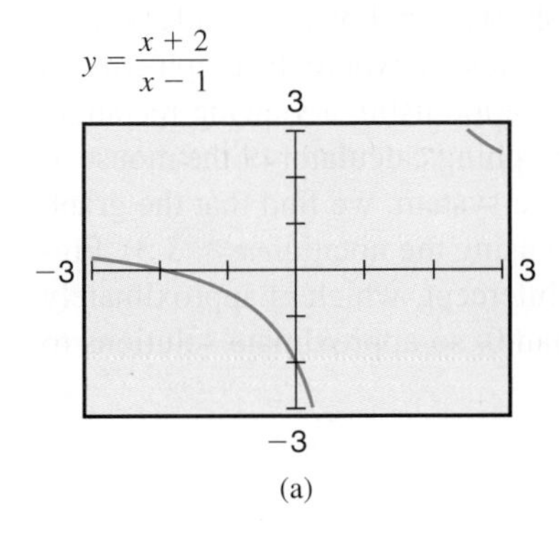

(a)

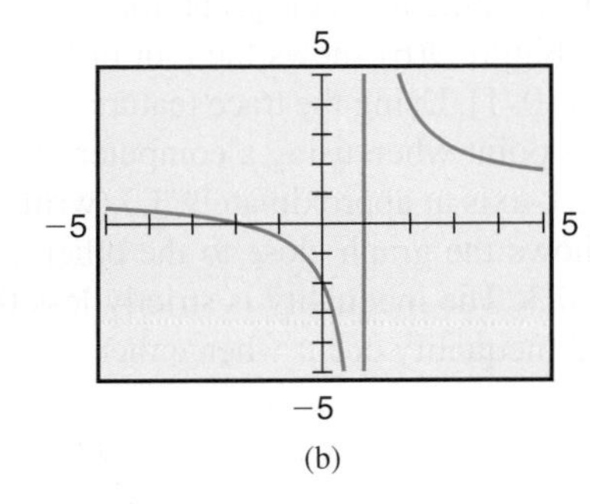

(b)

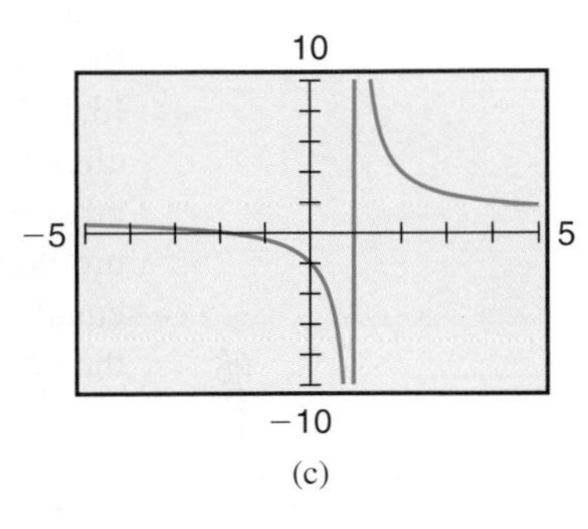

(c)

FIGURE 7

EXAMPLE 7 Graphically solve each of the inequalities.

a. $x^2 + x - 2 \geq 0$ **b.** $x^2 + 2 \geq 0$ **c.** $-x^2 + 3x + 1 < 0$

Solution To solve an inequality graphically, we plot a good representative for the curve determined by the inequality and then locate, from the graph, the x-intercepts. This allows us to find the intervals of x for which the inequality is satisfied by observing where the graph lies above or below the x-axis.

a. The graph of the equation $y = x^2 + x - 2$ is shown in Figure 8(a). The curve appears to cross the x-axis at $x = 1$ and $x = -2$. This can be seen algebraically by factoring the quadratic as

$$y = x^2 + x - 2 = (x - 1)(x + 2).$$

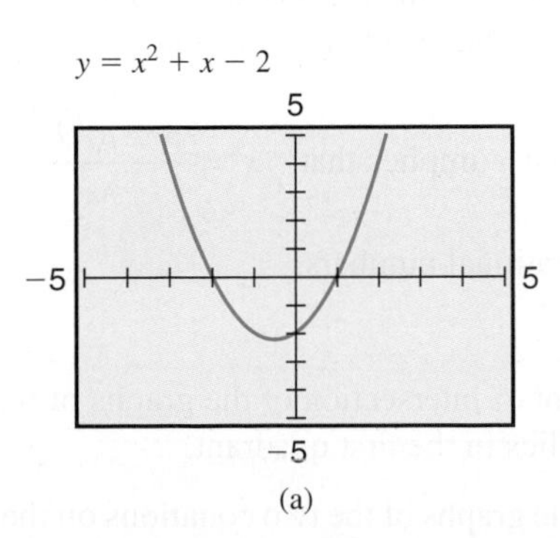

(a)

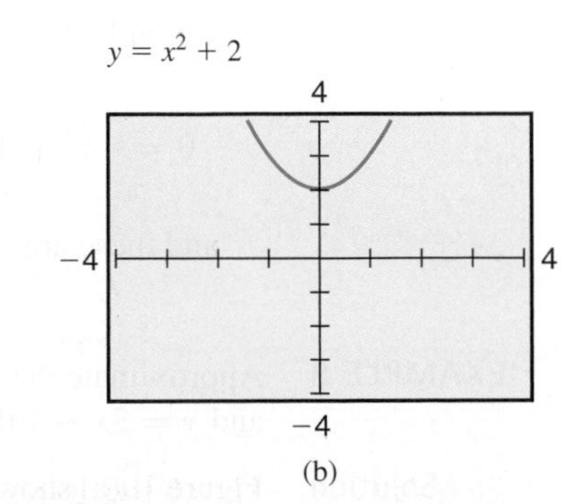

(b)

FIGURE 8

So $x^2 + x - 2 = 0$ exactly when $x = 1$ or $x = -2$. By observing where the graph is on or above the x-axis, we have

$$x^2 + x - 2 \geq 0 \quad \text{for } x \geq 1 \quad \text{or for } x \leq -2.$$

b. The graph of the equation $y = x^2 + 2$, in Figure 8(b), shows that the expression $x^2 + 2$ is always greater than 2. This is evident without plotting the curve since $x^2 \geq 0$ for all x, which implies that $x^2 + 2 \geq 2 > 0$ for all x.

c. Figure 9(a) shows the graph of $y = -x^2 + 3x + 1$. In this example we cannot determine the x-intercepts exactly from the graph. We can, however, find approximations by zooming in on the graph for x-values close to where the graph crosses the x-axis. Figure 9(b) shows the plot of the equation using a viewing rectangle $[3.1, 3.4] \times [0, 1]$. Using the trace feature of a graphing calculator or the mouse to click on the point when using a computer algebra system, we find that the graph crosses the x-axis at approximately 3.3 (written using the notation $x \approx 3.3$). Figure 9(c) shows the graph close to the other x-intercept, which is approximately equal to -0.3. The inequality is strictly less than 0, so approximate solutions to the original inequality occur when either

$$x < -0.3 \qquad \text{or} \qquad x > 3.3.$$

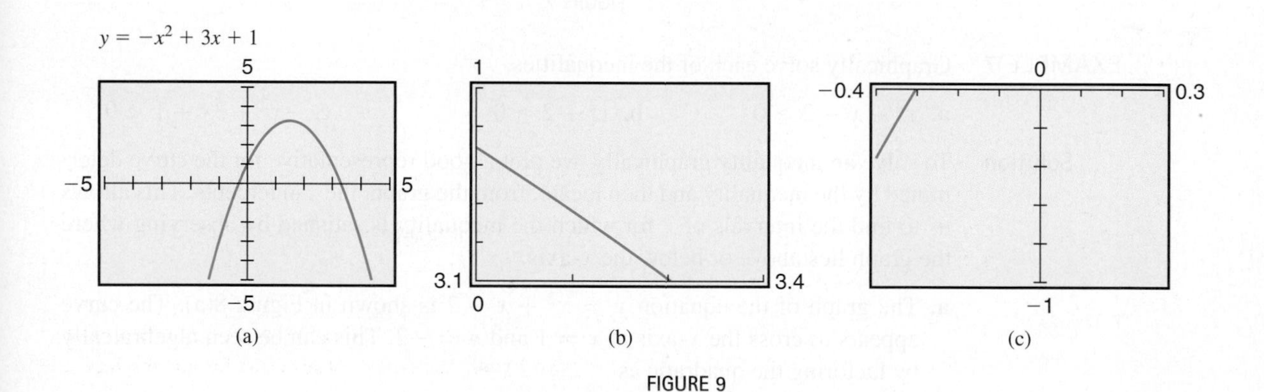

FIGURE 9

We cannot find the x-intercepts exactly using this graphing technique in this example, because by the Quadratic Formula

$$0 = -x^2 + 3x + 1 \quad \text{implies that} \quad x = \frac{3 \pm \sqrt{9 - 4(1)(-1)}}{-2} = -\frac{3}{2} \pm \frac{\sqrt{13}}{2},$$

and these are not rational numbers. ■

EXAMPLE 8 Approximate the point of intersection of the graphs of the equations $y = x^2 + x - 2$ and $y = 2x - 1$ that lies in the first quadrant.

Solution Figure 10(a) shows the graphs of the two equations on the same set of axes. The point of intersection in the first quadrant appears to have its x-coordinate between 1 and 2.

$y = 2x - 1$
4
−4
4
−4
$y = x^2 + x - 2$
(a)

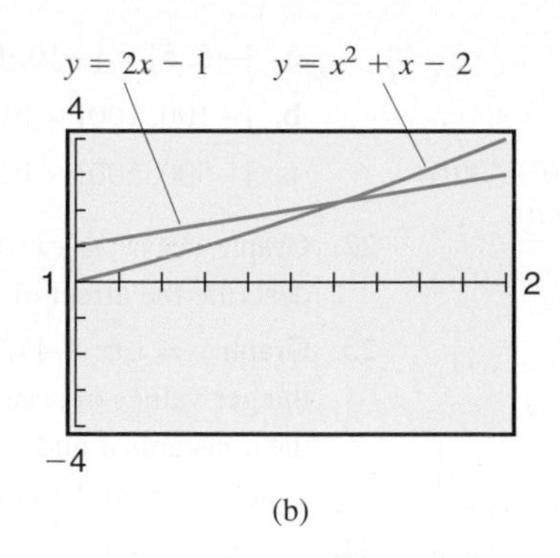

(b)

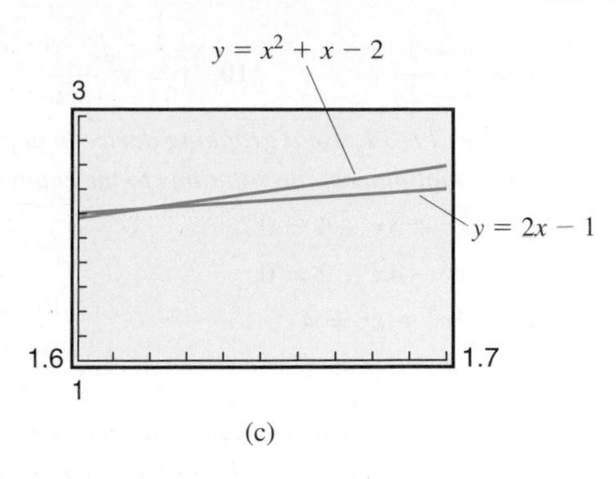

(c)

FIGURE 10

Calculus applications require finding the region between two curves. First find where the curves intersect.

Figure 10(b) shows close-ups of the two curves for x between 1 and 2. In Figure 10(c), the y-range has also been reduced using a viewing rectangle $[1.6, 1.7] \times [1, 3]$ to get a better view of the curves. The point of intersection is approximately $(1.62, 2.2)$. ■

We have presented only a few of the valuable ways that graphing devices can be used to help us graph equations and solve algebraic problems. We will describe more of these techniques in subsequent sections.

EXERCISE SET 1.5

1. Use a graphing device to sketch a graph of $y = x^2 - 4x + 9$ with the following viewing rectangles, and determine which gives the best representation for the graph of the equation.

a. $[-3, 3] \times [-3, 3]$

b. $[-7, 7] \times [-7, 7]$

c. $[-100, 100] \times [-100, 100]$

d. $[-3, 20] \times [-3, 20]$

2. Use a graphing device to sketch a graph of $y = x^2 + 10x + 15$ with the following viewing rectangles, and determine which gives the best representation for the graph of the equation.

a. $[-7, 7] \times [-7, 7]$

b. $[-10, 10] \times [-10, 10]$

c. $[-100, 100] \times [-500, 500]$

d. $[0, 10] \times [-10, 300]$

3. Use a graphing device to sketch a graph of $y = x^3 - 20x + 25$ with the following viewing rectangles, and determine which gives the best representation for the graph of the equation.

a. $[-2, 2] \times [-2, 2]$

b. $[-5, 5] \times [-5, 5]$

c. $[-10, 10] \times [-70, 70]$

d. $[-100, 100] \times [-200, 200]$

4. Use a graphing device to sketch a graph of

$$y = \frac{x^2 - 10}{x^2 - 1}$$

with the following viewing rectangles, and determine which gives the best representation for the graph of the equation.

a. $[-5, 5] \times [-5, 5]$

b. $[-10, 10] \times [-10, 10]$

c. $[-10, 10] \times [-100, 100]$

d. $[-5, 5] \times [-25, 25]$

In Exercises 5–10, determine an appropriate viewing rectangle for the graph of each equation, and use it to sketch the graph.

5. $y = x^2 - 10x + 18$

6. $y = x^2 + 14x + 59$

7. $y = \sqrt{3x - 8}$

8. $y = \sqrt{x^2 - 2x - 15}$

9. $y = \dfrac{x+3}{x-2}$ **10.** $y = x^2 + \dfrac{1}{x}$

In Exercises 11–14, use a graphing device to approximate, to two decimal places, the solutions to the equations.

11. $x^3 + x^2 + 3x - 4 = 0$

12. $x^3 - x^2 - 4x - 2 = 0$

13. $x^4 - 3x^3 + x^2 = 4$

14. $x^4 + x^3 - 2x^2 + 1 = x$

In Exercises 15–18, use a graphing device to approximate, to two decimal places, the values of x where the graphs of the equations intersect.

15. $y = x^2 + 3x + 2, \quad y = x^3 + 2x^2 + 2x + 1$

16. $y = x^3 + 5, \quad y = \frac{1}{2}x^2 + 7x$

17. $y = \sqrt{x+1} - 2, \quad y = x^2 - 1$

18. $y = \sqrt{x^2+1} - 3, \quad y = \sqrt[3]{1-x} - 2$

19. Graphically approximate the solutions to the inequalities.

a. $x^2 + 3x - 2 \ge 0$

b. $x^3 - 2x^2 - 6x + 9 < 0$

20. Graph the equations $y = -x^2 + 2x + 1$ and $y = x^3 - 2x^2 - x + 2$, and approximate the interval(s) on which

$$-x^2 + 2x + 1 \ge x^3 - 2x^2 - x + 2.$$

21. Graph the equations $y = x^4$ and $y = x^4 - 4x^3 + 3x^2$ on the same set of axes using the viewing rectangles, and compare their value for x large in magnitude.

a. $[-5, 5] \times [-10, 10]$

b. $[-100, 100] \times [0, 10^8]$

c. $[-500, 500] \times [0, 6 \times 10^9]$

22. Graph $y = x^4 + cx^2$ for different values of c, and describe the effect of the constant c on the curve.

23. Graph $y = (ax+1)/(bx-1)$ for different positive integer values of a and b, and describe the effect of the constants a and b on the curve. What can you conclude from the graphs?

24. Use a graphing device to plot a variety of curves of the form $y = ax + b$, where a and b are real numbers. Describe the effect that both positive and negative values of a and b have on the graph.

25. Use a graphing device to plot a variety of curves of the form $y = (x-a)^2 + b$, where a and b are real numbers. Describe the effect that both positive and negative values of a and b have on the graph.

26. Consider the family of curves given by

$$y = \frac{1}{x^n},$$

where $x \ne 0$ and n is a positive integer. Plot the graph for $n = 2, 4, 6,$ and 8 on the same set of axes. On another set of axes plot the graph for $n = 1, 3,$ 5, and 7. For a given value of n, compare the sizes of $1/x^n$ and $1/x^{n+2}$ on each of the intervals.

a. $0 < x < 1$ **b.** $x > 1$

c. $-1 < x < 0$ **d.** $x < -1$

1.6 FUNCTIONS

Calculus studies functions, one of the fundamental tools of mathematics.

There are a variety of ways to express a functional relationship between two quantities. One common way is by a formula expressed as an equation. For example, the formula $A = \pi r^2$ expresses how the area A of a circle depends on its radius r. Another way to express a functional relationship between two quantities is through a table of values. Table 1 shows the rapid increase in the public debt in the United States after 1940.

The information from Table 1 is shown graphically in Figure 1. The data have been plotted as year vs. public debt in billions of dollars, so years are marked on the horizontal, or x-axis, and debt is marked on the vertical, or y-axis. Straight lines have been used to join the data points so that the trend is clearer.

In a *functional* relationship, whether given by a formula, a table, or a graph, for each value of one quantity, called the *independent variable*, there is associated a unique value of another quantity, called the *dependent variable*.

TABLE 1

U.S. Gross Public Debt (in billions)

Year	Amount
1940	61
1950	257
1960	291
1970	381
1980	909
1990	3206
2000	5629
2010 (Est.)	14456

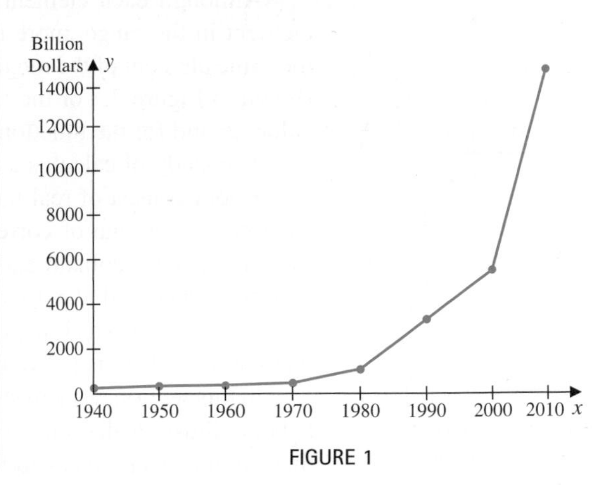

FIGURE 1

Functions

A **function** from a set $\mathbb{X}$ into a set $\mathbb{Y}$ is a rule that assigns each element in $\mathbb{X}$ to precisely one element in $\mathbb{Y}$.

Figure 2 illustrates the function concept. For the function f, illustrated on the left, each value of the independent variable x is associated with only one value of the dependent variable y. Note, that it is permissible for x_3 and x_4 to be associated with the same value, y_3. However, for g, illustrated on the right, both y_3 and y_4 are associated with the same value x_3. As a consequence, g is not a function.

The basis of the word *domain* is the Latin *domus*, meaning "home". So a function's domain is the set where it "feels at home," or is defined.

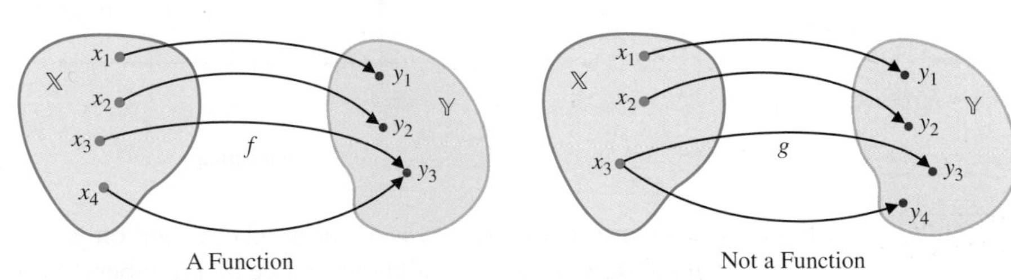

FIGURE 2

If f is used to denote a function from a set $\mathbb{X}$ into a set $\mathbb{Y}$, written $f : \mathbb{X} \to \mathbb{Y}$, then the *image* of x under f, denoted $f(x)$, is the unique value in $\mathbb{Y}$ associated with x. The set $\mathbb{X}$ is called the **domain** of f, and consists of all numbers that can be input to the function.

The word *range* means "the region over which you can roam, or wander". So a function's range is the set it can "wander over," or take on.

The **range** of f is the set of elements in $\mathbb{Y}$ that are associated with some element in $\mathbb{X}$. The domain of the function must be the entire set $\mathbb{X}$, but the range need not be all of $\mathbb{Y}$. For example, consider the function that takes the set of real numbers, $\mathbb{R}$, into the set $\mathbb{R}$ that is described by

$$f(x) = x^2.$$

This function has domain $\mathbb{R}$ but, since $x^2 \geq 0$ for all x, its range is $[0, \infty)$, a subset of $\mathbb{R}$.

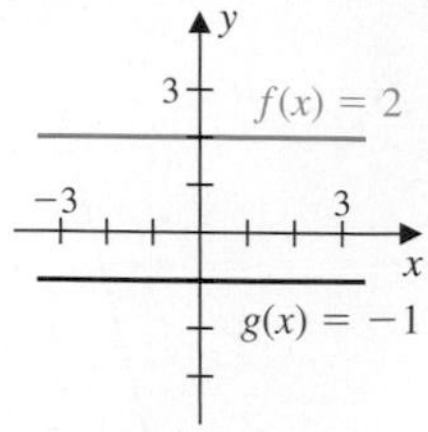

FIGURE 3

Although each element in the domain of a function corresponds to a unique element in the range, more than one element in the domain can be associated with the same element of the range. The most extreme examples are constant functions, as shown in Figure 3. For the function f, every number x is associated with the same value, 2, and for the function g, every number x is associated with -1.

The study of calculus is concerned primarily with functions whose domain and range both consist of real numbers. Functions of this type can be described simply by giving their rule of correspondence. For example, $f(x) = x^3 + 1$ describes a function f with domain and range both consisting of all the real numbers. Unless otherwise specified, the domain of the function is assumed to be the largest subset of real numbers for which the correspondence produces a real number. The domain of $f(x) = 1/x$, then, is $(-\infty, 0) \cup (0, \infty)$ because $1/x$ is defined except when $x = 0$.

Use a graphing device to help visualize problems having graphical representations.

In this section, we will show graphs of the functions we are using for the examples to better illustrate the concepts we wish to explore. Later in this chapter we will see that these graphs were obtained by simple modifications of the graphs of some common functions.

EXAMPLE 1 Use the graph in Figure 4 to find the domain and range of the function f defined by $f(x) = \sqrt{x} + 1$.

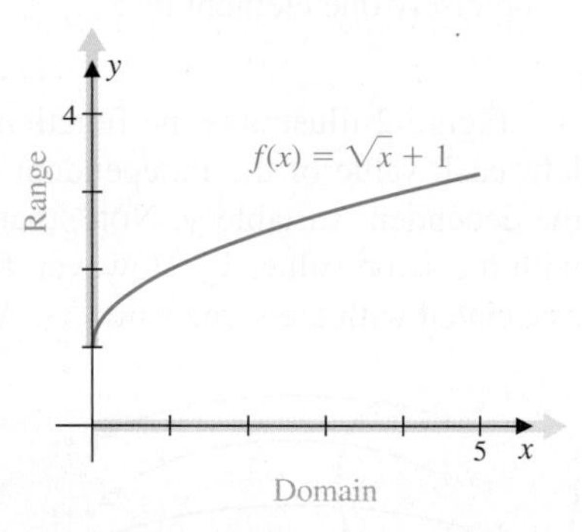

FIGURE 4

Solution The square root of x, $\sqrt{x}$, is a real number if and only if $x \geq 0$. Consequently, the domain of f is the set of all nonnegative real numbers. Notice that a value of x is in the domain if the vertical line through x intersects the graph.

The symbol $\sqrt{}$ is called a *radical* and indicates the *principal*, or nonnegative, square root. Hence we have $f(x) = \sqrt{x} + 1 \geq 1$ for all $x \geq 0$. The range of f is the set of real numbers greater than or equal to 1. A value y is in the range if the horizontal line through y intersects the graph. ■

There is no special significance to the variable x that is used to describe the function; in fact, it might be better to describe the function in Example 1 by writing $f(\square) = \sqrt{\square} + 1$. This form indicates more clearly that whatever value is used to fill the box on the left side of the equation, must also be used to fill the corresponding box on the right side. For example, 9 is in the domain of f and

$$f\left(\boxed{9}\right) = \sqrt{\boxed{9}} + 1 = 3 + 1 = 4.$$

EXAMPLE 2 Evaluate the function $f(x) = x^2 - 2x + 1$ at the given values.

a. 2 **b.** $\sqrt{5}$ **c.** $-1 + \sqrt{2}$ **d.** $2w + 1$

Solution The function can be thought of as

$$f(\square) = (\square)^2 - 2(\square) + 1$$

and each input replaces the boxes. Substituting each of the inputs for x into the function gives

a. $f(2) = 2^2 - 2(2) + 1 = 1$

b. $f(\sqrt{5}) = (\sqrt{5})^2 - 2(\sqrt{5}) + 1 = 5 - 2\sqrt{5} + 1 = 6 - 2\sqrt{5}$

c. $f(-1 + \sqrt{2}) = (-1 + \sqrt{2})^2 - 2(-1 + \sqrt{2}) + 1$
$= (1 - 2\sqrt{2} + 2) + 2 - 2\sqrt{2} + 1 = 6 - 4\sqrt{2}$

d. $f(2w + 1) = (2w + 1)^2 - 2(2w + 1) + 1$
$= 4w^2 + 4w + 1 - 4w - 2 + 1 = 4w^2$ ■

In Section 1.4, we considered the graphs of equations and discussed some symmetry properties of graphs of equations. In Example 3 of that section we saw the graphs of the equations $y = x^2$, $x = y^2$, and $x^2 + y^2 = 1$. These graphs, reproduced in Figure 5, illustrate a test to distinguish equations that represent functions from those that do not.

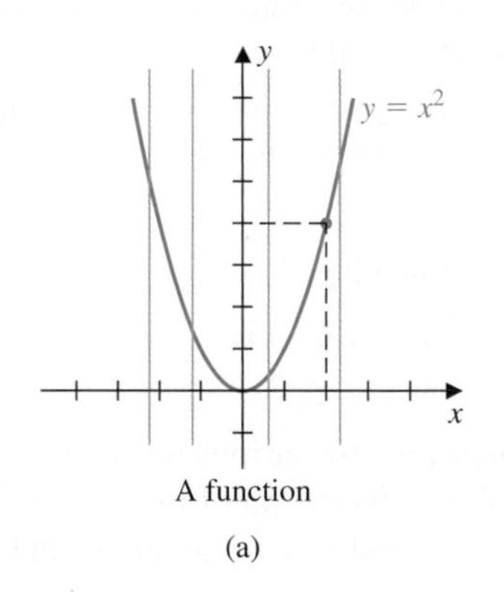

A function

(a)

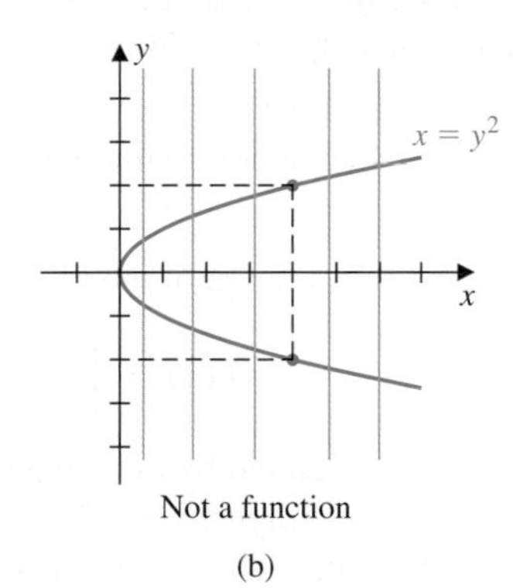

Not a function

(b)

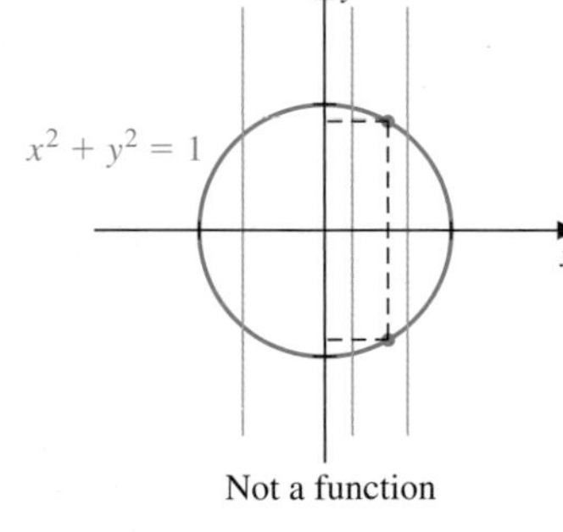

Not a function

(c)

FIGURE 5

Every vertical line intersects the graph of $y = x^2$, as shown in Figure 5(a), in exactly one place, so for each real number x there is precisely one point $(x, y) = (x, x^2)$ on the graph. This means that the graph of $y = x^2$ is the graph of a function, and the domain of this function is $(-\infty, \infty)$. Horizontal lines intersect the graph only on or above the x-axis, so the range is the interval $[0, \infty)$.

On the other hand, there are vertical lines that intersect the graph of $x = y^2$ twice. For example, the line $x = 4$ in Figure 5(b) intersects the graph of $x = y^2$ at both $y = 2$ and $y = -2$, so the graph of $x = y^2$ is not the graph of a function.

Similarly, the dashed line $x = \frac{1}{2}$ in Figure 5(c) intersects the graph of $x^2 + y^2 = 1$ at both $y = \sqrt{3}/2$ and $y = -\sqrt{3}/2$, so the graph of $x^2 + y^2 = 1$ is also not the graph of a function.

Vertical Line Test for Functions

An equation describes y as a function of x if and only if every vertical line intersects the graph of the equation at most once.

A general illustration of this test is shown in Figure 6.

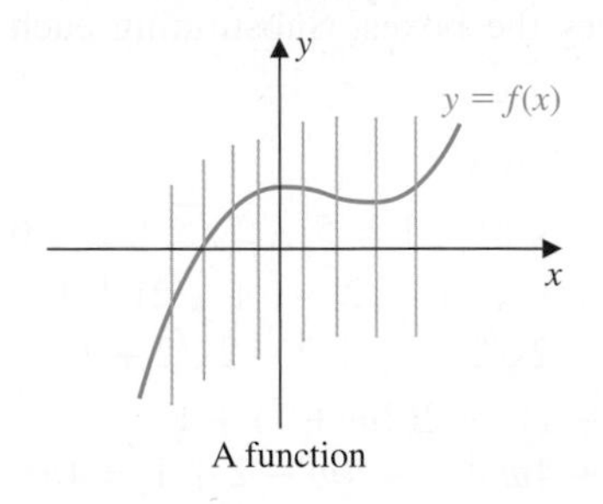

A function

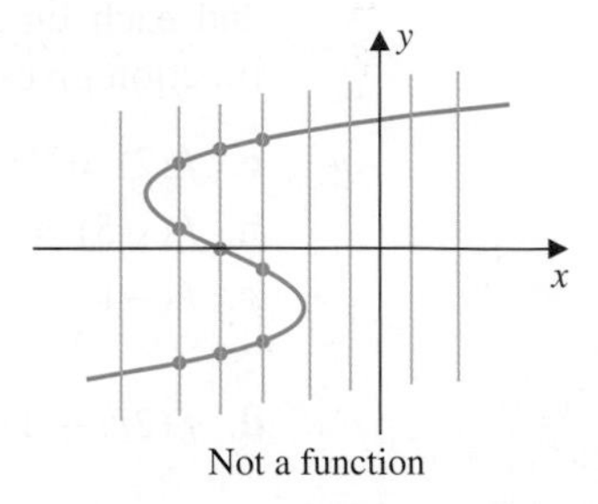

Not a function

FIGURE 6

When a graph is that of a function we have the following. (See Figure 7.)

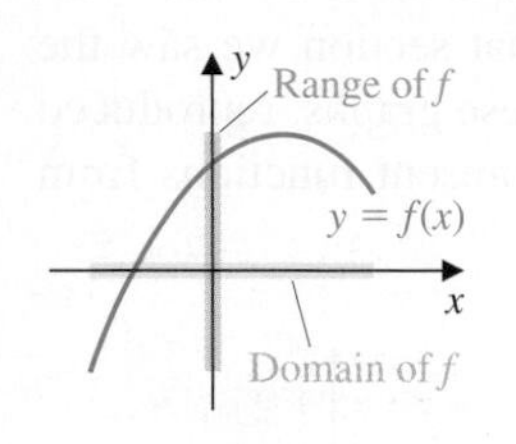

FIGURE 7

Finding the Domain and Range of a Function from a Graph

- The **domain** of a function is described by those values on the horizontal axis through which a vertical line intersects the graph.
- The **range** of a function is described by those values on the vertical axis through which a horizontal line intersects the graph.

EXAMPLE 3 Find the domain and range of the function defined by

$$f(x) = \begin{cases} x - 1, & \text{if } -3 \le x < 0 \\ x^2, & \text{if } 0 \le x \le 2. \end{cases}$$

Solution A function that is defined by differing expressions on various portions of its domain is called a *piecewise-defined* function. To sketch the graph of this piecewise-defined function, we first sketch the graphs of $y = x - 1$ and $y = x^2$, as shown in Figure 8(a).

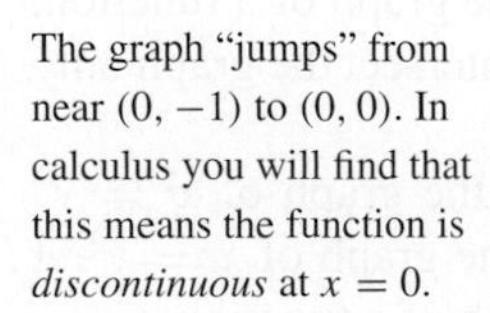
The graph "jumps" from near $(0, -1)$ to $(0, 0)$. In calculus you will find that this means the function is *discontinuous* at $x = 0$.

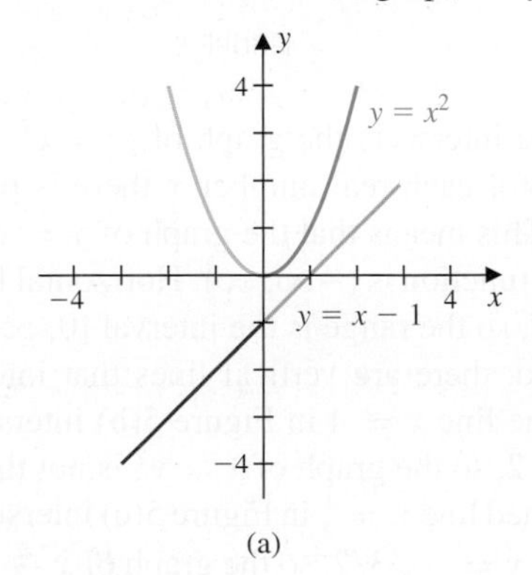

(a)

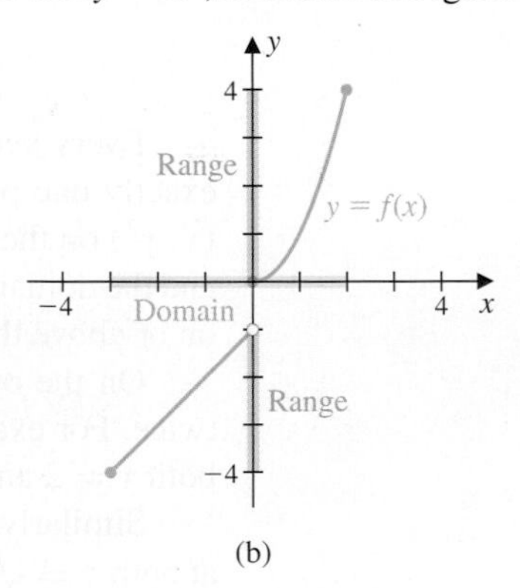

(b)

FIGURE 8

The graph of $y = f(x)$ switches from the graph of $y = x - 1$ to the graph of $y = x^2$ when $x = 0$ and is shown in Figure 8(b). Note the open circle at the point $(0, -1)$ and the solid circle at $(0, 1)$. This implies that $f(0)$ takes on the single value 1.

Vertical lines intersect the graph when $-3 \le x \le 2$, so the domain of f is $[-3, 2]$. Horizontal lines intersect the graph when $-4 \le y < -1$ or when $0 \le y \le 4$, so the range of f is $[-4, -1) \cup [0, 4]$. ■

It is important to be completely comfortable with function notation because it is used extensively in calculus. Example 4 illustrates the first steps of a common calculus problem.

EXAMPLE 4 Consider the function f described by $f(x) = x^2 + x$, whose graph is shown in Figure 9(a). Determine, for arbitrary x and $h \neq 0$, simplified expressions for

a. $f(x + h)$, **b.** $f(x + h) - f(x)$, and **c.** $\dfrac{f(x + h) - f(x)}{h}$.

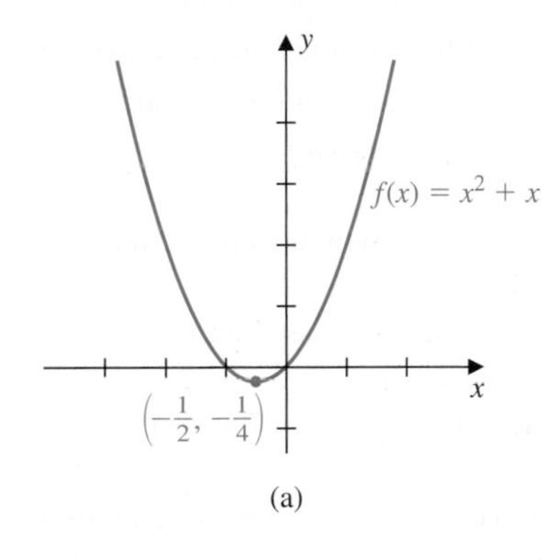

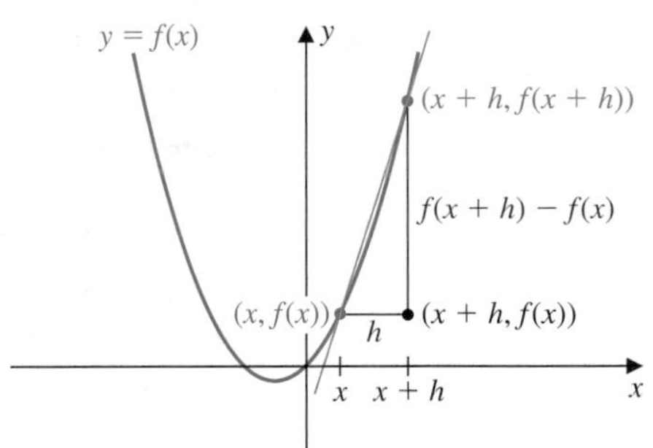

Average rate of change of f: $\dfrac{f(x + h) - f(x)}{h}$

(b)

FIGURE 9

Solution The solution to this problem is easier if we think of the function written as

$$f(\square) = \square^2 + \square,$$

where replacing the box on the left by something in the domain of f means the same value is placed in the boxes on the right. This implies, for example, that

a. $f(x + h) = (x + h)^2 + (x + h) = x^2 + 2xh + h^2 + x + h$.

b. Using part (a) gives

$$f(x + h) - f(x) = x^2 + 2xh + h^2 + x + h - (x^2 + x) = 2xh + h^2 + h.$$

c. For $h \neq 0$,

$$\frac{f(x + h) - f(x)}{h} = \frac{2xh + h^2 + h}{h} = \frac{h(2x + h + 1)}{h} = 2x + h + 1.$$ ■

The difference quotient is the basis for the definition of the *derivative* in calculus. Be sure that the notation is clear, and that you are able to simplify this type of expression.

The expression

$$\frac{f(x+h)-f(x)}{h}$$

is called a *difference quotient* of the function f. It describes the *average* rate of change of the values of the function as the variable changes from x to $x+h$, as shown in Figure 9(b). One of the primary concepts of calculus, the derivative, considers the limiting case of this difference quotient as h approaches zero. This is called the *instantaneous* rate of change. The difference quotient in Example 5 approaches $2x+1$ as h approaches 0. We write this as

$$\frac{f(x+h)-f(x)}{h} \to 2x+1 \quad \text{as} \quad h \to 0.$$

EXAMPLE 5 Let $f(x) = 1/(x-1)+2$.

a. Find the domain of f.

b. Find a value of x for which $f(x) = 5$.

Solution **a.** The quotient $1/(x-1)$ is defined except when the denominator is 0, which occurs at $x = 1$. The domain of the function is consequently $(-\infty, 1) \cup (1, \infty)$.

b. To determine the value x in the domain corresponding to $f(x) = 5$, we solve for x in the equation $5 = f(x)$. This gives

$$5 = \frac{1}{x-1}+2, \quad \text{which implies that} \quad \frac{1}{x-1} = 3.$$

Inverting both sides of this equation, assuming $x \neq 1$, gives

$$x - 1 = \frac{1}{3}, \quad \text{so} \quad x = 1 + \frac{1}{3} = \frac{4}{3}.$$

■

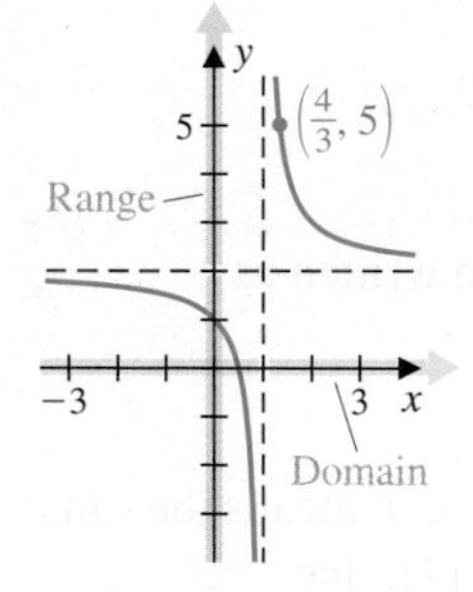

FIGURE 10

The procedure in part (b) of Example 5 can be used to determine the entire range of this function. Suppose that y is a value in the range. Then a number $x \neq 1$ must exist with

$$y = f(x) = \frac{1}{x-1}+2, \quad \text{which implies that} \quad y - 2 = \frac{1}{x-1}.$$

When $y \neq 2$ we can solve for x in terms of y to obtain

$$x - 1 = \frac{1}{y-2}, \quad \text{so} \quad x = \frac{1}{y-2}+1 = \frac{y-1}{y-2}.$$

The range of f consists of the set of numbers y for which this relation is defined, that is, the set of all real numbers y with $y \neq 2$, as shown in Figure 10.

We will always be interested in the domain of a given function since this tells us the values for which the function is defined. We do not always need to determine the range, which is fortunate because finding the exact range is often more difficult, and frequently impossible.

EXAMPLE 6 Find the domain of each function.

a. $f(x) = x^2 - 4x + 3$ **b.** $g(x) = \dfrac{1}{x^2 - 4x + 3}$

c. $h(x) = \sqrt{x^2 - 4x + 3}$ **d.** $k(x) = \dfrac{1}{\sqrt{x^2 - 4x + 3}}$

Solution

a. The expression $x^2 - 4x + 3$ is defined for all real numbers x, so the domain of the function f is $(-\infty, \infty)$.

b. The function g is defined for all real numbers except those where the denominator is zero. Factoring and solving for zero gives

$$0 = x^2 - 4x + 3 = (x - 1)(x - 3).$$

So the domain, which does not contain $x = 1$ and $x = 3$, is $(-\infty, 1) \cup (1, 3) \cup (3, \infty)$.

c. In this case the domain of h is the set of all real numbers x for which the expression under the radical is nonnegative—that is, all values of x for which

$$x^2 - 4x + 3 = (x - 1)(x - 3) \geq 0.$$

Use charts to determine the behavior of functions on intervals. It is easier than solving inequalities using algebra.

The sign chart in Figure 11 shows the domain is $(-\infty, 1] \cup [3, \infty)$.

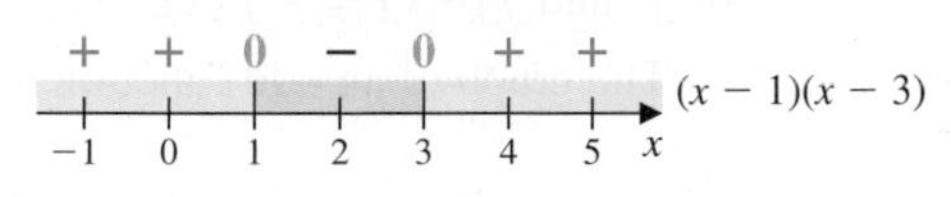

FIGURE 11

d. If x is in the domain, then

$$0 < x^2 - 4x + 3 = (x - 1)(x - 3).$$

The sign chart in part (c) tells us that the domain is $(-\infty, 1) \cup (3, \infty)$. ■

Finding the range of a function can be difficult. A graphing device can be useful.

Graphing devices are helpful in determining the range of a function that has a complicated representation.

EXAMPLE 7 Use the graph of the function $f(x) = \dfrac{x^2 + 4}{x^2 - 4}$ to determine its range.

Solution The graph is shown in Figure 12. The range includes all values except those in the interval $(-1, 1]$, since a horizontal line through any point in $(-\infty, -1] \cup (1, \infty)$ on the y-axis crosses the curve at least one time.

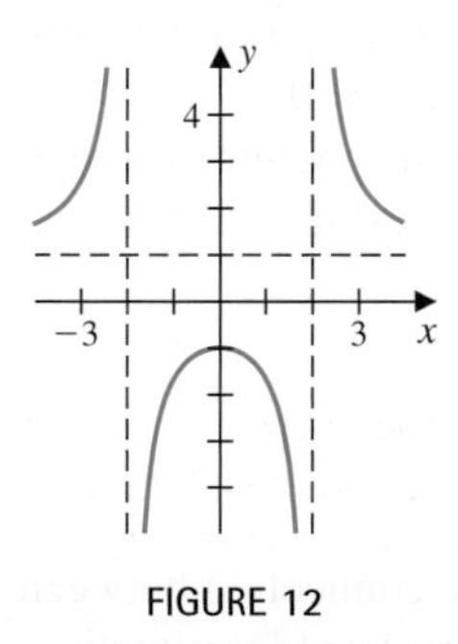

FIGURE 12

Suppose that for some number $x \neq \pm 2$ we have

$$y = \frac{x^2 + 4}{x^2 - 4}.$$

We can solve this equation for x in terms of y by writing

$$x^2 - 4y = x^2 + 4 \quad \text{and then} \quad x^2(y - 1) = 4(y + 1).$$

This implies that

$$x = \pm\sqrt{\frac{y + 1}{y - 1}}.$$

The range of f consists of those values of y with $(y + 1)/(y - 1) \geq 0$. Using the sign chart in Figure 13 we see that $y \leq -1$ or $y > 1$, which tells us, as before, that the range is $(-\infty, -1] \cup (1, \infty)$. ■

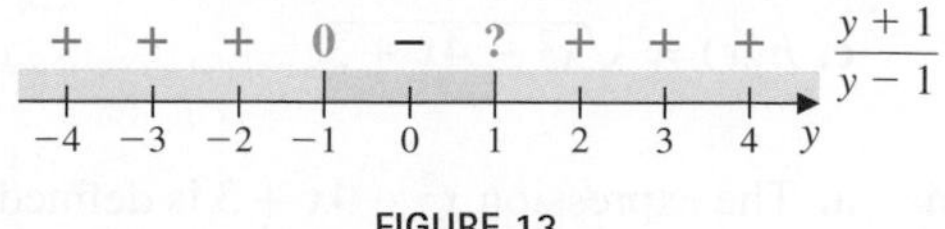

FIGURE 13

In Section 1.4 we discussed the symmetry properties of graphs. The following definition characterizes those functions whose graphs have y-axis and origin symmetry.

Odd and Even Functions

- A function f is **even** if $-x$ is in the domain of f whenever x is in the domain of f, and $f(-x) = f(x)$.
  - The graph of an even function has y-axis symmetry.
- A function f is **odd** if $-x$ is in the domain of f whenever x is in the domain of f, and $f(-x) = -f(x)$.
  - The graph of an odd function has origin symmetry.

Knowing that a graph has y-axis or origin symmetry cuts by half the work required to sketch the graph.

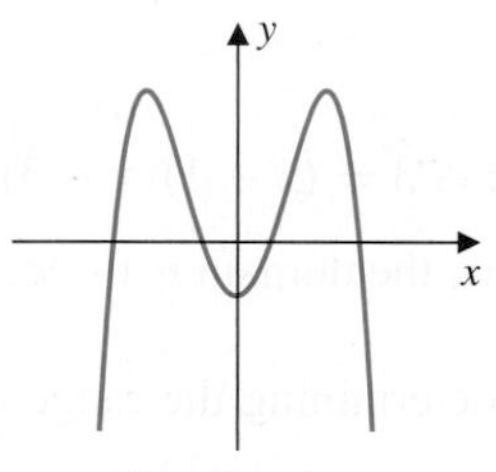

Even function
y-axis symmetry

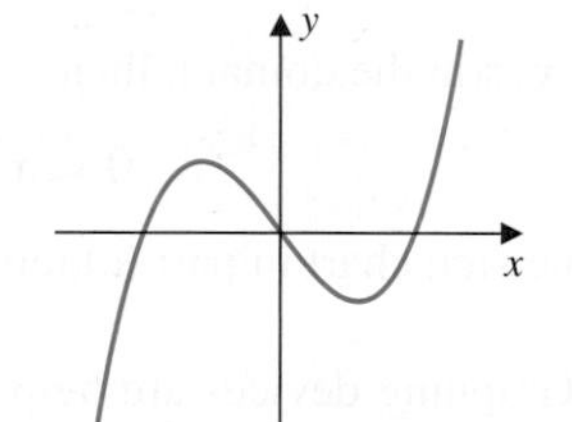

Odd function
Origin symmetry

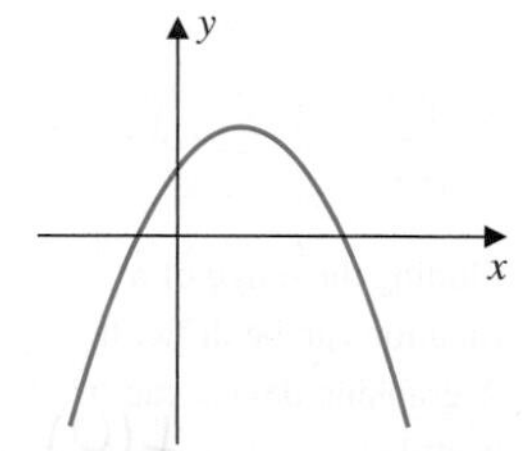

Neither odd nor even
No axis or origin symmetry

FIGURE 14

The graph of a *function* cannot have x-axis symmetry unless the function is the constant zero function. This follows from the fact that (x, y) and $(x, -y)$ cannot both be on the graph of a function unless $y = -y$, which implies that $y = 0$.

Figure 14 shows the graph of an even function has y-axis symmetry, and the graph of an odd function has origin symmetry. If a graph has no axis or origin symmetry, it represents a function that is neither odd nor even.

Function Graphing Summary

Graphs of functions provide an important tool for analyzing the connection between variables. Not all the techniques in this section are useful in every graphing situation, but the collection gives an excellent starting point for observing trends in data that are described by the variables. Listed on the following page is a summary that can be consulted when graphing functions. This will ensure that you do not inadvertently omit information that is important.

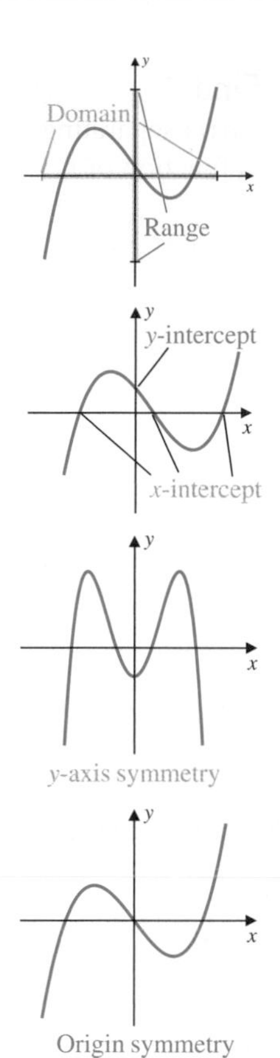

Graphing Procedures for Functions

- **Domain**: Unless specified otherwise, this is the largest set of numbers for which the function is defined. It consists of the values on the horizontal axis where vertical lines intersect the graph.
- **Range**: These are the numbers that are produced when all the domain values are substituted into the function. It consists of the values on the vertical axis where horizontal lines intersect the graph. The range is generally more difficult to determine than the domain.
- **The y-intercept**: This is the y-value that results when $x = 0$ is substituted into the function, provided that $x = 0$ is in the domain.
- **The x-intercepts**: These are the values of x that produce a y-value of 0. The x-intercepts are generally more difficult to determine than the y-intercept, and are often impossible to determine.
- **Symmetry to the y-axis**: This occurs when the function is even, that is, when $f(-x) = f(x)$ for every value in the domain.
- **Symmetry to the origin**: This occurs when the function is odd, that is, when $f(-x) = -f(x)$ for every value in the domain.
- **Local Maximums and Minimums**: These are the relative high and low points on the graph.

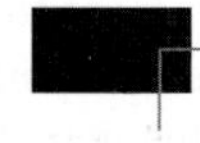

Applications

Our final examples use function notation to place physical problems in a mathematical form. The most difficult part of a problem is translating the written description of the problem into a form that uses mathematical equations to construct a mathematical model. The "Know–Find" outline used in these examples can be helpful in making this translation.

EXAMPLE 8 Two ships sail from the same port. The first ship leaves port at 1:00 A.M. and travels eastward at 15 knots (nautical miles per hour). The second ship leaves port at 2:00 A.M. and travels northward at 10 knots. Find the distance between the ships as a function of the time after 2:00 A.M.

Solution We first introduce a time variable so that we can express the distances using equations. Let t denote the time in hours after 2:00 A.M. Since the second ship travels 10 nautical miles per hour, it is $10t$ nautical miles from port at time t.

A nautical mile is the length of one minute of longitude on a great circle. It is equivalent to 1852 meters, or about 1.13 standard miles.

The first ship is traveling 15 nautical miles per hour and has traveled for $(1 + t)$ hours at t hours after 2:00 A.M. Hence its distance from port is $15(1 + t) = 15 + 15t$ nautical miles at time t. Figure 15 shows the position of the ships at a given time t.

The following is a concise description of the problem.

Know

1. The distance of the first ship from the port: $x = 15 + 15t, t \geq 0$.
2. The distance of the second ship from the port: $y = 10t, t \geq 0$.
3. The paths of the ships are perpendicular.

Find

a. The distance $d(t)$ separating the ships as a function of t.

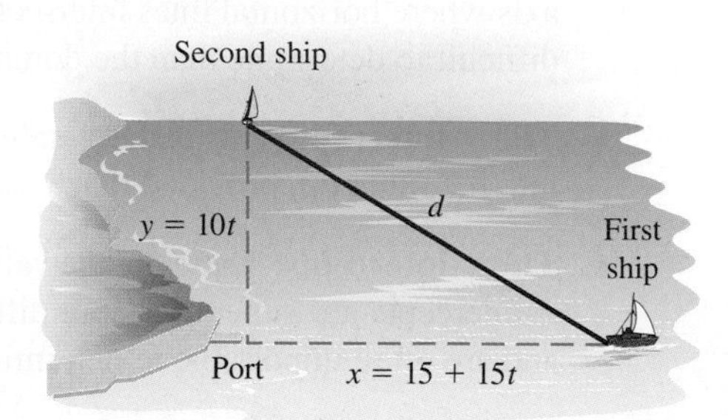

FIGURE 15

The paths of the ships are perpendicular, so we use the Pythagorean Theorem to determine the answer. The ships are a distance of

$$d(t) = \sqrt{x^2 + y^2} = \sqrt{(15 + 15t)^2 + (10t)^2} = \sqrt{325t^2 + 450t + 225}$$

nautical miles apart t hours after 2:00 A.M. ■

Note that the domain of the function described in Example 8 has been restricted to the interval $[0, \infty)$ by the physical conditions, or *constraints*, of the problem. In fact, a reasonable approximation to the actual distance between the ships is given by $d(t)$ only for small values of t. This is due to variations in the actual speed of the ships, deviation from the assumed course, the curvature of the earth, and many other factors. Keep in mind that when mathematical expressions are used to solve physical problems, the answer to the mathematical problem can vary from the answer to the physical problem due to such neglected technicalities.

EXAMPLE 9 A box without a top is to be constructed from a square piece of cardboard with sides of length 3 ft by cutting out squares of equal size at each corner and bending up the flaps. Approximate the size of the square that should be removed in order to produce the maximum volume.

Solution We will write the volume V as a function of x, the size of the flap removed, which is also the height of the constructed box.

Know

1. The height of the box is x.
2. The length of the side of the base of the box is $3 - 2x$.

Find

a. The volume $V(x)$ of the box.
b. An approximate value for x when $V(x)$ is a maximum.

The volume of a box is the area of the base times the height x. Because x units are cut from each corner and the sheet of cardboard is a square of side 3, the base of the constructed box is a square of side $3 - 2x$, as shown in Figure 16. The volume of the box is therefore

$$V(x) = x(3 - 2x)^2.$$

The derivative in calculus can determine maximums and minimums. Setting up the problem is often the difficult part.

The side of the flap must be positive and less than half the width of the side of the original cardboard piece. This implies that the domain of V is the interval $(0, 1.5)$.

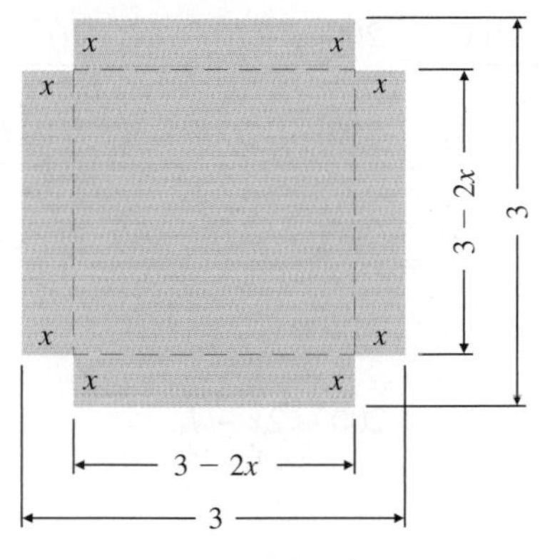

FIGURE 16

To approximate the value of x that maximizes the volume, we use a graphing device to plot the graph of $y = V(x)$, as shown in Figure 17. Moving to the peak of the curve we see that the maximum volume occurs when x is near 0.5, so the volume is approximately $V(0.5) = 0.5(3 - 2(0.5))^2 = 2.0 \text{ ft}^3$. Calculus can be used to show that these values are correct. ■

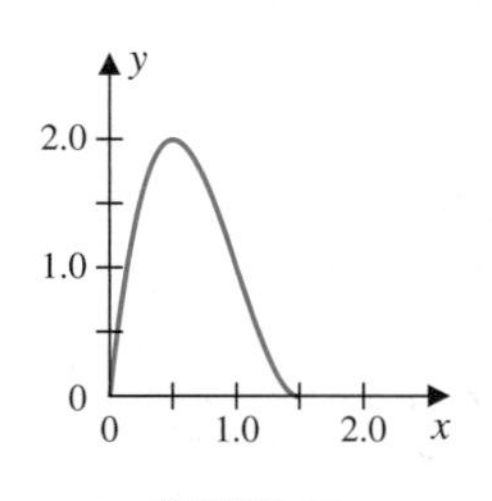

FIGURE 17

The difference quotient for a function f,

$$\frac{f(x+h) - f(x)}{h},$$

discussed in Example 4, can be interpreted as the average rate of change in the function as the input changes from an initial value x to a final value $x + h$. This interpretation has many applications, as illustrated in the next two examples.

EXAMPLE 10 In economics, the profit function P describes the profit $P(x)$ when x units of a commodity are sold. From experience, a manufacturer estimates that the profit function for a product is approximately

$$P(x) = 200x - x^2.$$

a. What is the average rate of change in the profit as the number of units changes from x to $x + h$?

b. What is the instantaneous rate of change at x?

c. Use the result in part (a) to find the average rate of change in the profit if the number of units produced changes from 50 to 75.

d. Find the instantaneous rate of change when the number of units is 50.

e. Plot the graph of $y = P(x)$ along with the line that passes through the points $(50, P(50))$ and $(75, P(75))$.

Solution The graph of the profit function shown in Figure 18(a) indicates that there is no profit when no items are sold, that a maximum profit is produced when 100 units are sold, and that beyond 100 units profit begins to decline.

a. The average rate of change is determined by computing the difference quotient of the function P. That is,

$$\begin{aligned}\frac{P(x+h)-P(x)}{h} &= \frac{200(x+h)-(x+h)^2-(200x-x^2)}{h}\\ &= \frac{200x+200h-x^2-2xh-h^2-200x+x^2}{h}\\ &= \frac{200h-2xh-h^2}{h}\\ &= \frac{h(200-2x-h)}{h}\\ &= 200-2x-h.\end{aligned}$$

In this part of the example we are determining the *average* rate of change of the profit. Calculus is concerned with the *instantaneous* rate of change.

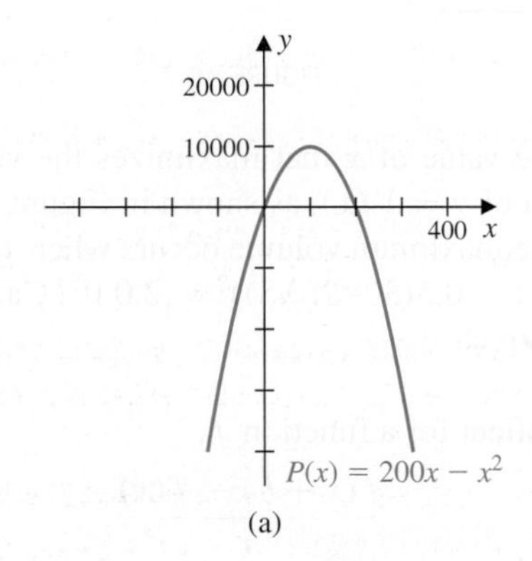

(a)

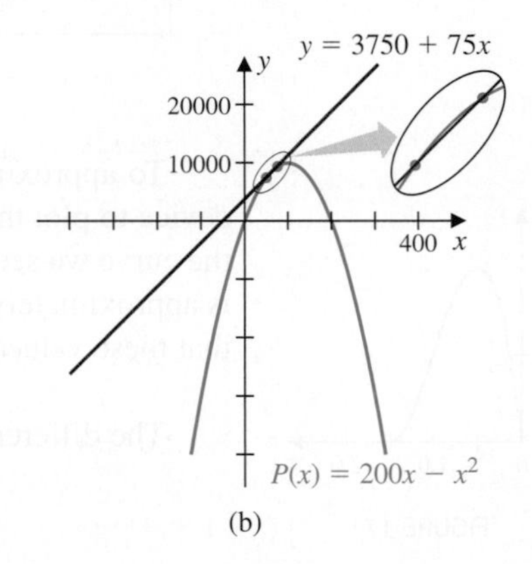

(b)

FIGURE 18

b. The instantaneous rate of change is the value of the difference quotient as h approaches 0. That is,

$$\frac{P(x+h)-P(x)}{h} \to 200-2x \quad \text{as} \quad h \to 0.$$

This rate of change of profit with respect to the number of units produced is called *marginal* profit.

c. If the number of units starts at $x = 50$ and changes by $h = 25$, the average rate of change is given by

$$200-2(50)-25 = 75.$$

For each additional unit manufactured and sold, the profit will increase on average by \$75.

d. The instantaneous rate of change when 50 units are sold is

$$200 - 2(50) = 100.$$

This indicates that the next unit beyond 50 that is sold will give an approximate profit of \$100.

e. The graph of $y = P(x)$ is shown in Figure 18(b). The inclination of the line indicates the average rate of change in the profit as the number of units produced increases from 50 to 75. ■

EXAMPLE 11 When a ball is dropped from the top of a building 200 meters high, Newton's law of motion tells us that the distance of the ball from the ground at time t is approximately

$$s(t) = 200 - 4.9t^2 \text{ m}.$$

a. Find the average velocity of the ball over the time interval from $t = 2$ to $t = 4$ seconds.

b. Find the *instantaneous velocity* of the ball at time $t = 2$.

Solution **a.** The average velocity is the change in the distance divided by the change in the time for the trip. The change in the distance is

$$s(4) - s(2) = (200 - 4.9(4)^2) - (200 - (4.9)(2)^2) = -58.8 \text{ m}$$

and the change in the time for the trip is $4 - 2 = 2$ seconds. The average velocity is

$$\text{average velocity} = \frac{\text{change in distance}}{\text{change in time}} = \frac{s(4) - s(2)}{2} = -29.4 \text{ m/s}.$$

The magnitude of the velocity, 29.4, is its speed, and the negative indicates that the distance from the ground is decreasing with time, which is certainly true since the ball is falling.

b. For any specific time t and elapsed time h, the average velocity over the time interval $[t, t + h]$ is the difference quotient

$$\begin{aligned}\frac{s(t+h) - s(t)}{h} &= \frac{(200 - 4.9(t+h)^2) - (200 - 4.9t^2)}{h}\\ &= \frac{200 - 4.9t^2 - 9.8th - 4.9h^2 - 200 + 4.9t^2}{h}\\ &= \frac{-9.8th - 4.9h^2}{h} = \frac{-h(9.8t + 4.9h)}{h} = -9.8t - 4.9h \text{ m/s}.\end{aligned}$$

The *instantaneous* velocity is the value of the difference quotient as h approaches 0. That is,

$$\frac{s(t+h) - s(t)}{h} \to -9.8t \text{ m/s} \quad \text{as} \quad h \to 0.$$

So when $t = 2$ the instantaneous velocity is $-9.8(2) = -19.6$ meters per second. ■

EXERCISE SET 1.6

1. If $f(x) = 2x^2 + 3$, find

   a. $f(2)$ b. $f(\sqrt{3})$
   c. $f(2+\sqrt{3})$ d. $f(2) + f(\sqrt{3})$
   e. $f(2x)$ f. $f(1-x)$
   g. $f(x+h)$ h. $f(x+h) - f(x)$

2. If $f(x) = \sqrt{x+4}$, find

   a. $f(-1)$ b. $f(0)$ c. $f(4)$
   d. $f(5)$ e. $f(a)$ f. $f(2a-1)$
   g. $f(x+h)$ h. $f(x+h) - f(x)$

3. If $f(t) = |t-2|$, find

   a. $f(4)$ b. $f(1)$
   c. $f(0)$ d. $f(t+2)$
   e. $f(2-t^2)$ f. $f(-t)$

4. If

$$f(x) = \begin{cases} x^2, & \text{if } x \geq 0 \\ 2x-1, & \text{if } x < 0 \end{cases},$$

   find

   a. $f(1)$ b. $f(-1)$
   c. $f(-2)$ d. $f(3)$
   e. For $a \geq 1$, find $f(1-a)$.
   f. For $0 < a < 1$, find $f(1+a)$.

5. The graph of the function f is given in the figure.

   a. Determine the values $f(-1)$, $f(0)$, $f(1)$, $f(3)$.
   b. Determine the domain and range of the function.

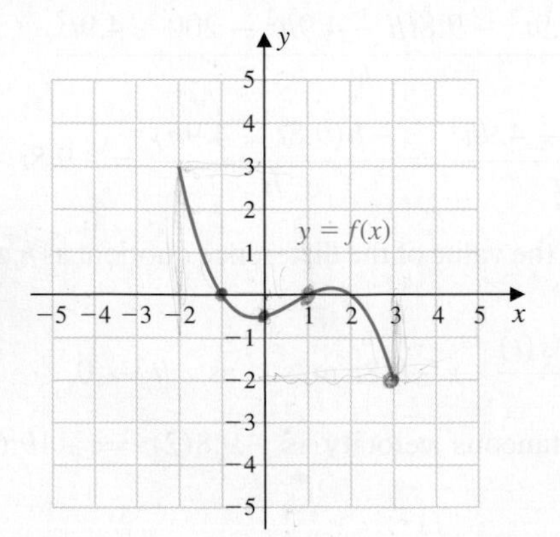

6. The graph of the function f is given in the figure.

   a. Determine the values $f(-1)$, $f(0)$, $f(1)$, $f(3)$.
   b. Determine the domain and range of the function.

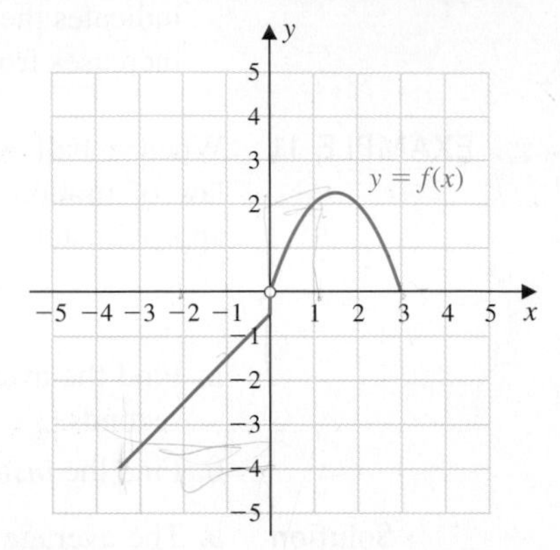

In Exercises 7–10, determine which of the curves are graphs of functions.

7.

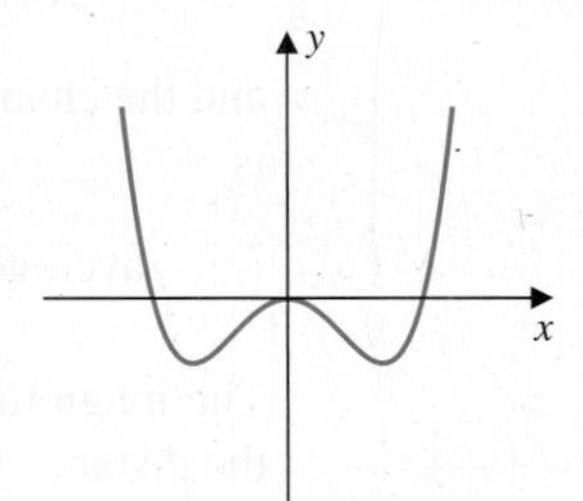

8.

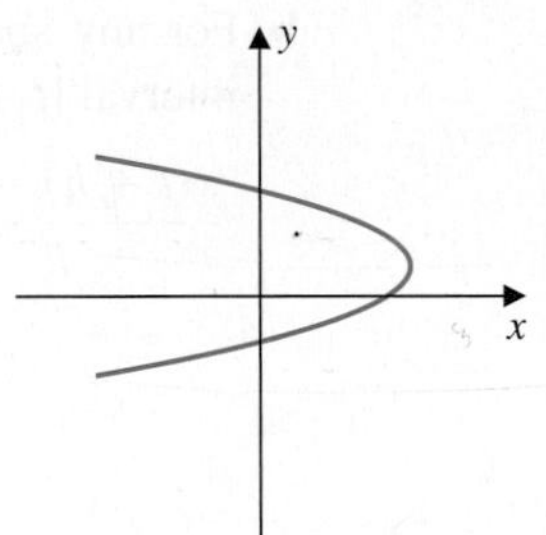

9.

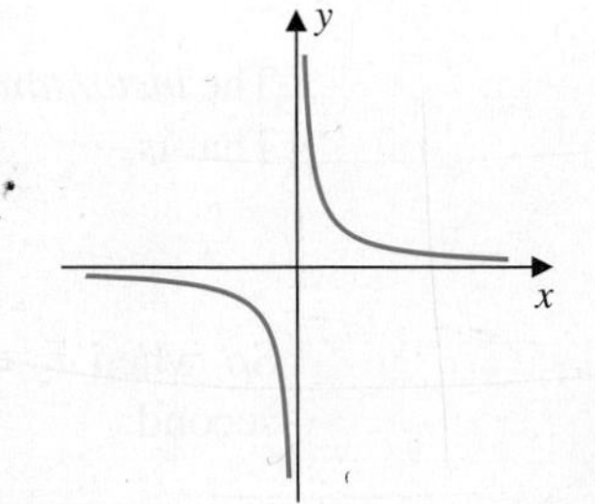

10.

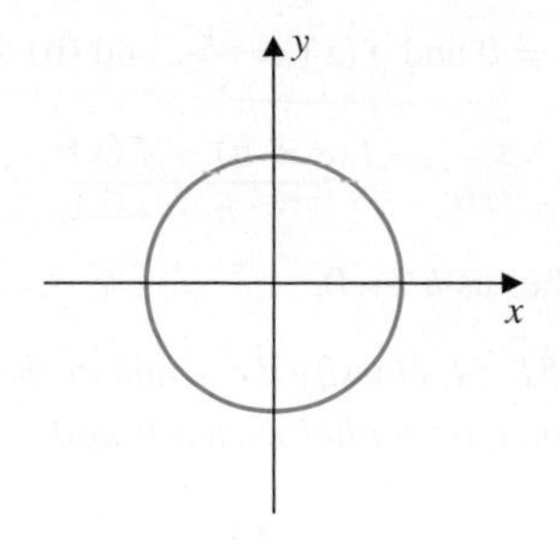

*In Exercises 11–14, use the graphs to determine **(a)** the domain and **(b)** the range of the function.*

11.

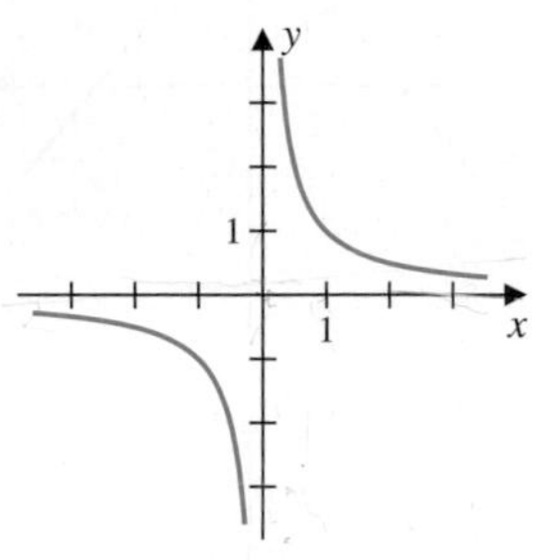

12.

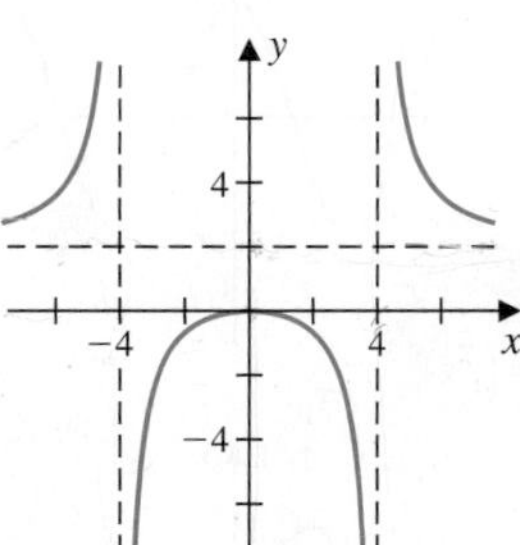

13.

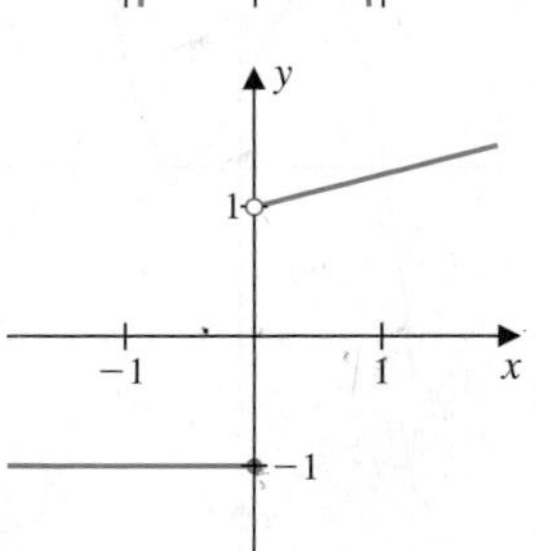

14.

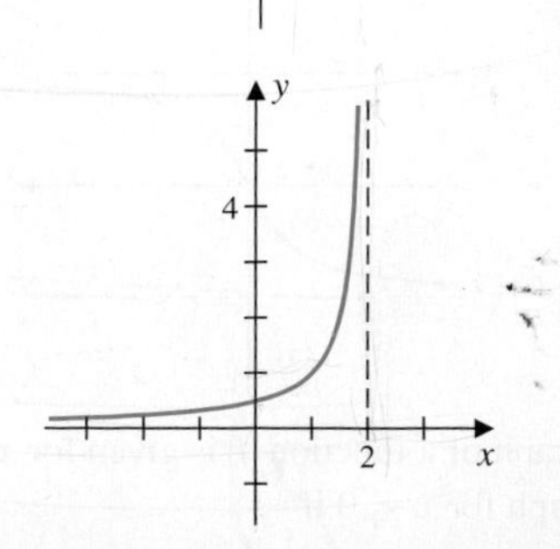

*In Exercises 15–22, find **(a)** the domain and **(b)** the*
of the function.

15. $f(x) = x^2 - 1$ **16.** $f(x) = x^2 + 1$

17. $f(x) = \sqrt{x} + 4$ **18.** $f(x) = \sqrt{x} - 2$

19. $f(x) = x^2 - 2x + 1$

20. $f(x) = x^2 - 2x + 2$

21. $f(x) = \begin{cases} 1, & \text{if } x \geq 0 \\ -1, & \text{if } x < 0 \end{cases}$

22. $f(x) = \begin{cases} x^3, & \text{if } x \geq 0 \\ -2x, & \text{if } x < 0 \end{cases}$

In Exercises 23–26, determine a number a in the domain of the function f with $f(a) = b$ for the given value of b.

23. $f(x) = x^2 - 1; b = 0$

24. $f(x) = x^2 - 1; b = 2$

25. $f(x) = \sqrt{x - 1}; b = \frac{1}{2}$

26. $f(x) = \sqrt{x} - 1; b = \frac{1}{2}$

In Exercises 27–32, find the domain of each function.

27. **a.** $f(x) = 2 - x$ **b.** $f(x) = \dfrac{1}{2 - x}$

c. $f(x) = \sqrt{2 - x}$ **d.** $f(x) = \dfrac{1}{\sqrt{2 - x}}$

28. **a.** $f(x) = 3x + 1$ **b.** $f(x) = \dfrac{1}{3x + 1}$

c. $f(x) = \sqrt{3x + 1}$ **d.** $f(x) = \dfrac{1}{\sqrt{3x + 1}}$

29. **a.** $f(x) = \dfrac{2x}{x^2 - 2}$ **b.** $f(x) = \dfrac{x - 2}{x^2 - 2}$

c. $f(x) = \sqrt{\dfrac{2x^2}{x^2 - 2}}$

30. **a.** $f(x) = \dfrac{x - 5}{x^2 + 3x - 10}$

b. $f(x) = \dfrac{x + 5}{x^2 + 3x - 10}$

c. $f(x) = \sqrt{\dfrac{(x - 5)^2}{x^2 + 3x - 10}}$

31. **a.** $f(x) = \sqrt{x(x - 2)}$

b. $f(x) = \sqrt{x(2 - x)}$

c. $f(x) = \dfrac{x^2}{x^2 - 2x}$

d. $f(x) = \sqrt{\dfrac{x^2}{x^2 - 2x}}$

32. a. $f(x) = \sqrt{(x+1)(x-1)}$

b. $f(x) = \sqrt{(x+1)(1-x)}$

c. $f(x) = \dfrac{x^4 - x^2}{x^2 - 1}$

d. $f(x) = \sqrt{\dfrac{x^4 - x^2}{x^2 - 1}}$

In Exercises 33–36, determine formulas for $f(-x)$, $-f(x)$, $f(1/x)$, $1/f(x)$, $f(\sqrt{x})$, and $\sqrt{f(x)}$.

33. $f(x) = x^2 + 2$

34. $f(x) = x^2 + 4x$

35. $f(x) = 1/x$

36. $f(x) = \sqrt{x}$

In Exercises 37–48, find

(a) $f(x+h)$,

(b) $f(x+h) - f(x)$,

(c) $\dfrac{f(x+h) - f(x)}{h}$, where $h \neq 0$, and

(d) the value $\dfrac{f(x+h) - f(x)}{h}$ approaches as $h \to 0$.

37. $f(x) = 3x - 2$

38. $f(x) = \dfrac{3}{2}x + \dfrac{1}{4}$

39. $f(x) = x^2$

40. $f(x) = -x^2$

41. $f(x) = 2 - x - x^2$

42. $f(x) = 3x^2 + 2x + 1$

43. $f(x) = \dfrac{1}{x}$

44. $f(x) = x + \dfrac{1}{x}$

45. $f(x) = \dfrac{x}{x-3}$

46. $f(x) = \dfrac{3 - 5x}{2x}$

47. $f(x) = x^3$

48. $f(x) = 2x - x^3$

49. Find **(a)**

$$\frac{f(x+h) - f(x)}{h},$$

where $h \neq 0$, and $f(x) = \sqrt{x}$, and **(b)** the value

$$\frac{f(x+h) - f(x)}{h}$$

approaches as $h \to 0$. (*Hint*: Rationalize the expression by multiplying the numerator and the denominator by the quantity $\sqrt{x+h} + \sqrt{x}$.)

50. Use a technique similar to that in Exercise 49 to find **(a)**

$$\frac{f(x+h) - f(x)}{h},$$

where $h \neq 0$ and $f(x) = \frac{1}{\sqrt{x}}$, and **(b)** the value

$$\frac{f(x+h) - f(x)}{h}$$

approaches as $h \to 0$.

In Exercises 51–54, classify the graph as that of a function that is even, odd, or neither even nor odd.

51.

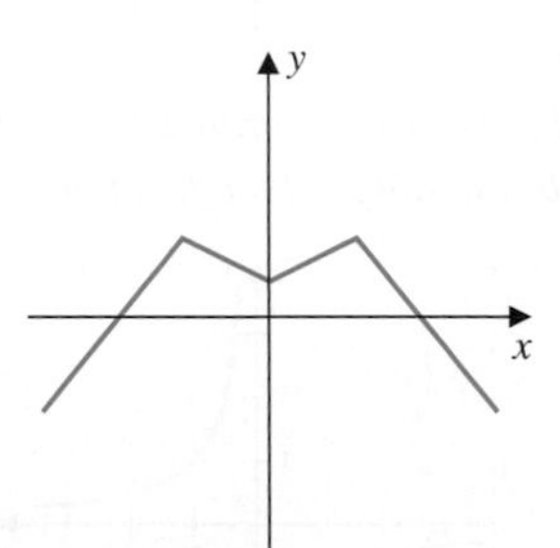

52.

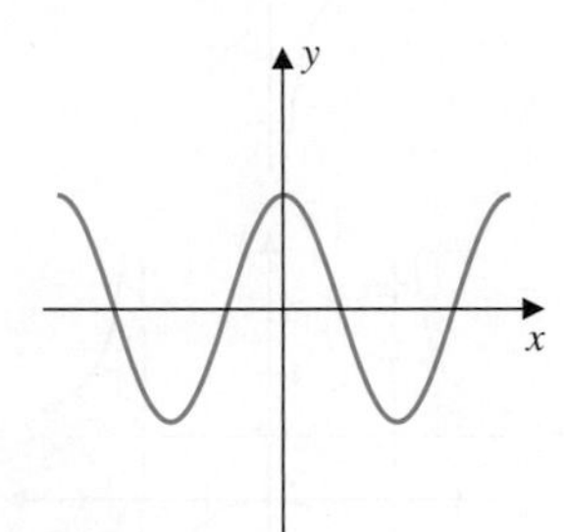

53.

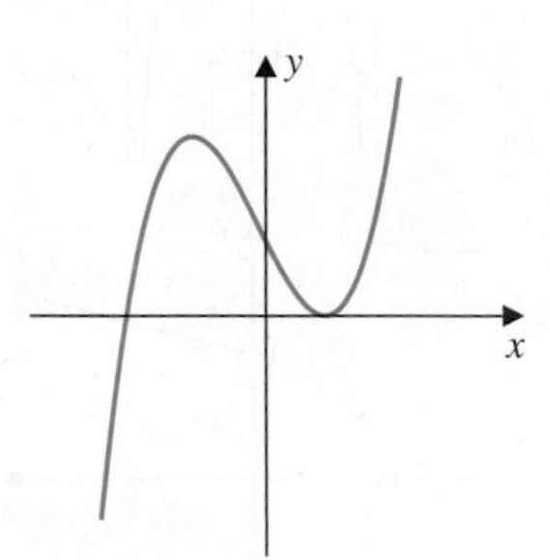

54.

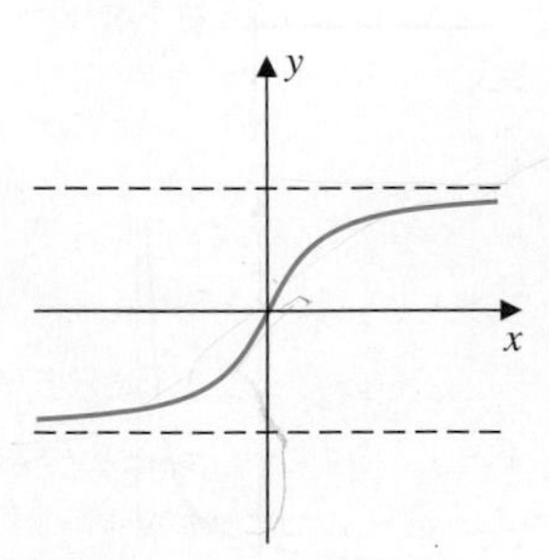

55. The graph of a function f is given for $x \geq 0$. Extend the graph for $x < 0$ if

a. f is even. **b.** f is odd.

56. The graph of a function f is given for $x \geq 0$. Extend the graph for $x < 0$ if

a. f is even. **b.** f is odd.

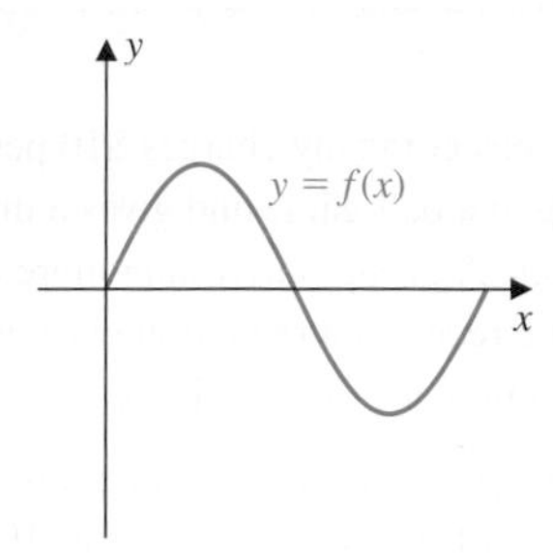

57. Redraw the following graph, and draw each graph on the same set of axes.

a. $y = -f(x)$ **b.** $y = f(-x)$ **c.** $y = |f(x)|$

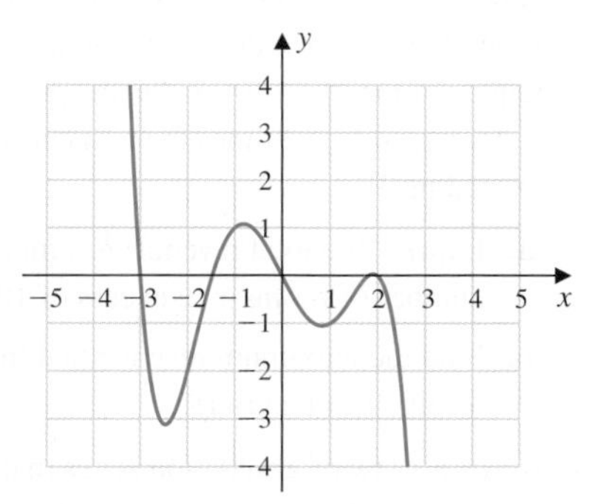

58. Sketch a possible graph for a function that satisfies all the following conditions.

a. $f(0) = 1$.

b. $f(x)$ is increasing on the interval $(0, 2)$.

c. $f(x)$ is decreasing on the interval $(2, 5)$.

d. $f(x)$ is increasing on the interval $(5, \infty)$.

e. $f(x)$ approaches 3 as x becomes large and approaches -2 as x becomes small.

59. Two ships sail from the same port. The first ship leaves at noon and travels eastward at 10 knots. The second ship leaves at 3:00 P.M. and travels southward at 15 knots. Find the distance d between the ships as a function of the time after 3:00 P.M.

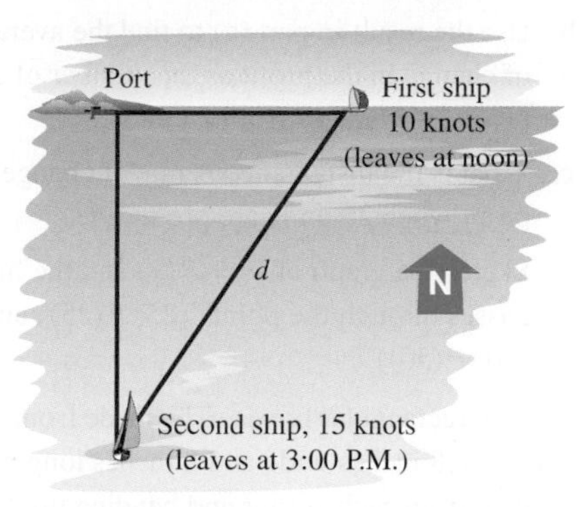

60. If the height h of a right circular cylinder is twice the radius r, express the volume V as a function of r.

61. A rectangle has an area of 64 m^2. Express the perimeter P of the rectangle as a function of the length s of one of the sides.

62. Express the area A of an equilateral triangle as a function of the length x of a side.

63. Express the surface area S of a cube as a function of its volume.

64. Express the area of a circle as a function of its circumference.

65. A rectangle is placed inside a circle of radius r so the center of the rectangle and the center of the circle coincide, and the corners of the rectangle are on the circle. Express the area of the rectangle as a function of the length x of one of its sides.

66. A rectangular plot of ground containing 432 ft^2 is to be fenced within a large lot.

a. Express the perimeter of the plot as a function of the width. What is the domain of this function?

b. Use a graphing device to approximate the dimensions of the plot that requires the least amount of fence.

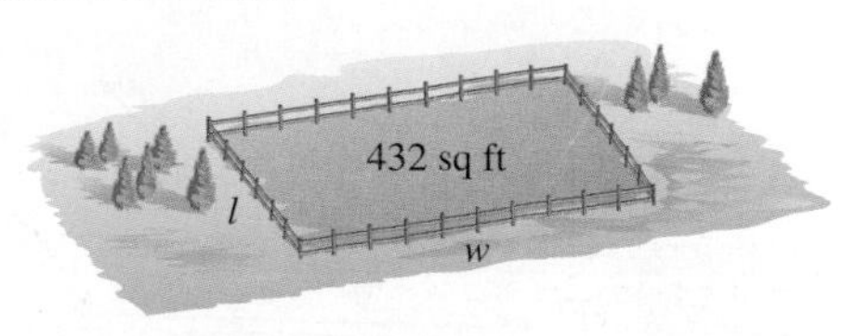

67. A manufacturer estimates the profit on producing x units of its product at

$$P(x) = 300x - 2x^2.$$

a. What is the average rate of change in the profit as the number of units changes from x to $x + h$?

b. Use the result in part (a) to find the average rate of change in the profit as the number of units produced changes from 25 to 50.

c. What is the instantaneous rate of change in the profit when 25 units are produced?

d. Sketch the graph of $y = P(x)$ and the line that passes through the points $(25, P(25))$ and $(50, P(50))$.

68. An open rectangular box is to be made from a piece of cardboard 8 inches wide and 8 inches long by cutting a square from each corner and bending up the sides.

a. Express the volume of the box as a function of the size x of the cutout.

b. Approximate the dimensions of the box with the largest volume.

69. A can has a volume of 900 cm^3. The can is in the shape of a right circular cylinder with a top and a bottom.

a. Express the amount of material needed to construct the can as a function of the radius of the top of the can.

b. Approximate the dimensions of the can that minimizes the amount of material needed to construct the can.

70. In constructing a can with a volume of 900 cm^3 in the shape of a right circular cylinder, no waste is produced when the side of the can is cut, but the top and bottom are each stamped from a square sheet and the remainder is wasted as shown in the figure.

a. Express the area of the material used when stamping the top and bottom as a function of the radius r.

b. Approximate the dimensions of the can that uses the least amount of material with this construction method.

71. A 1-mile race track is to be built with two straight sides of length l and semicircles at the ends of radius r.

a. Express the area enclosed by the oval as a function of the radius r.

b. Approximate the maximum amount of area needed to construct the track.

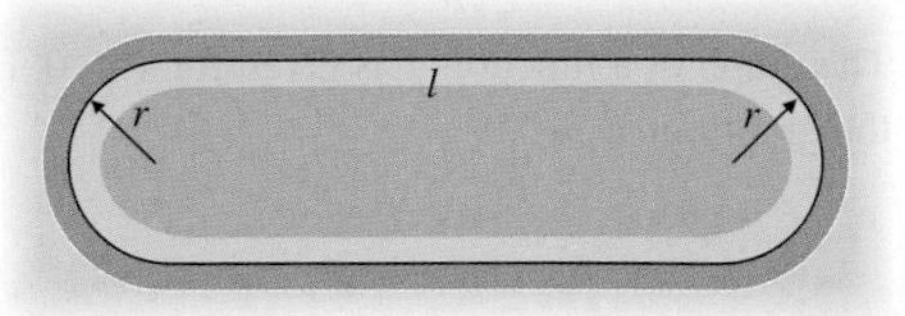

72. A charter bus company charges \$10 per person for a round trip to a ball game and gives a discount for group fares. A group purchasing more than 10 tickets at one time receives a reduction per ticket of \$0.25 times the number of tickets in excess of 10.

a. Express the total revenue as a function of the number of tickets x in excess of 10.

b. What is the maximum revenue that can be received by the bus company from a group?

73. A hotel with 125 rooms normally charges \$80 for a room, but will give discounts for groups. If the group requires more than 10 rooms, the price for each room is decreased by \$2 times the number of rooms exceeding 10.

a. Express the total revenue as a function of the number of rooms x in excess of 10.

b. Find the maximum revenue that the hotel can receive from a group.

74. Newton's law of gravitation states that the attraction between an object of mass m_1 and an object of mass m_2 is directly proportional to the product of the masses m_1 and m_2 of the objects and inversely proportional to the square of the distance r between the centers of mass of the objects.

a. Write a functional relationship expressing this force in terms of the distance r, assuming that the masses remain constant.

b. What restrictions must be put on the domain of this function if it is to describe the physical situation?

c. Sketch the graph of the function.

1.7 LINEAR FUNCTIONS

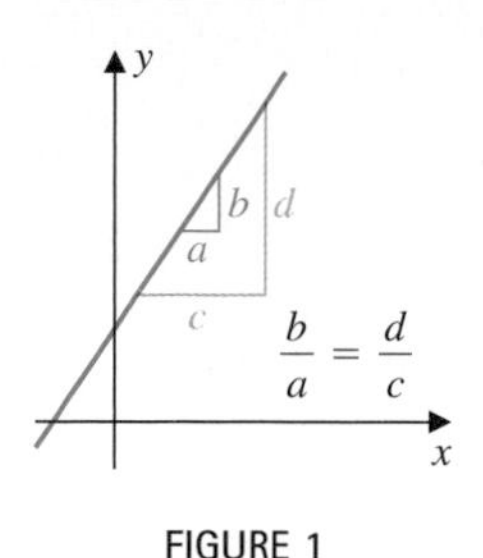

FIGURE 1

A *linear* relationship between two variables occurs when there is a constant increase or a constant decrease in one variable with respect to the other. So linear functions have the property that any change in the independent variable results in a proportional change in the dependent variable, as shown in Figure 1.

Many physical situations can be modeled using a linear relationship. Table 1 shows the approximately linear increasing trend in the global atmospheric concentrations of carbon dioxide in parts per million (ppm) from 1970 to 2010. Scientists are concerned that a continuation of this trend will lead to a "greenhouse effect," with the result that the average temperature of the earth will increase. This could produce catastrophic effects such as widespread drought and the flooding of coastal areas.

The data in Table 1 do not form an exact linear relationship because the carbon dioxide levels do not always increase by a fixed amount. However, a plot of the data, called a *scatter plot*, shown in Figure 2(a), shows a steady, increasing pattern that is approximately linear.

TABLE 1

Year	Carbon Dioxide (ppm)
1970	325
1980	339
1990	354
2000	369
2010	387

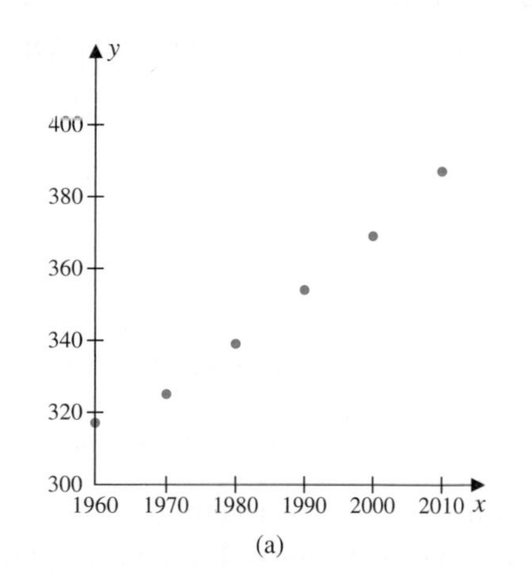

(a)

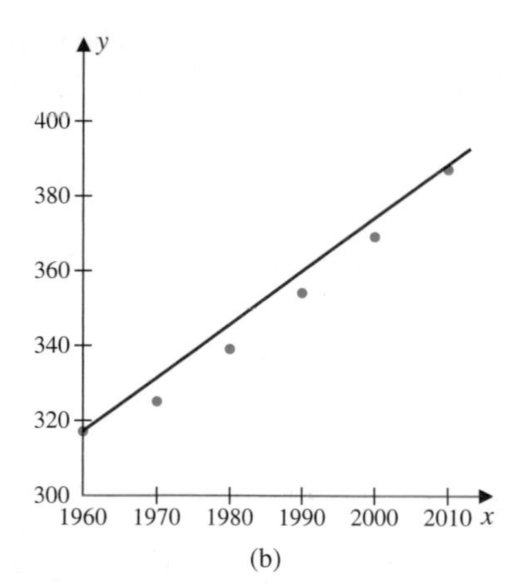

(b)

FIGURE 2

Calculus permits us to find the equation of the regression line that "best" fits all the data.

The line shown in Figure 2(b) passes through the two endpoints of the data and does a reasonable job of approximating the remaining points. Later in this section we will determine the equation of this line and use it to approximate the concentration in the year 2020.

A Linear Function

A function f defined by a linear equation of the form

$$y = f(x) = ax + b, \quad \text{where } a \text{ and } b \text{ are constants,}$$

is called a **linear function.**

Linear functions have graphs that are straight lines, and any nonvertical straight line is the graph of a linear function.

In calculus, the slope of a line is used to describe the slope of the graph of a function.

Suppose that l is a nonvertical straight line and that $P(x_1, y_1)$ and $Q(x_2, y_2)$ are two distinct points lying on l, as illustrated in Figure 3.

The quotient of the difference of the y-coordinates over the difference of the x-coordinates is called the *slope* of the line.

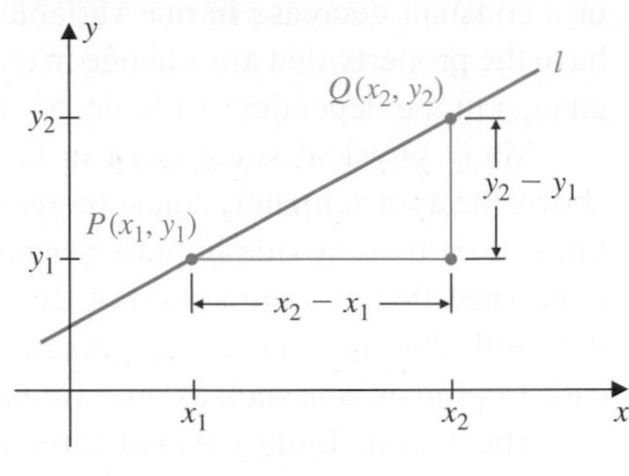

FIGURE 3

The Slope of a Line

The **slope** of the line passing through (x_1, y_1) and (x_2, y_2), when $x_1 \neq x_2$, is

$$m = \frac{y_2 - y_1}{x_2 - x_1}.$$

EXAMPLE 1 **a.** Find the slope of the line $2x - 3y = 3$.

b. Find the y- and x-intercepts of $2x - 3y = 3$, and sketch the graph of this line.

Solution **a.** To find the slope, place the equation in the form $y = mx + b$. Since

$$2x - 3y = 3, \quad \text{we have} \quad -3y = -2x + 3 \quad \text{and} \quad y = \frac{2}{3}x - 1.$$

The slope of the line is consequently $\frac{2}{3}$.

b. The graph is shown in Figure 4. The y-intercept occurs when $x = 0$, so

$$y = \frac{2}{3}(0) - 1 = -1.$$

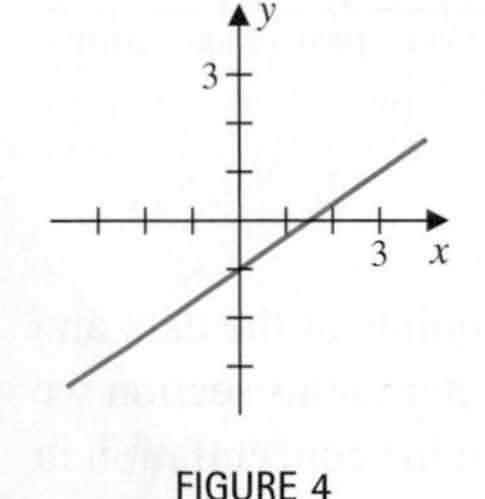

FIGURE 4

The x-intercept occurs when $y = 0$, so

$$0 = \frac{2}{3}x - 1, \quad \text{which implies that} \quad x = \frac{3}{2}.$$ ■

Any two distinct points on a line will give the same value for the slope. So, if we replace one of the pairs of points, say, (x_2, y_2), with an arbitrary pair (x, y) lying on the line, we have the *point-slope* form of the equation of the line.

Point-Slope Equation of a Line

The line with slope m that passes through the point (x_1, y_1) has the **point-slope** equation

$$y - y_1 = m(x - x_1).$$

Using a graphing device to plot a variety of lines quickly in the same viewing rectangle provides a visualization of the slope of a line.

The slope of a line determines its direction, in the following sense (see Figure 5).

Slopes of Lines

- A line with a **positive slope** is directed upward (from left to right), and y increases as x increases. The larger the positive slope, the faster the values of y increase.
- A line with a **negative slope** is directed downward (from left to right), and y decreases as x increases. The values of y decrease more rapidly on a line with a negative slope that is large in magnitude than on a line with a negative slope that is small in magnitude.
- A horizontal line has **zero slope**, and a vertical line has **no slope.**

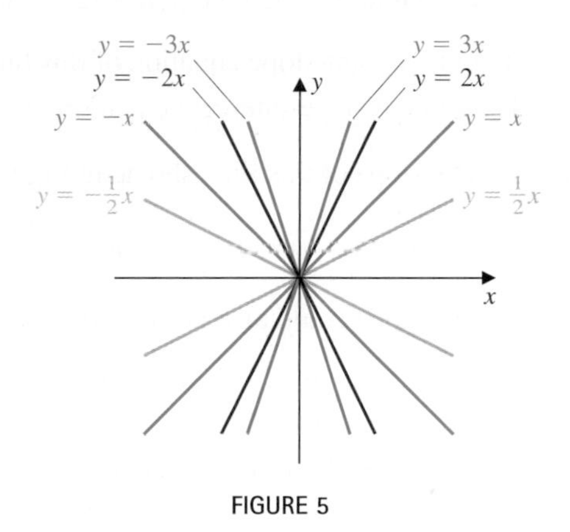

FIGURE 5

The point-slope equation $y - y_1 = m(x - x_1)$ can be rewritten as

$$y = m(x - x_1) + y_1 = mx + (y_1 - mx_1).$$

This gives the *slope-intercept* form of the equation of the line.

Slope-Intercept Equation of a Line

The line with slope m passing through the point (x_1, y_1) has the **slope-intercept** equation

$$y = mx + b, \quad \text{where} \quad b = y_1 - mx_1.$$

The name *slope-intercept* comes from the fact that it involves m, the *slope*, and b, the *y-intercept* of the line. The linear function that is described by this equation has the form $f(x) = mx + b$.

When $m = 0$ the linear function $f(x) = b$ is constant and the graph is a horizontal line, as shown in Figure 6(a). Points lying on vertical lines have the same x-coordinates, so they have equations of the form $x = c$ for some constant c. Vertical lines have no slope, so they cannot be graphs of functions. (See Figure 6(b).)

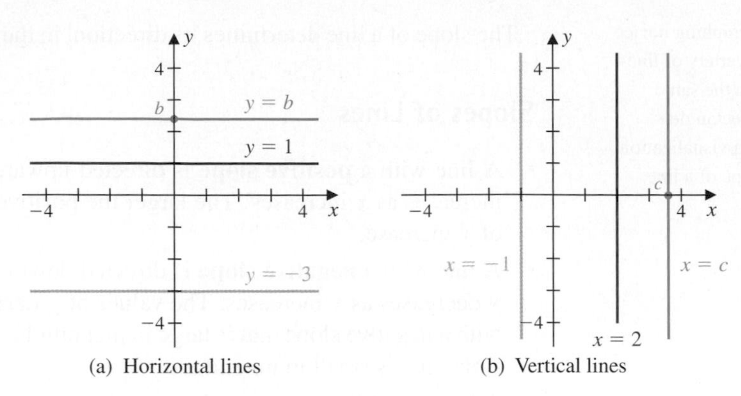

FIGURE 6

EXAMPLE 2 A line passes through the points (2, 3) and (−4, 0).

a. Find a point-slope equation of this line.

b. Find the slope-intercept equation of this line.

Solution **a.** The slope of this line, shown in Figure 7, is

$$m = \frac{0-3}{-4-2} = \frac{-3}{-6} = \frac{1}{2}.$$

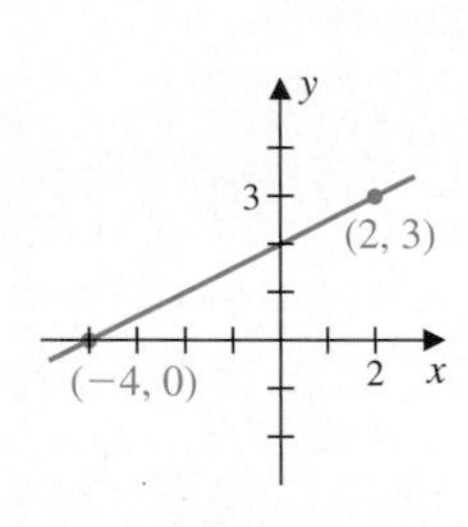

FIGURE 7

The slope could also be computed, reversing the roles of the two points, as

$$m = \frac{3-0}{2-(-4)} = \frac{3}{6} = \frac{1}{2}.$$

The point-slope equation obtained by using the point (2, 3) is

$$y - 3 = \frac{1}{2}(x-2).$$

The point-slope equation obtained by using the point (−4, 0) is

$$y - 0 = \frac{1}{2}(x-(-4)), \quad \text{or} \quad y = \frac{1}{2}(x+4).$$

b. Both point-slope equations reduce to the same slope-intercept equation,

$$y = \frac{1}{2}x + 2.$$ ■

In part (a) of Example 2, we found two different point-slope formulas, but the two equations reduce to the same slope-intercept formula in part (b).

- Each point on a line with nonzero slope generates a different point-slope equation, but they all reduce to the same slope-intercept equation.

EXAMPLE 3 Sketch the graph of the linear function described by $f(x) = 2 - \frac{2}{3}x$.

Solution The equation

$$y = 2 - \frac{2}{3}x = -\frac{2}{3}x + 2$$

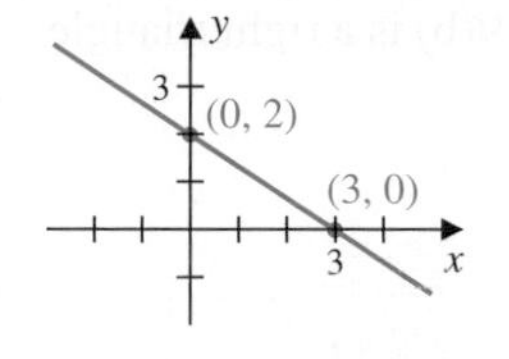

FIGURE 8

describes the line with slope $-\frac{2}{3}$ and y-intercept 2, so the point (0, 2) lies on the graph of the line. To find another point on the line we can set $y = 0$ to find the x-intercept of the line. This gives

$$0 = 2 - \frac{2}{3}x, \quad \text{and} \quad \frac{2}{3}x = 2, \quad \text{so} \quad x = 3.$$

The graph of f is the line through (0, 2) and (3, 0) shown in Figure 8. ■

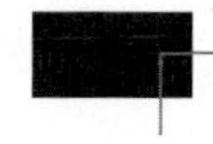

Parallel and Perpendicular Lines

In calculus you frequently need to find equations of lines as part of a larger problem. Tangent lines to curves are particularly important.

Straight lines that never intersect are *parallel*, and lines that intersect at right angles are *perpendicular*. The slopes of lines provide an easy way to determine when two lines are parallel or perpendicular.

Slopes of Parallel and Perpendicular Lines

Suppose the nonvertical lines l_1 and l_2 have the slopes m_1 and m_2, respectively.

- Lines l_1 and l_2 are **parallel** if and only if $m_1 = m_2$.
- Lines l_1 and l_2 are **perpendicular** if and only if $m_1 m_2 = -1$.

We will first show the perpendicular line result. Consider the lines shown in Figure 9(a), where the origin of the coordinate system is at the point of intersection. Line l_1 has y-intercept 0 and slope m_1, so it has the equation $y = m_1 x$ and it passes through the point $(1, m_1)$. Similarly, line l_2 has the equation $y = m_2 x$ and passes through $(1, m_2)$.

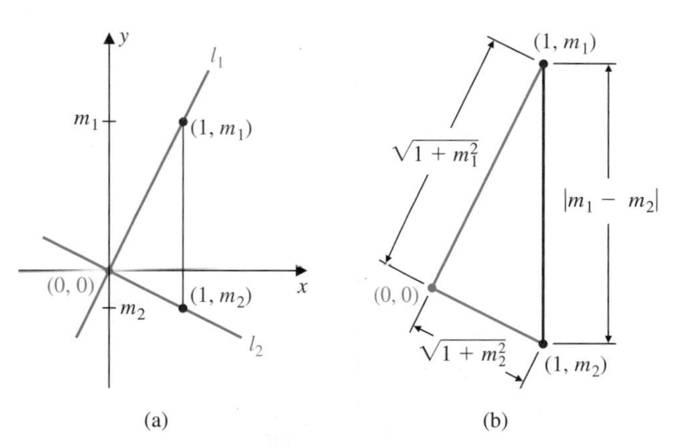

FIGURE 9

A line perpendicular to a curve or surface is called a *normal line.* Finding normal lines to surfaces is important in calculus.

Figure 9(b) shows the lengths of the sides of the triangle with vertices (0, 0), $(1, m_1)$, and $(1, m_2)$. To find the lengths of the sides use the distance formula between points in the plane. For example, the side with ends (0, 0) and $(1, m_1)$ has length

$$\sqrt{(1-0)^2 + (m_1 - 0)^2} = \sqrt{1 + m_1^2}.$$

The lines are perpendicular if and only if the triangle in Figure 9(b) is a right triangle, for which the Pythagorean Theorem implies that

$$(1 + m_1^2) + (1 + m_2^2) = |m_1 - m_2|^2.$$

Expanding the right side of the last equation gives

$$2 + m_1^2 + m_2^2 = m_1^2 - 2m_1m_2 + m_2^2,$$

which is equivalent to

$$-2 = 2m_1m_2, \quad \text{or to} \quad m_1m_2 = -1.$$

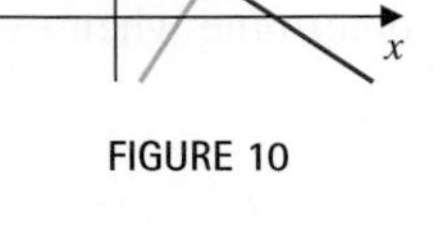

FIGURE 10

Hence, the two lines are perpendicular precisely when $m_1m_2 = -1$.

The result about parallel lines follows from the perpendicular result. Two lines l_1 and l_2, with slopes m_1 and m_2, are parallel if and only if they are perpendicular to a common line l_3 with slope $m_3 \neq 0$, as shown in Figure 10. This means that both

$$m_1m_3 = -1 \quad \text{and} \quad m_2m_3 = -1, \quad \text{so} \quad m_1m_3 = m_2m_3.$$

Since $m_3 \neq 0$, we have $m_1 = m_2$.

EXAMPLE 4 Find an equation of the line:

a. Passing through (3, 1) and parallel to the line with equation $y = 2x - 1$;

b. Passing through (3, 1) and perpendicular to the line with equation $y = 2x - 1$.

Solution **a.** The slope of the line with equation $y = 2x - 1$ is 2, which is also the slope of any line parallel to $y = 2x - 1$. The required line passes through (3, 1), as shown in Figure 11(a), so it has equation

$$y - 1 = 2(x - 3), \quad \text{or} \quad y = 2x - 5.$$

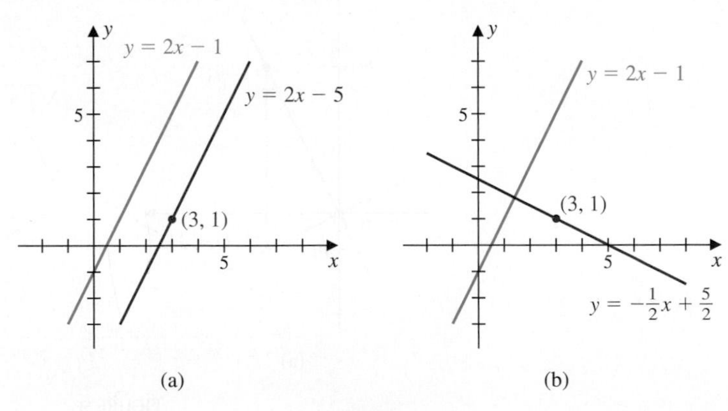

FIGURE 11

b. Since $y = 2x - 1$ has slope 2, any line perpendicular to $y = 2x - 1$ has slope $-\frac{1}{2}$. So the perpendicular line through (3, 1), which is shown in Figure 11(b), has equation

$$y - 1 = -\frac{1}{2}(x - 3), \quad \text{or} \quad y = -\frac{1}{2}x + \frac{5}{2}. \qquad \blacksquare$$

We have been careful to exclude vertical lines to this point in the section because these lines cannot be the graphs of linear *functions*. Vertical lines are, however, included in the general form of linear equations. These are equations of the form

$$Ax + By + C = 0,$$

where A, B, and C are constants.

General Linear Equations $Ax + By + C = 0$

- When $B = 0$ and $A \neq 0$, the equation describes a vertical line. Vertical lines have no slope.
- When $B \neq 0$, the equation describes a line with slope $-A/B$ and y-intercept $-C/B$. (When $A = 0$, the line is horizontal.)

Applications

EXAMPLE 5 The carbon dioxide data given in Table 1 at the beginning of this section state that there were 325 ppm in the atmosphere in 1970 and 387 ppm in 2010. Determine a linear function to predict the level of carbon dioxide in the atmosphere in the year 2020.

Solution Let t represent time and $f(t)$ the number of parts per million at time t. Then (1970, 325) and (2010, 387) are points on the graph, and the slope of the line joining the points, shown in Figure 12, is

$$m = \frac{387 - 325}{2010 - 1970} = \frac{52}{40} = \frac{13}{10}.$$

The linear function satisfies

$$f(t) - 325 = \frac{13}{10}(t - 1970), \quad \text{so} \quad f(t) = 325 + \frac{13}{10}(t - 1970).$$

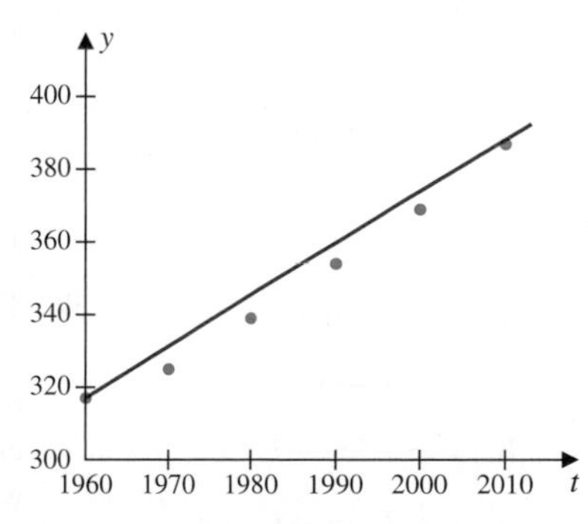

FIGURE 12

The predicted concentration in the year 2020 is

$$f(2020) = 325 + \frac{13}{10}(2020 - 1970) = 390 \text{ ppm}.$$

If different data points in Table 1 had been used to determine the equation, the approximation for the carbon dioxide levels in the atmosphere would differ slightly. ■

EXERCISE SET 1.7

In Exercises 1–8, sketch the straight line determined by the points, and find the slope-intercept equation of the line.

1. $(1, 1), (2, 4)$
2. $(1, 2), (3, -2)$
3. $(-2, 1), (2, -4)$
4. $(1, -3), (5, 1)$
5. $(-1, 3), (2, 3)$
6. $(3, 5), (3, -2)$
7. $(-1, -3), (-1, 6)$
8. $(-2, -4), (2, -4)$

In Exercises 9–16, find the slope of the line, and sketch its graph.

9. $y = x - 3$
10. $x + y = 3$
11. $3x - y = 2$
12. $2x - y = -1$
13. $-3x - 4y = 2$
14. $-2x - 3y = 1$
15. $y = 2$
16. $x = 2$
17. Find equations of the lines that pass through the point $(1, 4)$ and have the given slope.
    a. 1 **b.** -1 **c.** 3 **d.** $\frac{1}{3}$
18. Find equations of the lines that pass through the point $(-1, -2)$ and have the given slope.
    a. 2 **b.** -2 **c.** 3 **d.** $\frac{1}{3}$
19. Group the following lines into sets that are parallel.
    a. $y = 2$ **b.** $y = x + 1$
    c. $y = -x + 2$ **d.** $y = x + 4$
    e. $y = 2x - 5$ **f.** $y = -4$
    g. $x + y = 0$ **h.** $y - 2x = 2$
    i. $x - y = 4$ **j.** $2x - 2y = 1$
20. Group the following lines into sets that are perpendicular.
    a. $y = -1$ **b.** $x = -2$
    c. $y = 2x + 3$ **d.** $y = 3x + 5$
    e. $y = \frac{1}{3}x - \frac{1}{3}$ **f.** $y = x - 3$
    g. $y = -x + 1$ **h.** $y = -3x - 7$
    i. $x + 3y = -5$ **j.** $x + 2y = -1$

*In Exercises 21–24, the equation of a line is given, together with a point that is not on the line. Find the slope-intercept form of the equation of the line that passes through the given point and is **(a)** parallel to the given line and **(b)** perpendicular to the given line.*

21. $y = 2x + 1; (0, 0)$
22. $x + y = -2; (0, 0)$
23. $y = -2x + 3; (-1, 2)$
24. $y = 3x - 2; (1, 2)$

In Exercises 25–36, find the slope-intercept equation of the line that satisfies the given conditions.

25. Has slope -1 and y-intercept 2.
26. Has slope -4 and y-intercept 1.
27. Passes through $(1, -2)$ with slope 3.
28. Passes through $(-1, -3)$ with slope -2.
29. Has x-intercept 1 and y-intercept 3.
30. Has x-intercept -2 and y-intercept -3.
31. Passes through $(2, -1)$ and is parallel to the x-axis.
32. Passes through $(-2, 3)$ and is parallel to the y-axis.
33. Passes through $(4, 3)$ and is parallel to $2x - 3y = 2$.
34. Passes through $(3, -2)$ and is parallel to $2x + y = 3$.
35. Is perpendicular to $y = 2 - x$ at the point $(1, 1)$.
36. Passes through $(-3, 5)$ and is perpendicular to $x - 2y = 4$.
37. **a.** Find an equation of the line that is tangent to the circle $x^2 + y^2 = 3$ at the point $(1, \sqrt{2})$.
    b. At what other point on the circle will the tangent line be parallel to the line in part (a)?
38. Show that the line with x-intercept at $a \neq 0$ and y-intercept at $b \neq 0$ has equation

$$\frac{x}{a} + \frac{y}{b} = 1.$$

This is called the *intercept* equation of a line.

39. The function defined by $v(t) = -32t$ describes the velocity of a rock t seconds after it has been dropped from the top of a 784-foot-high building. Sketch the graph of v, and determine the velocity of the rock when it hits the ground, 7 seconds after it has been dropped.
40. Determine the linear function that relates the temperature in degrees Celsius to the temperature in degrees Fahrenheit, and use this function to determine the Fahrenheit temperature corresponding to 30°C. [*Note*: At sea level, water freezes at 32°F (0°C) and boils at 212°F (100°C).]

41. The average weight W, in grams, of a fish in a particular pond depends on the total number n of fish in the pond according to the model

$$W(n) = 500 - 0.5n.$$

   a. Sketch the graph of the function W.
   b. Express the total fish weight production in grams as a function of the number of fish in the pond.
   c. What happens when $n \geq 1000$?

42. A new computer workstation costs \$10,000. Its useful lifetime is 5 years, at which time it will be worth an estimated \$2000. The company calculates its depreciation using the linear decline method that is an option in U.S. tax law.

   a. Find the linear equation that expresses the value V of the equipment as a function of time t, for $0 \leq t \leq 5$.
   b. How much will the equipment be worth after 2.5 years?
   c. What is the average rate of change in the value of the equipment from 1 to 3 years?

43. A new car cost \$28,000. After 3 years its value is \$16,000. The depreciation of the car is assumed to be linear.

   a. Find an equation that expresses the value of the car in terms of the age.
   b. What is the value of the car after 5 years?
   c. When will the car have no value?

44. A piece of equipment costs a company \$250,000 when new and 7 years later is worth \$180,000. The company uses a linear depreciation model to value the equipment.

   a. Express the value of the equipment as a function of the age and sketch a graph of the function.
   b. The company has decided to keep the equipment as long as the value remains above \$70,000. When should the company sell or trade in the equipment?

1.8 QUADRATIC FUNCTIONS

The motion of a falling object is described using a quadratic function.

A function f defined by a quadratic equation of the form

$$y = f(x) = ax^2 + bx + c, \quad \text{where} \quad a \neq 0,$$

is a **quadratic function**, and its graph is a **parabola**. The most basic parabola is the graph of the *squaring* function, defined by $y = f(x) = x^2$. This is the quadratic equation with $a = 1$, $b = 0$, and $c = 0$, whose graph is shown in Figure 1.

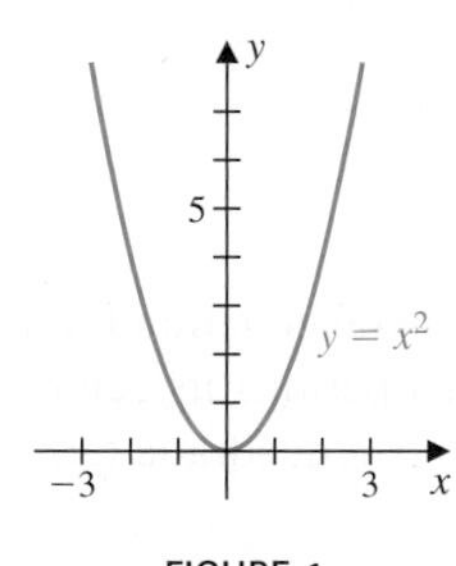

FIGURE 1

For $x < 0$, the values of $f(x) = x^2$ decrease as x increases, so we say that f *decreases* when $x < 0$. For $x > 0$, the values of $f(x)$ increase as x increases, so f *increases* when $x > 0$. Notice that the function f has a minimum value of 0 at $x = 0$, but it has no maximum value, since

$$x^2 \to \infty \text{ as } x \to \infty \quad \text{and} \quad x^2 \to \infty \text{ as } x \to -\infty.$$

The notions of increasing and decreasing help to describe the features of arbitrary curves.

The graph of $f(x) = x^2$ is used to determine the graphs of other quadratic functions by using a basic transformation technique.

The next example begins a study of how the graphs of many functions can be generated systematically from the graphs of a few common functions. Mastering these techniques is essential because you will frequently use them in this course and in calculus.

EXAMPLE 1 Sketch the graph of $g(x) = (x - 1)^2$.

Solution The correspondence described by $g(x) = (x - 1)^2$ is similar to that of $f(x) = x^2$ except that 1 unit is subtracted *before* the squaring operation is performed. Consequently,

$$g(1) = (1 - 1)^2 = 0^2 = f(0), \quad g(2) = 1^2 = f(1), \quad g(3) = f(2),$$

and so on. Because of this, the graph of $y = g(x)$ is the same as the graph of $y = x^2$ except that it is moved to the right 1 unit, as shown in Figure 2. The function g is decreasing when $x < 1$, is increasing when $x > 1$, and has a minimum value of 0 at $x = 1$. ■

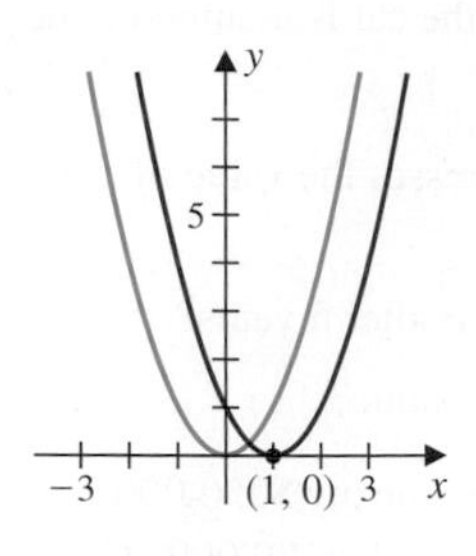

FIGURE 2

The horizontal shifting, or translation, technique described in Example 1 can be applied in general, as illustrated in Figure 3.

Horizontal Shifts of Graphs

Suppose that $c > 0$.

- The graph of $y = f(x - c)$ is the graph of $y = f(x)$ shifted right c units.
- The graph of $y = f(x + c)$ is the graph of $y = f(x)$ shifted left c units.

Knowing how to graph functions can help distinguish between a true representative of a function and an incomplete graph.

y
c
y = f(x)
x
y = f(x + c)

Horizontal translation
c units to the left

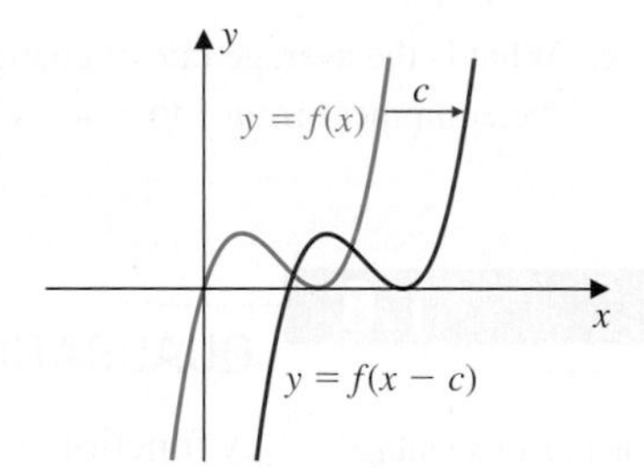

Horizontal translation
c units to the right

FIGURE 3

The Standard Form of a Quadratic Equation

In Section 1.3 we found that *completing the square* changes a general equation for a circle in the form $ax^2 + by^2 + cx + dy + e = 0$ into a form that permits its center and radius to be determined. This technique is also used to change a quadratic term of the form

$$ax^2 + bx + c, \quad \text{with} \quad a \neq 0,$$

into a *standard form*, one that is the sum of a constant and a square term that involves the variable x.

The coefficient of x^2 must be 1 in order to complete the square, so we first factor a from the terms involving x to produce

$$ax^2 + bx + c = a\left(x^2 + \frac{b}{a}x\right) + c.$$

Now we add and subtract the term $(b/2a)^2$ inside the parentheses to obtain

$$ax^2 + bx + c = a\left(x^2 + \frac{b}{a}x + \left(\frac{b}{2a}\right)^2 - \left(\frac{b}{2a}\right)^2\right) + c$$

$$= a\left(x^2 + \frac{b}{a}x + \left(\frac{b}{2a}\right)^2\right) - a\left(\frac{b}{2a}\right)^2 + c.$$

The term inside the first set of parentheses is a perfect square, so

Determining maximal and minimal values for arbitrary functions is often difficult. This is one of the important applications you will see in calculus.

$$ax^2 + bx + c = a\left(x + \frac{b}{2a}\right)^2 + c - \frac{b^2}{4a} = a\left(x + \frac{b}{2a}\right)^2 + \frac{4ac - b^2}{4a}.$$

The squared term is 0 precisely when $x = -b/(2a)$, so the *vertex* of the parabola with equation $y = ax^2 + bx + c$ occurs at the point

$$\left(-\frac{b}{2a}, \frac{4ac - b^2}{4a}\right).$$

This vertex gives the minimal y-value when $a > 0$ and the maximal y-value when $a < 0$, as illustrated in Figure 4.

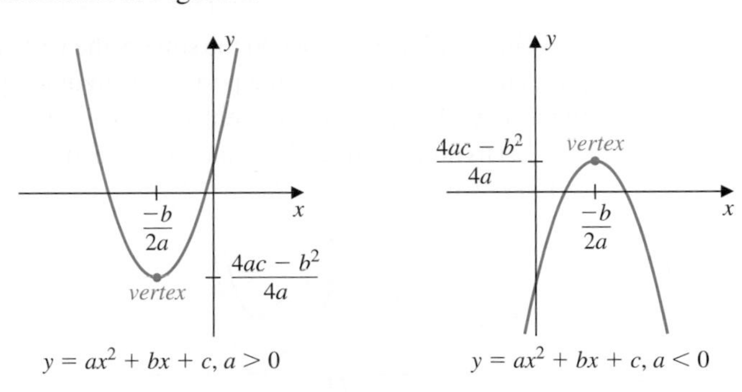

FIGURE 4

Completing the square on a quadratic function permits its graph to be determined by simply shifting the graph of a basic quadratic function of the form $f(x) = ax^2$. Example 2 illustrates the technique when the coefficient of the x^2 term is 1. Later in the section we consider the more general situation when $a \neq 1$.

EXAMPLE 2 Sketch the graph of $h(x) = x^2 - 2x + 3$.

Solution We first complete the square on the quadratic by adding and subtracting the term

$$\left(\frac{b}{2a}\right)^2 = \left(\frac{-2}{2 \cdot 1}\right)^2 = (-1)^2 = 1,$$

to obtain

$$h(x) = x^2 - 2x + 3 = (x^2 - 2x + 1) - 1 + 3 = (x - 1)^2 + 2.$$

In Example 1 we saw that the graph of $g(x) = (x - 1)^2$ is the same as the graph of $f(x) = x^2$, except that it is shifted right 1 unit. Each y-coordinate of the graph of

$$h(x) = (x - 1)^2 + 2 = g(x) + 2$$

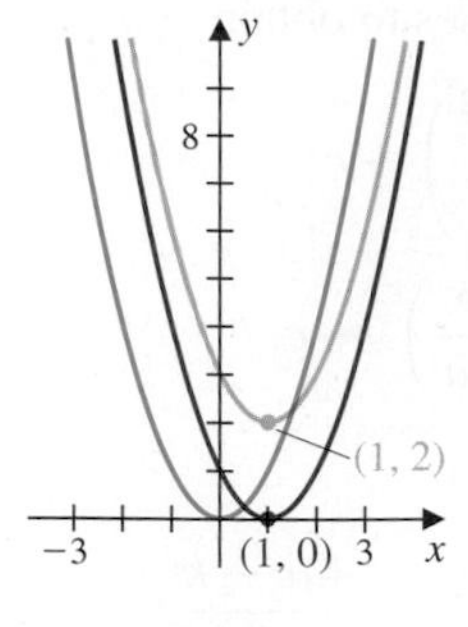

FIGURE 5

is 2 units greater than the corresponding y-coordinate on the graph of $y = g(x)$. So the graph of $y = h(x) = (x - 1)^2 + 2$ is obtained by shifting the graph of $y = g(x) = (x - 1)^2$ upward 2 units. Consequently, the graph of

$$h(x) = x^2 - 2x + 3 = (x - 1)^2 + 2,$$

shown in yellow in Figure 5, is the graph of $y = x^2$ shifted to the right 1 unit and upward 2 units. The function h is decreasing when $x < 1$, is increasing when $x > 1$, and has a minimum value of $y = 2$ at $x = 1$. The vertex of the parabola is $(1, 2)$. ■

The general rule associated with the vertical shifting, or translation, technique described in Example 2 is illustrated in Figure 6.

Vertical Shifts of Graphs

Suppose that $c > 0$.

- The graph of $y = f(x) + c$ is the graph of $y = f(x)$ shifted up c units.
- The graph of $y = f(x) - c$ is the graph of $y = f(x)$ shifted down c units.

Once we have completed the square, the vertical shifting result combined with the horizontal shifting technique permits us to quickly sketch the graph of any quadratic function with leading coefficient 1.

Let us now consider the situation when the quadratic equation has an x^2-coefficient different from 1.

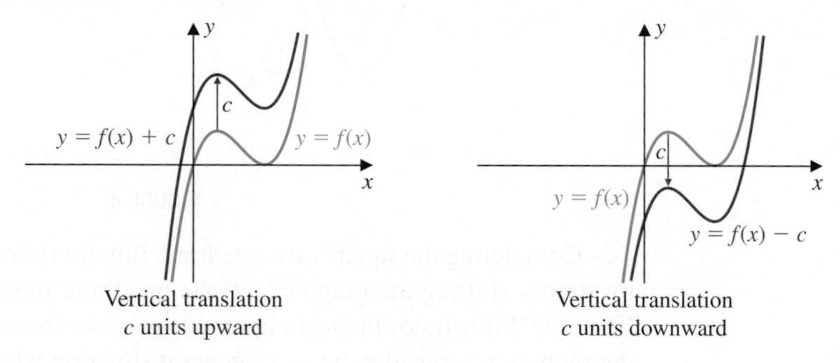

FIGURE 6

We will first consider the most basic quadratics, those with $b = 0$ and $c = 0$. The graph of a quadratic function with equation $y = ax^2$, for an arbitrary constant $a \neq 0$, has a shape similar to the graph of $y = x^2$, but is scaled vertically. (See Figure 7.)

The Graph of the Quadratic Equation $f(x) = ax^2$

- The graph opens upward when $a > 0$, and opens downward when $a < 0$.
- The greater the magnitude of a, the steeper the graph and the narrower the opening.
- If $a > 0$, then $f(x) \to \infty$ as $x \to -\infty$ and $f(x) \to \infty$ as $x \to \infty$.
- If $a < 0$, then $f(x) \to -\infty$ as $x \to -\infty$ and $f(x) \to -\infty$ as $x \to \infty$.

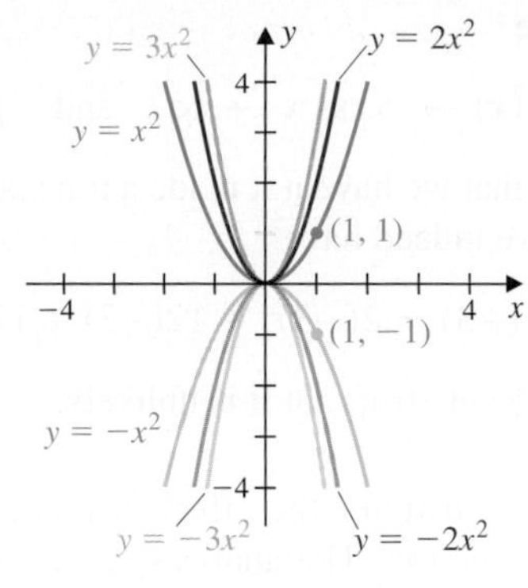

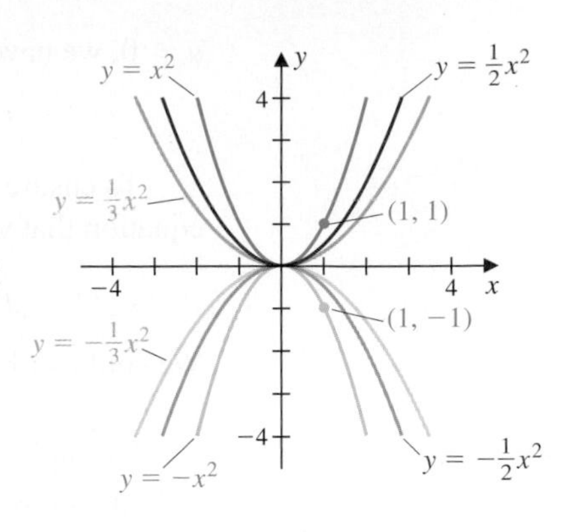

FIGURE 7

EXAMPLE 3 Sketch the graph of $f(x) = 2x^2 + 12x + 17$.

Solution We first complete the square on the quadratic. Then we use horizontal and vertical shifting to determine the graph.

Factor 2, the coefficient of x^2, from the first two terms of the expression, and then complete the square. This gives

$$f(x) = 2(x^2 + 6x) + 17 = 2(x^2 + 6x + 9) - 2 \cdot 9 + 17 = 2(x + 3)^2 - 1.$$

The graph of $y = 2x^2$ has the same shape as the graph of $y = x^2$, except that it is vertically compressed by a factor of 2, as shown in Figure 8(a).

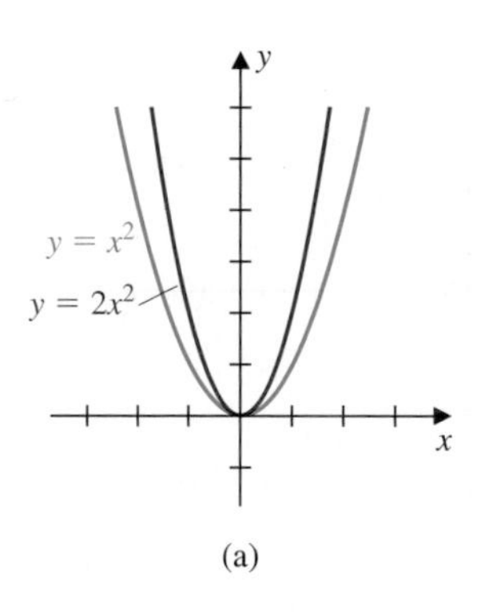

(a)

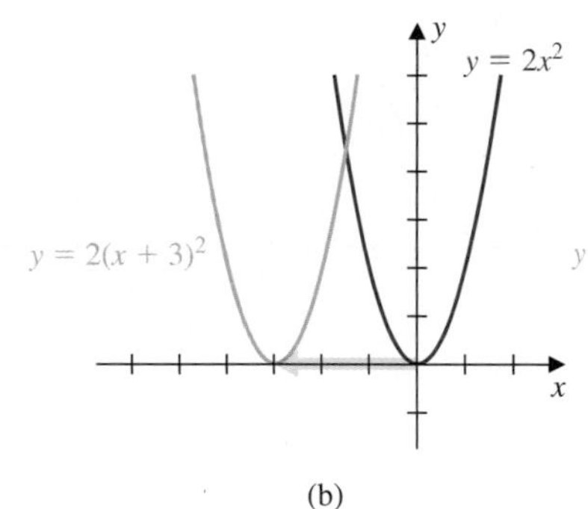

(b)

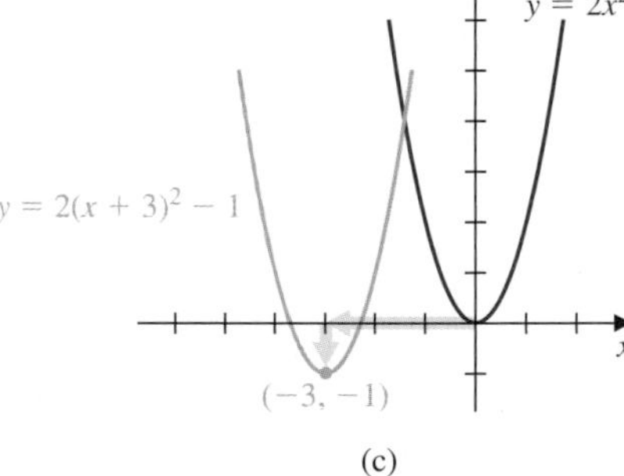

(c)

FIGURE 8

To sketch the graph of $y = 2(x+3)^2 - 1$, sketch in sequence

- $y = x^2$
- $y = 2x^2$
- $y = 2(x+3)^2$
- $y = 2(x+3)^2 - 1$

The graph of $y = 2(x+3)^2$ is the graph of $y = 2x^2$ shifted to the left 3 units, as shown in Figure 8(b).

Then the graph of $y = 2(x + 3)^2 - 1$ is the graph of $y = 2(x + 3)^2$ shifted downward 1 unit. Consequently, the graph of

$$y = f(x) = 2x^2 + 12x + 17 = 2(x + 3)^2 - 1$$

has the same shape as the graph of $y = 2x^2$, but it is shifted to the left 3 units and downward 1 unit, as shown in Figure 8(c).

The function f is decreasing when $x < -3$, is increasing when $x > -3$, and has a minimum value of -1 at $x = -3$. The vertex of the parabola is $(-3, -1)$. Since

$a > 0$, we have

$$f(x) \to \infty \text{ as } x \to \infty \quad \text{and} \quad f(x) \to \infty \text{ as } x \to -\infty.$$

To ensure that we have not made a translation error we can check in the original equation that we indeed have

$$f(-3) = 2(-3)^2 + 12(-3) + 17 = 18 - 36 + 17 = -1.$$

We could still be in error, but it is unlikely. ■

Example 4 illustrates the effect of a negative leading coefficient on the graph of a quadratic function. The analysis is the same as when the leading coefficient is positive, but care is required to ensure that the algebra is performed correctly.

EXAMPLE 4 Sketch the graph of $f(x) = -\frac{1}{3}x^2 + 2x + 3$.

Solution Factor $-\frac{1}{3}$ from the x^2- and x-terms of the expression and then complete the square. This gives

$$\begin{aligned} f(x) &= -\tfrac{1}{3}(x^2 - 6x) + 3 \\ &= -\tfrac{1}{3}(x^2 - 6x + 9) - \left(-\tfrac{1}{3}\right) \cdot 9 + 3 = -\tfrac{1}{3}(x - 3)^2 + 6. \end{aligned}$$

First sketch the graph of $y = \frac{1}{3}x^2$. Then reflect the graph of $y = \frac{1}{3}x^2$ about the x-axis to produce the graph of $y = -\frac{1}{3}x^2$, as is shown in Figure 9(a).

Principles of Problem Solving:
Reduce a hard problem to a series of easier problems. This shows the heart of the problem.

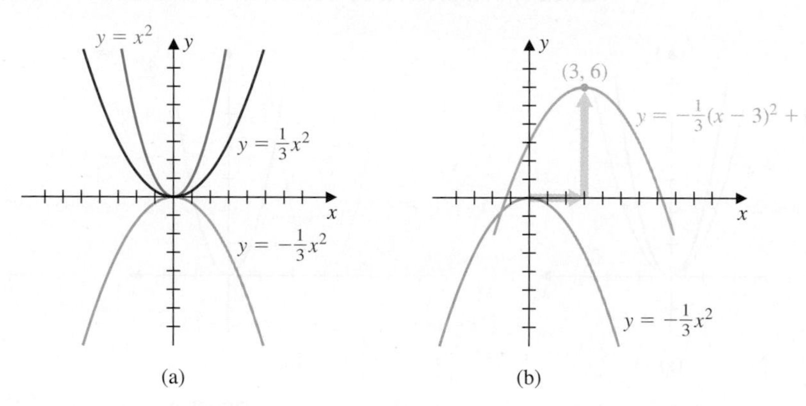

FIGURE 9

Now apply the shifting technique to $y = -\frac{1}{3}x^2$ to obtain the graph of

$$y = -\tfrac{1}{3}x^2 + 2x + 3 = -\tfrac{1}{3}(x - 3)^2 + 6.$$

It has the same shape as $y = -\frac{1}{3}x^2$, but it is shifted to the right 3 units and upward 6 units, as shown in Figure 9(b).

The function f is increasing when $x < 3$, is decreasing when $x > 3$, and has a maximum value of 6 at $x = 3$. The vertex of the parabola is (3, 6). Checking in the original equation gives

$$f(3) = -\tfrac{1}{3}(3)^2 + 2(3) + 3 = -3 + 6 + 3 = 6.$$

Since $-1/3 = a < 0$, we have

$$f(x) \to -\infty \text{ as } x \to \infty \quad \text{and} \quad f(x) \to -\infty \text{ as } x \to -\infty. \qquad \blacksquare$$

The graph in Figure 9(b) shows x-axis intercepts near -1 and 7. The completed square of the quadratic can also be used to determine the exact values of these x-intercepts. Setting y to 0 and solving for x gives

$$0 = -\tfrac{1}{3}x^2 + 2x + 3 = -\tfrac{1}{3}(x-3)^2 + 6,$$

so

$$\tfrac{1}{3}(x-3)^2 = 6, \quad \text{and} \quad (x-3)^2 = 18.$$

We have the two solutions

$$x - 3 = \pm\sqrt{18} = \pm 3\sqrt{2},$$

which simplify to

$$x = 3 + 3\sqrt{2} \quad \text{and} \quad x = 3 - 3\sqrt{2}.$$

Since $\sqrt{2}$ is approximately 1.4 (written $\sqrt{2} \approx 1.4$), the solutions are

$$x = 3 - 3\sqrt{2} \approx 3 - 3(1.4) \approx -1.2 \quad \text{and} \quad x = 3 + 3\sqrt{2} \approx 7.2,$$

which appear to agree with our graph.

EXAMPLE 5 Find the function whose graph is the parabola with vertex $(2, -1)$ and that contains the point $(4, 2)$.

Solution The graph is a parabola with vertex $(2, -1)$, so the function has the form

$$f(x) = a(x-2)^2 - 1$$

for some number a. Since $(4, 2)$ is on the parabola,

$$2 = f(4) = a(4-2)^2 - 1, \quad \text{so} \quad 4a = 3, \quad \text{which simplifies to } a = \tfrac{3}{4}.$$

Hence $f(x) = \frac{3}{4}(x-2)^2 - 1$. ■

EXAMPLE 6 Use a graphing device to approximate the point on the parabola $y = (x-4)^2$ that is nearest the origin.

Solution Figure 10(a) shows the parabola and the distance $d(x)$ that must be minimized. An arbitrary point on the parabola has coordinates $(x, (x-4)^2)$ and the distance from $(0, 0)$ to $(x, (x-4)^2)$ is

$$d(x) = \sqrt{(x-0)^2 + ((x-4)^2 - 0)^2} = \sqrt{x^2 + (x-4)^4}.$$

This is a common calculus problem. Although we do not have the power of calculus, a graphing device can be used to approximate the solution.

The value of x that minimizes $d(x)$ is the same as the value that minimizes its square $\hat{d}(x) = x^2 + (x-4)^4$, and $\hat{d}$ is a simpler function than d.

A plot of $y = \hat{d}(x)$ is shown in Figure 10(b) using a viewing rectangle $[0, 6] \times [0, 20]$. Clicking on the curve gives the minimum point on $y = \hat{d}(x)$ as approximately $(2.9, 9.9)$. Hence the closest point to the origin occurs when $x \approx 2.9$, $y \approx (2.9-4)^2 = 1.21$, and the minimal distance to the origin is approximately

$$d(2.9) = \sqrt{(2.9)^2 + (2.9-4)^4} \approx 3. \qquad \blacksquare$$

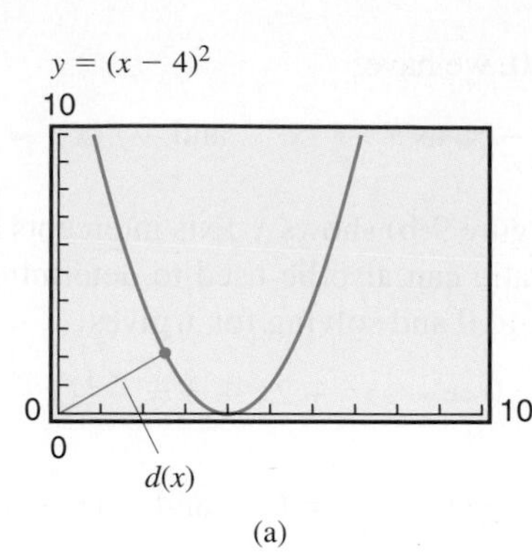

(a)

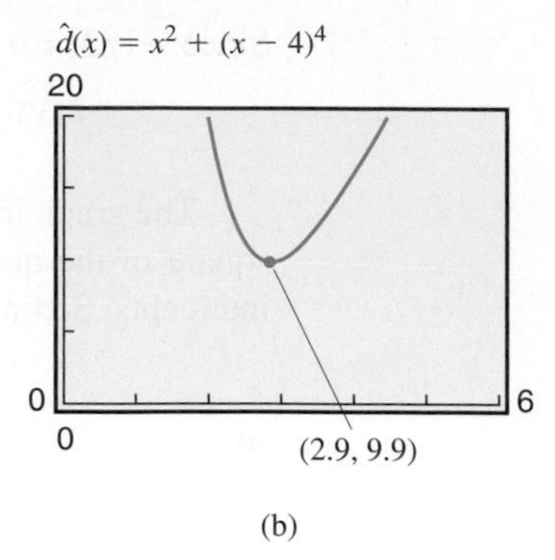

(b)

FIGURE 10

The Quadratic Formula

The technique of completing the square is used to derive a formula for finding solutions of the general quadratic equation

$$ax^2 + bx + c = 0, \quad \text{where } a \neq 0.$$

Since

$$ax^2 + bx + c = a\left(x^2 + \frac{b}{a}x + \left(\frac{b}{2a}\right)^2\right) - a\left(\frac{b}{2a}\right)^2 + c,$$

we have

$$0 = ax^2 + bx + c = a\left(x + \frac{b}{2a}\right)^2 + \frac{4ac - b^2}{4a}.$$

This implies that

$$a\left(x + \frac{b}{2a}\right)^2 = -\frac{4ac - b^2}{4a} = \frac{b^2 - 4ac}{4a}.$$

Hence

$$\left(x + \frac{b}{2a}\right)^2 = \frac{b^2 - 4ac}{4a^2}, \quad \text{and} \quad x + \frac{b}{2a} = \pm\frac{\sqrt{b^2 - 4ac}}{2a}.$$

Solving for x in this expression gives the frequently used *Quadratic Formula.*

The Quadratic Formula

The solutions to $ax^2 + bx + c = 0$, when $a \neq 0$, are

$$x = -\frac{b}{2a} \pm \frac{\sqrt{b^2 - 4ac}}{2a} = \frac{-b \pm \sqrt{b^2 - 4ac}}{2a}.$$

The *discriminant* of the quadratic equation, $b^2 - 4ac$, determines the number of real solutions of the equation.

- When $b^2 - 4ac > 0$, there are two distinct real number solutions.
- When $b^2 - 4ac = 0$, there is one real number solution.
- When $b^2 - 4ac < 0$, there are no real number solutions.

EXAMPLE 7 Sketch the graph of the parabola $y = x^2 + 2x - 1$.

Solution Completing the square gives

$$y = x^2 + 2x - 1 = x^2 + 2x + 1 - 1 - 1 = (x + 1)^2 - 2$$

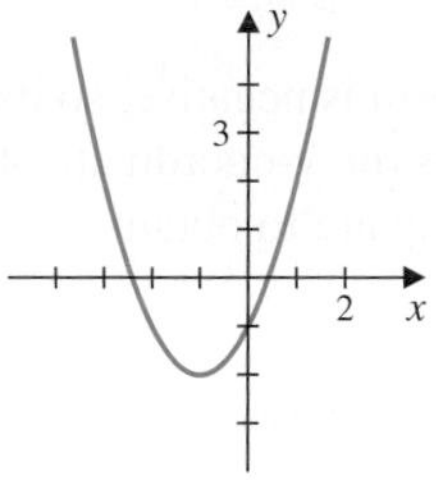

and the vertex of the parabola is $(-1, -2)$. To find the y-intercept, set $x = 0$ so that $y = -1$, and the y-intercept is the point $(0, -1)$.

We can use the Quadratic Formula to find the x-intercepts. We have $x^2 + 2x - 1 = 0$ precisely when

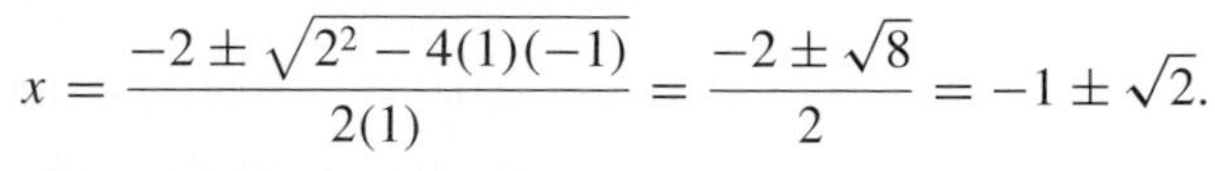

$$x = \frac{-2 \pm \sqrt{2^2 - 4(1)(-1)}}{2(1)} = \frac{-2 \pm \sqrt{8}}{2} = -1 \pm \sqrt{2}.$$

The graph crosses the x-axis when $x = -1 + \sqrt{2} \approx 0.4$ and $x = -1 - \sqrt{2} \approx -2.4$, as shown in Figure 11. ■

FIGURE 11

The linear and quadratic functions we have seen in this and the previous section are special cases of polynomial functions. A *polynomial of degree n* has the form

$$a_n x^n + a_{n-1} x^{n-1} + \cdots + a_1 x + a_0,$$

where n is a nonnegative integer and the $a_0, a_1, \ldots, a_n$ are all constants, with $a_n \neq 0$.

Linear functions that are not constant are first-degree polynomial functions because they can be expressed in the form

$$y = a_1 x + a_0,$$

where a_1 is the slope of the line and a_0 is the y-intercept. Quadratic functions are second-degree polynomial functions because they have the form

$$y = a_2 x^2 + a_1 x + a_0.$$

General polynomial functions are considered in Chapter 3.

Applications

EXAMPLE 8 A company finds that it can sell x units of a product per day when the price is $p(x) = 100 - 0.05x$, for x between 250 and 800. The cost of operating the company, regardless of the number of units produced, is \$4000 per day, and the production cost when x units are produced is $c(x) = 60 - 0.01x$ dollars per unit. Determine the number of units that should be produced each day to maximize the profit, and find this maximum profit.

Solution The revenue $R(x)$ is the product of x, the number of units sold, and the price $p(x)$ per unit. So the revenue in dollars is

$$R(x) = xp(x) = x(100 - 0.05x) = 100x - 0.05x^2.$$

The cost $C(x)$ is the sum of the fixed cost, 4000, and the variable cost that depends on the units sold. So

$$C(x) = 4000 + xc(x) = 4000 + x(60 - 0.01x) = 4000 + 60x - 0.01x^2.$$

The profit function is the revenue minus the cost, so

$$\begin{aligned} P(x) &= R(x) - C(x) \\ &= (100x - 0.05x^2) - (4000 + 60x - 0.01x^2) \\ &= -0.04x^2 + 40x - 4000. \end{aligned}$$

The profit function is a quadratic and the coefficient of the x^2-term is negative, so its graph is a parabola that opens downward. The maximum profit is the y-coordinate of the vertex of the parabola. To find the vertex, we complete the square to obtain

$$\begin{aligned} P(x) &= -0.04(x^2 - 1000x) - 4000 \\ &= -0.04(x^2 - 1000x + (500)^2) + 0.04(500)^2 - 4000 \\ &= -0.04(x - 500)^2 + 10000 - 4000 \\ &= -0.04(x - 500)^2 + 6000. \end{aligned}$$

The vertex of the parabola is (500, 6000), so the maximum profit is \$6000, which occurs when 500 units are sold, as shown in Figure 12. ■

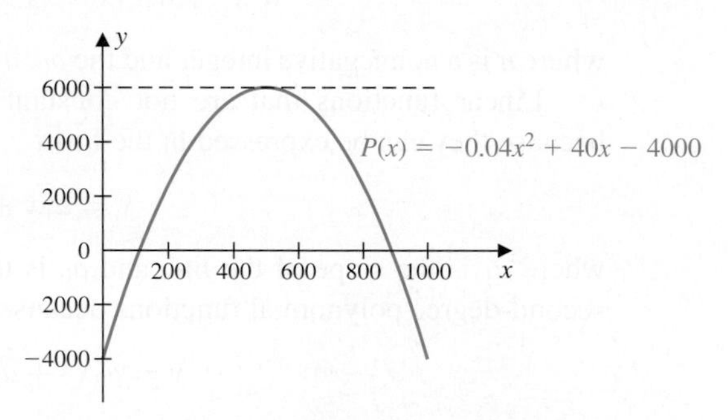

FIGURE 12

EXAMPLE 9 Show that the rectangle with a fixed perimeter that encloses the largest possible area is a square.

Solution Let l denote the length and w the width. Let P denote the constant (but arbitrary) perimeter. Then

$$2l + 2w = P, \quad \text{so} \quad l + w = \frac{P}{2}.$$

The area of the rectangle is $A = lw$, and $w = \frac{P}{2} - l$. So a function for the area in terms of just the variable l is (see Figure 13)

$$A(l) = lw = l\left(\frac{P}{2} - l\right) = \frac{P}{2}l - l^2.$$

The parabola $y = ax^2 + bx + c$ is symmetric about $x = -\frac{b}{2a}$.

The parabola is symmetric with respect to the l-intercepts, so the maximum area occurs when

$$l = \frac{P}{4} \quad \text{and} \quad w = \frac{P}{2} - l = \frac{P}{4}.$$

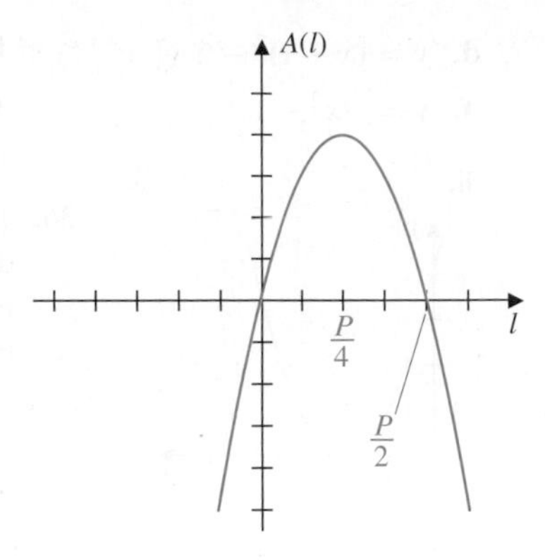

FIGURE 13

This shows that the rectangle with maximum area and fixed perimeter is a square, as you probably suspected. ■

EXERCISE SET 1.8

In Exercises 1–18, sketch the graph of the quadratic equation.

1. $y = x^2 + 1$
2. $y = x^2 - 3$
3. $y = -x^2 + 1$
4. $y = -x^2 - 1$
5. $y = (x - 3)^2$
6. $y = (x + 2)^2$
7. $y = (x + 1)^2 - 1$
8. $y = -(x - 1)^2 - 1$
9. $y = x^2 - 4x + 3$
10. $y = x^2 + 2x + 2$
11. $y = -x^2 - 2x$
12. $y = -x^2 + 2x$
13. $y = 3x^2 + 6x$
14. $y = 2x^2 - 4x + 8$
15. $y = \frac{1}{2}x^2 - 1$
16. $y = \frac{1}{2}x^2 + 2$
17. $y = \frac{1}{2}x^2 - x + 3$
18. $y = \frac{1}{3}x^2 - 2x + 1$

In Exercises 19–22, a quadratic function is given.

a. *Express the quadratic in standard form.*

b. *Find any axis intercepts.*

c. *Find the maximum or minimum value of the function.*

19. $f(x) = x^2 - 6x + 7$
20. $f(x) = x^2 + 4x + 5$
21. $f(x) = -x^2 + 4x + 6$
22. $f(x) = -x^2 - 4x - 4$

In Exercises 23–28, use the Quadratic Formula to find any x-intercepts of the parabola.

23. $y = 6x^2 - 5x + 1$
24. $y = 12x^2 + x - 1$
25. $y = 2x^2 - 4x + 1$
26. $y = 2x^2 - 6x + 3$
27. $y = 2x^2 + 4x + 3$
28. $y = 4x^2 - 8x + 7$
29. Use the graph of the function shown in the accompanying figure to sketch the graph.
    a. $y = f(x - 1)$
    b. $y = f(x - 1) + 2$
    c. $y = f(x + 2)$
    d. $y = f(x + 2) + 1$
    e. $y = f(x + 2) - 1$
    f. $y = f(x - 1) - 2$

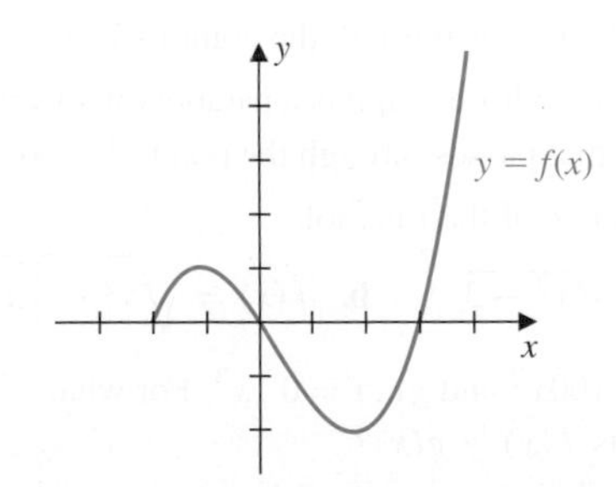

30. Match the equation with the curve in the figure.
    a. $y = x^2 + 2$
    b. $y = x^2 - 4x + 4$

c. $y = (x + 2)^2$ **d.** $y = (x + 1)^2 - 2$

e. $y = x^2 - 2x + 3$ **f.** $y = -x^2 - 1$

i.

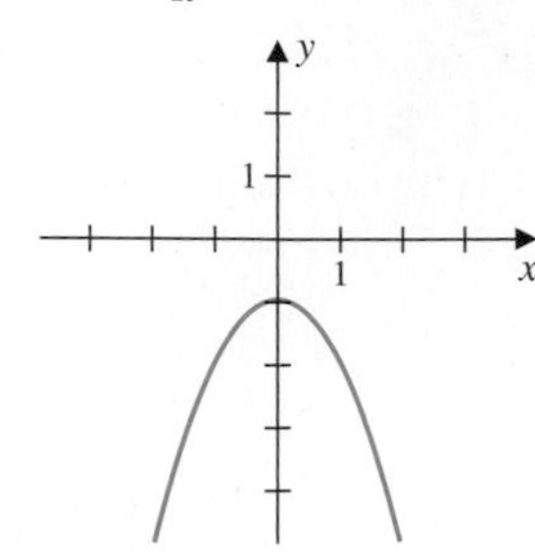

ii.

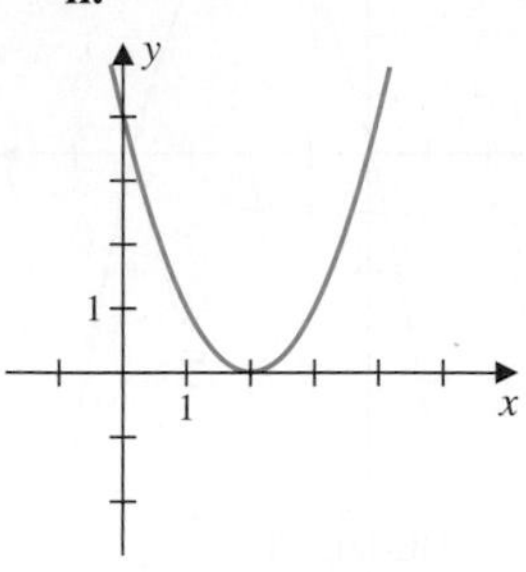

iii.

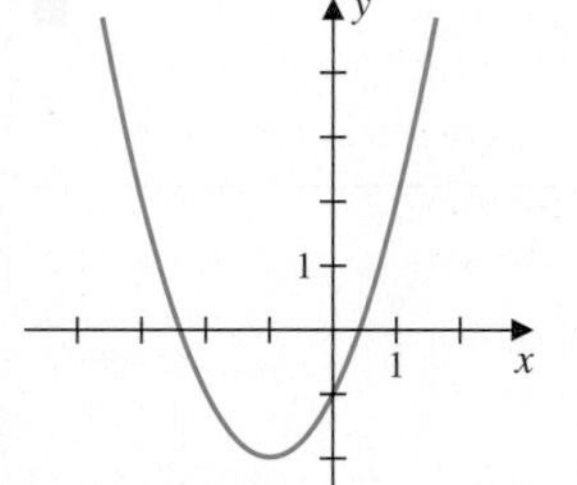

iv.

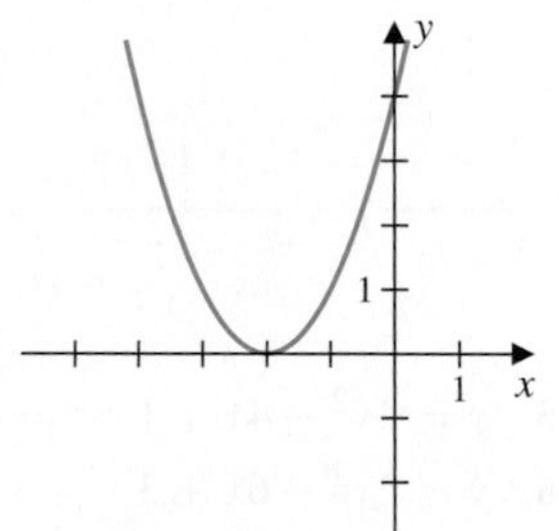

v.

vi.

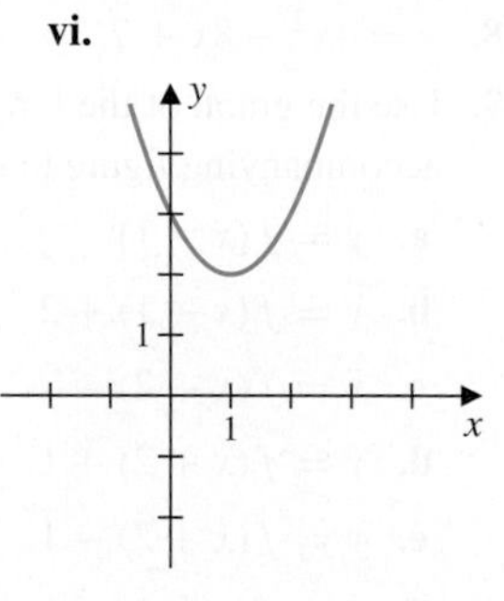

31. Find a function whose graph is a parabola with vertex $(1, 3)$ and that passes through the point $(-2, 5)$.

32. Find a function whose graph is a parabola with vertex $(-2, 2)$ and that passes through the point $(1, -4)$.

33. Find the domain of the function.

a. $f(x) = \sqrt{x^2 - 3}$ **b.** $f(x) = \sqrt{x^2 - \frac{1}{2}x}$

34. Let $f(x) = 100x^2$ and $g(x) = 0.1x^3$. For what values of x is $f(x) \geq g(x)$?

35. If $f(x) = ax^2 + bx + c$ and the following conditions are satisfied, what can you say about the values of a, b, and c?

a. The point $(1, 1)$ is on the graph of f.

b. The y-intercept is 6.

c. The vertex is $(1, 1)$.

d. Conditions (a), (b), and (c) are all satisfied.

36. The function defined by $s(t) = 576 + 144t - 16t^2$ describes the height, in feet, of a rock t seconds after it has been thrown upward at 144 feet per second from the top of a 50-story building, as shown in the figure.

a. Sketch the graph of s.

b. How long does it take the rock to hit the ground?

c. Determine the maximum height and the time it takes the rock to reach it.

d. Determine physically reasonable definitions for the domain and range of s.

37. The function defined by $v(t) = 144 - 32t$ describes the velocity in feet per second of the rock t seconds after it is thrown from the building described in Exercise 36.

a. Sketch the graph of v.

b. What is the velocity of the rock when it strikes the ground?

c. Determine physically reasonable definitions for the domain and range of v.

38. For a small manufacturing firm, the unit cost $C(x)$ in dollars of producing x units per day is given by

$$C(x) = x^2 - 120x + 4000.$$

How many items should be produced per day to minimize the unit cost, and what is the minimum unit cost?

39. A company that produces computer terminals analyzes production and finds that it should make a profit $P(x)$, in dollars, for selling x terminals per month, where

$$P(x) = -0.1x^2 + 160x - 20000.$$

a. How many terminals should be sold per month to produce the maximum profit?

b. What is the maximum profit?

40. Let (x, y) be a point on the graph of the parabola $y = f(x) = x^2 + 2$.

a. Express the distance from (x, y) to the point $(5, 1)$ as a function of x.

b. Approximate the point on the parabola $y = f(x) = x^2 + 2$ that is closest to $(5, 1)$.

41. A rectangle is inscribed under the parabola $y = 4 - x^2$ as shown in the figure.

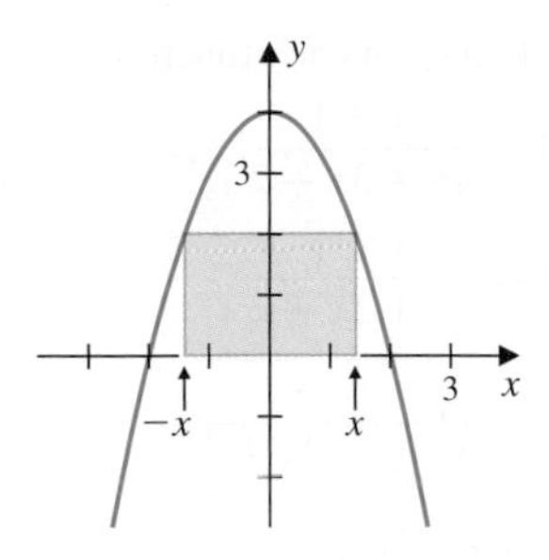

a. Express the area of the rectangle as a function of x.

b. Approximate the dimensions of the rectangle with largest area that can be inscribed under the rectangle in this way.

42. National health-care spending, in billions of dollars, has taken the shape of a parabola over the last few decades, increasing at an alarming rate. (See the table.)

a. Using only the 1980, 2000, and 2010 data, fit a parabola to the data of the form

$$y = a(x - 1980)^2 + b(x - 1980) + c.$$

b. What does the parabola predict will be the spending in the year 2020?

Year	Dollars (in billions)
1970	80
1980	250
1990	690
2000	1299
2010	2078

43. It has been determined that the maximum population that a certain region can support is a constant M. In addition, the rate R at which a population changes is proportional to the size of the population P and to the difference between M and the current population. Express R as a function of P and the constant M.

44. The profit function of a manufacturer when x units of a commodity are produced and sold is given by

$$P(x) = 200x - x^2.$$

a. Sketch a graph of the profit function.

b. What is the maximum profit, and how many units should be produced to yield it?

c. Compute the difference quotient

$$\frac{P(x+h) - P(x)}{h},$$

and determine the value the difference quotient approaches as h approaches 0. (This quantity is called the *marginal* profit when x units are sold. It approximates the change in the profit when one additional unit is produced.)

45. A driver traveling on an interstate highway sees traffic coming to a stop ahead and applies the brakes. The car's speed, in feet per second, is given by $v(t) = -4t^2 - 4t + 80$.

a. Make a sketch of the speed curve for $t \geq 0$.

b. How long does it take for the car to come to rest?

c. Make a table of speeds starting from time $t = 0$, the instant when the driver applies the brakes, including every half-second until the car comes to rest.

d. What is the slowest the car is traveling in the first half-second after the brakes are applied? What is the fastest the car is traveling in the first half-second?

e. Repeat part (d) for the second half-second and the third half-second.

f. What is the minimum distance the car can travel in the first half-second? In the second half-second? What is the maximum distance the car can travel in the first half-second? In the second half-second?

g. Use the table from part (c) to estimate a lower and an upper bound on the distance it takes for the car to come to rest.

REVIEW EXERCISES FOR CHAPTER 1

*In Exercises 1–4, (**a**) express the given interval using inequalities, and (**b**) sketch the numbers in the interval.*

1. $[-1, 7]$ **2.** $(1, \sqrt{3})$

3. $(-\infty, 7)$ **4.** $[-5, \infty)$

*In Exercises 5–8, (**a**) express the given inequalities using interval notation, and (**b**) sketch the numbers in the interval.*

5. $x > -4$ **6.** $-3 < x \le 3$

7. $2 \le x < 10$ **8.** $x \le 3$

In Exercises 9–18, find all values of x that satisfy the inequality.

9. $2x + 3 \ge 4$ **10.** $-(2x + 1) \le 4$

11. $x^2 + 2x + 1 \ge 1$ **12.** $x^2 - 4x > -3$

13. $(x - 1)(x + 2)(x - 2) \ge 0$

14. $(x - 1)(x - 4)(x + 2) < 0$

15. $x^2 + 3x > 0$ **16.** $x^3 - 4x^2 \le 0$

17. $\dfrac{2x - 1}{x + 1} \le -2$ **18.** $\dfrac{x^2 - 16}{x^2 - 1} \le 0$

*In Exercises 19–22, (**a**) solve the given inequality, and (**b**) show the solution graphically.*

19. $|2x - 3| < 5$ **20.** $|4x - 2| < 0.01$

21. $|3 - x| \le 4$ **22.** $|x^2 - 4| \le 1$

In Exercises 23–30, indicate in an xy-plane those points for which the statement holds.

23. $2 < y \le 3$ **24.** $|y| < 2$

25. $|x - 1| < 2$

26. $|x| < 3$ and $|y| \le 1$

27. $|x| + |y| = 1$

28. $1 \le |x| + |y| \le 4$

29. $|x| \le |y|$

30. $|x + 1| \le |y|$

*In Exercises 31–34, a function is described. Find (**a**) the domain, and (**b**) the range of the function.*

31. $f(x) = x^2 - 3$

32. $f(x) = \dfrac{1}{2x - 5}$

33. $f(x) = \sqrt{x - 2} + 2$

34. $f(x) = \begin{cases} -x + 3, & \text{if } x < 0 \\ -2x^2, & \text{if } x \ge 0 \end{cases}$

35. Find the domain of each function.

a. $f(x) = \dfrac{1}{x^2 - 6x + 8}$

b. $f(x) = \dfrac{x - 2}{x^2 - 6x + 8}$

c. $f(x) = \sqrt{\dfrac{x^2}{x^2 - 6x + 8}}$

36. Find the domain of each function.

a. $f(x) = \dfrac{x - 1}{x^2 + 3x + 2}$

b. $f(x) = \dfrac{x + 1}{x^2 + 3x + 2}$

c. $f(x) = \sqrt{\dfrac{x - 1}{x^2 + 3x + 2}}$

In Exercises 37–42, find

a. $f(x + h)$,

b. $\dfrac{f(x + h) - f(x)}{h}$, where $h \ne 0$

37. $f(x) = 5x + 3$ **38.** $f(x) = -\frac{3}{2}x + 5$

39. $f(x) = x^2 - 1$ **40.** $f(x) = 2x^2 - x$

41. $f(x) = \dfrac{1}{x - 1}$ **42.** $f(x) = \dfrac{2}{1 - x}$

43. Match the equations with the graphs.

a. $y = x + 3$ **b.** $x = -2$

c. $y = 2x/3$ **d.** $y = -4x - 3$

e. $y = -2x + 1$ **f.** $y = 5$

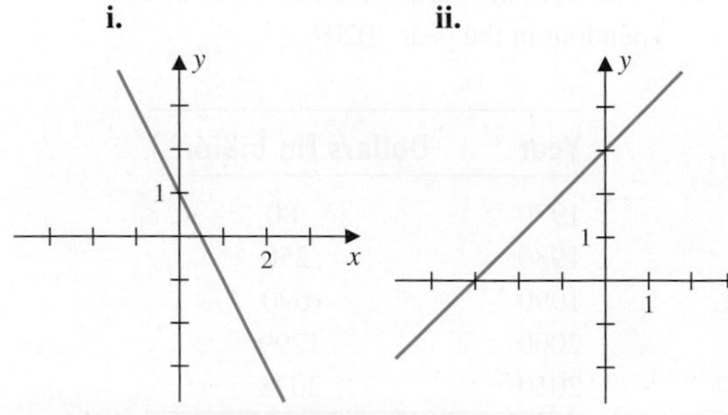

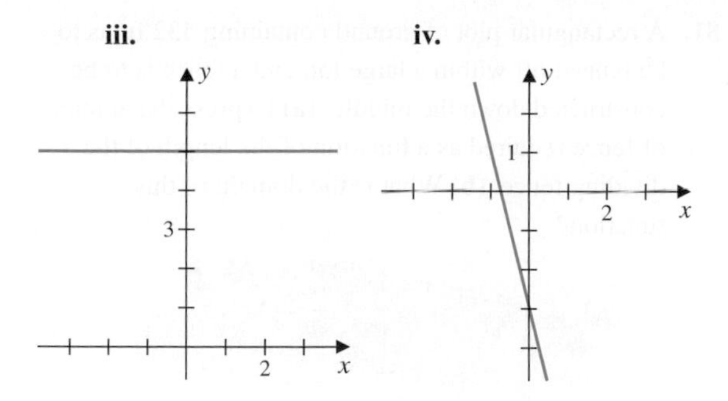

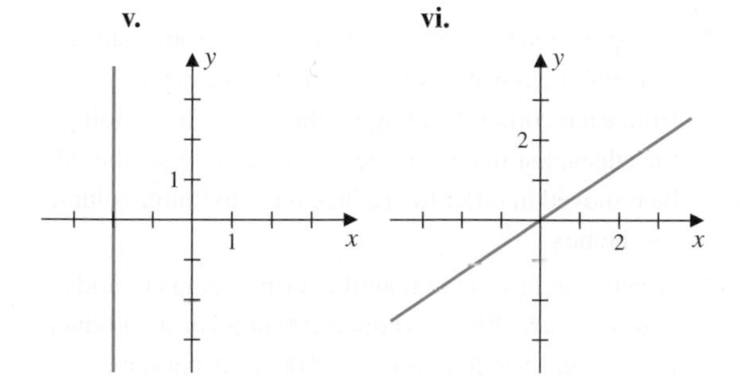

44. Match the equations with the graphs.

a. $y = -(x-1)^2 + 1$ **b.** $y = x^2 + x$

c. $y = x^2 - 2x - 1$ **d.** $y = x^2 + 2x + 3$

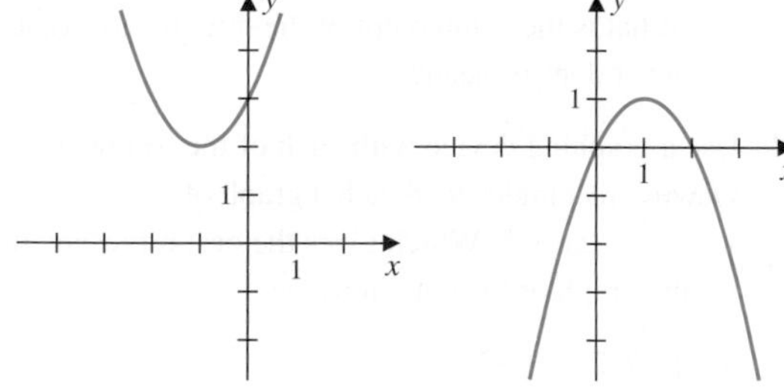

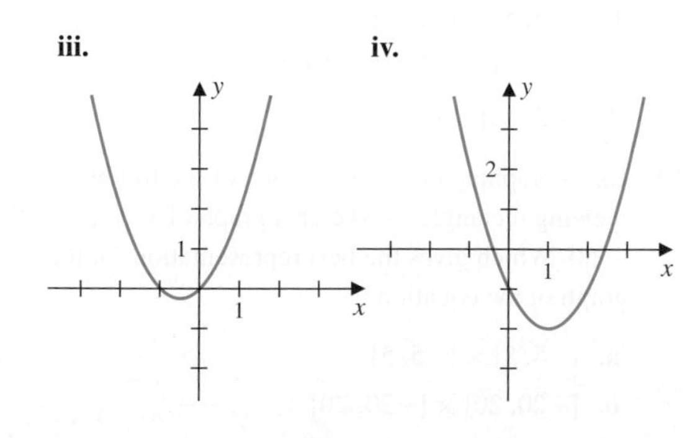

*In Exercises 45–48, **(a)** plot the pair of points, **(b)** determine the distance between the points, **(c)** find the midpoint of the line segment connecting the points, **(d)** sketch the straight line determined by these points, and **(e)** find the slope-intercept equation of the line.*

45. $(4, -2), (1, 1)$ **46.** $(3, 1), (-1, -2)$

47. $(-1, -2), (2, -3)$ **48.** $(1, -1), (4, -1)$

*In Exercises 49–52, the equation of a line is given together with a point that is not on the line. Find the slope-intercept form of the equation of the line that passes through the given point and is **(a)** parallel to the given line, and **(b)** perpendicular to the given line.*

49. $y = 4x + 3$; $(0, 0)$

50. $y = \frac{1}{3}x + 5$; $(1, 2)$

51. $-7x - 5y = -1$; $(-1, -3)$

52. $2x - 3y = 4$; $(1, -1)$

*In Exercises 53–62, a quadratic equation is given. **(a)** Sketch the graph of the equation, **(b)** find the range of the function described by the equation, **(c)** find the axis intercepts, and **(d)** find the maximum or minimum value on the curve.*

53. $y = (x-1)^2 - 2$ **54.** $y = -(x+1)^2 + 1$

55. $y = -(x+1)^2 - 1$ **56.** $y = (x-1)^2 + 1$

57. $y = x^2 - 4x$ **58.** $y = -x^2 + 4x - 5$

59. $y = 2x^2 - 12x + 18$ **60.** $y = 2x^2 + 5x + 10$

61. $y = -\frac{1}{2}x^2 + 3x - 3$ **62.** $y = -\frac{1}{3}x^2 + 2x - 1$

In Exercises 63–68, the equation of a circle is given. Find the center and radius of each circle, and sketch its graph.

63. $x^2 + y^2 = 16$

64. $(x-1)^2 + y^2 = 1$

65. $(x+2)^2 + (y-1)^2 = 9$

66. $(x-2)^2 + (y-3)^2 = 4$

67. $x^2 + y^2 + 4x + 2y = 4$

68. $x^2 + 6x + y^2 - 2y - 6 = 0$

69. Show that the points $(-3, 3)$, $(3, -5)$, and $(7, -2)$ are vertices of a rectangle, and find the fourth vertex.

70. Find the equation of the line with x-intercept 3 and y-intercept 2.

71. Find the shortest distance from the point $(-2, -1)$ to the line $y - 2x = -2$.

72. Find an equation of a circle with center $(-3, -1)$ that passes through $(-5, -3)$.

73. Find an equation of a circle with center $(4, 2)$ that is tangent to the x-axis.

74. Find the equation of the line passing through (3, 4) and the center of the circle

$$x^2 - 4x + y^2 - 2y - 1 = 0.$$

75. Find a point on the x-axis equidistant from the points $(-2, -3)$ and $(3, 5)$.

76. Sketch the graph of

$$g(x) = \begin{cases} 0, & \text{if } x < 0 \\ x^2, & \text{if } 0 \le x \le 2 \\ -x + 6, & \text{if } x > 2. \end{cases}$$

77. From each of the data sets given, sketch a likely graph of f and determine a likely formula for $f(x)$.

a.

x	$f(x)$
0	−26
1	−6
2	6
3	10
4	6
5	−6
6	−26

b.

x	$f(x)$
−3	7
−2	5
−1	3
0	1
1	−1
2	−3
3	−5

78. A baseball is thrown upward from the ground with an initial velocity of 57.6 m/s. If air resistance is neglected, its distance $s(t)$ above the ground t seconds later is $s(t) = -4.8t^2 + 57.6t$ m.

a. Estimate the maximum height of the ball.

b. Estimate when the ball is at least 120 m above the ground.

c. Estimate when the ball strikes the ground.

79. A restaurant has a fixed price of \$36 for a complete dinner. The average number of customers per night is 200. The owner estimates that for each dollar increase in the price of the dinner, there will be, on average, four fewer customers per night. Estimate the price for the dinner that will produce the maximum amount the owner can expect to receive.

80. The daily car rental charge for agency A is \$21, plus 21 cents per mile. For agency B the rate is \$32, plus 18 cents per mile. A driver plans to rent from one of these agencies and take a 320-mile trip. **(a)** From which agency should the driver rent to get the better rate? **(b)** Estimate the minimum number of miles beyond which agency B's cost is less than that of agency A.

81. A rectangular plot of ground containing 432 ft^2 is to be fenced off within a large lot, and a fence is to be constructed down the middle. **(a)** Express the amount of fence required as a function of the length of the dividing fence. **(b)** What is the domain of this function?

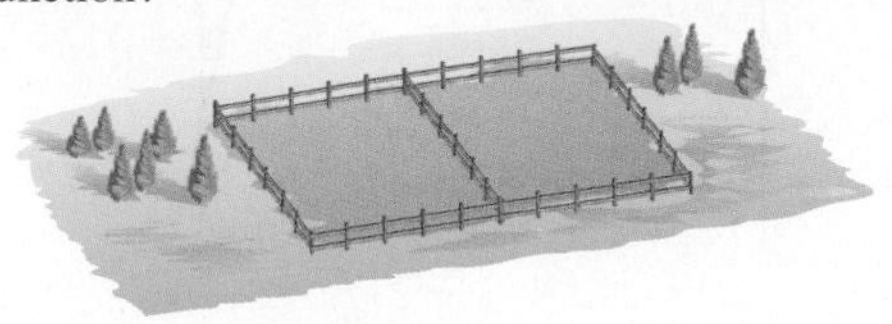

82. An open rectangular box is to be made from a sheet of metal 5 cm wide by 9 cm long by cutting a square from each corner, bending up the sides, and welding the edges. Estimate the size of the square that should be removed in order to produce the maximum volume for the box.

83. A manager of a small manufacturing company finds that it costs \$3200 to produce 100 units of its product per day and \$9600 to produce 500 units per day.

a. Assuming the relationship between cost and number of units produced per day is linear, find the equation that expresses the relationship and graph the equation.

b. What does the slope of the line in part (a) mean?

c. What is the y-intercept of the line in part (a), and what does it mean?

84. Use a graphing device with each of the following viewing rectangles to sketch a graph of $y = x^2 - 2x + 5$. Which gives the best representation for the graph of the equation?

a. $[-2, 2] \times [-2, 2]$

b. $[-5, 5] \times [-5, 5]$

c. $[-10, 10] \times [-100, 100]$

d. $[-5, 20] \times [-5, 20]$

85. Use a graphing device with each of the following viewing rectangles to sketch a graph of $y = x^3 - 25x + 50$. Which gives the best representation for the graph of the equation?

a. $[-5, 5] \times [-5, 5]$

b. $[-20, 20] \times [-20, 20]$

c. $[-100, 100] \times [-100, 100]$

d. $[-10, 10] \times [-100, 100]$

86. Determine an approximate viewing rectangle for the graph of the equation and use it to sketch the graph.

a. $y = x^2 - 16x + 61$

b. $y = x^2 + 10x + 38$

c. $y = \sqrt{x^2 - 10x + 29}$

d. $y = \dfrac{x+4}{x-3}$

87. Use the figure to answer the questions about the function f. Between which pairs of points labeled on the graph is the function

a. increasing?

b. decreasing?

c. having a local maximum?

d. having a local minimum?

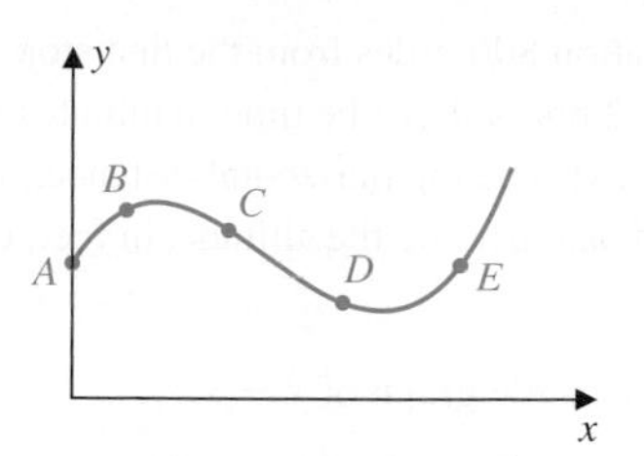

88. Sketch the graph of a function f that satisfies all of the conditions. The function

a. satisfies $f(0) = 1$

b. is decreasing on the interval $(0, 2)$

c. is increasing on the interval $(2, 3)$

d. is decreasing on the interval $(3, \infty)$

e. has values approach 0 as x becomes large and approach 2 as x becomes small.

CHAPTER 1: EXERCISES FOR CALCULUS

1. One of the functions f or g in the accompanying table is linear and one is quadratic of the form ax^2. Find equations for $f(x)$ and $g(x)$.

x	$f(x)$	$g(x)$
-2	2	2
2	2	14
6	18	26

2. Determine any values of the constant a so that the line with equation $3x + ay = -2a$ satisfies the specified condition.

a. Has slope 2; **b.** Is horizontal; **c.** Is vertical;

d. Passes through the point $(0, -2)$.

3. a. The sum of the first n consecutive natural numbers is given by

$$1 + 2 + 3 + \cdots + n = \frac{n(n+1)}{2}.$$

Find all values of n with the property that this sum is greater than 100 and less than 150.

b. The sum of the squares of the first n consecutive natural numbers is given by

$$1^2 + 2^2 + 3^2 + \cdots + n^2 = \frac{n(n+1)(2n+1)}{6}.$$

Find all values of n with the property that this sum is greater than 200 and less than 300.

4. a. Sketch the graph of the equations $y = ax^2 + bx$, where a and b are arbitrary and $a \neq 0$.

b. Show that the graph of $y = ax^2 + bx$ has symmetry to the vertical line $x = -b/(2a)$.

c. Show that for any number c, the graph of $y = ax^2 + bx + c$ also has symmetry to $x = -b/(2a)$.

5. Ice cubes are added to a glass of room temperature (20°C) water and the water is allowed to stand until the ice cubes melt 30 minutes later. Sketch a possible graph of the temperature of the water in the glass with respect to the elapsed time.

6. A child with a fever is given an antibiotic. The fever continues to rise for another hour, but not as rapidly as it did for the hour before the antibiotic was given. The child's temperature then begins to slowly drop until it stabilizes. Sketch a possible graph of the child's temperature as a function of time.

7. An airplane makes one stop before reaching its final destination. This stop is 400 miles away and takes 1 hour to reach. The layover time at this stop is 1 hour and 30 minutes. It then completes its flight, landing at its final destination 800 miles from the first stop, taking another 2 hours. Let t be time in minutes after the flight starts, $s(t)$ be the horizontal distance, in miles, traveled, and $a(t)$ be the altitude, in feet, of the plane.

a. Sketch a plausible graph of $y = s(t)$.

b. Sketch a plausible graph of $y = a(t)$.

c. Sketch a plausible graph of the ground speed.

d. Sketch a plausible graph of the vertical velocity.

8. A standard can in the shape of a right circular cylinder with a top and a bottom has volume 900 cm^3.

a. Express the surface area of the can as a function of the radius.

b. Estimate the dimensions of the can that will minimize the metal needed to manufacture the can.

9. To encourage volume purchases of a certain product the price per unit is decreased as the number purchased increases. For orders up to 100 units the price per unit is \$300. On orders of more than 100 units but not in excess of 150 units the price per unit is reduced by \$1 for each additional unit purchased. For all orders of more than 150 units the price per unit is \$225.

a. Write a function $P(x)$ for the price per unit when x units are purchased.

b. Sketch the graph of $y = P(x)$.

10. Agricultural research on a certain variety of fruit has shown that if 30 trees are planted per acre, then each tree will yield an average of 300 pounds of fruit per season. For each additional tree planted per acre, the average yield per tree is reduced by 5 pounds.

a. Find an equation for the yield, Y, of a 1-acre plot as a function of the number of trees planted.

b. How many trees should be planted per acre to produce the maximum yield? What is the maximum yield per acre?

11. When a solid rod is heated, its length increases depending on its coefficient of linear expansion a. The amount of increase in length is the product of the length, the change in temperature, and a. Suppose a steel rod has a length of 2 m at 0°C and that the coefficient of linear expansion for this material is $a = 11 \times 10^{-6}$.

a. Find a function that describes the length of the rod in terms of its temperature above 0°C.

b. Determine its length when the temperature is 1000°C.

12. A central concept in calculus is the extension of the tangent line to a circle to an arbitrary curve. In this exercise you will investigate finding the tangent line to a circle using geometry and essentially using calculus.

a. Find the equation of the tangent line to the unit circle at the point $(1/2, \sqrt{3}/2)$ by finding the slope of the line that coincides with the radius through the point as shown in the figure.

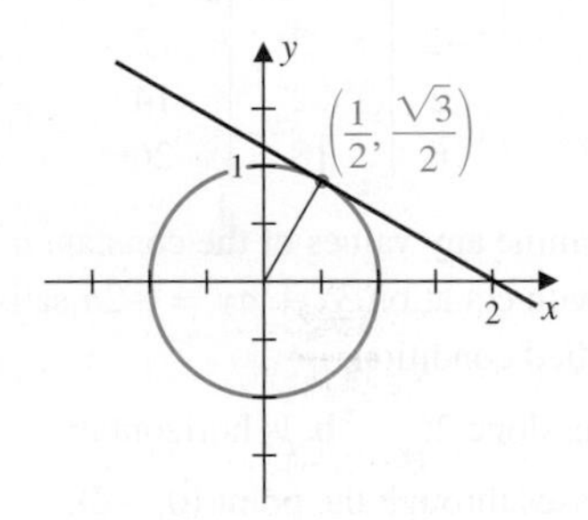

b. The slope of a tangent line can also be computed using the difference quotient.

- Write the upper half semi-circle as $y = f(x)$.
- Write the difference quotient for $f(x)$.
- Simplify the difference quotient using

$$\frac{\sqrt{a} - \sqrt{b}}{h} = \frac{\left(\sqrt{a} - \sqrt{b}\right)\left(\sqrt{a} + \sqrt{b}\right)}{h\left(\sqrt{a} + \sqrt{b}\right)}.$$

- What happens to the simplified difference quotient as $h \to 0$? Compare this result with what you found in part (a).

CHAPTER 1: CHAPTER TEST

Determine whether the statement is true or false. If false, describe how the statement might be changed to make it true.

1. The solution to the equation $3x - 2 = 4$ is $x = \frac{7}{2}$.
2. The solutions to the equation $x^2 - 3x + 2 = 0$ are $x = 2$ and $x = -1$.
3. The zeros of $f(x) = x^2 + 2x - 4$ are irrational numbers.
4. The zeros of $f(x) = 2x^3 - x^2 - x$ are rational numbers.
5. The solution set of the inequality $2x + 1 < 3x - 4$ is the interval $(5, \infty)$.
6. The solution set of the inequality $-3x + 4 \geq 10$ is the interval $(-\infty, -2]$.
7. The only solution to $|3x - 4| = 2$ is $x = \frac{2}{3}$.
8. The solution set of the inequality $|x - 4| \leq 3$ is the interval $[1, 7]$.
9. If $|x - 5| = 3$, then the distance from x to 5 is 3.
10. If $|2x - 5| = 3$, then the distance from x to $\frac{5}{2}$ is $\frac{3}{2}$.
11. The domain of the function $f(x) = \sqrt{x - 3}$ is $[-3, \infty)$.
12. The domain of the function $f(x) = \dfrac{x}{\sqrt{x - 3}}$ is $[3, \infty)$.
13. The domain of the function $f(x) = \sqrt{(x + 1)(x - 2)}$ is $(-\infty, -1] \cup [2, \infty)$.
14. The domain of the function $f(x) = \sqrt{x^2 + x - 2}$ is $(-\infty, -2] \cup [1, \infty)$.
15. The center of the circle $(x - 2)^2 + y^2 = 4$ is $(2, 0)$ and the radius is 4.
16. The center of the circle $x^2 + 2x + y^2 - 4y = 4$ is $(1, 2)$ and the radius is 3.
17. The slope of the line $2x - 3y = 4$ is $-\frac{2}{3}$.
18. The lines $x + y = 2$ and $3x - 2y = 1$ intersect at the point $(2, 1)$.
19. The lines $-3x + 2y = 5$ and $4y = 6x + 7$ are perpendicular.
20. The lines $x - 3y = 3$ and $4x - 6y = 5$ are parallel.
21. The lines $2x + y = 2$ and $2y + x = -1$ are perpendicular.
22. The lines $x + 2y = 1$ and $-2x + y = 3$ are parallel.
23. The equation of the line that has slope -3 and passes through the point $(0, 1)$ is $y = -3x - 1$.
24. The equation of the line that passes through the two points $(2, 1)$ and $(-3, 2)$ is $5y - x = 7$.

In Exercises 25 and 26, use the figure.

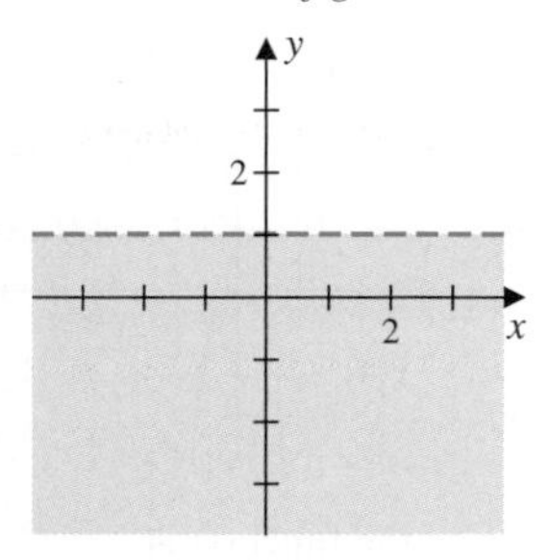

25. The shaded region is described by $y \leq 1$.
26. The region outside the shaded region is described by $y > 1$.

In Exercises 27 and 28, use the figure.

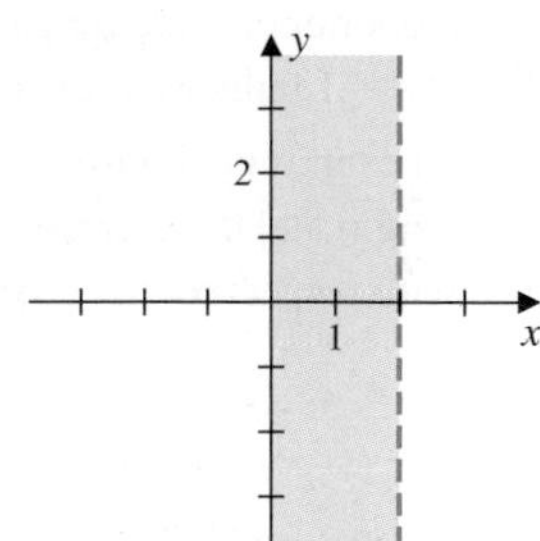

27. The shaded region is described by $0 \leq x < 2$.
28. The region outside the shaded region is described by $x < 0$ or $x \geq 2$.

In Exercises 29 and 30, use the figure.

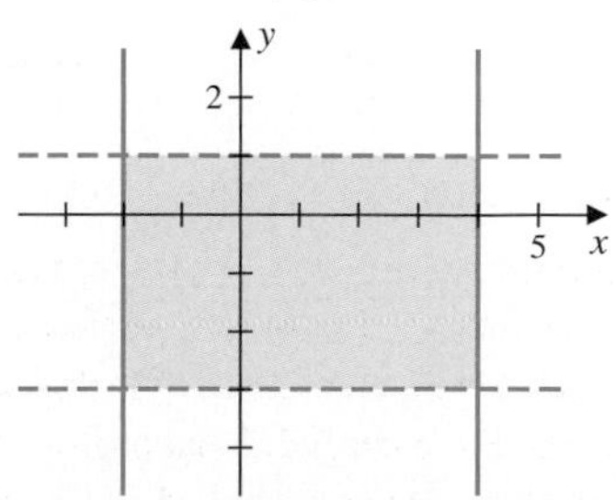

29. The shaded region is described by $|x - 1| \leq 3$ and $|y + 1| < 2$.

30. The region outside the shaded region is described by $|x - 1| > 3$ and $|y + 1| \geq 2$.

31. The vertex of the parabola $y = x^2 + 4x + 3$ is at $(-2, -1)$.

32. The parabola $y = -(x + 1)^2 - 2$ has a maximum point at $(-1, 2)$.

In Exercises 33 and 34, use the figure.

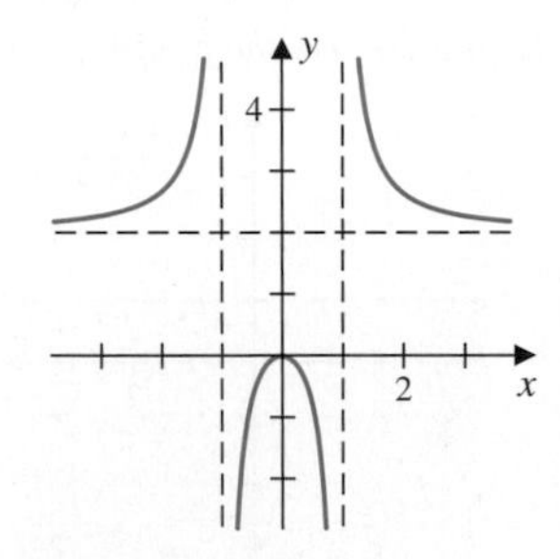

33. The domain of the function is $(-\infty, -1) \cup (-1, 1) \cup (1, \infty)$.

34. The range of the function is $(-\infty, 0] \cup (2, \infty)$.

35. The difference quotient for the function $f(x) = -2x + 3$ reduces to -2.

36. The difference quotient for the function $f(x) = x^2 + 2x - 1$ reduces to $2x + 2 + h$.

37. The graph of an even function is symmetric with respect to the origin and the graph of an odd function is symmetric with respect to the y-axis.

38. The graph of $y = -f(x)$ is the reflection of the graph of $y = f(x)$ about the y-axis.

39. A curve will describe the graph of a function provided every horizontal line crosses the curve at most one time.

40. The graph of $y = (x - 1)^2 + 2$ is obtained by shifting the graph of $y = x^2$ to the left 1 unit and upward 2 units.

41. The graph of $y = (x + 2)^2 - 1$ is obtained by shifting the graph of $y = x^2$ to the left 2 units and downward 1 unit.

In Exercises 42–45, use the figure.

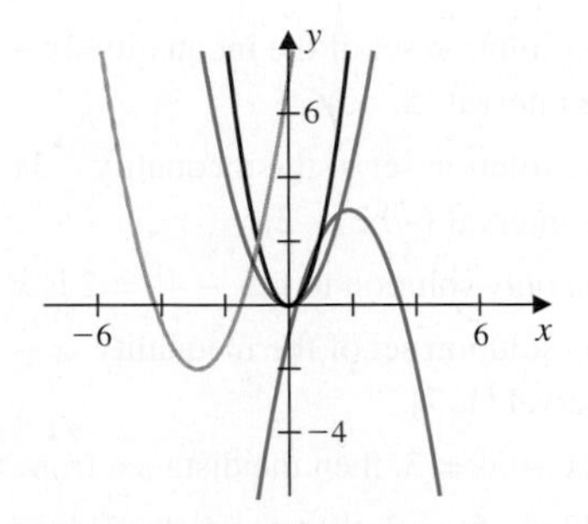

42. The leading coefficient of the red parabola is more than the leading coefficient of the green parabola.

43. The leading coefficient of the yellow parabola is positive.

44. The blue parabola has equation $y = -(x - 2)^2 + 3$.

45. The yellow parabola has equation $y = x^2 + 6x + 7$.

New Functions from Old

2

Calculus Connections

Human immunodeficiency virus (HIV) is the virus that causes the disease acquired ammunodeficiency syndrome (AIDS). The table of data estimates the cumulative HIV infections and deaths due to AIDS in millions worldwide from the 2008 report of the Joint United Nations Programme on HIV/AIDS (UNAIDS). It is estimated that more than 25 million people have died of AIDS since 1980. In recent years the number of deaths has been on the decline, even though the number of people contracting HIV is increasing. Despite the fact that more people are living with HIV, the World Health Organization predicts that AIDS will remain a leading cause of death for decades to come.

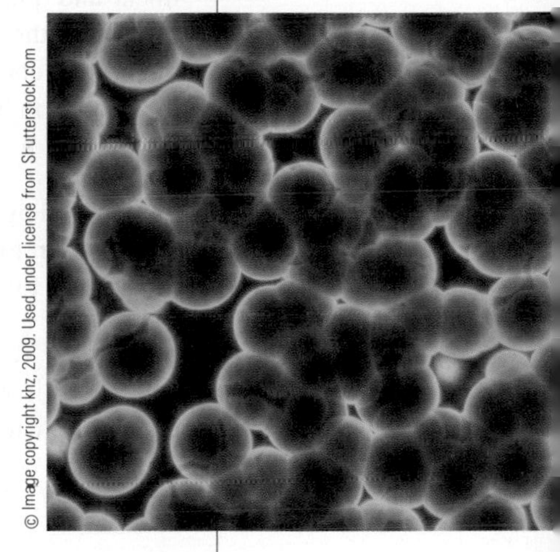
© Image copyright khz, 2009. Used under license from Shutterstock.com

The table gives HIV infections and AIDS deaths as functions of time. Plotting the data points and connecting successive points by straight line segments shows the increasing trends in HIV infections and AIDS deaths. The data are also used to produce functions $H(x)$ for the approximate number of HIV infections, and $A(x)$ for the approximate number of AIDS deaths. A function $S(x)$ describing the number of survivors of AIDS is defined by

$$S(x) = H(x) - A(x).$$

On the graph, $S(x)$ is determined by subtracting the y-coordinates on the graph of $A(x)$ from the corresponding y-coordinates on the graph of $H(x)$.

Year	HIV Infections	AIDS Deaths
1990	8	0.2
1995	19	2
2000	28	8.7
2002	31	12.3
2005	32	18.8
2007	33	23.1
2008	33.4	25.1

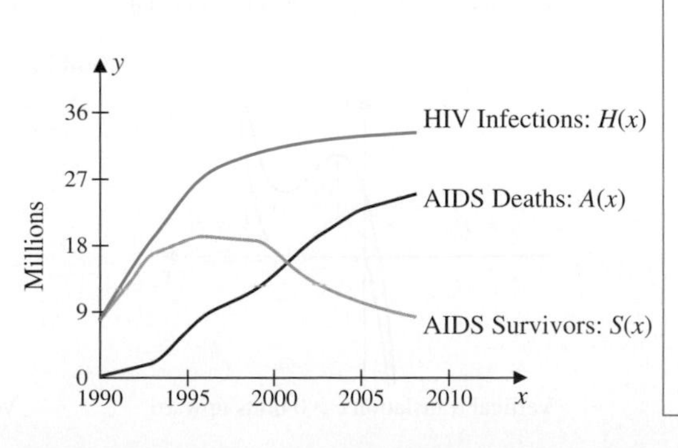

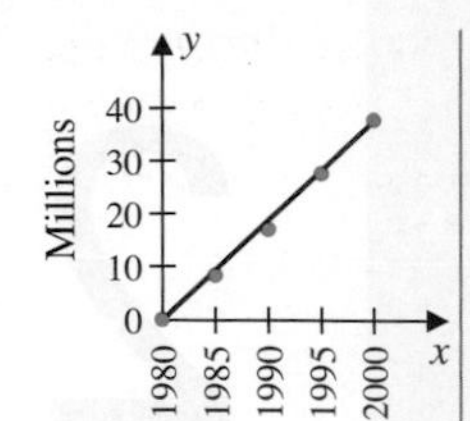

A function describing the cumulative death rates as a percentage of the cumulative number of infections is

$$R(x) = \frac{A(x)}{H(x)} \cdot 100$$

and plotting the data shows the steady increase in the death rate. The figure in the margin shows the graphical representation of the data.

The functions $S(x)$ and $R(x)$ are examples of combining functions; of constructing new functions from old. In Section 2.3 we examine a number of algebraic combinations of functions by adding, subtracting, multiplying, and dividing functions. In Section 2.4 we consider the composition of functions, which is an important operation in calculus. Inverse functions are considered in Section 2.5.

2.1 INTRODUCTION

In Chapter 1 we introduced the important concept of a function and considered the linear and quadratic functions that have applications in calculus. In this chapter we discuss methods for building new functions from those that are already familiar. One method for constructing new functions from old uses the graphical shifting techniques introduced in Section 1.8. For example, the function

$$f(x) = (x-1)^2 + 2,$$

represented by the red parabola in Figure 1, can be realized from the function $g(x) = x^2$ by a shift to the right 1 unit followed by a shift upward of 2 units. In terms of the function $g(x) = x^2$ this is the same as writing

$$f(x) = g(x-1) + 2.$$

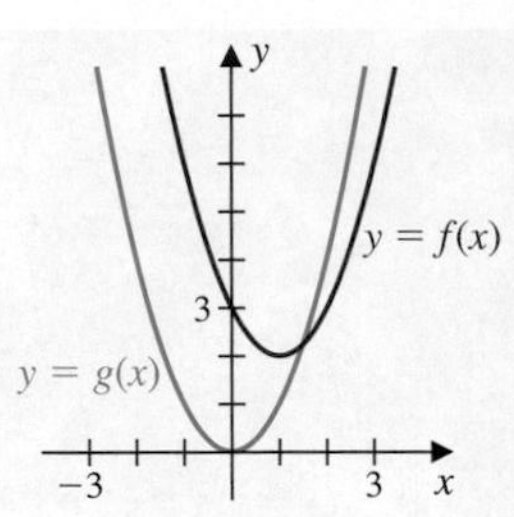

FIGURE 1

There is nothing special about shifting the quadratic functions; these shifting techniques can be used with any function, as is shown in Figures 2 and 3.

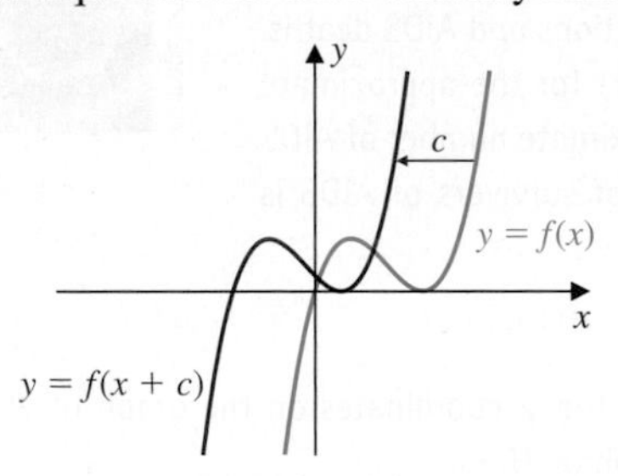

Horizontal translation $c > 0$ units to the left

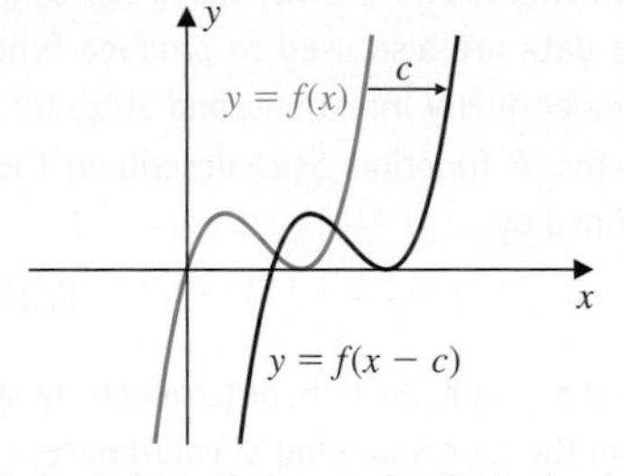

Horizontal translation $c > 0$ units to the right

FIGURE 2

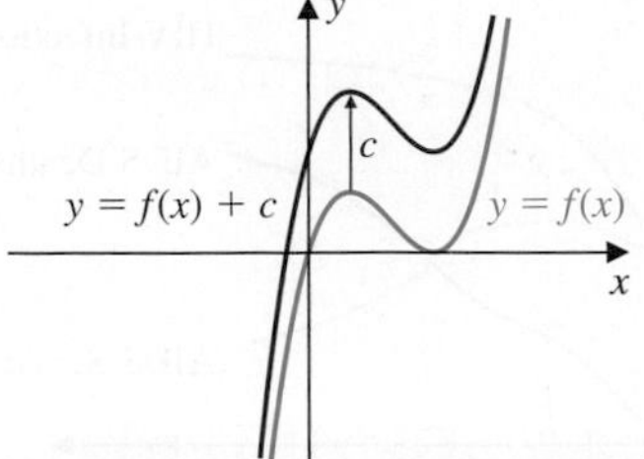

Vertical translation $c > 0$ units upward

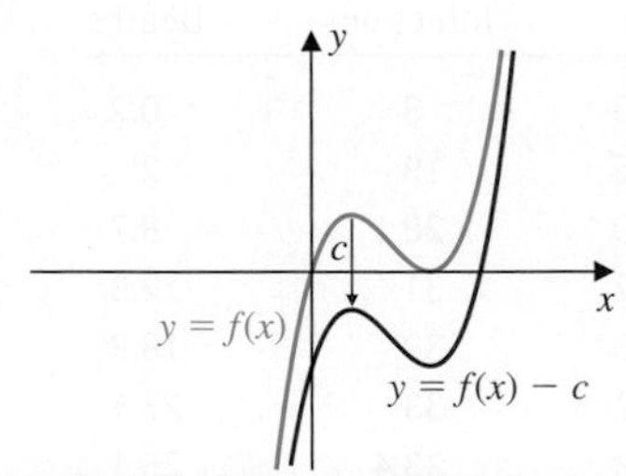

Vertical translation $c > 0$ units downward

FIGURE 3

Throughout this chapter and the remainder of the text we will use the graphical shifting techniques. Constructing a graph is often an important first step in solving a problem. The more functions you can picture, the better problem solver you will become.

2.2 OTHER COMMON FUNCTIONS

In this section we consider some of the most common algebraic functions. These will be used as building blocks for constructing many other useful functions.

The Absolute Value Function

We saw in Section 1.2 that the absolute value of a real number x, defined by

$$|x| = \begin{cases} x, & \text{if } x \geq 0, \\ -x, & \text{if } x < 0, \end{cases}$$

describes the distance from x to the origin. The **absolute value function** is defined by $f(x) = |x|$. When $x \geq 0$, the graph coincides with the line $y = x$. When $x < 0$, the graph coincides with the line $y = -x$. Consequently,

$$|x| \to \infty \text{ as } x \to \infty \quad \text{and} \quad |x| \to \infty \text{ as } x \to -\infty.$$

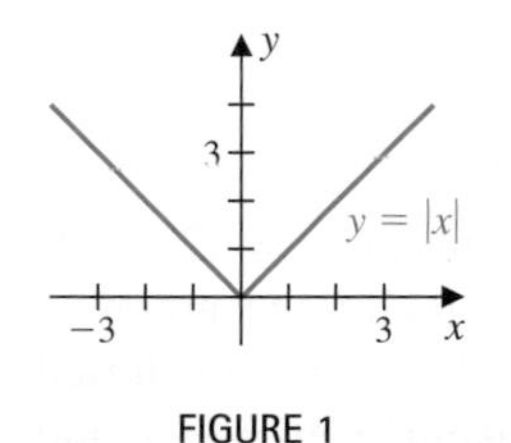

FIGURE 1

For any real number $|x| = |-x|$, so the absolute value function is an even function, and its graph has y-axis symmetry. The absolute value function is decreasing when $x < 0$, increasing when $x > 0$, and has the minimal value of 0 at $x = 0$. Its graph is shown in Figure 1.

EXAMPLE 1 Use the graph of $y = |x|$ to sketch the graph of $g(x) = |x - 1| - 2$.

Solution Replacing x by $x - 1$ in $y = |x|$ results in a horizontal shift of the graph of $y = |x|$ to the right 1 unit. Then applying vertical shifting, the graph of $y = g(x) = |x - 1| - 2$ is the graph of $y = |x - 1|$ shifted downward 2 units. Putting these results together gives the graph shown in Figure 2. Notice that g is decreasing when $x < 1$, is increasing when $x > 1$, and has minimum value of -2 at $x = 1$. ■

The absolute value function is not "smooth" at (0, 0), since the graph turns to form a right angle. This non-smooth behavior is of particular interest in calculus.

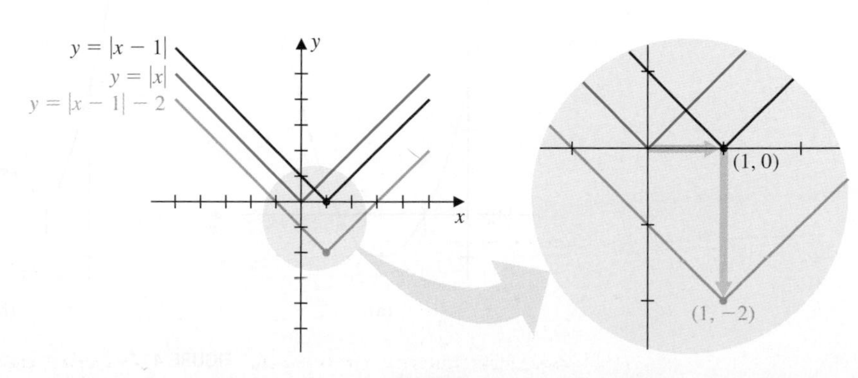

FIGURE 2

EXAMPLE 2 Sketch the graph of $h(x) = -|2x - 1| + 2$.

Solution The graph of $y = |2x| = 2|x|$ is similar to the graph of $y = |x|$ but is horizontally compressed, as shown in Figure 3(a), just as the graph of $y = 2x^2$ is compressed when compared to the graph of $y = x^2$. The graph of

$$y = |2x - 1| = 2\left|x - \tfrac{1}{2}\right|$$

is the graph of $y = 2|x|$ shifted to the right $\frac{1}{2}$ unit. These graphs are shown in Figure 3(b).

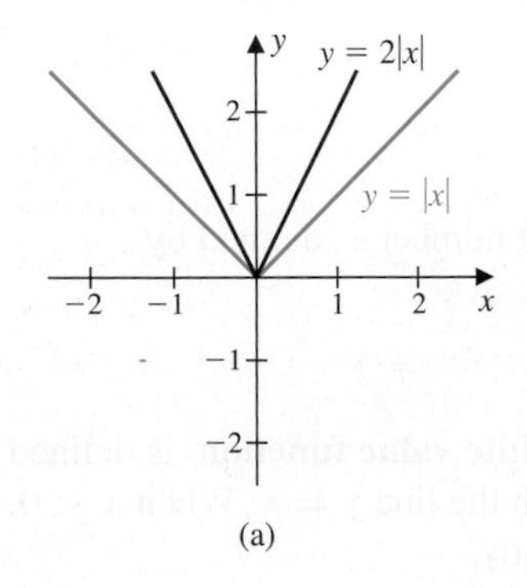

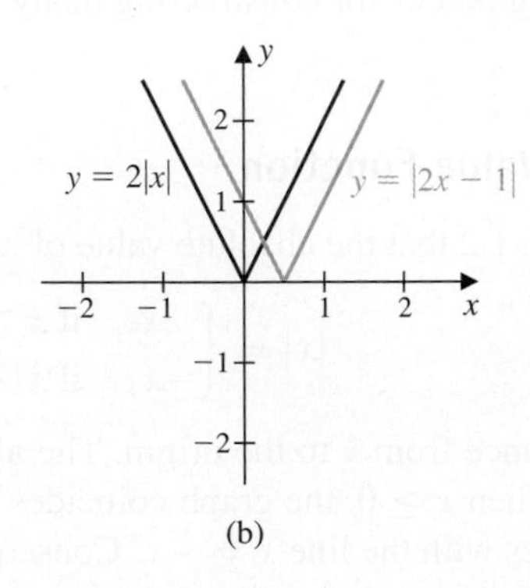

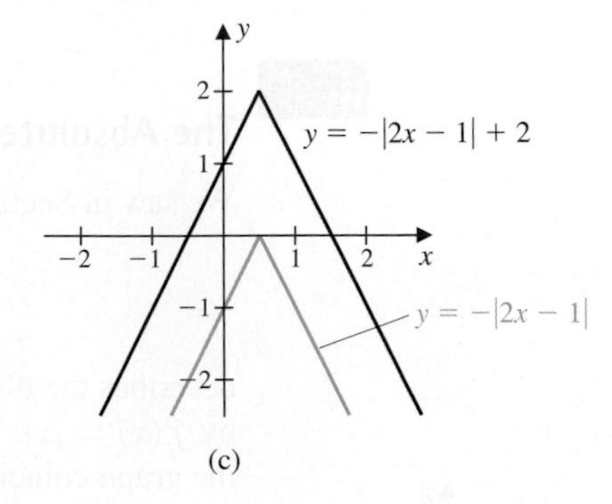

FIGURE 3

The graph of $y = -|2x - 1|$ is the reflection of the graph of $y = |2x - 1|$ about the x-axis, so it opens downward instead of upward. Finally, the graph of $y = h(x) = -|2x - 1| + 2$ is the same as the graph of $y = -|2x - 1|$, but shifted upward 2 units, as shown in Figure 3(c). Notice that h is increasing when $x < \frac{1}{2}$, and is decreasing when $x > \frac{1}{2}$. It has a maximum value of 2 at $x = \frac{1}{2}$. ■

EXAMPLE 3 Sketch the graph of $f(x) = |x^2 - 4x + 3|$.

Solution Figure 4(a) shows the graph of the parabola

$$y = x^2 - 4x + 3 = (x^2 - 4x + 4) - 1 = (x - 2)^2 - 1.$$

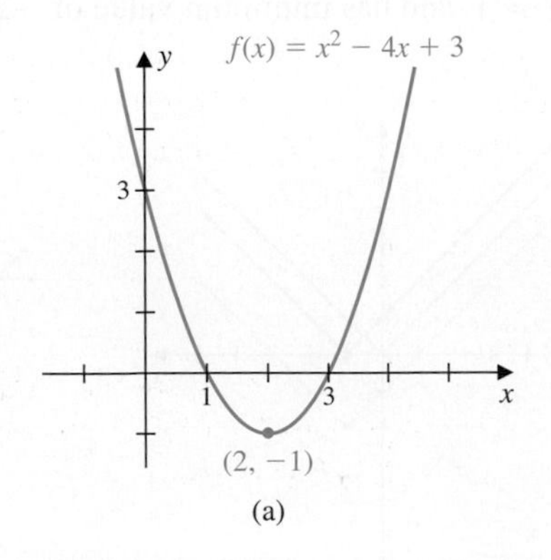

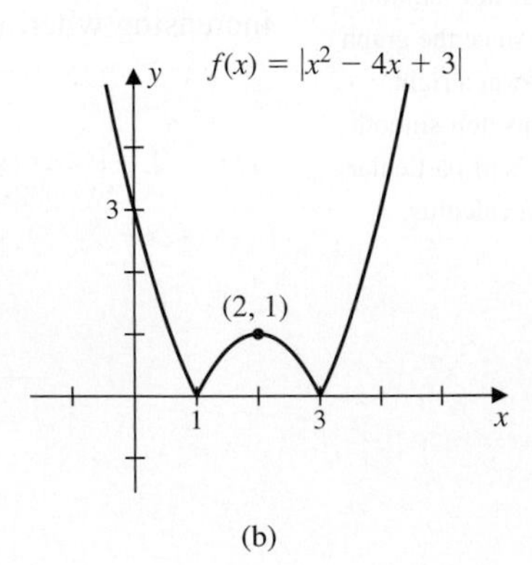

FIGURE 4

The graph of $y = |x^2 - 4x + 3|$ is not "smooth" at the points $(1, 0)$ and $(3, 0)$.

The absolute value of a number is the positive magnitude of the number, so the portion of the graph in Figure 4(a) below the x-axis is reflected above the x-axis. The portion already above the x-axis remains the same. The graph of $f(x) = |x^2 - 4x + 3|$ is shown in Figure 4(b). The function has a minimum value of 0 at $x = 1$ and $x = 3$. The function does not have a maximum value since

$$f(x) \to \infty \text{ as } x \to \infty \quad \text{and} \quad f(x) \to \infty \text{ as } x \to -\infty.$$

However, for $1 \leq x \leq 3$, the maximum value is 1, which occurs when $x = 2$. This is said to be a *local* maximum of f. ■

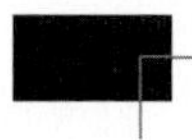

The Square Root Function

The graph of $f(x) = \sqrt{x}$ appears to approach the y-axis vertically as x approaches zero. This feature will be important in calculus.

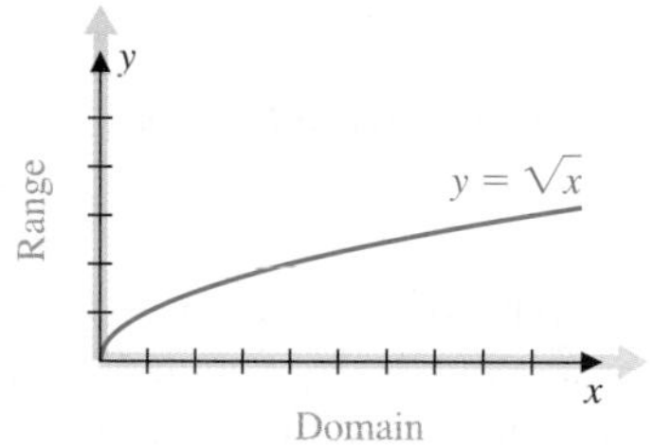

The graph of the *square root function*, $f(x) = \sqrt{x}$, is shown in Figure 5. Square roots are defined only for nonnegative numbers, so $f(x) = \sqrt{x}$ is defined only for $x \geq 0$ and the domain of the square root function is the set of all nonnegative real numbers. The range is also $[0, \infty)$, because by definition $\sqrt{}$ is the principal, or nonnegative, *square root*. The square root function is increasing on its entire domain, has a minimum value of zero at $x = 0$, and

$$\sqrt{x} \to \infty \quad \text{as} \quad x \to \infty.$$

FIGURE 5

EXAMPLE 4 Sketch the graph of $g(x) = \sqrt{x - 2} - 1$ and determine the domain and range of g.

Solution The graph of $y = \sqrt{x - 2}$ is the graph of $y = \sqrt{x}$ shifted to the right 2 units. As a consequence, the graph of $y = \sqrt{x - 2} - 1$ is the graph of $y = \sqrt{x}$ shifted to the right 2 units and downward 1 unit, as shown in Figure 6. The function g is always increasing and has a minimum value of -1 at $x = 2$.

For x to be in the domain of

$$g(x) = \sqrt{x - 2} - 1$$

we need $x - 2 \geq 0$, that is, $x \geq 2$.

The domain and range of g can easily be determined by using the horizontal and vertical line tests on the graph. As illustrated in Figure 6, the domain of g is $[2, \infty)$,

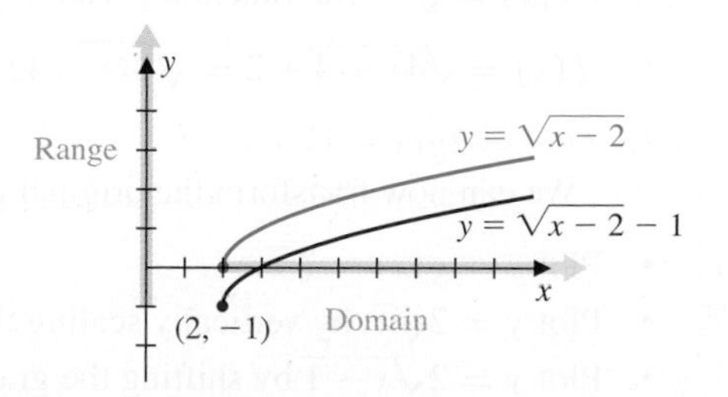

FIGURE 6

because the numbers in this interval of the x-axis are those through which a vertical line intersects the graph. In a similar manner, the range of g is $[-1, \infty)$, because the numbers in this interval of the y-axis are those through which a horizontal line intersects the graph. ■

EXAMPLE 5 Sketch the graph of $h(x) = \sqrt{-x-2} - 1$ and determine the domain and range of h.

Solution The graph of h involves shifting the graph of $y = \sqrt{-x}$. Because $\sqrt{x}$ is defined only for $x \geq 0$, $\sqrt{-x}$ is defined only when $x \leq 0$, and the graph of $y = \sqrt{-x}$ is the reflection about the y-axis of the graph of $y = \sqrt{x}$, as shown in red in Figure 7.

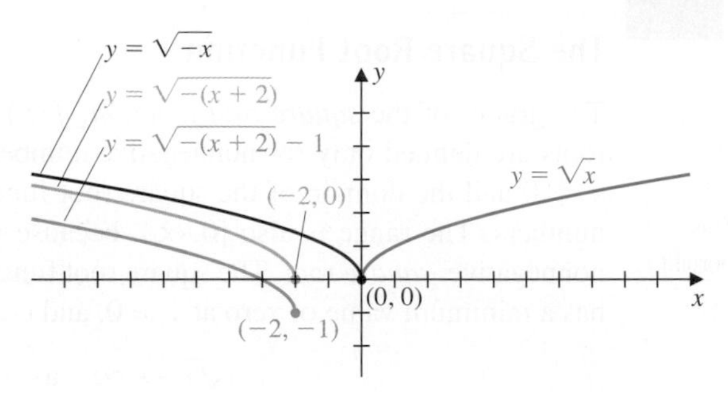

FIGURE 7

We can rewrite $h(x)$ as

$$h(x) = \sqrt{-x-2} - 1 = \sqrt{-(x+2)} - 1.$$

The graph of $y = \sqrt{-x-2} = \sqrt{-(x+2)}$ is obtained by shifting the graph of $y = \sqrt{-x}$ to the left 2 units. The final graph of $y = h(x) = \sqrt{-x-2} - 1$ is found by shifting the graph of $y = \sqrt{-x-2}$ downward 1 unit, as shown in green in Figure 7.

Vertical lines intersect the graph whenever $x \leq -2$, so the domain of h is $(-\infty, -2]$. The range of h is $[-1, \infty)$, since the horizontal lines that intersect the graph have $y \geq -1$.

Notice that h is decreasing on its entire domain and has a minimum value of -1 at $x = -2$. ■

EXAMPLE 6 Sketch the graph of the function $f(x) = \sqrt{4x-4} + 2$, and find the domain and range.

Solution We will use transformations to simplify the process, starting with the basic graph of $y = g(x) = \sqrt{x}$. The function f can be rewritten as

$$f(x) = \sqrt{4x-4} + 2 = \sqrt{4(x-1)} + 2 = \sqrt{4}\sqrt{x-1} + 2 = 2\sqrt{x-1} + 2,$$

so $f(x) = 2g(x-1) + 2$.

We can now transform the original problem into four easier ones.

Principles of Problem Solving:

- Reduce hard problems to a sequence of easier problems.

- Plot $y = g(x) = \sqrt{x}$.
- Plot $y = 2\sqrt{x}$, by vertically scaling the graph of $y = \sqrt{x}$ by a factor of 2.
- Plot $y = 2\sqrt{x-1}$ by shifting the graph of $y = 2\sqrt{x}$ to the right 1 unit.
- Plot $y = f(x) = 2\sqrt{x-1} + 2$ by shifting $y = 2\sqrt{x-1}$ upward 2 units.

The graph of the intermediate curves and the graph of the function are shown in Figure 8.

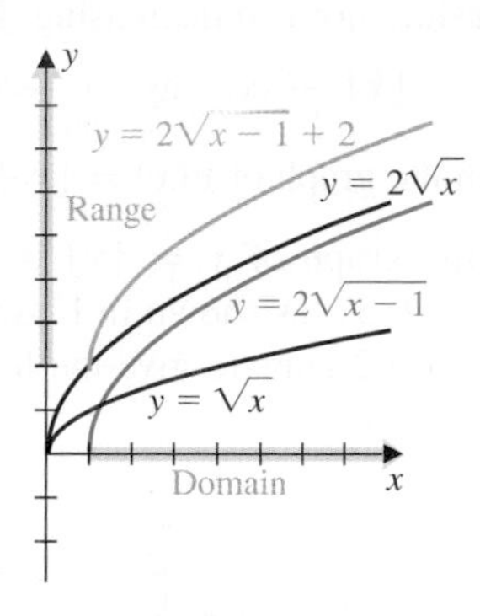

FIGURE 8

The graph can now be used to determine the domain and range of the function. Since the graph lies to the right of the vertical line $x = 1$ and includes the point $(1, 2)$, the domain is $[1, \infty)$. Since the graph lies above the horizontal line $y = 2$ and includes the point $(1, 2)$, the range is $[2, \infty)$. ■

The Greatest Integer Function

The **greatest integer function** (also called the **floor function**) is denoted $f(x) = \lfloor x \rfloor$, and defined for a given real number x to be the largest integer that is less than or equal to x. So

$$\lfloor x \rfloor = m, \quad \text{where } m \text{ is the integer satisfying } m \le x < m + 1.$$

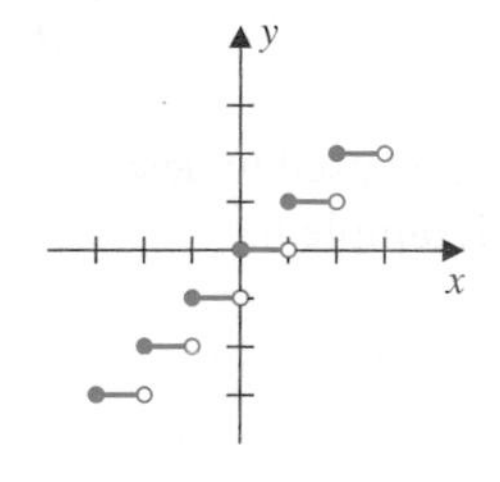

FIGURE 9

The greatest integer function has a range with gaps and its graph "jumps." The function is *discontinuous* at each integer.

In essence, the greatest integer function "rounds down" a real number to the next lowest integer. For example,

$$\lfloor 1.2 \rfloor = 1, \quad \lfloor -1.51 \rfloor = -2, \quad \lfloor 2 \rfloor = 2, \quad \text{and} \quad \lfloor 0.33 \rfloor = 0.$$

The greatest integer function has wide application in computer science, where this *floor* function has a complementary *ceiling* function, denoted $f(x) = \lceil x \rceil$, which represents the least integer that is greater than or equal to x.

The greatest integer function assumes the constant integer value m on the interval $[m, m + 1)$, so its graph consists of horizontal line segments beginning at each of the integers. For example, if x is in the interval $[-2, -1)$, then $-2 \le x < -1$ and $f(x) = \lfloor x \rfloor = -2$; if $0 \le x < 1$, then $f(x) = \lfloor x \rfloor = 0$; and so on, as shown in Figure 9 and Table 1.

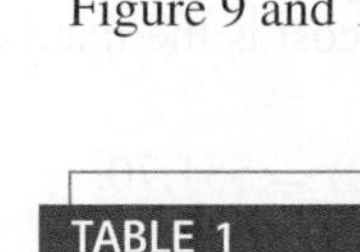

TABLE 1

When	$-2 \le x < -1$	$-1 \le x < 0$	$0 \le x < 1$	$1 \le x < 2$	$2 \le x < 3$
$\lfloor x \rfloor$ is	-2	-1	0	1	2

The domain of the greatest integer function is the set of real numbers, and its range is the set of integers. Because the function is constant between integers, it is not increasing, nor is it decreasing. In addition, it has no maximum or minimum since

$$\lfloor x \rfloor \to \infty \quad \text{as} \quad x \to \infty, \qquad \text{and} \qquad \lfloor x \rfloor \to -\infty \quad \text{as} \quad x \to -\infty.$$

EXAMPLE 7 Sketch the graph of $g(x) = \lfloor x + 3 \rfloor - 2$.

Solution First the graph of $y = \lfloor x \rfloor$ is shifted to the left 3 units to produce the graph of $y = \lfloor x + 3 \rfloor$, as shown in Figure 10(a). The graph of $y = \lfloor x + 3 \rfloor$ is then shifted downward 2 units to give the final graph of $y = g(x) = \lfloor x + 3 \rfloor - 2$, as shown in Figure 10(b).

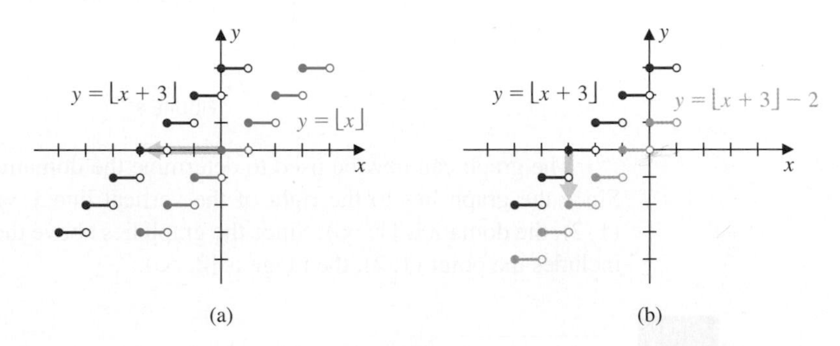

FIGURE 10

Notice that the graph is the same as the graph obtained by shifting the greatest integer function to the left 1 unit or upward 1 unit. This implies that $\lfloor x + 3 \rfloor - 2 = \lfloor x \rfloor + 1$, and that $\lfloor x + 1 \rfloor = \lfloor x \rfloor + 1$. ■

The greatest integer function jumps at each integer, so it can be used to describe and approximate certain real-world situations, as illustrated in Example 8.

Applications

EXAMPLE 8 A power company charges a monthly service charge of \$8.50, a rate of 5.8 cents per kilowatt-hour (kWh) for the first 500 kWh, and a rate of 8.3 cents per kWh for usage above 500.

a. What is the charge for 400 kWh?

b. What is the charge for 850 kWh?

c. Give a function that describes the cost in terms of kilowatt-hours.

Solution **a.** Not more than 500 kWh were used, so the cost is the fixed charge plus 5.8 cents per kWh. That is,

$$C = 8.5 + 0.058(400) = \$31.70.$$

b. The usage is above 500 kWh, so the charge is the fixed charge plus 0.058(500) cents for the first 500 kWh and 0.083(350) cents for the number of hours that exceeds 500. So

$$C = 8.5 + 0.058(500) + 0.083(350) = \$66.55.$$

c. Let x denote the number of kilowatt-hours used. Then for $x > 500$ we have

$$C(x) = \underbrace{8.5}_{\text{fixed cost}} + \underbrace{0.058(500)}_{\text{first 500 kWh}} + \underbrace{0.083(x-500)}_{\text{kWh over 500}}$$

$$= -4.0 + 0.083x,$$

so the cost is a piecewise-defined function given by

$$C(x) = \begin{cases} 8.5 + 0.058x, & \text{if } 0 \le x \le 500 \\ -4.0 + 0.083x, & \text{if } x > 500. \end{cases}$$

■

EXERCISE SET 2.2

In Exercises 1–8, use the graph of $y = |x|$ to sketch the graph of each of the functions.

1. $f(x) = |x - 4|$ **2.** $f(x) = |x + 1| - 1$

3. $f(x) = |x + 2| - 2$ **4.** $f(x) = |x - 2| + 2$

5. $f(x) = -2|x|$ **6.** $f(x) = |3x - 2|$

7. $f(x) = -3|x - 1| + 1$

8. $f(x) = -2|x + 1| - 3$

*In Exercises 9–16, **(a)** use the graph of $y = \sqrt{x}$ to sketch the graph of each of the functions, and **(b)** find the domain and range of each of the functions.*

9. $g(x) = \sqrt{x} + 3$ **10.** $g(x) = \sqrt{x} - 1$

11. $g(x) = \sqrt{x+2} - 2$ **12.** $g(x) = \sqrt{x-1} + 2$

13. $g(x) = -\sqrt{x+2}$ **14.** $g(x) = 3 - \sqrt{x+3}$

15. $g(x) = \sqrt{2-x} - 1$ **16.** $g(x) = \sqrt{3-x} + 2$

17. Use the graph of $f(x) = x^3$, shown in the figure, to sketch the graph of g defined by each equation.

a. $g(x) = x^3 + 2$ **b.** $g(x) = x^3 - 2$

c. $g(x) = (x + 2)^3$ **d.** $g(x) = (x - 2)^3$

e. $g(x) = 3x^3$ **f.** $g(x) = -3x^3$

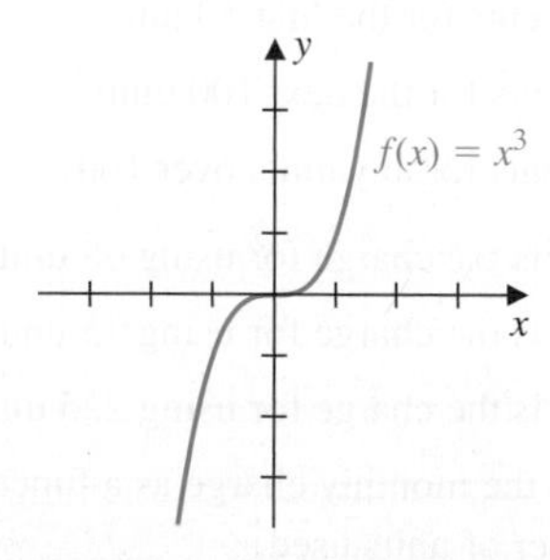

18. Use the graph of the cube root function, $f(x) = \sqrt[3]{x}$, shown in the figure, to sketch the graph of g defined by each equation.

a. $g(x) = \sqrt[3]{x} + 1$ **b.** $g(x) = \sqrt[3]{x+1}$

c. $g(x) = \sqrt[3]{x} - 1$ **d.** $g(x) = \sqrt[3]{x-1}$

e. $g(x) = 2\sqrt[3]{x}$ **f.** $g(x) = -2\sqrt[3]{x}$

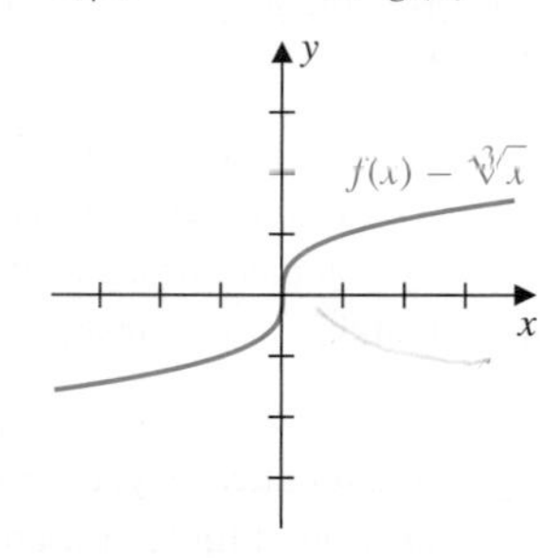

In Exercises 19–26, sketch $y = f(x)$ and $y = |f(x)|$.

19. $f(x) = 2x - 3$ **20.** $f(x) = -x + 2$

21. $f(x) = -x^2 - 4$ **22.** $f(x) = -x^2 + 2$

23. $f(x) = -(x - 1)^2 + 1$

24. $f(x) = (x + 2)^2 - 3$

25. $f(x) = x^2 - 6x + 7$

26. $f(x) = x^2 - 3x + 2$

In Exercises 27–32, use the graph of $y = \lfloor x \rfloor$ to sketch the graph of each function.

27. $f(x) = \lfloor x - 2 \rfloor$ **28.** $f(x) = \lfloor x \rfloor - 2$

29. $f(x) = \lfloor x + 1 \rfloor - 2$ **30.** $f(x) = \lfloor x - 2 \rfloor + 1$

31. $f(x) = 2 - \lfloor x \rfloor$ **32.** $f(x) = 2 - \lfloor x - 1 \rfloor$

In Exercises 33 and 34, use the graph of $y = f(x)$ to sketch the graph of each of the following.

a. $y = f(x) + 1$ **b.** $y = f(x - 2)$

c. $y = 2f(x)$ **d.** $y = f(2x)$

e. $y = -f(x)$ **f.** $y = f(-x)$

g. $y = |f(x)|$ **h.** $y = f(|x|)$

33. **34.**

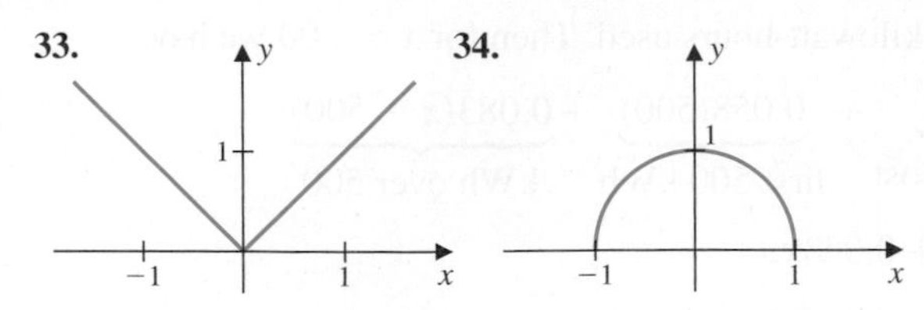

35. Let $f(x) = x + 1$.

a. Sketch the graph of $y = f(x)$ on the interval $[0, 4]$.

b. Let $A(t)$ denote the area of the region bounded by the x-axis, the y-axis, the curve $y = f(x)$, and the vertical line $x = t$. Find an expression for $A(t)$ and sketch the graph of $y = A(t)$ on the interval $[0, 4]$.

c. Find an expression $d(x)$ for the distance from the origin to the point $(x, f(x))$. Use a graphing device to sketch the graph of $y = d(x)$ on the interval $[0, 4]$.

36. You are driving at a constant 50 miles/hour along a straight road at a constant speed between cities A and B. Along the way you pass through city C. The cities B and C are 100 and 25 miles, respectively, from city A.

a. Draw a graph that indicates your distance from city C as a function of time, measured from the time you leave city A.

b. Use absolute value functions to express this distance as a function of t.

37. In 2010, the cost of mailing a first-class large envelope letter in the United States was 88 cents up to the first ounce and an additional 17 cents for each ounce beyond the first ounce up to a total of 13 ounces. We can use the greatest integer function to express this rate as

$$P(w) = \begin{cases} 0.88 + 0.17\lfloor w \rfloor, & \text{if } w > 0 \text{ is not a positive integer,} \\ 0.39 + 0.24(w - 1), & \text{if } w \text{ is a positive integer.} \end{cases}$$

Sketch the graph of $y = P(w)$ and determine the domain and range of P.

38. The Ohio Turnpike is 241 miles in length and has service plazas located 75 and 160 miles from Eastgate, the entrance to the turnpike at the Pennsylvania line. Express the distance of a car from the nearest service plaza as a function of the car's distance from Eastgate, and sketch the graph of this function.

39. The ancient Babylonians discovered a rule for approximating the square root of a positive number p. Begin by making a reasonable estimate a of $\sqrt{p}$ and construct the improved estimate

$$b = \frac{1}{2}\left(a + \frac{p}{a}\right).$$

Continue the process with b as the new value of a.

a. Use the method to approximate $\sqrt{2}$ to three decimal places.

b. Repeat the process to approximate $\sqrt{13}$ to three decimal places.

40. Show algebraically that $\lfloor x + 3 \rfloor - 2 = \lfloor x + 1 \rfloor$. (See Example 6.)

41. Calculators commonly permit you to enter the number of decimal places you would like to display, and to round all displayed results when appropriate.

a. Show that the function $f_0(x) = \lfloor x + 0.5 \rfloor$ rounds a number to its nearest integer.

b. Show that the function $f_1(x) = (0.1)\lfloor 10x + 0.5 \rfloor$ rounds a number to one decimal point, with multiples.

c. Construct a function f_n that rounds a number to n decimal points.

42. The functions in the preceding exercise rounded all midpoint values upward. Redo the parts of that exercise but construct the functions so that the midpoints round downward.

43. A natural gas provider charges customers based on the following fee schedule:

- Monthly service charge of \$6.58
- 3.65 cents for the first 50 units
- 2.8 cents for the next 100 units
- 1.4 cents for any units over 100

a. What is the charge for using 38 units in a month?

b. What is the charge for using 75 units in a month?

c. What is the charge for using 225 units in a month?

d. Write the monthly charge as a function of the number of units used.

44. Sketch on the same set of axes the graph of the function defined by $f(x) = -\lfloor -x \rfloor$ and the graph of $g(x) = \lceil x \rceil$.

2.3 ARITHMETIC COMBINATIONS OF FUNCTIONS

Functions are combined in calculus to form functions that are more complex. It is important to feel comfortable with the notation introduced in this section.

In this section we show how combinations of familiar functions together with some function operations can be used to construct more complex functions. The rules involving the arithmetic combinations of functions are quite natural; the only minor difficulty involves the domain of the quotient of two functions.

Arithmetic Combinations of Functions

If f and g are functions, then the functions $f+g$, $f-g$, $f \cdot g$, and f/g are defined by

$$(f+g)(x) = f(x) + g(x), \qquad (f-g)(x) = f(x) - g(x),$$
$$(f \cdot g)(x) = f(x) \cdot g(x), \qquad (f/g)(x) = \left(\frac{f}{g}\right)(x) = \frac{f(x)}{g(x)}.$$

Both $f(x)$ and $g(x)$ must be defined for any of these arithmetic operations to be defined at x. This means that the domains of $f+g$, $f-g$, and $f \cdot g$ consist of those real numbers that are common to both the domain of f and the domain of g. The domain of the quotient f/g consists of those real numbers x that are in both the domain of f and the domain of g, and that also satisfy $g(x) \neq 0$.

EXAMPLE 1 Let f and g be defined for the integers from -3 to 3 by the values in Table 1. Show that f and g are functions, and find $f+g$, $f-g$, $f \cdot g$, and f/g.

TABLE 1

x	-3	-2	-1	0	1	2	3
$f(x)$	-1	1	2	-3	0	7	5
$g(x)$	0	-2	4	8	-1	3	5

Solution Both f and g are functions since for each value of x there are unique image values. The new functions are shown in Table 2. To find the new values, either add, subtract, multiply, or divide the entries in the second and third columns of Table 2.

Since $g(-3) = 0$, $x = -3$ is not in the domain of f/g. ■

TABLE 2

x	$f(x)$	$g(x)$	$(f+g)(x)$	$(f-g)(x)$	$(f \cdot g)(x)$	$(f/g)(x)$
-3	-1	0	-1	-1	0	Not defined
-2	1	-2	-1	3	-2	$-\frac{1}{2}$
-1	2	4	6	-2	8	$\frac{1}{2}$
0	-3	8	5	-11	-24	$-\frac{3}{8}$
1	0	-1	-1	1	0	0
2	7	3	10	4	21	$\frac{7}{3}$
3	5	5	10	0	25	1

EXAMPLE 2 Let $f(x) = 1/(x^2 - 1)$ and $g(x) = x/\sqrt{x+2}$. Find $f + g$, $f - g$, $f \cdot g$, and f/g, and give their domains.

Solution By the definitions we have

$$(f+g)(x) = f(x) + g(x) = \frac{1}{x^2-1} + \frac{x}{\sqrt{x+2}},$$

$$(f-g)(x) = f(x) - g(x) = \frac{1}{x^2-1} - \frac{x}{\sqrt{x+2}},$$

$$(f \cdot g)(x) = f(x) \cdot g(x) = \frac{1}{x^2-1} \cdot \frac{x}{\sqrt{x+2}}, \quad \text{and}$$

$$\left(\frac{f}{g}\right)(x) = \frac{f(x)}{g(x)} = \frac{\frac{1}{x^2-1}}{\frac{x}{\sqrt{x+2}}} = \frac{\sqrt{x+2}}{x(x^2-1)}.$$

The domain of f is the set of all real numbers except 1 and -1, the values of x that would produce a 0 in the denominator. The domain of g is the interval $(-2, \infty)$ because the square root in the denominator requires that $x \geq -2$, but $x = -2$ would produce a 0 in the denominator so it is also eliminated.

Figure 1(a) shows that the common domain of $f + g$, $f - g$, and $f \cdot g$ is the set of real numbers x that satisfy $x > -2$, excluding $x = \pm 1$. That is, the domains of $f + g$, $f - g$, and $f \cdot g$ are $(-2, -1) \cup (-1, 1) \cup (1, \infty)$.

For x to be in the domain of the quotient f/g, we also need $g(x) = x/\sqrt{x+2} \neq 0$, so we cannot have $x = 0$. Therefore, the domain of f/g is the set of real numbers x that satisfy $x > -2$, excluding $x = \pm 1$ and $x = 0$. This can be expressed as $(-2, -1) \cup (-1, 0) \cup (0, 1) \cup (1, \infty)$ and is shown in Figure 1(b). ■

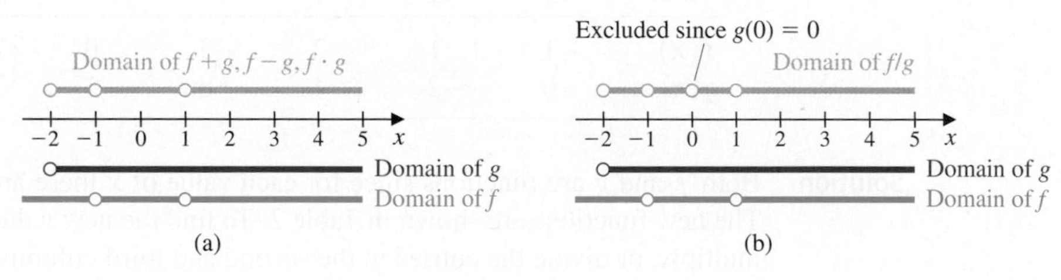

FIGURE 1

The simplified form on the right side of the equation describing the function f/g in Example 2 is

$$\left(\frac{f}{g}\right)(x) = \frac{f(x)}{g(x)} = \frac{\frac{1}{x^2-1}}{\frac{x}{\sqrt{x+2}}} = \frac{\sqrt{x+2}}{x(x^2-1)}.$$

This seems to imply that $x = -2$ is in the domain of f/g. However, this is *not* the case. Since $x = -2$ is not in the domain of g, it is not a candidate for the domain of any arithmetic combination involving the function g. When computing the quotient or product of two functions, always be careful to exclude from the domain those values that are not in the common domain of the original functions, particularly when the final form has been simplified.

The graph of $y = (f + g)(x)$ is the graphical addition of the curves $y = f(x)$ and $y = g(x)$. For each x, the y-coordinate of the point on the graph of the sum is the sum of the corresponding y-coordinates of the points on $y = f(x)$ and $y = g(x)$.

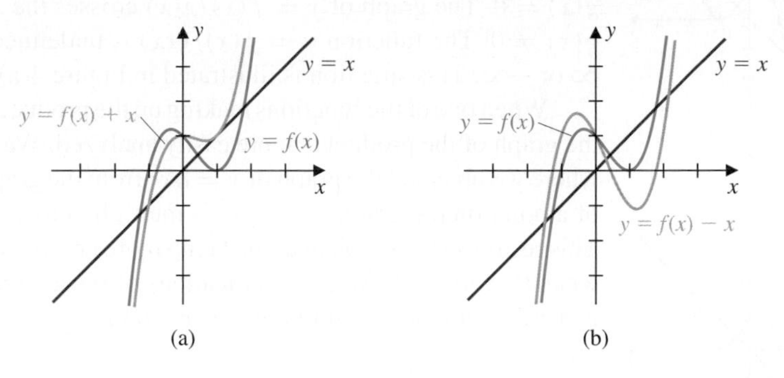

FIGURE 2

Similarly, the graph of $y = (f - g)(x)$ is the graphical subtraction of the graphs of $y = f(x)$ and $y = g(x)$. Figure 2(a) shows the graph of the sum of the arbitrary function $y = f(x)$ and $y = x$, and Figure 2(b) shows the difference. The x-intercepts for the graph of $y = (f + g)(x)$ are those x for which $f(x) = -g(x)$, and the x-intercepts for the graph of $y = (f - g)(x)$ are those x for which $f(x) = g(x)$.

EXAMPLE 3 Sketch the graphs of $f(x) = \lfloor x \rfloor + |x - 1|$ and $g(x) = \lfloor x \rfloor - |x - 1|$.

Solution The graph of $y = |x - 1|$ is the graph of $y = |x|$ shifted to the right 1 unit, as shown in red in Figure 3(a). Summing the vertical components of the graphs of $y = \lfloor x \rfloor$ and $y = |x - 1|$ on several intervals gives

$$f(x) = \lfloor x \rfloor + |x - 1| = \begin{cases} -1 - (x - 1) = -x, & \text{if } -1 \le x < 0 \\ 0 - (x - 1) = -x + 1, & \text{if } 0 \le x < 1 \\ 1 + (x - 1) = x, & \text{if } 1 \le x < 2 \\ 2 + (x - 1) = x + 1, & \text{if } 2 \le x < 3. \end{cases}$$

The graph of $f(x) = \lfloor x \rfloor + |x - 1|$ is shown in Figure 3(b). By contrast, the graph of $g(x) = \lfloor x \rfloor - |x - 1|$ is shown in Figure 3(c). ■

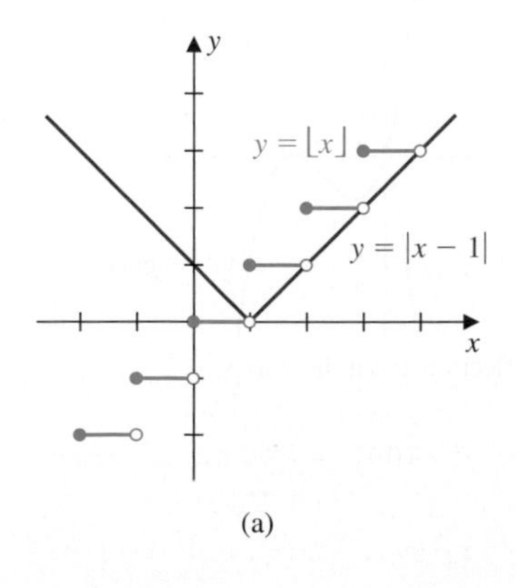

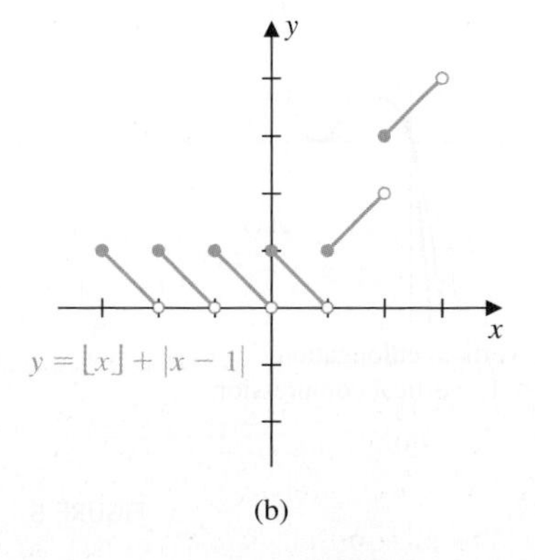

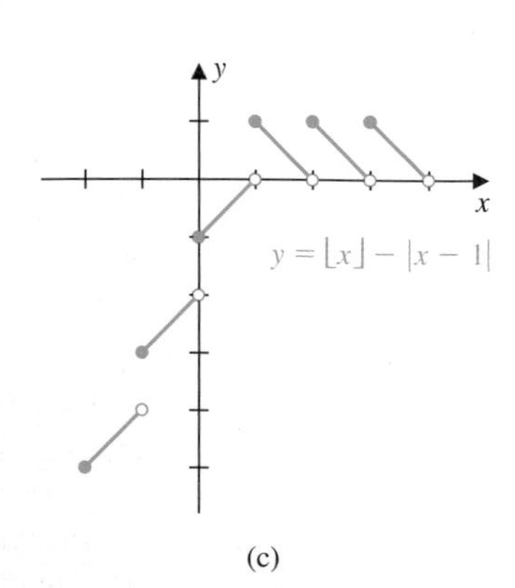

FIGURE 3

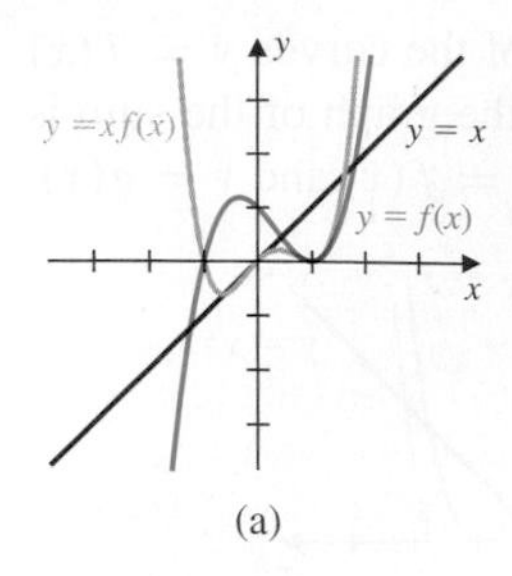

(a)

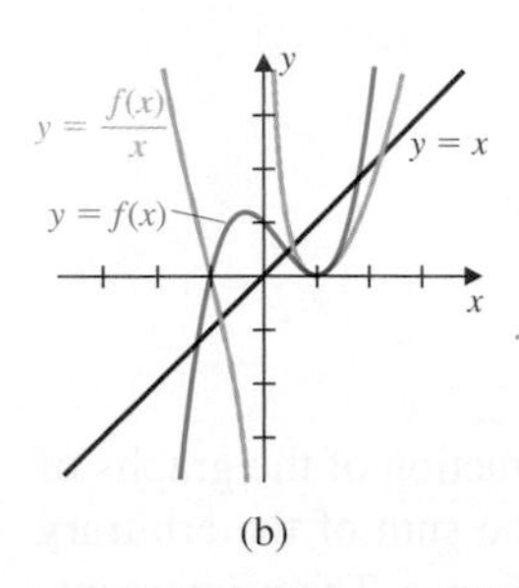

(b)

FIGURE 4

It is generally difficult to determine the graphs for the product $f \cdot g$ and quotient f/g, but some valuable information about these graphs can often be obtained. For example, the graph of $y = f(x)g(x)$ crosses the x-axis when either $f(x) = 0$ or $g(x) = 0$. The graph of $y = f(x)/g(x)$ crosses the x-axis when both $f(x) = 0$ and $g(x) \neq 0$. The function $y = f(x)/g(x)$ is undefined when $g(x) = 0$ and can go to ∞ or $-\infty$. This situation is illustrated in Figure 4(a) and (b).

When one of the functions making up the product $f \cdot g$ is a constant, say $f(x) = c$, the graph of the product is more easily analyzed. We encountered this in Section 1.8, where we obtained the graph of $y = ax^2$ from the graph of $y = x^2$. Each y-coordinate of a point on the graph of $y = x^2$ is multiplied by a to obtain the graph of $y = ax^2$. This results in the graph opening being more compressed when $|a| > 1$, and widening when $0 < |a| < 1$. Some modifications of those guidelines are needed to reflect the general nature of the graph of $y = c \cdot g(x)$.

Vertical Compression and Elongation of Graphs

- For any constant $c > 0$, the basic shape of the graph of $y = c \cdot g(x)$ is the same as the graph of $y = g(x)$ but with a change in the vertical scale. The domains of cg and g are the same and the graphs have the same x-intercepts.
- For $c > 1$, the graph of $y = c \cdot g(x)$ is a **vertical elongation**, and for $0 < c < 1$ it is a **vertical compression** of the graph of $y = g(x)$. (See Figure 5(a).)
- The graph of $y = -g(x)$ is the **reflection about the x-axis** of the graph of $y = g(x)$, as shown in Figure 5(b).
  - For $c < -1$, the graph of $y = c \cdot g(x)$ is a vertical elongation and reflection of the graph of $y = g(x)$. (See Figure 6(a).)
  - For $-1 < c < 0$, the graph of $y = c \cdot g(x)$ is a vertical compression and reflection of the graph of $y = g(x)$. (See Figure 6(b)).

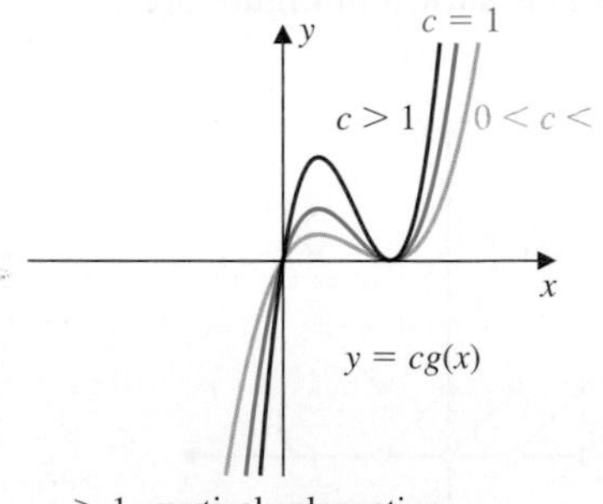

$c > 1$: vertical enlongation
$0 < c < 1$: vertical compression

(a)

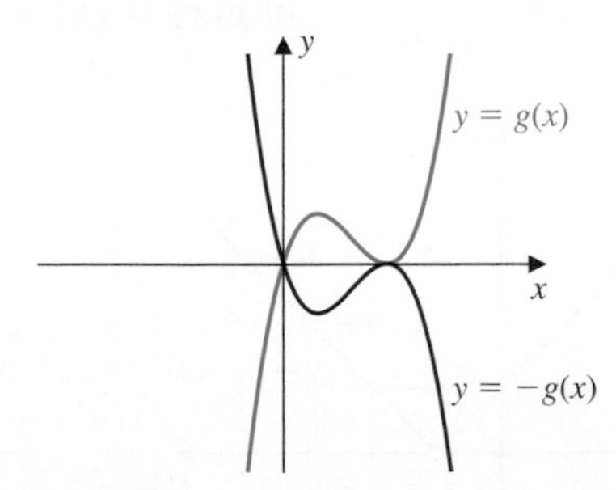

Reflection about the x-axis

(b)

FIGURE 5

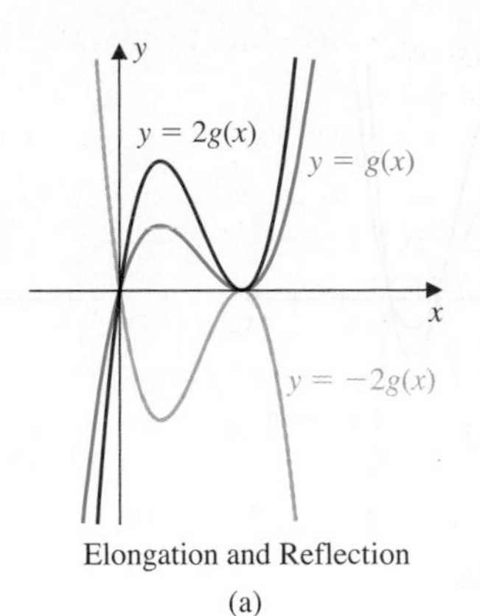

Elongation and Reflection

(a)

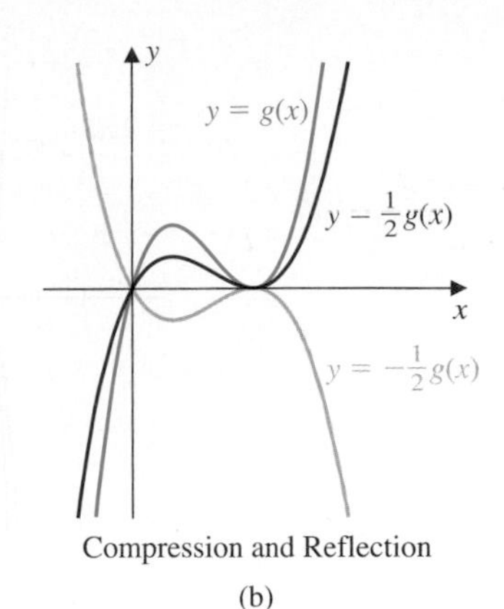

Compression and Reflection

(b)

FIGURE 6

EXAMPLE 4 Consider the graph shown in Figure 7 of

$$g(x) = x^3 - 2x^2 - x + 2 = (x - 1)(x - 2)(x + 1).$$

Compare the graph of $y = g(x)$ with the graphs of $y = 2g(x)$, $y = \frac{1}{2}g(x)$, $y = -2g(x)$, and $y = -\frac{1}{2}g(x)$.

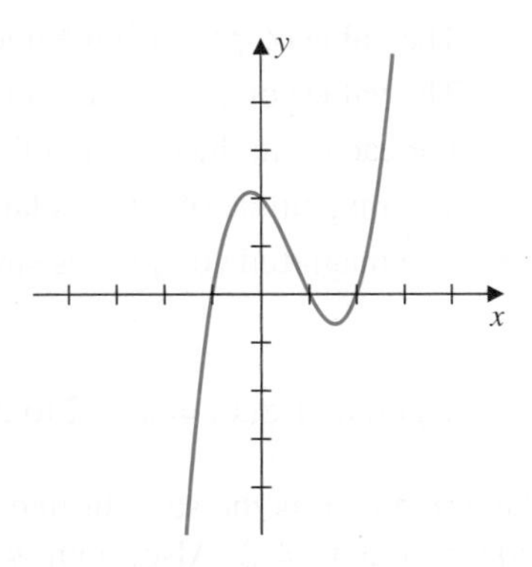

FIGURE 7

Solution These graphs are vertical elongations or compressions of the graph of $y = g(x)$. If the scaling factor is $c = 2$, then the graph is an elongation since the y-coordinates are twice as large, and if $c = \frac{1}{2}$, then the graph is a compression. A negative scaling factor will in addition reflect the graph of $y = g(x)$ about the x-axis. Figure 8 shows the various graphs. Notice that they all have the same x-intercepts, and that the y-intercepts are directly proportional to the scaling factor c. ■

A special case of the quotient of two functions is both useful and quite elementary. The *reciprocal* of a function g, defined by $(1/g)(x) = 1/g(x)$, is a quotient that occurs when the function in the numerator is the constant 1. The following relationships

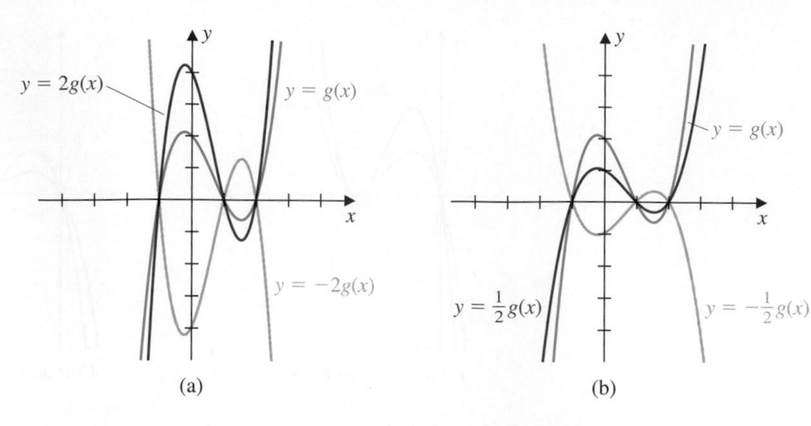

FIGURE 8

between a function and its reciprocal can be used to produce a graph of the reciprocal function that shows its most interesting features. (See Figure 9.)

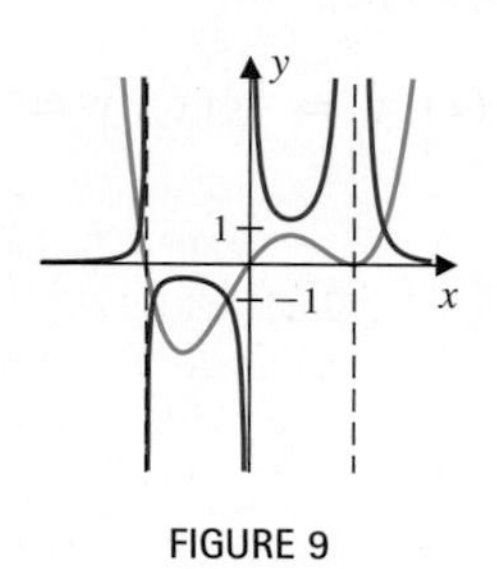

FIGURE 9

The Graph of the Reciprocal of a Function

Let $f(x) = \dfrac{1}{g(x)}$ be the reciprocal of $g(x)$.

- The value of $f(x)$ is undefined when $g(x) = 0$.
- The values of $f(x)$ and $g(x)$ are the same when $g(x) = 1$ or $g(x) = -1$.
- For each x in the domain of f, $f(x)$ and $g(x)$ have the same sign.
- The magnitude of $f(x)$ is large when the magnitude of $g(x)$ is small.
- The magnitude of $f(x)$ is small when the magnitude of $g(x)$ is large.

EXAMPLE 5 Use the graph of $g(x) = x - 2$ to determine the graph of $f(x) = \dfrac{1}{x-2}$.

Solution The graph of g is the straight line through $(2, 0)$ with slope 1, so the function f is undefined at $x = 2$. Also, both graphs pass through the points $(1, -1)$ and $(3, 1)$. Since $g(x) > 0$ on $(2, \infty)$ and $g(x) < 0$ on $(-\infty, 2)$, we have $f(x) > 0$ on $(2, \infty)$ and $f(x) < 0$ on $(-\infty, 2)$. In addition,

$$f(x) = \frac{1}{x-2} \to \infty \quad \text{as } x \to 2 \text{ with } x > 2 \text{ (denoted by } x \to 2^+)$$

and

$$f(x) = \frac{1}{x-2} \to -\infty \quad \text{as } x \to 2 \text{ with } x < 2 \text{ (denoted by } x \to 2^-).$$

Finally,

$$g(x) = x - 2 \to \infty \quad \text{as} \quad x \to \infty$$

and

$$g(x) = x - 2 \to -\infty \quad \text{as} \quad x \to -\infty,$$

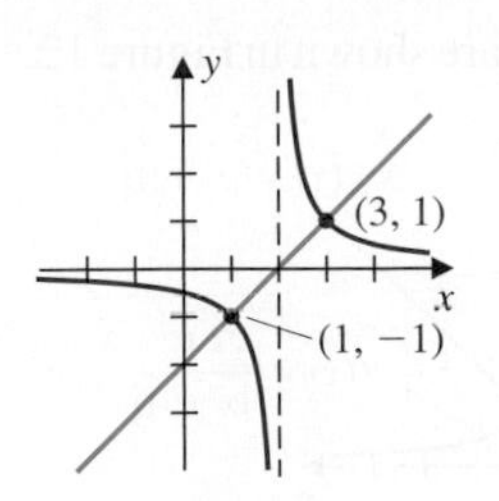

FIGURE 10

so

$$f(x) = \frac{1}{x-2} \to 0 \quad \text{as } x \to \infty$$

and

$$f(x) = \frac{1}{x-2} \to 0 \quad \text{as } x \to -\infty.$$

This leads to the graph of f shown in Figure 10. The dashed line indicates that the graph of f approaches but does not touch the vertical line $x = 2$. ■

EXAMPLE 6 Sketch the graph of $f(x) = \dfrac{1}{x^2 + 2x - 3}$.

Solution The graph of the reciprocal of $y = f(x)$ is the parabola $y = x^2 + 2x - 3$. The essential points of the parabola are the vertex and the x-intercepts. To find the x-intercepts, solve

$$0 = x^2 + 2x - 3 = (x-1)(x+3) \quad \text{to give} \quad x = 1 \text{ and } x = -3.$$

To find the vertex, complete the square to obtain

$$y = x^2 + 2x - 3 = x^2 + 2x + 1 - 1 - 3 = (x+1)^2 - 4.$$

The vertex is $(-1, -4)$. The graph of the parabola is shown in Figure 11.

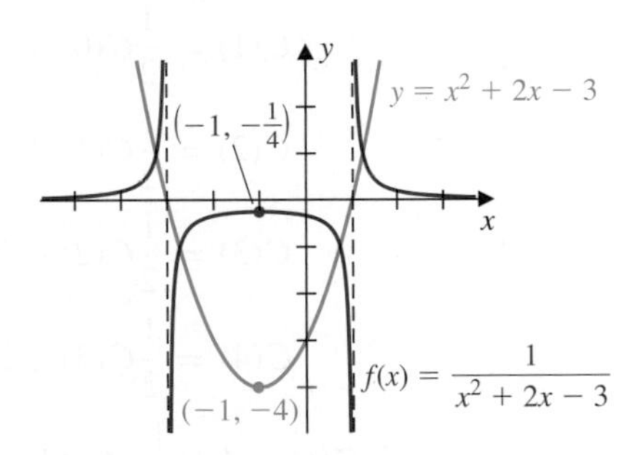

FIGURE 11

The graph of the reciprocal of the parabola is shown in red in Figure 11. Note that

$$\text{when } x > 1 : f(x) \to \infty \text{ as } x \to 1,$$
$$\text{when } x < 1 : f(x) \to -\infty \text{ as } x \to 1,$$
$$\text{when } x > -3 : f(x) \to -\infty \text{ as } x \to -3,$$
$$\text{when } x < -3 : f(x) \to \infty \text{ as } x \to -3.$$

Also

$$x^2 + 2x - 3 \to \infty \text{ as } x \to \infty \quad \text{and} \quad x^2 + 2x - 3 \to \infty \text{ as } x \to -\infty$$

so

$$f(x) \to 0 \text{ as } x \to \infty \quad \text{and} \quad f(x) \to 0 \text{ as } x \to -\infty.$$

Finally, the vertex of the parabola is $(-1, -4)$, so we have $f(-1) = -\frac{1}{4}$. ■

Some other examples of the reciprocal graphing technique are shown in Figure 12.

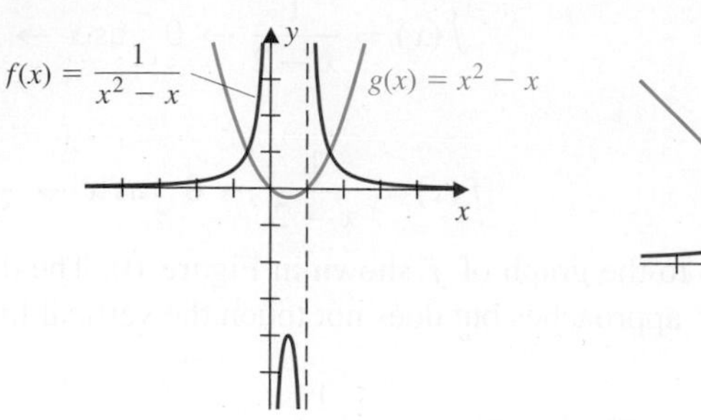

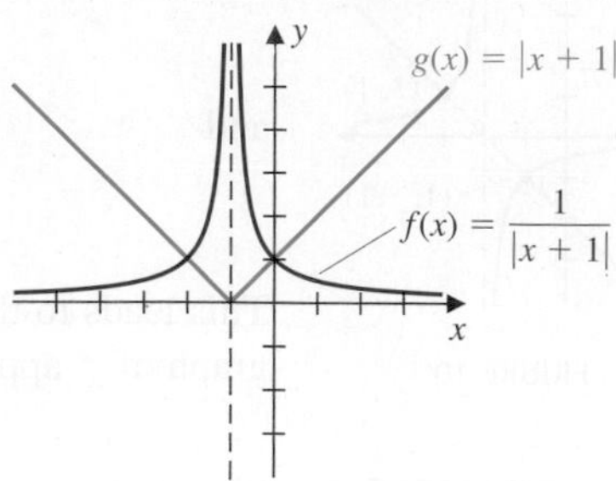

FIGURE 12

EXAMPLE 7 A drug is being administered to a patient on a daily basis, and an accurate amount of the drug in the patient's bloodstream is needed during the treatment phase. The amount must decrease over time but also never drop below a specified level, about 2 mg. In a 24-hour period, half the dose will be eliminated. Assume that 5 mg is given the first day, and 2 mg is given each succeeding day. Find a function for the amount of the drug in the patient in terms of time.

Solution Let $C(t)$ denote the amount of the drug in mg in the patient after the tth amount has been administered. Then $C(0) = 5$,

$$C(1) = \frac{1}{2}C(0) + 2 = \frac{5}{2} + 2 = 4 + \frac{1}{2},$$

$$C(2) = \frac{1}{2}C(1) + 2 = \frac{5}{4} + 3 = 4 + \frac{1}{4},$$

$$C(3) = \frac{1}{2}C(2) + 2 = \frac{5}{8} + \frac{7}{2} = 4 + \frac{1}{8},$$

$$C(4) = \frac{1}{2}C(3) + 2 = \frac{5}{16} + \frac{15}{4} = 4 + \frac{1}{16},$$

and, in general, $C(t) = 4 + \frac{1}{2^t}$. As t becomes increasingly large, the dosage becomes closer to 4 mg. ■

EXERCISE SET 2.3

In Exercises 1–8, find $f + g$, $f - g$, $f \cdot g$, and f/g, and give the domain of each new function.

1. $f(x) = 2x$; $g(x) = x^2 + 1$
2. $f(x) = x^2$; $g(x) = x + 1$
3. $f(x) = \frac{1}{x}$; $g(x) = \frac{x}{x - 2}$
4. $f(x) = \frac{1}{x - 1}$; $g(x) = \frac{1}{x + 1}$
5. $f(x) = \sqrt{x + 1}$; $g(x) = \sqrt{3 - x}$
6. $f(x) = \frac{1}{x}$; $g(x) = \sqrt{x - 1}$
7. $f(x) = \begin{cases} -1, & \text{if } x < 0, \\ 1, & \text{if } \geq 0; \end{cases}$

   $g(x) = \begin{cases} 1, & \text{if } x < 0, \\ 0, & \text{if } x \geq 0 \end{cases}$
8. $f(x) = \begin{cases} 0, & \text{if } x < 0, \\ x, & \text{if } x \geq 0; \end{cases}$

   $g(x) = \begin{cases} -x, & \text{if } x < 0, \\ 0, & \text{if } x \geq 0 \end{cases}$

In Exercises 9–16, use the results about the graph of the reciprocal of a function to sketch the graph of the function $h(x) = 1/g(x)$.

9. $g(x) = x - 3$ **10.** $g(x) = 2 - x$

11. $g(x) = |x|$ **12.** $g(x) = |x - 1|$

13. $g(x) = x^2 - 1$ **14.** $g(x) = x^2 - 3$

15. $g(x) = x^2 - 4x + 3$ **16.** $g(x) = x^2 - x - 2$

In Exercises 17–20, the graphs of functions f and g are given. Sketch the graphs of $f + g$ and $f - g$.

17.

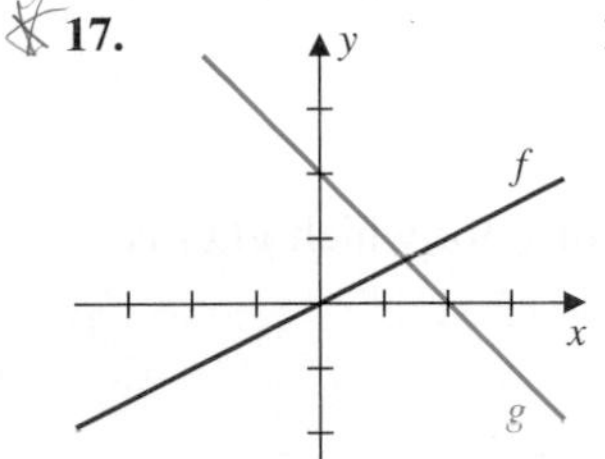

18.

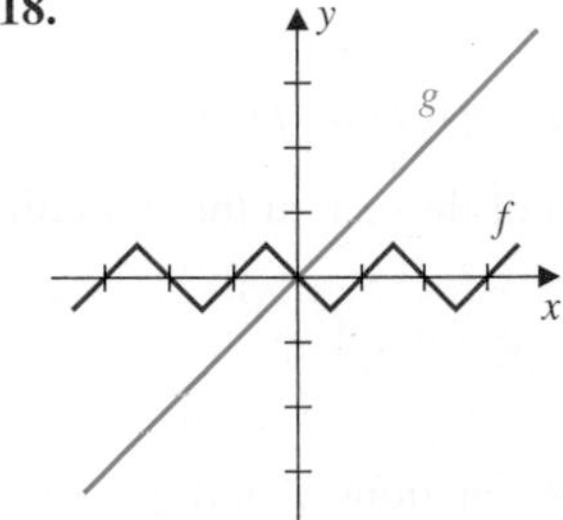

19.

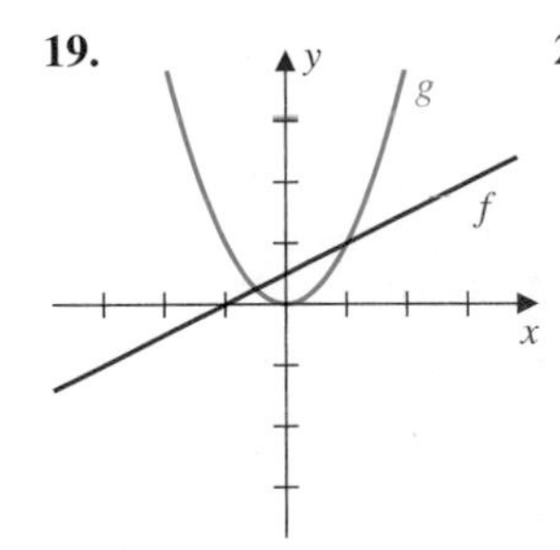

20.

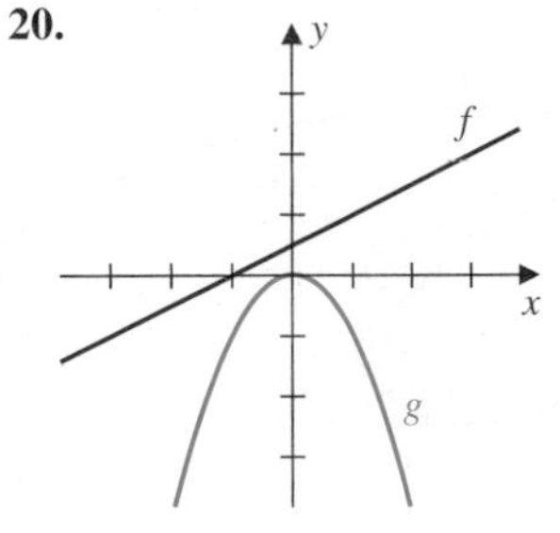

21. Functions f and g are defined by

$$f(x) = \frac{x^2 - 4}{x - 2} \text{ and } g(x) = x + 2.$$

How do the graphs of f and g differ?

22. Functions f and g are defined by

$$f(x) = \frac{x^3 - 2x^2}{x} \text{ and } g(x) = x^2 - 2x.$$

How do the graphs of f and g differ?

In Exercises 23–26, use a graphing device to sketch the graphs of f, g, and $f + g$ on the same set of axes to illustrate graphical addition.

23. $f(x) = \sqrt{x - 1}$; $g(x) = \sqrt{x + 2}$

24. $f(x) = x^2 - 1$; $g(x) = x^3$

25. $f(x) = x^3 - 2x^2 - x + 2$; $g(x) = x^3 - 7x + 6$

26. $f(x) = -x^3 - x^2 + 2x$; $g(x) = x^3 - x^2 - 2x$

27. The rate of the spread of an infectious disease is proportional to the number of infected individuals and to the number of healthy individuals in the population. Suppose that a community has 5000 people, and that if 30 people have the disease, then 2 additional people per day will contract the disease. What is the rate of infection when 70 people have been infected?

28. A culture of bacteria normally doubles every hour. However, after each hour an amount A of bacteria are harvested. Assume the initial population contains I bacteria.

a. Find the population in terms of I and A after each of the first 4 hours.

b. Find $P(t)$ the number of bacteria present at time t.

c. Find $H(t)$ the number of bacteria harvested after t hours.

2.4 COMPOSITION OF FUNCTIONS

In Section 2.3 we saw how functions can be combined using the arithmetic operations of addition, subtraction, multiplication, and division. *Composition* creates a new function by taking the output of one function and applying it as input to another function.

To illustrate, suppose that $f(x) = x^2 + 1$ and $g(x) = x - 2$. If the input to g is x, then the output is $g(x) = x - 2$. For example, the illustration

$$2 \longrightarrow g(2) = 0 \longrightarrow f(0) = 1$$

describes the composition $f(g(x))$ when $x = 2$.

If we use $x - 2$ as the input to f, then we have

$$f(g(x)) = f(x - 2) = (x - 2)^2 + 1.$$

This is called the *composition* of f with g and denoted $f \circ g$. It results in a building process that is used to make successively more complicated functions from a sequence of elementary functions.

Composition of Functions

The **composition** of the function f with the function g, denoted $f \circ g$, is defined by

$$(f \circ g)(x) = f(g(x)).$$

The domain of $f \circ g$ consists of those x in the domain of g for which $g(x)$ is in the domain of f.

The composition of a pair of functions f and g is illustrated in Figure 1.

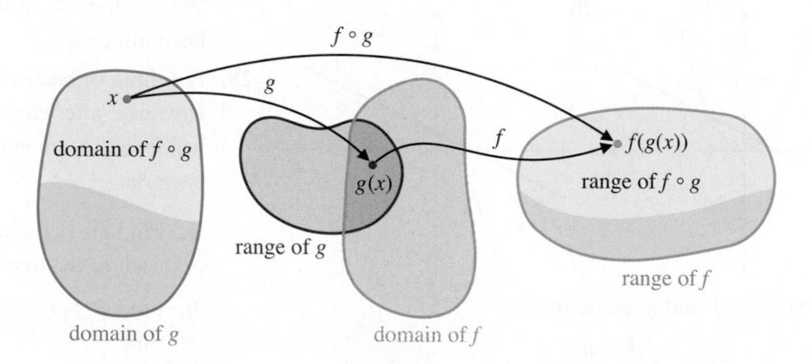

FIGURE 1

The diagram in Figure 2 shows how the value $(f \circ g)(x) = f(g(x))$ is computed. First x is input to the function g, producing the output $g(x)$. Then $g(x)$ is used as the input to the function f. This gives the final value of the composition. The composition is defined provided x is a valid input for the function g, that is, x is in the domain of g, and, in addition, $g(x)$ is in the domain of f.

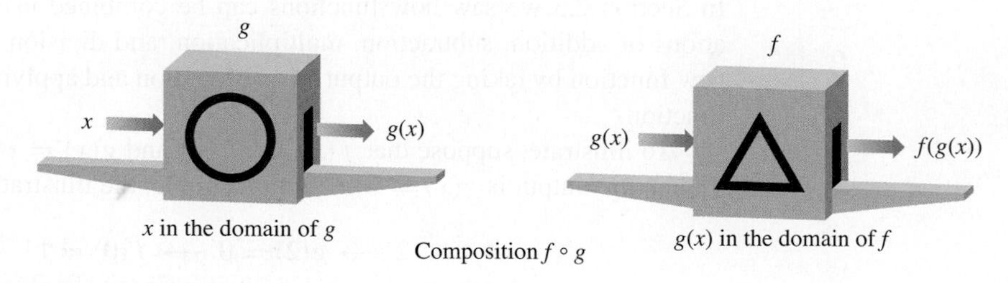

FIGURE 2

EXAMPLE 1 Find $(f \circ g)(2)$ and $(g \circ f)(2)$ for $f(x) = x - 2$ and $g(x) = x^2 - 1$.

Solution As shown in Figure 3,

$$(f \circ g)(2) = f(g(2)) = f(3) = 1$$

and

$$(g \circ f)(2) = g(f(2)) = g(0) = -1.$$

Notice that $(f \circ g)(2) \neq (g \circ f)(2)$. ■

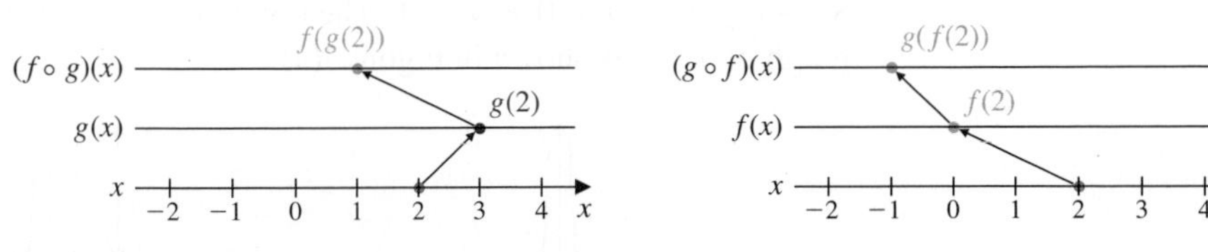

FIGURE 3

Example 1 illustrates that even when $f \circ g$ and $g \circ f$ have a common value x in their domains, it is generally the case that

$$(f \circ g)(x) \neq (g \circ f)(x).$$

Keep in mind that function notation such as $f(x) = x^2 - 1$ means that for any element x in the domain, the value $f(x)$ is obtained by squaring x and subtracting 1. In particular, keep in mind that there is nothing special about the variable used in the description of f. Notation of the type

$$f(\square) = (\square)^2 - 1$$

is helpful when considering the composition since it better emphasizes that

$$f\left(\boxed{g(x)}\right) = \left(\boxed{g(x)}\right)^2 - 1.$$

EXAMPLE 2 Let $f(x) = \sqrt{x - 1}$ and $g(x) = 1/x^2$.

a. Is $(f \circ g)(0)$ defined?

b. Is $(f \circ g)(2)$ defined?

c. Find $(f \circ g)(x)$ and the domain of $f \circ g$.

d. Find $(g \circ f)(x)$ and the domain of $g \circ f$.

Solution **a.** To find the composition $(f \circ g)(0)$ we first need to compute $g(0)$. However, $g(0)$ is undefined so $(f \circ g)(0)$ is also undefined.

b. To find the composition

$$(f \circ g)(2) = f(g(2))$$

we first compute $g(2) = \frac{1}{4}$. Although $g(2)$ is defined, $(f \circ g)(2)$ is undefined because $\frac{1}{4} - 1 = -\frac{3}{4}$, and negative numbers are not in the domain of the square root function f.

c. The composition is

$$(f \circ g)(x) = f(g(x)) = f\left(\frac{1}{x^2}\right) = \sqrt{\frac{1}{x^2} - 1} = \sqrt{\frac{1 - x^2}{x^2}}.$$

The domain of g is the set of all nonzero real numbers, and the domain of f is the interval $[1, \infty)$. To find the domain of $f \circ g$, we need to determine the numbers x in the domain of g with $g(x)$ in the domain of f. That is, we need to find $x \neq 0$ so that

$$\frac{1}{x^2} - 1 = \frac{1 - x^2}{x} \geq 0, \quad \text{which implies that} \quad 0 < x^2 \leq 1.$$

So $-1 \leq x < 0$ or $0 < x \leq 1$. The domain of $f \circ g$ is $[-1, 0) \cup (0, 1]$, and the graph of $f \circ g$ is shown in Figure 4(a).

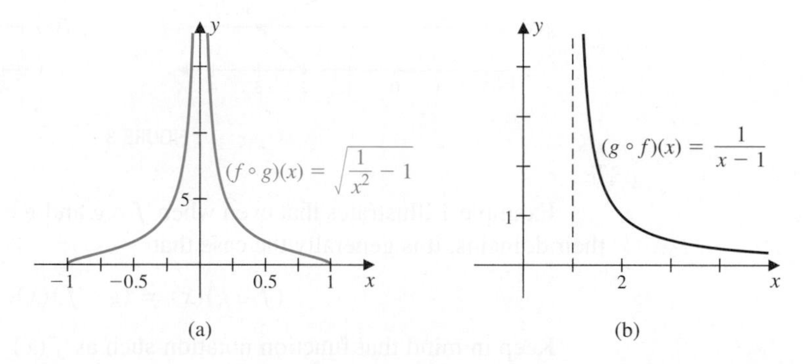

FIGURE 4

d. The composition is

$$(g \circ f)(x) = g(f(x)) = g(\sqrt{x - 1}) = \frac{1}{\left(\sqrt{x - 1}\right)^2} = \frac{1}{x - 1}.$$

The domain of f is the interval $[1, \infty)$ because the term under the square root cannot be negative. We can apply g to any of the resulting values of $f(x)$ except when $x = 1$. This value is not permitted because $f(1) = 0$ is not in the domain of g. The domain of $g \circ f$ is the interval $(1, \infty)$. and the graph of $g \circ f$ is shown in Figure 4(b). ■

Notice that the simplified form of $(g \circ f)(x) = 1/(x - 1)$ in Example 2 cannot be used to determine the domain of $g \circ f$, because $\left(\sqrt{x - 1}\right)^2 = x - 1$ only when both of these terms are defined. The left side of the equation is not defined when $x < 1$.

Decomposing a complicated function into a sequence of more familiar functions is a common technique in calculus. The following examples illustrate how this is done.

EXAMPLE 3 **a.** Write the function $f(x) = \sqrt{x^2 + 1}$ as the composition of two functions.

b. Write the function

$$k(x) = \frac{2}{\sqrt{x^2 + 1}}$$

as the composition of three functions.

Solution **a.** The function f can be visualized by the chain of transformations:

$$x \to x^2+1 \to \sqrt{x^2+1} = f(x).$$

This leads us to define a pair of functions g and h as

$$g(x) = x^2+1 \quad \text{and} \quad h(x) = \sqrt{x},$$

so

$$f(x) = h(g(x)) = \sqrt{g(x)} = \sqrt{x^2+1}.$$

Being able to unravel a complex composition of functions is valuable in calculus when finding derivatives and integrals.

This representation in terms of a composition is not unique. We can alternatively define, for example,

$$\hat{g}(x) = x^2 \quad \text{and} \quad \hat{h}(x) = \sqrt{x+1},$$

and have

$$f(x) = \hat{h}(\hat{g}(x)) = \sqrt{\hat{g}(x)+1} = \sqrt{x^2+1}.$$

This second decomposition does not, however, seem to be quite as natural as the first.

b. Most of the work has already been done in part (a). Consider the chain

$$x \to x^2+1 \to \sqrt{x^2+1} \to \frac{2}{\sqrt{x^2+1}} = k(x).$$

Define $j(x) = 2/x$ and use the same functions, $g(x) = x^2+1$ and $h(x) = \sqrt{x}$, as in the decomposition in part (a). Then

$$k(x) = j(f(x)) = j(h(g(x))) = \frac{2}{h(g(x))} = \frac{2}{\sqrt{x^2+1}}.$$ ■

It is often useful first to illustrate a composition by a chain, as we did in Example 3.

Using composition to decompose a complicated function into several basic pieces can lead to a better understanding of the graph, as illustrated in the next example.

EXAMPLE 4 Write $f(x) = \dfrac{1}{|x^2-4x+1|}$ as a composition of functions, and sketch the graph of f.

Solution First, we reemphasize that there is no single method of writing a function as a composition. There are, however, ways that are more natural than others.

Since $f(x)$ includes a quadratic term in the denominator, we first complete the square to write the quadratic in standard form. This gives

$$x^2-4x+1 = x^2-4x+4-4+1 = (x-2)^2-3,$$

which allows us to write

$$f(x) = \frac{1}{|(x-2)^2-3|}.$$

To construct $f(x)$ in a systematic manner involving only common operations, we use the "chain"

$$x \to (x-2)^2 \to (x-2)^2 - 3 \to |(x-2)^2 - 3| \to \frac{1}{|(x-2)^2 - 3|}.$$

The sequence of functions that produces each of these operations is

$$g_1(x) = (x-2)^2, \quad g_2(x) = x - 3, \quad g_3(x) = |x|, \quad \text{and} \quad g_4(x) = \frac{1}{x}.$$

Then

$$\begin{aligned} g_4(g_3(g_2(g_1(x)))) &= g_4\left(g_3\left(g_2\left((x-2)^2\right)\right)\right) \\ &= g_4\left(g_3\left((x-2)^2 - 3\right)\right) \\ &= g_4\left(\left|(x-2)^2 - 3\right|\right) \\ &= \frac{1}{\left|(x-2)^2 - 3\right|} = f(x). \end{aligned}$$

In this way, $f(x)$ is written as a composition of familiar functions.

To use the decomposition of $f(x)$ to sketch its graph, we first sketch in sequence the graphs of

$$y = g_1(x) = (x-2)^2,$$
$$y = g_2(g_1(x)) = (x-2)^2 - 3,$$

and

$$y = g_3(g_2(g_1(x))) = |(x-2)^2 - 3|.$$

These are shown in Figure 5.

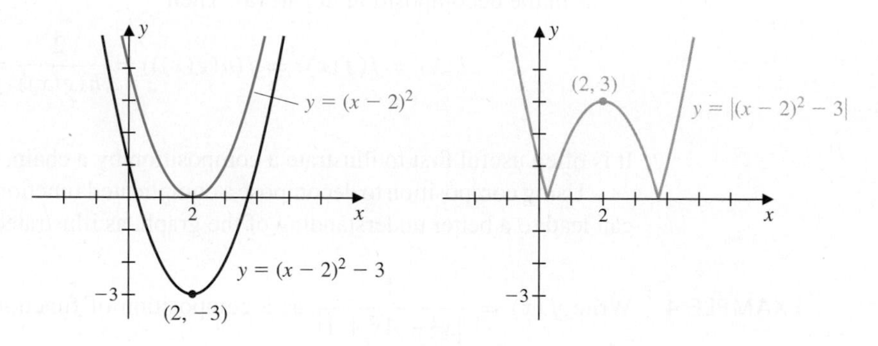

FIGURE 5

We can now use the reciprocal graphing technique to obtain the graph of

$$f(x) = \frac{1}{g_3(g_2(g_1(x)))} = \frac{1}{|(x-2)^2 - 3|}.$$

When applied to the graph in Figure 5, it produces the graph of $y = f(x)$, shown in red in Figure 6. ■

In Section 2.3 we found that the graph of $y = c \cdot f(x)$ is a vertical compression or elongation of the graph of $y = f(x)$, together with a reflection about the x-axis in the

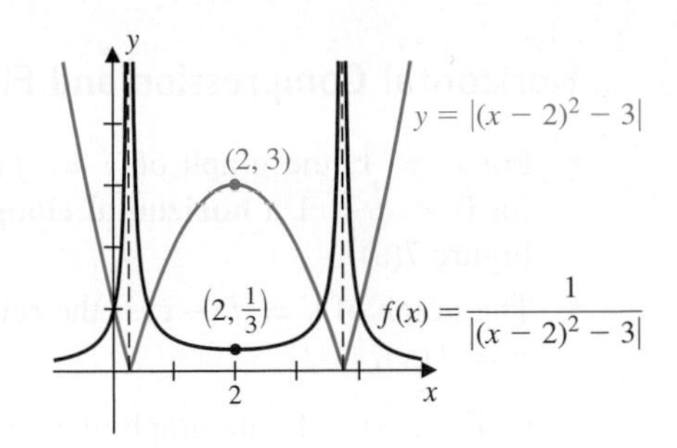

FIGURE 6

case that the constant is negative. The composition of a function f with $g(x) = cx$ is $(f \circ g)(x) = f(g(x)) = f(cx)$. Multiplying the independent variable, or *argument*, of a function by a constant produces an effect similar to that of multiplying the function by a constant, but the scaling of the graph occurs in the horizontal rather than the vertical direction.

For example, the graph of $y = f(2x)$ is a horizontal compression of the graph of $y = f(x)$, and the graph of $y = f(\frac{1}{2}x)$ is a horizontal elongation. The y-coordinate of the point on the graph of $y = f(2x)$ at x is the same as the y-coordinate of the point on the graph of $y = f(x)$ at $2x$, so the graph is horizontally compressed. The y-coordinate of the point on $y = f(\frac{1}{2}x)$ at x is the same as the y-coordinate of the point on $y = f(x)$ at $\frac{1}{2}x$, so the graph is elongated.

EXAMPLE 5 The graph of

$$f(x) = x^3 - 2x^2 - x + 2 = (x - 1)(x - 2)(x + 1)$$

is given in Figure 7. Compare the graphs of $f(x)$, $f(2x)$, $f(\frac{1}{2}x)$, $f(-2x)$, and $f(-\frac{1}{2}x)$.

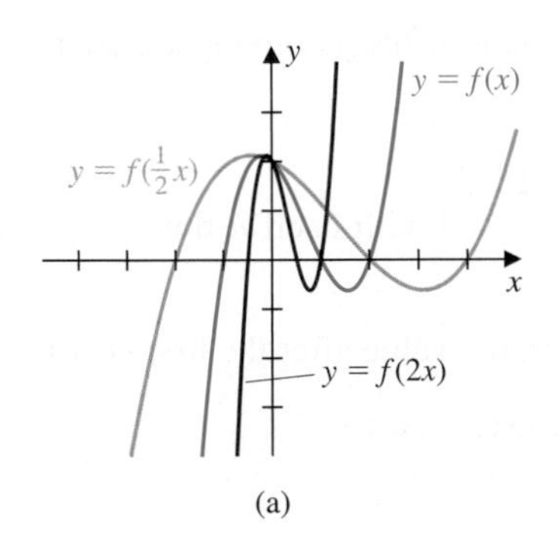

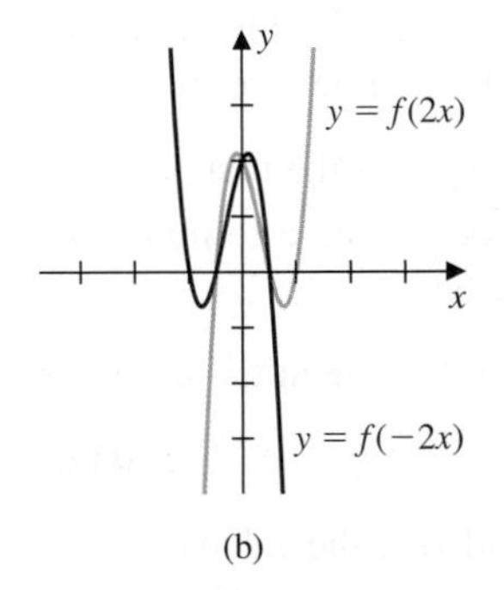

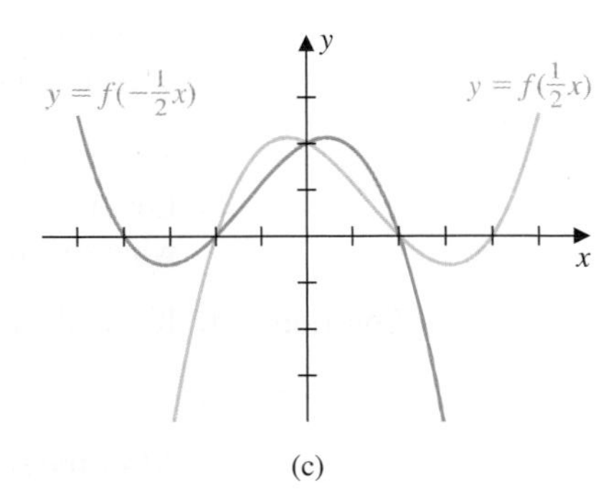

FIGURE 7

Solution Figure 7 shows the various graphs. Notice that the y-intercepts of the graphs agree. However, the horizontal scale has changed, so the graphs do not have the same x-intercepts. The x-intercepts are inversely proportional to the scaling factor. ■

Horizontal Compression and Elongation of Graphs

- For $c > 1$, the graph of $y = f(cx)$ is a **horizontal compression**, and for $0 < c < 1$ a **horizontal elongation**, of the graph of $y = f(x)$. (See Figure 7(a).)
- The graph of $y = f(-x)$ is the **reflection about the y-axis** of the graph of $y = f(x)$.
  - For $c < -1$, the graph of $y = f(cx)$ is a horizontal compression and reflection of the graph of $y = f(x)$. (See Figure 7(b).)
  - For $-1 < c < 0$, the graph of $y = f(cx)$ is a horizontal elongation and reflection of the graph of $y = f(x)$. (See Figure 7(c).)

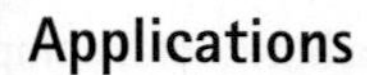

Applications

Example 6 illustrates the use of the composition of functions to describe a problem involving variation in time. These applications involving related rates form an important part of any calculus course.

EXAMPLE 6 Oil leaks from a tanker into a lake. Suppose the shape of the oil spill is approximately circular, and at any time t minutes after the leak has begun, the radius of the circle is $r(t) = \sqrt{t} + \sqrt[3]{t}$. Find the area of the spill at any time t after the leak has begun.

Solution The area of a circle of radius r is $A(r) = \pi r^2$. The area of the spill depends on the radius, and the radius depends on time. So A is the function of t given by the composition

In calculus you will want to know how the area of the circle changes as the radius changes relative to time.

$$(A \circ r)(t) = A(r(t)) = A\left(\sqrt{t} + \sqrt[3]{t}\right) = \pi\left(\sqrt{t} + \sqrt[3]{t}\right)^2.$$ ∎

EXAMPLE 7 You have the opportunity to invest an initial amount at 6% compounded annually (interest computed one time a year).

a. What is the value after the first, second, and third years?

b. Let $A(x) = 1.06x$. Find a formula for A composed with itself n times, and tell what this represents.

Solution **a.** If x is the amount of the initial investment, then the value after the first year is

$$x + 0.06x = 1.06x.$$

After the second year the value is

$$1.06x + 0.06(1.06x) = 1.06x(1 + 0.06) = (1.06)^2x.$$

After three years the value is

$$(1.06)^2x + 0.06(1.06)^2x = 1.06x^2(1 + 0.06) = (1.06)^3x,$$

and so on.

b. The value after n years is $(1.06)^n x$. In terms of $A(x) = 1.06x$ we have

$$(A \circ A)(x) = A(A(x)) = A(1.06x) = 1.06(1.06x) = (1.06)^2 x,$$
$$(A \circ A \circ A)(x) = A(A(A(x))) = A((1.06)^2 x) = 1.06((1.06)^2 x) = (1.06)^3 x,$$

and, in general, the composition of A with itself n times gives $(1.06)^n x$, the value after n years. ■

EXERCISE SET 2.4

In Exercises 1–6, let $f(x) = 2x - 3$ and $g(x) = x^2 + 2$, and evaluate the composition.

1. $(f \circ g)(3)$ **2.** $(g \circ f)(-1)$

3. $f(g(-3))$ **4.** $g(f(5))$

5. $(f \circ f)(-2)$ **6.** $g(g(\frac{1}{2}))$

In Exercises 7–14, find $f \circ g$ and $g \circ f$, and give the domain of each composition.

7. $f(x) = 2x + 1$; $g(x) = 3x - 1$

8. $f(x) = x^2 + 1$; $g(x) = x - 1$

9. $f(x) = \dfrac{1}{x}$; $g(x) = x^2 + 2x$

10. $f(x) = \dfrac{2}{x-5}$; $g(x) = x^2 + 4x$

11. $f(x) = \sqrt{x-1}$; $g(x) = x^2 - 3$

12. $f(x) = \sqrt{x-9}$; $g(x) = x^2$

13. $f(x) = \dfrac{1}{x}$; $g(x) = \dfrac{1}{x+1}$

14. $f(x) = \dfrac{1}{x-1}$; $g(x) = \dfrac{x+1}{x-1}$

In Exercises 15–20, find functions f and g such that $h = f \circ g$.

15. $h(x) = (3x^2 - 2)^4$ **16.** $h(x) = (x^2 - x + 1)^8$

17. $h(x) = \sqrt[3]{x-4}$ **18.** $h(x) = \sqrt[3]{x + \sqrt{x}}$

19. $h(x) = \dfrac{1}{x+2}$ **20.** $h(x) = |x^2 + x + 1|$

21. a. Use the graphs of f and g in the figure to evaluate each expression.

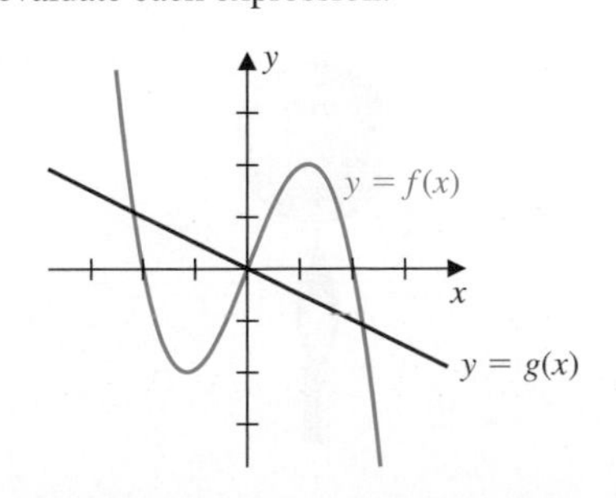

i. $(f \circ g)(-2)$ **ii.** $(g \circ f)(-2)$

iii. $(g \circ f)(2)$ **iv.** $(f \circ g)(2)$

b. Determine all values of x that satisfy the equation.

i. $(f \circ g)(x) = 0$ **ii.** $(g \circ f)(x) = 0$

22. In the following table f is even and g is odd. Complete the table.

x	$f(x)$	$g(x)$	$(f \circ g)(x)$	$(g \circ f)(x)$
-3	-2	-2		
-2	3	3		
-1	0	0		
0	1	1		
1				
2				
3				

23. The graph of $f(x) = x^3 - x$ is shown in the figure. Use the concepts of horizontal compression and elongation to sketch the graphs of $f(2x)$, $f(-2x)$, $f(\frac{1}{2}x)$, and $f(-\frac{1}{2}x)$.

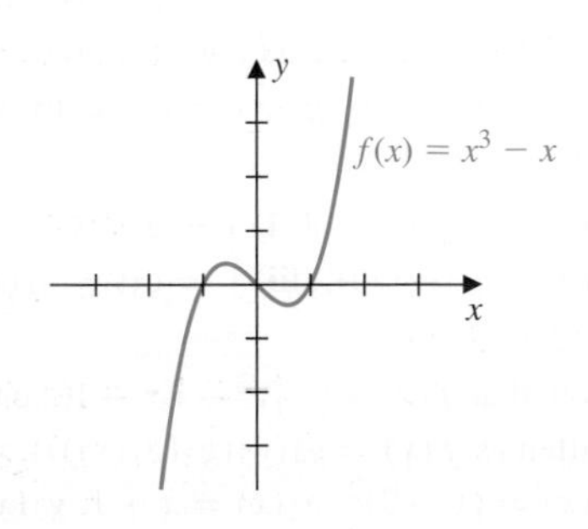

24. The graph of $f(x) = x^3 - 4x^2 + x + 4$ is shown in the figure. Use the concepts of horizontal compression and elongation to sketch the graphs of $f(2x)$, $f(-2x)$, $f(\frac{1}{2}x)$, and $f(-\frac{1}{2}x)$.

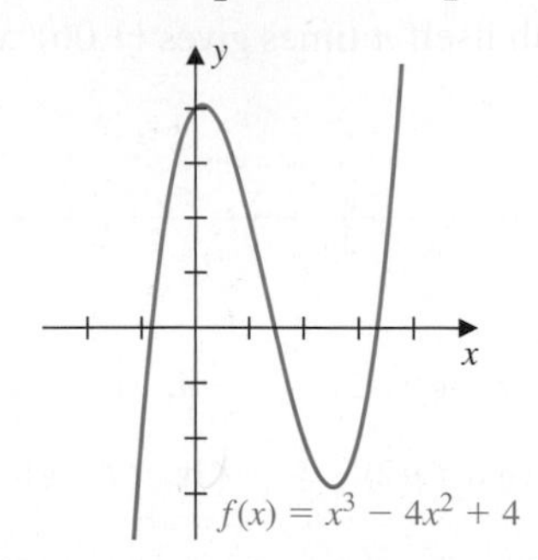

25. Sketch the graphs of the functions in the order given, and observe the difference in the graph that each successive complication introduces.

a. $f_1(x) = x - 1$ **b.** $f_2(x) = (x - 1)^2$

c. $f_3(x) = x^2 - 2x$ **d.** $f_4(x) = |x^2 - 2x|$

e. $f_5(x) = \dfrac{1}{|x^2 - 2x|}$ **f.** $f_6(x) = \dfrac{-1}{|x^2 - 2x|}$

26. Sketch the graphs of the functions in the order given, and observe the difference in the graph that each successive complication introduces.

a. $f_1(x) = x - 4$

b. $f_2(x) = (x - 4)^2$

c. $f_3(x) = x^2 - 8x + 13$

d. $f_4(x) = |x^2 - 8x + 13|$

e. $f_5(x) = \dfrac{1}{|x^2 - 8x + 13|}$

f. $f_6(x) = \dfrac{2}{|x^2 - 8x + 13|}$

27. **a.** Show that $f(x) = 1/|x^2 + 2x - 1|$ can be written as $f(x) = g_4(g_3(g_2(g_1(x))))$, where $g_1(x) = (x + 1)^2$, $g_2(x) = x - 2$, $g_3(x) = |x|$, and $g_4(x) = 1/x$.

b. Sketch the graphs of **(i)** $y = g_1(x)$, **(ii)** $y = g_2(g_1(x))$, **(iii)** $y = g_3(g_2(g_1(x)))$, and **(iv)** $y = f(x)$.

28. **a.** Show that $f(x) = 1/|x^2 - 6x + 10|$ can be written as $f(x) = g_4(g_3(g_2(g_1(x))))$, where $g_1(x) = (x - 3)^2$, $g_2(x) = x + 1$, $g_3(x) = |x|$, and $g_4(x) = 1/x$.

b. Sketch the graphs of **(i)** $y = g_1(x)$, **(ii)** $y = g_2(g_1(x))$, **(iii)** $y = g_3(g_2(g_1(x)))$, and **(iv)** $y = f(x)$.

29. We have seen that, in general, $f \circ g$ and $g \circ f$ are different. Find examples of f and g when these two compositions are the same.

30. Show that the composition of two odd functions is an odd function.

31. Show that the composition of an odd and an even function, in either order, is an even function.

32. Find a linear function f such that $(f \circ f)(x) = 4x + 3$.

33. Let f and g be linear functions with $f(x) = ax + b$ and $g(x) = cx + d$. When does

a. $(f \circ g)(x) = (g \circ f)(x)$?

b. $(f \circ g)(x) = f(x)$?

c. $(f \circ g)(x) = g(x)$?

34. What type of function is **(a)** the composition of two linear functions? **(b)** the composition of a linear and a quadratic function? **(c)** the composition of two quadratic functions?

35. Describe a function $g(x)$ in terms of $f(x)$, if the graph of g is obtained by shifting the graph of f to the right 1 unit and upward 2 units, and if it is vertically stretched by a factor of 3 when compared to the graph of f.

36. Describe a function $g(x)$ in terms of $f(x)$, if the graph of g is obtained by reflecting the graph of f about the x-axis and shifting it to the left 2 units, and if it is horizontally stretched by a factor of 2 when compared to the graph of f.

37. A metal sphere is heated so that t seconds after the heat has been applied, the radius $r(t)$ is given by $r(t) = 3 + 0.01t$ cm. Express the volume of the sphere as a function of t.

38. Sand is poured onto a conical pile whose radius and height are always equal, although both increase with time. The height of the pile t seconds after pouring begins is given by $h(t) = 10 + 0.25t$ ft. Express the volume of the pile as a function of t.

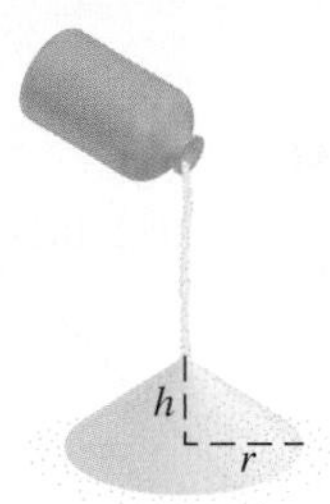

39. A spherical balloon is inflated so that its radius at the end of t seconds is $r(t) = 3\sqrt{t} + 5$ cm, $0 \le t \le 4$. Express **(a)** the volume and **(b)** the surface area as a function of time. What are the units of these quantities?

40. In Exercise 36 of Exercise Set 1.8, the height, in feet, of a rock above the ground t seconds after it has been thrown upward from the top of a building was given as $s(t) = 576 + 144t - 16t^2$. The domain of the function s is the closed interval whose left endpoint is zero and whose right endpoint is the time it takes the rock to reach the ground. Define a function $\bar{s}$ that describes the height of the rock above the ground, assuming that the rock is thrown at $t = 2$ instead of $t = 0$. What are the domain and range of $\bar{s}$?

41. An initial investment of x dollars is made at a fixed rate of interest of 4% compounded annually, and at the end of each year P additional dollars are invested.

a. What is the value of the investment at the end of the second and third years?

b. Let $V(x)$ be the value of the account at the end of the first year after the new investment is made. Use compositions of $V(x)$ to find a formula for the value of the investment after n years.

2.5 INVERSE FUNCTIONS

The relations of functions to their inverses is an important topic in calculus. Some of the most important functions cannot be approached directly.

The word *inverse* brings to mind a reversal of some process or operation. In the case of functions, this reversal involves the interchange of the domain with the range and a corresponding reversal of the operation describing the function.

As an illustration, there is a linear function that converts Celsius temperature to Fahrenheit temperature. The relationship between the scales can be determined by two facts:

- Water freezes at 0°C and at 32°F.
- Water boils at 100°C and at 212°F.

The graph of the function that converts Celsius temperature into Fahrenheit temperature, expressed as $F(C)$, is a line that passes through the points $(0, 32)$ and $(100, 212)$, and this linear function has slope

$$\frac{F(100) - F(0)}{100 - 0} = \frac{212 - 32}{100} = \frac{9}{5}.$$

Because $F(0) = 32$, its equation is

$$F \equiv F(C) = \tfrac{9}{5}C + 32.$$

It is equally useful to know the *inverse* relationship, the one that converts Fahrenheit temperature into Celsius. This is found by solving for C in terms of F in the Celsius-to-Fahrenheit equation,

$$F - 32 = \tfrac{9}{5}C, \quad \text{so} \quad C \equiv C(F) = \tfrac{5}{9}(F - 32).$$

Graphs of these equations are shown in Figure 1.

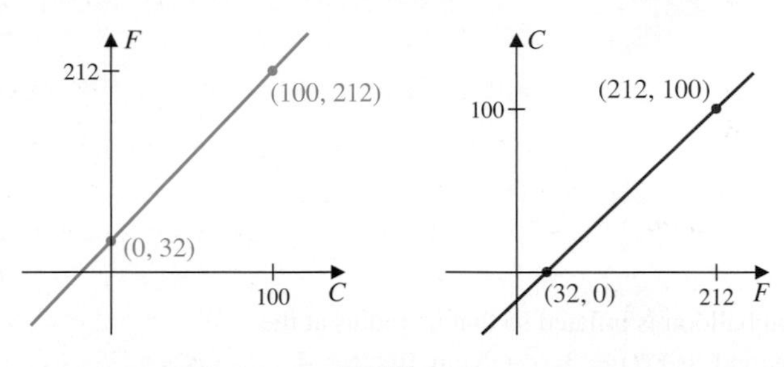

FIGURE 1

If we start with a temperature given in Celsius, convert it to Fahrenheit, and then convert it back to Celsius, the result is the starting temperature. A similar result occurs for a Fahrenheit temperature converted to Celsius and then back to Fahrenheit. So

$$F \equiv F(C) = \tfrac{9}{5}C + 32 \quad \text{if and only if} \quad C \equiv C(F) = \tfrac{5}{9}(F - 32).$$

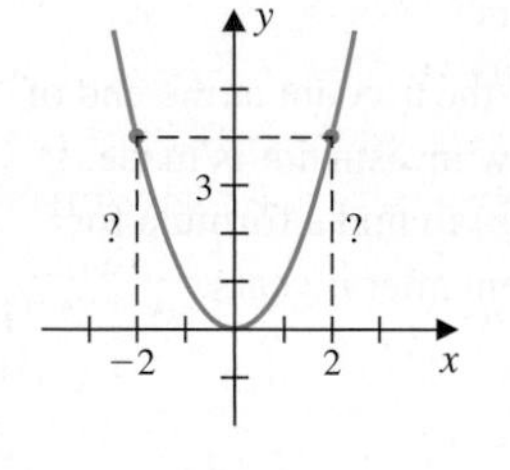

FIGURE 2

This is the essence of the inverse function relationship. The inverse function reverses the process and returns unambiguously to the starting point. Not all functions are reversible. The function $f(x) = x^2$ cannot be reversed since we have no way of knowing, for example, whether the number 4 in the range of f originated at $x = 2$ or at $x = -2$. The same quandary occurs at every number in the range of f except at 0, as shown in Figure 2.

Before we can study inverse functions, then, we need to consider precisely which functions can be reversed. These functions are called *one-to-one*. (See Figure 3.)

One-to-One Functions

A function f is **one-to-one** if for all x_1 and x_2 in its domain,

$$f(x_1) = f(x_2) \quad \text{implies that} \quad x_1 = x_2.$$

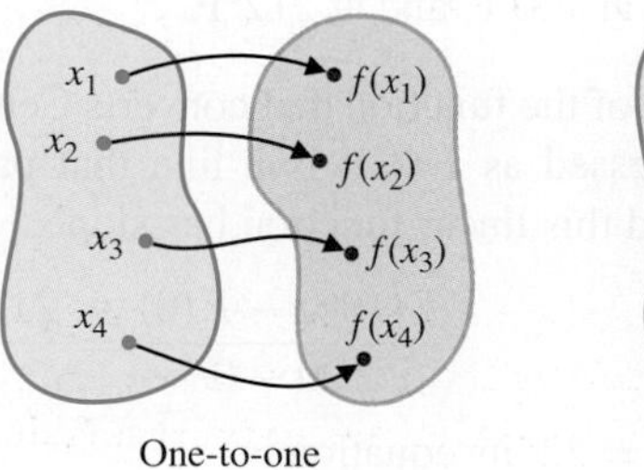

One-to-one

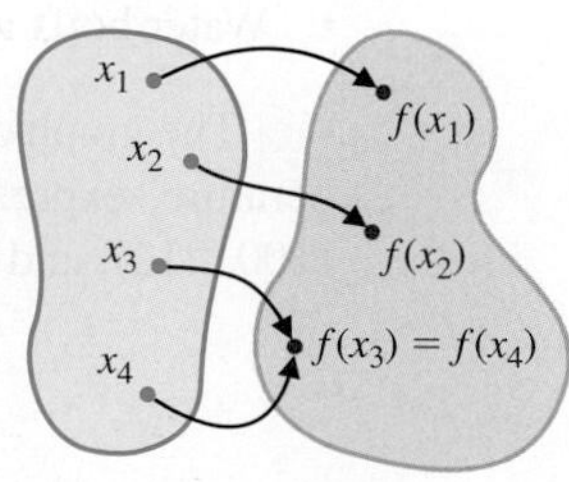

Not one-to-one

FIGURE 3

An alternative to the statement in the definition is called the *contrapositive* statement. This states that a function is one-to-one provided that

$$x_1 \neq x_2 \quad \text{implies} \quad f(x_1) \neq f(x_2).$$

The contrapositive statement is logically equivalent to the statement in the definition. The terminology *one-to-one* refers to the fact that we want these functions to have the property that

- each *one* of the range elements corresponds *to* precisely *one* domain element.

The Celsius-to-Fahrenheit conversion function is one-to-one because different Celsius temperatures always correspond to different Fahrenheit temperatures. The squaring function is not one-to-one because, for example, both $2^2 = 4$ and $(-2)^2 = 4$.

Given the graph of a function, it is easy to determine if the function is one-to-one.

Horizontal Line Test for One-to-One Functions

A function is one-to-one precisely when every horizontal line intersects its graph at most once.

Figure 4(a) illustrates that the graphs of the one-to-one functions $f(x) = x^3$ and $g(x) = 1/x$ satisfy the Horizontal Line Test. However, the graphs of the functions $f(x) = x^2$ and $g(x) = |x|$ in Figure 4(b) fail the Horizontal Line Test, since these functions are not one-to-one.

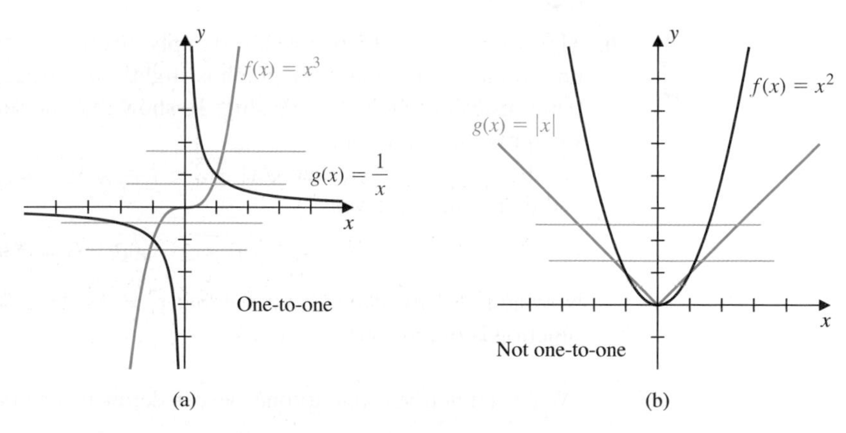

FIGURE 4

Although the Horizontal Line Test is a valuable tool when the graph of a function is known, it is generally necessary to use algebra to verify that a function is one-to-one. On the other hand, to verify that a function f is not one-to-one, we need only find a pair of numbers $x_1 \neq x_2$ with $f(x_1) = f(x_2)$. The next example illustrates a typical situation.

EXAMPLE 1 **a.** Verify that the function $f(x) = -|2x - 1| + 2$ is not one-to-one.
b. Verify that the function $f(x) = \sqrt{x - 2} - 1$ is one-to-one.

Solution The graphs of these functions were considered in Section 2.2 and are reproduced in Figure 5.

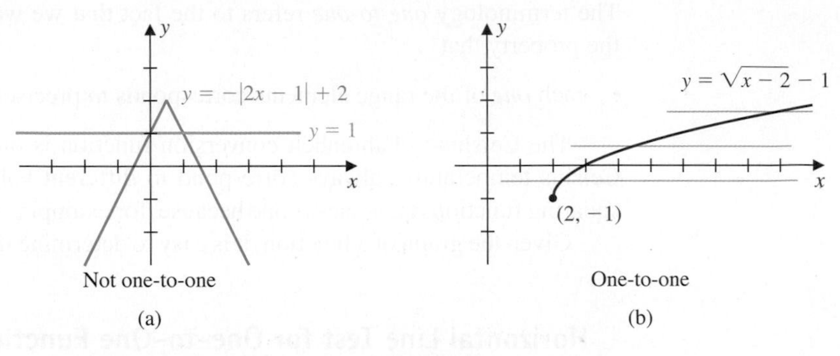

FIGURE 5

a. The graph of $f(x) = -|2x - 1| + 2$ in Figure 5(a) shows quite clearly that the function is not one-to-one and gives us a good indication of how to choose numbers to demonstrate it. For example, the horizontal line $y = 1$ crosses the graph at (0, 1) and at (1, 1). To verify this, note that

$$f(0) = -|0 - 1| + 2 = 1 \quad \text{and} \quad f(1) = -|2 - 1| + 2 = 1.$$

This shows that the function is not one-to-one, because $f(0) = f(1)$, even though $0 \neq 1$.

b. The graph in Figure 5(b) appears to imply that the function $f(x) = \sqrt{x - 2} - 1$ is one-to-one. But perhaps the graph is slightly incorrect, and it dips down at some point instead of always increasing. To show that the function is truly one-to-one, we apply the definition.

If $f(x_1) = f(x_2)$, then $\sqrt{x_1 - 2} - 1 = \sqrt{x_2 - 2} - 1$. Adding 1 to each side and then squaring gives

$$\sqrt{x_1 - 2} = \sqrt{x_2 - 2} \quad \text{and} \quad x_1 - 2 = x_2 - 2.$$

Now adding 2 to each side we see that $x_1 = x_2$. This shows conclusively that the function is one-to-one. ■

When a function is one-to-one, we can define its inverse function. (See Figure 6.)

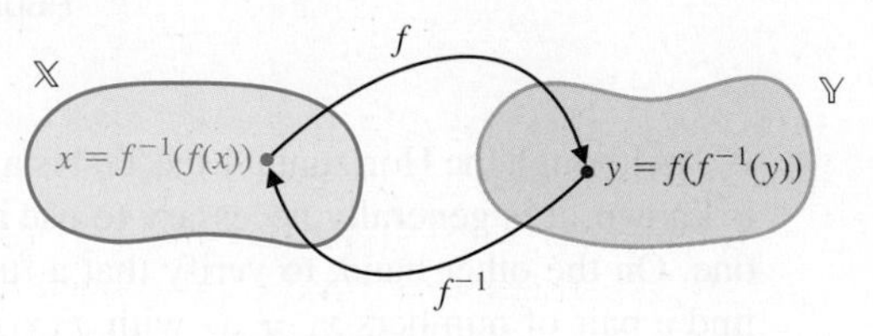

FIGURE 6

Inverse Functions

Suppose that f is a one-to-one function with domain $\mathbb{X}$ and range $\mathbb{Y}$. The **inverse function** for the function f is the function denoted f^{-1} with domain $\mathbb{Y}$ and range $\mathbb{X}$ and defined for all values of $x \in \mathbb{X}$ by

$$f^{-1}(f(x)) = x.$$

Although the inverse function of an arbitrary function f is denoted f^{-1}, many of the inverse functions that are particularly important have special notations. These will be introduced when we study these functions.

Some important properties follow directly from the definition of an inverse function, and the fact that the original function is one-to-one.

Properties of Inverse Functions

Suppose that f is a one-to-one function with domain $\mathbb{X}$ and range $\mathbb{Y}$.

- The inverse function f^{-1} is unique.
- The domain of f^{-1} is the range of f.
- The range of f^{-1} is the domain of f.
- The statement $f(x) = y$ is equivalent to $f^{-1}(y) = x$.
- Applying f followed by f^{-1} gives $f^{-1}(f(x)) = x$ and reversing the order gives $f(f^{-1}(y)) = y$.

EXAMPLE 2 Consider the function defined by $f(x) = 2x - 3$ with the domain restricted to the set $[-1, 4]$.

a. Show that the function f is one-to-one.
b. Find the inverse function f^{-1}.

Solution We have restricted the domain of the function in this example to better distinguish the domains and ranges of the function and its inverse. The graph of $f(x) = 2x - 3$ is a straight-line segment with positive slope. So the function is increasing and, when restricted to the interval $[-1, 4]$, the range of f is the interval $[f(-1), f(4)] = [-5, 5]$. (See Figure 7.)

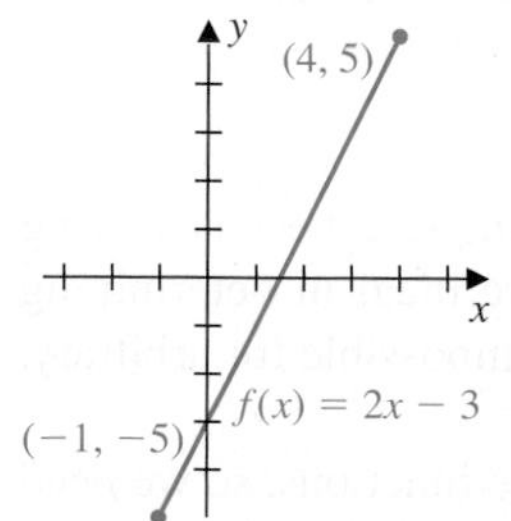

FIGURE 7

a. The function f is one-to-one since

$$f(x_1) = 2x_1 - 3 = 2x_2 - 3 = f(x_2) \quad \text{if and only if} \quad 2x_1 = 2x_2.$$

But this holds precisely when $x_1 = x_2$.

b. The domain of the inverse function, f^{-1}, is $[-5, 5]$, the range of f. The range of f^{-1} is $[-1, 4]$, the domain of f. To determine the correspondence relation for f^{-1}, we let $y = f(x) = 2x - 3$ and solve for the variable x in terms of y. Since

$$y = 2x - 3, \quad \text{we have} \quad f^{-1}(y) = x = \tfrac{1}{2}(y + 3).$$

We can verify that this relationship is correct by noting that, indeed,

$$f^{-1}(f(x)) = f^{-1}(2x-3) = \tfrac{1}{2}((2x-3)+3) = x.$$

Although the relationship $x = f^{-1}(y) = \frac{1}{2}(y+3)$ is expressed as x in terms of y, there is no significance to the defining variable for a function. It is the relationship that is important.

It is customary to reexpress the relationship for f^{-1} using the variable x to represent the values in the domain of f^{-1}. In this way we can give the graph of f^{-1} in the same framework as the graph of f, that is, with its domain along the x-axis. Hence, we have

$$f^{-1}(x) = \tfrac{1}{2}(x+3).$$

The graph of f^{-1} is the straight-line segment with slope $\frac{1}{2}$ and y-intercept $\frac{3}{2}$ shown with the graph of f in Figure 8. ■

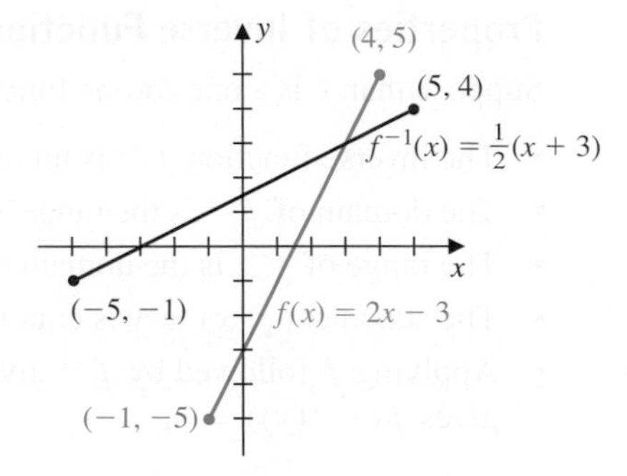

FIGURE 8

In Example 2 three steps were used to determine the inverse function.

Determining the Inverse of a One-to-One Function f

- Set $y = f(x)$.
- Solve for x in terms of y.
- Interchange the variables x and y so that the domain of f^{-1} is represented by numbers on the x-axis.

We are able to find the inverse for the function in Example 2 because, being linear, it is easy to solve for x in terms of y. This step is critical in determining inverse functions, but one that is generally difficult or often impossible for arbitrary, nonlinear, functions.

We cannot always find explicit representations for inverse functions, so we need to gain information about them in other ways. The next observation helps considerably by describing the relationship between the graph of a one-to-one function and the graph of its inverse.

Notice in Figure 8 that the graphs of f and f^{-1} are quite similar. There is a symmetry property that holds for the graph of a function and its inverse. If the point

(a, b) is on the graph of the one-to-one function f, then $f(a) = b$. So $f^{-1}(b) = a$, which means that (b, a) is on the graph of f^{-1}. Geometrically, this means that the graphs of $y = f(x)$ and $y = f^{-1}(x)$ have symmetry with respect to the line $y = x$. (See Figure 9.)

The graphs of a function and its inverse both increase or decrease together, but the steepness of the graph of the inverse function is inversely related to the steepness of the graph of the function.

The Graph of the Inverse of a One-to-One Function

The graph of $y = f^{-1}(x)$ is the reflection of the graph of $y = f(x)$ about the line $y = x$.

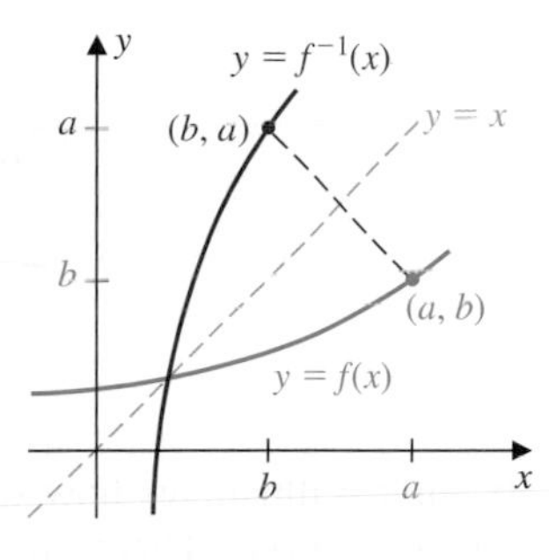

FIGURE 9

This result will frequently permit us to determine many important features of the graph of an inverse function, even when we do not have an explicit expression for the inverse function.

EXAMPLE 3 Sketch the graph of the inverse function f^{-1} for the function defined by $f(x) = x + 2x^3$.

Solution Although it is possible to solve for x in terms of y in the equation

$$y = x + 2x^3,$$

it is not done easily. Instead, we first show that the inverse function exists by showing that f is one-to-one. Then we will use the graph of f to produce the graph of f^{-1}. The graph of f in Figure 10 appears to satisfy the horizontal line test, so we suspect that f is one-to-one. To show that this is true we will use the contrapositive of the definition of one-to-one. Suppose that $x_1 \neq x_2$. Then either $x_1 < x_2$ or $x_2 < x_1$. We will assume that $x_1 < x_2$.

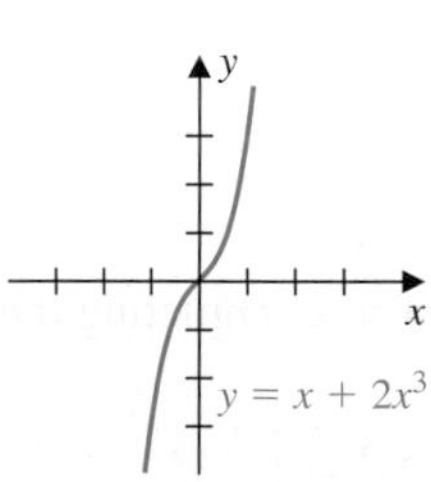

FIGURE 10

The cubing function is always increasing, so

$$x_1^3 < x_2^3, \quad \text{which implies that} \quad 2x_1^3 < 2x_2^3.$$

Adding the inequalities $x_1 < x_2$ and $2x_1^3 < 2x_2^3$ gives

$$f(x_1) = x_1 + 2x_1^3 < x_2 + 2x_2^3 = f(x_2).$$

In calculus the derivative is also used to show that a function is one-to-one.

So $x_1 < x_2$ implies that $f(x_1) < f(x_2)$. If $x_2 < x_1$ we have a similar result. Together these imply that if $x_1 \neq x_2$, we have $f(x_1) \neq f(x_2)$, so f is one-to-one.

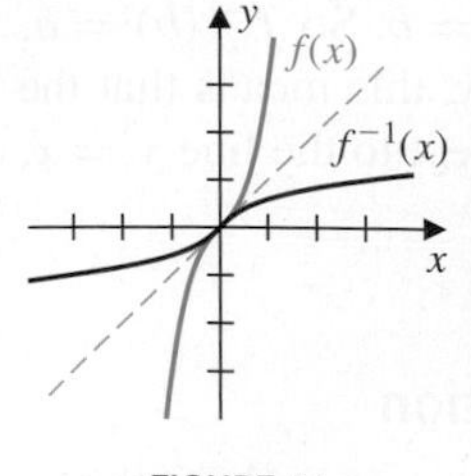

FIGURE 11

The graph of f^{-1} is simply the reflection about $y = x$ of the graph of f, so the graph of f^{-1} is as shown in Figure 11. ■

The inverse function graphing technique is so useful that many graphing devices have a built-in operation to reflect a graph about the line $y = x$. However, this is generally done whether or not the function is one-to-one. When it is not, the resulting reflected graph will not be the graph of a function. An example of this is shown in Figure 12, where the graph of $y = x^3 - x$ is reflected about the line $y = x$. The reflected graph is certainly not the graph of a function, because (0, 0), (0, 1), and (0, −1) are all points on the graph.

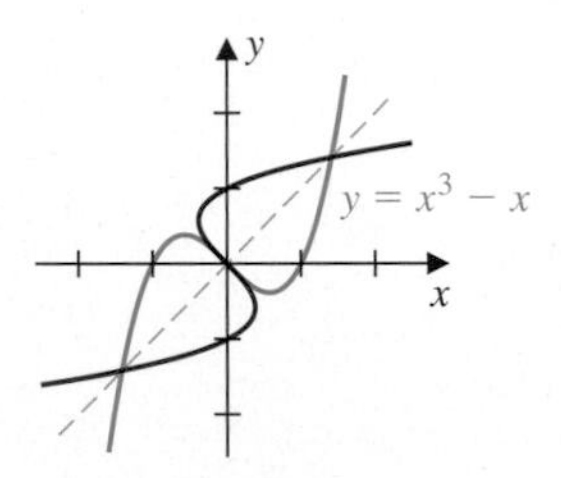

FIGURE 12

Example 4 illustrates how we can at times find a partial inverse for a function that is not one-to-one. This technique will be needed in Chapter 4 when we consider inverses of the trigonometric functions.

EXAMPLE 4 Figure 13 shows the graph of $f(x) = x^2 + 4x + 3$. Since f is not one-to-one, it does not have an inverse. However, if we restricted the domain of f to the interval $[-2, \infty)$, then f would be one-to-one. Determine f^{-1} after assuming that this domain restraint has been made.

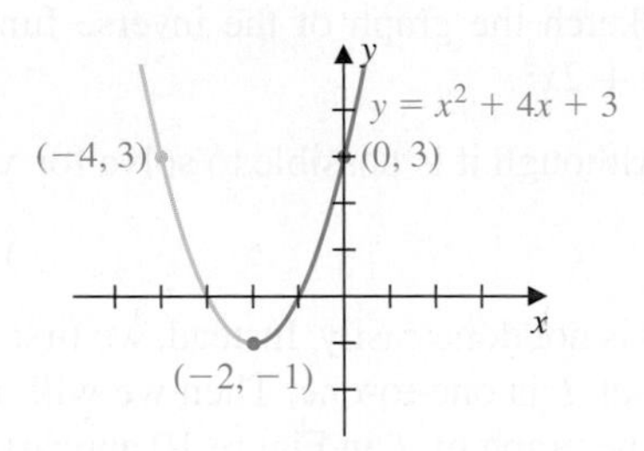

FIGURE 13

Solution To find the inverse, first set $y = f(x)$ and solve for x in terms of y. Completing the square of the quadratic gives

$$y = x^2 + 4x + 3 = (x^2 + 4x + 4) - 4 + 3 = (x + 2)^2 - 1.$$

Then solving for x we have

$$y + 1 = (x + 2)^2, \quad \text{so} \quad x = -2 \pm \sqrt{y + 1}.$$

There are two possibilities for writing x in terms of y,

$$x = -2 + \sqrt{y + 1} \quad \text{and} \quad x = -2 - \sqrt{y + 1}.$$

If x was restricted to $x \geq -2$, we would need to choose

$$x = -2 + \sqrt{y + 1}.$$

Interchanging the roles of x and y then gives the inverse function with domain $[-1, \infty)$ on the x-axis as

$$f^{-1}(x) = -2 + \sqrt{x + 1}.$$

The graph of f^{-1} with the domain of f restricted to $[-2, \infty)$ is shown in Figure 14(a). ■

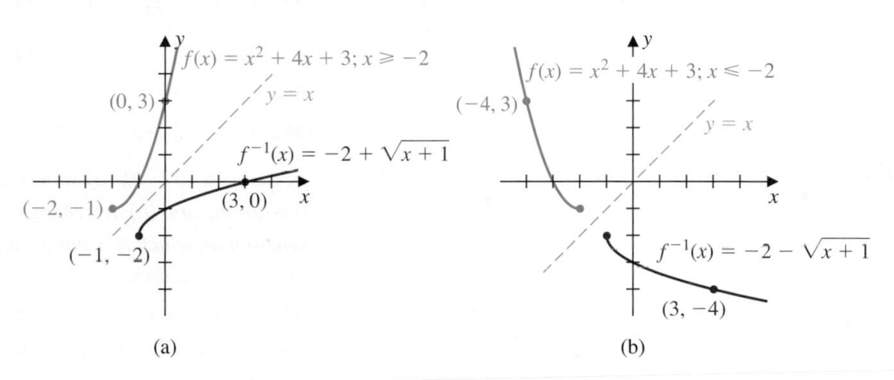

FIGURE 14

If the domain of f had instead been restricted to $(-\infty, -2]$, then the inverse function would have been $f^{-1}(x) = -2 - \sqrt{x + 1}$, as shown in Figure 14(b).

EXERCISE SET 2.5

In Exercises 1–6, determine if the function whose graph is given is one-to-one.

1. **2.**

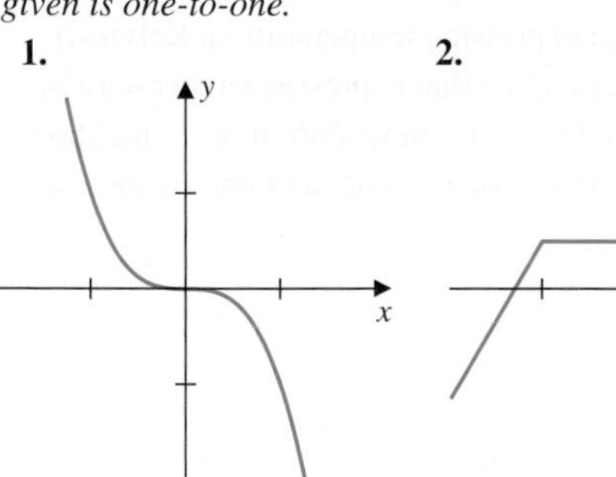

3. **4.**

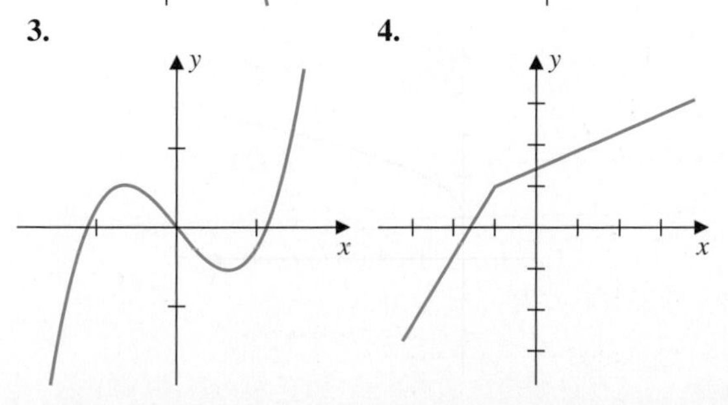

5. **6.**

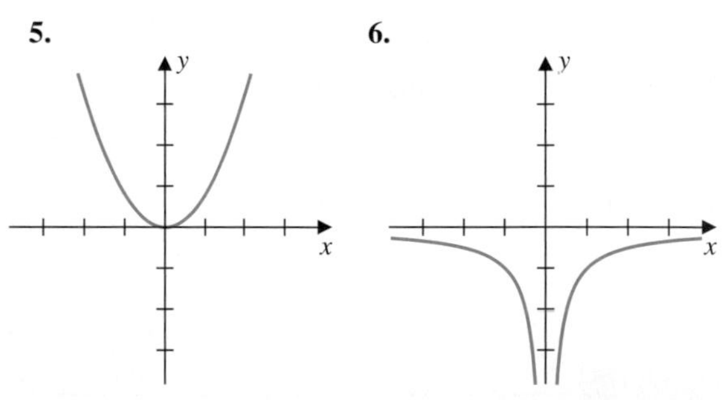

In Exercises 7–14, determine if the given function is one-to-one.

7. $f(x) = 3x - 4$

8. $f(x) = \sqrt{x - 1}$

9. $f(x) = |x - 1| + 1$

10. $f(x) = x^2 - 6x + 5$

11. $f(x) = x^4 + 1$

12. $f(x) = x^6 - 2$

13. $f(x) = \begin{cases} 2x^2, & \text{if } x \geq 0 \\ x - 2, & \text{if } x < 0 \end{cases}$

14. $f(x) = \begin{cases} \sqrt{x}, & \text{if } x \geq 0 \\ x + 1, & \text{if } x < 0 \end{cases}$

In Exercises 15–18, the graphs of one-to-one functions are given. Sketch the graph of the inverse of each function.

15.

16.

17.

18.

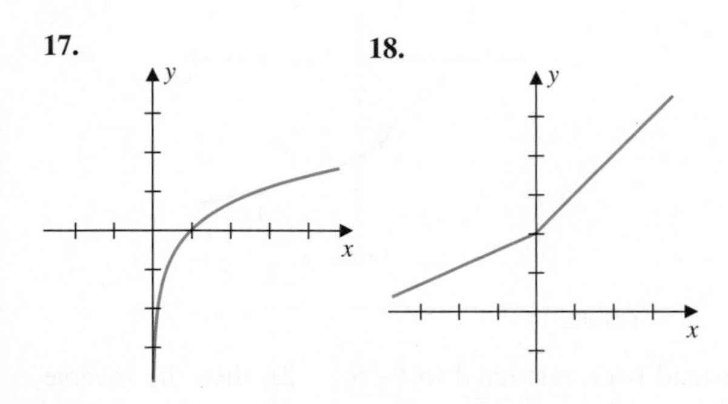

In Exercises 19–24, find the indicated value.

19. If $f(x) = x + 1$, find $f^{-1}(2)$.

20. If $f(x) = x - 2$, find $f^{-1}(0)$.

21. If $f(x) = 2x - 2$, find $f^{-1}(1)$.

22. If $f(x) = 3x - 2$, find $f^{-1}(2)$.

23. If $f(x) = \dfrac{1}{x}$, find $f^{-1}(2)$.

24. If $f(x) = x^3$, find $f^{-1}(-8)$.

In Exercises 25–36, find $f^{-1}(x)$ for the one-to-one function f, and graph f and f^{-1} on the same coordinate axes.

25. $f(x) = 2x - 1$

26. $f(x) = \dfrac{2x - 3}{4}$

27. $f(x) = \sqrt{x - 3}$

28. $f(x) = \sqrt{x + 2}$

29. $f(x) = \dfrac{1}{2x}$

30. $f(x) = \dfrac{1}{x + 1}$

31. $f(x) = \dfrac{1}{\sqrt{x}}$

32. $f(x) = \dfrac{1}{\sqrt{x}} + 1$

33. $f(x) = 1 + x^3$

34. $f(x) = 2 - x^3$

35. $f(x) = x^2 + 1, x \geq 0$

36. $f(x) = 1 - x^2, x \leq 0$

*In Exercises 37–40, **(a)** show that the given function is not one-to-one and **(b)** determine a subset of the domain of the function on which it is one-to-one, and find its inverse on this restricted domain.*

37. $f(x) = |2 - x|$

38. $f(x) = -|x - 3| - 2$

39. $f(x) = x^2 - 2x$

40. $f(x) = x^2 - 2x + 3$

41. For which values of m is the linear function $f(x) = mx + b$ one-to-one? For these values of m find $f^{-1}(x)$.

42. No object can assume a temperature lower than $-273°$C, a temperature called *absolute zero*. The Kelvin temperature scale, named for the Irish physicist William Thomson (Lord Kelvin), is defined as a translation of the Celsius scale so that $-273°$C corresponds to 0 K. (The degree symbol is generally omitted when expressing temperature in Kelvins.) Determine a function that expresses temperature in Fahrenheit in terms of temperature in Kelvins, show that this function is one-to-one, and find its inverse.

REVIEW EXERCISES FOR CHAPTER 2

1. Match the equation with the graph.

a. $y = \sqrt{x - 2} + 1$

b. $y = \sqrt{x} + 1$

c. $y = -\sqrt{x + 1} + 3$

d. $y = -2\sqrt{x}$

i.

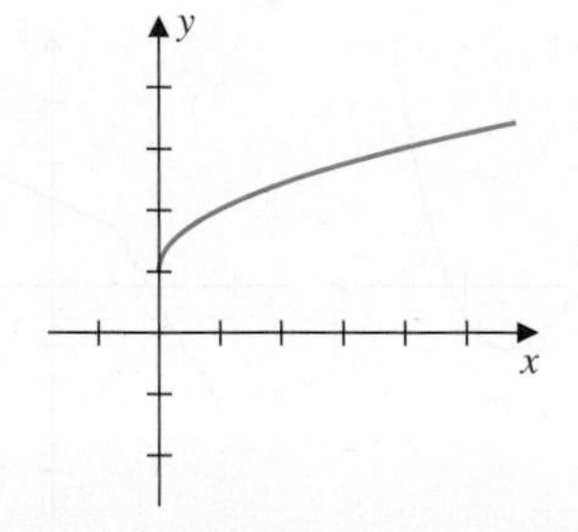

ii.

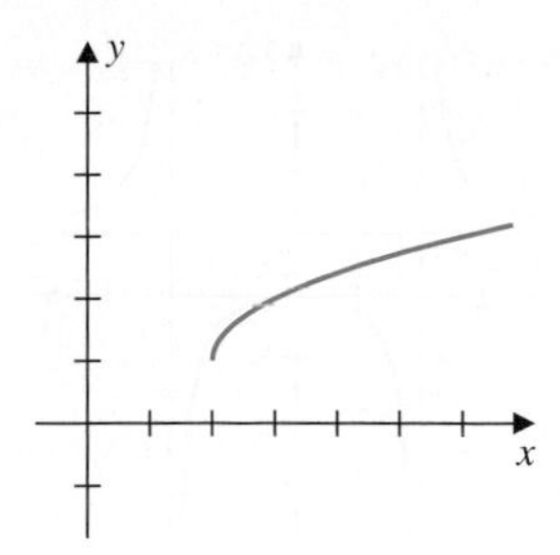

iii.

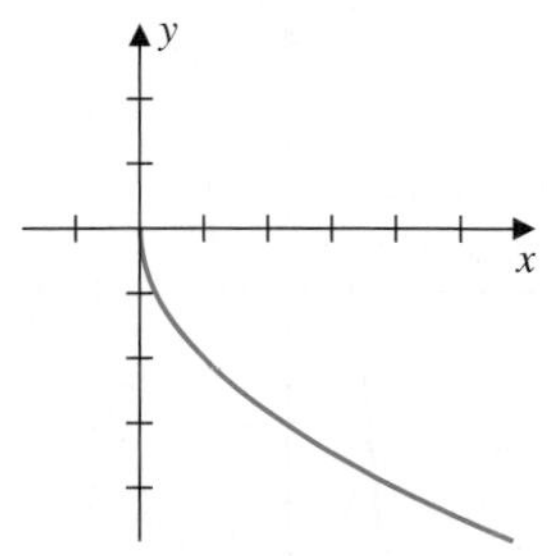

iv.

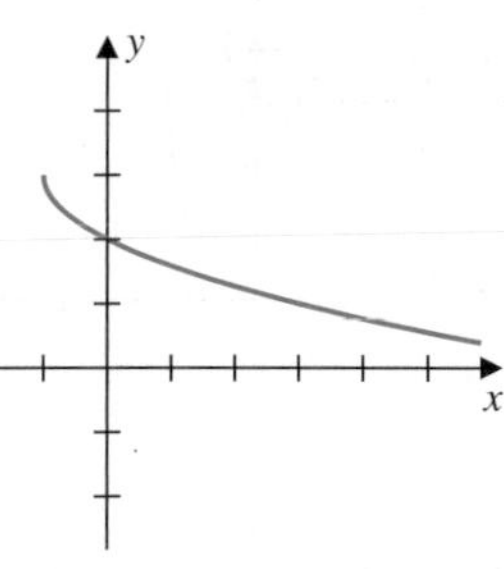

2. Match the equation with the graph.

a. $y = -|x - 1| + 2$ **b.** $y = 2|x| - 1$

c. $y = |x + 1| - 1$ **d.** $y = |x| + 2$

i.

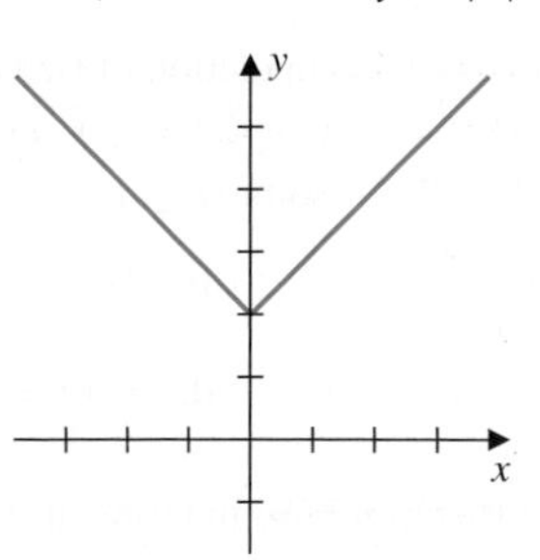

ii.

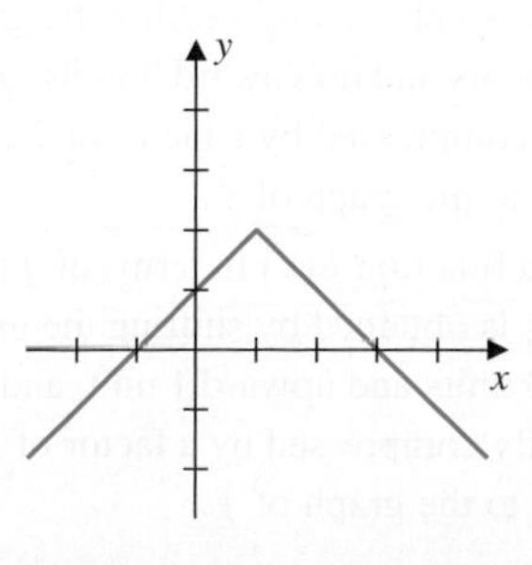

iii.

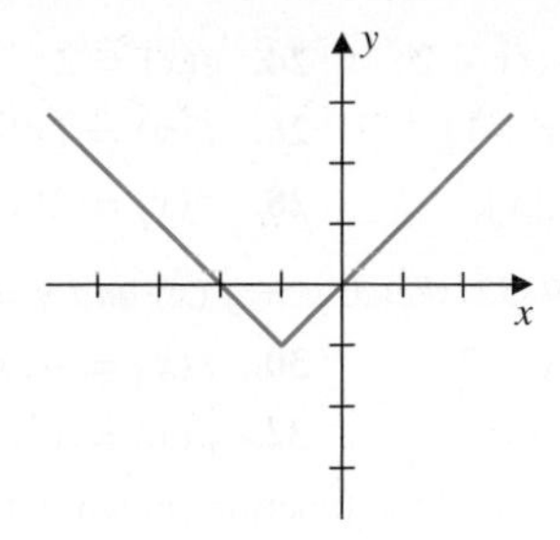

iv.

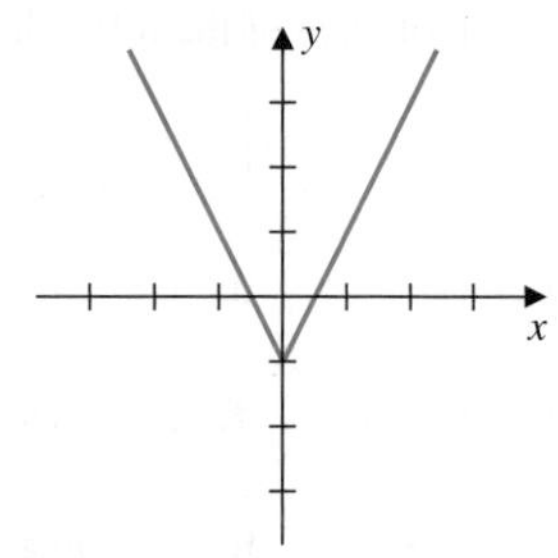

In Exercises 3–6, (a) find $f + g$, $f - g$, $f \cdot g$, f/g, $f \circ g$, $g \circ f$, $f \circ f$, $g \circ g$. (b) Give the domain of each new function.

3. $f(x) = 2x - 1; g(x) = x + 2$

4. $f(x) = x^2; g(x) = x - 2$

5. $f(x) = x + 1; g(x) = \sqrt{x - 1}$

6. $f(x) = 2x + 5; g(x) = 4/x$

In Exercises 7–12, find functions f and g such that $h = f \circ g$.

7. $h(x) = (x^3 - 2x + 1)^6$

8. $h(x) = \dfrac{1}{(x^4 - 2x + 1)^3}$

9. $h(x) = \sqrt{3x + 3}$

10. $h(x) = \sqrt{(x + 1)^3}$

11. $h(x) = |x^2 - 2x + 1|$

12. $h(x) = |x^2 - 2x| + 1$

In Exercises 13–16, use the results about the graph of the reciprocal of a function to sketch the graph of $h(x) = 1/f(x)$.

13. $f(x) = 2x + 1$ **14.** $f(x) = x^2 - 1$

15. $f(x) = x^2 - 2x - 3$ **16.** $f(x) = |x + 2|$

In Exercises 17–28, sketch the graph of each of the functions.

17. $f(x) = |x + 1|$ **18.** $f(x) = |x - 1| + 2$

19. $f(x) = |x + 2| - 2$ **20.** $f(x) = |2x + 4| - 3$

21. $g(x) = \sqrt{x + 1}$ **22.** $g(x) = \sqrt{x - 2} + 2$

23. $g(x) = -\sqrt{x} - 2$ **24.** $g(x) = 2 - \sqrt{x-2}$

25. $f(x) = \lfloor x+1 \rfloor$ **26.** $f(x) = \lfloor x \rfloor + 1$

27. $f(x) = -\lfloor x \rfloor$ **28.** $f(x) = 2\lfloor x \rfloor + 1$

In Exercises 29–32, sketch $y = f(x)$ *and* $y = |f(x)|$.

29. $f(x) = 3x - 2$ **30.** $f(x) = -2x + 1$

31. $f(x) = -x^2 + 3$ **32.** $f(x) = x^2 - 6x + 8$

33. Use the graph of the function shown in the figure to sketch the graph of each of the following.

a. $y = f(x+1)$ **b.** $y = f(x+1) + 1$

c. $y = f(x-1)$ **d.** $y = f(x-1) + 2$

e. $y = f(x+1) - 1$ **f.** $y = f(x-1) - 2$

g. $y = f(2x)$ **h.** $y = f(x/2)$

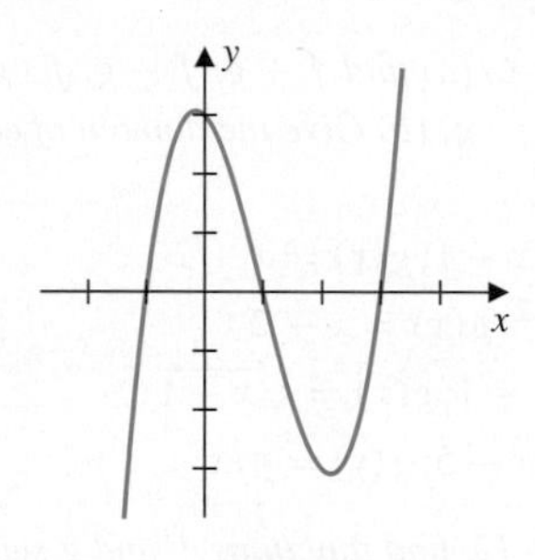

34. For each of the graphs, specify

a. any origin or axis symmetries, and

b. whether the graph describes a function. If the graph is the graph of a function,

c. find the domain and range,

d. specify whether the function is even, odd, or neither, and

e. determine if the function is one-one. If the function is one-to-one,

f. sketch the graph of the inverse function.

i.

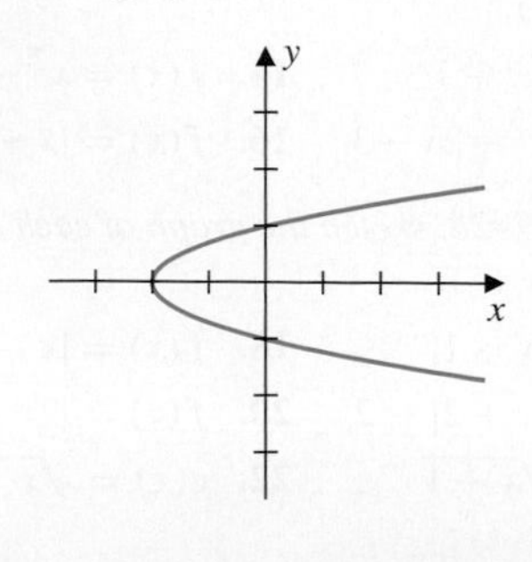

ii.

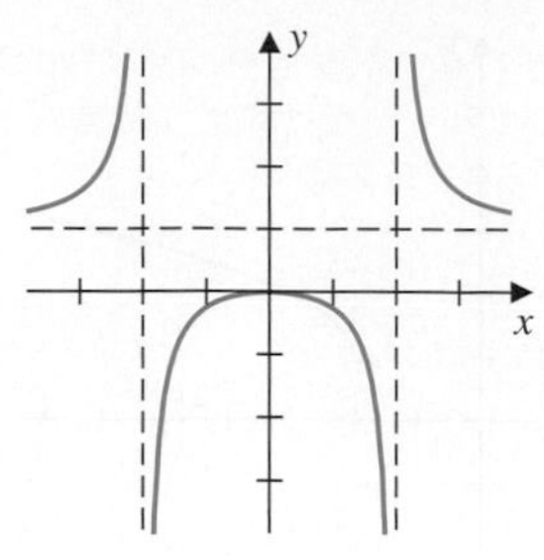

iii.

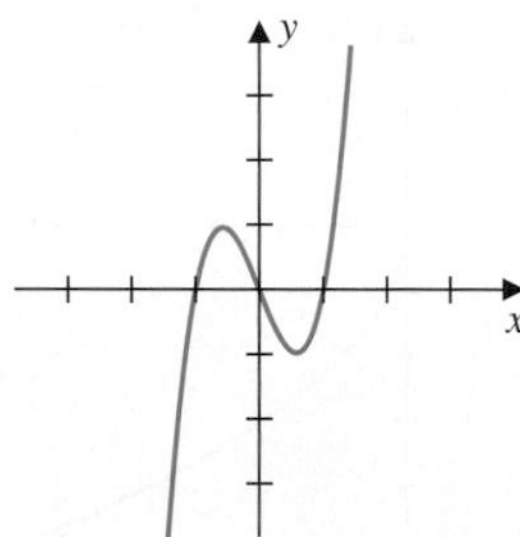

iv.

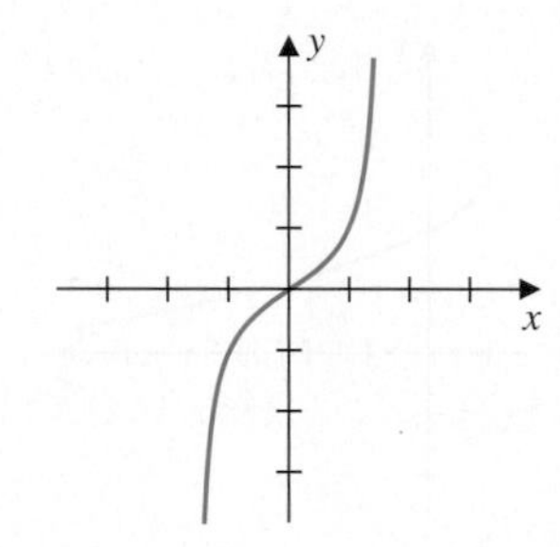

35. Determine formulas for $f(-x)$, $-f(x)$, $f(1/x)$, $1/f(x)$, $f(\sqrt{x})$, and $\sqrt{f(x)}$ for the given $f(x)$.

a. $f(x) = 2x^2 - 3$ **b.** $f(x) = 1/x^2$

36. Write $f(x)$ as the composition of two or more of the functions $g(x) = 1/x$, $w(x) = \sqrt{x}$, $v(x) = x - 2$, and $h_n(x) = x^n$, for some $n \geq 1$.

a. $f(x) = \dfrac{1}{x^5}$ **b.** $f(x) = \sqrt{x^4 - 2}$

c. $f(x) = (x^2 - 2)^6$ **d.** $f(x) = \dfrac{1}{\sqrt{x^3 - 2}}$

37. Describe a function $g(x)$ in terms of $f(x)$ if the graph of g is obtained by shifting the graph of f to the left 2 units and downward 3 units, and if it is vertically compressed by a factor of 2 when compared to the graph of f.

38. Describe a function $g(x)$ in terms of $f(x)$ if the graph of g is obtained by shifting the graph of f to the right 3 units and upward 1 unit, and if it is horizontally compressed by a factor of 3 when compared to the graph of f.

39. Determine whether the given function is one-to-one. If it is one-to-one, find its inverse and sketch both $y = f(x)$ and $y = f^{-1}(x)$.

a. $f(x) = 2x + 1$ **b.** $f(x) = |x - 1| + 1$

c. $f(x) = 2 + x^3$ **d.** $f(x) = x^2 + 2x - 2$

e. $f(x) = 1/x^2$ **f.** $f(x) = \sqrt{x - 1} + 2$

40. Let $f(x) = x^2 - 2$.

a. Sketch the graph of $y = f(x)$ for $x \geq 0$.

b. Use the graph from (a) to sketch the graph of $y = f^{-1}(x)$.

c. Find an equation for f^{-1}.

41. Let $f(x) = x^2 - 4x$.

a. Determine a subset of the function domain on which the function is one-to-one. Sketch the function restricted to the subset.

b. Use the graph in (a) to draw the graph of f^{-1}.

c. Find an equation for f^{-1} and specify the domain of f^{-1}.

42. A public works department is to construct a new road between towns A and B. Town A lies on an abandoned road running east–west. Town B lies 20 miles north of this road and 40 miles east of town A. An engineer proposes that the road be constructed by restoring a section of the old road leaving town A and joining it to an entirely new section of road at a point to be determined, connecting that point with town B (see the figure).

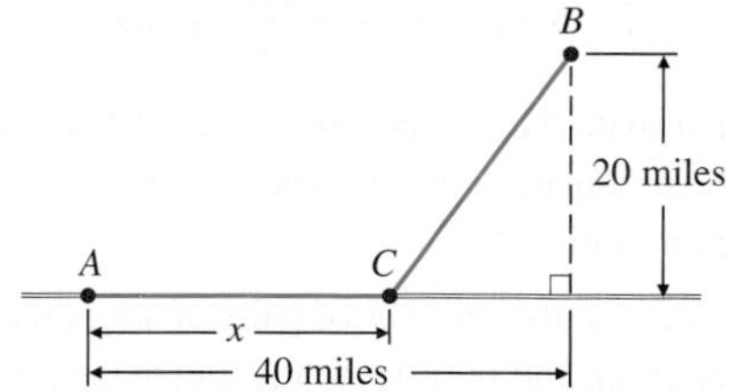

a. If the cost of restoring the old road is \$200,000 per mile and the cost of new construction is \$400,000 per mile, determine the function describing the total cost.

b. Estimate how much of the old road should be restored in order to minimize the department's cost.

43. A manufacturer of solar energy cells finds that the cost, in dollars, of producing x cells per month is $30x + 1500$, and sets the selling price per unit at $120 - 0.1x$.

a. Find the profit function $P(x)$.

b. How many units should be sold for $P(x)$ to be a maximum?

c. What is the approximate maximum profit?

44. Use the figure to answer each of the following.

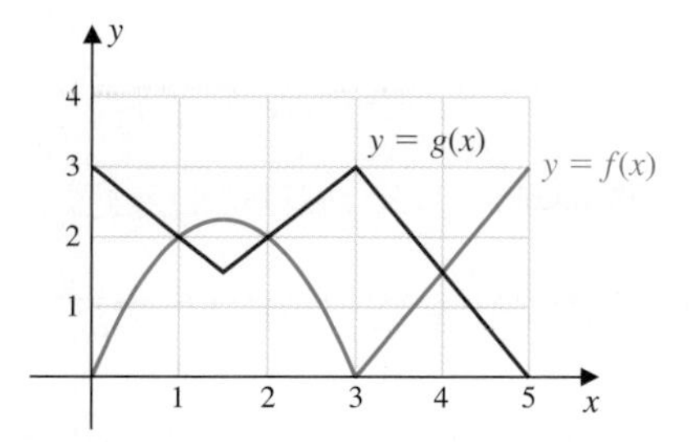

a. For which values of x is $f(x) = g(x)$?

b. For which values of x is $f(x) \leq g(x)$?

c. For which values of x is $f(x) = 0$?

d. For which values of x is $g(x)$ decreasing?

e. Evaluate $(f + g)(3)$, $(f + g)(4)$, $(f - g)(3)$, and $(f - g)(4)$.

f. Evaluate $(f \circ g)(4)$ and $(g \circ f)(5)$.

g. What are the ranges of f and g?

h. What is the largest value of $f(x)$?

CHAPTER 2: EXERCISES FOR CALCULUS

1. Use the graph of $y = f(x)$ given in the figure to sketch the graph of $y = 1/f(x)$. Which properties of f are the most essential in sketching $y = 1/f(x)$? Describe how the properties of f are used.

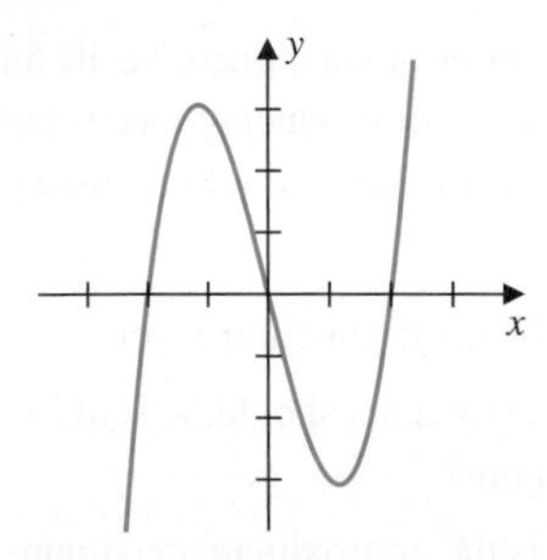

2. An automobile is traveling between points A and B at 60 miles/hour.

a. Express the distance the car is from the point C in terms of the distance d the car is from point A.

b. Express the distance the car is from the point C in terms of the amount of time the car has been traveling.

c. Write the distance the car is from point C with respect to the time t as a composition of functions.

3. The table gives data for the world population and the world urban population in billions. Let $W(x)$ and $U(x)$ denote functions describing the world population and world urban population, respectively, at time x.

Year	World	Urban
1950	2.555	0.750
1960	3.039	1.017
1970	3.708	1.357
1980	4.457	1.754
1990	5.284	2.280
2000	6.080	2.860

a. Use the data in the table to sketch approximate graphs of $y = W(x)$ and $y = U(x)$.

b. Use $W(x)$ and $U(x)$ to describe two new functions representing the world nonurban population $N(x)$ and $S(x)$ the share of the world population that is urban as a percentage of the total world population. Use the data in the table to sketch approximate graphs of $y = N(x)$ and $y = S(x)$.

4. The rate R at which a population increases is proportional to the product of the current population P and the difference between the maximum population M that can be sustained and the current population.

a. Determine an expression for R as a function of P. Use a proportionality constant k in the expression.

b. Sketch the graph of the function R when $k = 2$ and $M = 10{,}000$.

5. A small shoe company can produce between 100 and 1000 pairs of shoes per day. The cost of producing x pairs of shoes, for $100 \le x \le 1000$, is given by

$$C(x) = 295 + 3.28x + 0.003x^2 \text{ dollars.}$$

If x pairs of shoes are sold, then the price per pair that the company will receive is given by

$$p(x) = 7.47 + \frac{321}{x} \text{ dollars.}$$

a. Estimate the number of pairs of shoes that should be produced if the average cost per unit is to be a minimum.

b. Estimate the number of pairs of shoes that should be produced in order to maximize the profit.

6. According to Boyle's Law, if the temperature of a confined gas is held fixed, then the product of the pressure P (in pounds/inch^2) and the volume V (in inches^3) is a constant. Suppose for a certain gas, $PV = 8000$. Determine **(a)** the average rate of change of P as V increases from 200 to 250, and **(b)** the instantaneous rate of change of P when $V = 200$.

7. Define a function f by $f(1) = 1$, $f(2) = 1$, and $f(n) = f(n-1) + f(n-2)$ for all positive integers $n \geq 2$. The function f generates the Fibonacci sequence of numbers.

   a. Find the first 10 Fibonacci numbers.

   b. Show that for any positive integer n, the only positive integer d that divides both F_n and F_{n+1} is $d = 1$.

8. For positive integers $n \geq 1$ the iterates of a function f are

$$\begin{aligned} f^1(x) &= f(x) \\ f^2(x) &= (f \circ f)(x) \\ f^3(x) &= (f \circ f \circ f)(x) \\ &\vdots \\ f^n(x) &= \underbrace{(f \circ f \circ f \circ \cdots \circ f)}_{n\text{-times}}(x). \end{aligned}$$

   For fixed x, the values $f^1(x), f^2(x), \ldots, f^n(x), \ldots$ are called the *orbit* of x under f. Describe the orbits under the function $f(x) = -x + b$, where b is a fixed number.

9. Let

$$f(x) = \begin{cases} 2x, & \text{if } 0 \leq x \leq \frac{1}{2} \\ 2 - 2x, & \text{if } \frac{1}{2} < x \leq 1. \end{cases}$$

   a. Find an expression for $(f \circ f)(x)$.

   b. Sketch the graphs of f and $(f \circ f)$.

10. Suppose that $f(x) = ax + b$ and $g(x) = bx + c$. What must be true about the constants a and b if for all x we have $(f \circ g)(x) = (g \circ f)(x)$?

11. Let $f(x) = \dfrac{x+a}{x+b}$, where $a \neq b$.

   a. Show that f^{-1} exists, and find $f^{-1}(x)$.

   b. Give the domains and ranges of f and f^{-1}.

   c. Verify that $(f^{-1} \circ f)(x) = x$ for all x in the domain of f, and $(f \circ f^{-1})(x) = x$ for all x in the domain of f^{-1}.

CHAPTER 2: CHAPTER TEST

Determine whether the statement is true or false. If false, describe how the statement might be changed to make it true.

1. The function $f(x) = \sqrt{x-2} + 1$ is always increasing with the minimum value 1.
2. The domain of the function $f(x) = \sqrt{x-2} + 1$ is $[2, \infty)$ and the range is $[1, \infty)$.
3. The function $f(x) = -|x+1| + 2$ is increasing on $(-\infty, -1)$ and decreasing on $(-1, \infty)$.
4. The domain of the function $f(x) = -|x+1| + 2$ is $[-1, \infty)$ and the range is $[2, \infty)$.
5. The function $f(x) = \sqrt{x^2 + 2x - 3}$ has domain $(-\infty, -3] \cup [1, \infty)$.

In Exercises 6–9, use $f(x) = \lfloor x - 2 \rfloor + 1$.

6. The value of $f(-2)$ is -3.
7. The value of $f(2)$ is 3.
8. The function defined by $g(x) = f(x+1)$ is constantly 1 on the interval $[2, 3)$.
9. The function defined by $g(x) = f(x-1) + 2$ is constantly -2 on the interval $[-1, 0)$.
10. A function will have an inverse function provided every vertical line crosses the graph of the function at most one time.
11. The graph of $y = f(x-1) + 2$ is obtained by shifting the graph of $y = f(x)$ to the right 1 unit and upward 2 units.
12. The graph of $y = f(x+2) - 1$ is obtained by shifting the graph of $y = f(x)$ to the left 2 units and upward 1 unit.

In Exercises 13–20, use $f(x) = x + 1$ *and* $g(x) = x^2 + 2x - 3$.

13. The value of $(f + g)(2) = 8$.
14. The value of $(g - f)(1) = -2$.
15. The value of $(f/g)(3) = \frac{1}{3}$.
16. The domain of the function f/g is $\{x | x \neq 3 \text{ and } x \neq -1\}$.
17. The value of $(f \circ g)(-1) = -4$.
18. The value of $(g \circ f)(-1) = -3$.
19. The value of $(f \circ f)(2) = 0$.
20. The expression $(g \circ f)(x) = x^2 + 4x$.
21. The evaluation of $(f \circ g)(x)$ is equivalent to the evaluation of $f(g(x))$.

22. For functions f and g it is always the case that $(f \circ g)(x) - (g \circ f)(x) = 0$.

23. If $f(x) = x^2 - 2x + 1$ and $g(x) = \sqrt{x}$, then $(f \circ g)(4) = -3$.

24. The function $f(x) = (x^3 + 2x - 1)^8$ can be written as $(h \circ g)(x)$ where $h(x) = x^8$ and $g(x) = x^3 + 2x - 1$.

In Exercises 25–28, use the figure.

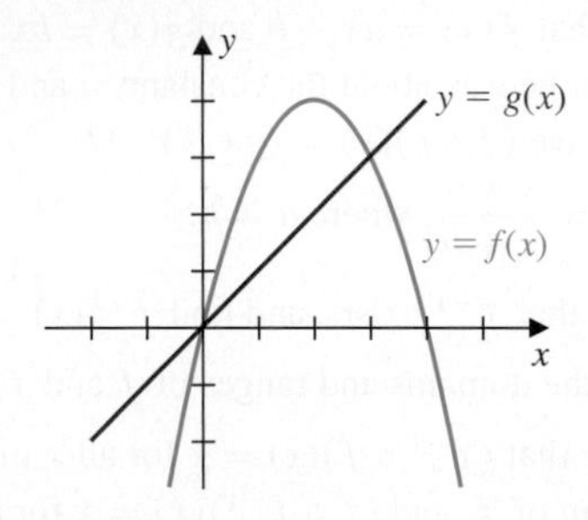

25. The value of $(f + g)(2) = 4$.

26. The value of $(f - g)(3) = 0$.

27. The value of $(f \circ g)(2) = 4$.

28. The value of $(g \circ f)(4) = 0$.

29. The inverse function for $f(x) = 3x - 2$ is $f^{-1}(x) = \frac{1}{3}x + \frac{2}{3}$.

30. The inverse function for $f(x) = 8x^3 - 1$ is $f^{-1}(x) = \frac{1}{2}\sqrt[3]{x + 1}$.

In Exercises 31–34, use the figure.

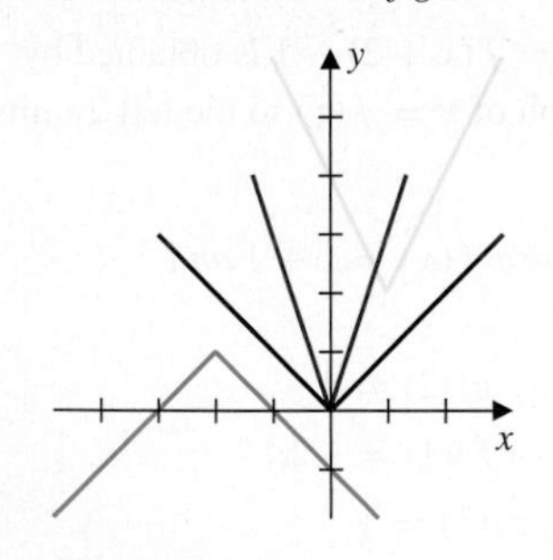

31. The red curve is a scaling of the black curve by a factor greater than 1.

32. The blue curve is a horizontal shift to the left 2 units of the black curve followed by a vertical shift upward 1 unit.

33. The blue curve is $y = -|x + 2| + 1$.

34. The yellow curve is $y = 2|x + 1| - 2$.

In Exercises 35–38, use the figures.

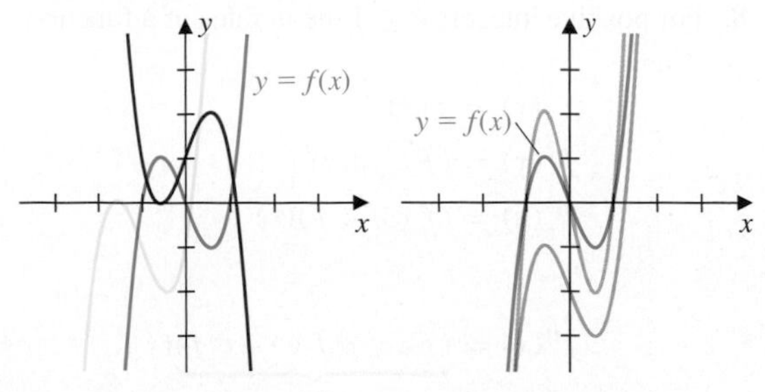

35. The red curve can be described by $y = -f(x) + 1$.

36. The yellow curve can be described by $y = f(x + 1) - 1$.

37. The green curve is $y = f(x) + 2$.

38. The orange curve is $y = 2f(x)$.

39. As $x \to 2$, the function $f(x) = \frac{1}{x-2} \to \infty$.

40. As $x \to 1^-$, the function $f(x) = \frac{x}{x-1} \to -\infty$.

41. As $x \to 3$, the function $f(x) = \frac{2x+1}{(x-3)^2} \to 2$.

42. As $x \to \infty$, the function $f(x) = \frac{1}{x} + 2 \to -2$.

Algebraic Functions

3

Calculus Connections

The functions introduced in the first two chapters are used to construct mathematical *models* to describe the behavior of a system or process. The phenomena being modeled might represent a physical, biological, economic, or other quantitative problem. Simplifying assumptions are generally made to make the modeling possible, so the resulting representation will not describe the physical situation exactly. Instead, the model describes the important features of the problem. The fewer simplifying assumptions made, the more likely the model will describe the actual situation, but the more difficult it is to solve the modeling problem.

Take, as an example, the problem of determining the path of water coming out of a garden hose. Experience shows that the path the water takes depends on both the velocity of the water and the angle at which we hold the nozzle of the hose. Place an xy-coordinate system with the end of the nozzle at the origin with the positive y-axis pointing upward, as shown in the figure.

© Image copyright Andrey Yurlov, 2009. Used under license from Shutterstock.com

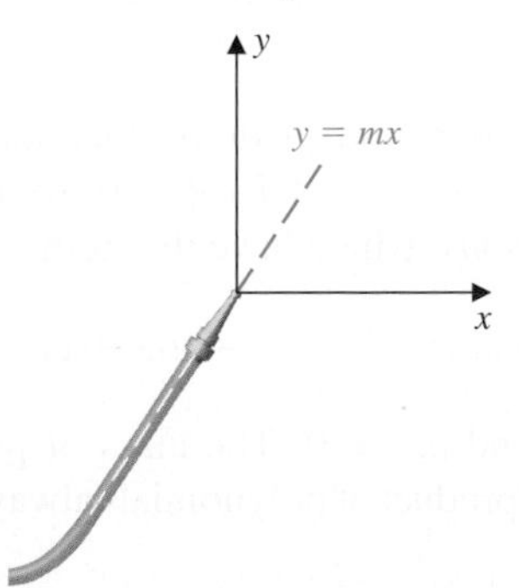

Suppose the velocity of the water as it leaves the origin is the constant v_0 and that the nozzle is pointed in the direction of the line with equation $y = mx$. Wind, the friction of

the water in the air, the spread of the water as it moves through the air, and various other things affect the path of the water. But our model ignores all these minor (we hope) influences. We assume that the only significant influence on the water is the constant acceleration of the gravity of the earth, which is approximately -32 ft/s^2 (or, in metric units, -9.8 m/s^2). Physics (actually applied calculus) tells us that the path of the water is described by

$$y = mx - \frac{16}{v_0^2}(1 + m^2)x^2,$$

whose graph is a parabola. Because y indicates the height above the ground, if we set y to zero in this polynomial equation and solve for x we will know approximately where the water hits the ground.

Determining zeros of polynomials is one of the topics considered in this chapter. In this example we have a quadratic polynomial to solve, and $y = 0$ precisely when either

$$x = 0 \quad \text{or} \quad x = \frac{mv_0^2}{16(m^2 + 1)}.$$

The figure shows how the trajectory of the water varies as the angle of the nozzle is changed. We can determine from the equation the distance the water travels, how high it reaches, and the velocity when it hits the ground.

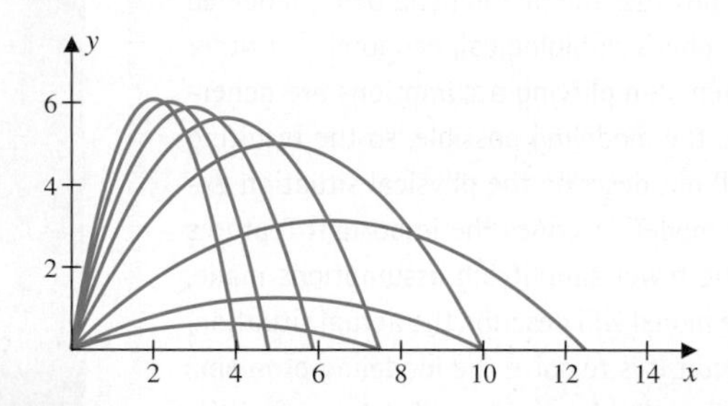

To improve the mathematical model would require taking into consideration some of the other factors that influence the path of the water; for example, the friction from the air or the effect of wind. But then the problem would likely be more difficult, or even impossible, to solve.

3.1 INTRODUCTION

In Chapter 1 we discussed constant, linear, and quadratic functions. They are defined by $f(x) = c$, $f(x) = mx + b$, and $f(x) = ax^2 + bx + c$, respectively. These are all special cases of the *polynomial functions*, which have the form

The applications generally require calculus, but to understand the application you need to recognize the behavior of the polynomial.

$$P(x) = a_n x^n + a_{n-1}x^{n-1} + \cdots + a_1 x + a_0,$$

where $a_0, a_1, \ldots, a_n$ are constants and $a_n \neq 0$. The class of polynomials has the property that the sum, difference, and product of polynomials always produces another polynomial.

Functions that arise in applications in the sciences, economics, and other areas are often polynomial functions, or are approximated using them. This is due in part to the fact that the values of polynomial functions can be computed using only the operations of addition, subtraction, and multiplication, so their evaluation can be done quickly, particularly by computer.

In this chapter we see how to systematically sketch the graphs of polynomial functions. We also look at some common *rational functions*, which are obtained by dividing polynomials, and some more general *algebraic functions*. Algebraic functions are obtained from polynomials by a finite combination of the operations of addition, subtraction, multiplication, division, and extracting integral roots. For example,

$$f(x) = \sqrt{x^4 + 2x + 1} \quad \text{and} \quad g(x) = \frac{x^3 + 2x^2 - x + 5}{x^4 + 3x + 2}$$

are algebraic functions. The function g is also rational, since $g(x)$ is the quotient of the polynomials $x^3 + 2x^2 - x + 5$ and $x^4 + 3x + 2$. On the other hand, the trigonometric functions that we study in Chapter 4 are not algebraic functions, nor are the exponential and logarithmic functions considered in Chapter 5.

3.2 POLYNOMIAL FUNCTIONS

A **polynomial of degree n**, where n is a nonnegative integer, has the form

$$a_n x^n + a_{n-1} x^{n-1} + \cdots + a_1 x + a_0,$$

where $a_0, a_1, \ldots, a_n$ are constants, and $a_n \neq 0$. The numbers $a_0, a_1, \ldots, a_n$ are called the **coefficients** of the polynomial; a_n is the **leading coefficient**, and a_0 is the **constant term**.

From this definition we see, for example, that:

- A constant function $f(x) = c$ is a polynomial function of degree 0.
- A linear function $f(x) = mx + b$, with $m \neq 0$, is a polynomial function of degree 1.
- A quadratic function $f(x) = ax^2 + bx + c$, with $a \neq 0$, is a polynomial function of degree 2.

Some other examples of polynomials are shown in Table 1.

TABLE 1

Polynomial	Degree	Leading Coefficient	Constant Term
$-3x^6 + 2x^4 + x^2 - x + 1$	6	-3	1
17	0	17	17
$2x^3 - 2x^2 + x - 5$	3	2	-5
x^{10}	10	1	0

We saw in Section 1.7 that the graphs of polynomials of degrees 0 and 1 are lines and, in Section 1.8, that the graph of a polynomial of degree 2 is a parabola. (See Figure 1.)

Graphing devices, such as graphing calculators and computer algebra systems, can be helpful for generating graphs of polynomials, but the graph produced by one of

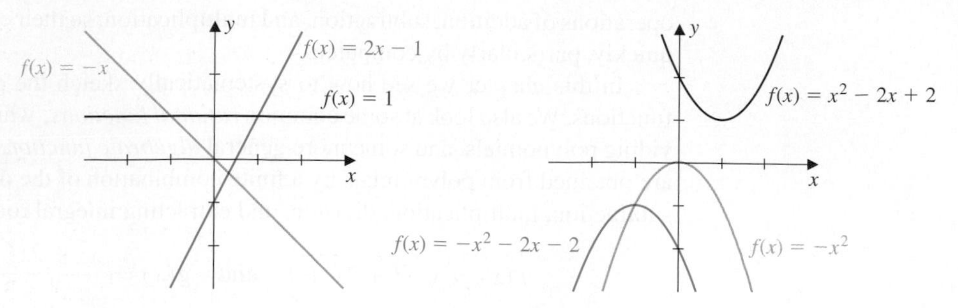

FIGURE 1

Graphing devices are a powerful tool when you understand the possibilities for the shape of the graph.

these devices is found by simply sketching a large number of points. It is more useful to develop a logical way to approach graphing problems that relies only minimally on plotting points. We concentrate more on the overall behavior of the graph than on the fine detail.

EXAMPLE 1 Use the graphing techniques of Chapter 1 to sketch the graph of $y = -3(x-2)^3 + 1$.

Solution The graph of $y = 3x^3$ is just a vertical stretching of the graph of $y = x^3$, as shown in Figure 2(a). To graph $y = -3x^3$, we reflect the graph of $y = 3x^3$ about the x-axis, as shown in Figure 2(b). Then the graph of $y = -3(x-2)^3 + 1$ is a horizontal shift right 2 units, followed by a vertical shift upward 1 unit, as seen in Figure 2(c). ■

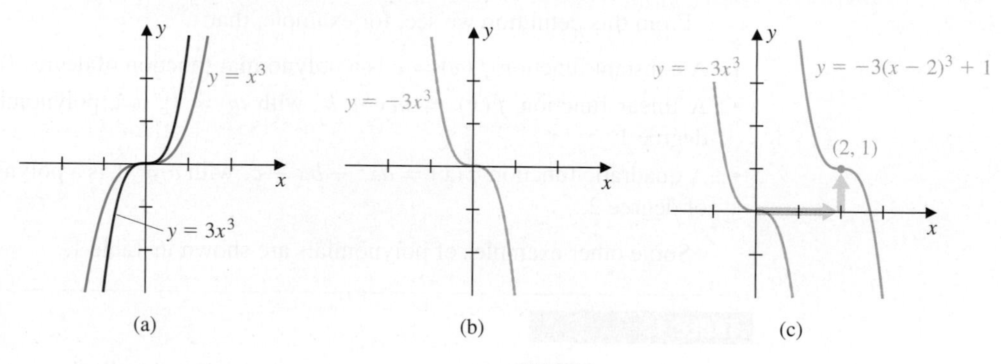

FIGURE 2

EXAMPLE 2 Sketch the graph of $f(x) = x^3 - 4x = x(x-2)(x+2)$.

Solution First notice that as x becomes large the term x^3 dominates the term $4x$, as shown in Table 2.

TABLE 2

x	x^3	$4x$
10	10^3	40
100	10^6	400
1000	10^9	4000

This implies that the graph of $y = x^3 - 4x$ approaches the graph of $y = x^3$. The graph has x-intercepts at -2, 0, and 2, so it has to pass through the points $(-2, 0)$, $(0, 0)$, and $(2, 0)$. The graph consequently has the features shown in Figure 3(a).

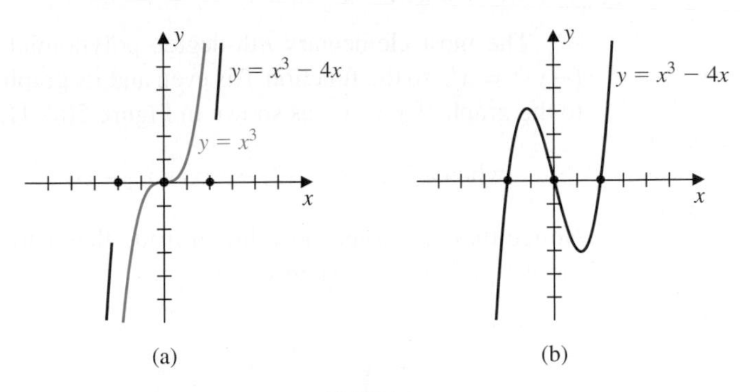

FIGURE 3

Moreover, f is an odd function because

$$f(-x) = (-x)^3 - 4(-x) = -x^3 + 4x = -(x^3 - 4x) = -f(x),$$

so the graph has origin symmetry.

Producing accurate graphs of functions is an application of the derivative that you will study in your first calculus course.

To determine the behavior of the graph near the origin, note that $f(1) = -3$. So the graph is below the x-axis between 0 and 1. But $f(3) = 15$, so the graph is above the x-axis beyond 2. This information indicates that the graph likely appears as shown in Figure 3(b).

To fine-tune the graph we would need to know the location of the local high and low points, but this requires calculus. An approximation to these points can come from a graphing device, but for most purposes the graph in Figure 3(b) will suffice. ■

EXAMPLE 3 Sketch the graph of $y = f(x) = (x - 2)^3 - 4(x - 2) + 5$.

Solution The graph of $y = x^3 - 4x$ is given in Example 2. Shifting this graph to the right 2 units produces the graph of $y = (x - 2)^3 - 4(x - 2)$ shown in Figure 4(a). The graph of $f(x) = (x - 2)^3 - 4(x - 2) + 5$ is then obtained by shifting the graph upward 5 units, as shown in Figure 4(b).

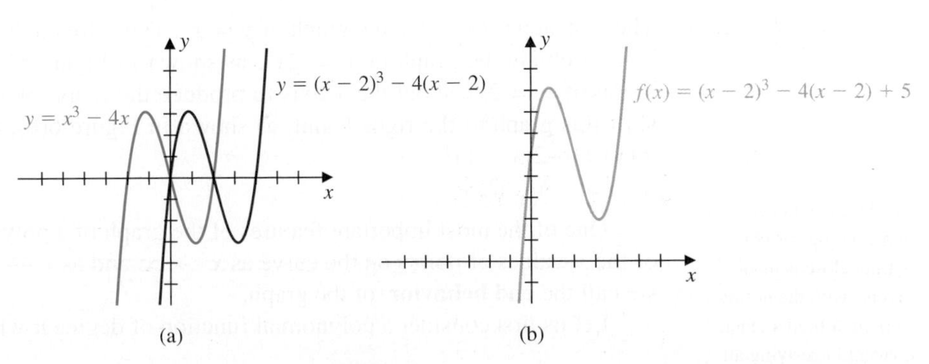

FIGURE 4

Notice that by shifting the graph to the right 2 units and then upward 5 units, the origin symmetry that was true for the graph of $y = x^3 - 4x$ is now lost. It is replaced by a symmetry with respect to the point (2, 5). ■

The most elementary nth-degree polynomial is $f(x) = x^n$. When n is even, $(-x)^n = x^n$, so the function f is even and its graph has y-axis symmetry. It is similar to the graph of $y = x^2$, as shown in Figure 5(a). Hence

Comparing the rates of growth of different functions is an important topic in calculus.

$$\text{when } n \text{ is even:} \quad x^n \to \infty \text{ as } x \to \infty \quad \text{and} \quad x^n \to \infty \text{ as } x \to -\infty.$$

Notice that the larger the value of n the flatter the graph is near the origin and the more rapidly it rises when $x > 1$.

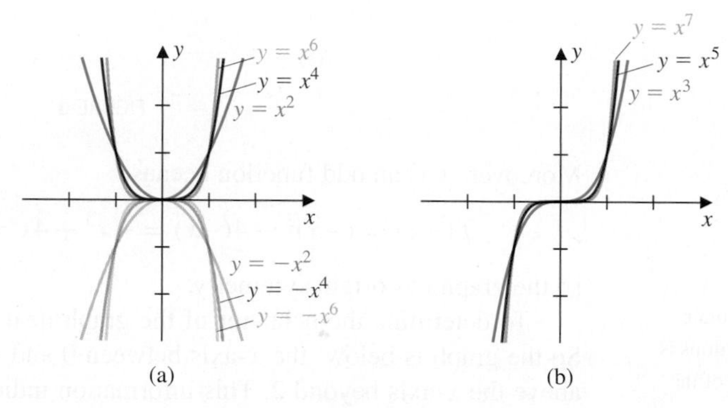

FIGURE 5

When n is odd, $(-x)^n = -x^n$, so the function f is odd and its graph has origin symmetry, as shown in Figure 5(b). For $n > 3$, the graph is similar to the graph of $y = x^3$. Hence,

$$\text{when } n \text{ is odd:} \quad x^n \to \infty \text{ as } x \to \infty \quad \text{and} \quad x^n \to -\infty \text{ as } x \to -\infty.$$

The larger the value of n, the flatter the graph near the origin and the more rapidly it rises when $x > 1$.

EXAMPLE 4 Sketch the graph of $f(x) = -2(x - 1)^4$.

Solution The first step is to sketch the graph of $y = x^4$. Then stretch this vertically by a factor of 2 to obtain the graph of $y = 2x^4$, as shown in Figure 6(a). Next we reflect the graph of $y = 2x^4$ about the x-axis to produce the graph of $y = -2x^4$. Finally we shift this graph to the right 1 unit, as shown in Figure 6(b), to obtain the graph of $f(x) = -2(x - 1)^4$. ■

The intuitive notion of approaching zero or getting close is made precise with the notion of *limit*, a fundamental concept underlying all of calculus.

One of the most important features of the graph of a polynomial is the behavior of the y-values of points on the curve as $x \to \infty$ and as $x \to -\infty$. This gives what we call the **end behavior** of the graph.

Let us first consider a polynomial function of degree n with the form

$$P(x) = x^n + a_{n-1}x^{n-1} + a_{n-2}x^{n-2} + \cdots + a_1x + a_0.$$

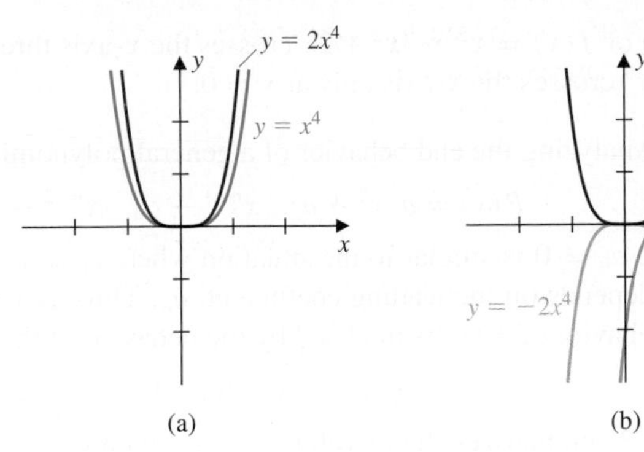

(a) (b)

FIGURE 6

The end behavior of $P(x)$ depends only on the behavior of the leading term x^n. To see this, first factor x^n from each term to rewrite $P(x)$ as

$$P(x) = x^n\left(1 + \frac{a_{n-1}}{x} + \frac{a_{n-2}}{x^2} + \cdots + \frac{a_1}{x^{n-1}} + \frac{a_0}{x^n}\right).$$

As the values of x become large in magnitude, either positively or negatively, the terms

$$\frac{1}{x}, \quad \frac{1}{x^2}, \quad \ldots, \quad \frac{1}{x^{n-1}}, \quad \text{and} \quad \frac{1}{x^n}$$

all approach zero. So as x becomes large in magnitude, the behavior of $P(x)$ is modeled by the behavior of the equation $y = x^n$.

EXAMPLE 5 Use the end behavior to give a sketch of the graph of $P(x) = x^3 - 3x^2 + 2x$.

Solution The graph of $y = P(x)$ has the same end behavior as the graph of $y = x^3$. So, as Figure 7(a) shows,

$$x^3 - 3x^2 + 2x \to \infty \text{ as } x \to \infty \quad \text{and} \quad x^3 - 3x^2 + 2x \to -\infty \text{ as } x \to -\infty.$$

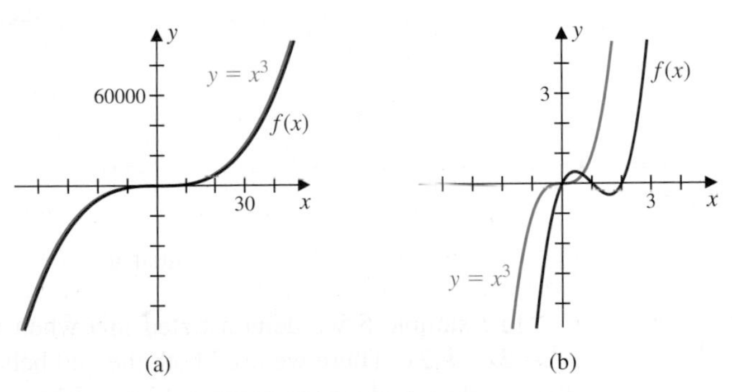

(a) (b)

FIGURE 7

Now factor $P(x)$ to determine any x-intercepts. Since

$$P(x) = x^3 - 3x^2 + 2x = x(x^2 - 3x + 2) = x(x-1)(x-2),$$

$P(x) = 0$ when $x = 0$, $x = 1$, and $x = 2$. In Figure 7(b) we see that for small values of x, the polynomial does not behave like $y = x^3$. In particular, for $0 \le x \le 2$ the graph of $f(x) = x^3 - 3x^2 + 2x$ crosses the x-axis three times, whereas the graph of $y = x^3$ crosses the x-axis only at $x = 0$. ■

Analyzing the end behavior of a general polynomial function

$$P(x) = a_n x^n + a_{n-1}x^{n-1} + a_{n-2}x^{n-2} + \cdots + a_1 x + a_0$$

where $a_n \neq 0$ is similar to the situation when $a_n = 1$, except that the behavior now also depends on the leading coefficient, a_n. Thus as x becomes large in magnitude, the behavior of $P(x)$ is modeled by the behavior of the equation $y = a_n x^n$. That is,

$$P(x) \approx a_n x^n, \text{ for } x \text{ large in magnitude.}$$

Table 3 summarizes the possibilities for the end behavior of polynomials, and Figure 8 illustrates the various possibilities. Of course, the end behavior only applies for x-values that become large in magnitude.

Don't memorize the table. Polynomials of degree $n \geq 1$ go to $\pm\infty$ as x goes to $\pm\infty$. Adjust the sign based on the term of highest degree.

TABLE 3

n	$a_n > 0$	$a_n < 0$
Even	$P(x) \to \infty$ as $x \to \infty$ $P(x) \to \infty$ as $x \to -\infty$ (See Figure 8(a))	$P(x) \to -\infty$ as $x \to \infty$ $P(x) \to -\infty$ as $x \to -\infty$ (See Figure 8(b))
Odd	$P(x) \to \infty$ as $x \to \infty$ $P(x) \to -\infty$ as $x \to -\infty$ (See Figure 8(c))	$P(x) \to -\infty$ as $x \to \infty$ $P(x) \to \infty$ as $x \to -\infty$ (See Figure 8(d))

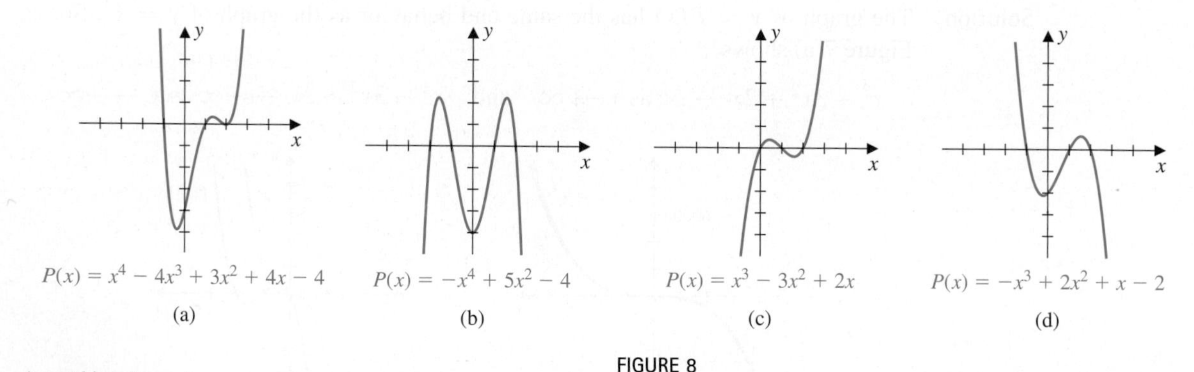

$P(x) = x^4 - 4x^3 + 3x^2 + 4x - 4$ (a)

$P(x) = -x^4 + 5x^2 - 4$ (b)

$P(x) = x^3 - 3x^2 + 2x$ (c)

$P(x) = -x^3 + 2x^2 + x - 2$ (d)

FIGURE 8

A graphing device is useful in visualizing the end behavior of a polynomial.

In Example 5 we demonstrated this when we sketched the graph of $P(x) = x^3 - 3x^2 + 2x$. There we used both the end behavior and the values of x that make $P(x) = 0$ to produce the graph in Figure 7(b).

Calculus is concerned with *continuous* functions. The definition of continuous requires a precise definition of the limit of a function.

Zeros of Functions

A number c is called a **zero** of the function f if $f(c) = 0$. The graph of $y = f(x)$ crosses the x-axis at $x = c$ precisely when c is a zero of f.

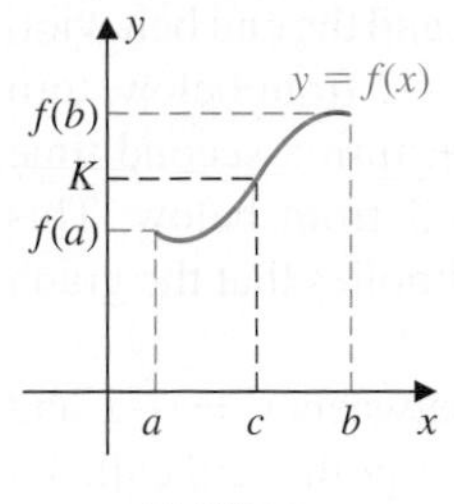

FIGURE 9

When c is a zero of the function f, it is also common to say that c is a **root** of the equation $f(x) = 0$. Determining—or even approximating—the zeros of a polynomial can greatly aid in sketching its graph since every polynomial function has the property of *continuity*. Continuity for a function on an interval $[a, b]$ essentially means that its graph has no breaks or interruptions on $[a, b]$. This is the essence of the *Intermediate Value Theorem* (see Figure 9). The Intermediate Value Theorem was used implicitly in the sketch of the graph of $P(x) = x^3 - 3x^2 + 2x$ in Example 5.

Intermediate Value Theorem

Suppose that f is continuous on $[a, b]$ and K is a number between $f(a)$ and $f(b)$. Then at least one number c in (a, b) exists with $f(c) = K$.

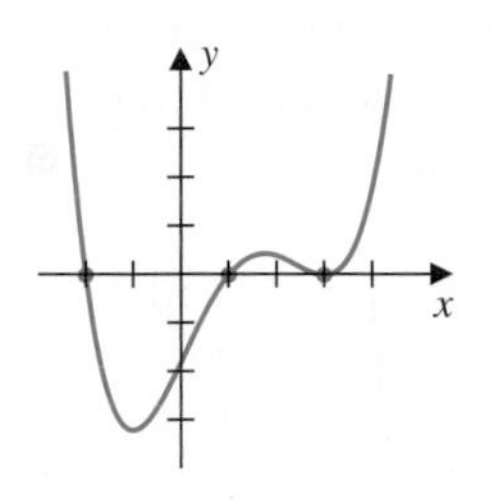

FIGURE 10

The Intermediate Value Theorem tells us that continuous functions do not skip over values in the range. In particular, a polynomial that is both positive and negative in some interval must take on the value zero somewhere between the positive and negative values. This means that the graph of a polynomial between successive zeros is either always positive or always negative. So the graph will either lie entirely above or entirely below the x-axis between successive zeros. (See Figure 10.)

In the next example we again show how knowledge of the zeros and end behavior of a polynomial is used to produce a graph.

EXAMPLE 6 Sketch the graph of $y = f(x) = x^3 - x^2 - 6x = x(x+2)(x-3)$.

Solution It is easy to determine the points where the graph crosses the x-axis because

$$f(x) = x(x+2)(x-3) = 0$$

is satisfied precisely when one of the factors is 0. The zeros of f are $x = -2$, $x = 0$, and $x = 3$, and the points in the plane where the graph crosses the x-axis are $(-2, 0)$, $(0, 0)$, and $(3, 0)$. To determine whether the graph is above or below the x-axis between two successive zeros, we select an arbitrary x-value between the two zeros and determine the sign of $f(x)$. The sign chart in Figure 11 indicates that the graph of $f(x)$ is above the x-axis for x in the intervals $(-2, 0)$ and $(3, \infty)$ and below the x-axis for x in the intervals $(-\infty, -2)$ and $(0, 3)$.

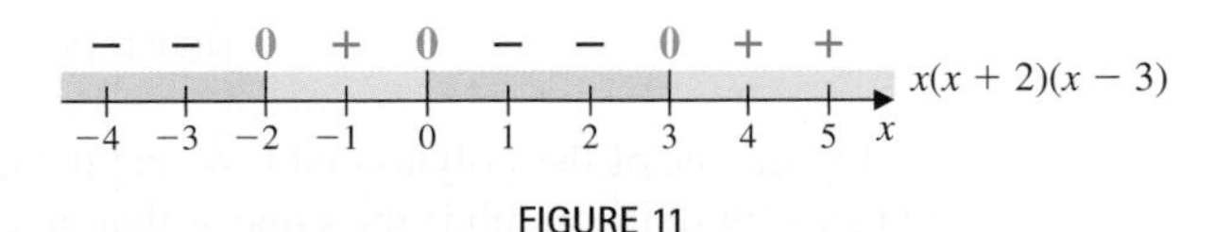

FIGURE 11

To determine the end behavior of the graph, we consider $f(x)$ in its original form:

$$f(x) = x^3 - x^2 - 6x.$$

For x-values large in magnitude, $f(x) \approx x^3$, so

$$x^3 - x^2 - 6x \to \infty \text{ as } x \to \infty \quad \text{and} \quad x^3 - x^2 - 6x \to -\infty \text{ as } x \to -\infty.$$

Figure 12(a) shows the information provided from the zeros and the end behavior. To satisfy the end behavior, the graph has to pass through $x = -2$ from below, turn between $x = -2$ and $x = 0$, pass through $x = 0$ from above, turn a second time between $x = 0$ and $x = 3$, and finally pass through $x = 3$ from below. This information, together with the results in the chart in Figure 11, implies that the graph is as shown in Figure 12(b).

The graph in Figure 12(b) has a high point somewhere between $x = -2$ and $x = -1$ and has a low point between $x = 1$ and $x = 2$. These points are called a *local maximum* and *local minimum*, respectively. Collectively they are called *local extrema*.

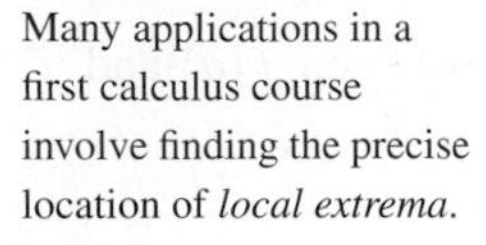

Many applications in a first calculus course involve finding the precise location of *local extrema*.

One of the most valuable applications of calculus is determining the locations of local extrema. Without calculus these points generally cannot be determined exactly, but by zooming in with a graphing device we can approximate them. ■

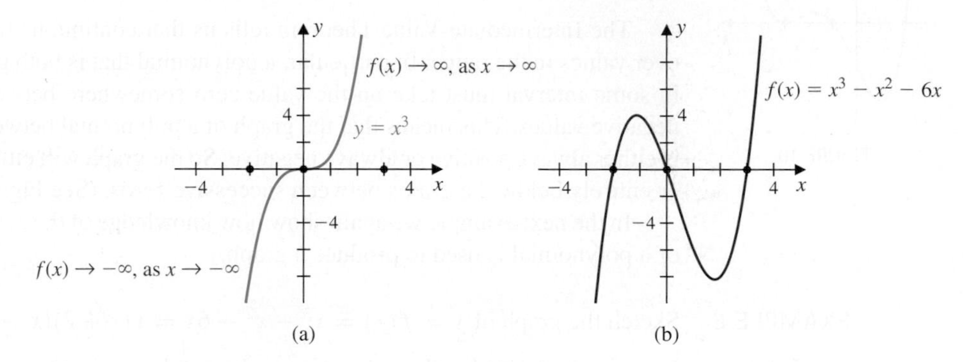

FIGURE 12

EXAMPLE 7 Sketch the graph of $f(x) = x^4 - 6x^3 + 8x^2 = x^2(x-2)(x-4)$.

Solution The zeros of f are $x = 0$, $x = 2$, and $x = 4$. Be careful to include the zero $x = 0$ from the term x^2 in the factored form $f(x) = x^2(x-2)(x-4)$. The sign chart in Figure 13 indicates there are only two sign changes. Because the factor x^2 is raised to an even power, the sign of the product does not change at $x = 0$.

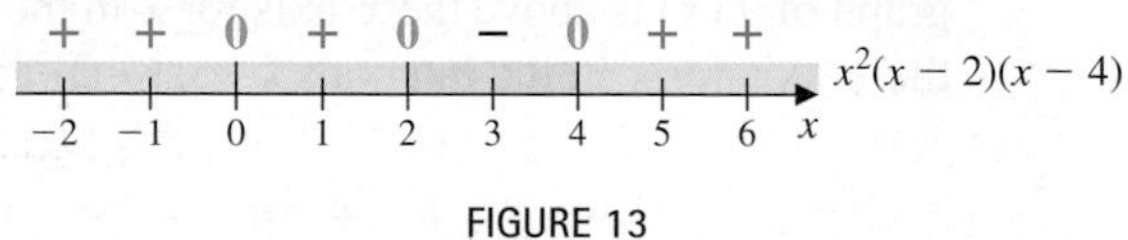

FIGURE 13

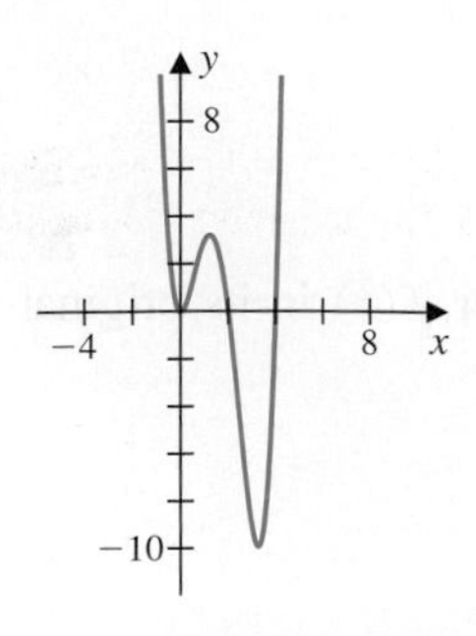

FIGURE 14

The degree of the polynomial is 4, and the leading coefficient is positive, so the end behavior of the graph is the same as that of $y = x^4$. For x large in magnitude, we have $f(x) \approx x^4$. So

$$f(x) \to \infty \text{ as } x \to \infty \quad \text{and} \quad f(x) \to \infty \text{ as } x \to -\infty.$$

Plotting the points (0, 0), (2, 0), and (4, 0) helps us infer that the graph is similar to that shown in Figure 14.

The sign chart in Figure 13 indicates that a local minimum occurs at $x = 0$, since the graph passes through (0, 0), but does not change sign there. The curve has to just touch the x-axis at the point (0, 0), as shown in Figure 14.

The graph remains positive and passes through $(2, 0)$, where it then changes sign. By zooming in with a graphing device on the intervals $[0, 2]$ and $[2, 4]$, we can approximate the other local extrema. However, we cannot determine them precisely in this way, because graphing devices only give rational numbers and the extrema happen to occur at the irrational numbers $\frac{1}{4}(9 \pm \sqrt{17})$. Notice also that the graph has both a zero and a local minimum at $x = 0$. This is due to the factor x^2, which caused $x = 0$ to be a *repeated zero* of *multiplicity* 2. ■

Multiplicity of a Zero

Suppose that a polynomial $f(x)$ has a factor of the form $(x - c)^k$, for a positive integer k, but that $(x - c)^{k+1}$ is not a factor. Then $x = c$ is a **zero** of f of **multiplicity** k. A zero of multiplicity 1 is called a *simple* zero.

The graph of a function has different features at a zero depending on its multiplicity.

- If the multiplicity k of the zero is even, then the graph flattens and just touches the x-axis at $x = c$, as shown with $c = 0$ in Figure 15(a).
- If the multiplicity $k > 1$ of the zero is odd, the graph flattens and then crosses the x-axis at $x = c$, as shown with $c = 0$ in Figure 15(b).

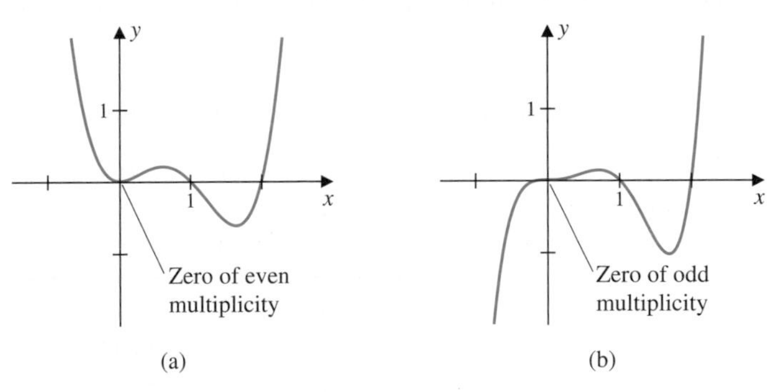

FIGURE 15

Suppose we change the function in Example 7 slightly, to

$$g(x) = (x^2 + 1)(x - 2)(x - 4).$$

Then g has only two zeros, at $x = 2$ and $x = 4$, and there are still two sign changes. As shown in Figure 16(a), the graph of g now has only one local extreme point, a local minimum between $x = 2$ and $x = 4$.

Minor changes in a polynomial can produce significant changes in the shape of the graph.

On the other hand, if the function in Example 7 is changed to

$$h(x) = (x^2 - 1)(x - 2)(x - 4),$$

we can factor the quadratic term as

$$x^2 - 1 = (x - 1)(x + 1).$$

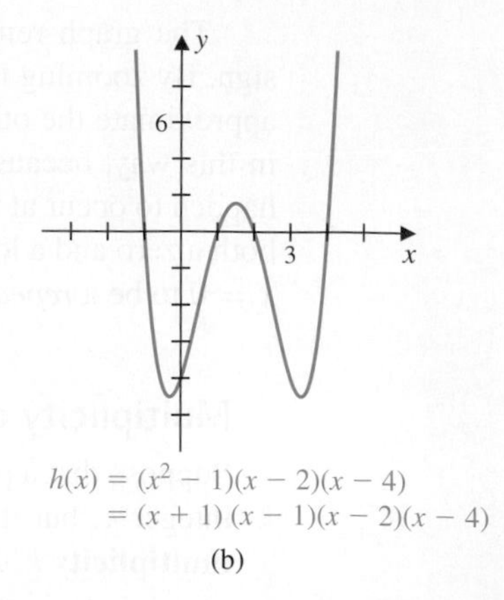

$g(x) = (x^2 + 1)(x - 2)(x - 4)$

(a)

$h(x) = (x^2 - 1)(x - 2)(x - 4)$
$= (x + 1)(x - 1)(x - 2)(x - 4)$

(b)

FIGURE 16

Now there are four zeros, at $-1, 1, 2$, and 4, and the graph appears as shown in Figure 16(b). Local minimums occur between -1 and 1 and between 2 and 4. A local maximum occurs between 1 and 2. Note that

- Small changes in the function can result in significant changes in the graph.

EXAMPLE 8 Sketch the graph of $f(x) = x^3 - x^2 - 9x + 9$.

Solution We begin by grouping the terms of f and factoring:

$$f(x) = x^3 - x^2 - 9x + 9 = x^2(x - 1) - 9(x - 1).$$

The factor $x - 1$ is common to both terms, so we have

$$f(x) = (x - 1)(x^2 - 9) = (x + 3)(x - 1)(x - 3).$$

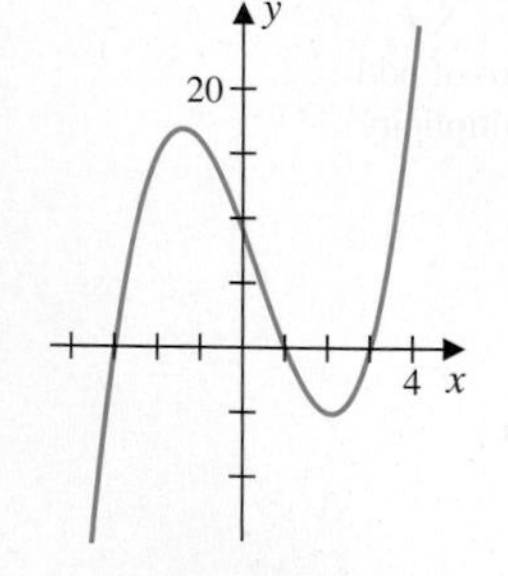

FIGURE 18

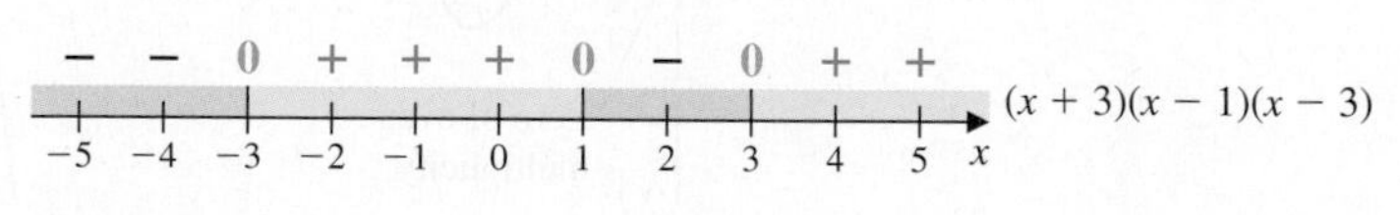

FIGURE 17

This implies that f has three zeros, at $x = -3$, $x = 1$, and $x = 3$, and the chart in Figure 17 shows that there are three sign changes. The graph shown in Figure 18 has a local maximum near $x = -1.5$ and a local minimum near $x = 2$. Note that the graph has the same end behavior as $y = x^3$. ■

EXAMPLE 9 A fifth-degree polynomial $P(x)$ has a zero of multiplicity 3 at $x = 2$, simple zeros at $x = -1$ and $x = 0$, and passes through the point $(1, 4)$. Determine $P(x)$ and sketch its graph.

Solution The polynomial has simple zeros at $x = -1$ and $x = 0$, so it has factors $(x + 1)$ and x. It has a zero of multiplicity 3 at $x = 2$, so it also has a factor $(x - 2)^3$. The fifth-degree polynomial

$$Q(x) = x(x + 1)(x - 2)^3$$

has the correct zeros, but $Q(1) = -2$. The polynomial $P(x)$ we need has the form

$$P(x) = aQ(x) = ax(x+1)(x-2)^3$$

for some nonzero constant a. The requirement that $P(1) = 4$ implies that

$$4 = a(1)(2)(-1), \quad \text{so} \quad a = -2.$$

Hence

$$P(x) = -2x(x+1)(x-2)^3.$$

For x large in magnitude

$$P(x) \approx -2x^3,$$

so the end behavior is

$$P(x) \to -\infty \text{ as } x \to \infty \quad \text{and} \quad P(x) \to \infty \text{ as } x \to -\infty.$$

The graph passes directly through the zero at $x = -1$ from above, turns somewhere between $x = -1$ and $x = 0$, passes directly through the zero at $x = 0$ from below, and turns somewhere between $x = 0$ and $x = 2$. It passes through the zero at $x = 2$ from above but also flattens there because the zero at $x = 2$ is of odd multiplicity greater than 1. The graph has the shape shown in Figure 19. ■

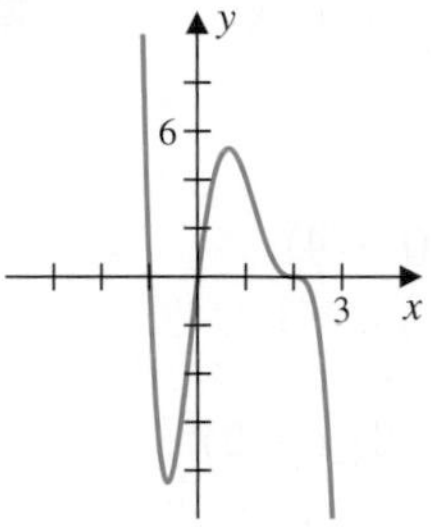

FIGURE 19

Graphing higher-degree polynomials without a graphing device can be difficult unless the zeros of the polynomial are easy to determine. Only by determining the zeros can we find intervals where the polynomial is positive and negative. We examine this way of locating zeros in the next section.

Applications

EXAMPLE 10 A container without a lid is constructed from a square sheet of material with sides 40 inches long cutting a square of size x from each corner, as shown in Figure 20.

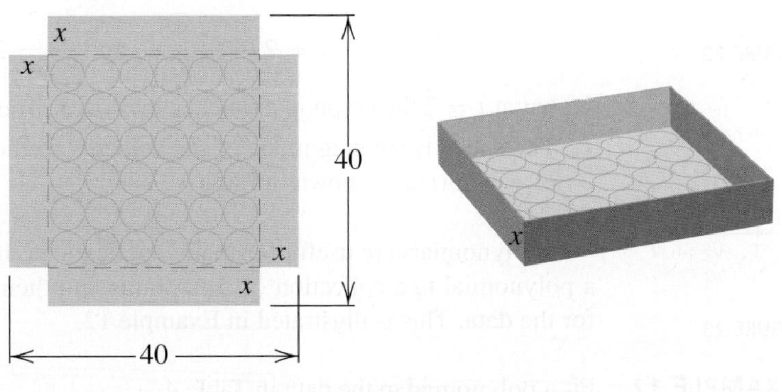

FIGURE 20

a. Express the volume of the container as a function of x.

b. Express the surface area of the container as a function of x.

Solution **a.** The side of the cutout is restricted to half the side of the sheet, so $0 \le x \le 20$. When squares of side x are cut from each corner, the area of the base of the container is

$$(40 - 2x)(40 - 2x),$$

so the volume is

$$V(x) = x(40 - 2x)(40 - 2x) = 4x^3 - 160x^2 + 1600x.$$

b. The surface area, $S(x)$, is the area of the base plus the area of the four vertical sides. So

$$S(x) = (40 - 2x)(40 - 2x) + 4x(40 - 2x) = 4x^2 + 1600.$$ ■

EXAMPLE 11 The velocity of a particle is given by

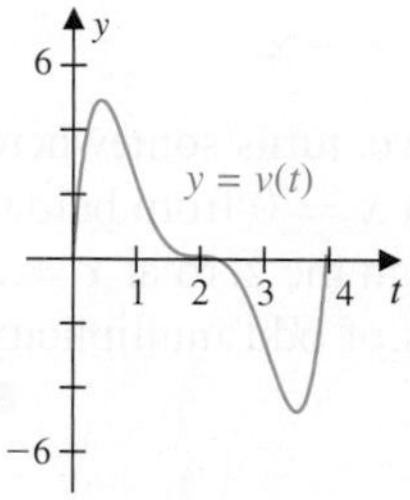

$$v(t) = t^5 - 10t^4 + 36t^3 - 56t^2 + 32t = t(t - 2)^3(t - 4),$$

and its acceleration is given by

$$a(t) = 5t^4 - 40t^3 + 108t^2 - 112t + 32 = (5t^2 - 20t + 8)(t - 2)^2.$$

Its speed is defined to be $s(t) = |v(t)|$. In each case, $0 \le t \le 4$ denotes time in seconds. Sketch the graphs of

a. $y = v(t)$; **b.** $y = a(t)$; **c.** $y = s(t)$.

FIGURE 21

Solution **a.** The function v has simple zeros at $t = 0$ and $t = 4$, and a zero of multiplicity 3 at $t = 2$, so the graph passes through the three t-intercepts. At $t = 2$ the graph will be instantaneously flat, as shown in Figure 21.

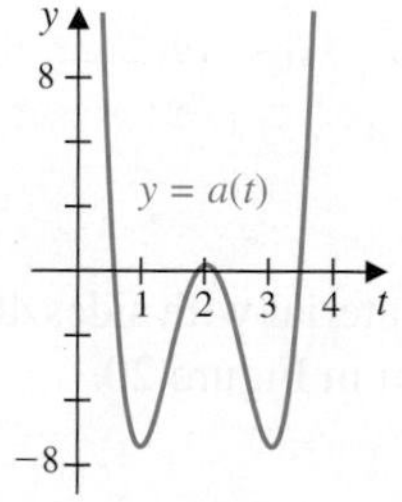

b. The quadratic term in the polynomial $a(t)$ can be factored using the quadratic formula. If $5t^2 - 20t + 8 = 0$, then

$$t = \frac{20 \pm \sqrt{400 - 4(5)(8)}}{10} = 2 \pm \frac{2\sqrt{15}}{5}, \quad \text{so} \quad t \approx 3.5 \text{ and } t \approx 0.5.$$

The graph of $y = a(t)$ passes through the t-intercepts at

$$t = 2 - \frac{2\sqrt{15}}{5} \quad \text{and} \quad t = 2 + \frac{2\sqrt{15}}{5},$$

but at $t = 2$ the graph just touches the t-axis. The graph is shown in Figure 22.

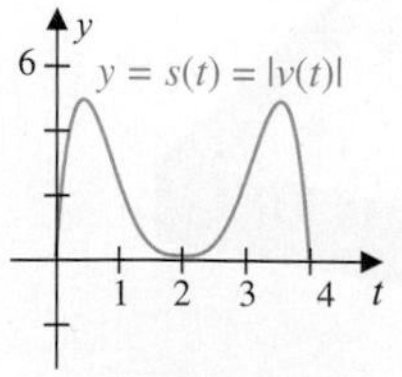

FIGURE 22

c. The speed is the magnitude of the velocity, so the graph of $y = s(t)$ is the graph of $y = |v(t)|$, as shown in Figure 23. ■

FIGURE 23

Polynomials are useful for many applications. It is common, for example, to fit a polynomial to a collection of data points and then use the polynomial as a model for the data. This is illustrated in Example 12.

EXAMPLE 12 Fit a polynomial to the data in Table 4.

Solution A plot of the data points, shown in Figure 24, suggests fitting the data with a cubic polynomial. Note that the points are approximately symmetric about the point with coordinates (2, 1.5), which is marked with the cross in Figure 24.

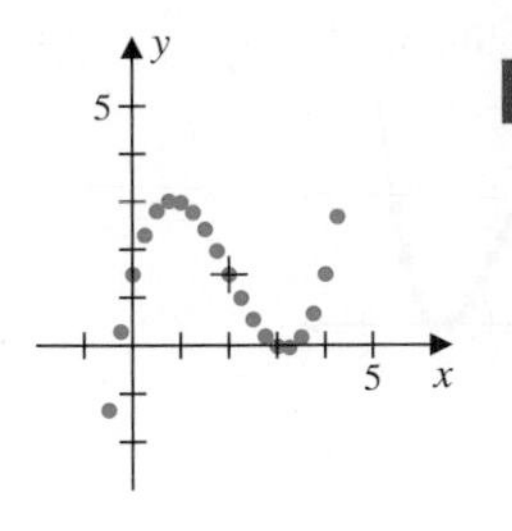

FIGURE 24

TABLE 4

x	y	x	y	x	y	x	y
−0.5	−1.32	0.75	3.03	2.00	1.50	3.25	−0.03
−0.25	0.31	1.0	3.0	2.25	1.01	3.5	1.19
0	1.5	1.25	2.76	2.50	0.56	3.75	0.68
0.25	2.32	1.5	2.42	2.75	0.21	4.0	1.50
0.5	2.82	1.75	1.99	3.00	0	4.25	2.70

We will first shift the data points by 2 units to the left on the x-axis and 1.5 units downward on the y-axis, as shown in Table 5, so that the graph of the shifted data will be symmetric about the origin.

TABLE 5

x	y	x	y	x	y	x	y
−2.5	−2.82	−1.25	1.53	0.0	0.0	1.25	−1.58
−2.25	−1.19	−1.0	1.5	0.25	−0.49	1.5	−0.31
−2	0.0	−0.75	1.26	0.5	−0.94	1.75	−0.82
−1.75	0.82	−0.5	0.92	0.75	−1.29	2.0	0.0
−1.5	1.32	−0.25	0.49	1.0	−1.5	2.25	1.2

The cubic equation approximating the shifted data crosses the x-axis three times, as shown in Figure 25(a). The x-intercepts of the shifted cubic polynomial are approximately $x = -2$, $x = 0$, and $x = 2$.

The cubic polynomial function $g(x) = x(x-2)(x+2)$ has the correct x-intercepts, but it needs to be vertically scaled by a factor of $1/2$ to fit the shifted data.

To determine a cubic polynomial to fit the original data, we define the function f as

- a vertical scaling of $y = g(x)$ by a factor of 1/2, to $y = \frac{1}{2}g(x)$, followed by
- a shift to the right 2 units, to $y = \frac{1}{2}g(x-2)$, followed by
- an upward shift of 1.5 units, to $y = \frac{1}{2}g(x-2) + 1.5$.

This produces

$$f(x) = \frac{1}{2}g(x-2) + 1.5 = \frac{1}{2}(x-2)((x-2)-2)((x-2)+2) + 1.5,$$

which simplifies to

$$f(x) = \frac{1}{2}x^3 - 3x^2 + 4x + 1.5$$

and gives the curve shown in Figure 25(b). ■

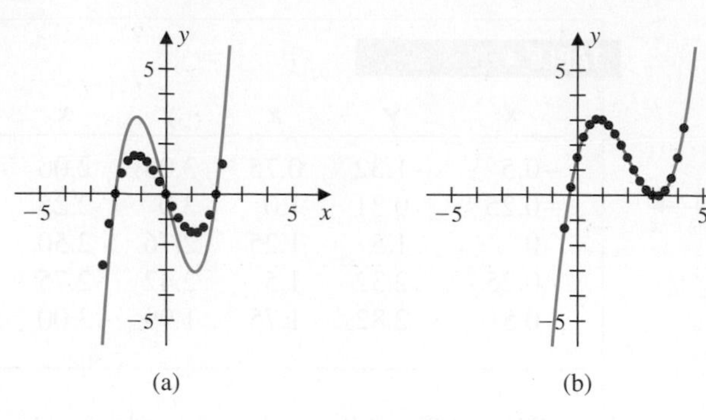

FIGURE 25

EXERCISE SET 3.2

1. Match the equation with the graph.

a. $y = x(x + 2)^2(x - 2)$

b. $y = x(x + 2)(x - 2)^2$

c. $y = x(x + 2)(x - 2)$

d. $y = x^2(x + 2)(x - 2)$

i.

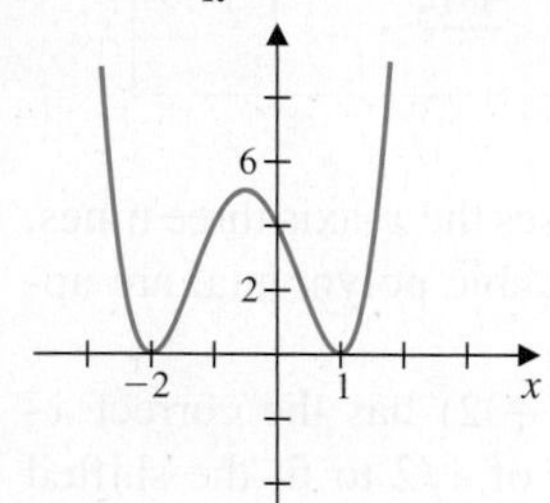

ii.

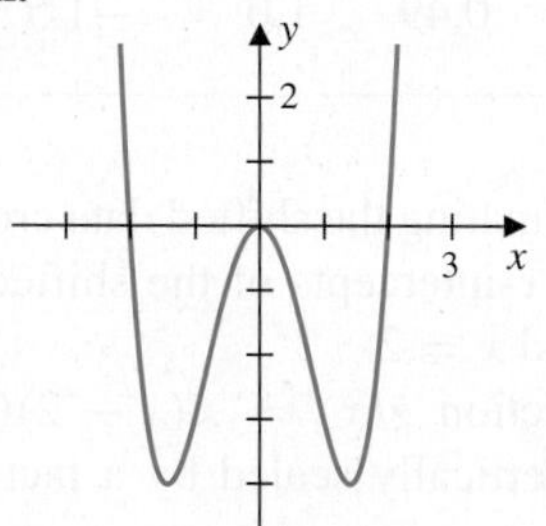

iii.

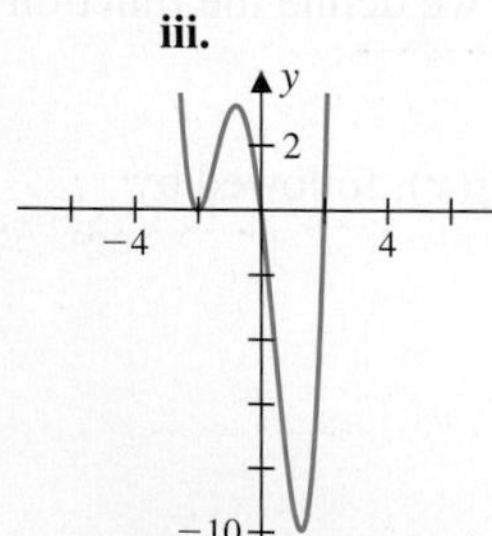

iv.

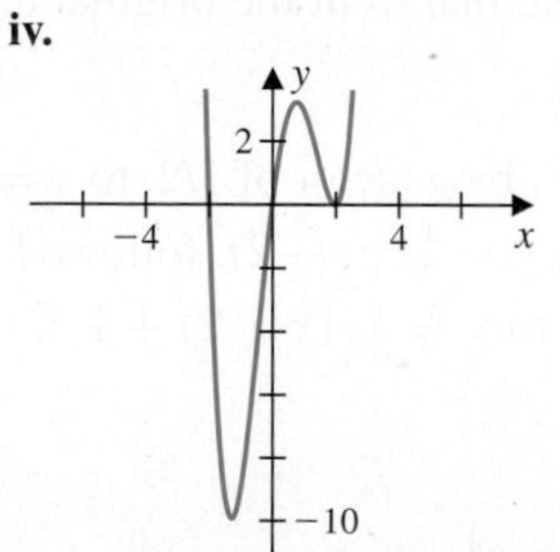

2. Match the equation with the graph.

a. $y = (x - 1)(x + 2)(x - 2)$

b. $y = x^2(x - 1)$

c. $y = (x + 1)(x - 2)^2$

d. $y = x(x + 1)(2 - x)$

i.

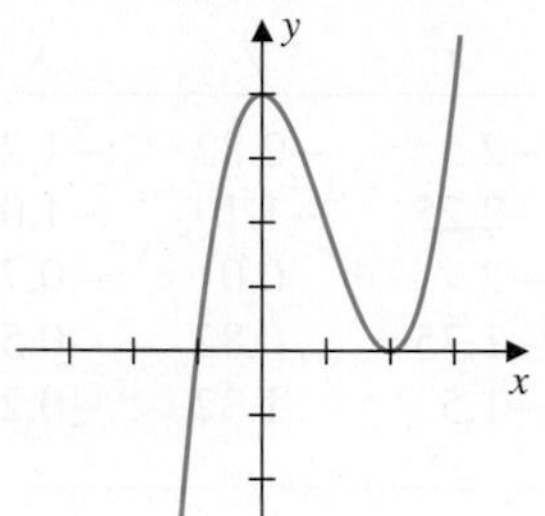

ii.

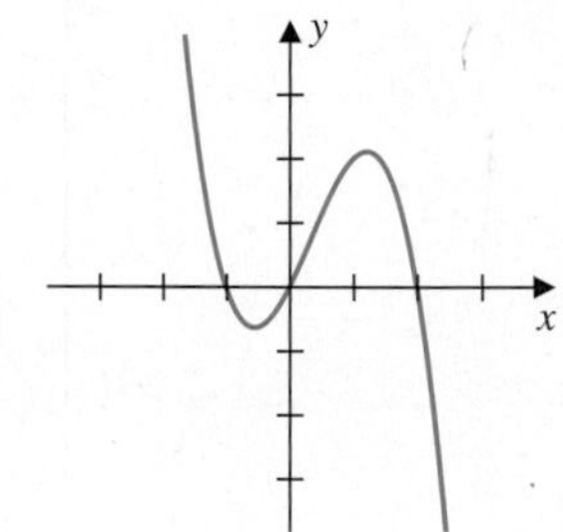

iii.

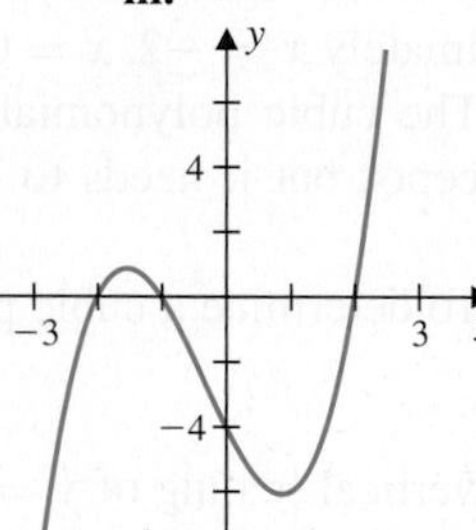

iv.

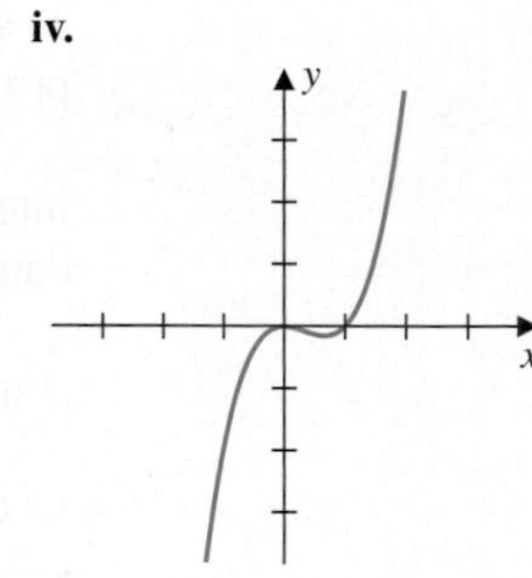

In Exercises 3–6, determine the lowest possible degree for the polynomial whose graph is shown.

3.

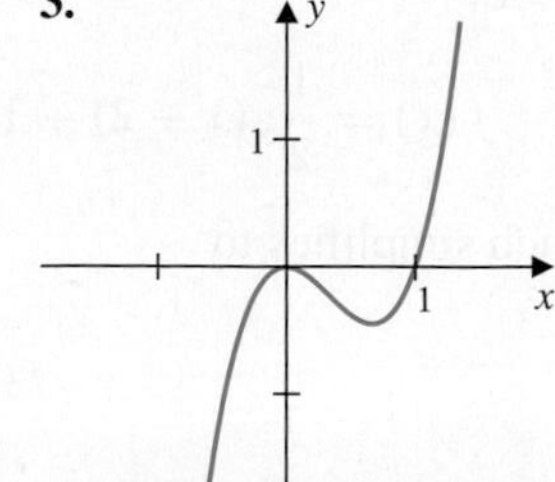

4.

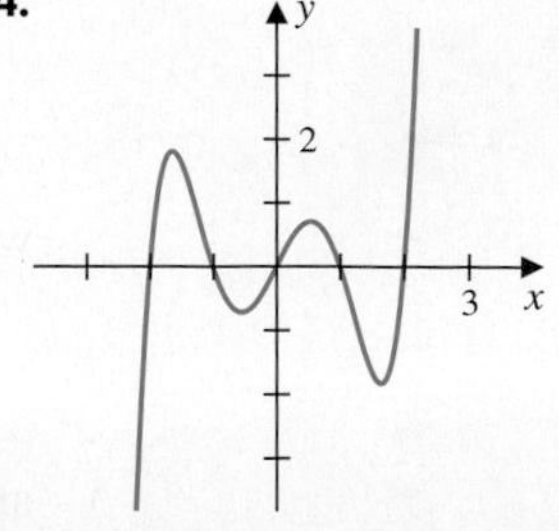

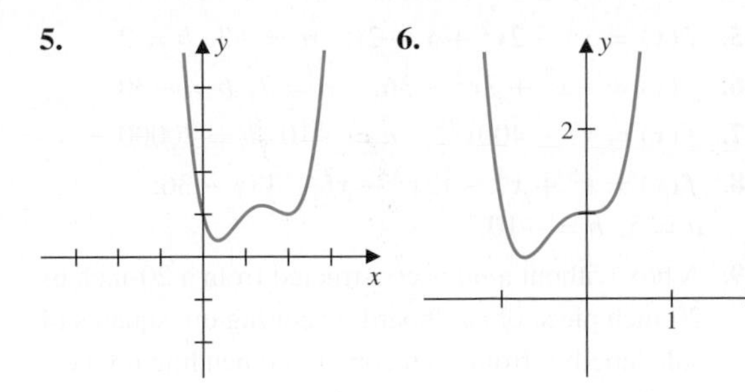

In Exercises 7–12, sketch the graph of the function by transforming the graph of a function of the form $g(x) = x^n$.

7. $f(x) = 3x^3 - 2$
8. $f(x) = \frac{1}{2}x^3 + 2$
9. $f(x) = (x-2)^4 + 1$
10. $f(x) = (x-1)^3 + 3$
11. $f(x) = -\frac{1}{2}(x-3)^3 - 3$
12. $f(x) = -2(x+2)^4 + 3$

In Exercises 13–22, use the zeros and the end behavior of the polynomial function to sketch an approximation to the graph of the function.

13. $f(x) = (x-2)(x+2)(x-3)$
14. $f(x) = (x+3)(x-2)(x-1)$
15. $f(x) = x(x+1)(x-1)(x-2)$
16. $f(x) = -x(x+1)(x+2)(x-3)$
17. $f(x) = -x^3(x-1)$
18. $f(x) = x^2(x+1)(x-2)$
19. $f(x) = -x^2(x+1)^2$
20. $f(x) = (x-1)^3(x+1)^3$
21. $f(x) = x^3 + x^2 - 2x$
22. $f(x) = x^4 - x^3 - 2x^2$

In Exercises 23–30, sketch each of the following curves by shifting the graph in the accompanying figure, and determine the coordinates of the labeled points.

23. $y = f(x-2)$
24. $y = f(x+1)$

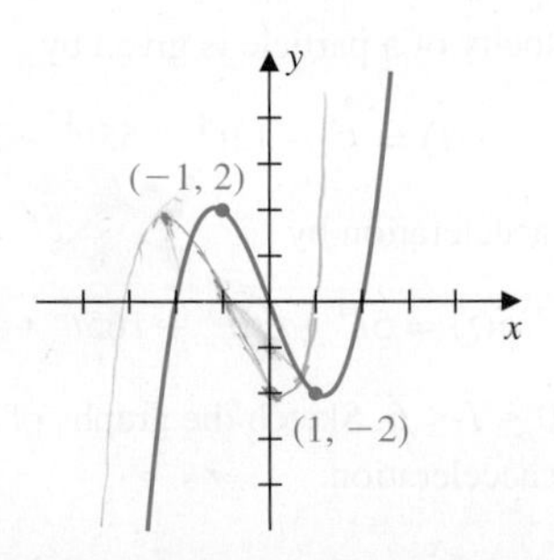

25. $y = f(x-2) + 1$
26. $y = -f(x)$
27. $y = f(-x)$
28. $y = -f(-x-2)$
29. $y = |f(x)|$
30. $y = f(|x|) - 1$
31. **a.** Sketch the graph of a polynomial $P(x)$ that has zeros of multiplicity 1 at $x = 1$, $x = -1$, and $x = 2$ and that satisfies $P(x) \to \infty$ as $x \to \infty$ and $P(x) \to -\infty$ as $x \to -\infty$.
    b. What is the least possible degree of this polynomial?
32. **a.** Sketch the graph of a polynomial $P(x)$ that has zeros of multiplicity 1 at $x = 0$ and $x = 1$, has a zero of multiplicity 3 at $x = 3$, and that satisfies $P(x) \to -\infty$ as $x \to \infty$ and $P(x) \to \infty$ as $x \to -\infty$.
    b. What is the least possible degree of this polynomial?
33. The cubic polynomial $P(x)$ has a zero of multiplicity 1 at $x = 2$ and a zero of multiplicity 2 at $x = 1$, and $P(-1) = 4$. Determine $P(x)$ and sketch its graph.
34. The cubic polynomial $P(x)$ has zeros at $x = 1$, $x = 2$, and $x = -2$ and y-intercept at 2. Determine $P(x)$ and sketch its graph.
35. Find a cubic polynomial that approximates the data in the table.

x	y	x	y
−1	−10	1	2
−0.5	−1.75	1.5	1.25
0	2	2	2
0.5	2.75	2.5	5.75

36. Find a cubic polynomial that approximates the data in the table.

x	y	x	y
−3.5	4.6	−1	−1
−3	−1	−0.5	0.9
−2.5	−3.6	0	2
−2	−4	0.5	1.6
−1.5	2.9	1	−1

37. Determine the polynomial of degree 4 whose graph is shown in the figure.

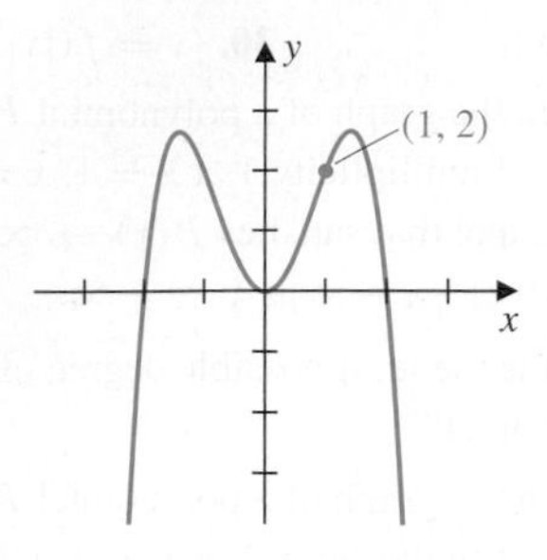

38. Determine the polynomial of degree 5 whose graph is shown in the figure.

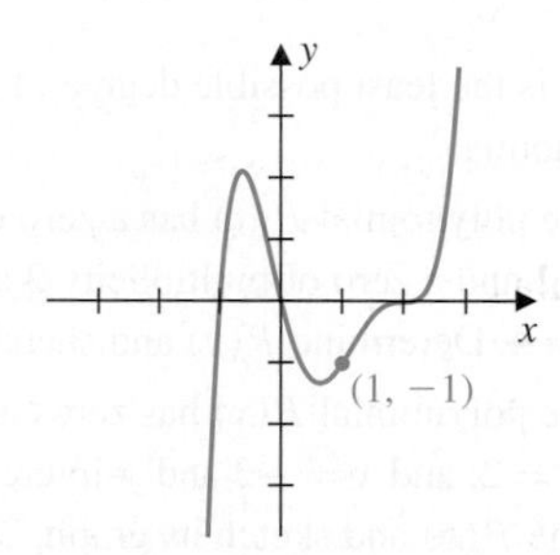

39. For each positive integer n, let $f_n(x) = x^n$. Compare the sizes of $f_n(x)$ and $f_{n+1}(x)$ on the intervals $(0, 1)$ and $(1, \infty)$.

40. Use a graphing device to sketch the graphs of the polynomials $P(x) = x^4 + x^3 - x^2 - x$ and $Q(x) = 4 - x^2$ and determine all the values of x for which $P(x) < Q(x)$.

In Exercises 41–44, use a graphing device to sketch the graph of the given function and approximate the following:

a. *Intervals on which the function is increasing.*

b. *Local maximum and minimum values.*

41. $f(x) = x^4 + x^3 - 2x^2$

42. $f(x) = x^4 - 2x^3$

43. $f(x) = \frac{1}{5}x^5 - \frac{5}{4}x^4 + \frac{5}{3}x^3 + \frac{5}{2}x^2 - 6x + 1$

44. $f(x) = x^5 - 2x^4 + 2x^2 - x$

In Exercises 45–48, use a graphing device to sketch the graph of the given function. Then sketch the graphs of $y = -f(x)$, $y = f(-x)$, $y = |f(x)|$, and $y = f(x + a) + b$ for the given values of a and b.

45. $f(x) = x^3 - 2x^2 + x + 2; \quad a = -3,\ b = 2$

46. $f(x) = -x^3 + 7x^2 - 36; \quad a = 7,\ b = -30$

47. $f(x) = x^4 - 400x^2; \quad a = -10,\ b = 30000$

48. $f(x) = x^5 + x^4 - 13x^3 - x^2 + 48x - 36;$
$a = 5,\ b = -10$

49. A box without a lid is constructed from a 20-inch by 20-inch piece of cardboard by cutting out squares of side length x from each corner and bending up the resulting flaps, as shown in the figure.

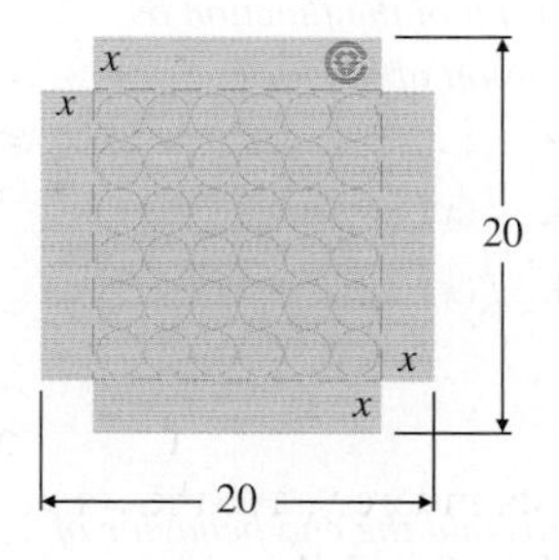

a. Determine the volume of the box as a function of the variable x.

b. Use a graphing device to approximate the value of x that produces a volume of 500 inch^3.

50. A fuel tank is in the shape of a right circular cylinder of length 10 m capped on either end with a hemisphere, as shown in the figure. Determine the radius r of the tank if the volume of the tank is 50 m^3.

51. The velocity of a particle is given by

$$v(t) = t^5 - 13t^4 + 55t^3 - 75t^2$$

and its acceleration by

$$a(t) = 5t^4 - 52t^3 + 165t^2 - 150t,$$

where $0 \le t \le 6$. Sketch the graphs of the velocity and the acceleration.

3.3 FINDING FACTORS AND ZEROS OF POLYNOMIALS

We saw in Section 3.2 that the zeros of a polynomial give important graphing information. Finding the zero of a linear polynomial is easy, and the quadratic formula

$$x = \frac{-b \pm \sqrt{b^2 - 4ac}}{2a}$$

for finding the zeros of the quadratic polynomial $ax^2 + bx + c$ has been known for over 2000 years. But finding the zeros of polynomials of higher degree is generally difficult.

There are formulas for finding the exact values of the zeros for polynomials of degrees 3 and 4. These were discovered in the middle of the 16th century by mathematicians in the north of Italy. This area was the birthplace of the Renaissance and produced many of the artists, musicians, and scientists of the period. A great deal of scandal and intrigue surrounded the development of these formulas, but the third-degree formula is generally credited to Nicolo Fontana (known as Tartaglia—the stammerer), and the fourth-degree formula to his rival Ludovico Ferrari. Like the quadratic formula, these are *algebraic solutions*, formulas that require only common arithmetic operations together with the extraction of roots. But unlike the quadratic formula, they are difficult to apply.

In the 16th through the early 19th centuries, efforts were made by many outstanding mathematicians to determine formulas for finding the zeros of higher-degree polynomials. However, in 1824 the 22-year-old Norwegian Niels Abel proved that there is no general algebraic solution for finding the zeros of fifth-degree polynomials. Certain fifth-degree polynomials can be solved exactly. For example, $x^5 - x^4$ has the zeros 0 and 1. But no formula will work for all fifth-degree polynomials.

Just a few years after Abel's proof appeared, a brilliant teenage mathematician in France, Evariste Galois, developed results that show precisely which polynomials of degree 5 and higher have an algebraic solution. Unfortunately, he was killed in a duel soon afterward, and his work was so complicated that its full significance was not appreciated for nearly 40 years.

In this section we will look at ways to find zeros of polynomials and see how this information is useful. A major role in this discussion is played by the technique of polynomial division, which is where we begin.

EXAMPLE 1 Find the zeros of $P(x) = x^3 + 2x^2 - x - 2$.

Solution By inspection we can see that $P(1) = 0$, so $x = 1$ is a zero of the polynomial P. This implies that $(x - 1)$ is a factor of $P(x)$, and

$$P(x) = (x - 1)Q(x)$$

for some quadratic polynomial $Q(x)$. We will use polynomial division to determine $Q(x)$.

Division of polynomials is similar to division of numbers. The first step in determining the quotient $Q(x)$ is to divide the leading term, x, in the divisor $x - 1$

into the leading term, x^3, of the dividend $x^3 + 2x^2 - x - 2$. This gives the leading term, x^2, in the *quotient*. This term, x^2, is then multiplied by the divisor $x - 1$ to get $x^3 - x^2$. We subtract this term, just like in the division of numbers, from the dividend $x^3 + 2x^2 - x - 2$ to obtain

$$(x^3 + 2x^2 - x - 2) - (x^3 - x^2) = 3x^2 - x - 2.$$

So at this stage we have

$$\begin{array}{llrl}
 & & x^2 + 3x + 2 & \text{Quotient} \\ \cline{3-3}
\text{Divisor} & x - 1 & \big) \; x^3 + 2x^2 - x - 2 & \text{Dividend} \\
 & & x^3 - x^2 \qquad\quad & \\ \cline{3-3}
 & & 3x^2 - x - 2 &
\end{array}$$

The process is repeated for a second and third iteration using the divisor and the reduced dividend at the bottom of the calculations until this line has a degree that is less than the degree of the divisor. In our example the calculations stop when the bottom line is a constant, because the degree of the divisor is 1.

$$\begin{array}{llrl}
 & & x^2 + 3x + 2 & \text{Quotient} \\ \cline{3-3}
\text{Divisor} & x - 1 & \big) \; x^3 + 2x^2 - x - 2 & \text{Dividend} \\
 & & x^3 - x^2 \qquad\quad & \\ \cline{3-3}
 & & 3x^2 - x - 2 & \text{Start Second Iteration} \\
 & & 3x^2 - 3x \qquad & \\ \cline{3-3}
 & & 2x - 2 & \text{Start Third Iteration} \\
 & & 2x - 2 & \\ \cline{3-3}
 & & 0 & \text{Remainder}
\end{array}$$

The constant in the *remainder* of our example is 0 since the linear factor $x - 1$ divides evenly into the polynomial $P(x)$, and we have factored $P(x)$ as

$$P(x) = x^3 + 2x^2 - x - 2 = (x - 1)(x^2 + 3x + 2).$$

The quadratic on the right of this equation factors as

$$x^2 + 3x + 2 = (x + 1)(x + 2),$$

so the complete factorization of $P(x)$ is

$$P(x) = (x - 1)(x^2 + 3x + 2) = (x - 1)(x + 1)(x + 2).$$

When a polynomial is written as a product of linear factors, the zeros are the numbers that make these factors zero. So the polynomial $P(x)$ has three distinct zeros, at $x = 1$, $x = -1$, and $x = -2$. Using the facts that $P(0) = -2$ and that the end behavior is similar to that of $f(x) = x^3$ gives the graph shown in Figure 1. ■

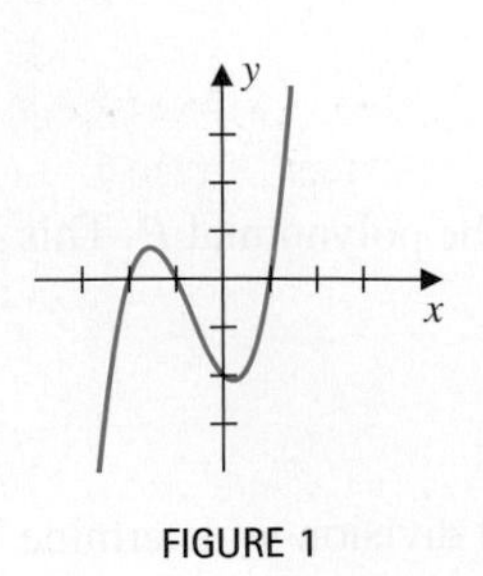

FIGURE 1

In Example 1 we used the following result in the special case where the divisor was the linear term $D(x) = x - 1$.

The Division Algorithm

Suppose $D(x)$ and $P(x)$ are polynomials with $D(x) \neq 0$, and the degree of $D(x)$ is less than the degree of $P(x)$. There exist unique polynomials $Q(x)$ and $R(x)$, where $R(x)$ is either 0 or has degree less than the degree of $D(x)$, such that

$$P(x) = Q(x) \cdot D(x) + R(x) \quad \text{or, equivalently,} \quad \frac{P(x)}{D(x)} = Q(x) + \frac{R(x)}{D(x)}.$$

EXAMPLE 2 Find the quotient and remainder when the polynomial $2x^5 + x^4 - x^3 - x - 1$ is divided by the polynomial $x^2 - 2x + 1$.

Solution Let $P(x) = 2x^5 + x^4 - x^3 - x - 1$ and $D(x) = x^2 - 2x + 1$. It is usually easier to perform the division if a position is reserved in $P(x)$ for every power of x starting with the highest-power term (in this case $2x^5$). So before performing the division, we insert a $0x^2$ term into $P(x)$. Sincc

$$\begin{array}{r|rrrrrr}
 & & & 2x^3 & +\,5x^2 & +\,7x & +\,9 \\
\hline
x^2 - 2x + 1 & 2x^5 & +\,x^4 & -\,x^3 & +\,0x^2 & -\,x & -\,1 \\
 & 2x^5 & -\,4x^4 & +\,2x^3 & & & \\
\hline
 & & 5x^4 & -\,3x^3 & +\,0x^2 & -\,x & -\,1 \\
 & & 5x^4 & -\,10x^3 & +\,5x^2 & & \\
\hline
 & & & 7x^3 & -\,5x^2 & -\,x & -\,1 \\
 & & & 7x^3 & -\,14x^2 & +\,7x & \\
\hline
 & & & & 9x^2 & -\,8x & -\,1 \\
 & & & & 9x^2 & -\,18x & +\,9 \\
\hline
 & & & & & 10x & -\,10,
\end{array}$$

wc havc

$$2x^5 + x^4 - x^3 - x - 1 = (2x^3 + 5x^2 + 7x + 9)(x^2 - 2x + 1) + (10x - 10),$$

or, equivalently,

$$\frac{2x^5 + x^4 - x^3 - x - 1}{x^2 - 2x + 1} = 2x^3 + 5x^2 + 7x + 9 + \frac{10x - 10}{x^2 - 2x + 1}. \quad \blacksquare$$

Synthetic division can simplify the calculation when the divisor has the form $x - c$. This technique is discussed in the Student Study Guide.

If the polynomial $P(x)$ is divided by a linear term $x - c$, then the remainder produced by the Division Algorithm is a constant. Calling this constant R gives

$$P(x) = (x - c)Q(x) + R,$$

and setting $x = c$ gives

$$P(c) = (c - c)Q(c) + R = 0 + R = R.$$

Hence the remainder R in the division of $P(x)$ by $x - c$ is the value of the polynomial $P(x)$ at $x = c$. In particular, this implies that the constant R will be zero precisely when $x - c$ is a factor of $P(x)$ (c is a zero of P).

Review the Division Algorithm and Factor Theorem carefully. They are needed frequently in calculus.

The Factor Theorem

If the polynomial $P(x)$ is divided by the linear factor $x - c$, then the remainder is $P(c)$. The linear term $x - c$ is a factor of the polynomial $P(x)$ if and only if $P(c) = 0$.

EXAMPLE 3 Factor $P(x) = x^4 + 5x^3 + 5x^2 - 5x - 6$, and sketch the graph of P.

Solution It is not difficult to see that both $P(1) = 0$ and $P(-1) = 0$. So the Factor Theorem implies that $x - 1$ and $x - (-1) = x + 1$ are factors of $P(x)$. To find any remaining factors, divide $P(x)$ by the product $(x - 1)(x + 1) = x^2 - 1$ to produce

$$P(x) = (x^2 - 1)(x^2 + 5x + 6) = (x - 1)(x + 1)(x^2 + 5x + 6).$$

Finally, factor the quadratic as

$$x^2 + 5x + 6 = (x + 2)(x + 3)$$

and write

$$P(x) = (x - 1)(x + 1)(x + 2)(x + 3).$$

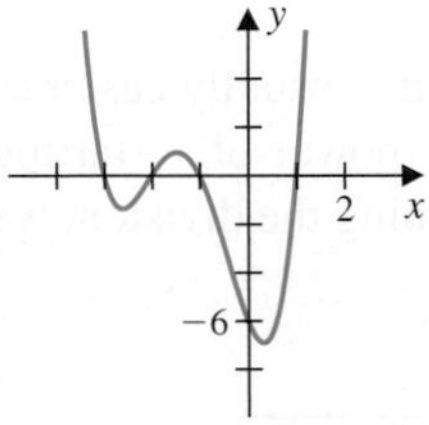

FIGURE 2

For x large in magnitude, $P(x) \approx x^4$, so

$$P(x) \to \infty \text{ as } x \to \infty \quad \text{and} \quad P(x) \to \infty \text{ as } x \to -\infty.$$

This information, together with the fact that $P(0) = -6$, implies that the graph of $y = P(x)$ is similar to that shown in Figure 2. ■

Notice that if the problem in Example 3 is changed slightly, to $P(x) = x^4 + 5x^3 + 6x^2 - 5x - 7$, we still have $P(1) = 0$ and $P(-1) = 0$, so $P(x)$ still has factors $x - 1$ and $x + 1$. But division by $x^2 - 1$ for this polynomial produces

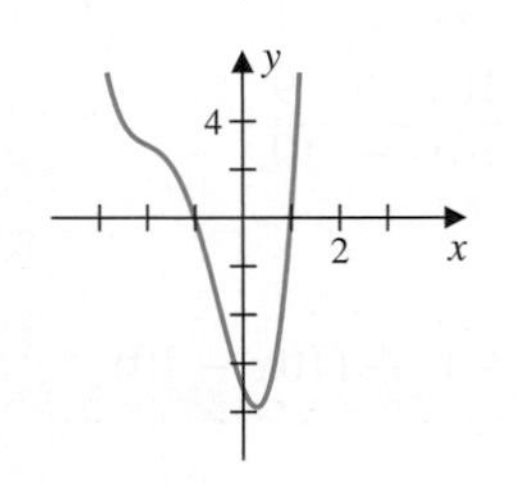

FIGURE 3

$$P(x) = (x - 1)(x + 1)(x^2 + 5x + 7).$$

Applying the Quadratic Formula to the quadratic equation $x^2 + 5x + 7 = 0$, we find that

$$x = \frac{-5}{2(1)} \pm \frac{\sqrt{5^2 - 4(1)(7)}}{2(1)} = -\frac{5}{2} \pm \frac{\sqrt{-3}}{2}.$$

Since $\sqrt{-3}$ is not a real number, the quadratic has no real zeros. The Factor Theorem implies that the quadratic has no real factors. Hence the factorization

$$P(x) = (x - 1)(x + 1)(x^2 + 5x + 7)$$

is complete. A computer-generated graph of $P(x) = x^4 + 5x^3 + 6x^2 - 5x - 7$ is shown in Figure 3. Notice that the graph crosses the x-axis at only two points, compared to the four x-intercepts for the graph in Example 3.

In Example 4 we consider the special case of finding a polynomial that has specific zeros.

EXAMPLE 4 Find a fourth-degree polynomial that has zeros at $x = 1, -1, \frac{1}{2}$, and 2, and has all integer coefficients.

Solution Since $1, -1, \frac{1}{2}$, and 2 are zeros of the polynomial, the Factor Theorem implies that the linear expressions $x-1, x+1, x-\frac{1}{2}$, and $x-2$ are factors of the polynomial. So a fourth-degree polynomial that has the given zeros is

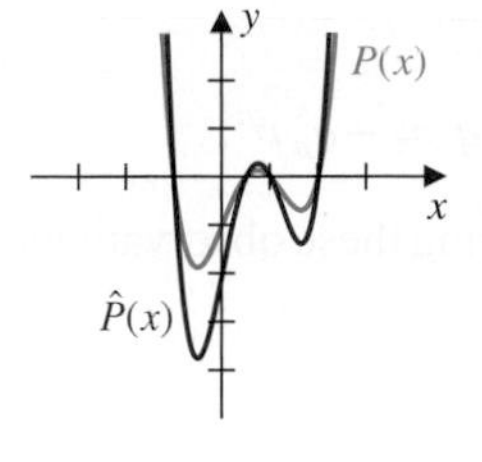

FIGURE 4

$$P(x) = (x-1)(x+1)(x-\tfrac{1}{2})(x-2) = x^4 - \tfrac{5}{2}x^3 + \tfrac{5}{2}x - 1.$$

We want a polynomial with *integer* coefficients having the same zeros, so we multiply $P(x)$ by 2 to produce the polynomial

$$\hat{P}(x) = 2x^4 - 5x^3 + 5x - 2.$$

The graphs of these two polynomials are shown in Figure 4. The graph of $y = \hat{P}(x)$ is a vertical stretching, by a factor of 2, of the graph of $y = P(x)$. ■

The Rational Zero Test

In our examples to this point we have been able to determine quite easily the factors and zeros of the polynomials. When the factors are not so readily determined, there are tests to help isolate the zeros. One test determines all the rational numbers that can be zeros when a polynomial has rational coefficients.

The Rational Zero Test applies only to polynomials whose coefficients are rational numbers, that is, numbers of the form p/q, where p and q are integers with $q \neq 0$. (So, for example, we cannot apply this test to $P(x) = \sqrt{2}x^2 - x + 1$ or to $P(x) = x^5 - \pi x^3 - x + 1$.) In fact, we need only consider polynomials with integer coefficients. This is because when we multiply a polynomial with rational coefficients by the common denominator of the coefficients, as was done in Example 4, the result is a polynomial having the same zeros and all integer coefficients.

For example, the polynomial

$$\tfrac{1}{2}x^4 - \tfrac{2}{3}x^2 - x + 2$$

has the same zeros as the polynomial with integer coefficients

$$6(\tfrac{1}{2}x^4 - \tfrac{2}{3}x^2 - x + 2) = 3x^4 - 4x^2 - 6x + 12.$$

As a consequence, we can concentrate on polynomials with only integer coefficients.

To develop the Rational Zero Test, let

$$P(x) = a_n x^n + a_{n-1}x^{n-1} + \cdots + a_1 x + a_0$$

be a polynomial, where $a_0, a_1, \ldots, a_n$ are all integers and $a_n \neq 0$. Suppose $x = p/q$ is a rational zero of $P(x)$, where all possible cancelation has been done so that p and q have no common factors.

Consider the possibilities for p and q. Because p/q is a zero of $P(x)$, we have

$$a_n \frac{p^n}{q^n} + a_{n-1}\frac{p^{n-1}}{q^{n-1}} + \cdots + a_1 \frac{p}{q} + a_0 = 0.$$

Multiplying both sides of this equation by q^n gives

$$a_n p^n + a_{n-1}p^{n-1}q + a_{n-2}p^{n-2}q^2 + \cdots + a_1 pq^{n-1} + a_0 q^n = 0$$

and

$$p\left(a_n p^{n-1} + a_{n-1}p^{n-2}q + a_{n-2}p^{n-3}q^2 + \cdots + a_1 q^{n-1}\right) = -a_0 q^n.$$

However, p divides the left side of this last equation, so it must also divide the right side. But p has no factors in common with q, so p cannot divide q^n. Hence p must divide the constant term a_0.

In a similar manner, we can rewrite the equation as

$$\left(a_{n-1}p^{n-1} + a_{n-2}p^{n-2}q + \cdots + a_1 pq^{n-2} + a_0 q^{n-1}\right) q = -a_n p^n,$$

and deduce that q must divide the leading coefficient a_n. Combining these observations gives the *Rational Zero Test.*

The Rational Zero Test

Suppose p/q is a rational zero of

$$P(x) = a_n x^n + a_{n-1}x^{n-1} + \cdots + a_1 x + a_0,$$

where $a_0, \ldots, a_n$ are integers and $a_n \neq 0$. Then p divides a_0 and q divides a_n. The Rational Zero Test provides a list of possible rational zeros.

The Rational Zero Test tells us which numbers *cannot* be rational zeros, because it describes which numbers are possible candidates for zeros. The Test does not tell us which of these possibilities are truly zeros. To determine this requires additional analysis, as illustrated in the next examples.

EXAMPLE 5 Find all the rational numbers that are possibilities for zeros of the polynomial $P(x) = x^3 - \frac{13}{4}x^2 + \frac{11}{4}x - \frac{1}{2}$.

Solution Finding the zeros of $P(x) = x^3 - \frac{13}{4}x^2 + \frac{11}{4}x - \frac{1}{2}$ is equivalent to finding the values of x for which

$$Q(x) = 4P(x) = 4x^3 - 13x^2 + 11x - 2 = 0.$$

When finding exact values for zeros of a polynomial, a graphing device can eliminate many possibilities.

The only possible rational zeros of $Q(x)$ are the factors of -2 divided by the factors of 4. That is,

$$\pm\frac{\text{divisors of 2}}{\text{divisors of 4}} = \pm\frac{1, 2}{1, 2, 4} = \pm\frac{1}{1}, \quad \pm\frac{1}{2}, \quad \pm\frac{1}{4}, \quad \pm\frac{2}{1}, \quad \pm\frac{2}{2}, \quad \pm\frac{2}{4}.$$

Simplifying these fractions and eliminating duplication gives the possibilities

$$\pm 1, \quad \pm\tfrac{1}{2}, \quad \pm\tfrac{1}{4}, \quad \pm 2.$$

Table 1 shows $Q(x) = 4x^3 - 13x^2 + 11x - 2$ evaluated at each of the possible rational zeros.

TABLE 1

x	1	-1	2	-2	$\frac{1}{4}$	$-\frac{1}{4}$	$\frac{1}{2}$	$-\frac{1}{2}$
$Q(x)$	0	-30	0	-108	0	$-\frac{45}{8}$	$\frac{3}{4}$	$-\frac{45}{4}$

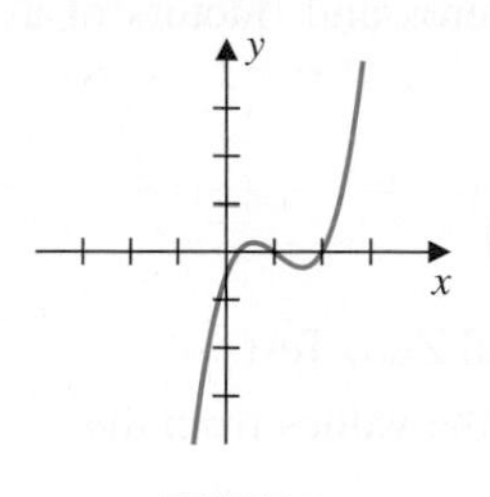

FIGURE 5

The polynomial $Q(x)$, and hence also $P(x)$, has zeros $x = 1, x = 2$, and $x = \frac{1}{4}$. The leading coefficient of $P(x)$ is 1, so it is completely factored as

$$P(x) = x^3 - \tfrac{13}{4}x^2 + \tfrac{11}{4}x - \tfrac{1}{2} = (x-1)(x-2)(x-\tfrac{1}{4}).$$

A graphing device can be used to eliminate many of the possible rational numbers that could be zeros. Figure 5 shows a graph of $P(x) = x^3 - \frac{13}{4}x^2 + \frac{11}{4}x - \frac{1}{2}$. It is clear from this graph, for example, that no negative numbers are zeros. ■

EXAMPLE 6 Find all the zeros of the polynomial $P(x) = x^4 - 3x^3 + x^2 + 3x - 2$ and factor the polynomial completely.

Solution The possible rational zeros of $P(x)$ are $\pm 1, \pm 2$, and

$$P(1) = 1 - 3 + 1 + 3 - 2 = 0.$$

So $x = 1$ is a zero and $x - 1$ is a factor of the polynomial. Dividing $P(x)$ by $x - 1$ gives

$$P(x) = (x-1)(x^3 - 2x^2 - x + 2).$$

Let $Q(x) = x^3 - 2x^2 - x + 2$. Then

$$P(x) = (x-1)Q(x)$$

and additional zeros of $Q(x)$ will also be zeros of $P(x)$. The possible rational zeros of $Q(x)$ are once again ± 1 and ± 2. The possibility that $x = 1$ cannot be ignored, because it might be a zero for $P(x)$ of multiplicity 2 or higher. In fact,

$$Q(1) = 1 - 2 - 1 + 2 = 0,$$

so $x = 1$ is a zero of $Q(x)$, and hence is a zero of multiplicity at least 2 of $P(x)$.

Dividing $Q(x)$ by $x - 1$ gives

$$P(x) = (x-1)Q(x) = (x-1)(x-1)(x^2 - x - 2) = (x-1)^2(x^2 - x - 2).$$

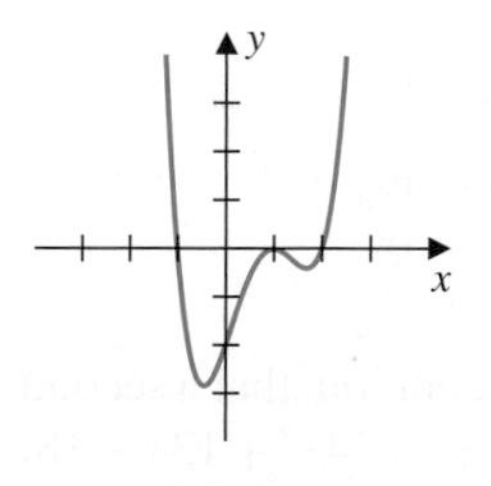

FIGURE 6

The quadratic factors as $(x-2)(x+1)$, so the zeros of the polynomial $P(x)$ are $x = 1$ of multiplicity 2, $x = 2$, $x = -1$, and

$$P(x) = (x-1)^2(x-2)(x+1).$$

Because there is a double root at $x = 1$, the graph just touches the x-axis at $(1, 0)$, as shown in Figure 6. ■

In Example 6 the polynomial $P(x)$ was written as $P(x) = (x-1)Q(x)$ so that to find additional zeros of $P(x)$ it is enough to find zeros of $Q(x)$. We cannot ignore any possible rational zeros of $Q(x)$ that were already zeros of $P(x)$, because a zero can have multiplicity greater than 1. However, once a possible rational zero for $P(x)$ is eliminated it does not have to be considered for $Q(x)$.

The following list summarizes a procedure for finding zeros and factors of a polynomial.

Principles of Problem Solving:

- Create a summary list when solving a complicated problem. It ensures that you fully understand the problem.

Finding Zeros and Factors of a Polynomial $P(x)$

- List all possible rational zeros of $P(x)$ using the Rational Zero Test.
- Check the candidates. It is generally easier to substitute the values from the smallest in magnitude to the largest.
- When a zero is found, factor $P(x)$ and repeat the process on the quotient. There is no need to check possible zeros of the quotient that have already been eliminated from the list of zeros. But check any zeros that have been found, because they may have a multiplicity greater than 1.
- Once $P(x)$ has been factored to linear terms and quadratic terms, use the Quadratic Formula, if necessary, to factor the quadratics.

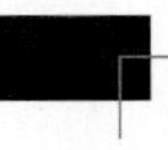

Applications

One of the many applications of calculus involves finding the area between regions in the plane. The regions lie between the graphs of functions, and the intersections of the graphs determine the horizontal boundaries of the regions. The boundaries are needed before calculus can be applied to determine the area. Example 7 illustrates this when the region is bounded by the graphs of polynomials.

EXAMPLE 7 A region is bounded between the graphs of $f(x) = x^4 + 2x^3 - 15x^2 + 9x + 12$ and $g(x) = -x^4 + x^3 + 19x^2 - 4x - 6$. Determine the values of x that give the horizontal boundaries of the region.

Solution We need to determine the x-values when the graphs intersect, that is, those values of x with

$$f(x) = x^4 + 2x^3 - 15x^2 + 9x + 12 = -x^4 + x^3 + 19x^2 - 4x - 6 = g(x).$$

Hence, we need all the solutions to the polynomial

$$0 = f(x) - g(x) = 2x^4 + x^3 - 34x^2 + 13x + 18.$$

The Rational Zero Test implies that the rational zeros can occur only at

$$\pm 1,\ \pm 2,\ \pm 3,\ \pm 6,\ \pm 9,\ \pm 18,\ \pm\frac{1}{2},\ \pm\frac{3}{2} \text{ and } \pm\frac{9}{2}.$$

It is easily verified that $x = 1$ is a zero, and, after some experimentation, that a second zero is $x = -9/2$. So both $x - 1$ and $2x + 9$ are factors of $2x^4 + x^3 - 34x^2 + 13x + 18$. Dividing this fourth-degree polynomial by $(x - 1)(2x + 9) = 2x^2 + 7x + 18$ gives

$$2x^4 + x^3 - 34x^2 + 13x + 18 = (2x^2 + 7x + 18)(x^2 - 3x - 2) = (x - 1)(2x + 9)(x^2 - 3x - 2).$$

We can use the Quadratic Formula on the remaining quadratic term to show that its zeros are

$$x = \frac{3}{2} + \frac{\sqrt{17}}{2} \approx 3.56 \quad \text{and} \quad x = \frac{3}{2} - \frac{\sqrt{17}}{2} \approx -0.56.$$

The intersection points of the graphs of $f(x)$ and $g(x)$ consequently occur when

$$x = -\frac{9}{2}, \quad x = \frac{1}{2} - \frac{\sqrt{17}}{2}, \quad x = 1, \quad \text{and} \quad x = \frac{1}{2} + \frac{\sqrt{17}}{2}.$$

Evaluating $f(x) - g(x) = 2x^4 + x^3 - 34x^2 + 13x + 18$ at values of x between the intersection points tells us when $f(x) > g(x)$ and when $g(x) > f(x)$.

- For values of x that are large in magnitude and negative, we have $f(x) > g(x)$, so $f(x) > g(x)$ on $\left(-\infty, -\frac{9}{2}\right)$.
- For $x = -1$, we have $f(-1) - g(-1) < 0$, so $g(x) > f(x)$ on $\left(-\frac{9}{2}, \frac{1}{2} - \frac{\sqrt{17}}{2}\right)$.
- For $x = 0$, we have $f(0) - g(0) > 0$, so $f(x) > g(x)$ on $\left(\frac{1}{2} - \frac{\sqrt{17}}{2}, 1\right)$.
- For $x = 2$, we have $f(2) - g(2) < 0$, so $g(x) > f(x)$ on $\left(1, \frac{1}{2} + \frac{\sqrt{17}}{2}\right)$.
- For values of x that are large in magnitude and positive, we have $f(x) > g(x)$, so $f(x) > g(x)$ on $\left(\frac{1}{2} + \frac{\sqrt{17}}{2}, \infty\right)$.

The graphs of the functions shown in Figure 7 indicate that this is indeed the case. ■

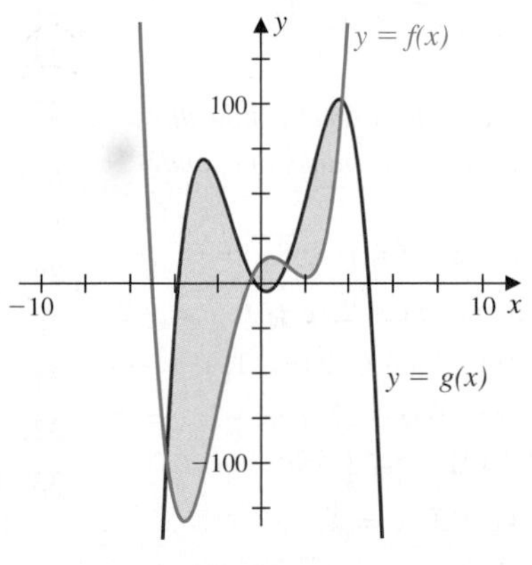

FIGURE 7

Polynomials are basic functions that are useful for many applications, and it is common to use a polynomial to approximate another function within a small interval. For example, the polynomial

$$P(x) = 1 + \tfrac{1}{2}(x-1) - \tfrac{1}{8}(x-1)^2 + \tfrac{1}{16}(x-1)^3$$

provides good approximations to the values of $f(x) = \sqrt{x}$, provided that we are only interested in the values of $\sqrt{x}$ near $x = 1$, as shown in Figure 8.

The polynomial $P(x)$ shown in Figure 8 is a *Taylor polynomial* for the function f at $x = 1$. In calculus, Taylor polynomials are used to approximate various functions.

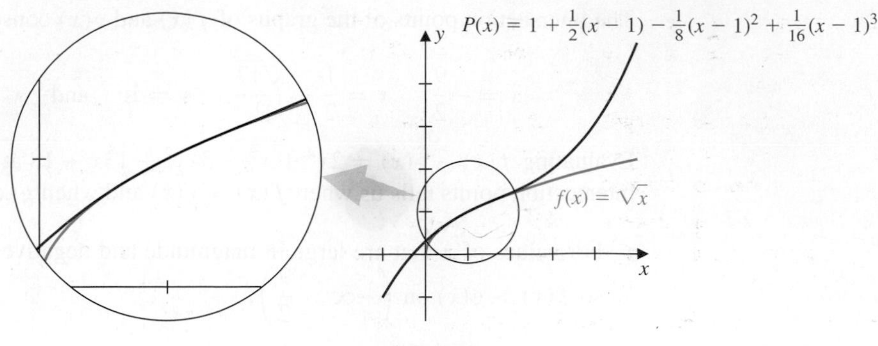

FIGURE 8

EXERCISE SET 3.3

In Exercises 1–6, find the quotient $Q(x)$ and remainder $R(x)$ when the polynomial $P(x)$ is divided by the polynomial $D(x)$.

1. $P(x) = 2x^2 + 3x - 3,\ D(x) = x - 2$
2. $P(x) = 2x^2 - 3x + 4,\ D(x) = x + 2$
3. $P(x) = x^3 + x^2 - 2,\ D(x) = x - 1$
4. $P(x) = 3x^4 + 2x^3 - x + 2,\ D(x) = x^2 + 2x - 1$
5. $P(x) = 2x^4 - 2x^3 + 4x^2 + x - 2,\ D(x) = x^2 - x - 1$
6. $P(x) = 2x^5 + 3x^4 - 2x^3 + x^2 + x - 1,$ $D(x) = x^4 + x^3 - 2x - 1$

In Exercises 7–12, use the Factor Theorem to show that $x - c$ is a factor of $P(x)$ for the given values of c, and factor $P(x)$ completely.

7. $P(x) = x^3 - 5x^2 + 8x - 4,\ c = 1$
8. $P(x) = 3x^4 + 5x^3 - 5x^2 - 5x + 2,\ c = 1,\ c = -2$
9. $P(x) = 2x^4 + 3x^3 - 12x^2 - 7x + 6,\ c = -1,\ c = -3$
10. $P(x) = 4x^4 - 7x^3 - 33x^2 + 63x - 27,\ c = -3,\ c = 3$
11. $P(x) = 3x^3 + x^2 - 8x + 4,\ c = \frac{2}{3}$
12. $P(x) = 2x^3 - 5x^2 - 4x + 3,\ c = \frac{1}{2}$

In Exercises 13–18, determine all the possibilities for rational zeros.

13. $x^4 + 2x^3 - x^2 + 3x + 4 = 0$
14. $x^5 + 2x^4 - 8x^3 + 4x^2 + 12x - 16 = 0$
15. $10x^5 - 14x^3 + 18x^2 + 6x - 4 = 0$
16. $6x^5 - 24x^4 + 40x^3 - 48x^2 + 24x - 4 = 0$
17. $\frac{3}{2}x^3 - \frac{5}{2}x^2 + 7x - 4 = 0$
18. $\frac{3}{4}x^3 + \frac{1}{8}x^2 - \frac{9}{4}x + 1 = 0$

In Exercises 19–28, find all rational zeros of the polynomial. Then determine any irrational zeros, and factor the polynomial completely.

19. $x^3 - 3x^2 + 4$
20. $3x^3 - 8x^2 - 5x + 6$
21. $2x^4 - 3x^3 - 3x^2 + 2x$
22. $3x^4 - 11x^3 + 5x^2 + 3x$
23. $2x^5 + x^4 - 12x^3 + 10x^2 + 2x - 3$
24. $x^4 - 4x^3 + 3x^2 + 4x - 4$
25. $x^3 + 2x^2 - 4x + 1$
26. $x^4 - x^3 - 6x^2 + 8x$
27. $4x^3 - 12x^2 + 9x - 1$
28. $9x^3 - 3x^2 - 7x + 1$

In Exercises 29–32, sketch the graphs of $y = f(x)$ and $y = g(x)$, and determine all the points of intersection.

29. $f(x) = x^3;\ g(x) = 2x^2 + x - 2$
30. $f(x) = x^3 + 3;\ g(x) = 2x^2 + 5x - 3$
31. $f(x) = x^3 - 1;\ g(x) = 6x + 3$
32. $f(x) = x^3 - x^2;\ g(x) = 3x - 2$
33. Determine the polynomial $P(x)$ of least degree whose graph is shown in the figure.

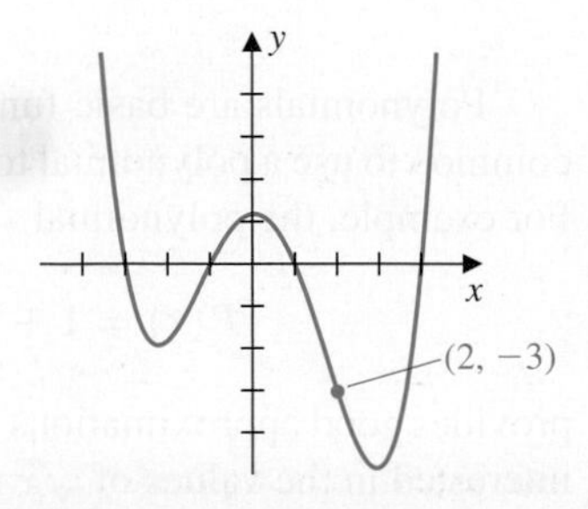

34. Find a third-degree polynomial $P(x)$ that has zeros at $x = -1$, $x = 1$, and $x = 2$, and whose x-term has coefficient 3.

35. Find a fourth-degree polynomial $P(x)$ that has zeros at $x = 0$, $x = 1$, $x = -1$, and $x = -2$, and whose x^2-term has coefficient 5.

36. Find a fifth-degree polynomial that has a zero of multiplicity 2 at $x = 1$, a zero at $x = 4$, and the factor $x^2 + x + 1$.

37. Find a sixth-degree polynomial that has a zero of multiplicity 3 at $x = 2$, a zero of multiplicity 2 at $x = 0$, and a zero at $x = 1$.

38. **a.** The polynomial $P(x) = x^4 - 4x^3 + 4x^2 - 1$ has a local maximum at $(1, 0)$ and local minima at $(0, -1)$ and $(2, -1)$. Factor the polynomial completely and sketch its graph.

b. Determine how many real zeros the polynomial $Q(x) = P(x) + c$ has for each constant c.

c. Show that $x - 1$ is a factor of $x^n - 1$ for all positive integers n.

d. Show that $x + 1$ is a factor of $x^n + 1$ for all odd, but no even, positive integers n.

39. A region is bounded between the graphs of the polynomials $f(x) = x^3 - 2x$ and $g(x) = x^3 - x^2$.

a. Determine the horizontal boundaries of the region.

b. Determine any intervals within the boundaries when $f(x) \geq g(x)$.

40. A region is bounded between the graphs of the polynomials $f(x) = x^3 - 3x^2 - x + 3$ and $g(x) = x^2 - 1$.

a. Determine the horizontal boundaries of the region.

b. Determine any intervals within the boundaries when $f(x) \geq g(x)$.

41. A region is bounded between the graphs of the polynomials $f(x) = x^5 - 3x^4 + 27x^2 - 32x + 10$ and $g(x) = 5x^3 - 2$.

a. Determine the horizontal boundaries of the region.

b. Determine any intervals within the boundaries when $f(x) \geq g(x)$.

3.4 RATIONAL FUNCTIONS

A rational number is the quotient of two integers, which are the most basic numbers. In a similar manner, a *rational function* is the quotient of polynomials, which are the most basic functions.

Rational Functions

A **rational function** has the form

$$f(x) = \frac{P(x)}{Q(x)},$$

where $P(x)$ and $Q(x)$ are polynomials. The domain of f is the set of all real numbers x with $Q(x) \neq 0$.

The functions described by

$$f(x) = \frac{x+2}{x^2-1}, \quad g(x) = \frac{1}{x}, \quad \text{and} \quad h(x) = \frac{x^2 - 2x + 1}{x^4 + 2x^2 + 5}$$

are all rational. The domain of f is the set of all real numbers except $x = 1$ and $x = -1$. The domain of g is the set of all nonzero real numbers. The domain of h is the set of all real numbers, $\mathbb{R}$, because the denominator is never zero.

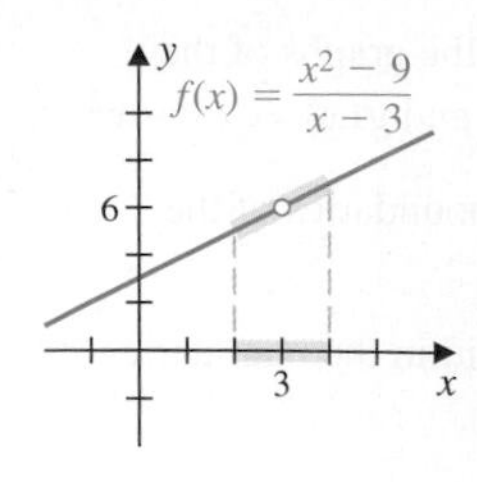

FIGURE 1

If the numerator and denominator of a rational function have common factors, we can cancel these factors and write the fraction in a simpler form. For example,

$$f(x) = \frac{x^2 - 9}{x - 3} = \frac{(x - 3)(x + 3)}{x - 3} = x + 3 \quad \text{when} \quad x \neq 3.$$

Notice that $x = 3$ is excluded from the domain of the function f, since f is undefined at $x = 3$ in the original form. The graph of f is shown in Figure 1. The open circle at the point $(3, 6)$ indicates that $x = 3$ is not in the domain of f, but that $f(x)$ approaches 6 as x approaches 3. That is,

$$f(x) \to 6 \quad \text{as} \quad x \to 3.$$

The open circle will not likely be displayed when using a graphing device.

EXAMPLE 1 Find the domain and axis intercepts of the rational function f given by

$$f(x) = \frac{x^2 + x - 2}{x - 2}.$$

Solution The denominator of $f(x)$ is zero precisely when $x = 2$, so the domain of f consists of all real numbers except $x = 2$.

The y-intercept of the graph occurs at $(0, f(0)) = (0, 1)$. To find the x-intercepts we need to determine when the numerator of $f(x)$ is 0. Factoring gives

$$\frac{x^2 + x - 2}{x - 2} = \frac{(x + 2)(x - 1)}{x - 2},$$

and $f(x) = 0$ precisely when $x = -2$ or $x = 1$. The graph crosses the y- and x-axes at the points $(0, 1)$, $(-2, 0)$, and $(1, 0)$.

TABLE 1

x	$f(x)$	x	$f(x)$
1.9	−35.1	2.1	45.1
1.99	−395.01	2.01	405.01
1.999	−3995.001	2.001	4005.001
1.9999	−39995.0001	2.0001	40005.0001

Table 1 shows $f(x)$ for values of x that are close but unequal to $x = 2$. As x gets close to 2 from the left side (as x gets close to 2 but remains less than 2), the values of the function become negative and large in magnitude. As x gets close to 2 from the right side (as x gets close to 2 but remains greater than 2), the values of the function become positive and large. This suggests that the graph is as shown in Figure 2. ■

An **asymptote** of a graph is a line that the graph approaches. The vertical line $x = a$ is a *vertical asymptote* of the graph of f when $f(x)$ approaches either ∞ or $-\infty$ as x approaches a from either the left side or from the right side.

Asymptotes are closely connected to the concept of *limit*, a basic notion in calculus.

For example, Figure 2 shows the dashed vertical line at $x = 2$ as a vertical asymptote of the graph in Example 1. In Figure 3 we see the various possibilities that

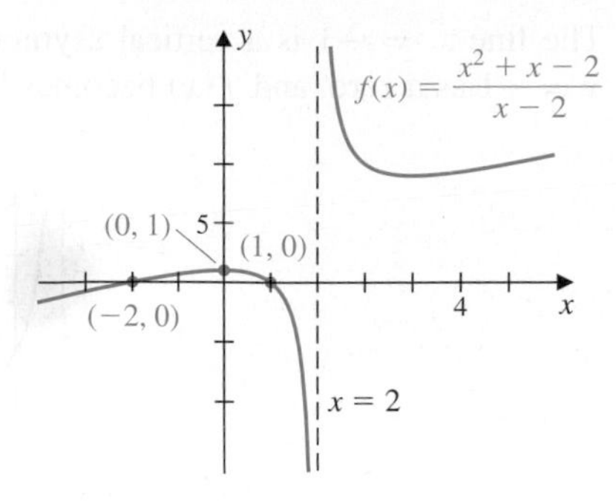

FIGURE 2

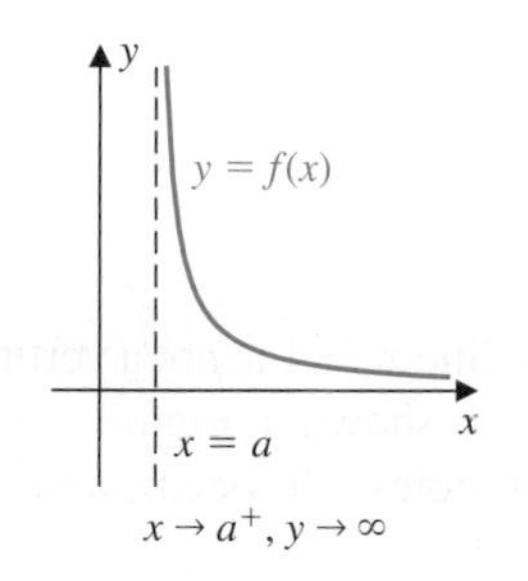

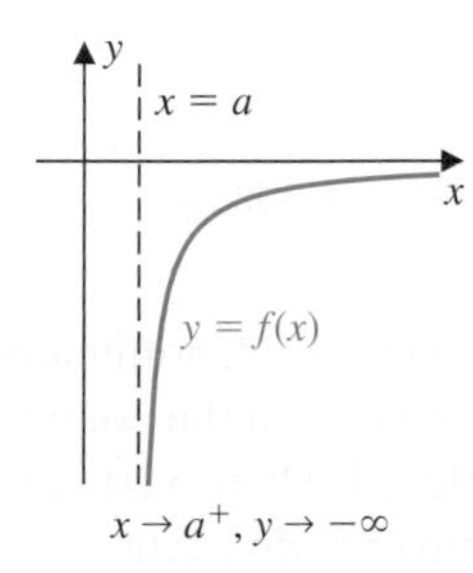

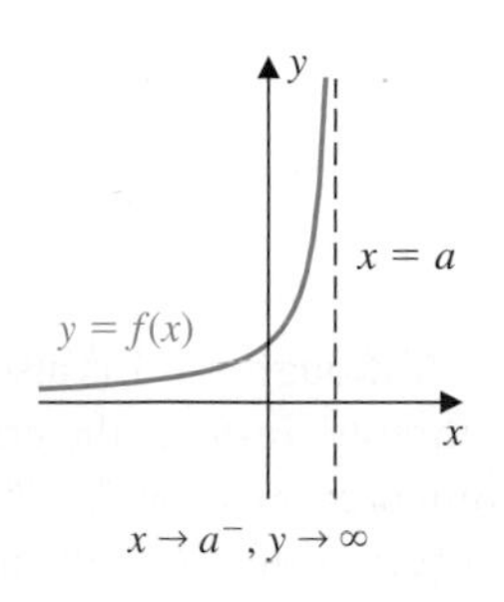

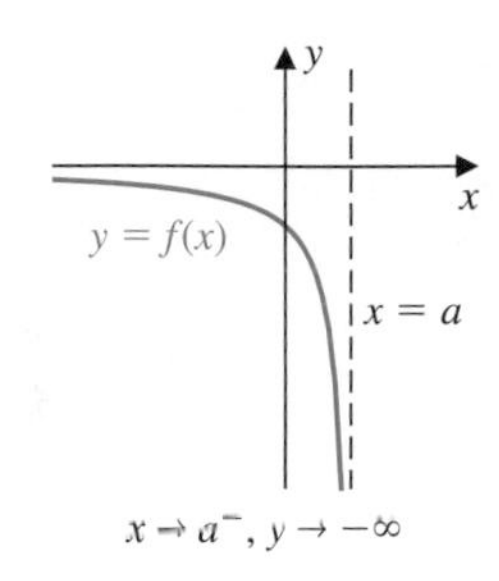

FIGURE 3

can produce vertical asymptotes, together with some notation that is used to describe the situations.

When used properly, graphing devices will display the behavior of a function near a vertical asymptote.

Vertical Asymptotes

The vertical line $x = a$ is a **vertical asymptote** for the graph of f if

$$|f(x)| \to \infty \quad \text{as} \quad x \to a^-, \quad \text{or} \quad |f(x)| \to \infty \quad \text{as} \quad x \to a^+.$$

- The symbol a^+ means that x *approaches a from the right*, that is, $x > a$.
- The symbol a^- means that x *approaches a from the left*, that is, $x < a$.

A vertical asymptote of a rational function can occur at $x = a$ only if $x - a$ is a factor of its denominator. However, $x - a$ can be a factor of the denominator without $x = a$ being a vertical asymptote. This can occur when it is also a factor of the numerator, as we saw in Figure 1.

As another example, consider the rational function defined by

$$f(x) = \frac{x^2 + 2x - 3}{x^2 - 1}.$$

Its domain is the set of all real numbers except $x = -1$ and $x = 1$, the zeros of the denominator. But factoring the numerator and denominator gives

$$f(x) = \frac{x^2 + 2x - 3}{x^2 - 1} = \frac{(x - 1)(x + 3)}{(x - 1)(x + 1)} = \frac{x + 3}{x + 1} \quad \text{when } x \neq 1.$$

The line $x = -1$ is a vertical asymptote, because the simplified denominator has $x = -1$ as a zero, and $f(x)$ becomes large in magnitude as x approaches -1.

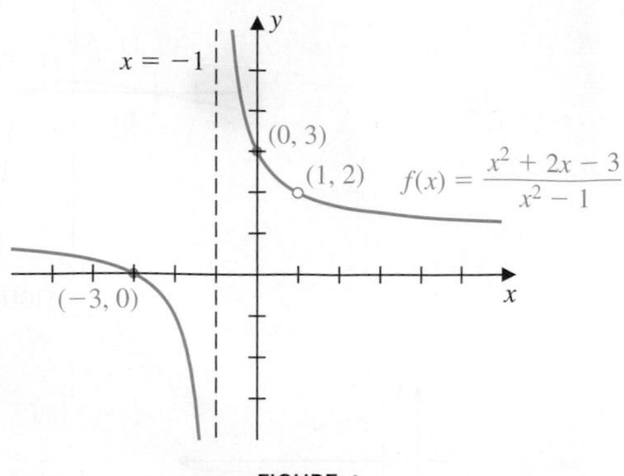

FIGURE 4

Although $x - 1$ is also a factor of the denominator, the line $x = 1$ is not a vertical asymptote. Instead, the graph has a hole at the point $(1, 2)$, as shown in Figure 4. The original expression for $f(x)$ tells us where vertical asymptotes *can* occur, and the reduced expression tells us where they *do* occur.

A rational function can change sign only when the function is zero or when passing over points that are not in the domain. Hence the sign of $f(x)$ remains the same in intervals that are bracketed by these values. By determining the sign of $f(x)$ at a single value in each of these intervals, we determine the sign of $f(x)$ on its entire domain. This is illustrated in Example 2.

EXAMPLE 2 Sketch the graph of

$$f(x) = \frac{x^2 - 6x + 8}{x^3 - 4x} \qquad \text{for } x \text{ in } [-4, 6].$$

Solution First we factor the numerator and the denominator to obtain

$$f(x) = \frac{(x-2)(x-4)}{x(x-2)(x+2)} = \frac{x-4}{x(x+2)}, \qquad \text{if } x \neq 2.$$

The domain of f excludes the numbers that make the original denominator zero, namely, $x = 0$, $x = 2$, and $x = -2$. Vertical asymptotes occur at $x = 0$ and at $x = -2$. Because the simplified form of the function does not contain the factor $x - 2$, a hole appears in the graph when $x = 2$.

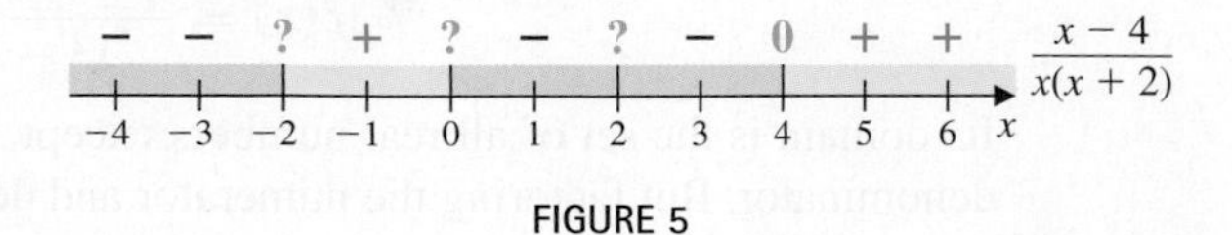

FIGURE 5

To determine how the graph approaches the vertical asymptotes $x = 0$ and $x = -2$, we need to know if the function is positive or negative near these values.

The chart in Figure 5 shows the situation for our function. It implies that

$$\begin{array}{rlll} \text{For } -4 < x < -2: & f(x) < 0 \text{ so } f(x) \to -\infty & \text{as} & x \to -2^-; \\ \text{For } -2 < x < -1: & f(x) > 0 \text{ so } f(x) \to \infty & \text{as} & x \to -2^+; \\ \text{For } -1 < x < 0: & f(x) > 0 \text{ so } f(x) \to \infty & \text{as} & x \to 0^-; \\ \text{For } 0 < x < 1: & f(x) < 0 \text{ so } f(x) \to \infty & \text{as} & x \to 0^+. \end{array}$$

The y-coordinate of the hole is the value of the reduced fraction at $x = 2$, that is,

$$y = \frac{2-4}{2(2+2)} = -\frac{1}{4}.$$

We cannot analyze the graph of the rational function completely without calculus, but the graph in the figure is reasonable.

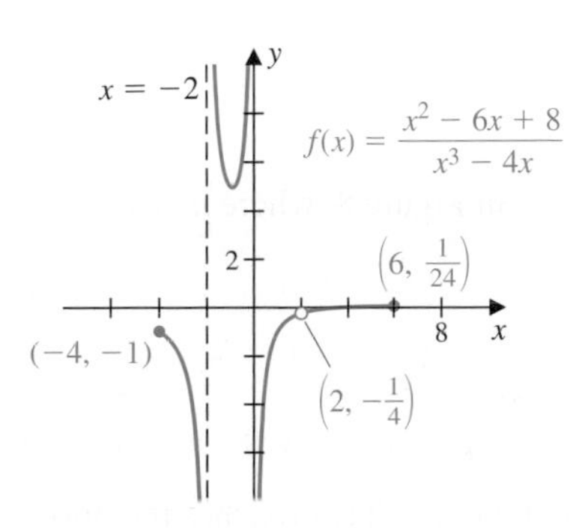

FIGURE 6

Using this information and evaluating $f(x)$ at the endpoints -4 and 6 verifies the features of the graph in Figure 6. ■

In Section 3.2 we discussed the end behavior of polynomials and showed that the degree and leading coefficient of the polynomial (the term with the highest exponent) determines the behavior of the polynomial as $x \to \infty$ and as $x \to -\infty$. In a similar manner, the leading terms of the polynomials in the numerator and denominator of a rational function determine the behavior of the rational function as $x \to \infty$ and as $x \to -\infty$.

Suppose that the rational function f has the form

$$f(x) = \frac{P(x)}{Q(x)} = \frac{a_n x^n + a_{n-1}x^{n-1} + \cdots + a_1 x + a_0}{b_m x^m + b_{m-1}x^{m-1} + \cdots + b_1 x + b_0},$$

where a_n and b_m are nonzero. Then

$$f(x) \approx \frac{a_n x^n}{b_m x^m} = \frac{a_n}{b_m}x^{n-m} \text{ as } x \to \infty \quad \text{and} \quad f(x) \approx \frac{a_n}{b_m}x^{n-m} \text{ as } x \to -\infty.$$

When $n = m$ the graph approaches the horizontal line $y = a_n/b_m$, which is a *horizontal asymptote* to the graph, as shown on the graph in Figure 7. For x large in magnitude

$$f(x) = \frac{4x+1}{2x-1} \approx \frac{4x}{2x} = 2,$$

so

$$f(x) \to 2 \quad \text{as} \quad x \to \infty \quad \text{and} \quad f(x) \to 2 \quad \text{as} \quad x \to -\infty.$$

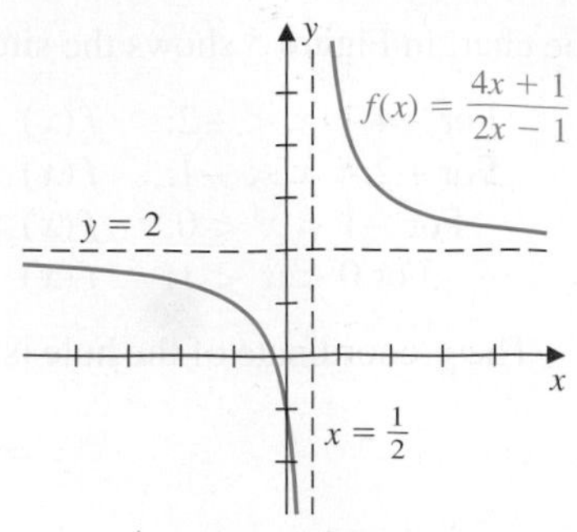

As $x \to \infty$, $f(x) \to 2$ from above.
As $x \to -\infty$, $f(x) \to 2$ from below.

FIGURE 7

In Figure 8, where $n < m$, when x is large in magnitude we have

$$f(x) = -\frac{8}{x^2+4} \approx -\frac{8}{x^2},$$

so

$$f(x) \to 0 \quad \text{as} \quad x \to \infty \quad \text{and} \quad f(x) \to 0 \quad \text{as} \quad x \to -\infty$$

and the graph approaches the horizontal asymptote $y = 0$.

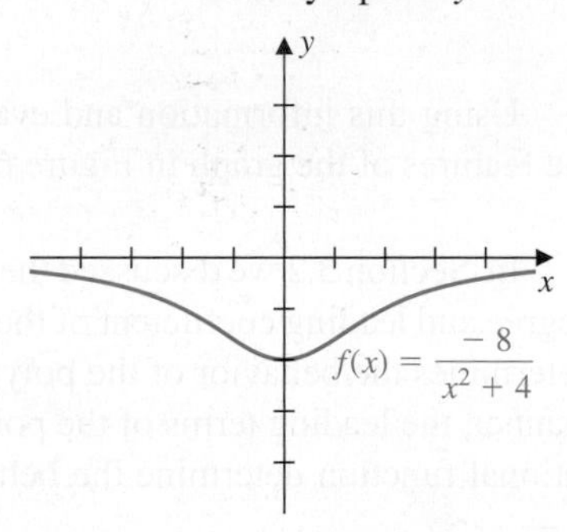

As $x \to \infty$, $f(x) \to 0$ from below.
As $x \to -\infty$, $f(x) \to 0$ from below.

FIGURE 8

Figure 9 illustrates some of the possibilities that can occur in the case $n > m$, when the graph has no horizontal asymptote.

The Greek word *horizein* means "to divide or separate." The horizon separates earth from sky.

Horizontal Asymptotes

The horizontal line $y = a$ is a **horizontal asymptote** to the graph of f if

$$f(x) \to a \text{ as } x \to \infty \quad \text{or} \quad f(x) \to a \text{ as } x \to -\infty.$$

A rational function can have at most one horizontal asymptote. If $f(x)$ approaches a finite value as x approaches ∞, it approaches that same value as x approaches $-\infty$.

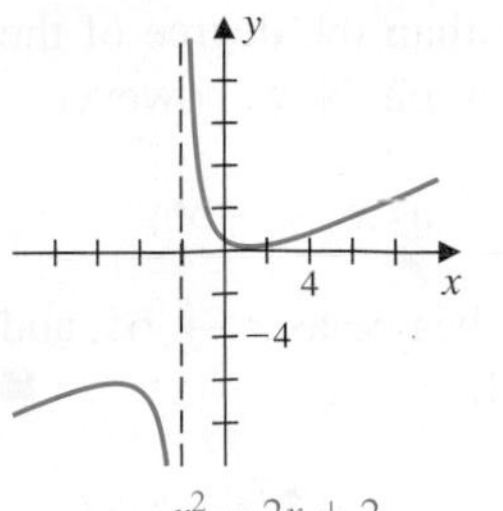

$f(x) = \dfrac{x^2 - 2x + 2}{2x + 4}$

As $x \to \infty, f(x) \to \infty$.
As $x \to -\infty, f(x) \to -\infty$.

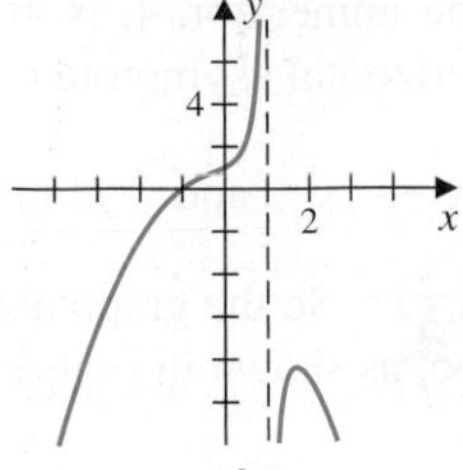

$f(x) = \dfrac{x^3 + 1}{1 - x}$

As $x \to \infty, f(x) \to -\infty$.
As $x \to -\infty, f(x) \to -\infty$.

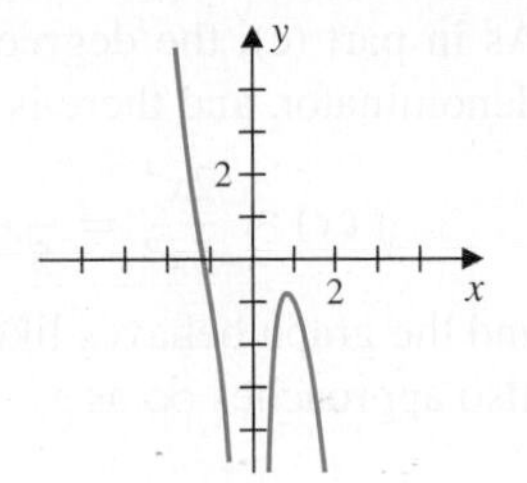

$f(x) = \dfrac{-x^5 + x - 1}{x^2}$

As $x \to \infty, f(x) \to -\infty$.
As $x \to -\infty, f(x) \to \infty$.

$f(x) = \dfrac{(x-1)^3(x-4)^2(x+1)}{5x^2 + 5}$

As $x \to \infty, f(x) \to \infty$.
As $x \to -\infty, f(x) \to \infty$.

FIGURE 9

In the next section we will see examples of nonrational functions that have two distinct horizontal asymptotes, one occurring as $x \to \infty$ and the other as $x \to -\infty$.

EXAMPLE 3 Determine if the graph of the rational function has a horizontal asymptote.

a. $f(x) = \dfrac{3x^4 - 2x + 1}{-7x^4 + 5x^3 - 2x^2}$ **b.** $f(x) = \dfrac{x^3 - 2x + 2}{x^4 - x^3 + 5x^2 - 3}$

c. $f(x) = \dfrac{2x^4 - 3x^3 + 4x - 1}{5x^3 + 7x + 2}$ **d.** $f(x) = \dfrac{2x^4 - 3x^3 + 4x - 1}{5x^2 + 7x + 2}$

Solution **a.** The degree, 4, of the polynomial in the numerator agrees with the degree of the polynomial in the denominator, so

$$f(x) \approx \frac{3x^4}{-7x^4} = -\frac{3}{7} \text{ as } x \to \infty \quad \text{and} \quad f(x) \approx -\frac{3}{7} \text{ as } x \to -\infty.$$

The graph, shown in Figure 10(a), has the horizontal asymptote $y = -\frac{3}{7}$.

b. The degree of the polynomial in the numerator, 3, is less than the degree, 4, of the polynomial in the denominator, so

$$f(x) \approx \frac{x^3}{x^4} = \frac{1}{x} \to 0 \text{ as } x \to \infty \quad \text{and} \quad f(x) \approx \frac{1}{x} \to 0 \text{ as } x \to -\infty.$$

As a consequence, $y = 0$ is a horizontal asymptote to the graph, as shown in Figure 10(b). Notice that the graph crosses this asymptote.

c. The degree of the numerator, 4, is greater than the degree, 3, of the denominator, so there is no horizontal asymptote to the graph. Also, when $|x|$ is large,

$$f(x) \approx \frac{2x^4}{5x^3} = \frac{2}{5}x.$$

So the graph, when $|x|$ is large, is similar to the graph of $y = \frac{2}{5}x$, and

$$f(x) \to \infty \text{ as } x \to \infty \quad \text{and} \quad f(x) \to -\infty \text{ as } x \to -\infty.$$

The graph approaches a line parallel to $y = \frac{2}{5}x$, as shown in Figure 10(c).

d. As in part (c), the degree of the numerator, 4, is greater than the degree of the denominator, and there is no horizontal asymptote to the graph. Now, however,

$$f(x) \approx \frac{2x^4}{5x^2} = \frac{2}{5}x^2 \text{ as } x \to \infty \quad \text{and} \quad f(x) \approx \frac{2}{5}x^2 \text{ as } x \to -\infty,$$

and the graph behaves like $y = \frac{2}{5}x^2$. So the graph approaches ∞ as $x \to \infty$, and also approaches ∞ as $x \to -\infty$, as shown in Figure 10(d). ■

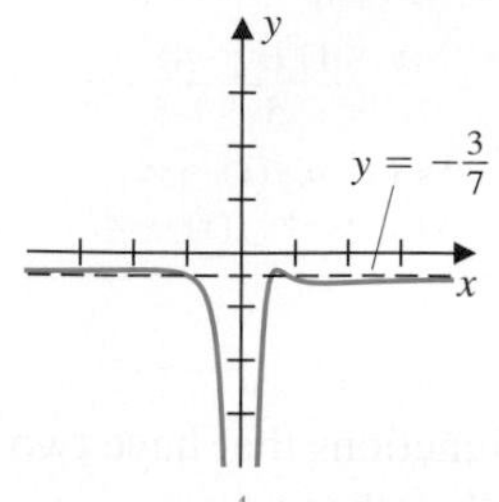

$f(x) = \dfrac{3x^4 - 2x + 1}{-7x^4 + 5x^3 - 2x^2}$

(a)

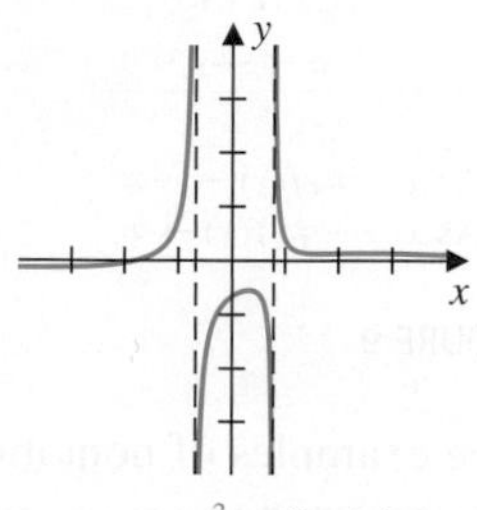

$f(x) = \dfrac{x^3 - 2x + 2}{x^4 - x^3 + 5x^2 - 3}$

(b)

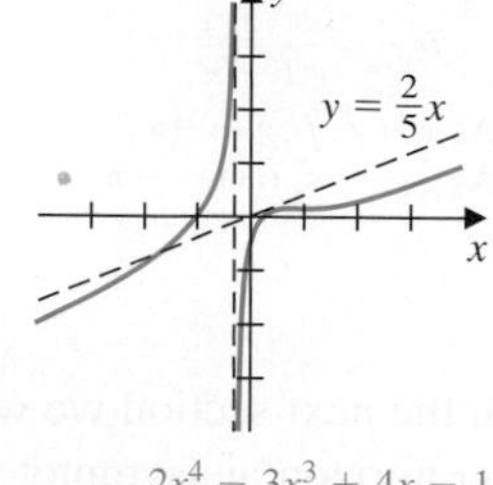

$f(x) = \dfrac{2x^4 - 3x^3 + 4x - 1}{5x^3 + 7x + 2}$

(c)

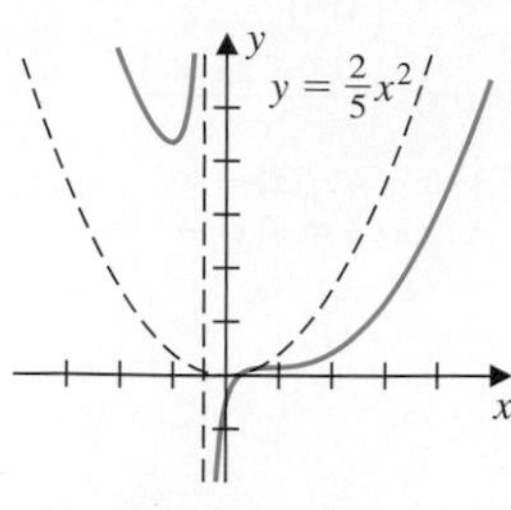

$f(x) = \dfrac{2x^4 - 3x^3 + 4x - 1}{5x^2 + 7x + 2}$

(d)

FIGURE 10

EXAMPLE 4 Sketch the graph of $f(x) = \dfrac{x^2 - 1}{x^2 - 9}$.

Solution The numerator and denominator both have degree 2 and leading coefficient 1, so

$$f(x) \approx \frac{x^2}{x^2} = 1 \text{ as } x \to \infty \quad \text{and} \quad f(x) \approx \frac{x^2}{x^2} = 1 \text{ as } x \to -\infty.$$

Hence $y = 1$ is a horizontal asymptote to the graph.

The y-intercept for the graph occurs when $x = 0$, which gives

$$y = f(0) = \frac{0 - 1}{0 - 9} = \frac{-1}{-9} = \frac{1}{9}.$$

To determine any x-intercepts and vertical asymptotes, we factor $f(x)$ as

$$f(x) = \frac{(x + 1)(x - 1)}{(x + 3)(x - 3)}.$$

The fraction has no common factors, so the graph has x-intercepts at $x = -1$ and $x = 1$, and vertical asymptotes $x = -3$ and $x = 3$. The sign chart given in Figure 11 describes how the graph approaches the vertical asymptotes. For example, to determine the sign of $f(x)$ on the interval $(-1, 1)$ choose any number inside the interval and substitute the value in the function. We have already found that $f(0) = \frac{1}{9} > 0$, which implies that $f(x)$ is positive on the entire interval $(-1, 1)$.

+ + ? − 0 + 0 − ? + + $\dfrac{(x + 1)(x - 1)}{(x + 3)(x - 3)}$

−5 −4 −3 −2 −1 0 1 2 3 4 5 x

FIGURE 11

The graph is symmetric with respect to the y-axis because

$$f(-x) = \frac{(-x)^2 - 1}{(-x)^2 - 9} = \frac{x^2 - 1}{x^2 - 9} = f(x).$$

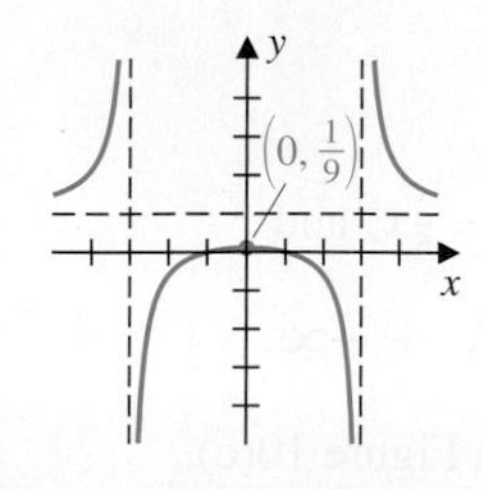

FIGURE 12

So the graph likely appears as shown in Figure 12. ■

EXAMPLE 5 Sketch the graph of the function defined by $f(x) = \dfrac{x^2 + 2x - 3}{2x^2 - 5x - 3}$.

Solution The degrees of the numerator and denominator are the same, so

$$f(x) \to \tfrac{1}{2} \text{ as } x \to \infty \quad \text{and} \quad f(x) \to \tfrac{1}{2} \text{ as } x \to -\infty,$$

and $y = \frac{1}{2}$ is a horizontal asymptote.

The y-intercept to the graph of f is $(0, f(0)) = (0, 1)$. To find the x-intercepts and vertical asymptotes, we first factor the numerator and denominator of $f(x)$ as

$$f(x) = \frac{x^2 + 2x - 3}{2x^2 - 5x - 3} = \frac{(x+3)(x-1)}{(x-3)(2x+1)}.$$

From this factored form we see that the x-intercepts are $(1, 0)$ and $(-3, 0)$, and that the graph has vertical asymptotes $x = 3$ and $x = -\frac{1}{2}$. The chart in Figure 13 shows the sign of $f(x)$ on the regions of the domain, lying between the zeros of the numerator and denominator. It tells how the graph approaches the vertical asymptotes, and gives us the sketch shown in Figure 14.

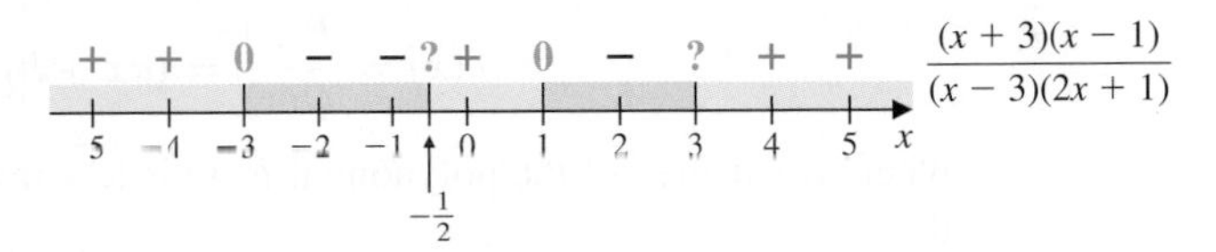

FIGURE 13

Notice that the graph crosses the horizontal asymptote somewhere in the interval $(0, 1)$. To find this intersection point we need to determine the value of x with

$$\frac{1}{2} = \frac{x^2 + 2x - 3}{2x^2 - 5x - 3}.$$

This implies that

$$2x^2 - 5x - 3 = 2x^2 + 4x - 6.$$

Simplifying this equation gives

$$3 = 9x, \quad \text{so} \quad x = \tfrac{3}{9} = \tfrac{1}{3}.$$

■

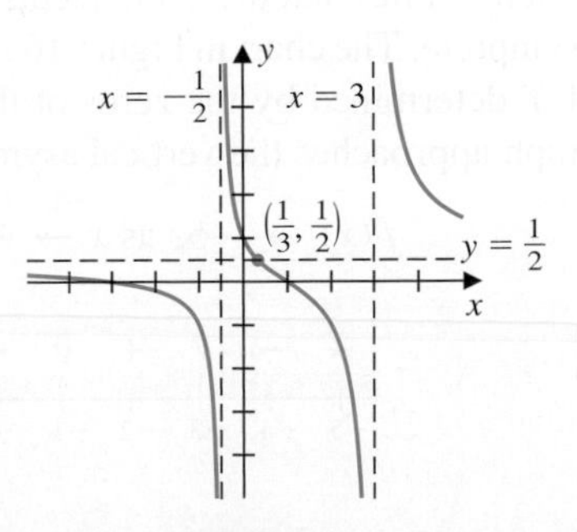

FIGURE 14

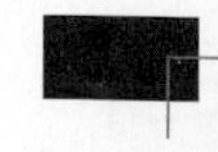

Slant Asymptotes

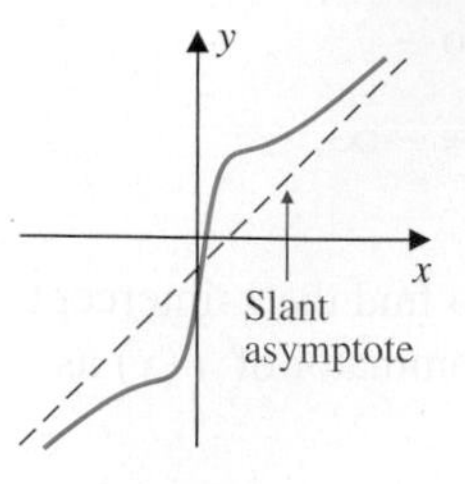

FIGURE 15

We have seen that a rational function does not have a horizontal asymptote when the degree of the numerator is greater than the degree of the denominator. However, in the special case where the degree of the numerator is just 1 more than the degree of the denominator, the graph approaches a nonhorizontal line as $x \to \infty$ and as $x \to -\infty$. This line is called a **slant**, or *oblique*, **asymptote** to the graph. (See Figure 15.)

To determine the equation of the slant asymptote, suppose that

$$f(x) = \frac{P(x)}{Q(x)},$$

where the degree of the polynomial $P(x)$ is exactly 1 greater than the degree of the polynomial $Q(x)$. By the Division Algorithm there exists a linear quotient, say $ax + b$, such that

$$P(x) = (ax + b)Q(x) + R(x).$$

So

$$f(x) = \frac{P(x)}{Q(x)} = (ax + b) + \frac{R(x)}{Q(x)},$$

where the degree of the polynomial $R(x)$ is less than the degree of $Q(x)$. Because of this,

$$\frac{R(x)}{Q(x)} \to 0 \text{ as } x \to \infty \quad \text{and} \quad \frac{R(x)}{Q(x)} \to 0 \text{ as } x \to -\infty.$$

As a consequence,

$$f(x) \to ax + b \text{ as } x \to \infty \quad \text{and} \quad f(x) \to ax + b \text{ as } x \to -\infty,$$

so $y = ax + b$ is a slant asymptote for the graph.

EXAMPLE 6 Sketch the graph of $f(x) = \dfrac{x^2 - 2x - 3}{x + 3}$.

Solution The y-intercept for the graph is $(0, f(0)) = (0, -1)$. Factoring the numerator of $f(x)$ gives

$$f(x) = \frac{(x + 1)(x - 3)}{x + 3},$$

which implies that the x-intercepts are at $(-1, 0)$ and $(3, 0)$, and $x = -3$ is a vertical asymptote. The chart in Figure 16 shows the sign of $f(x)$ in the regions of the domain of f determined by the zeros of the numerator and denominator, and tells how the graph approaches the vertical asymptote $x = -3$:

$$f(x) \to -\infty \text{ as } x \to -3^- \quad \text{and} \quad f(x) \to \infty \text{ as } x \to -3^+.$$

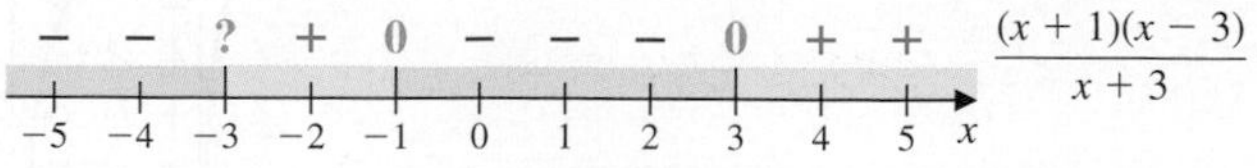

− − ? + 0 − − − 0 + + $\dfrac{(x+1)(x-3)}{x+3}$

−5 −4 −3 −2 −1 0 1 2 3 4 5 x

FIGURE 16

There are no horizontal asymptotes to the graph because the degree of the numerator is 2 and the degree of the denominator is 1. However, if we divide the denominator, $x+3$, into the original numerator, x^2-2x-3, we find that

$$f(x) = x - 5 + \frac{12}{x+3}.$$

As x becomes large in magnitude the term $\frac{12}{x+3}$ approaches 0, so

$$f(x) \to x - 5 \text{ as } x \to \pm\infty.$$

The line $y = x - 5$ is a slant asymptote to the graph, as shown in Figure 17. ■

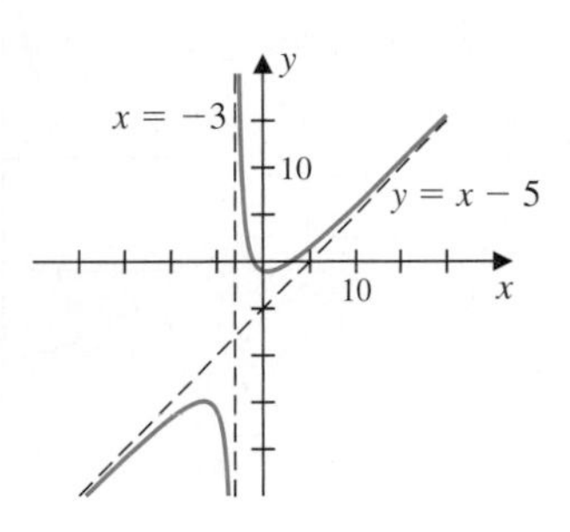

FIGURE 17

EXAMPLE 7 Determine a rational function f that has

i. vertical asymptotes $x = -2$ and $x = 1$,
ii. a horizontal asymptote $y = 2$, and
iii. a graph that passes through the point $(0, 3)$.

Solution A function with vertical asymptotes $x = -2$ and $x = 1$ has factors $x + 2$ and $x - 1$ in the denominator, but not in the numerator. The graph of

$$y = \frac{1}{(x+2)(x-1)}$$

has vertical asymptotes $x = 1$ and $x = -2$. But as shown in Figure 18, $y = 2$ is not a horizontal asymptote and the graph does not contain the point $(0, 3)$.

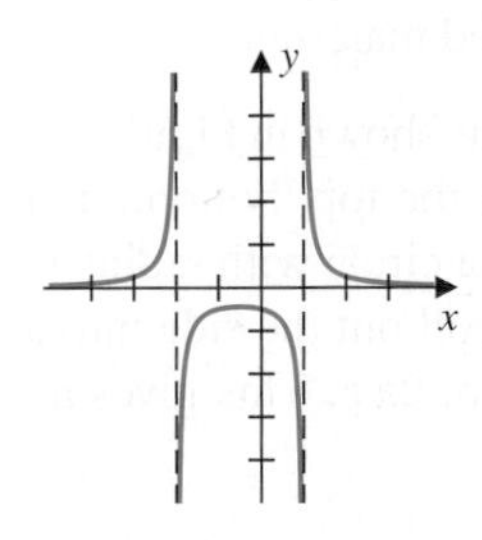

FIGURE 18

The graph of a rational function has a horizontal asymptote when the degree of the numerator is the same as the degree of the denominator. The graph of

$$y = \frac{2x^2}{(x+2)(x-1)}$$

has a horizontal asymptote $y = 2$, but still the graph does not contain the point $(0, 3)$, as shown in Figure 19. We need to adjust the curve without changing the horizontal asymptote. We can do this by adding a constant to the numerator. Consider

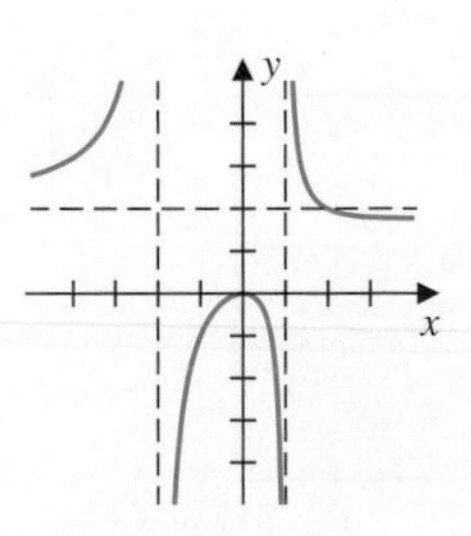

FIGURE 19

$$y = \frac{2x^2 + a}{(x+2)(x-1)}$$

for some nonzero constant a. The graph contains $(0, 3)$ when

$$3 = \frac{2 \cdot 0^2 + a}{(0+2)(0-1)} = \frac{a}{-2}, \quad \text{so} \quad a = -6.$$

A function satisfying all the requirements is

$$f(x) = \frac{2x^2 - 6}{(x+2)(x-1)} = \frac{2x^2 - 6}{x^2 + x - 2}.$$

This graph is shown in Figure 20. ■

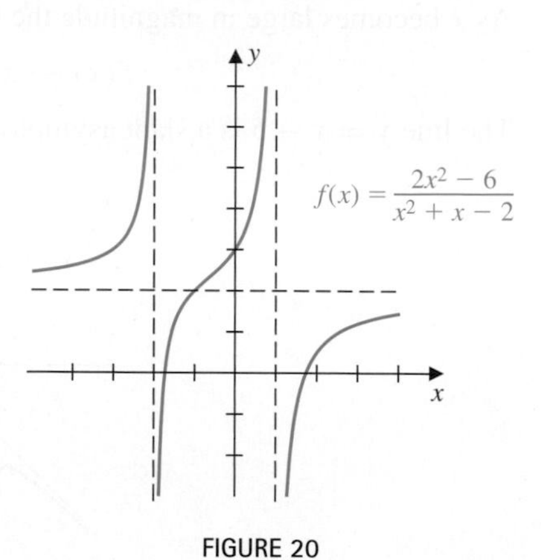

FIGURE 20

The function found in the previous example is not uniquely determined by the stated conditions. For example, replacing the numerator $2x^2 - 6$ by $2x^2 + x - 6$ would also satisfy the conditions.

Applications

EXAMPLE 8 An aluminum can is to be constructed to contain 318 cm³ of liquid. Approximate the dimensions of the can that will minimize the amount of required material.

Solution Let r represent the radius of the can and h represent the height, as shown in Figure 21. The total surface area A of the can is the sum of the areas of the top, bottom, and side. The areas of the top and bottom are both πr^2, the area of a circle with radius r. To determine the area of the side, cut the can vertically. Then roll out the side into a rectangle with length h and width the circumference of the top, $2\pi r$. This gives an area of $h(2\pi r)$. So

$$A = \pi r^2 + \pi r^2 + 2\pi rh = 2\pi r^2 + 2\pi rh.$$

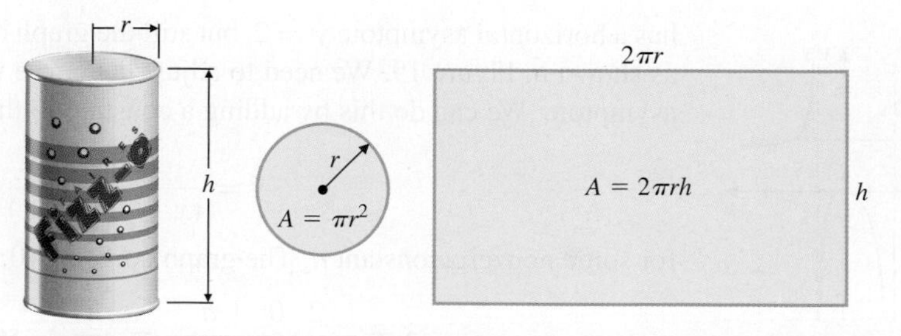

FIGURE 21

We can summarize the problem as follows:

Know	Find
1. The surface area of the can is $A = 2\pi r^2 + 2\pi rh$.	a. The area of the can, $A(r)$, in terms of the radius.
2. The volume of the can is $V = \pi r^2 h = 318\text{cm}^3$.	b. An approximate minimal value for $A(r)$.
	c. Values of r and h that give the result in (b).

Optimization problems like this often involve rational functions. Calculus is used to find the exact minimum value.

Using what we know in (2) gives

$$318 = \pi r^2 h, \quad \text{so} \quad h = \frac{318}{\pi r^2}.$$

Adding this to what we know in (1) gives the area in terms of the radius as

$$A(r) = 2\pi r^2 + 2\pi r\left(\frac{318}{\pi r^2}\right) = 2\pi r^2 + \frac{636}{r}.$$

When r is close to 0, the graph of $A(r)$ behaves like the graph of $636/r$, so $r = 0$ is a vertical asymptote. When r is large the graph is similar to that of r^2.

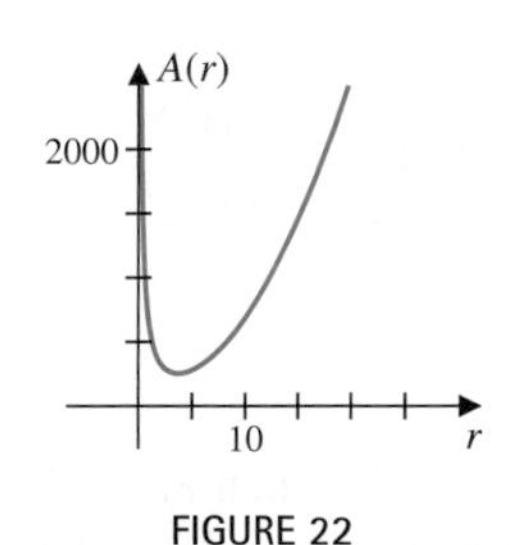

FIGURE 22

The graph of $A(r)$ in Figure 22 indicates that the minimum value of $A(r)$ occurs when r is approximately 3.7. As a consequence, the minimum amount of material needed to construct the can is approximately

$$A(3.7) = 2\pi(3.7)^2 + \frac{636}{3.7} \approx 258 \text{ cm}^2.$$

Methods of calculus can be used to show that the true value of r that minimizes the area is $r = (159/\pi)^{1/3} \approx 3.7$. This value gives an approximate area of 257.91, which does not significantly differ from our approximate value. The height of the can is

$$h = \frac{318}{\pi(3.7)^2} \approx 7.4 \text{ cm}.$$

■

EXAMPLE 9 The Calculus Connection in the opening to this chapter described the parabolic path of water coming out of a hose. This was given by

$$p(x) = mx - \frac{16}{v_0^2}(1 + m^2)x^2,$$

where v_0 is the initial velocity of the water, m is the slope of the nozzle, and x is the distance, in feet, along the x-axis. Suppose that the initial velocity $v_0 = 32$ ft/sec.

a. Find a function $r(m)$ for the maximum horizontal distance of the water.

b. Find the function $h(m)$ for the maximum height of the water.

c. For what value of m is $r(m)$ a maximum?

Solution **a.** The height of the water when it is x units from the nozzle is

$$p(x) = mx - \frac{16}{32^2}(1 + m^2)x^2 = mx - \frac{1}{64}(1 + m^2)x^2.$$

So the water hits the ground when

$$mx - \frac{1}{64}(1+m^2)x^2 = 0, \quad \text{that is, when} \quad x\left(m - \frac{1}{64}(1+m^2)x\right) = 0.$$

This means that the water is on the ground when $x = 0$ and when

$$r(m) = x = \frac{64m}{1+m^2}.$$

b. Because $y = p(x)$ is a parabola, it is symmetric about the horizontal line that lies halfway between its zeros. This is the line $x = 32m/(1+m^2)$. The height at this value is

$$h(m) = p\left(\frac{32m}{1+m^2}\right) = m\left(\frac{32m}{1+m^2}\right) - \frac{1}{64}(1+m^2)\left(\frac{32m}{1+m^2}\right)^2$$

$$= \frac{32m^2 - 16m^2}{1+m^2} = \frac{16m^2}{1+m^2}.$$

c. The graph of $y = r(m)$ shown in Figure 23 would lead one to suspect that the value of m that makes $r(m)$ a maximum is $m = 1$ and that this maximum would be

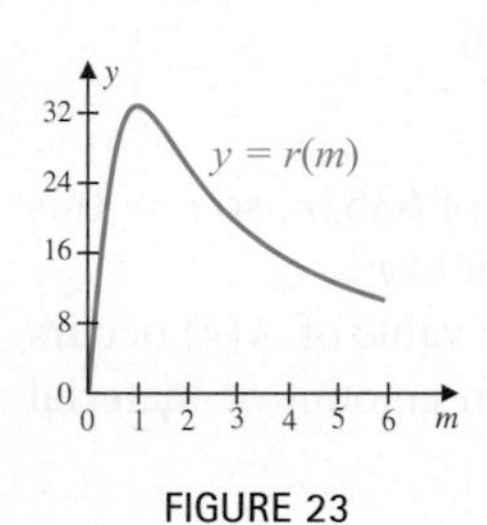

FIGURE 23

$$\frac{64 \cdot 1}{1+1^2} = 32.$$

To show that this is true, first consider the difference

$$r(1) - r(m) = 32 - \frac{64m}{1+m^2} = 32\left(\frac{1+m^2-2m}{1+m^2}\right) = 32\left(\frac{(1-m)^2}{1+m^2}\right).$$

Thus

$$r(1) = r(m) + 32\left(\frac{(1-m)^2}{1+m^2}\right).$$

Because the second term on the right is positive unless $m = 1$, when $m \neq 1$ we must add a positive quantity to $r(m)$ to obtain $r(1)$. Hence $r(1) = 32$ is the maximum value for the range. So if the nozzle is aimed in the direction of the line $y = x$, that is, at an angle of 45°, the water will hit the ground as far away as possible, a distance of 32 feet. ■

EXERCISE SET 3.4

In Exercises 1–6, the graph of a rational function is given. Specify **(a)** *the domain,* **(b)** *the range, and* **(c)** *the vertical and horizontal asymptotes of the function.*

1. **2.**

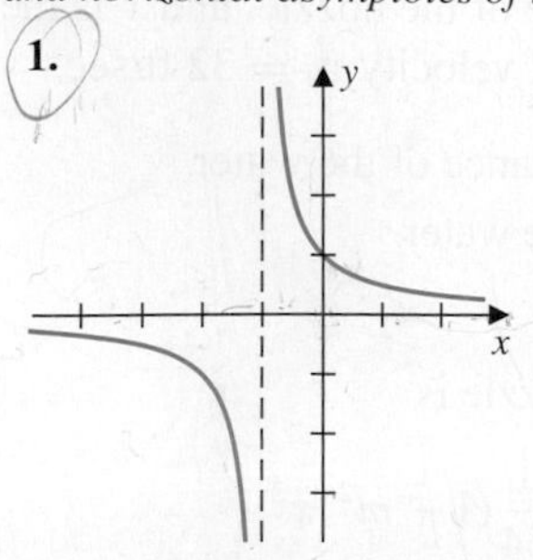

3. **4.**

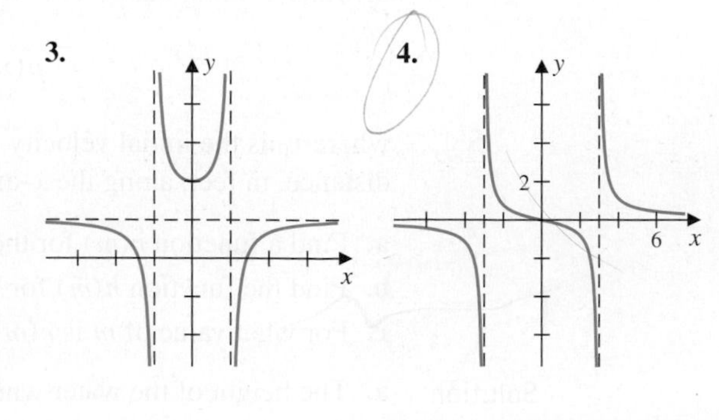

5.

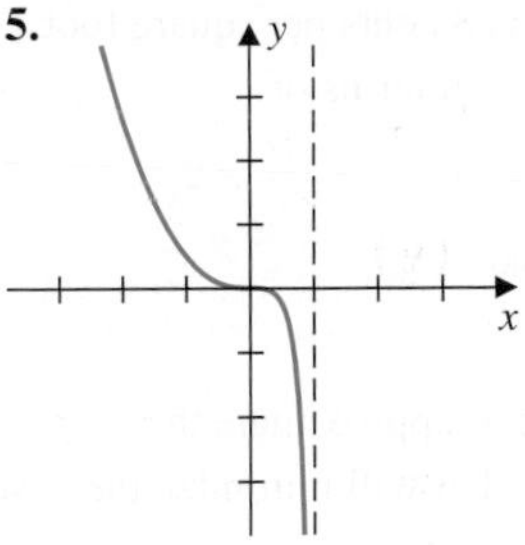

6.

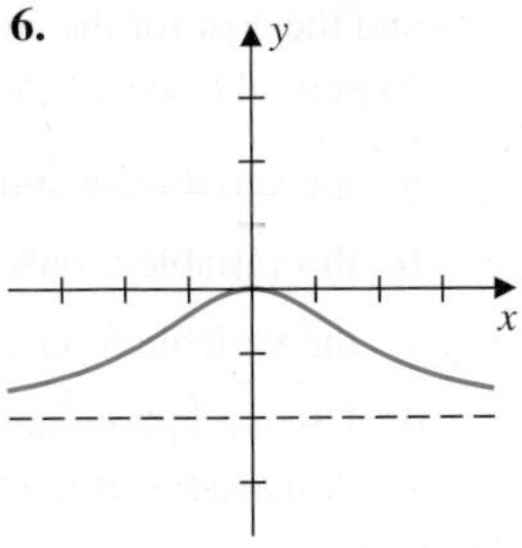

*In Exercises 7–10, complete the table to determine **(a)** the vertical and **(b)** the horizontal asymptotes of the rational function.*

7. $f(x) = \dfrac{(x+1)(x-1)}{(x-2)(x+2)}$

Behavior of x	Sign of $f(x)$	Behavior of $f(x)$
$x \to -2^-$	$\frac{(\ )(\ )}{(\ )(\ )}$	$f(x) \to$
$x \to -2^+$	$\frac{(\ )(\ )}{(\ )(\ )}$	$f(x) \to$
$x \to 2^-$	$\frac{(\ )(\ )}{(\ )(\ )}$	$f(x) \to$
$x \to 2^+$	$\frac{(\ )(\ )}{(\ )(\ )}$	$f(x) \to$
$x \to -\infty$	$\frac{(\ )(\ )}{(\ )(\ )}$	$f(x) \to$
$x \to \infty$	$\frac{(\ )(\ )}{(\ )(\ )}$	$f(x) \to$

8. $f(x) = \dfrac{x}{(x+1)(x-3)}$

Behavior of x	Sign of $f(x)$	Behavior of $f(x)$
$x \to -1^-$	$\frac{(\ )(\ )}{(\ )(\ )}$	$f(x) \to$
$x \to -1^+$	$\frac{(\ )(\ )}{(\ )(\ )}$	$f(x) \to$
$x \to 3^-$	$\frac{(\ )(\ )}{(\ )(\ )}$	$f(x) \to$
$x \to 3^+$	$\frac{(\ )(\ )}{(\ )(\ )}$	$f(x) \to$
$x \to \infty$	$\frac{(\ )(\ )}{(\ )(\ )}$	$f(x) \to$
$x \to -\infty$	$\frac{(\ )(\ )}{(\ )(\ )}$	$f(x) \to$

9. $f(x) = \dfrac{x^2-1}{x^2-2x+1}$

Behavior of x	Sign of $f(x)$	Behavior of $f(x)$
$x \to -1^-$	$\frac{(\ )(\ )}{(\ )(\ )}$	$f(x) \to$
$x \to -1^+$	$\frac{(\ )(\ )}{(\ )(\ )}$	$f(x) \to$
$x \to 1^-$	$\frac{(\ )(\ )}{(\ )(\ )}$	$f(x) \to$
$x \to 1^+$	$\frac{(\ )(\ )}{(\ )(\ )}$	$f(x) \to$
$x \to -\infty$	$\frac{(\ )(\ )}{(\ )(\ )}$	$f(x) \to$
$x \to \infty$	$\frac{(\ )(\ )}{(\ )(\ )}$	$f(x) \to$

10. $f(x) = \dfrac{x^2+4x+4}{x^2-4}$

Behavior of x	Sign of $f(x)$	Behavior of $f(x)$
$x \to -2^-$	$\frac{(\ )(\ )}{(\ )(\ )}$	$f(x) \to$
$x \to -2^+$	$\frac{(\ )(\ )}{(\ )(\ )}$	$f(x) \to$
$x \to 2^-$	$\frac{(\ )(\ )}{(\ )(\ )}$	$f(x) \to$
$x \to 2^+$	$\frac{(\ )(\ )}{(\ )(\ )}$	$f(x) \to$
$x \to \infty$	$\frac{(\ )(\ )}{(\ )(\ )}$	$f(x) \to$
$x \to -\infty$	$\frac{(\ )(\ )}{(\ )(\ )}$	$f(x) \to$

*In Exercises 11–16, find **(a)** the domain, **(b)** the axis intercepts, and **(c)** the vertical and horizontal asymptotes of the rational function.*

11. $f(x) = \dfrac{x-3}{x+1}$

12. $f(x) = \dfrac{(x-1)(2x+2)}{(x+3)(x-4)}$

13. $f(x) = \dfrac{3x^2-11x-4}{x^2-1}$

14. $f(x) = \dfrac{x^2+3x-10}{x^2-4x+4}$

15. $f(x) = \dfrac{x^3-2x^2-x+2}{x^2}$

16. $f(x) = \dfrac{x^5-2x^4-x+2}{x^3-1}$

In Exercises 17–30, sketch the graph, labeling any horizontal and vertical asymptotes and axis intercepts.

17. $f(x) = \dfrac{3}{x-2}$

18. $f(x) = \dfrac{4}{x+3}$

19. $f(x) = \dfrac{x}{x-3}$

20. $f(x) = \dfrac{2x}{3x-1}$

21. $f(x) = \dfrac{(2x-1)(x+2)}{(3x+1)(x-3)}$

22. $f(x) = \dfrac{x(x+4)}{(x-1)(x+2)}$

23. $f(x) = \dfrac{x^2-9}{x^2-16}$

24. $f(x) = \dfrac{x^2-16}{x^2-9}$

25. $f(x) = \dfrac{x^2+2x-3}{x-1}$

26. $f(x) = \dfrac{x^2-2x-8}{x+2}$

27. $f(x) = \dfrac{x^2-x-2}{x^2-2x-3}$

28. $f(x) = \dfrac{x^2 - 1}{x^2 - 3x + 2}$

29. $f(x) = \dfrac{3x - 2}{x^2 + x - 6}$ **30.** $f(x) = \dfrac{-2x}{x^2 - 9}$

In Exercises 31–36, determine a rational function f satisfying all of the given conditions.

31. Has a vertical asymptote at $x = 1$, a horizontal asymptote at $y = 0$, a y-intercept at $(0, -1)$, and never crosses the x-axis.

32. Has a vertical asymptote at $x = -2$, a horizontal asymptote at $y = 0$, a y-intercept at $(0, 2)$, and never crosses the x-axis.

33. Has vertical asymptotes at $x = 2$ and $x = -3$, a horizontal asymptote at $y = 1$, and x-intercepts at 3 and -4.

34. Has vertical asymptotes at $x = 2$ and $x = -3$, a horizontal asymptote at $y = 1$, and x-intercepts at 3 and 4.

35. Has $f(2) = 0$, $f(x) \to 3$ as $x \to \infty$, $f(x) \to 3$ as $x \to -\infty$, $f(x) \to -\infty$ as $x \to 0^+$, and $f(x) \to \infty$ as $x \to 0^-$.

36. Has $f(4) = 0$, $f(x) \to 2$ as $x \to \infty$, $f(x) \to -\infty$ as $x \to 3^+$, and $f(x) \to \infty$ as $x \to 3^-$, and has y-axis symmetry.

In Exercises 37–40, sketch the graph of the function and label the slant asymptote.

37. $f(x) = \dfrac{x^2}{x - 1}$ **38.** $f(x) = \dfrac{x^2 - x}{x - 2}$

39. $f(x) = \dfrac{x^2 + x - 2}{x + 1}$ **40.** $f(x) = \dfrac{x^2 - 2x - 3}{x - 1}$

In Exercises 41–44, the graph of the rational function crosses the horizontal asymptote once. Sketch each graph and label the point of intersection of the curve with the horizontal asymptote.

41. $f(x) = \dfrac{x}{x^2 - 3x + 2}$ **42.** $f(x) = \dfrac{x^2 - 6x + 8}{x^2 - 4x + 3}$

43. $f(x) = \dfrac{x^2}{x^2 - 5x + 6}$ **44.** $f(x) = \dfrac{3 - x^2}{2x^2 + x - 3}$

45. Rework the problem posed in Example 8 of this section with the change that the can to be constructed has a bottom but no top.

46. A rectangular box with a square base of length x and height h is to have a volume of 20 ft^3. The cost of the top and bottom of the box is 20 cents per square foot and the cost for the sides is 8 cents per square foot. Express the cost of the box in terms of

a. the variables x and h;

b. the variable x only; and

c. the variable h only.

d. Use a graphing device to approximate the dimensions of the box that will minimize the cost.

47. The number of bacteria in a culture at time t is given by

$$n = 10{,}000\left(\frac{3t^2 + 1}{t^2 + 1}\right).$$

a. As time increases, does the size of the bacteria colony become stable?

b. If the answer to part (a) is yes, what is the stabilizing level?

c. How long does it take for the number of bacteria to exceed 22,000?

48. A drug in the bloodstream has a concentration of

$$c(t) = \frac{at}{t^2 + b}$$

at time $t \geq 0$ hours. The constants a and b depend on the particular drug.

a. Use a graphing device to sketch the graph of $y = c(t)$ for several choices of a and b. How do a and b affect the graph?

b. For $a = 3$ and $b = 1$, approximate the highest concentration of the drug reached in the bloodstream.

c. What happens to the drug concentration as time t becomes large?

d. For $a = 3$ and $b = 1$, determine how long it takes for the drug concentration to drop below 0.2.

49. A population at time t is given by

$$P(t) = \frac{at^2}{t^2 + 1},$$

where a is a constant determined by the particular species.

a. Graph $y = P(t)$ for different choices of a and describe the effect of the parameter a on the curve.

b. What happens to the population as t increases indefinitely?

50. An oil drum in the shape of a right circular cylinder holds a volume of 3000 cubic inches.

a. Express the surface area in terms of the radius r and the height h.

b. Express the surface area as a function of r only.

c. Express the the surface area as a function of h only.

3.5 OTHER ALGEBRAIC FUNCTIONS

Polynomial and rational functions are examples of a larger class of functions, called **algebraic**, that are obtained from polynomials by a finite combination of the operations of addition, subtraction, multiplication, division, raising to integral powers, and extracting roots.

The simplest algebraic functions that are not polynomials or rational functions are the *rational power functions*. These have the form

$$f(x) = x^{m/n},$$

where m/n is a rational number, n is an integer greater than 1, and the integers m and n have no common factors. In this case we can alternatively write $f(x)$ as

$$x^{m/n} = \left(\sqrt[n]{x}\right)^m = \sqrt[n]{x^m}.$$

The graphs of some rational power functions for these cases are shown in Figure 1. The graphs in parts (c) and (d) can be obtained by applying the reciprocal graphing techniques to the graphs in parts (a) and (b).

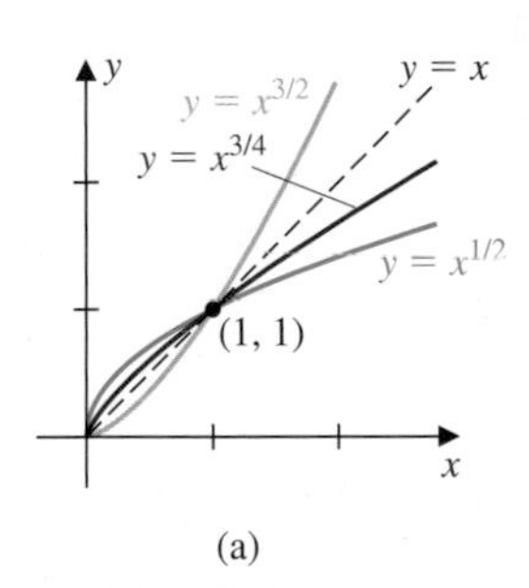

(a)

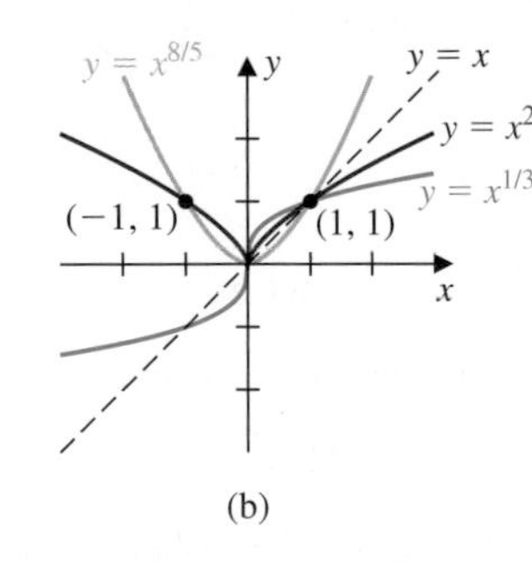

(b)

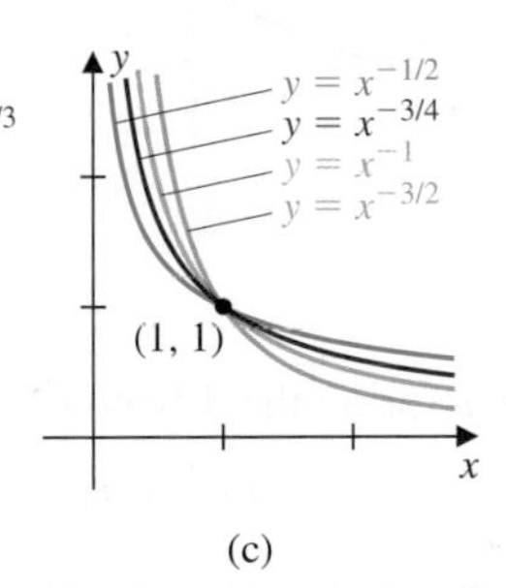

(c)

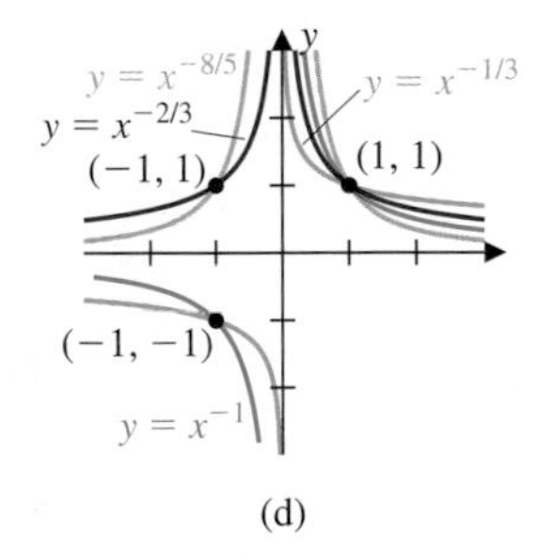

(d)

FIGURE 1

Symmetry for Rational Power Functions

- Rational power functions whose exponents have odd denominators have symmetry.
  - If the numerator of the exponent is even, it is y-axis symmetry.
  - If the numerator of the exponent is odd, it is origin symmetry.
- Rational power functions whose exponents have even denominators have no symmetry, since the domain consists of only positive values.

In Section 2.2 we examined the graphs of functions that occur as translations and changes of scale of the square root function,

$$f(x) = \sqrt{x} = x^{1/2}.$$

Example 1 shows how this translation technique can be applied to any of the rational power functions.

EXAMPLE 1 Sketch the graph of the function.

a. $f(x) = 2\sqrt[3]{x+2} - 1$ **b.** $f(x) = (4x-3)^{3/2}$

Solution **a.** First graph the rational power function $y = x^{1/3}$. Then stretch the vertical scale by 2 units to give the graph of $y = 2x^{1/3}$ shown in Figure 2(a). The graph of

$$f(x) = 2\sqrt[3]{x+2} - 1$$

is obtained by shifting the graph of $y = 2\sqrt[3]{x}$ left 2 units and downward 1 unit, as shown in Figure 2(b). The graph of $y = 2\sqrt[3]{x}$ is symmetric with respect to the origin. Because of the translations, the new graph is symmetric with respect to the point $(-2, -1)$.

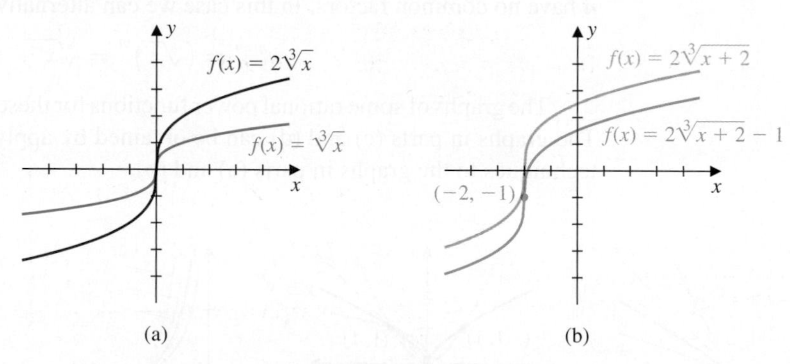

FIGURE 2

b. Factoring the 4 from within the parentheses gives

$$f(x) = (4x-3)^{3/2} = \left(4\left(x - \tfrac{3}{4}\right)\right)^{3/2} = 4^{3/2}\left(x - \tfrac{3}{4}\right)^{3/2} = 8\left(x - \tfrac{3}{4}\right)^{3/2}.$$

The graphs of $y = x^{3/2}$ and $y = 8x^{3/2} = (4x)^{3/2}$ are shown in Figure 3(a). The final graph of $f(x) = (4x-3)^{3/2}$ is shown in Figure 3(b). ■

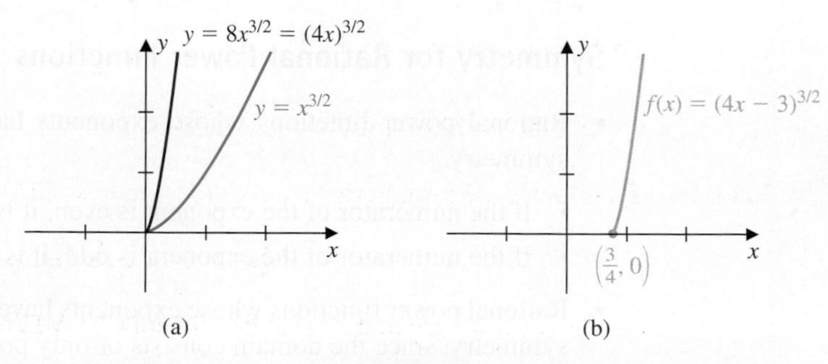

FIGURE 3

In Example 2 we consider the effect of composing the square root function with a rational function.

EXAMPLE 2 Sketch the graph of $f(x) = \sqrt{\dfrac{x-2}{x+2}}$.

Solution The domain consists of all real numbers x with

$$\frac{x-2}{x+2} \geq 0.$$

The numerator is zero when $x = 2$, so the x-intercept is $(2, 0)$. The denominator is zero when $x = -2$, so $x = -2$ is not in the domain. The sign chart in Figure 4 indicates that the domain is $(-\infty, -2) \cup [2, \infty)$.

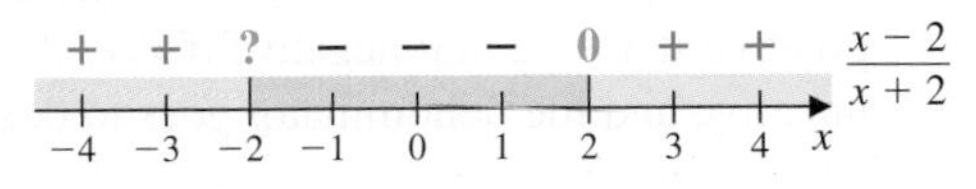

FIGURE 4

The graph does not cross the y-axis, because $x = 0$ is not in the domain. Hence there is no y-intercept.

The graph has the vertical asymptote $x = -2$, because the denominator of the function is zero at this value but the numerator is not. The square root function is always positive, so

$$f(x) \to \infty \quad \text{as} \quad x \to -2^-.$$

There is no limit from the right at -2 because the interval $(-2, 2)$ is not in the domain of f.

Finally,

$$\frac{x-2}{x+2} \to 1 \text{ as } x \to \infty \quad \text{and} \quad \frac{x-2}{x+2} \to 1 \text{ as } x \to -\infty,$$

so $f(x)$, the square root of this quotient, also approaches 1. This implies that $y = 1$ is a horizontal asymptote. This analysis gives the graph shown in Figure 5. ■

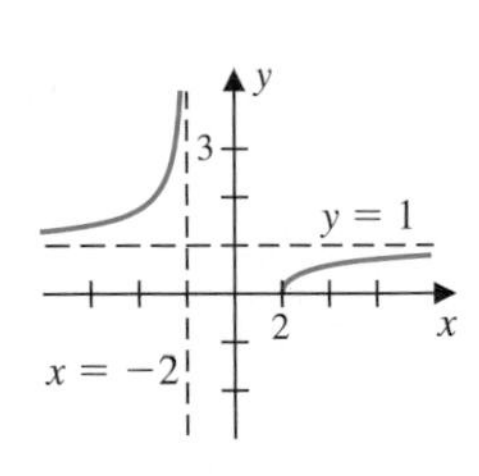

FIGURE 5

In Section 3.4 we saw that the graph of a rational function f can have at most one horizontal asymptote. This follows from the fact that for a rational function

- $f(x) \to L$ as $x \to \infty$ exactly when $f(x) \to L$ as $x \to -\infty$.

Example 3 considers a function (by necessity not a rational function) that has two distinct horizontal asymptotes.

EXAMPLE 3 Sketch the graph of $f(x) = \dfrac{x-2}{\sqrt{x^2+x-2}}$.

Solution First factor the term in the denominator and rewrite $f(x)$ as

$$f(x) = \frac{x-2}{\sqrt{(x+2)(x-1)}}.$$

In this form we see that x is in the domain of f whenever

$$(x+2)(x-1) > 0.$$

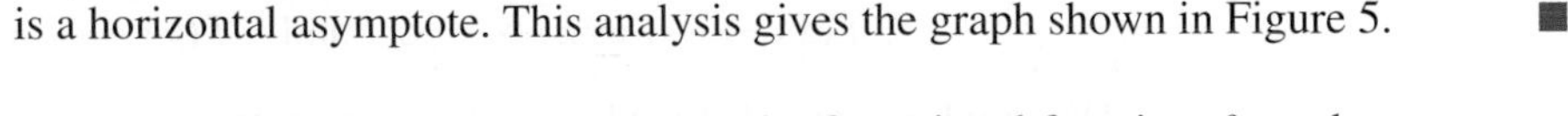

The sign chart for $(x+2)(x-1)$ in Figure 6 implies that the domain is $(-\infty, -2) \cup (1, \infty)$.

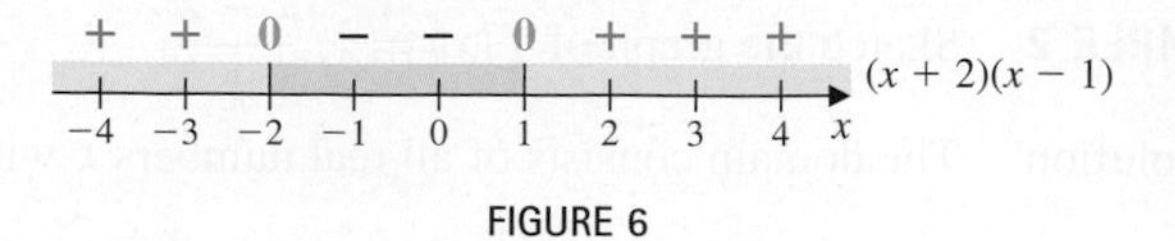

FIGURE 6

The numerator is zero precisely when $x = 2$, so the only x-intercept is $(2, 0)$. There is no y-intercept, because 0 is not in the domain of f.

The graph has vertical asymptotes $x = -2$ and $x = 1$. The denominator of $f(x)$ is always positive, so the sign of $f(x)$ depends only on the sign of $x - 2$, which is positive if $x > 2$ and negative if $x < 2$. As x approaches -2 from the left, $x - 2$ is negative and the denominator goes to 0, so

$$f(x) \to -\infty \quad \text{as} \quad x \to -2^-.$$

As x approaches 1 from the right, $x - 2$ is negative and the denominator goes to 0, so

$$f(x) \to -\infty \quad \text{as} \quad x \to 1^+.$$

When x is large in magnitude—that is, for $|x|$ large—the numerator of

$$f(x) = \frac{x-2}{\sqrt{(x+2)(x-1)}}$$

is approximately x, and the denominator is approximately $\sqrt{x^2} = |x|$. So when $|x|$ is large,

$$f(x) = \frac{x-2}{\sqrt{(x+2)(x-1)}} \approx \frac{x}{\sqrt{x^2}} = \frac{x}{|x|} = \begin{cases} 1, & \text{if } x > 0, \\ -1, & \text{if } x < 0. \end{cases}$$

This implies that

$$f(x) \to -1 \text{ as } x \to -\infty \quad \text{and} \quad f(x) \to 1 \text{ as } x \to \infty.$$

The graph has two distinct horizontal asymptotes,

$$y = -1 \text{ as } x \to -\infty \quad \text{and} \quad y = 1 \text{ as } x \to \infty,$$

as shown in Figure 7. ■

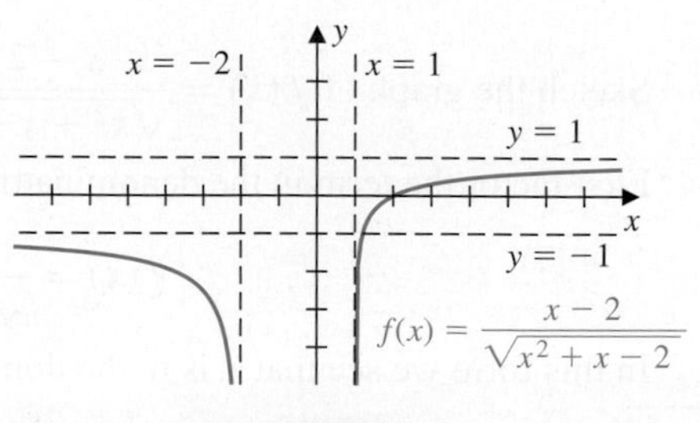

FIGURE 7

Applications

Allometry is the study of the relationships between different characteristics of an organism, such as the relationship between its volume and its weight. An organism grows as it matures; its size changes but its shape remains similar. Often these relationships can be modeled using power functions with rational exponents of the form

$$y = ax^b.$$

This idea can be applied to many scenarios, one of which is considered in Example 4.

EXAMPLE 4 A model for the growth rate of a tumor is given by a function of the form

$$g(x) = ax^\alpha - bx^\beta,$$

where x is the volume of the tumor (size), the term ax^α describes the growth of the tumor, and the term bx^β, the shrinking of the tumor. The parameters a, b, α, and β are positive. For which values of x is the tumor growing? For which values is the tumor shrinking?

Solution The function g gives the growth rate, so the tumor is growing when $g(x) > 0$ and is shrinking when $g(x) < 0$. To solve $g(x) > 0$, we have

$$ax^\alpha - bx^\beta > 0 \quad \text{so} \quad \frac{x^\alpha}{x^\beta} > \frac{b}{a}.$$

Combining the exponents on x and then taking roots gives

$$x^{\alpha-\beta} > \frac{b}{a}, \quad \text{so} \quad x > \left(\frac{b}{a}\right)^{\frac{1}{\alpha-\beta}} = \left(\frac{b}{a}\right)^{\beta-\alpha}.$$

In a similar manner, the tumor is shrinking when $x < \left(\frac{b}{a}\right)^{\beta-\alpha}$. ■

EXERCISE SET 3.5

In Exercises 1–10, determine the domain of the function.

1. $f(x) = \sqrt{x^2 + 2x - 15}$

2. $f(x) = \sqrt{-6 + 5x - x^2}$

3. $f(x) = \sqrt{(x-1)^2(x+2)}$

4. $f(x) = \sqrt{x^2(x-3)}$

5. $f(x) = \sqrt{(x+1)(x-3)(x+4)}$

6. $f(x) = \sqrt{(x-1)(x+2)(x-4)}$

7. $f(x) = \sqrt{\dfrac{1-x}{x+3}}$

8. $f(x) = \sqrt{\dfrac{7-2x}{x-5}}$

9. $f(x) = \sqrt{\dfrac{x+1}{x^2+2x-3}}$

10. $f(x) = \sqrt{\dfrac{x^2+2x-24}{x^2-9}}$

11. Match the function with its graph.

a. $f(x) = (x+1)^{2/3} - 1$

b. $f(x) = (x+1)^{3/2} - 1$

c. $f(x) = \sqrt{\dfrac{x-1}{x+3}}$

d. $f(x) = \sqrt{\dfrac{4x+4}{x-1}}$

e. $f(x) = \dfrac{x+1}{\sqrt{x^2-x-2}}$

f. $f(x) = \dfrac{2x+1}{\sqrt{x^2+3x+2}}$

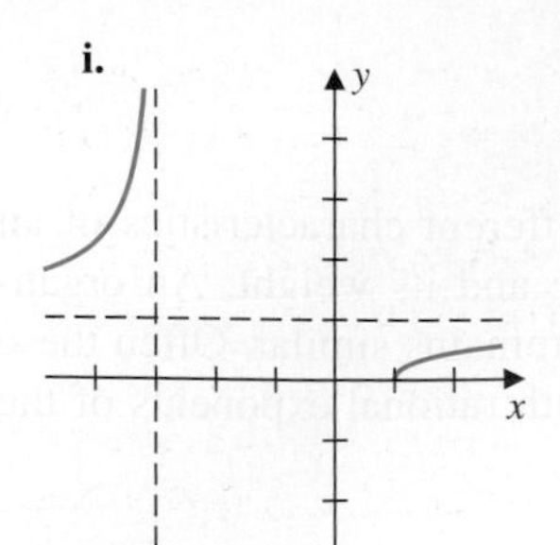

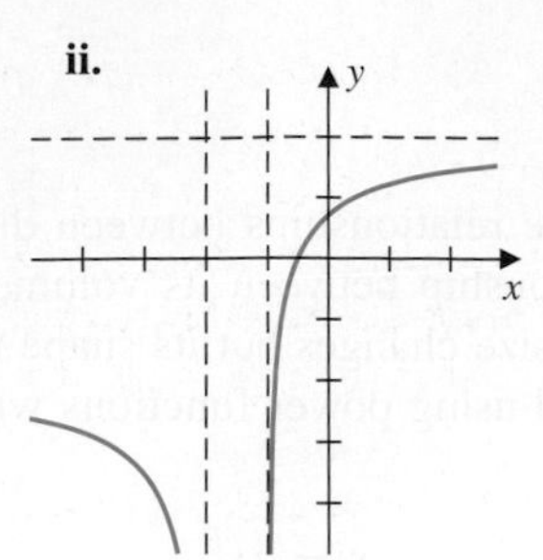

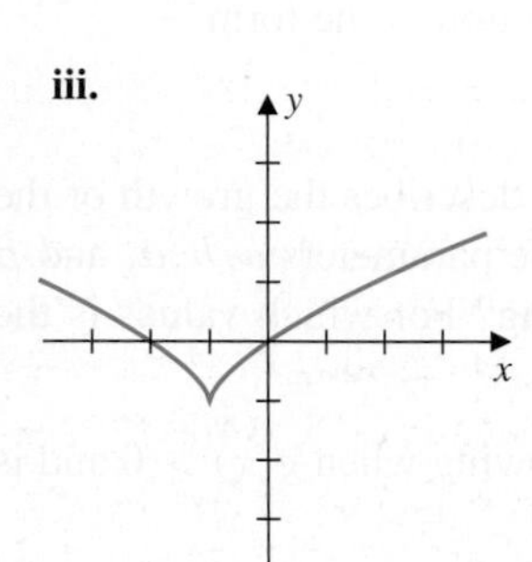

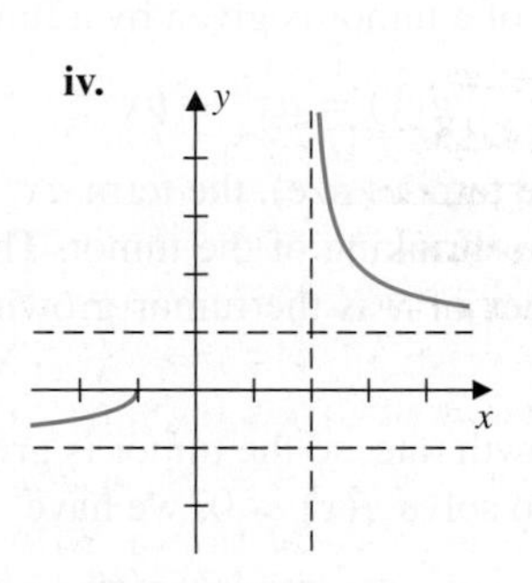

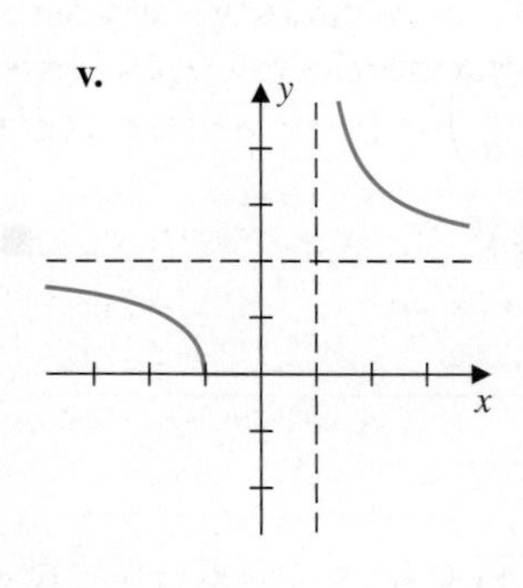

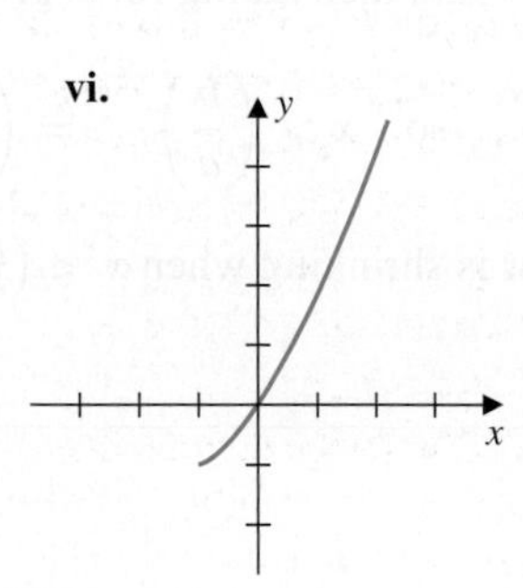

12. Match the function with its graph.

a. $f(x) = (x - 1)^{1/5} + 1$

b. $f(x) = (x - 1)^{3/4} + 1$

c. $f(x) = -(x - 1)^{1/7} + 1$

d. $f(x) = -(x + 1)^{1/6} - 1$

e. $f(x) = \sqrt{\dfrac{1 - x}{x + 2}}$

f. $f(x) = \sqrt{\dfrac{2 - x}{x + 3}}$

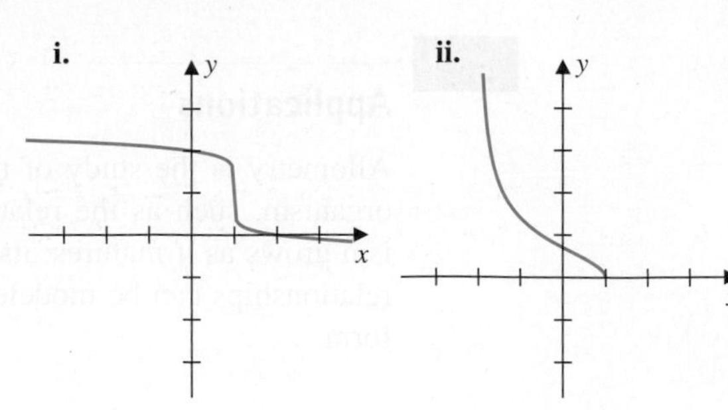

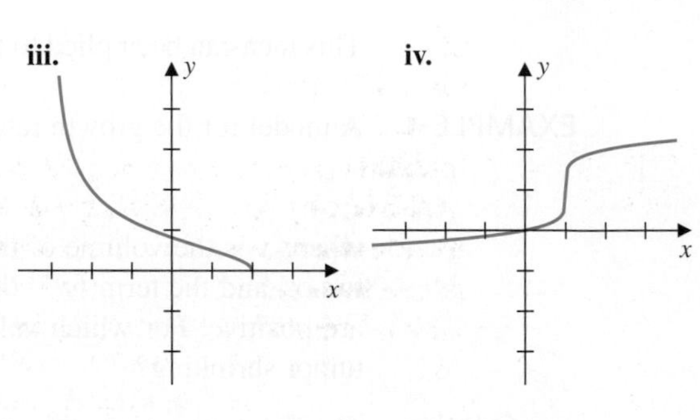

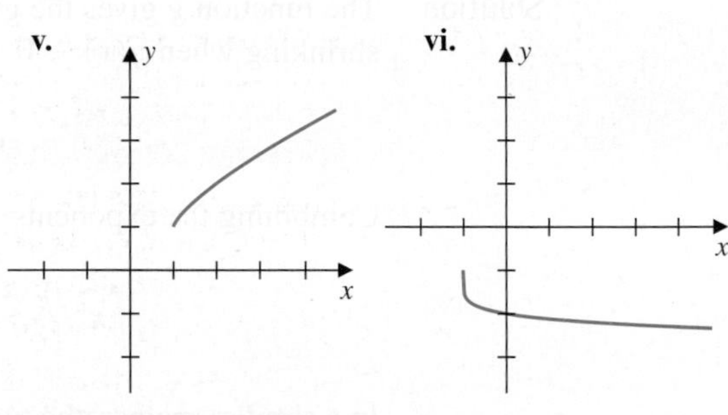

In Exercises 13–32, sketch the graph and label the axis intercepts and asymptotes.

13. $f(x) = x^{3/2} - 2$

14. $f(x) = x^{2/3} + 1$

15. $f(x) = (x - 1)^{1/3} + 1$

16. $f(x) = (x + 1)^{1/4} - 1$

17. $f(x) = -2(x + 1)^{2/3} - 2$

18. $f(x) = -3(x - 1)^{3/4} + 2$

19. $f(x) = (1 - x)^{4/3} - 1$

20. $f(x) = (4 - x)^{3/2} + 1$

21. $f(x) = x^{-1/2} - 2$

22. $f(x) = x^{-1/3} + 1$

23. $f(x) = \sqrt{\dfrac{x + 2}{x - 1}}$

24. $f(x) = \sqrt{\dfrac{x - 1}{x + 2}}$

25. $f(x) = \sqrt{\dfrac{2 - x}{x + 2}}$

26. $f(x) = \sqrt{\dfrac{1 - x}{x + 3}}$

27. $f(x) = \dfrac{x}{\sqrt{x^2+1}}$

28. $f(x) = \dfrac{3x}{\sqrt{4x^2+9}}$

29. $f(x) = \dfrac{x-1}{\sqrt{(x+1)(x-2)}}$

30. $f(x) = \dfrac{x}{\sqrt{x^2+3x+2}}$

31. $f(x) = \dfrac{x}{\sqrt{4-x^2}}$

32. $f(x) = -\dfrac{x}{\sqrt{9-x^2}}$

33. Heavier birds have greater surface areas and larger wingspans. If the relationship between surface area and weight of a certain species of bird is given by $S(w) = 0.3w^{2/3}$, estimate the weight, in kilograms, of a bird with a surface area of $0.35\ \text{m}^2$.

34. The basal metabolic rate (BMR) in humans is essentially a measure of the amount of energy expended at rest. The relationship between BMR and body weight is assumed to be $B(w) = 3.8w^{0.75}$, where w is given in pounds. Determine your own basal metabolic rate.

3.6 COMPLEX ROOTS OF POLYNOMIALS

We have seen throughout this chapter that determining where the graph of a function crosses the x-axis gives us a zero of the function. What we have avoided to this point is the fact that some of the most elementary functions have graphs that fail to cross the x-axis. Consider, for example, the graphs of

$$f(x) = x^2 - 1, \quad g(x) = x^2, \quad \text{and} \quad h(x) = x^2 + 1$$

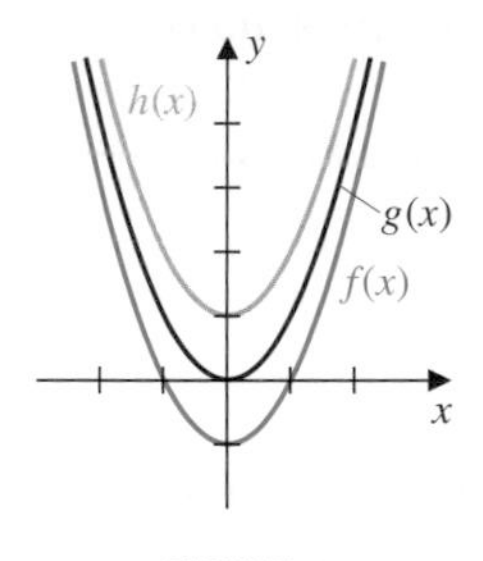

FIGURE 1

shown in Figure 1. The function f has two distinct zeros, one at $x = -1$ and the other at $x = 1$. The function g has a single zero of multiplicity 2 at $x = 0$. The graph of h does not cross the x-axis, so there is no real number that makes $h(x) = 0$.

To examine the zeros of functions like h, we extend our number system once more, this time to the set of *complex numbers*. To define the set of complex numbers we first introduce a number i whose square is -1.

The Square Root of −1

We define

$$i = \sqrt{-1} \quad \text{so that} \quad i^2 = -1.$$

The number i cannot be a real number, because there are no real numbers x that satisfy $x^2 + 1 = 0$, and by definition we have $i^2 + 1 = -1 + 1 = 0$. The *imaginary number* i is used to separate the ordinary real numbers we have used in the past from those new numbers constructed in the *complex number* system.

Complex Numbers

The set of **complex numbers** consists of all expressions of the form

$$z = a + bi,$$

where a and b are real numbers. This set is denoted $\mathbb{C}$. The number a is called the **real part** of the complex number z, and the number b is called the **imaginary part** of z.

When $b = 0$, the complex number is simply a real number, so all real numbers are also complex numbers. To simplify the language in this section, however, when we speak of complex numbers we generally assume that the imaginary part is not zero.

Two complex numbers are equal when their real and imaginary parts are the same. So

$$a + b\,i = c + d\,i \quad \text{precisely when} \quad a = c \quad \text{and} \quad b = d.$$

Addition and subtraction of complex numbers are defined in a natural way, if we keep in mind that the real and imaginary parts of a complex number are kept separate:

$$(a + b\,i) + (c + d\,i) = (a + c) + (b + d)\,i,$$

and

$$(a + b\,i) - (c + d\,i) = (a - c) + (b - d)\,i.$$

Multiplication is also straightforward if we use the standard rules of algebra and the fact that $i^2 = -1$. So

$$\begin{aligned}(a + b\,i)(c + d\,i) &= ac + bc\,i + ad\,i + bd\,i^2\\ &= ac + bc\,i + ad\,i - bd = (ac - bd) + (bc + ad)\,i.\end{aligned}$$

EXAMPLE 1 Determine the sum and product of the complex numbers $2 + 3\,i$ and $5 - 7\,i$.

Solution The definitions of complex addition and multiplication give

$$(2 + 3\,i) + (5 - 7\,i) = (2 + 5) + (3 + (-7))\,i = 7 - 4\,i$$

and

$$\begin{aligned}(2 + 3\,i) \cdot (5 - 7\,i) &= 2 \cdot 5 + 2 \cdot (-7\,i) + (3\,i) \cdot 5 + (3\,i) \cdot (-7\,i)\\ &= 10 - 14\,i + 15\,i - 21\,i^2\\ &= 10 + i - 21(-1) = 31 + i.\end{aligned}$$

■

The Quadratic Formula

The Quadratic Formula tells us that the solutions to the quadratic equation $ax^2 + bx + c = 0$ are

$$x = -\frac{b}{2a} \pm \frac{\sqrt{b^2 - 4ac}}{2a}.$$

The sign of $b^2 - 4ac$, called the *discriminant* of the quadratic, determines the nature of the solution.

When $b^2 - 4ac > 0$, there are two real solutions to this equation,

$$x_1 = -\frac{b}{2a} - \frac{\sqrt{b^2 - 4ac}}{2a} \quad \text{and} \quad x_2 = -\frac{b}{2a} + \frac{\sqrt{b^2 - 4ac}}{2a}.$$

When $b^2 - 4ac = 0$, there is a single solution of multiplicity 2,

$$x = -\frac{b}{2a}.$$

However, when $b^2 - 4ac < 0$, the equation has two complex solutions. To determine these in standard complex form, we first observe that

$$b^2 - 4ac < 0, \quad \text{implies that} \quad 4ac - b^2 > 0.$$

So $\sqrt{4ac - b^2}$ is a real number, and

$$\sqrt{b^2 - 4ac} = \sqrt{(4ac - b^2)(-1)} = \sqrt{4ac - b^2}\sqrt{-1} = \sqrt{4ac - b^2}\, i.$$

The two solutions to the equation $ax^2 + bx + c = 0$ when $b^2 - 4ac < 0$ are expressed in standard complex number form as

$$x = -\frac{b}{2a} - \frac{\sqrt{4ac - b^2}}{2a} i \quad \text{and} \quad x = -\frac{b}{2a} + \frac{\sqrt{4ac - b^2}}{2a} i.$$

These are commonly expressed together as

$$x = -\frac{b}{2a} \pm \frac{\sqrt{4ac - b^2}}{2a} i.$$

Solutions to the Quadratic Equation $ax^2 + bx + c = 0$

The solutions depend on the sign of the **discriminant** $b^2 - 4ac$.

- If $b^2 - 4ac > 0$, there are two real solutions,

$$x = -\frac{b}{2a} + \frac{\sqrt{b^2 - 4ac}}{2a} \quad \text{and} \quad x = -\frac{b}{2a} - \frac{\sqrt{b^2 - 4ac}}{2a}.$$

- If $b^2 - 4ac = 0$, there is a single real solution of multiplicity 2,

$$x = -\frac{b}{2a}.$$

- If $b^2 - 4ac < 0$, there are two complex solutions,

$$x = -\frac{b}{2a} + \frac{\sqrt{4ac - b^2}}{2a} i \quad \text{and} \quad x = -\frac{b}{2a} - \frac{\sqrt{4ac - b^2}}{2a} i.$$

EXAMPLE 2 Determine all the solutions to the equation $x^2 - 4x + 5 = 0$.

Solution The Quadratic Formula applied to this equation gives

$$x = \frac{-(-4)}{2(1)} \pm \frac{\sqrt{(-4)^2 - 4(1)(5)}}{2(1)} = 2 \pm \frac{1}{2}\sqrt{-4} = 2 \pm \frac{1}{2}(2i).$$

So the two solutions are

$$x = 2 + i \quad \text{and} \quad x = 2 - i.$$

The quadratic $x^2 - 4x + 5$ can consequently be factored in complex form as

$$x^2 - 4x + 5 = (x - (2 + i)) \cdot (x - (2 - i)).$$

The graph of $f(x) = x^2 - 4x + 5$ is easy to determine. Completing the square gives

$$f(x) = x^2 - 4x + 5 = (x^2 - 4x + 4) + 1 = (x - 2)^2 + 1.$$

Note that the graph, shown in red in Figure 2, does not cross the x-axis, so there are no x-axis intercepts to the graph. This implies that there are no *real* numbers x satisfying $x^2 - 4x + 5 = 0$. So the solutions must be complex numbers. ■

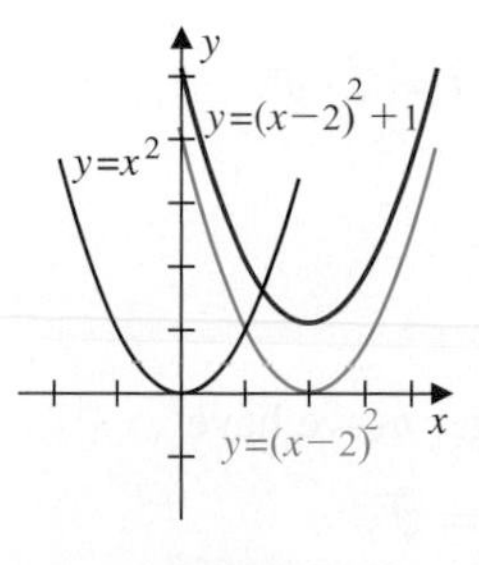

FIGURE 2

The two complex solutions to the quadratic equation in Example 2 have the same real part, and imaginary parts that differ only in sign. These two numbers are called *complex conjugates*.

Complex Conjugate

The **complex conjugate** of the complex number $z = a + bi$ is the complex number $\overline{z} = a - bi$. For any complex number z,

$$z + \overline{z} = (a + bi) + (a - bi) = 2a$$

and

$$z \cdot \overline{z} = (a + bi) \cdot (a - bi) = (a^2 - b^2 i^2) + (ba - ab)i = a^2 + b^2.$$

So $z + \overline{z}$ and $z \cdot \overline{z}$ are both real numbers.

This definition and the Quadratic Formula imply that a quadratic equation with real coefficients has complex zeros occurring in conjugate pairs.

EXAMPLE 3 Determine the complex conjugates of $z = 4 + 3i$ and $w = 2 - 5i$. Then show that $\overline{z} + \overline{w} = \overline{z + w}$, and that $\overline{z} \cdot \overline{w} = \overline{z \cdot w}$.

Solution The conjugates of z and w are

$$\overline{z} = 4 - 3i \quad \text{and} \quad \overline{w} = 2 + 5i,$$

so

$$\overline{z} + \overline{w} = 6 + 2i \quad \text{and} \quad \overline{z} \cdot \overline{w} = (8 + 15) + (20 - 6)i = 23 + 14i.$$

Since

$$z + w = (4 + 3i) + (2 - 5i) = 6 - 2i$$

and

$$z \cdot w = (4 + 3i) \cdot (2 - 5i) = (8 + 15) + (6 - 20)i = 23 - 14i,$$

we also have

$$\overline{z + w} = 6 + 2i \quad \text{and} \quad \overline{z \cdot w} = 23 + 14i.$$

■

In Example 3 we saw that

$$\overline{z + w} = 6 + 2i = \overline{z} + \overline{w} \quad \text{and} \quad \overline{z \cdot w} = 23 + 14i = \overline{z} \cdot \overline{w}.$$

Results of this type are true in general.

Conjugate Results

For any pair of complex numbers z and w and for any integer n, we have

$$\overline{z} + \overline{w} = \overline{z + w}, \quad \overline{z} \cdot \overline{w} = \overline{z \cdot w}, \quad \text{and} \quad \overline{z}^n = \overline{z^n}.$$

The fact that the product of a complex number and its conjugate is a real number gives us a way to define the quotient of complex numbers in standard complex form.

Let $z = a + bi$ and $w = c + di$, where $w \neq 0$. Then

$$\frac{z}{w} = \frac{z}{w} \cdot \frac{\overline{w}}{\overline{w}} = \frac{a + b\,i}{c + d\,i} \cdot \frac{c - d\,i}{c - d\,i} = \frac{ac + bd}{c^2 + d^2} + \frac{bc - ad}{c^2 + d^2}\,i.$$

For the pair of complex numbers we considered in Example 1, the quotient can be written in standard complex form as

$$\frac{2 + 3i}{5 - 7i} = \frac{2 + 3i}{5 - 7i} \cdot \frac{5 + 7i}{5 + 7i} = \frac{-11}{74} + \frac{29}{74}\,i = -\frac{11}{74} + \frac{29}{74}\,i.$$

We can now summarize all the results of elementary complex arithmetic.

Complex Arithmetic

For any pair of complex numbers $z = a + b\,i$ and $w = c + d\,i$, we have

- $z + w = (a + c) + (b + d)\,i$
- $z \cdot w = (ac - bd) + (bc + ad)\,i$
- $\dfrac{z}{w} = \dfrac{ac + bd}{c^2 + d^2} + \dfrac{bc - ad}{c^2 + b^2}\,i$, if $w \neq 0$

We have seen from the Quadratic Formula that complex zeros of quadratic equations with real coefficients occur in conjugate pairs. This result holds in a much broader situation. If z is a complex zero of any polynomial with real coefficients

$$P(x) = a_n x^n + a_{n-1} x^{n-1} + \cdots + a_1 x + a_0,$$

then

$$a_n z^n + a_{n-1} z^{n-1} + \cdots + a_1 z + a_0 = 0.$$

Since $0, a_n, a_{n-1}, \ldots, a_1$, and a_0 are all real numbers, their complex conjugates are simply themselves. The conjugate results imply that

$$\begin{aligned}
0 = \overline{0} &= \overline{a_n z^n + a_{n-1} z^{n-1} + \cdots + a_1 z + a_0} \\
&= \overline{a_n z^n} + \overline{a_{n-1} z^{n-1}} + \cdots + \overline{a_1 z} + \overline{a_0} \\
&= a_n \overline{z^n} + a_{n-1} \overline{z^{n-1}} + \cdots + a_1 \overline{z} + a_0 \\
&= a_n \overline{z}^n + a_{n-1} \overline{z}^{n-1} + \cdots + a_1 \overline{z} + a_0.
\end{aligned}$$

So if z is a complex zero of a real polynomial $P(x)$, then $\overline{z}$ is also a zero of $P(x)$.

EXAMPLE 4 Verify that $x = i$ is a zero of the polynomial $P(x) = x^3 - x^2 + x - 1$, and factor the polynomial into all linear terms.

Solution Substituting $x = i$ gives

$$P(i) = i^3 - i^2 + i - 1 = -i + 1 + i - 1 = 0,$$

so $x = i$ is a zero of the polynomial. Since $P(i) = 0$, the conjugate $\overline{i} = -i$ also satisfies $P(-i) = 0$. Thus, i and $-i$ are zeros of the polynomial, and $x - i$ and

$x - (-i) = x + i$ are factors. Dividing $(x - i)(x + i) = x^2 + 1$ into $P(x)$ gives

$$P(x) = (x^2 + 1)(x - 1) = (x - i)(x + i)(x - 1)$$

and the polynomial is factored into all linear terms, two of which involve complex conjugates. ■

EXAMPLE 5 One solution to the equation

$$P(x) = x^4 - 6x^3 + 14x^2 - 22x + 5 = 0$$

is the complex number $x = 1 + 2i$. Determine all the solutions to this equation.

Solution Since $x = 1 + 2i$ is a solution, its complex conjugate $x = 1 - 2i$ is also a solution. So the quadratic term

$$\begin{aligned}(x - (1 + 2i))\,(x - (1 - 2i)) &= x^2 - (1 + 2i + 1 - 2i)x + (1 + 2i)(1 - 2i) \\ &= x^2 - 2x + 5\end{aligned}$$

is a factor of the polynomial $x^4 - 6x^3 + 14x^2 - 22x + 5$. Division of $P(x)$ by $x^2 - 2x + 5$ gives

$$x^4 - 6x^3 + 14x^2 - 22x + 5 = (x^2 - 2x + 5)(x^2 - 4x + 1).$$

We can now apply the Quadratic Formula to find the solutions to $x^2 - 4x + 1 = 0$, which are the remaining solutions to the original equation. The Quadratic Formula gives

$$x = \frac{-(-4)}{2(1)} \pm \frac{\sqrt{(-4)^2 - 4(1)(1)}}{2(1)} = 2 \pm \sqrt{3}.$$

The four solutions to the equation

$$x^4 - 6x^3 + 14x^2 - 22x + 5 = 0$$

are

$$x = 1 + 2i, \quad x = 1 - 2i, \quad x = 2 + \sqrt{3}, \quad \text{and} \quad x = 2 - \sqrt{3},$$

and the polynomial is factored into all linear terms as

$$x^4 - 6x^3 + 14x^2 - 22x + 5 = (x - (2 + \sqrt{3}))(x - (2 - \sqrt{3}))(x - (1 + 2i))(x - (1 - 2i)).$$

■

We have seen many results in this chapter concerning the zeros of polynomials, but have not discussed a fundamental question. Which polynomials have zeros, and how many do they have? The answer is that all polynomials have zeros, and, in a sense, as many zeros as the degree of the polynomial. The first result in this direction is the Fundamental Theorem of Algebra.

The Fundamental Theorem of Algebra

If

$$P(x) = a_n x^n + a_{n-1} x^{n-1} + \cdots + a_1 x + a_0$$

is a polynomial of degree $n > 0$ with real or complex coefficients, then P has at least one real or complex zero. In fact, P has precisely n zeros, provided that a zero of multiplicity m is counted m times.

The Fundamental Theorem of Algebra was first stated by the French mathematician Albert Girard in 1629, but with no accompanying demonstration of truth. Most of the great mathematicians in the 17th and 18th centuries tried to prove this result, including the foremost scientist of all time, Isaac Newton, and the most prolific mathematician of record, Leonhard Euler. But it was Carl Friedrich Gauss, considered by many to be the greatest mathematician of all time, who first proved this Fundamental Theorem for his doctoral dissertation in 1799, when he was only 20. The proof is much more difficult than the statement of the theorem would lead you to believe. All four proofs that Gauss eventually constructed introduced new mathematical ideas.

In addition, we have the following result.

Conjugate Pairs of Zeros of Real Polynomials

Suppose the coefficients of

$$P(x) = a_n x^n + a_{n-1}x^{n-1} + \cdots + a_1 x + a_0$$

are all real numbers and z is a complex zero of P with multiplicity m. Then its complex conjugate $\bar{z}$ is also a zero of P with multiplicity m.

Because complex zeros of polynomials with real coefficients occur in conjugate pairs of the same multiplicity, a polynomial of even degree might have only complex zeros. But a polynomial of odd degree must have at least one zero that is real. If we add the information from the Intermediate Value Theorem, we can often determine the character of the zeros, if not their precise location. This is illustrated in Example 6.

EXAMPLE 6 Determine possibilities for the zeros of the polynomial.

a. $P(x) = x^3 - 5x^2 + 2x + 1$

b. $Q(x) = 2x^5 - 4x^4 + 3x^3 - x + 3$

Solution **a.** Since $P(0) > 0$ and $P(1) < 0$, there is at least one positive zero in the interval $(0, 1)$. In addition, $P(4) < 0$ and $P(5) > 0$, so a positive zero lies in the interval $(4, 5)$. Because $P(x)$ is of degree 3, it has only one additional zero. The third zero cannot be complex, because complex zeros come in conjugate pairs. So the remaining zero is also real. Knowing this, we can evaluate $P(x)$ at other values of x to determine where the zero lies. Since $P(-1) < 0$ and $P(0) > 0$, a negative zero lies in the interval $(-1, 0)$, as shown in Figure 3.

b. The degree of the polynomial $Q(x)$ is odd, so it must have at least one real zero. Since complex zeros occur in conjugate pairs, the number of real zeros must be either 1, 3, or 5. One zero lies in the interval $(-1, 0)$, because $Q(-1) < 0$ and $Q(0) > 0$.

The computer-generated graph of $y = Q(x)$ in Figure 4 shows only one real root, so $Q(x)$ has four complex roots in two conjugate pairs. ■

FIGURE 3

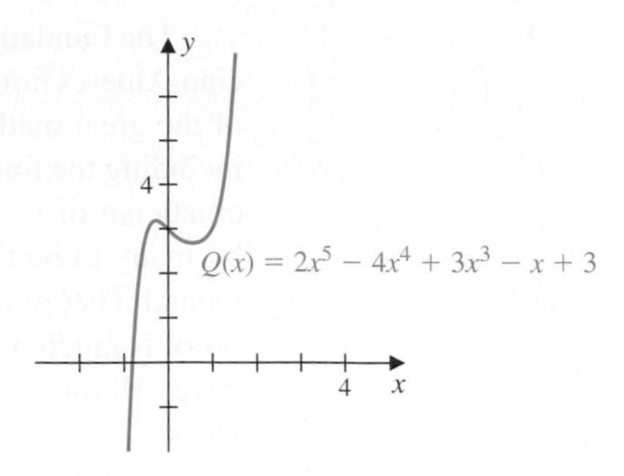

FIGURE 4

EXAMPLE 7 Find a polynomial of degree 5 that has

i. real coefficients,

ii. the leading coefficient 1, and

iii. zeros at i, $1+i$, and -2.

Solution The polynomial has real coefficients, so for each complex zero, the conjugate is also a zero. Hence the five zeros of the polynomial are $i, -i, 1+i, 1-i$, and -2. A polynomial with these zeros and leading coefficient 1 is

$$\begin{aligned} P(x) &= (x-i)(x-(-i))(x-(1+i))(x-(1-i))(x-(-2)) \\ &= (x-i)(x+i)(x-(1+i))(x-(1-i))(x+2) \\ &= (x^2+1)(x^2-2x+2)(x+2) = x^5 - x^3 + 4x^2 - 2x + 4. \end{aligned}$$

■

EXERCISE SET 3.6

In Exercises 1–30, write the complex number in the form $a + bi$.

1. $(-3+i)+(2-3i)$ **2.** $(3-i)+(2+4i)$

3. $(-3+5i)-(2-3i)$ **4.** $(5-7i)-(2-4i)$

5. $3i \cdot (2+i)$ **6.** $i \cdot (-2-i)$

7. $(2-i)\cdot(3+i)$ **8.** $(5-2i)\cdot(4-i)$

9. $(6+5i)\cdot(-3-2i)$ **10.** $(3-7i)\cdot(6-2i)$

11. $(2-3i)\cdot(2+3i)$ **12.** $(-3+4i)\cdot(-3-4i)$

13. $(3-8i)\cdot\overline{(2+i)}$ **14.** $(2-i)\cdot\overline{(2-i)}$

15. i^5 (Hint: $i^5 = i^4 \cdot i$.) **16.** i^6

17. i^{104} **18.** i^{101}

19. $\sqrt{-9}$ **20.** $\sqrt{-25}$

21. $(2+\sqrt{-5})\cdot(1+\sqrt{-1})$

22. $(1-\sqrt{-5})\cdot(1+\sqrt{-5})$

23. $\sqrt{-\frac{16}{9}}$ **24.** $\sqrt{-\frac{25}{4}}$

25. $\dfrac{1}{1-i}$ **26.** $\dfrac{3}{i}$

27. $\dfrac{1}{2+3i}$ **28.** $\dfrac{1}{3-4i}$

29. $\dfrac{1-4i}{1+4i}$ **30.** $\dfrac{5+6i}{5-6i}$

*In Exercises 31–36, **(a)** find the zeros of the quadratic function and **(b)** write the function in factored form.*

31. $f(x) = x^2 + 4$

32. $f(x) = 2x^2 + 18$

33. $f(x) = x^2 - 2x + 2$

34. $f(x) = x^2 - 3x + 3$

35. $f(x) = 2x^2 - x + 2$

36. $f(x) = 3x^2 + 2x + 1$

In Exercises 37–42, find all solutions of the equation.

37. $x^4 - 4 = 0$ **38.** $x^4 - 9 = 0$

39. $x^4 - x^2 - 6 = 0$

40. $x^4 + 2x^2 - 8 = 0$

41. $x^3 + 8 = 0$

42. $x^3 - 8 = 0$

In Exercises 43–46, given that the value of x is a solution of the equation, find all solutions.

43. $x^3 - 2x^2 + 9x - 18 = 0; \quad x = 3i$

44. $x^3 + 3x^2 + 16x - 20 = 0; \quad x = -2 + 4i$

45. $x^4 - 2x^3 - 2x^2 - 2x - 3 = 0; \quad x = i$

46. $x^5 - 2x^4 - 2x^3 + 8x^2 - 8x = 0; \quad x = 1 + i$

In Exercises 47–50, find all zeros of the given function.

47. $f(x) = x^3 - 3x^2 + 9x - 27$

48. $f(x) = x^3 + 2x^2 + 2x + 1$

49. $f(x) = x^4 - x^2 - 2x + 2$

50. $f(x) = x^4 + 3x^3 + 4x^2 + 3x + 1$

In Exercises 51–56, find the polynomial with leading coefficient 1 that satisfies the given conditions.

51. Degree 3, and zeros 2 and $2i$.

52. Degree 3, and zeros 1 and $1 - i$.

53. Degree 4, and zeros $\sqrt{3}i$ and $3i$.

54. Degree 4, a zero of multiplicity 2 at 1, and a zero at $2 + i$.

55. Degree 5, a zero of multiplicity 2 at -2, a zero at 0, and a zero at $1 + 2i$.

56. Degree 5, zeros i and $3 - i$, and passing through the origin.

57. A problem states that the sum of a number and its reciprocal is $\frac{b}{a}$, where a and b are positive real numbers. Find conditions on a and b that guarantee that this problem has a real number solution.

REVIEW EXERCISES FOR CHAPTER 3

1. Match the equation with the graph.

a. $y = (x + 1)^2(x + 2)(x - 1)$

b. $y = x(x - 3)(x - 2)^2$

c. $y = (x - 1)^3(x + 2)$

d. $y = (x - 1)^2(x + 1)^2$

i.

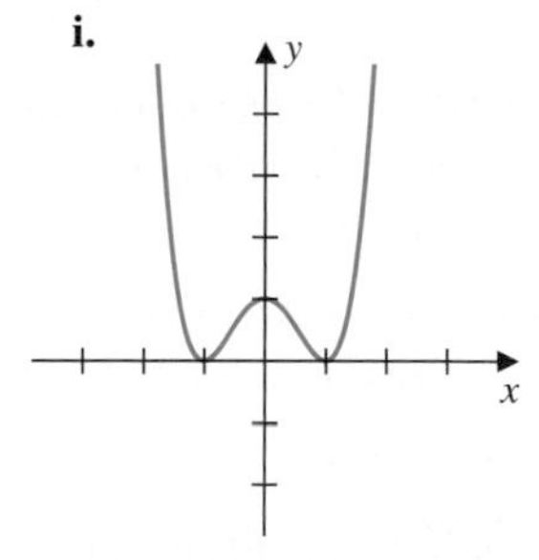

ii.

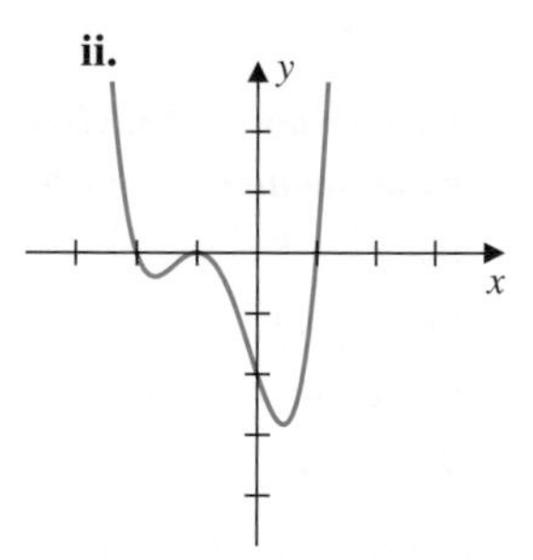

iii.

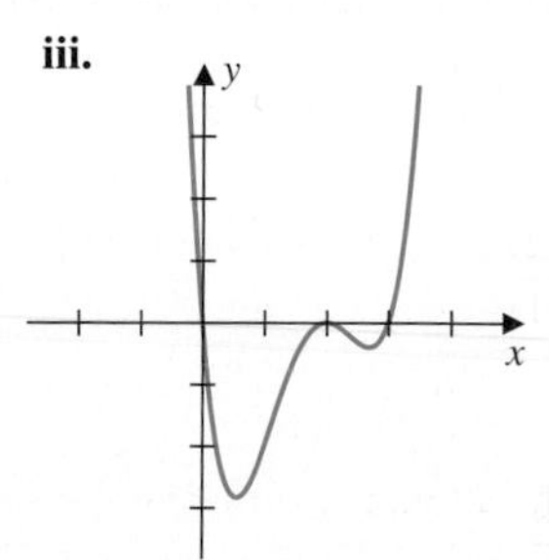

iv.

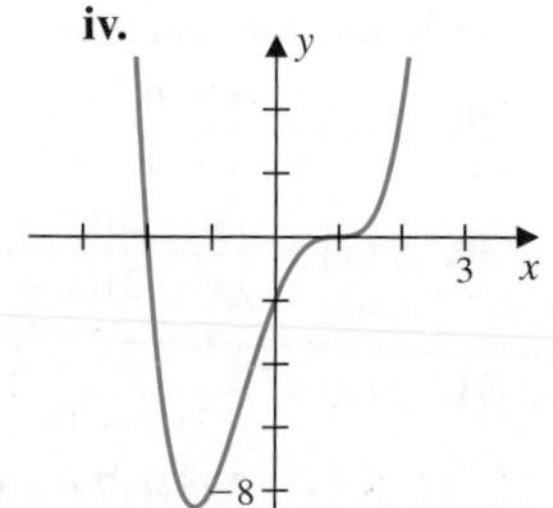

2. Match the equation with the graph.

a. $y = \dfrac{x}{x^2 - 1}$

b. $y = \dfrac{x^2}{x^2 - 1}$

c. $y = \dfrac{2x}{x + 2}$

d. $y = -\dfrac{x}{x + 2}$

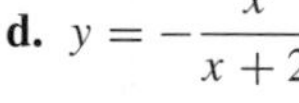

i.

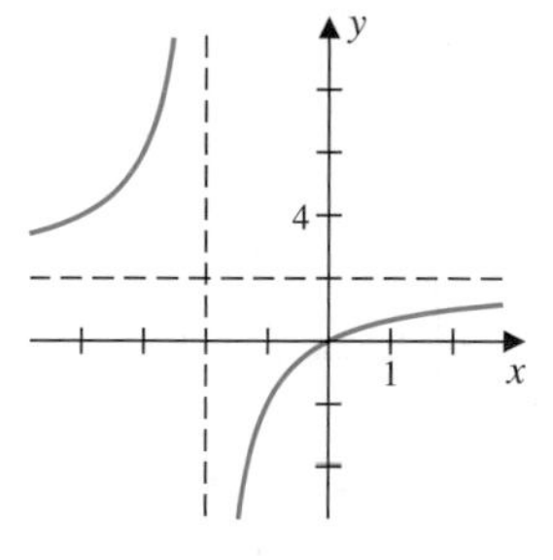

ii

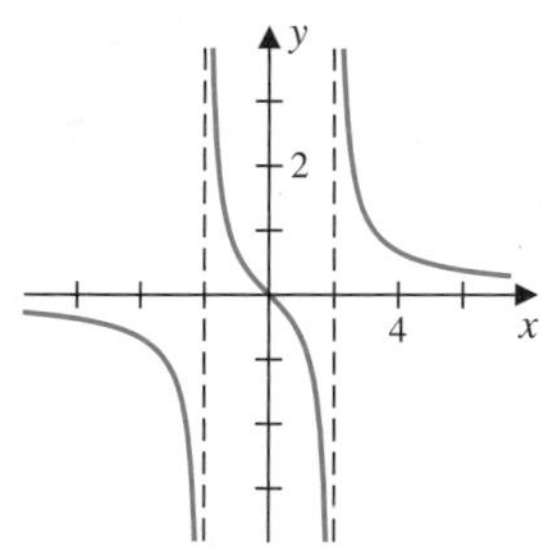

iii.

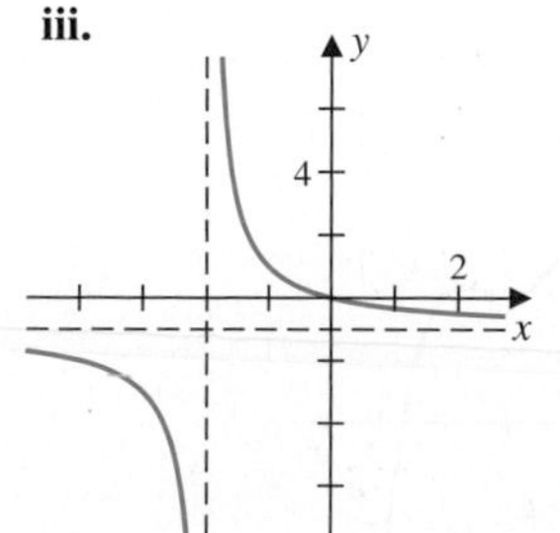

iv

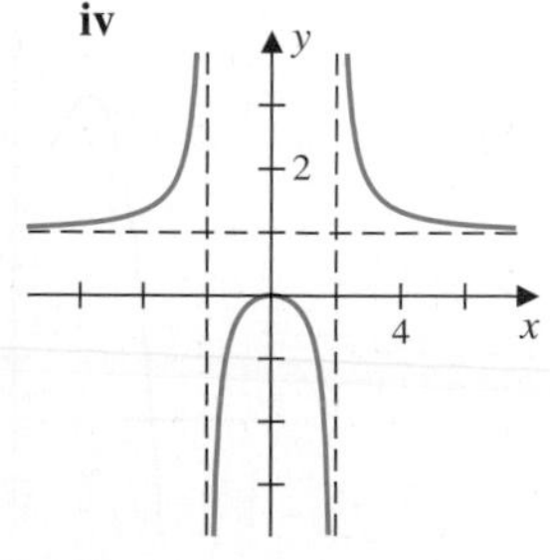

In Exercises 3–6, sketch the graph of the given function by transforming a curve of the form $y = x^n$.

3. $f(x) = -2(x-1)^2 + 2$

4. $f(x) = -2(x-2)^3 - 1$

5. $f(x) = -x^4 - 3$ **6.** $f(x) = (x+2)^4 - 1$

In Exercises 7–12, give a reasonable sketch of the graph of the function by plotting the axis intercepts and using the end behavior.

7. $f(x) = (x+1)(x+2)(x-3)$

8. $f(x) = x^2(x-1)$

9. $f(x) = \frac{1}{2}(x-2)^3(x+1)$

10. $f(x) = -\frac{1}{16}(x-2)^3(x+1)(x+2)$

11. $f(x) = x^3 - \frac{1}{2}x^2 - \frac{1}{2}x$

12. $f(x) = x^5 + 2x^4 + 4x + 8$

In Exercises 13–16, the graph of a polynomial is given. What is the lowest possible degree for the polynomial, and what is the sign of the leading coefficient?

13.

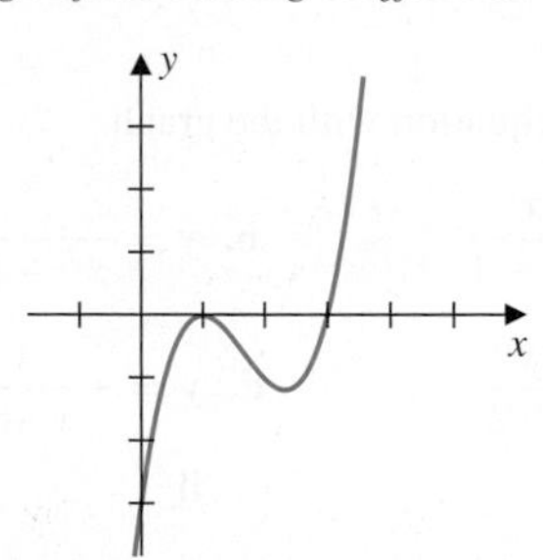

14.

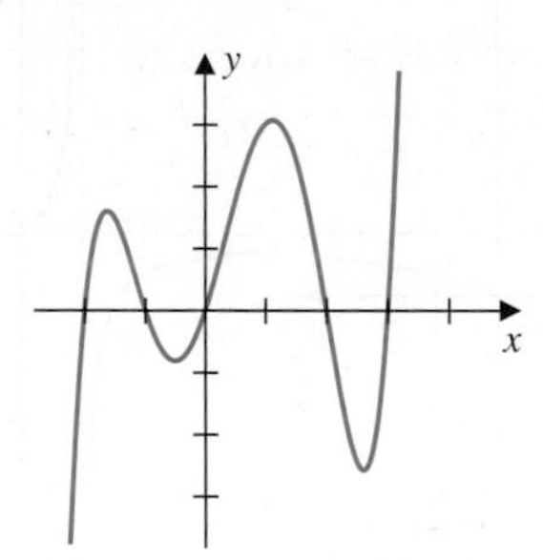

15.

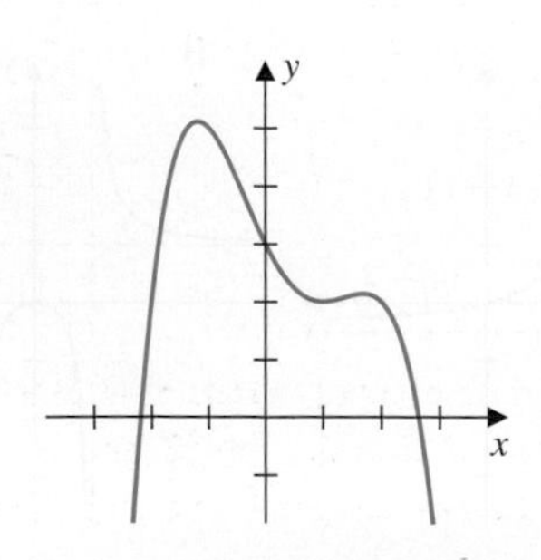

16.

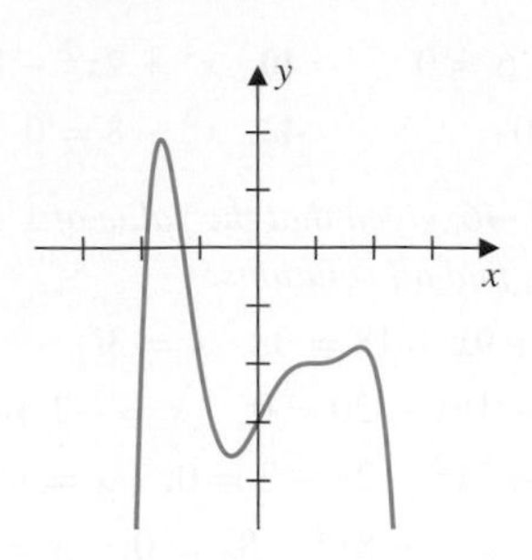

In Exercises 17–20, find the quotient $Q(x)$ and remainder $R(x)$ when $P(x)$ is divided by $D(x)$.

17. $P(x) = 2x^2 + 3x - 4$; $D(x) = x - 3$

18. $P(x) = 3x^2 - 4x + 7$; $D(x) = -2x + 1$

19. $P(x) = 3x^3 + 2x^2 - 2x + 1$; $D(x) = x + 2$

20. $P(x) = x^5 + x^4 - 5x^2 - 2x - 3$; $D(x) = x^4 - 2x^3 + x + 1$

In Exercises 21–24, ***(a)*** *list all the possibilities for rational zeros,* ***(b)*** *use the Factor Theorem to show that $x - c$ is a factor of the polynomial $P(x)$ for the given value of c, and* ***(c)*** *factor $P(x)$ completely in terms of real factors.*

21. $P(x) = 3x^4 - 9x^3 - 2x^2 + 5x + 3,\ c = 3$

22. $P(x) = x^4 + 2x^3 - 5x^2 - 6x + 8,\ c = -2$

23. $P(x) = x^5 - 3x^4 - 5x^3 + 27x^2 - 32x + 12,$ $c = -3$

24. $P(x) = x^6 - 5x^5 + 5x^4 + 9x^3 - 14x^2 - 4x + 8,$ $c = 1$

In Exercises 25–28, given that the value of x is a solution of the equation, factor the polynomial completely.

25. $x^4 - 5x^3 + 2x^2 + 22x - 20 = 0; x = 3 + i$

26. $x^4 - 4x^3 + 25x^2 - 36x + 144 = 0; x = 3i$

27. $x^5 - x^4 + 10x^3 - 10x^2 + 9x - 9 = 0; x = i$

28. $x^5 - x^4 + 4x - 4 = 0; x = 1 + i$

In Exercises 29–34, find ***(a)*** *the domain,* ***(b)*** *any axis intercepts, and* ***(c)*** *any vertical and horizontal asymptotes of the rational function.*

29. $f(x) = \dfrac{x-4}{x-1}$

30. $f(x) = \dfrac{(x-2)(x+2)}{(x-3)(x+1)}$

31. $f(x) = \dfrac{x^2 - 2x + 1}{2x^2 - 18}$

32. $f(x) = \dfrac{3x^2 + 7x - 6}{x^2 - x - 6}$

33. $f(x) = \dfrac{x^3 - x^2 - 4x + 4}{x^3}$

34. $f(x) = \dfrac{x^4 - 2x^3 + x^2}{x^3 - 1}$

In Exercises 35–42, sketch the graph showing any axis intercepts and asymptotes.

35. $f(x) = \dfrac{3}{x - 2}$ **36.** $f(x) = \dfrac{-4}{x + 2}$

37. $f(x) = \dfrac{-2}{(x - 1)(x - 2)}$

38. $f(x) = \dfrac{2}{(x + 2)(x - 4)}$

39. $f(x) = \dfrac{4}{x^2 - 4}$ **40.** $f(x) = \dfrac{x - 5}{x^2 - 2x - 3}$

41. $f(x) = \dfrac{x^2 - 4}{x^2 + 5x}$ **42.** $f(x) = \dfrac{4 - x^2}{x^2 - 9}$

In Exercises 43 and 44, each graph has a slant asymptote. Sketch the graph showing any vertical and slant asymptotes and the axis intercepts.

43. $f(x) = \dfrac{x^2 - 2x + 1}{x + 1}$

44. $f(x) = \dfrac{x^3 - 2x^2 + 4x - 3}{x^2 - 3x + 2}$

*In Exercises 45–50, **(a)** find the domain of the function, **(b)** draw a rough sketch of the graph showing axis intercepts and asymptotes, and **(c)** use a graphing device to check your work.*

45. $f(x) = \sqrt{\dfrac{x - 2}{x + 1}}$ **46.** $f(x) = \sqrt{\dfrac{x - 5}{x + 5}}$

47. $f(x) = \sqrt{\dfrac{4 - x}{x + 4}}$

48. $f(x) = \dfrac{x - 3}{\sqrt{(x - 1)(x + 2)}}$

49. $f(x) = \dfrac{x^2}{\sqrt{9 - x^2}}$ **50.** $f(x) = \dfrac{\sqrt{x^2 - 9}}{x - 2}$

In Exercises 51–54, use a graphing device to sketch the graph of the given function and estimate:

(a) Intervals where the function is increasing and where it is decreasing;

(b) Local maximum and local minimum points.

51. $f(x) = x^3 - 2x^2 - x + 2$

52. $f(x) = x^4 - 2x^3$

53. $f(x) = \dfrac{2}{3}x^3 + \dfrac{7}{2}x^2 - 12x$

54. $f(x) = x^5 + 4x^4 - 8x^2 + 4x$

In Exercises 55–64, write the complex number in the form $a + bi$.

55. $\dfrac{1}{4}(2 + i\sqrt{2})$ **56.** $\dfrac{1}{6}(-5 - \sqrt{-4})$

57. $(2 - i) - (3 - 2i)$ **58.** $(-2 + 6i) + (-3 + i)$

59. $(2 - i) \cdot \overline{(2 + i)}$ **60.** $(4 - 6i) \cdot \overline{(3 - 2i)}$

61. i^{20} **62.** i^{21}

63. $\dfrac{2 + 3i}{4 - 7i}$ **64.** $\dfrac{-5 + 3i}{2 - 3i}$

65. Sketch each of the following curves by shifting the graph in the accompanying figure and determine the coordinates of the labeled point.

a. $y = f(x - 1)$ **b.** $y = f(x - 1) - 1$

c. $y = -f(x + 1) + 1$ **d.** $y = |f(x)|$

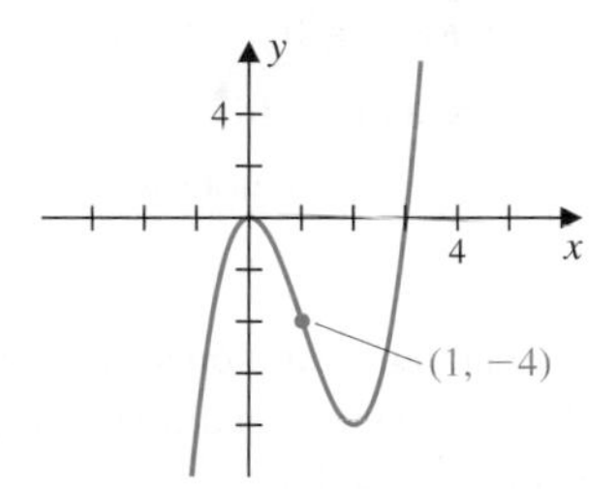

66. Sketch each of the following curves by shifting the graph in the accompanying figure, showing the horizontal and vertical asymptotes.

a. $y = f(x - 1) + 1$ **b.** $y = -f(x) + 1$

c. $y = f(-x) - 2$ **d.** $y = |f(x)|$

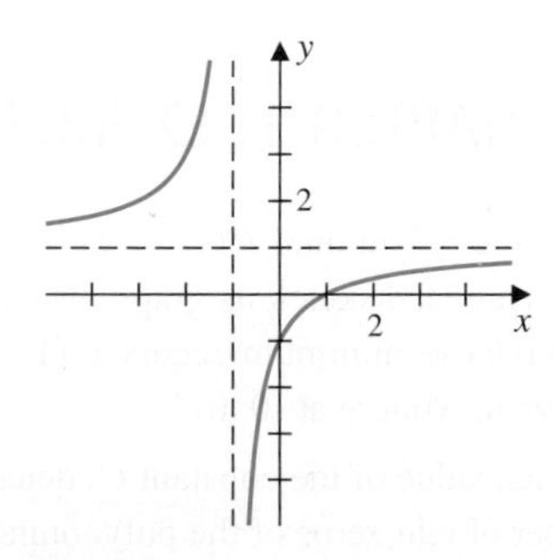

67. Sketch a graph of a polynomial $P(x)$ that has zeros of multiplicity 1 at $x = 2$, $x = -2$, and $x = 3$ and satisfies $P(x) \to \infty$ as $x \to \infty$ and $P(x) \to -\infty$ as $x \to -\infty$.

68. Sketch a graph of a polynomial $P(x)$ that has a zero of multiplicity 1 at $x = 0$, a zero of multiplicity 2 at $x = -2$, a zero of multiplicity 3 at $x = 3$, and satisfies $P(x) \to \infty$ as $x \to \infty$ and $P(x) \to \infty$ as $x \to -\infty$.

69. The cubic polynomial $P(x)$ has zeros at $x = 1$, $x = 3$, and $x = -1$, and $P(0) = 1$. Find $P(x)$ and sketch its graph.

70. The fourth-degree polynomial $P(x)$ has zeros of multiplicity 1 at $x = 1$ and $x = -1$, a zero of multiplicity 2 at $x = 2$, and $P(-2) = 2$. Find $P(x)$ and sketch its graph.

71. **(a)** Sketch the graph of a rational function that satisfies all the following conditions and **(b)** find an expression for the function.

i. $f(x) \to \infty$ as $x \to 2^+$

ii. $f(x) \to -\infty$ as $x \to 2^-$

iii. $f(x) \to \infty$ as $x \to 0^+$

iv. $f(x) \to -\infty$ as $x \to 0^-$

v. Has a horizontal asymptote $y = 0$.

vi. $f(1) = 0$

72. **(a)** Sketch the graph of a rational function that satisfies all the following conditions and **(b)** find an expression for the function.

i. $f(x) \to \infty$ as $x \to 1^+$

ii. $f(x) \to -\infty$ as $x \to 1^-$

iii. $f(x) \to -\infty$ as $x \to -2^+$

iv. $f(x) \to \infty$ as $x \to -2^-$

v. Has a horizontal asymptote $y = 0$.

vi. Never crosses the x-axis.

vii. Passes through $(-1, -2)$.

73. Sketch the curves $f(x) = x^3$ and $g(x) = 2x^2 + x - 2$ on the same set of axes and label the points of intersection.

74. Repeat Exercise 73 with $f(x) = x^3 - x$ and $g(x) = -x^2 + 1$.

75. Find a polynomial of degree 3 with integer coefficients and zeros at 1 and $-i$.

76. Find a polynomial of degree 4 with integer coefficients, $3 - i$ as a zero, and -2 as a zero of multiplicity 2.

77. Find a polynomial of degree 4 with integer coefficients, zeros at $\sqrt{2}i$ and $2i$, and constant term 8.

78. Find a polynomial of degree 5 with integer coefficients, zeros at i and $2 + i$, and a graph that passes through $(0, 5)$.

79. Find the dimensions of the rectangle of maximum area with a perimeter of 20 feet.

80. A box without a lid is constructed from a 4-foot by 2-foot piece of tin by cutting out equal squares of side x from each corner and bending up the flaps.

a. Express the volume of the box as a function of x.

b. What is the domain of the volume function?

c. Use a graphing device to approximate the value of x that produces the maximum volume.

CHAPTER 3: EXERCISES FOR CALCULUS

1. a. Factor the polynomial $P(x) = 2x^3 - 3x^2$ completely and sketch its graph, using the facts that a relative minimum occurs at $(1, -1)$ and a relative maximum at $(0, 0)$.

b. For each value of the constant C, determine the number of real zeros of the polynomial $Q(x) = P(x) + C$.

2. Suppose a polynomial has exactly two local maximums and one local minimum.

a. Sketch a possible graph of the polynomial.

b. What is the smallest possible degree of the polynomial?

c. What is the minimum number of real zeros the polynomial can have?

d. What is the maximum number of real zeros the polynomial can have?

3. a. Show that if $P(x)$ is a linear polynomial, then the composition $Q(x) = (P \circ P)(x)$ is also a linear polynomial and that $y = Q(x)$ has positive slope.

b. Suppose that $P(x)$ is a quadratic polynomial. Determine the degree of $Q(x) = (P \circ P)(x)$ and how the coefficients of this polynomial depend on the coefficients of $P(x)$.

4. The graph of $y = f(x)$ is shown in the figure.

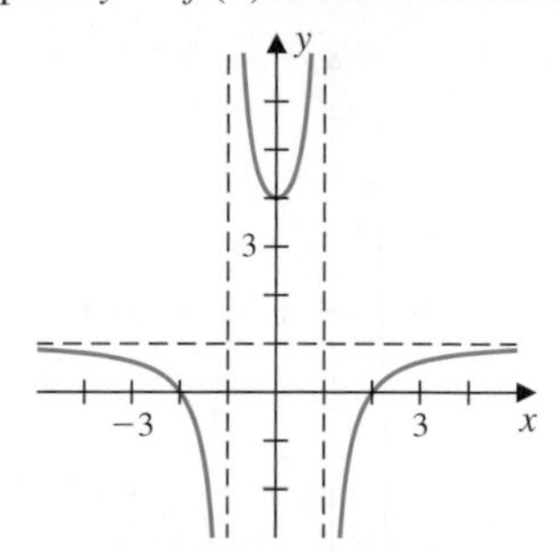

a. What is the domain of the function f?

b. What is the range of the function f?

c. On which interval or intervals is the function increasing?

d. On which interval or intervals is the function decreasing?

e. Specify any horizontal asymptotes for the graph.

f. Specify any vertical asymptotes for the graph.

g. Does the function have an inverse on the interval $(-1, 1)$?

h. Does the function have an inverse on the interval $(1, \infty)$?

5. For which rational numbers k does the equation

$$x^3 + x^2 + kx - 3 = 0$$

have at least one rational zero?

6. If $P(x) = x^2 + bx + c$ is a quadratic polynomial with zeros r_1 and r_2, then we can write

$$x^2 + bx + c = (x - r_1)(x - r_2),$$

where $b = -(r_1 + r_2)$ and $c = r_1 r_2$. Determine a similar relationship between the zeros r_1, r_2, and r_3 and the coefficients of the cubic polynomial $P(x) = x^3 + bx^2 + cx + d$.

7. a. Show that the linear polynomial defined by

$$P(x) = \frac{x - x_2}{x_1 - x_2} y_1 + \frac{x - x_1}{x_2 - x_1} y_2$$

passes through the points (x_1, y_1) and (x_2, y_2).

b. Use the fact in part (a) to determine an equation of the line that passes through $(-1, 6)$ and $(1, -2)$.

8. a. Expand the technique in Exercise 7 to determine a formula for the quadratic polynomial that passes through the points (x_1, y_1), (x_2, y_2), and (x_3, y_3).

b. Use the formula in part (a) to determine an equation of the quadratic polynomial that passes through the points $(-1, 6)$, $(1, -2)$, and $(2, 3)$.

9. The height of an object above the ground is $s(t) = v_0 t - (gt^2/2)$, where g represents the earth's gravity and v_0 is the initial velocity of the object.

a. What is the initial height of the object?

b. How long is the object in the air?

c. What is the maximum height reached by the object?

d. How long does it take the object to reach its maximum height?

10. You are asked to construct a rectangular box with volume 10 ft^3, a square base, and no top. The material for the bottom costs \$15 per ft^2 and that for the sides costs \$6 per ft^2. Estimate, to one decimal place, the dimensions of the most economical box, and its cost.

11. A pizza box with a lid is constructed from a 20-inch by 50-inch piece of cardboard, as shown in the figure. Determine the volume of the box as a function of the variable x.

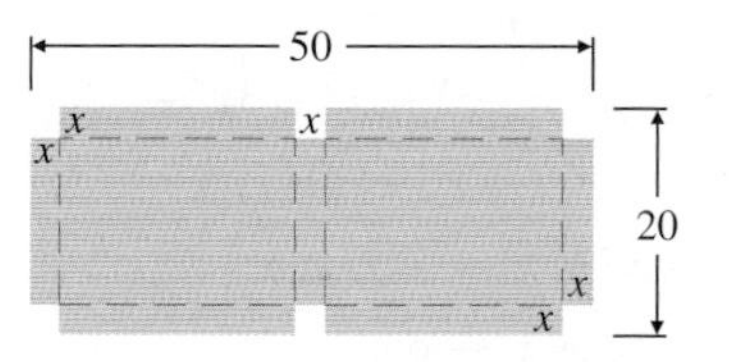

12. A rectangle is inscribed within a triangular region formed by the positive x-axis, the positive y-axis, and the graph of the line $x + y = 50$, as shown in the figure.

a. Determine the area as a function of the variable x.

b. Given the knowledge that there is a single value of x that produces a rectangle with maximum area, show that this rectangle must be a square.

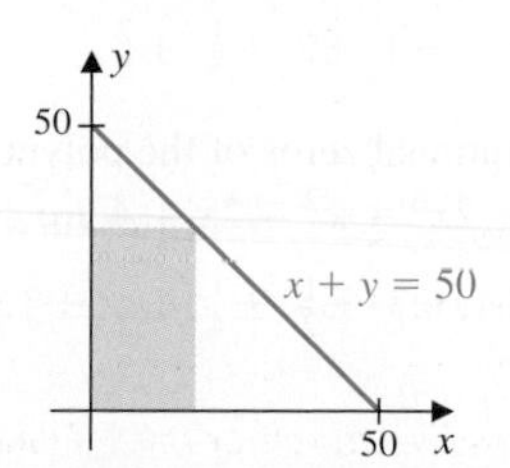

CHAPTER 3: CHAPTER TEST

Determine whether the statement is true or false. If false, describe how the statement might be changed to make it true.

1. If the number c is a zero of the polynomial $P(x)$, then $P(c) = 0$.
2. If the number c is a zero of the polynomial $P(x)$, then $x - c$ is a factor of the polynomial.
3. The polynomial $P(x) = x^3 - x^2 + x - 1$ has a factor $x - 1$.
4. The polynomial $P(x) = x^3 + x^2 - x + 2$ has a factor $x + 2$.
5. If a polynomial $P(x)$ has degree n, then there are n real numbers so that $P(x) = 0$.
6. The polynomial $P(x) = x^4 - 3x^3 + 2x^2$ has exactly three distinct real zeros.
7. The graph of $P(x) = x^3 + x^2 - 2x$ crosses the x-axis at $x = -1$, $x = 2$, and $x = 0$.
8. The graph of $P(x) = x(x - 1)^3 \times (x + 1)^2$ crosses the x-axis three times.
9. The end behavior of $P(x) = x^5 - 2x^4 + 10x^2 - 5$ satisfies

$$P(x) \to -\infty \text{ as } x \to -\infty \text{ and}$$
$$P(x) \to \infty \text{ as } x \to \infty.$$

10. The end behavior of $P(x) = -x^6 - x^5 + 3x^2 - 2x + 7$ satisfies

$$P(x) \to -\infty \text{ as } x \to -\infty \text{ and}$$
$$P(x) \to -\infty \text{ as } x \to \infty.$$

11. The polynomials $P(x) = 2x^4 - 3x^3 + 7x - 10$ and $Q(x) = 3x^4 - 5x^3 + 9x - 12$ have the same zeros.
12. The polynomials $P(x) = x^4 - 2x^3 + 6x - 9$ and $Q(x) = x^4 - 2x^3 + 6x - 5$ have the same zeros.
13. The possible rational zeros of the polynomial $P(x) = 2x^5 - 2x^3 + x^2 - 3x + 5$ are

$$\pm 1, \pm 2, \pm \tfrac{1}{5}, \pm \tfrac{2}{5}.$$

14. The possible rational zeros of the polynomial $P(x) = 3x^4 - 2x^3 + x^2 - 2x + 4$ are

$$\pm 1, \pm 3, \pm \tfrac{1}{2}, \pm \tfrac{1}{4}, \pm \tfrac{3}{2}, \pm \tfrac{3}{4}.$$

In Exercises 15–18, use the figure.

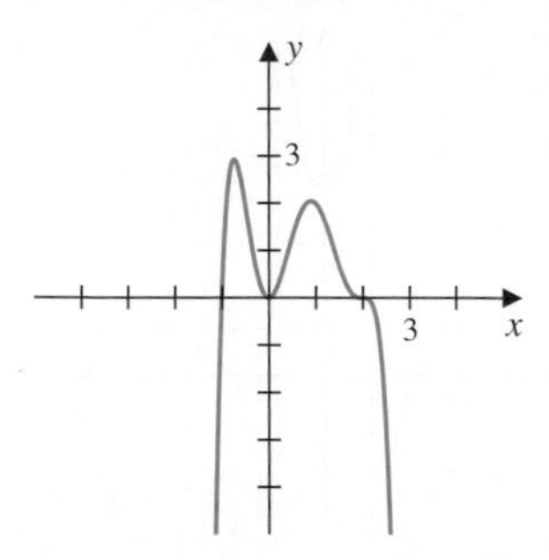

15. The polynomial shown in the figure has a zero of even multiplicity at $x = 0$.
16. The polynomial shown in the figure has a zero of even multiplicity at $x = 2$.
17. The polynomial $P(x) = x^2(x - 2)^3(x + 1)$ has the same zeros.
18. The degree of the polynomial shown in the figure is at least 8.

In Exercises 19–22, use the figure.

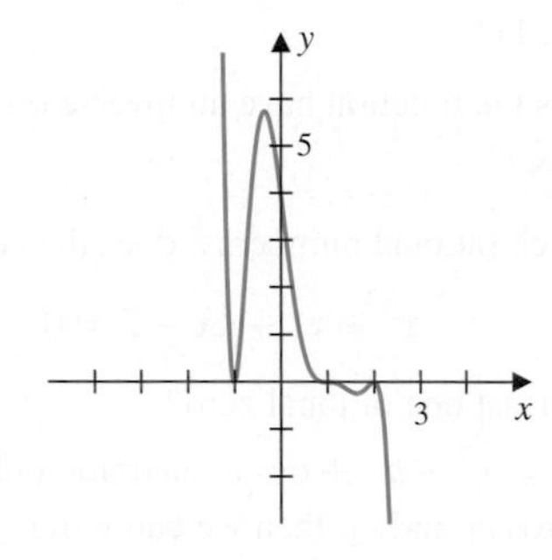

19. The polynomial shown in the figure has zeros at $x = -2$, $x = -1$, and $x = 1$.
20. The end behavior of the polynomial shown in the figure is the same as the end behavior of x^3.
21. A possible equation of the polynomial shown is

$$P(x) = -(x + 1)^2(x - 1)^3(x - 2)^2.$$

22. A polynomial with the same zeros but with y-intercept $(0, 1)$ is

$$P(x) = -\frac{1}{4}(x + 1)^2(x - 1)^3(x - 2)^2.$$

In Exercises 23–26, use the figure.

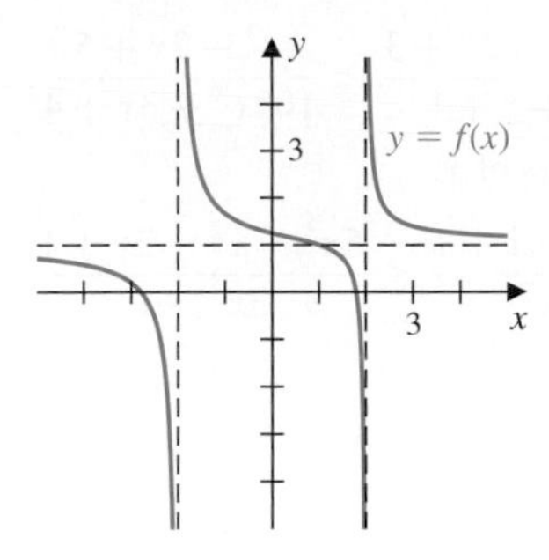

23. The graph has vertical asymptotes $x = 2$ and $x = -2$.

24. The graph has a horizontal asymptote $y = -1$.

25. The function satisfies

$$f(x) \to 1 \text{ as } x \to \infty,$$
$$f(x) \to 1 \text{ as } x \to -\infty,$$
$$f(x) \to \infty \text{ as } x \to -2^- \text{ and}$$
$$f(x) \to -\infty \text{ as } x \to -2^+.$$

26. A possible definition for the function is

$$f(x) = \frac{x-1}{x^2-4}.$$

In Exercises 27–32, use the figure.

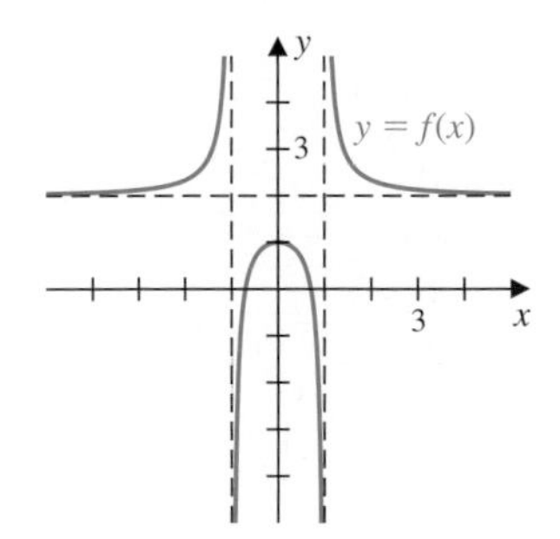

27. The graph of f has vertical asymptotes $x = 2$ and $x = -2$.

28. The graph of f has a horizontal asymptote $y = 2$.

29. The domain of f is $(-\infty, 1) \cup (1, \infty)$.

30. The range of f is $(-\infty, 1] \cup (2, \infty]$.

31. The function satisfies

$$f(x) \to 2, \text{ as } x \to \infty$$
$$f(x) \to 2, \text{ as } x \to -\infty$$
$$f(x) \to \infty, \text{ as } x \to 1^+$$
$$f(x) \to -\infty, \text{ as } x \to 1^-.$$

32. A possible definition for the function is

$$f(x) = \frac{2x^2-2}{x^2-1}.$$

33. The polynomial $P(x) = x^4 + x^3 - 3x^2 - x + 2$ has a zero of multiplicity 2 at $x = 1$ and zeros at $x = -1$ and $x = -2$.

34. The polynomial $P(x) = x^5 - 5x^4 + 6x^3 + 4x^2 - 8x$ has a zero of multiplicity 3 at $x = -2$.

35. The rational function

$$f(x) = \frac{1-x^2}{x^2+1}$$

has a vertical asymptote $x = -1$.

36. The rational function

$$f(x) = \frac{x^2-3x+2}{x-2}$$

has a hole in the graph at (2, 1)

37. The rational function

$$f(x) = \frac{x^2-1}{x^2-4}$$

has vertical asymptotes at $x = 1$ and $x = 4$.

38. The rational function

$$f(x) = \frac{x^2+5x+4}{x^2-2x-3}$$

has vertical asymptotes at $x = 1$ and $x = -3$.

39. The function

$$f(x) = \frac{x^2+1}{x^2-3x+2}$$

satisfies

$$f(x) \to \infty \text{ as } x \to 1^+, \text{ and}$$
$$f(x) \to -\infty \text{ as } x \to 1^-.$$

40. The function

$$f(x) = \frac{1-x^2}{x^2-4}$$

satisfies

$$f(x) \to \infty \text{ as } x \to 2^-, \text{ and}$$
$$f(x) \to \infty \text{ as } x \to -2^+.$$

41. The function

$$f(x) = \frac{3x^4-2x^3-x-1}{x^4+2x+1}$$

has a horizontal asymptote $y = 3$.

42. The function

$$f(x) = \frac{2x^3 - 3x^2 + x - 1}{3x^3 - 4x^2 - 2}$$

has a horizontal asymptote $y = \frac{3}{2}$.

43. For large values of x,

$$\frac{100x^2 - 2x + 3}{x^3 + x + 1} < \frac{x^3 - 2x + 5}{100x^3 + 3x + 4}.$$

44. For large values of x,

$$\frac{10x^2 - x + 1}{5x^2} \leq \frac{5x^3 - x^2 - 2x + 1}{3x^3 + 1}.$$

Trigonometric Functions

4

Calculus Connections

Many common phenomena have an oscillatory, or periodic, behavior. You can observe this when riding in a boat on a rough weather day, listening to an overloaded washer that shakes and rattles, or hearing the beat of your heart with a stethoscope. Electricity flowing through the wires in your house, the motion of your car after it hits a hole in the road, the music that you hear on your iPod, and your watch also exhibit this behavior. All these phenomena can be modeled using equations based on the familiar sine and cosine functions. These equations have the form

$$y = A + B\sin(Cx + D),$$

where A, B, C, and D are constants and sin represents the trigonometric sine function.

In this chapter we will see how to apply and construct functions that permit us to model behavior like that given in the figure.

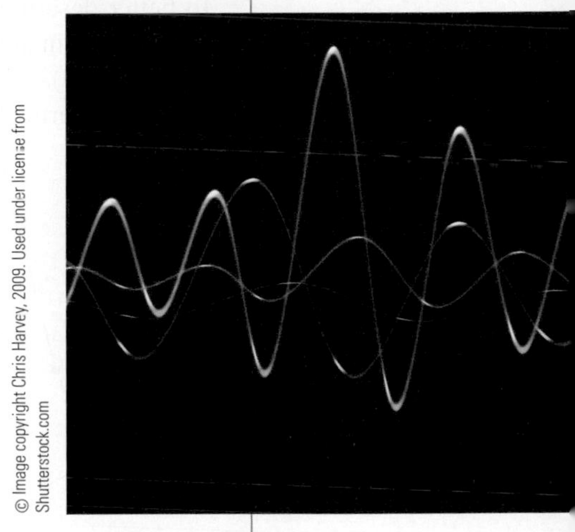

© Image copyright Chris Harvey, 2009. Used under license from Shutterstock.com

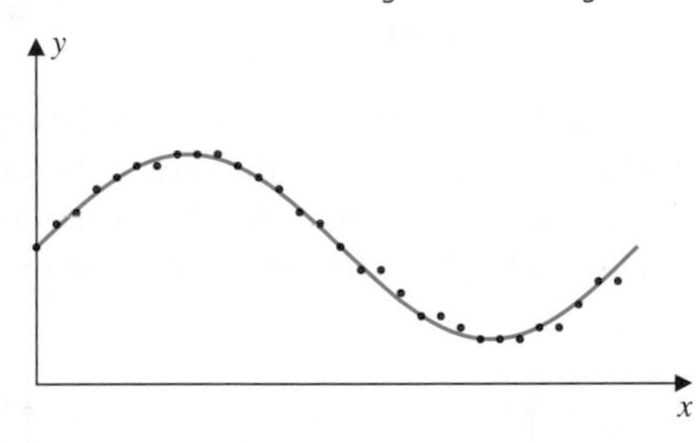

You have probably seen sines and cosines in your previous mathematics courses, but you may not have thought of them as functions. Instead, your study in these courses might have emphasized their relationships to the sides of a right triangle. We will review this approach in Section 4.3. However, to study problems with periodic and oscillatory behavior we need to extend trigonometry to functions on the real line, which we consider in Section 4.4. This is the approach you will need in your future work, so it is the one we use in the remainder of the chapter.

4.1 INTRODUCTION

The trigonometric functions have domains that are sets of real numbers, but the applications of trigonometry often involve triangles and the angles of their vertices. Our first step, then, is to develop a connection between angles and the set of real numbers.

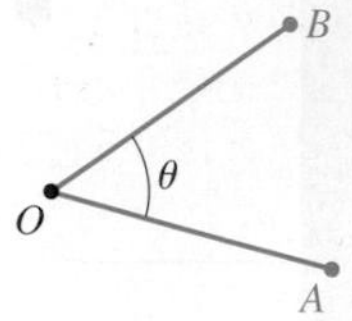

An *angle* consists of two **rays**, or half-lines, that originate at a common point called the **vertex**, which we will denote O. One of the rays is called the **initial side** of the angle, and the other ray is called the **terminal side**. The angle is generated by revolving the initial side about the point O until the terminal side is reached. We choose points A and B on the two rays and denote the angle as $\angle AOB$, as shown in Figure 1. Angles are commonly denoted using lowercase Greek letters, such as the θ shown in Figure 1.

FIGURE 1

The amount of rotation from the initial side to the terminal side is critical to the definition of an angle, but the particular points chosen on the rays to represent the angle are not. In general, the position of the angle in the plane is also unimportant. To better describe the rotation, superimpose an xy-coordinate system on the angle with the origin at the point O and the initial side along the positive x-axis, as shown in Figure 2(a). Then rotate the angle with its coordinate system to the more familiar horizontal-vertical position illustrated in Figure 2(b).

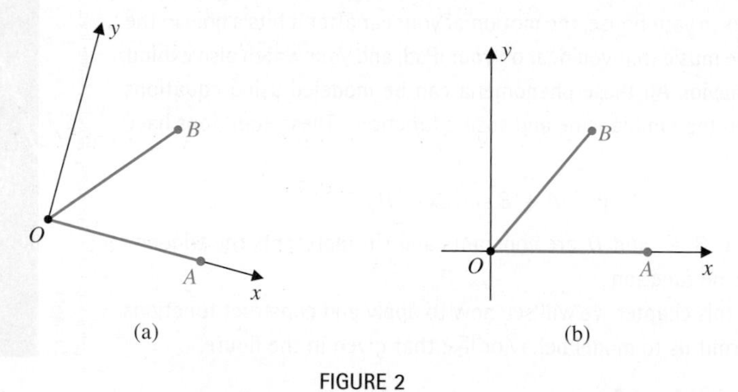

FIGURE 2

To provide a measure for an angle we need to know how the angle was generated. The angle in Figure 3 can be generated by any of the three rotations shown, or by an infinite number of other such rotations, one for each full revolution about the point O in the clockwise direction and another for each full rotation in the counterclockwise direction.

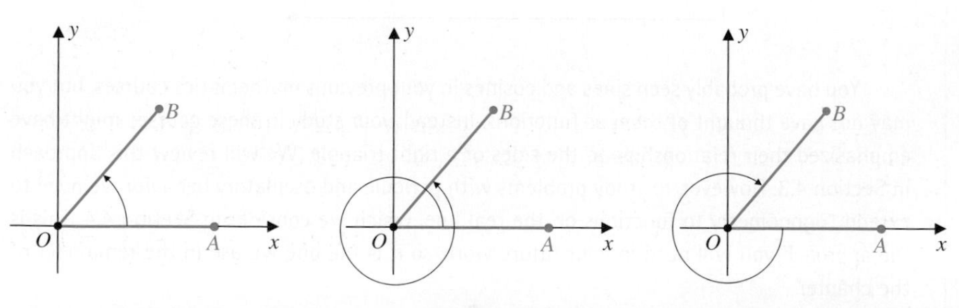

FIGURE 3

4.2 MEASURING ANGLES

There are several ways to define the measure of an angle. An angle formed by one complete rotation, by definition, consists of 360 equal parts called **degrees.** Degrees are denoted using the symbol °. An angle of 1° is $\frac{1}{360}$ of a revolution. A more natural method of measuring angles, one that is not based on this artificial value of 360 equal parts, is called *radian measure*. Radian measure is the method most often used in calculus. The measure of an angle in radians is determined by the length of the arc of a unit circle that is cut by the rays of the angle.

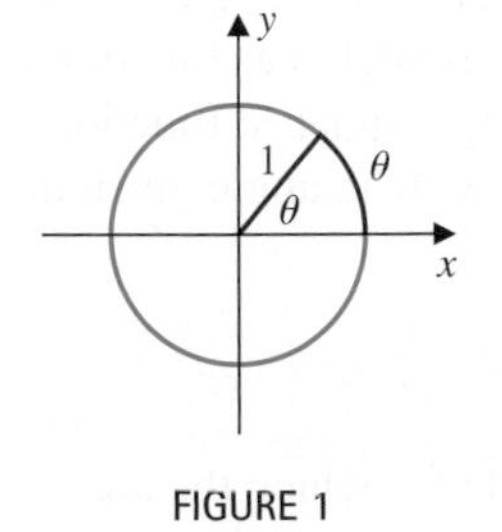

FIGURE 1

Radian Measure

The **radian measure** of an angle with vertex at the center of a unit circle is the length of the arc made by the angle. (See Figure 1.)

We begin the study of trigonometry by establishing the relationship between the two measurement systems. Since the circumference of a unit circle is 2π, an angle made by one complete revolution has measure 2π radians. It also has a measure of 360°. Half a revolution has measure π radians and 180°.

Conversion between Degrees and Radians

Since $180° = \pi$ radians,

$$1° = \frac{\pi}{180} \approx 0.01745 \text{ radians} \quad \text{and} \quad 1 \text{ radian} = \frac{180°}{\pi} \approx 57.296°.$$

EXAMPLE 1 Express 135° and 420° in radians, and express $5\pi/6$ radians in degrees.

Solution To convert from degrees to radians, multiply the number of degrees by $\pi/180$, the number of radians in one degree. This gives

$$135° = 135\left(\frac{\pi}{180}\right) = \frac{3\pi}{4} \text{ radians}$$

$$420° = 420\left(\frac{\pi}{180}\right) = \frac{7\pi}{3} \text{ radians.}$$

Radian measure of angles is the most useful system for precalculus and calculus, but we will also use degree measure.

To convert from radians to degrees, multiply the number of radians by $180/\pi$, the number of degrees in one radian. Thus

$$\frac{5\pi}{6} \text{ radians} = \frac{5\pi}{6}\left(\frac{180}{\pi}\right)^{\circ} = 150°.$$

The decimal equivalents are seldom required. ■

Table 1 gives degree and radian equivalents for some standard angles.

TABLE 1

Degrees	0	30	45	60	90
Radians	0	$\frac{\pi}{6}$	$\frac{\pi}{4}$	$\frac{\pi}{3}$	$\frac{\pi}{2}$

You have probably encountered trigonometric functions evaluated at angles given in degrees. In calculus the trigonometric functions are also important, but as functions defined for all real numbers. In preparation for defining the trigonometric functions for all real numbers we begin by associating each real number with an angle given in radians.

Real Numbers Related to Unit Circle Arcs

For any real number t, let $P(t)$ be the point on the unit circle for which the arc of the circle from $(1, 0)$ to $P(t)$ is $|t|$ units. The distance is measured in the counterclockwise direction when $t \geq 0$, and in the clockwise direction when $t < 0$. (See Figures 2(b) and 2(c).)

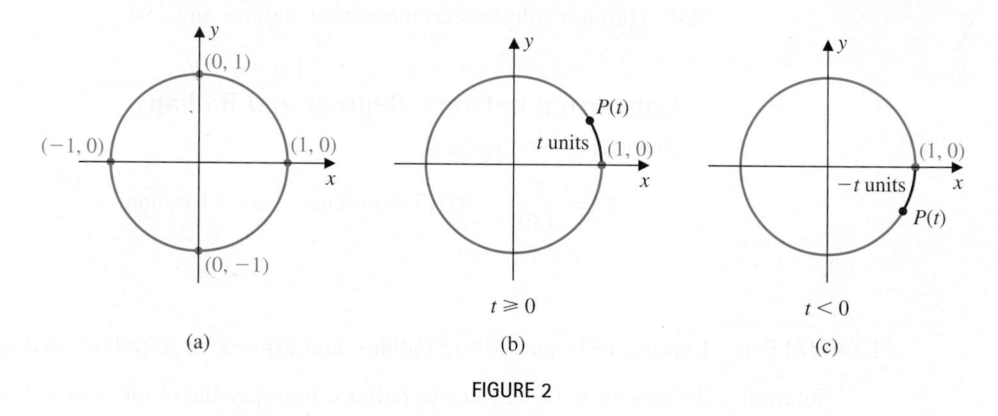

FIGURE 2

The mapping of t into $P(t)$ can be described by positioning a copy of the real line vertically with $t = 0$ coinciding with the point $(1, 0)$ on the unit circle. For any real number t, the point $P(t)$ is obtained by "wrapping" the line around the circle and marking as $P(t)$ the point on the circle that corresponds to the position of t. (See Figure 3.)

This wrapping corresponds to rotating the initial side of an angle around the point O until we reach the terminal side. The real number t that gives the point $P(t)$ on the terminal side of the angle is the radian measure of the angle. (See Figure 4.)

Radian Measure and Terminal Points

For any real number t, the angle generated by rotating counterclockwise from the positive x-axis to the point $P(t)$ on the unit circle has radian measure t.

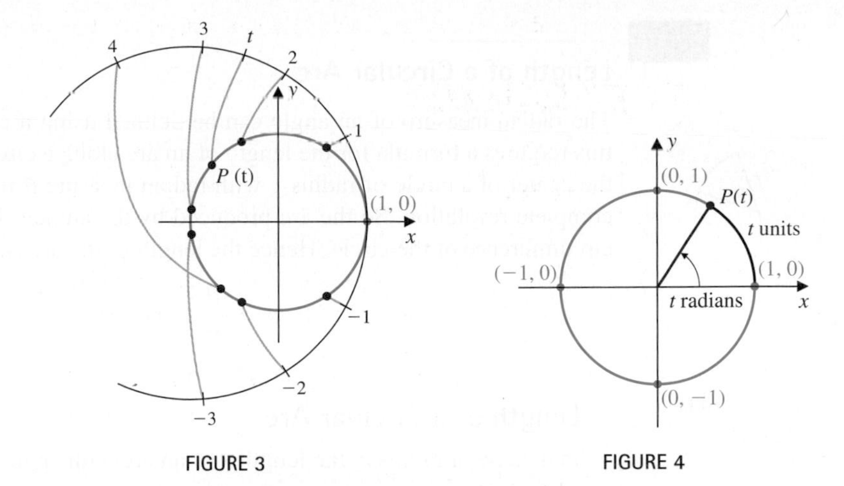

FIGURE 3 FIGURE 4

The smallest positive real number that is mapped onto $(-1, 0)$ is π, an irrational number whose value is approximately 3.14159. Because of the symmetry of the circle, the number $\pi/2$ is associated with the point $(0, 1)$, the number $3\pi/2$ with the point $(0, -1)$, the number 2π with the point $(1, 0)$, and so on, as illustrated in Figure 5.

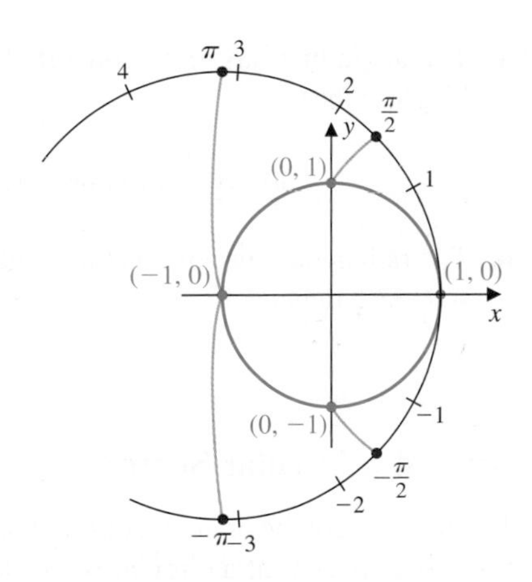

FIGURE 5

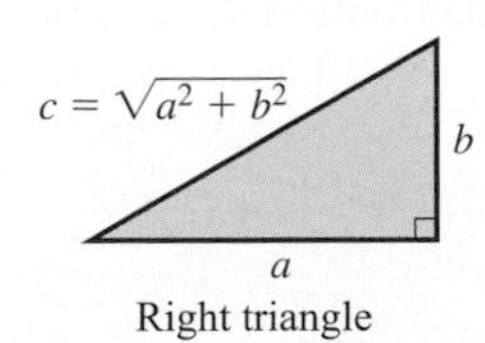

Right triangle

An angle with radian measure $\pi/2$ is called a *right angle*, and a triangle containing a right angle is a *right triangle*. The sides of this right triangle satisfy the Pythagorean relationship $a^2 + b^2 = c^2$. The angles of any triangle have radian measure greater than 0 but less than π. Angles with radian measure less than $\pi/2$ are called *acute* angles. So, for example, the nonright angles in a right triangle are acute. An angle with radian measure greater than $\pi/2$ is called an *obtuse* angle.

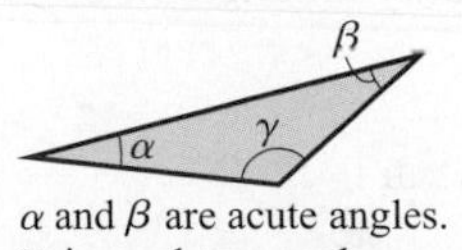

α and β are acute angles.
γ is an obtuse angle.

In Section 4.3 we will consider trigonometry as it applies to acute angles. For now we are only concerned with the connection between a real number t and the unique arc of the unit circle described by the point $P(t)$. The coordinates $(x(t), y(t))$ of $P(t)$ provide us with our basic trigonometric functions.

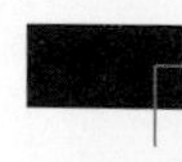

Length of a Circular Arc

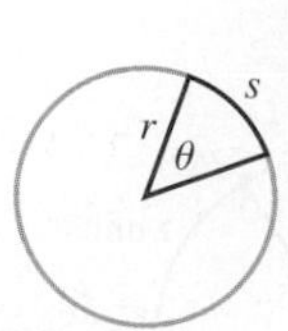

The radian measure of an angle can be defined using a circle of any radius. To do this requires a formula for the length of an arc along a circle. An angle positioned at the center of a circle of radius r with radian measure θ is the fraction $\theta/2\pi$ of one complete revolution. So the arc produced by the angle θ is the same fraction of the circumference of the circle. Hence the length of the arc is,

$$s = \frac{\theta}{2\pi}(2\pi r) = r\theta.$$

The formula $s = r\theta$ holds only for θ in radians.

Length of a Circular Arc

In a circle of radius r, the length s of an arc with angle θ radians is

$$s = r\theta.$$

EXAMPLE 2 **a.** Find the length of an arc of a circle with radius 5 and angle 30°.

b. The arc of a circle of radius 3 associated with an angle θ has length 5. What is the radian measure of θ?

Solution **a.** The angle first has to be converted to radians in order to use the formula for arc length. Since

$$30^\circ = \frac{\pi}{6} \text{ radians,} \quad \text{we have} \quad s = r\theta = 5 \cdot \frac{\pi}{6} = \frac{5\pi}{6}.$$

b. The radian measure of θ is the length of the arc divided by the radius of the circle, so $\theta = \frac{5}{3}$ radians. ■

Area of a Circular Sector

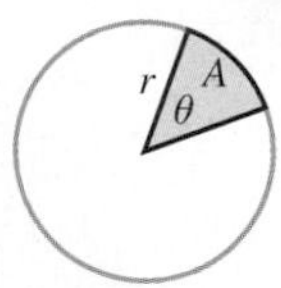

FIGURE 6

The formula for the area of a circular sector, shown in Figure 6, is also easy to derive. Since the area, A, of a circular sector formed by an angle θ is the fraction $\theta/(2\pi)$ of the total area of a circle, we have

$$A = \frac{\theta}{2\pi}(\pi r^2) = \frac{1}{2}r^2\theta.$$

The formula

$$A = \frac{1}{2}r^2\theta$$

holds only for θ in radians.

Area of a Circular Sector

In a circle of radius r, the area A of a circular sector formed by an angle of θ radians is

$$A = \frac{1}{2}r^2\theta.$$

EXAMPLE 3 Find the area A of a sector with angle $45°$ in a circle of radius 4.

Solution Since

$$45° = 45\left(\frac{\pi}{180}\right) = \frac{\pi}{4} \text{ radians}, \quad \text{the area is} \quad A = \frac{1}{2}4^2\left(\frac{\pi}{4}\right) = 2\pi. \quad ■$$

EXAMPLE 4 The shaded region in Figure 7 is bounded by the sides of the angle with radian measure θ and the arcs of the circles with radii r_1 and r_2. Determine the area of the region.

Solution The area inside the circle with radius r_2 is πr_2^2, and the area inside the circle with radius r_1 is πr_1^2. So the area in the ring between the circles, called an *annulus*, is

$$\pi r_2^2 - \pi r_1^2 = \pi(r_2^2 - r_1^2).$$

The radian measure of the entire circle is 2π and the radian measure of the angle producing the shaded portion is θ, so the shaded region has area

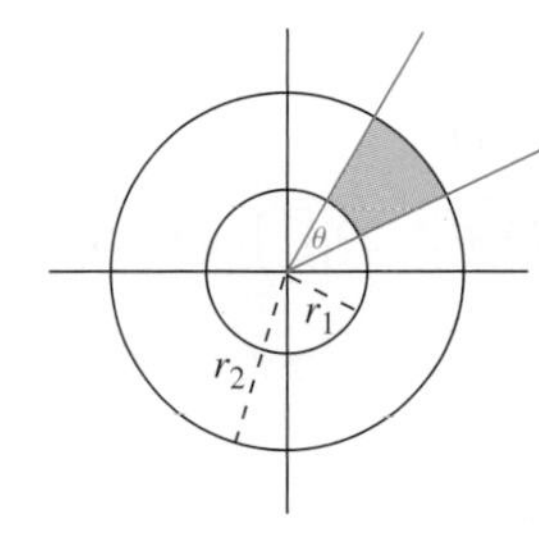

FIGURE 7

$$A = \left(\frac{\theta}{2\pi}\right)\pi(r_2^2 - r_1^2) = \frac{\theta}{2}(r_2^2 - r_1^2).$$

This area can also be expressed by factoring the difference of the two squares as

$$A = \frac{\theta}{2}(r_2 - r_1)(r_2 + r_1) = \frac{r_2 + r_1}{2}(r_2 - r_1)\theta.$$

Written in this form, we see that the area is the product of the average of the radii, the difference between the radii, and the radian measure of the angle. This representation is useful for applications in calculus. ■

The linear speed of an object moving at a constant rate is found by dividing the distance traveled by the time it takes to go this distance. The *angular speed*, commonly denoted as ω, of a point on a wheel turning at a constant speed is defined in a similar manner. It is the measure of the angle traveled divided by the time taken to rotate through that angle. Example 5 shows how linear and angular speeds can be related.

EXAMPLE 5 The tachometer in a car registers 2800 rpm, which indicates that the engine is turning at the rate of 2800 revolutions per minute. When this particular car is in sixth gear, the rear wheels turn once for every 3.005 revolutions of the engine. The diameter of the rear wheels is 24.5 inches. At what speed is the car traveling?

Solution The angular speed of the engine is

$$\omega_E = 2800 \frac{\text{revolutions}}{\text{minute}},$$

so the angular speed of the rear wheel is

$$\omega_w = \left(\frac{1}{3.005}\right) 2500 \frac{\text{revolutions}}{\text{minute}} \approx 931.8 \frac{\text{revolutions}}{\text{minute}}.$$

The diameter of the tire is 24.5 inches, so the circumference of the tire is

$$24.5\pi \text{ inches} \approx 76.969 \text{ inches} = 6.414 \text{ feet}.$$

Hence a point on the outside of the tire, which is also the theoretical speed of the car, is

$$931.8 \frac{\text{revolutions}}{\text{minute}} \cdot 6.414 \frac{\text{feet}}{\text{revolution}} \approx 5977 \frac{\text{feet}}{\text{minute}} = 99.61 \frac{\text{feet}}{\text{sec}}.$$

Because 88 ft/sec is equivalent to 60 miles per hour, the speed of the car is approximately 67.9 miles per hour. ■

EXERCISE SET 4.2

In Exercises 1–8, convert from degree measure to radian measure.

1. 30° **2.** 135° **3.** 150°
4. 40° **5.** 225° **6.** 315°
7. −72° **8.** −270°

In Exercises 9–14, convert from radian measure to degree measure.

9. $\frac{3\pi}{4}$ **10.** $-\frac{\pi}{4}$ **11.** $-\frac{11\pi}{6}$

12. $\frac{2\pi}{3}$ **13.** $\frac{9\pi}{2}$ **14.** 8π

In Exercises 15–20, the measure of two angles is given. Determine whether the angles coincide.

15. 45°, 405° **16.** −120°, 570°

17. $\frac{\pi}{6}, \frac{13\pi}{6}$ **18.** $\frac{5\pi}{6}, -\frac{11\pi}{6}$

19. 150°, −240° **20.** $\frac{2\pi}{3}, -\frac{4\pi}{3}$

In Exercises 21–24, show the approximate location on the unit circle of $P(t)$ for the given value of t.

21. a. $t = \frac{\pi}{2}$ **b.** $t = \pi$
c. $t = 2\pi$ **d.** $t = \frac{3\pi}{2}$

22. a. $t = \frac{2\pi}{3}$ **b.** $t = \frac{5\pi}{4}$
c. $t = \frac{\pi}{6}$ **d.** $t = \frac{11\pi}{6}$

23. a. $t = -\frac{\pi}{4}$ **b.** $t = -\frac{4\pi}{3}$
c. $t = -\frac{37\pi}{6}$ **d.** $t = -\frac{7\pi}{4}$

24. a. $t = \frac{21\pi}{2}$ **b.** $t = \frac{317\pi}{4}$
c. $t = -\frac{33\pi}{2}$ **d.** $t = -\frac{19\pi}{6}$

25. If $P(t)$ has coordinates $\left(\frac{3}{5}, \frac{4}{5}\right)$, find the coordinates of the point.

a. $P(t+\pi)$ **b.** $P(-t)$
c. $P(t-\pi)$ **d.** $P(-t-\pi)$

26. If $P(t)$ has coordinates $\left(\frac{4}{5}, -\frac{3}{5}\right)$, find the coordinates of the point.

a. $P(t+2\pi)$ **b.** $P(-t)$
c. $P(t-\pi)$ **d.** $P(t-3\pi)$

27. If $P(t)$ has coordinates $\left(-\frac{\sqrt{5}}{3}, \frac{2}{3}\right)$, find the coordinates of the point.

a. $P(t+\pi)$ **b.** $P(-t)$
c. $P(t-\pi)$ **d.** $P(-t-\pi)$

28. If $P(t)$ has coordinates $\left(-2\frac{\sqrt{13}}{13}, -3\frac{\sqrt{13}}{13}\right)$, find the coordinates of the point.

a. $P(t+2\pi)$ **b.** $P(-t)$
c. $P(t-\pi)$ **d.** $P(t-3\pi)$

29. Find the length of an arc with angle 45° in a circle with radius 8 meters.

30. Find the length of an arc with angle 45° in a circle with radius 2 miles.

31. In a circle of radius 3 inches an arc of length 6 inches has angle θ. Find the measure of the angle in degrees and radians.

32. In a circle of radius 5 miles an arc of length 3 miles has angle θ. Find the measure of the angle in degrees and radians.

33. Find the length of the arc in the figure.

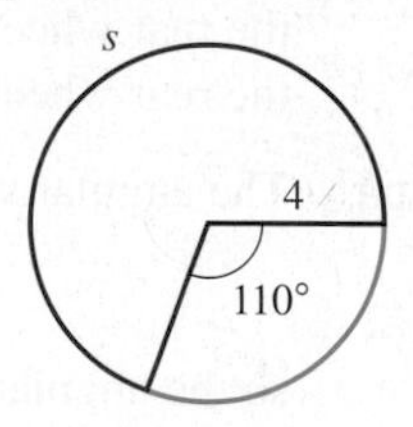

34. Find the radian measure of the angle in the figure.

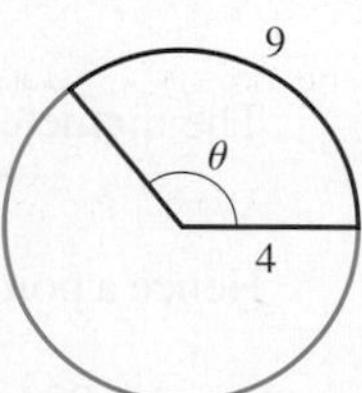

35. Find the area of the circular sector with angle $\frac{1}{2}$ radian in a circle of radius 10 meters.

36. Find the area of the circular sector with angle 30° in a circle of radius 4 inches.

37. The area of a sector of a circle with angle 45° is 2 square feet. Find the radius of the circle.

38. The area of a sector of a circle with radius 3 miles is 10 square miles. Find the angle of the sector.

39. Find the area of the sector in the figure.

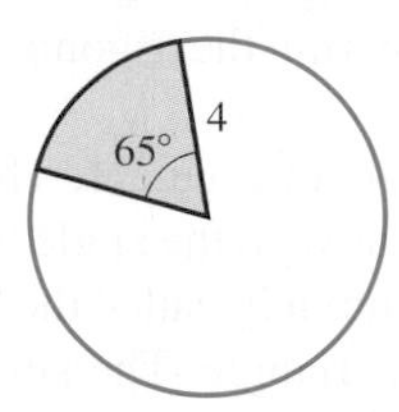

40. Find the area of the sector in the figure.

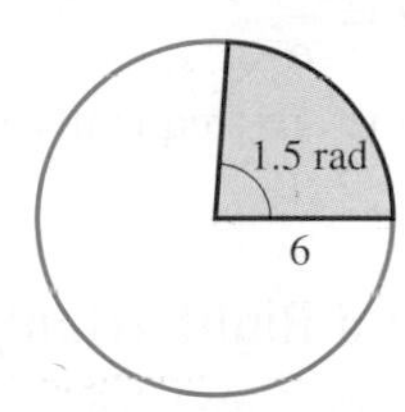

In Exercises 41 and 42, determine the distance between two cities that lie approximately on the same meridian but at different latitudes by finding the length of the arc between the two cities. The radius of the earth is approximately 3960 miles.

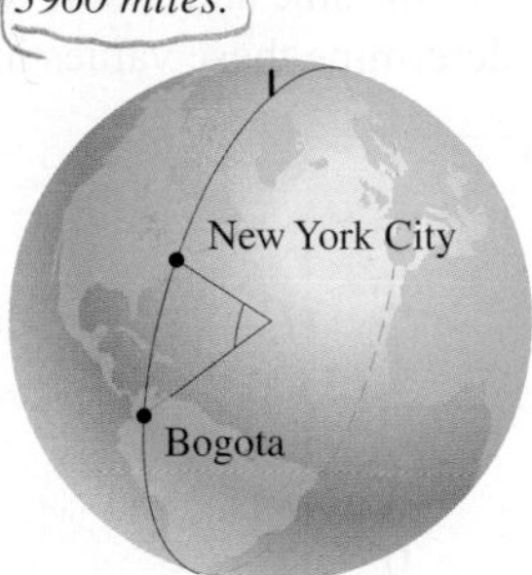

41. Syracuse, New York and Virginia Beach, Virginia lie on approximately the same meridian. Syracuse has latitude approximately 43° N (43°14′6″) and Virginia Beach has latitude approximately 36.5° N (36°30′57″). Find the distance between the two cities.

42. New York City and Bogota, Colombia lie on approximately the same meridian. New York City has latitude approximately 40.5° N (40°28′12″) and Bogota has latitude approximately 4.6° N (4°35′59″). Find the distance between the two cities.

43. At the equator the radius of the earth is approximately 3960 miles. Determine, in miles per hour, the speed at which a point on the equator is moving as a result of the earth's rotation about its axis.

44. The owner of the car in Example 5 replaces the 18-inch wheels with 19-inch wheels and tires. The profile of the new tires is lower, so the diameter increases by only 0.75 inch to 26.25 inches. What will the tachometer read when the car is traveling at the speed found in the example?

45. Two gears of different diameters are connected as shown in the figure. The smaller gear has diameter 4 inches and the larger gear has diameter 16 inches.

   a. Suppose that the red dot on the smaller gear travels a distance s when rotated through an angle θ, as shown in the figure. What is the relationship between θ and the angle φ swept by the dot on the larger gear?

   b. The smaller gear is rotating at a rate of 5 revolutions per second. At what rate is the larger gear is rotating?

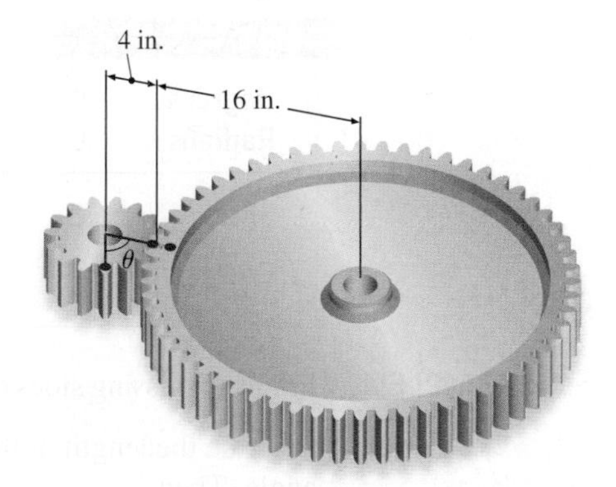

4.3 RIGHT-TRIANGLE TRIGONOMETRY

Your first introduction to trigonometry was likely through the use of right triangles, because this form is used for computational and geometric applications. The sine and cosine of an acute angle are defined as ratios of the sides of a right triangle. The remaining trigonometric functions are simply the quotients and reciprocals of the sine and cosine functions. In this section we review this process, and then in Section 4.4 we demonstrate how the definitions can be extended to permit the trigonometric functions to be defined for any real number and any angle.

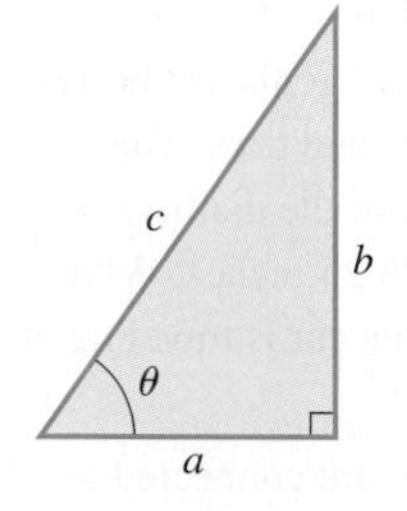

FIGURE 1

Consider the angle θ shown in the right triangle in Figure 1. The side a of the triangle is called the *adjacent* side of the triangle relative to the angle θ, and the side b is called the *opposite* side. Sides a and b are also commonly called the *legs* of the right triangle. The side c is called the *hypotenuse* of the triangle. The sine of θ, denoted $\sin\theta$, and the cosine of θ, denoted $\cos\theta$, are defined as

$$\sin\theta = \frac{b}{c} \quad \text{and} \quad \cos\theta = \frac{a}{c}.$$

The ratios and reciprocals of the sine and cosine are frequently used and given definitions as well.

Trigonometric Functions of an Angle in a Right Triangle

For the angle θ in the right triangle shown in Figure 1 we have

$$\sin\theta = \frac{b}{c}, \quad \cos\theta = \frac{a}{c}, \quad \tan\theta = \frac{b}{a}, \quad \csc\theta = \frac{c}{b}, \quad \sec\theta = \frac{c}{a}, \quad \cot\theta = \frac{a}{b}.$$

Table 1 has been provided to help you recall the conversions between degrees and radians. We have included in this table the values of the sine and cosine for some common angles between 0 and 90 degrees. We will determine these values in Section 4.4.

TABLE 1

Degrees	0	30	45	60	90
Radians	0	$\frac{\pi}{6}$	$\frac{\pi}{4}$	$\frac{\pi}{3}$	$\frac{\pi}{2}$
Sine	0	$\frac{1}{2}$	$\frac{\sqrt{2}}{2}$	$\frac{\sqrt{3}}{2}$	1
Cosine	1	$\frac{\sqrt{3}}{2}$	$\frac{\sqrt{2}}{2}$	$\frac{1}{2}$	0

EXAMPLE 1 Find any missing sides or angles of the right triangle from the information in Figure 2.

Solution Let x be the length of the hypotenuse and y the length of the adjacent side of the 30° angle. Then

$$\sin 30^\circ = \frac{4}{x} \quad \text{and} \quad x = \frac{4}{\sin 30^\circ} = \frac{4}{\frac{1}{2}} = 8.$$

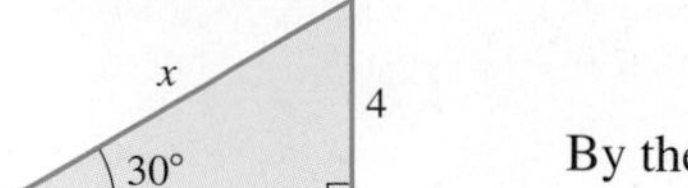

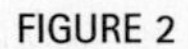
FIGURE 2

By the Pythagorean Theorem, we have

$$y^2 + 4^2 = 8^2 \quad \text{so} \quad y = \sqrt{48} = 4\sqrt{3}.$$

The angles of a triangle sum to 180°, so the missing angle is 60°. ■

Example 2 shows how to determine all the trigonometric values of an angle when one of these values is known.

EXAMPLE 2 Suppose that an acute angle θ is known to satisfy $\sin\theta = \frac{3}{5}$. Determine the other trigonometric functions of this angle.

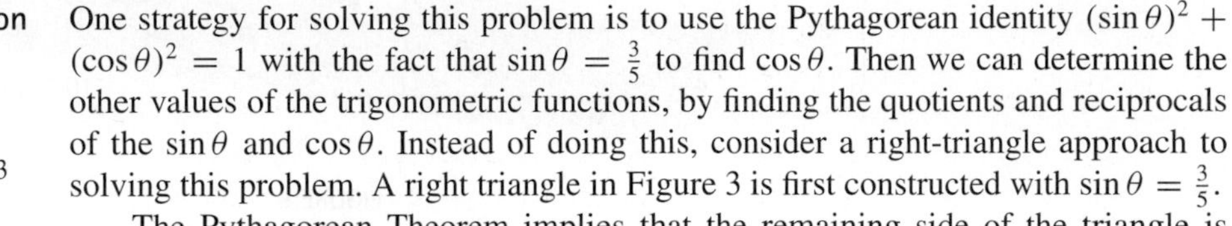

Solution One strategy for solving this problem is to use the Pythagorean identity $(\sin\theta)^2 + (\cos\theta)^2 = 1$ with the fact that $\sin\theta = \frac{3}{5}$ to find $\cos\theta$. Then we can determine the other values of the trigonometric functions, by finding the quotients and reciprocals of the $\sin\theta$ and $\cos\theta$. Instead of doing this, consider a right-triangle approach to solving this problem. A right triangle in Figure 3 is first constructed with $\sin\theta = \frac{3}{5}$.

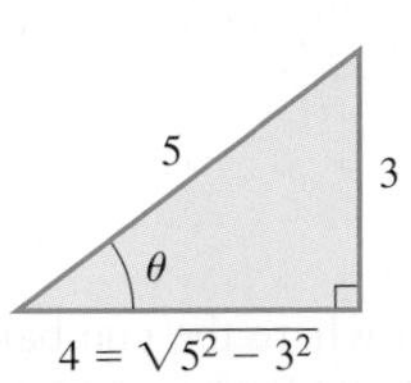

FIGURE 3

The Pythagorean Theorem implies that the remaining side of the triangle is $\sqrt{5^2 - 3^2} = 4$. Now that this is known, all the trigonometric functions of θ are read directly from the triangle:

$$\sin\theta = \frac{3}{5}, \quad \cos\theta = \frac{4}{5}, \quad \tan\theta = \frac{3}{4},$$
$$\csc\theta = \frac{5}{3}, \quad \sec\theta = \frac{5}{4}, \quad \text{and} \quad \cot\theta = \frac{4}{3}.$$

Applications

The following examples show some ways that the angle-side relationships of the trigonometric functions can be used in applications.

EXAMPLE 3 A climber who wants to measure the height of a cliff is standing 35 feet from the base of the cliff. An angle of approximately 60° is formed by the lines joining the climber's feet with the top and bottom of the cliff, as shown in Figure 4. Use this information to approximate the height of the cliff.

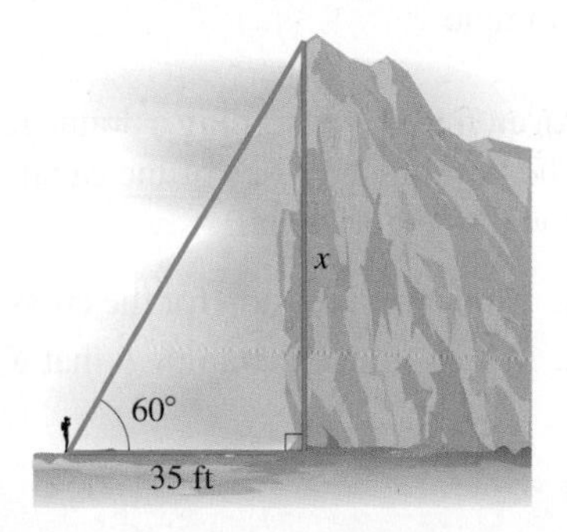

FIGURE 4

Solution Let x denote the height of the cliff. Since $60° = \pi/3$ radians, we have $\tan 60° = \sqrt{3}$. As a consequence,

$$\sqrt{3} = \tan 60° = \frac{x}{35}, \quad \text{and} \quad x = 35\sqrt{3} \approx 60.6 \text{ ft.}$$

EXAMPLE 4 Two balls are against the rail at opposite ends of a 10-foot billiard table, as shown in Figure 5. The player must hit the orange ball on the left with the white cue ball on the right without touching any of the other balls on the table. This is done by banking the cue ball off the bottom cushion. Where should the cue ball hit the bottom cushion, and what is the angle that its path makes with the bottom cushion?

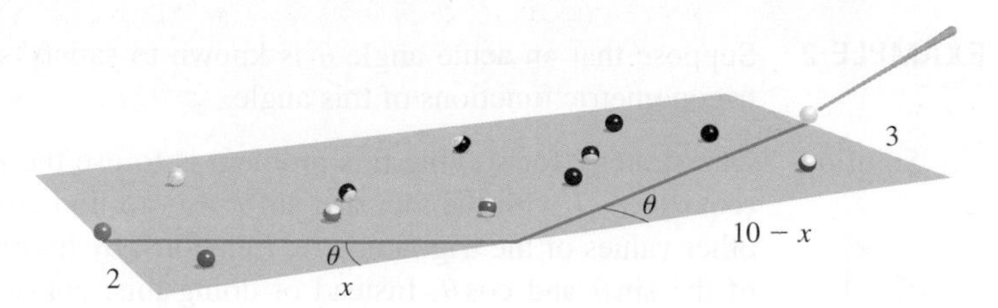

FIGURE 5

Solution Let x denote the distance from the left bottom corner to the spot where the cue ball hits the cushion. Billiard players instinctively realize that when a ball hits the cushion, its angle of reflection is the same as its angle of incidence (unless there is spin on the ball, which we will assume is not the case). This means that for the shot to be successful, the triangles shown in Figure 5 must be similar. Hence

$$\frac{2}{x} = \tan\theta = \frac{3}{10 - x}, \quad \text{and} \quad 2(10 - x) = 3x.$$

Solving for x gives

$$20 - 2x = 3x, \quad \text{and} \quad x = \frac{20}{5} = 4.$$

Since $\tan\theta = 2/x$, we also have

$$\tan\theta = \frac{2}{4} = \frac{1}{2}.$$

This is not quite the answer we want because we have not come upon any angles whose tangent is $\frac{1}{2}$. In Section 4.8 we will find a more satisfactory answer for this example. ■

EXAMPLE 5 An engineer is designing a drainage canal that has the cross section shown in Figure 6. The bottom and sides of the canal are each L feet long, and the side makes an angle θ with the horizontal.

This is an optimization problem similar to those in calculus.

a. Find an expression for the cross-sectional area of the canal in terms of the angle θ.

b. Approximate the angle θ that will maximize the capacity of a canal S feet long.

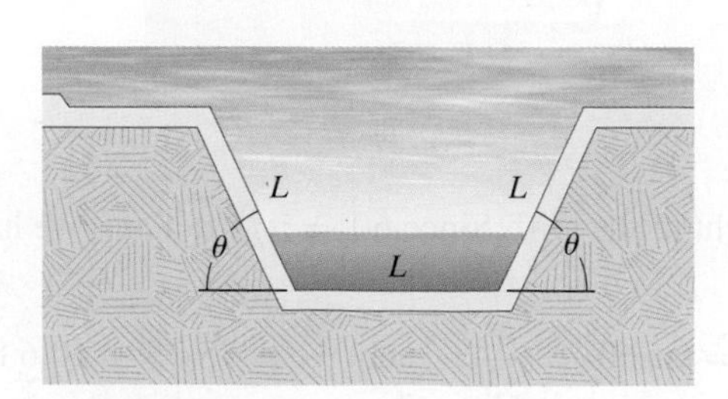

(a)

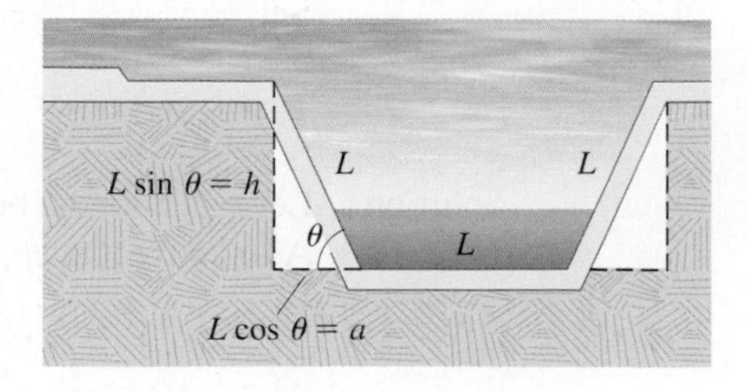

(b)

FIGURE 6

Solution **a.** The cross section of the canal is shown in Figure 6(b). The top and bottom sides are parallel, so the cross section is a *trapezoid*. The cross section has an area equal to the product of the height and the average of the widths of the top and bottom.

The width of the bottom is L, and Figure 6(b) shows that the width of the top is $L + 2(L\cos\theta)$ and the height is $L\sin\theta$, so the area of the cross section is

$$\begin{aligned} A &= \frac{1}{2}(L + (L + 2L\cos\theta)) \cdot (L\sin\theta) \\ &= \frac{1}{2}(2L + 2L\cos\theta)(L\sin\theta) \\ &= L^2(1 + \cos\theta)\sin\theta. \end{aligned}$$

We cannot find the exact maximum value, but a graphing device allows us to estimate it.

b. The capacity of the canal is the volume V of water the canal can hold, so a canal S feet long has a capacity

$$V = A \cdot S = SL^2(1 + \cos\theta)\sin\theta \text{ ft}^3,$$

and the capacity is maximized if θ maximizes

$$f(\theta) = (1 + \cos\theta)\sin\theta.$$

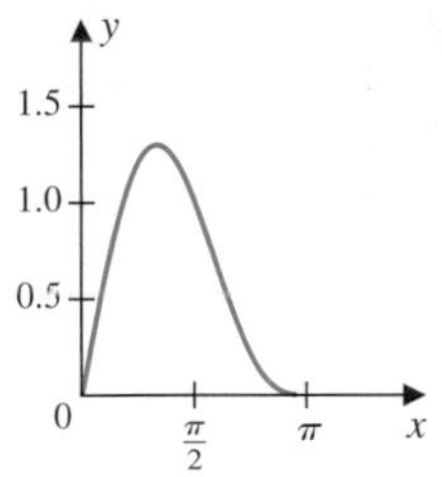

FIGURE 7

Figure 7 shows a computer-generated graph of $y = f(\theta)$. Using a graphing device to zoom in near the peak of the curve, you will find that near the maximum

$$\theta \approx 1.047 \text{ radians}.$$

So the maximum capacity of the canal is approximately

$$V \approx SL^2(1 + \cos(1.047))\sin(1.047) \approx 1.299\ SL^2 \text{ ft}^3.$$ ■

EXAMPLE 6 Two tracking stations x miles apart measure the angle of elevation of a weather balloon to be θ and φ, as shown in Figure 8. Find the altitude of the balloon in terms of the two angles.

Solution Let h denote the height of the balloon. Using the cotangent function and the two angles, we have

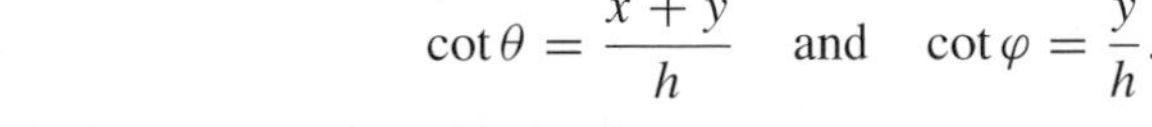

$$\cot\theta = \frac{x + y}{h} \quad \text{and} \quad \cot\varphi = \frac{y}{h}.$$

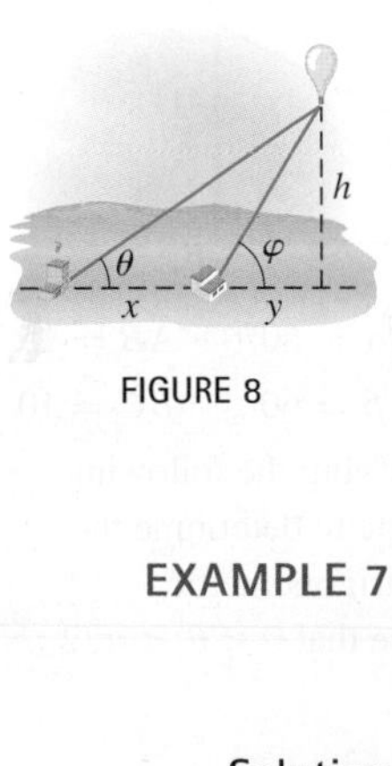

Next isolate y in each of the previous equations so

$$y = h\cot\theta - x \quad \text{and} \quad y = h\cot\varphi, \quad \text{hence} \quad h\cot\theta - x = h\cot\varphi.$$

Solving for h gives

$$h = \frac{x}{\cot\theta - \cot\varphi}.$$ ■

FIGURE 8

EXAMPLE 7 For many purposes the earth can be considered a perfect sphere with radius $R \approx 6400$ kilometers. Use this approximation to determine the distance you would travel around the earth if you stayed at the latitude of 30°.

Solution Let r denote the radius of the circle around the earth at the latitude of θ° degrees above the equator, and let E be the circumference at the equator. The right triangle

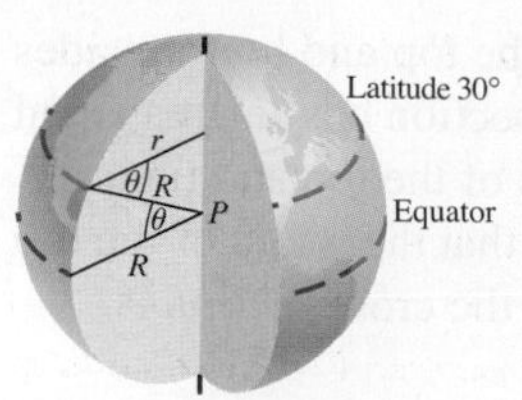

FIGURE 9

in Figure 9 illustrates the situation, where P is at the center of the earth. The radius at latitude θ is $r = R\cos\theta$, so the circumference of the latitude is

$$L = 2\pi r = 2\pi R\cos\theta = E\cos\theta.$$

At the latitude of 30°, the circumference, which is the distance traveled, is

$$L = 6400\cos 30^\circ = 6400\frac{\sqrt{3}}{2} \approx 5542.6 \text{ km.}$$

■

EXERCISE SET 4.3

In Exercises 1–6, find the value of the six trigonometric functions of the angle θ.

1.

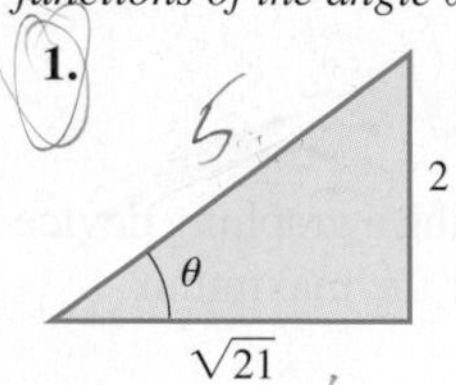

2.

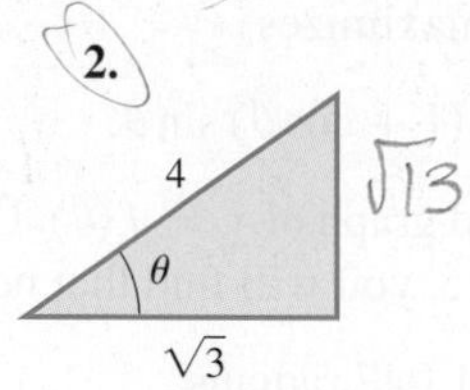

3.

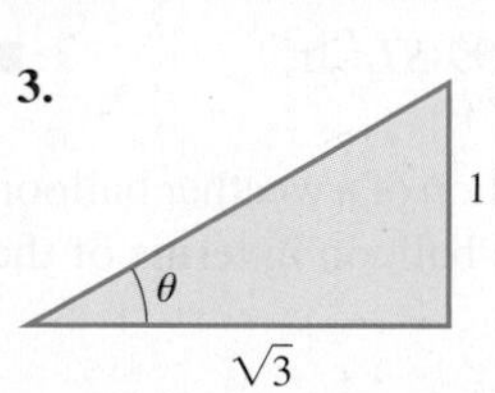

4.

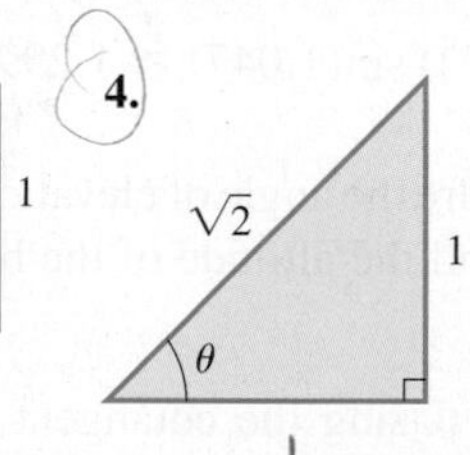

5.

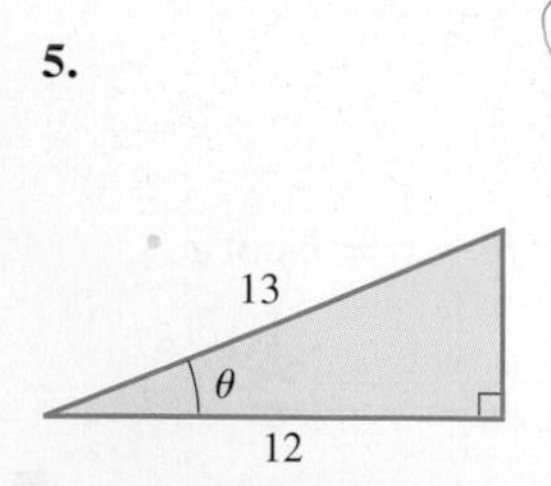

6.

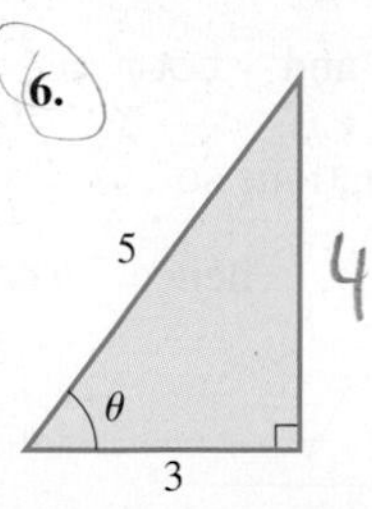

In Exercises 7–12, find the value of x.

7.

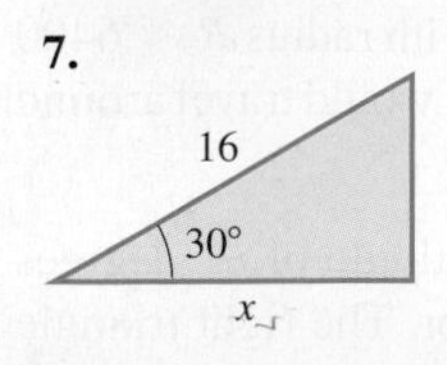

8.

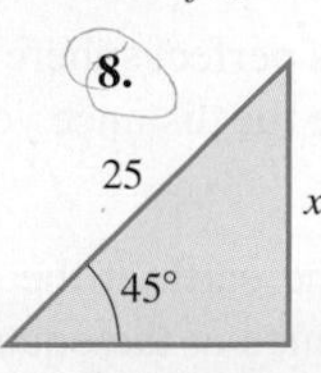

9.

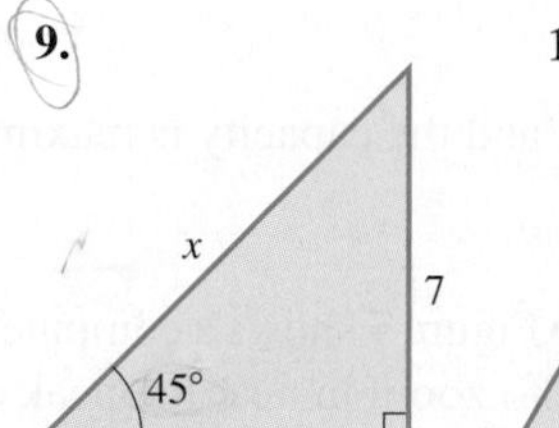

10.

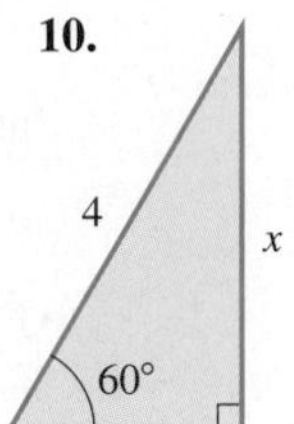

11.

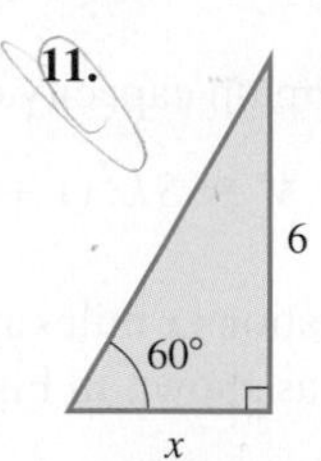

12.

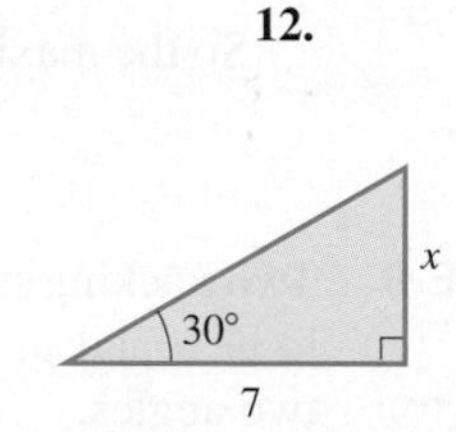

In Exercises 13–16, refer to the right triangle in the figure. From the information given, find any missing angles or sides.

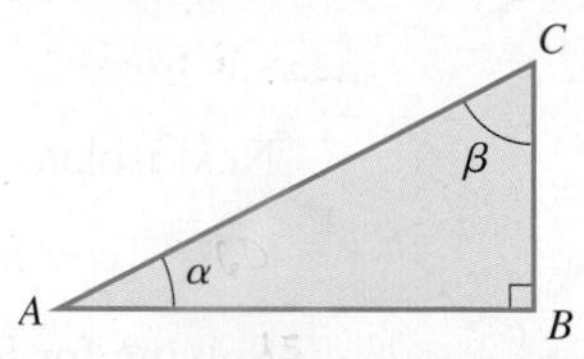

13. $\alpha = 30^\circ, \quad BC = 8$

14. $\alpha = 60^\circ, \quad AB = 15$

15. $\beta = 45^\circ, \quad AC = 5$

16. $\beta = 60^\circ, \quad BC = 10$

17. Construct a right triangle satisfying the following and use the sides of this triangle to determine the formulas of all the remaining trigonometric functions. Assume in each case that $0 < \theta < \pi/2$.

a. $\sin\theta = \frac{2}{3}$ **b.** $\cos\theta = \frac{1}{5}$

c. $\tan\theta = \frac{4}{3}$ **d.** $\csc\theta = \frac{3}{2}$

18. Use the results in Exercise 17 to determine the values of the remaining trigonometric functions.

a. $\sin\theta = \frac{2}{3}, \quad \frac{\pi}{2} < \theta < \pi$

b. $\cos\theta = \frac{1}{5}, \quad \frac{3\pi}{2} < \theta < 2\pi$

c. $\tan\theta = \frac{4}{3}, \quad \pi < \theta < \frac{3\pi}{2}$

d. $\csc\theta = \frac{3}{2}, \quad \frac{\pi}{2} < \theta < \pi$

19. Estimate the distance around the Earth at 45° degrees latitude north.

20. Two tracking stations 50 miles apart sight a weather balloon. The angle of the balloon from one of the stations is 45° and from the other is 60°. What is the height of the balloon?

21. The angle of elevation to the top of a building is 30° from the ground when viewed 1 mile from the building. Estimate the height of the building in feet.

22. The box in the figure has width 4 and height 3, and $\sin\theta = \frac{5}{13}$. What is the length of the box?

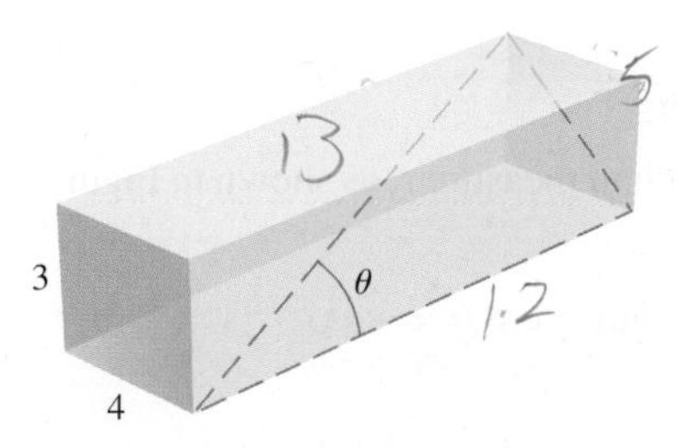

23. A climber who needs to estimate the height of a cliff stands at a point on the ground 40 feet from the base of the cliff and estimates that the angle from the ground to a line extending from the climber's feet to the top of the cliff is 60°. Assuming that the cliff is perpendicular to the ground, find the approximate height of the cliff.

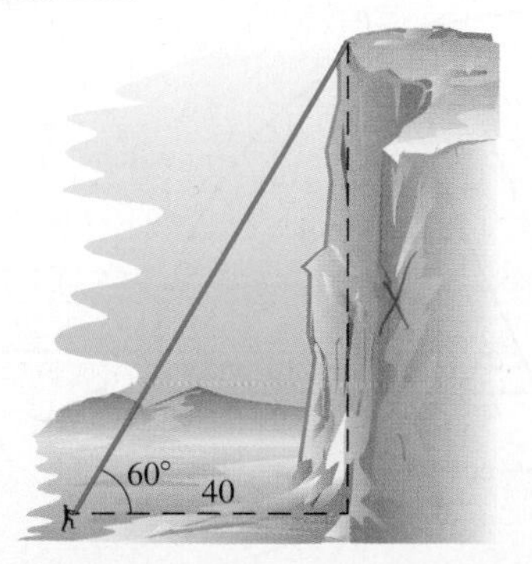

24. A pipeline is to be constructed between points A and B on opposite sides of a river, as shown in the figure. To determine the amount of pipe needed, using a theodolite at point A, a right triangle ACB is constructed with point C approximately 300 feet from A. At point A, a transit is used to determine that the angle θ is 45°. Determine the amount of pipe needed to connect points A and B.

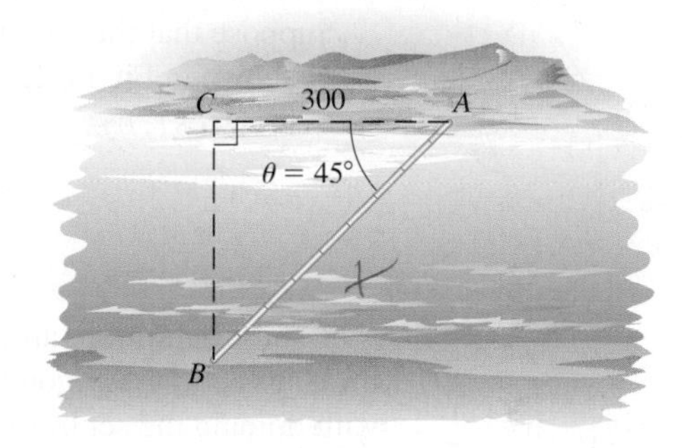

25. From the top of an observation tower, a forest ranger spots an illegal campfire. The angle of depression made by the ranger's line of sight and the campfire is 10.5°. If the tower is 80 feet high, how far is the campfire from the base of the tower? (Note: $\sin 10.5° \approx 0.182$.)

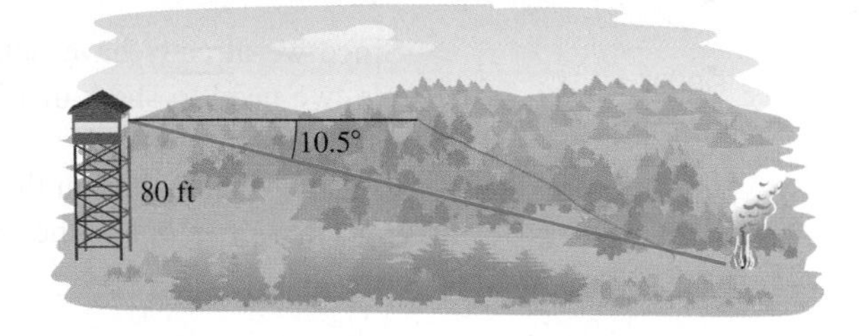

4.4 THE SINE AND COSINE FUNCTIONS

In Section 4.2 we saw that for each real number t, and hence for each angle with radian measure t, there is a pair $(x(t), y(t))$ of xy-coordinates describing the point $P(t)$ on the unit circle. These coordinates provide us with a way to extend the notion of the sine and cosine to functions whose domains are the set of real numbers.

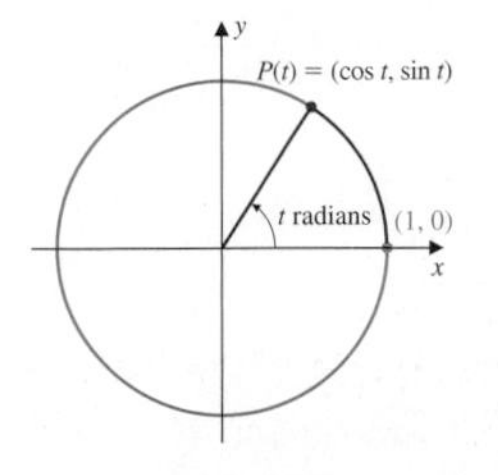

FIGURE 1

The Sine and Cosine Functions

Suppose that the coordinates of a point $P(t)$ on the unit circle are $(x(t), y(t))$ as shown in Figure 1. We define the **sine of** t, written $\sin t$, and the **cosine of** t, written $\cos t$, by

$$\sin t = y(t) \quad \text{and} \quad \cos t = x(t).$$

These definitions are also used to define the sine and cosine of an angle with radian measure t. So the trigonometric functions serve two purposes, directly as functions with domain the set of real numbers, and indirectly as functions whose domains are the set of all angles, where the angles are given in radian measure.

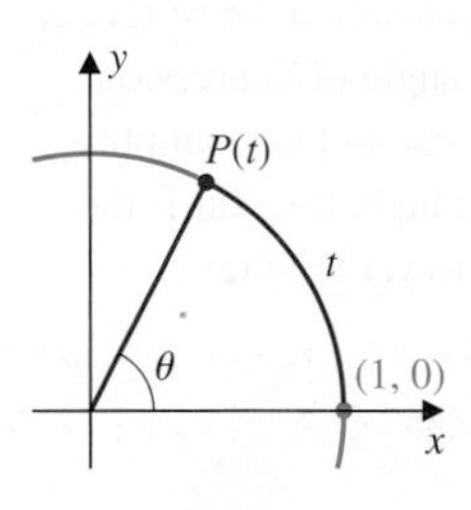

FIGURE 2

The Sine and Cosine Functions

Suppose that θ is an angle that has radian measure t, as shown in Figure 2. Then we define

$$\sin \theta = \sin t \quad \text{and} \quad \cos \theta = \cos t.$$

Since we already have definitions for the sine and cosine of angles in a right triangle, we need to show that in this situation we obtain the same results with the unit circle definitions.

Consider the angle θ in the right triangle in Figure 3(a). The adjacent side is a, the opposite side is b, and the hypotenuse of the triangle is c.

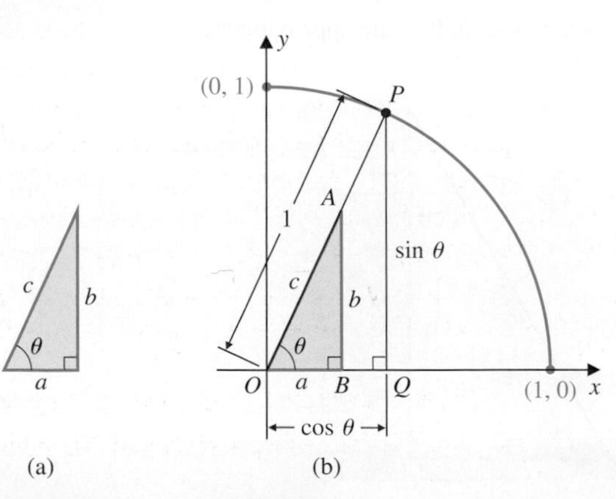

FIGURE 3

Superimpose an xy-coordinate system onto the triangle with the vertex of θ at the origin and positive x-axis along the side of the triangle with length a. Let P denote the intersection point of the unit circle with the terminal side of θ. Figure 3(b) shows the situation when $0 < c < 1$. The situation when $c > 1$ is similar, and the algebra does not depend on the magnitude of c.

Since $\triangle AOB$ is similar to $\triangle POQ$, relating the sides gives

$$\frac{\sin\theta}{1} = \frac{b}{c} \quad \text{and} \quad \frac{\cos\theta}{1} = \frac{a}{c},$$

so

$$\sin\theta = \frac{b}{c} \quad \text{and} \quad \cos\theta = \frac{a}{c},$$

which agrees with the right triangle definition. We generally consider the sine and cosine as functions whose domains are the set of real numbers, but you should keep in mind that there is an immediate angle relationship as well.

Some results follow quickly from the unit circle definitions of the sine and cosine. The first identity in trigonometry, and probably the most important, follows from the fact that for any number t the point $P(t) = (\cos t, \sin t)$ lies on the circle with equation $x^2 + y^2 = 1$.

It is also common to write

$$(\sin t)^2 = \sin^2 t$$

and

$$(\cos t)^2 = \cos^2 t.$$

The Pythagorean Identity

For all real numbers t, $\quad (\sin t)^2 + (\cos t)^2 = 1$.

In addition, since the points (x, y) on the circle $x^2 + y^2 = 1$ satisfy $-1 \le x \le 1$ and $-1 \le y \le 1$, these same bounds hold for the sine and cosine functions.

Bounds on the Sine and Cosine

For all real numbers t, $\quad -1 \le \sin t \le 1 \quad$ and $\quad -1 \le \cos t \le 1$.

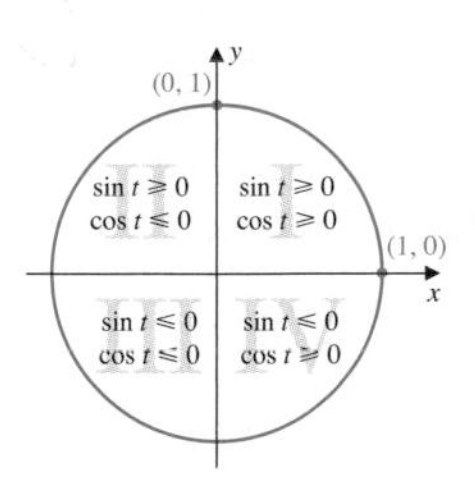

FIGURE 4

The signs of the sine and cosine functions are also easily determined once it is known in which quadrant of the plane $P(t)$ lies. For example, if $P(t)$ lies in quadrant II, the x-coordinate is negative and the y-coordinate is positive, so $\cos t \le 0$ and $\sin t \ge 0$. A summary of these results is shown in Figure 4.

The points $P(t) = (\cos t, \sin t)$ and $P(-t) = (\cos(-t), \sin(-t))$ are obtained in the same manner, except that in the first instance the rotation is counterclockwise from $(1, 0)$ and in the second the rotation is clockwise from $(1, 0)$. A typical situation is illustrated in Figure 5. Notice that

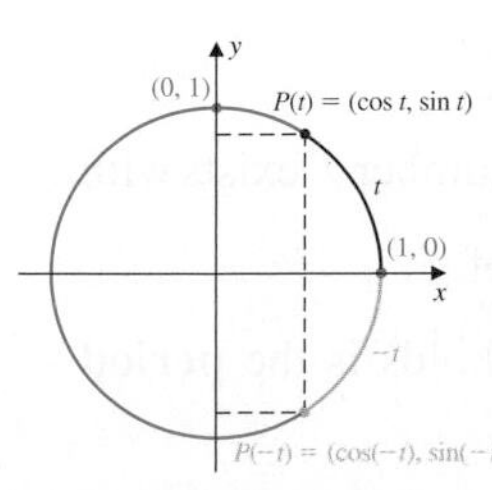

FIGURE 5

- The x-coordinates of $P(t)$ and $P(-t)$ are always the same.
- The y-coordinates of $P(t)$ and $P(-t)$ have the same magnitude but differ in sign.

This provides us with another important fact about the sine and cosine functions.

We will concentrate on the basic trigonometric identities that are most frequently used in calculus.

Cosine Function Is Even Sine Function Is Odd

For all real numbers t,

- $\cos(-t) = \cos t \qquad \sin(-t) = -\sin t$

Some specific values for the sine and cosine functions can be found using the geometric properties of the unit circle. The circumference of a circle with radius r is $2\pi r$, so the unit circle has circumference 2π. The axis intercepts for the unit circle shown in Figure 6 give the values of the sine and cosine for the multiples of $\pi/2$ listed in Table 1.

TABLE 1

t	0	$\frac{\pi}{2}$	π	$\frac{3\pi}{2}$	2π
$\cos t$	1	0	-1	0	1
$\sin t$	0	1	0	-1	0

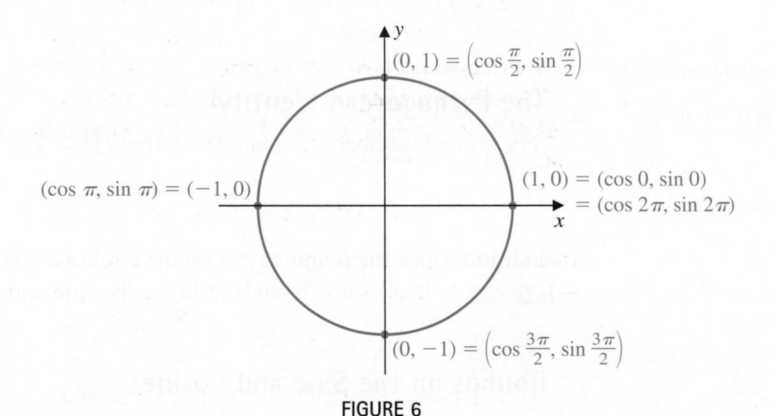

FIGURE 6

The unit circle has circumference 2π, so for any real number t we have

$$P(t + 2\pi) = P(t).$$

As a consequence, for any real number t,

$$\sin t = \sin(t + 2\pi) \quad \text{and} \quad \cos t = \cos(t + 2\pi).$$

Periodic Functions

A nonconstant function f is said to be **periodic** if a positive number T exists with

$$f(t + T) = f(t) \quad \text{for all } t \text{ in the domain of } f.$$

The smallest positive number T for which this equation holds is the **period** of f. Figure 7 shows some examples of periodic functions.

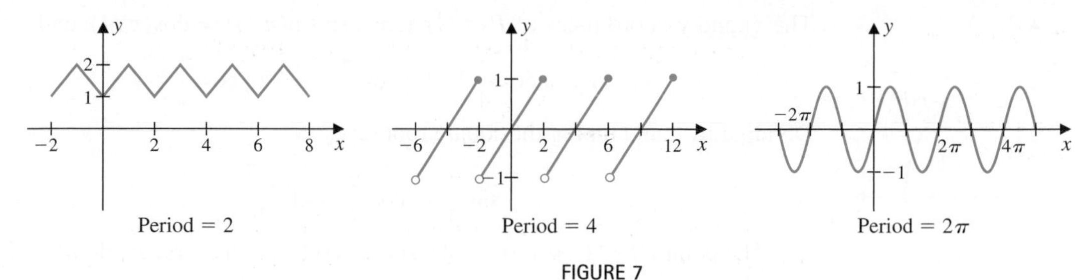

FIGURE 7

This definition implies that the sine and cosine functions have period 2π.

Period of the Sine and Cosine

The sine and cosine functions are periodic with **period** 2π. So for every real number t,

$$\sin(t + 2\pi) = \sin t \quad \text{and} \quad \cos(t + 2\pi) = \cos t.$$

To determine values of the sine and cosine functions for all values of t, then, we only need to determine the values of these functions for $0 \le t < 2\pi$. The other values of the sine and cosine are determined by finding a corresponding value in the interval $[0, 2\pi)$. For example, to find the sine and cosine of 17π, we note that the coordinates of $P(17\pi)$ agree with those of $P(17\pi - 2\pi) = P(15\pi)$, $P(15\pi - 2\pi) = P(13\pi), \ldots$, and, finally, $P(\pi)$. So

$$\sin 17\pi = \sin \pi = 0 \quad \text{and} \quad \cos 17\pi = \cos \pi = -1.$$

In a similar manner, the coordinates of $P(-3\pi/2)$ are the same as those of $P(\pi/2)$, since

$$P\left(-\frac{3\pi}{2}\right) = P\left(-\frac{3\pi}{2} + 2\pi\right) = P\left(\frac{\pi}{2}\right) = (0, 1).$$

So the cosine and sine of $-3\pi/2$ agree with those of $\pi/2$, which are 0 and 1, respectively.

In calculus you need the values of the trigonometric functions at the *standard* values, $0, \pi/6, \pi/4, \pi/3$, and $\pi/2$.

This periodic information is not of much use, of course, unless we can determine more values of $\sin t$ and $\cos t$ for t in the interval $[0, 2\pi)$. So our next step is to extend our knowledge to include other values of t that lie in the interval $[0, 2\pi)$. We will first determine the coordinates for $P(\pi/4)$, $P(\pi/3)$, and $P(\pi/6)$, and consequently the sines and cosines of $\pi/4$, $\pi/3$, and $\pi/6$. The symmetry of the unit circle will then permit us to determine the values of the sine and cosine functions for any integer multiple of these angles.

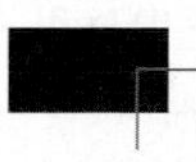

Determining $\sin\left(\frac{\pi}{4}\right)$ and $\cos\left(\frac{\pi}{4}\right)$

The point $P(\pi/4)$ lies on the unit circle midway between the points $P(0) = (1, 0)$ and $P(\pi/2) = (0, 1)$, as shown in Figure 8. The Pythagorean identity gives

$$1 = \left(\sin \frac{\pi}{4}\right)^2 + \left(\cos \frac{\pi}{4}\right)^2.$$

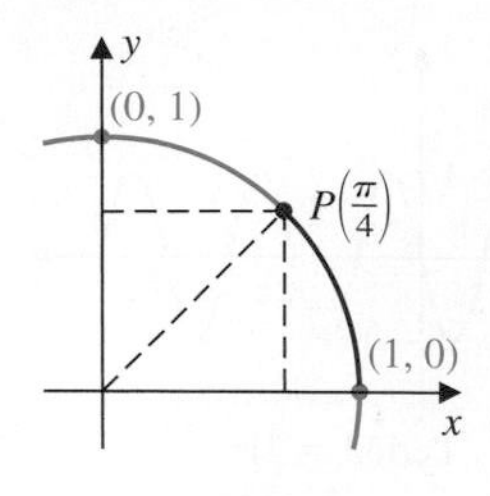

FIGURE 8

The x- and y-coordinates of $P(\pi/4)$ agree, so $\sin(\pi/4) = \cos(\pi/4)$, and

$$1 = \left(\cos\frac{\pi}{4}\right)^2 + \left(\cos\frac{\pi}{4}\right)^2 = 2\left(\cos\frac{\pi}{4}\right)^2.$$

Dividing by 2 and taking the square root gives

$$\sin\frac{\pi}{4} = \cos\frac{\pi}{4} = \pm\frac{\sqrt{2}}{2}.$$

The point $P(\pi/4) = (\cos(\pi/4), \sin(\pi/4))$ lies in the first quadrant of the plane, so both of its coordinates are positive. Hence

$$\sin\frac{\pi}{4} = \frac{\sqrt{2}}{2} \quad \text{and} \quad \cos\frac{\pi}{4} = \frac{\sqrt{2}}{2}.$$

The Sine and Cosine of $\frac{\pi}{4}$: $\sin\frac{\pi}{4} = \frac{\sqrt{2}}{2}$ and $\cos\frac{\pi}{4} = \frac{\sqrt{2}}{2}$.

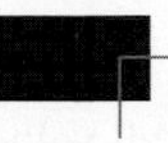

Determining $\sin\left(\frac{\pi}{3}\right)$ and $\cos\left(\frac{\pi}{3}\right)$

Consider $\triangle AOB$ shown in Figure 9(a), where O is at the origin $(0, 0)$ of the plane, A is at $P(0) = (1, 0)$, and B is at $P(\pi/3) = (\cos(\pi/3), \sin(\pi/3))$. The triangle is isosceles since OB and OA are both radii of the circle.

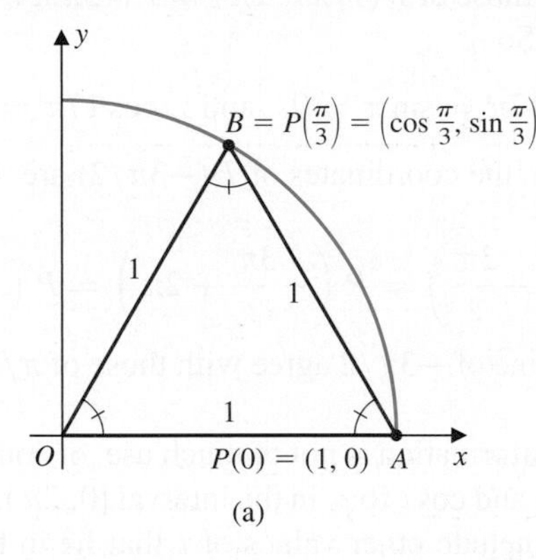

(a)

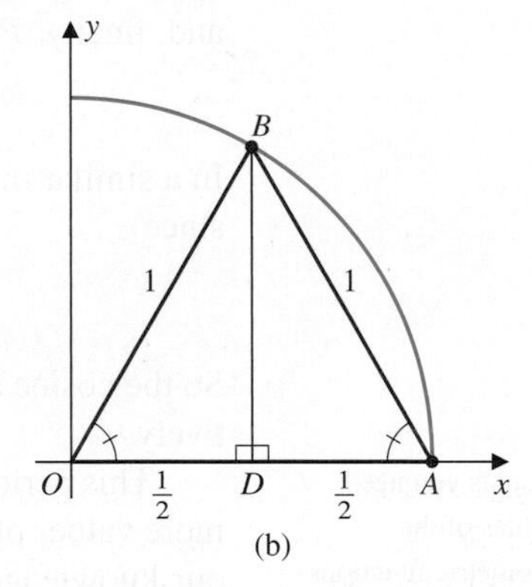

(b)

FIGURE 9

The base angles of an isosceles triangle are equal, so $\angle OAB = \angle OBA$. Since the sum of the angles in a triangle is π and $\angle AOB = \pi/3$, we must also have $\angle OAB = \angle OBA = \pi/3$. So $\triangle AOB$ is equilateral, with all its sides of length 1.

Drop a perpendicular from B to the x-axis meeting the axis at D, as shown in Figure 9(b). Then $\triangle ODB$ is congruent to $\triangle ADB$, and

$$OD = DA = \frac{1}{2}OA = \frac{1}{2}. \quad \text{Hence} \quad \cos\frac{\pi}{3} = OD = \frac{1}{2}.$$

The altitude of $\triangle ODB$ is $\sin(\pi/3)$, and by the Pythagorean Theorem

$$\left(\sin\frac{\pi}{3}\right)^2 + \left(\cos\frac{\pi}{3}\right)^2 = 1^2, \quad \text{so} \quad \left(\sin\frac{\pi}{3}\right)^2 + \left(\frac{1}{2}\right)^2 = 1.$$

Hence

$$\sin\frac{\pi}{3} = BD = \sqrt{1-\left(\frac{1}{2}\right)^2} = \sqrt{\frac{3}{4}} = \frac{\sqrt{3}}{2},$$

and $P(\pi/3) = (1/2, \sqrt{3}/2)$.

The Sine and Cosine of $\frac{\pi}{3}$: $\sin\frac{\pi}{3} = \frac{\sqrt{3}}{2}$ and $\cos\frac{\pi}{3} = \frac{1}{2}$.

Determining $\sin\left(\frac{\pi}{6}\right)$ and $\cos\left(\frac{\pi}{6}\right)$

In Figure 10, $\triangle AEO$ and $\triangle CDO$ are constructed so that $\angle COD$ and $\angle AOE$ are both $\pi/6$. Since $OA = OC = 1$, the right triangles $\triangle AEO$ and $\triangle CDO$ are congruent. But we have just found that $P(\pi/3) = (1/2, \sqrt{3}/2)$ so

$$\cos\frac{\pi}{6} = OD = OE = \frac{\sqrt{3}}{2}, \quad \text{and} \quad \sin\frac{\pi}{6} = DC = EA = \frac{1}{2}.$$

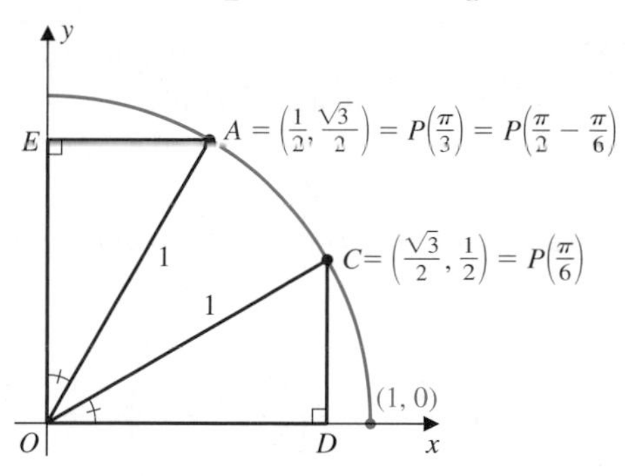

FIGURE 10

The Sine and Cosine of $\frac{\pi}{6}$: $\sin\frac{\pi}{6} = \frac{1}{2}$ and $\cos\frac{\pi}{6} = \frac{\sqrt{3}}{2}$.

Know the location of $P(\pi/6)$, $P(\pi/4)$, and $P(\pi/3)$ and remember that the numbers $1/2$, $\sqrt{2}/2$, and $\sqrt{3}/2$ are associated with the coordinates of these points.

We now have the values of the sine and cosine functions for five values of t in the interval $[0, \pi/2]$. These are shown in Figure 11 and in Table 2, and form the basis for our knowledge of the values of these trigonometric functions.

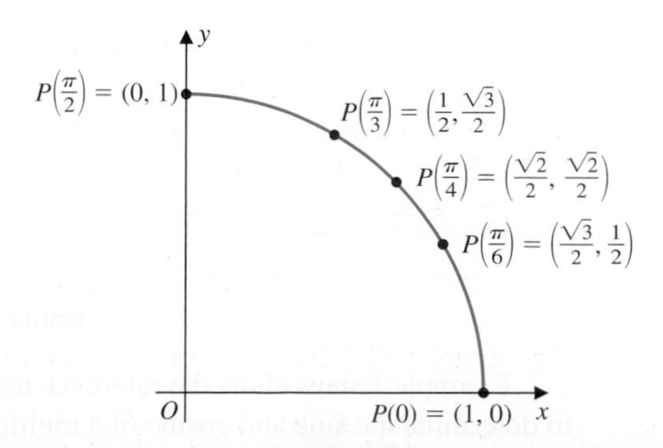

FIGURE 11

The values of any multiple of $\pi/6$ or $\pi/4$ are found using the entries in Table 2. To do this we introduce the notion of the *reference number*.

TABLE 2

t	0	$\frac{\pi}{6}$	$\frac{\pi}{4}$	$\frac{\pi}{3}$	$\frac{\pi}{2}$
$\cos t$	1	$\frac{\sqrt{3}}{2}$	$\frac{\sqrt{2}}{2}$	$\frac{1}{2}$	0
$\sin t$	0	$\frac{1}{2}$	$\frac{\sqrt{2}}{2}$	$\frac{\sqrt{3}}{2}$	1

Reference Number

For any real number t, the **reference number** r associated with t is the shortest distance along the unit circle from t to the x-axis. For any t, the reference number r is in $[0, \pi/2]$.

Figure 12 shows values of t in each of the four quadrants together with their reference numbers. Notice that in each case the coordinates of $P(t)$ are easily found from the coordinates of $P(r)$.

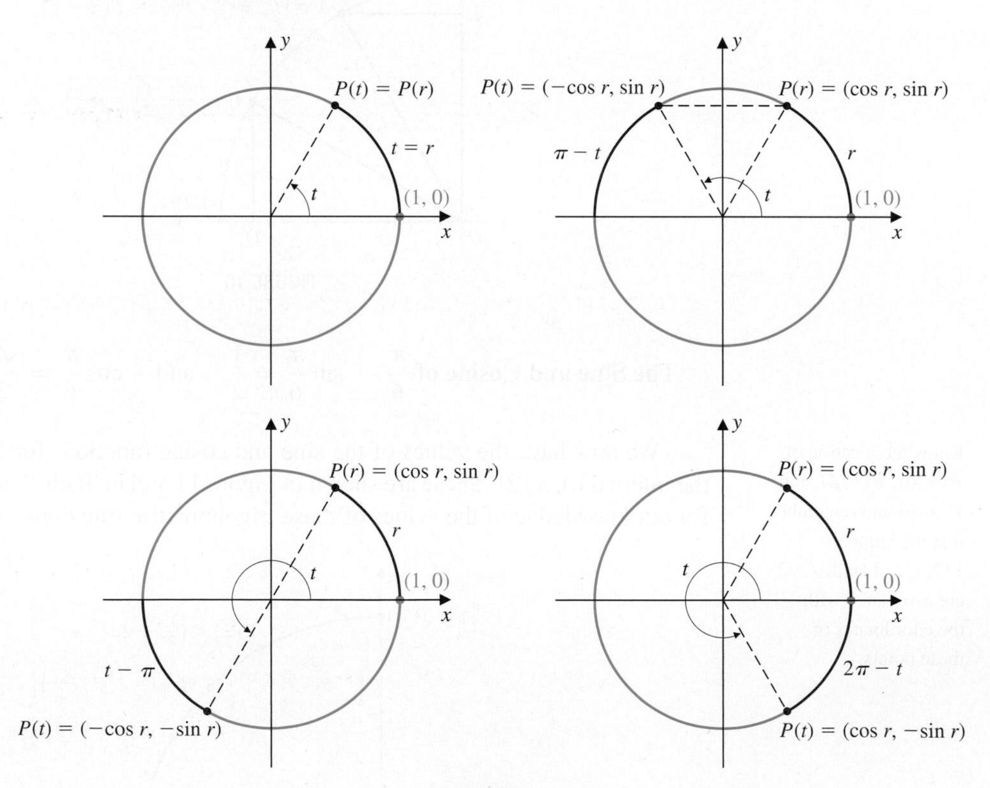

FIGURE 12

Example 1 shows how the reference numbers are used with the values in Table 2 to determine the sine and cosine of a multiple of $\pi/6$ or $\pi/4$.

EXAMPLE 1 Find the values of $\sin t$ and $\cos t$ for the value of t.

a. $t = \dfrac{5\pi}{6}$ **b.** $t = -\dfrac{3\pi}{4}$ **c.** $t = \dfrac{17\pi}{3}$

Solution We first determine the reference number in $[0, \pi/2]$ that corresponds to t. The coordinates of $P(t)$ are the same as those of the reference number, except for an adjustment of signs to reflect the quadrant in which $P(t)$ lies.

a. Since $5\pi/6$ lies in the second quadrant, its reference number is

$$\pi - \frac{5\pi}{6} = \frac{\pi}{6},$$

so the point $P(5\pi/6)$, shown in Figure 13, has the same y-coordinate as the point $P(\pi/6)$. The x-coordinates of these two points have the same magnitude, but differ in sign. So $P(5\pi/6) = (-\sqrt{3}/2, 1/2)$, which implies that

$$\sin\frac{5\pi}{6} = \sin\frac{\pi}{6} = \frac{1}{2} \quad \text{and} \quad \cos\frac{5\pi}{6} = -\cos\frac{\pi}{6} = -\frac{\sqrt{3}}{2}.$$

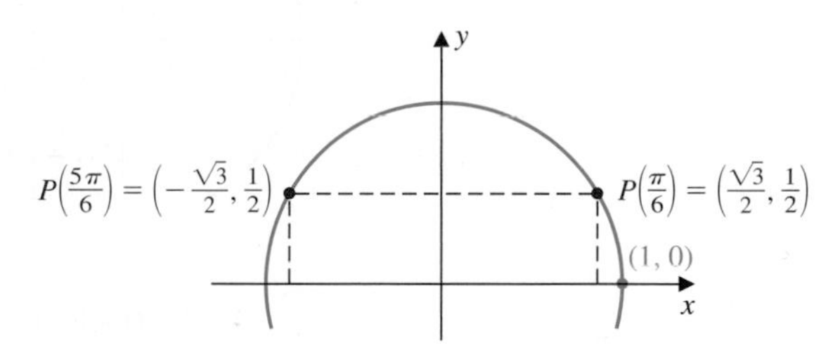

FIGURE 13

b. The point $P(-3\pi/4)$, shown in Figure 14, lies in the third quadrant with reference number $\pi/4$. Hence

$$\sin\left(-\frac{3\pi}{4}\right) = -\sin\left(\frac{\pi}{4}\right) = -\frac{\sqrt{2}}{2}$$

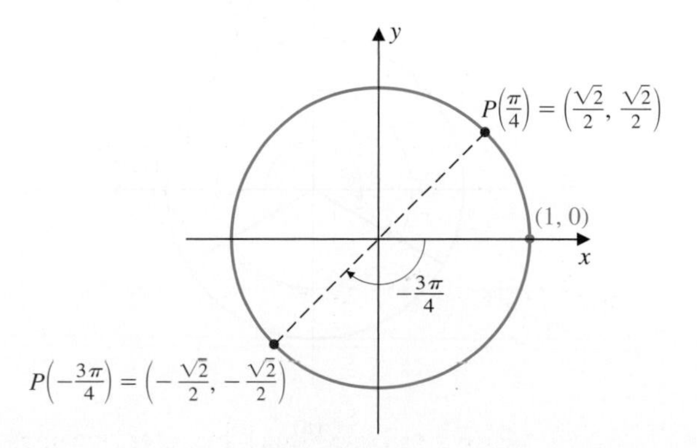

FIGURE 14

and

$$\cos\left(-\frac{3\pi}{4}\right) = -\cos\left(\frac{\pi}{4}\right) = -\frac{\sqrt{2}}{2}.$$

c. Although $17\pi/3$ exceeds 2π, notice that

$$\frac{17\pi}{3} = \frac{18\pi}{3} - \frac{\pi}{3} = 3(2\pi) - \frac{\pi}{3}.$$

So $17\pi/3$ is $\pi/3$ units less than three counterclockwise revolutions of the circle, beginning at $(1, 0)$. This implies that $P(17\pi/3)$ lies in the fourth quadrant and $17\pi/3$ has reference number $\pi/3$, as shown in Figure 15. Hence

$$\sin\frac{17\pi}{3} = -\sin\frac{\pi}{3} = -\frac{\sqrt{3}}{2} \quad \text{and} \quad \cos\frac{17\pi}{3} = \cos\frac{\pi}{3} = \frac{1}{2}.$$ ■

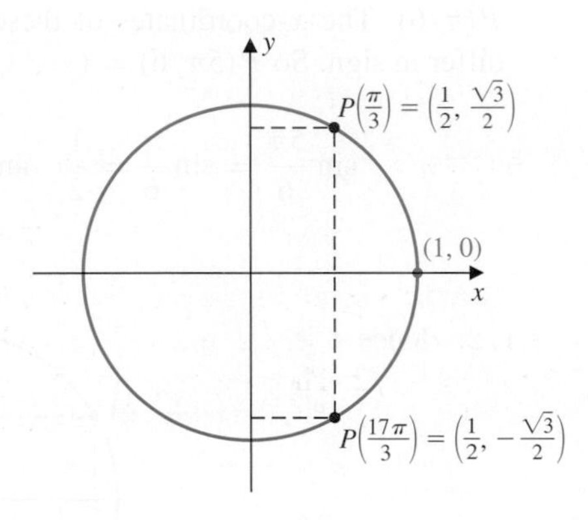

FIGURE 15

EXAMPLE 2 Find all values of t in the interval $[0, 2\pi]$ that satisfy $\sin 2t = -\frac{1}{2}$.

Solution First note that when t is in $[0, 2\pi]$, $2t$ is in $[0, 4\pi]$. The angle in quadrant I where the sine is $1/2$ is $\pi/6$. Since the sine is negative in quadrants III and IV, we need the angles in quadrants III and IV with reference angle $\pi/6$. So the solutions when $2t$ is in $[0, 2\pi]$ are $7\pi/6$ and $11\pi/6$, as shown in Figure 16(a).

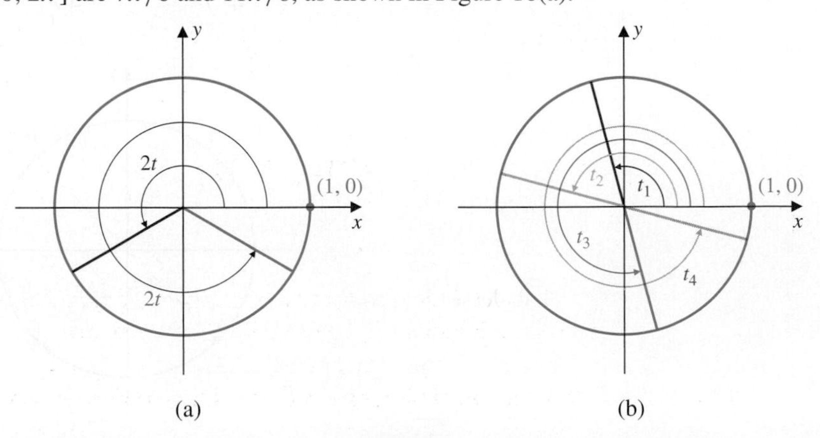

FIGURE 16

Additional answers for $2t$ in $[2\pi, 4\pi]$ are

$$\frac{7\pi}{6} + 2\pi = \frac{19\pi}{6} \quad \text{and} \quad \frac{11\pi}{6} + 2\pi = \frac{23\pi}{6}.$$

The values of t in $[0, 2\pi]$ with $\sin 2t = -1/2$ are as shown in Figure 16(b). They are

$$t_1 = \frac{1}{2}\left(\frac{7\pi}{6}\right) = \frac{7\pi}{12}, \quad t_2 = \frac{11\pi}{12}, \quad t_3 = \frac{19\pi}{12}, \quad \text{and} \quad t_4 = \frac{23\pi}{12}.$$

■

We have described in this section how to obtain the sine and cosine of the most frequently used values of t. Any scientific calculator will give approximations of the sine and cosine functions for arbitrary values of t.

EXAMPLE 3 Determine the values of t in $[0, 2\pi]$ that satisfy $2(\sin t)^2 + 3\cos t = 0$.

Solution We first use the Pythagorean identity to write the $(\sin t)^2$ term as a cosine term, giving

$$0 = 2(\sin t)^2 + 3\cos t = 2(1 - (\cos t)^2) + 3\cos t = 2 + 3\cos t - 2(\cos t)^2.$$

The term on the right is quadratic in the variable $\cos t$ and is factored to give

$$0 = 2 + 3\cos t - 2(\cos t)^2 = (2 - \cos t)(1 + 2\cos t).$$

For t to be a solution, we must have either $\cos t = 2$, which is impossible, or $\cos t = -1/2$. Hence t in the interval $[0, 2\pi]$ can be a solutions to the equation only if $\cos t = -1/2$. The reference number whose cosine is $1/2$ is $\pi/3$, and the cosine is negative in quadrants II and III as shown in Figure 17. Hence the two values of t in $[0, 2\pi]$ with $2(\sin t)^2 + 3\cos t = 0$ are

$$t_1 = \pi - \frac{\pi}{3} = \frac{2\pi}{3} \quad \text{and} \quad t_2 = \pi + \frac{\pi}{3} = \frac{4\pi}{3}.$$

■

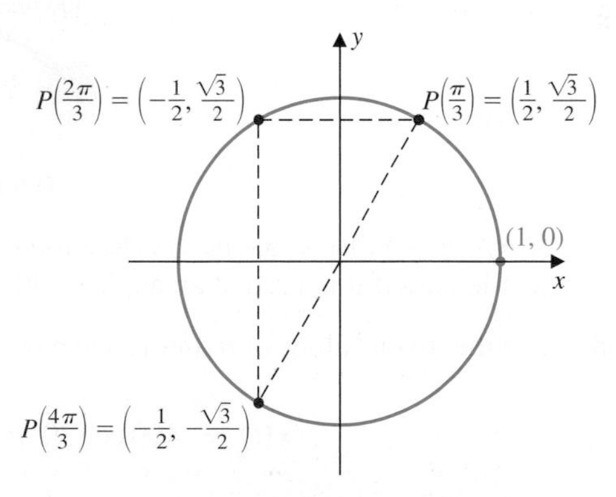

FIGURE 17

In calculus it is common to see trigonometric functions composed with algebraic functions. Example 4 gives two illustrations of this.

EXAMPLE 4 Write the function $h(x)$ as a composition $h(x) = f(g(x))$.

a. $h(x) = (\sin x)^3$ **b.** $h(x) = \cos\left(x^2 - 2x + 1\right)$

Solution **a.** The function $h(x) = (\sin x)^3$ first evaluates the sine of x and then cubes this result, so it is the composition of these operations. Let

$$g(x) = \sin x \quad \text{and} \quad f(x) = x^3.$$

Then

$$h(x) = f(g(x)) = \overbrace{\Big(\underbrace{\sin x}_{\text{first}}\Big)^3}^{\text{second}}.$$

b. Let

$$g(x) = x^2 - 2x + 1 \quad \text{and} \quad f(x) = \cos x.$$

Then

$$h(x) = f(g(x)) = f(x^2 - 2x + 1) = \cos(x^2 - 2x + 1).$$ ■

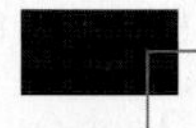

Applications

EXAMPLE 5 Suppose a projectile is fired at an angle of elevation θ with an initial velocity of v_0 ft/sec, as illustrated in Figure 18. In calculus and physics courses you will learn that points $(x(t), y(t))$ on the path of the projectile at time t are given by

$$x(t) = (v_0 \cos\theta)t \quad \text{and} \quad y(t) = (v_0 \sin\theta)t - \frac{1}{2}gt^2,$$

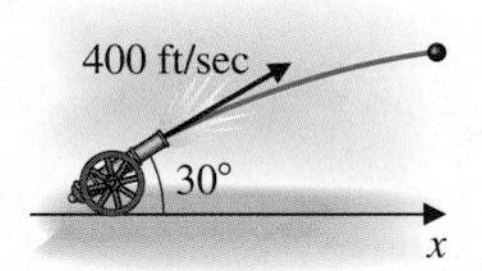

FIGURE 18

where $g \approx 32\text{ft/sec}^2$ is the acceleration due to the earth's gravity. Where will a projectile land if it is fired at an angle of 30° and an initial velocity of 400 ft/sec?

Solution For the given values of θ and v_0 we have

$$x(t) = (400 \cos 30^\circ)t = 400\frac{\sqrt{3}}{2}t = 200\sqrt{3}t,$$

and

$$y(t) = (400 \sin 30^\circ)t - \frac{1}{2}32t^2 = 400\left(\frac{1}{2}\right)t - 16t^2 = 200t - 16t^2.$$

The projectile hits the ground at the value of $t > 0$ for which $y(t) = 0$. This occurs when

$$0 = 200t - 16t^2 = 8t(25 - 2t), \quad \text{so} \quad t = 12.5 \text{ seconds}.$$

Hence the projectile hits the ground

$$x(12.5) = 200\sqrt{3}(12.5) \approx 200(1.732)(12.5) \approx 1169 \text{ feet}$$

from where it was fired. ■

EXERCISE SET 4.4

In Exercises 1–6, find the reference number r for the given value of t.

1. a. $t = \frac{\pi}{4}$ b. $t = \frac{\pi}{3}$ c. $t = \frac{2\pi}{3}$ d. $t = \frac{3\pi}{4}$
2. a. $t = \frac{9\pi}{4}$ b. $t = \frac{5\pi}{3}$ c. $t = \frac{11\pi}{3}$ d. $t = \frac{13\pi}{6}$
3. a. $t = -\frac{\pi}{3}$ b. $t = -\frac{5\pi}{6}$ c. $t = -\frac{4\pi}{3}$ d. $t = -\frac{5\pi}{4}$
4. a. $t = -\frac{11\pi}{6}$ b. $t = -\frac{19\pi}{3}$ c. $t = -\frac{11\pi}{3}$ d. $t = -\frac{23\pi}{4}$
5. a. $t = 45°$ b. $t = 150°$ c. $t = 240°$ d. $t = 330°$
6. a. $t = -45°$ b. $t = -30°$ c. $t = -150°$ d. $t = -240°$

In Exercises 7–38, find sin *t and* cos *t for the given value of t.*

7. $t = \frac{\pi}{6}$
8. $t = \frac{\pi}{4}$
9. $t = \frac{5\pi}{6}$
10. $t = \frac{2\pi}{3}$
11. $t = \frac{4\pi}{3}$
12. $t = \frac{7\pi}{6}$
13. $t = \frac{13\pi}{6}$
14. $t = \frac{7\pi}{4}$
15. $t = -\frac{\pi}{3}$
16. $t = -\frac{\pi}{6}$
17. $t = -\frac{5\pi}{6}$
18. $t = -\frac{7\pi}{6}$
19. $t = -\frac{5\pi}{4}$
20. $t = -\frac{5\pi}{3}$
21. $t = -\frac{3\pi}{2}$
22. $t = -\frac{7\pi}{2}$
23. $t = -7\pi$
24. $t = 8\pi$
25. $t = 45°$
26. $t = 90°$
27. $t = 150°$
28. $t = 135°$
29. $t = 240°$
30. $t = 120°$
31. $t = 330°$
32. $t = 315°$
33. $t = -60°$
34. $t = -150°$
35. $t = -240°$
36. $t = -300°$
37. $t = 540°$
38. $t = -900°$

In Exercises 39–46, find all values of t in the interval $[0, 2\pi]$ *that satisfy the given equation.*

39. $\cos t = \frac{\sqrt{2}}{2}$
40. $\sin t = \frac{\sqrt{3}}{2}$
41. $\sin t = -\frac{1}{2}$
42. $\cos t = -\frac{\sqrt{3}}{2}$
43. $\cos t = 1$
44. $\sin t = -1$
45. $\cos \frac{t}{2} = \frac{1}{2}$
46. $\sin(3t) = -\frac{\sqrt{2}}{2}$

In Exercises 47–50, determine whether the function is even, odd, or neither.

47. $f(x) = (\cos x)^2$
48. $f(x) = x^3 \sin x$
49. $f(x) = |x| \sin x$
50. $f(x) = \cos(\sin x)$
51. If $\sin t = -\frac{2\sqrt{2}}{3}$ and $\cos t = \frac{1}{3}$, find the sine and cosine of the given value.
    a. $t + \pi$ b. $-t$ c. $t + \frac{\pi}{2}$ d. $-t + \frac{\pi}{2}$
52. Find all t in the interval $[0, 2\pi]$ satisfying $(\cos t)^2 + \cos t - 2 = 0$.
53. Find all t in the interval $[0, 2\pi]$ satisfying $2(\sin t)^2 + \sin t - 1 = 0$.
54. Find all t in the interval $[0, 2\pi]$ satisfying $\sin t \cos t - \sin t - \cos t + 1 = 0$.
55. Find all t in the interval $[0, 2\pi]$ satisfying $\sin t + \cos t = 1$.

56. Suppose the function f is even and is periodic with period 2, that $f(-1) = 0$ and $f(0) = 1$, and that f is linear on the interval $[-1, 0]$ and on the interval $[0, 1]$.

a. Make a sketch of the graph of $y = f(x)$.

b. Determine the values of x for which $f(x) = 0$ and $f(x) = 1$, and the range of the function.

57. A projectile is fired with an initial velocity of 1600 ft/sec at an angle of 60°. Find the horizontal distance that the projectile travels.

58. Show that for a given angle of elevation θ and initial velocity v_0, the height y of a projectile can be expressed as a quadratic function of the horizontal distance x.

59. Use the result of Exercise 58 to determine the maximum height of the projectile described in Exercise 57.

4.5 GRAPHS OF THE SINE AND COSINE FUNCTIONS

In Section 4.4 the variable t is used to describe the domain of the sine and cosine functions so that an xy-coordinate system can be used to position the point $P(t)$. The coordinates of $P(t)$ simultaneously produced values of the sine and cosine of t.

Graphing devices give accurate graphs of the trigonometric functions. Remember that radian measure is used.

Now that we know the values of these functions, we can revert to the usual situation where the variable x represents the numbers in the domain of a function and the variable y represents the resulting values in the range. The information given in Section 4.4 is used to sketch the graphs of the sine and cosine functions.

We first consider the graph of the sine function, $f(x) = \sin x$. For x in the interval $[0, 2\pi]$ we have the values given in Table 1.

TABLE 1

x	0	$\frac{\pi}{6}$	$\frac{\pi}{4}$	$\frac{\pi}{3}$	$\frac{\pi}{2}$	$\frac{2\pi}{3}$	$\frac{3\pi}{4}$	$\frac{5\pi}{6}$	
$\sin x$	0	$\frac{1}{2}$	$\frac{\sqrt{2}}{2}$	$\frac{\sqrt{3}}{2}$	1	$\frac{\sqrt{3}}{2}$	$\frac{\sqrt{2}}{2}$	$\frac{1}{2}$	
x	π	$\frac{7\pi}{6}$	$\frac{5\pi}{4}$	$\frac{4\pi}{3}$	$\frac{3\pi}{2}$	$\frac{5\pi}{3}$	$\frac{7\pi}{4}$	$\frac{11\pi}{6}$	2π
$\sin x$	0	$-\frac{1}{2}$	$-\frac{\sqrt{2}}{2}$	$-\frac{\sqrt{3}}{2}$	-1	$-\frac{\sqrt{3}}{2}$	$-\frac{\sqrt{2}}{2}$	$-\frac{1}{2}$	0

Assuming a simple and smooth behavior based on these values gives the graph in Figure 1(a) for x in $[0, 2\pi]$. The graph has x-intercepts at $x = 0$, at $x = \pi$,

Recognizing the general shape and behavior of the graphs of the sine and cosine functions is essential in calculus.

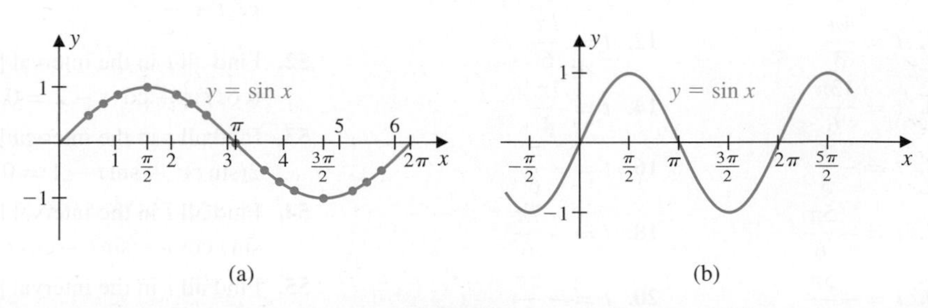

FIGURE 1

and at $x = 2\pi$. The sine function is increasing on $[0, \pi/2]$ and on $[3\pi/2, 2\pi]$, and is decreasing on $[\pi/2, 3\pi/2]$. A maximum value of 1 occurs at $x = \pi/2$ and a minimum value of -1 occurs at $x = 3\pi/2$.

Since $\sin(x \pm 2\pi) = \sin x$ for every real number x, the graph of the sine function repeats indefinitely to the right and to the left, as indicated in Figure 1(b).

In a similar manner, the values of the cosine function $f(x) = \cos x$, for x in $[0, 2\pi]$ given in Table 2, produces the graph shown in Figure 2(a).

The graph of the cosine function has x-intercepts at $x = \pi/2$ and at $x = 3\pi/2$. It is decreasing on $[0, \pi]$ and increasing on $[\pi, 2\pi]$. A maximum value of 1 occurs at $x = 0$ and $x = 2\pi$, and a minimum value of -1 occurs at $x = \pi$. Since for every real number x we have $\cos(x \pm 2\pi) = \cos x$, the graph repeats indefinitely to the right and to the left, as indicated in Figure 2(b).

TABLE 2

x	0	$\frac{\pi}{6}$	$\frac{\pi}{4}$	$\frac{\pi}{3}$	$\frac{\pi}{2}$	$\frac{2\pi}{3}$	$\frac{3\pi}{4}$	$\frac{5\pi}{6}$	
$\cos x$	1	$\frac{\sqrt{3}}{2}$	$\frac{\sqrt{2}}{2}$	$\frac{1}{2}$	0	$-\frac{1}{2}$	$-\frac{\sqrt{2}}{2}$	$-\frac{\sqrt{3}}{2}$	
x	π	$\frac{7\pi}{6}$	$\frac{5\pi}{4}$	$\frac{4\pi}{3}$	$\frac{3\pi}{2}$	$\frac{5\pi}{3}$	$\frac{7\pi}{4}$	$\frac{11\pi}{6}$	2π
$\cos x$	-1	$-\frac{\sqrt{3}}{2}$	$-\frac{\sqrt{2}}{2}$	$-\frac{1}{2}$	0	$\frac{1}{2}$	$\frac{\sqrt{2}}{2}$	$\frac{\sqrt{3}}{2}$	1

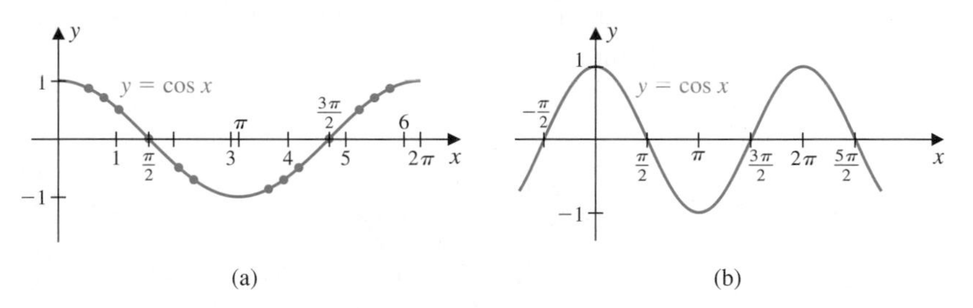

FIGURE 2

We found in Section 4.4 that the sine function is odd and the cosine is even. This implies that

- the graph of $y = \sin x$ has origin symmetry, and
- the graph of $y = \cos x$ has y-axis symmetry.

This relationship between the two basic trigonometric functions is a primary reason for their importance.

Notice also that the zeros of the sine function occur at integer multiples of π, that is, at $x = n\pi$ for all integers n, which is precisely where the maximum and minimum values of the cosine function occur. Moreover, the zeros of the cosine function occur when $x = n\pi + \pi/2$ for all integers n and coincide with the values producing the maximum and minimum values of the sine function. Hence

- extreme values of $\sin x$ occur when $\cos x = 0$, and
- extreme values of $\cos x$ occur when $\sin x = 0$.

The translation results used in Chapters 1 and 2 can, of course, also be applied to the trigonometric functions.

EXAMPLE 1 Use the graphs of $y = \sin x$ and $y = \cos x$ to sketch the graphs of the following.

a. $y = \sin\left(x - \frac{\pi}{2}\right)$ and $y = \sin\left(x + \frac{\pi}{2}\right)$,

b. $y = \cos\left(x - \frac{\pi}{2}\right)$ and $y = \cos\left(x + \frac{\pi}{2}\right)$.

Solution **a.** The graph of $y = \sin(x - \pi/2)$ is a horizontal shift of the graph of $y = \sin x$ to the right $\pi/2$ units, and the graph of $y = \sin(x + \pi/2)$ is a horizontal shift of the graph of $y = \sin x$ to the left $\pi/2$ units. The graphs of these functions are shown in Figure 3.

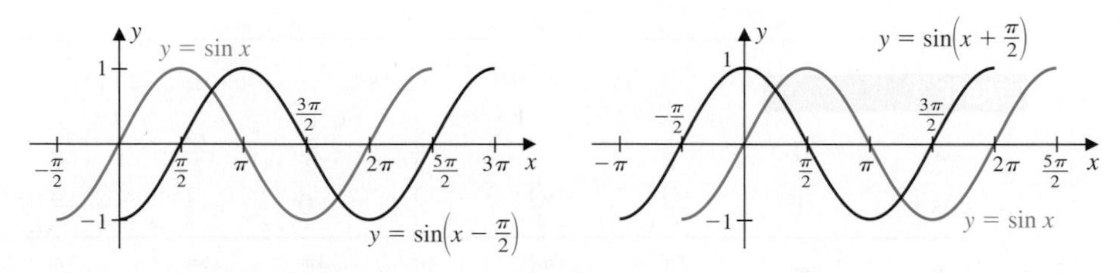

FIGURE 3

b. The graphs of $y = \cos(x - \pi/2)$ and $y = \cos(x + \pi/2)$ are obtained from the graph of $y = \cos x$ by a shift right and left $\pi/2$ units, respectively, as shown in Figure 4. ■

Identities involving the trigonometric functions can often be visualized using graphs.

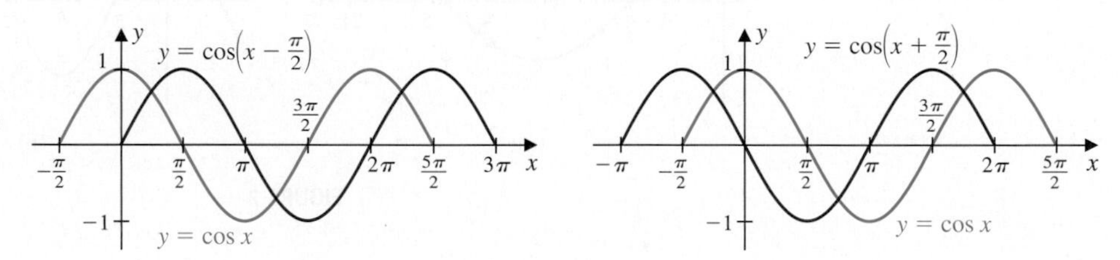

FIGURE 4

The graphs in Figure 3 and Figure 4 give four additional identities:

- $\sin\left(x + \dfrac{\pi}{2}\right) = \cos x, \quad \sin\left(x - \dfrac{\pi}{2}\right) = -\cos x,$
- $\cos\left(x - \dfrac{\pi}{2}\right) = \sin x, \quad \cos\left(x + \dfrac{\pi}{2}\right) = -\sin x.$

By shifting π units instead of $\pi/2$, as shown in Figure 5, we also have

- $\sin(x - \pi) = \sin(x + \pi) = -\sin x,$
- $\cos(x - \pi) = \cos(x + \pi) = -\cos x.$

These results are special cases of some general identities that we will consider in Section 4.7.

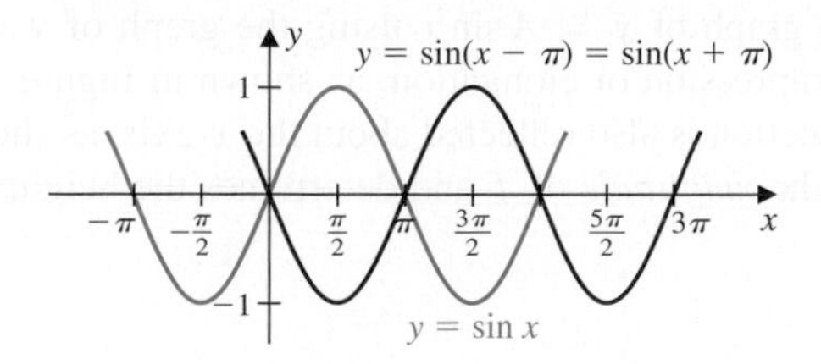

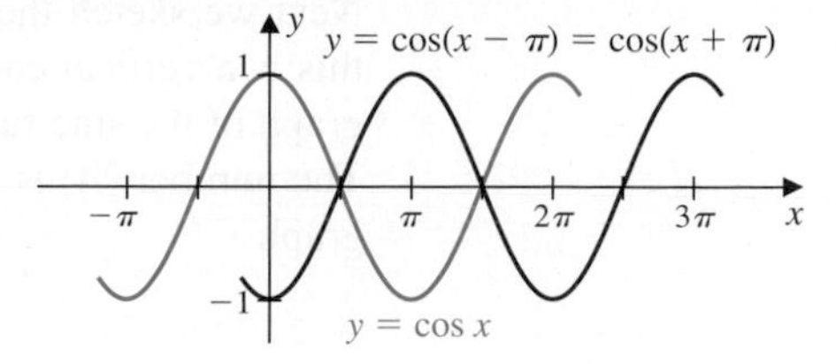

FIGURE 5

The remainder of this section considers some other transformations of the sine and cosine functions.

EXAMPLE 2 Sketch the graph of $y = -3\sin(x - \pi/4)$.

Solution As usual, we build the graph of this function by using the graphs of more elementary functions. We start with $y = \sin x$ and stretch this graph vertically by a factor of 3 to produce the graph of $y = 3\sin x$ shown in Figure 6(a). Then we shift this graph to the right $\pi/4$ units to produce the graph of $y = 3\sin(x - \pi/4)$ shown in yellow in Figure 6(b). The graph of $y = -3\sin(x - \pi/4)$, shown in green in Figure 6(b), is the reflection about the x-axis of the graph of $y = 3\sin(x - \pi/4)$. ■

You must be able to apply the basic graphing techniques to each new function you encounter.

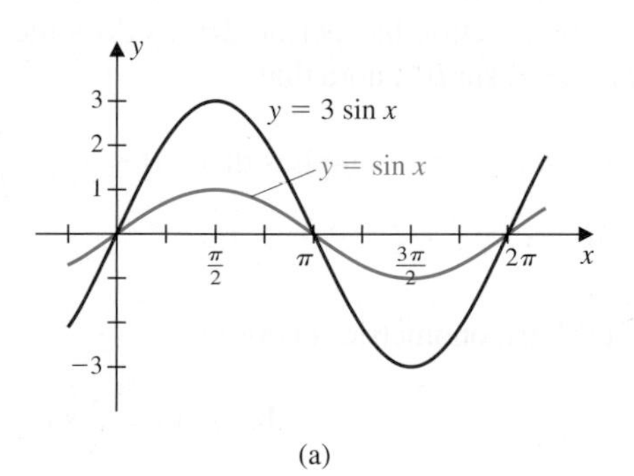

(a)

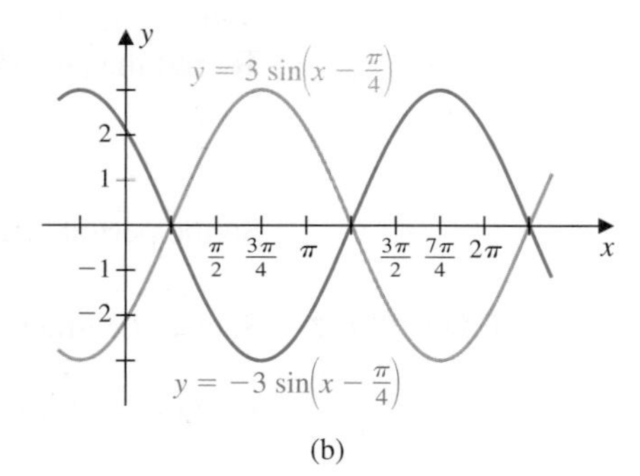

(b)

FIGURE 6

Examples 1 and 2 provide illustrations of the techniques needed to graph a member of the class of functions of the form

$$f(x) = A\sin(Bx + C) \quad \text{and} \quad f(x) = A\cos(Bx + C),$$

where $A \neq 0$ and $B > 0$. These functions are frequently seen in physics and engineering. The graph of any function of this type results in the compression or elongation, translation, and possible reflection of a sine or cosine graph. The cases of the sine and the cosine functions are similar, so we begin by concentrating on the graph of $f(x) = A\sin(Bx + C)$.

The first step is to factor the constant B from the terms in $Bx + C$ to produce

$$f(x) = A\sin B\left(x + \frac{C}{B}\right).$$

Next we sketch the graph of $y = A\sin x$ using the graph of $y = \sin x$. If $A > 0$, this is a vertical compression or elongation, as shown in Figure 7(a). If $A < 0$, the graph of the sine function is also reflected about the x-axis, as shown in Figure 7(b). The number $|A|$ is the *amplitude* of f and determines the height of a "wave" of the graph.

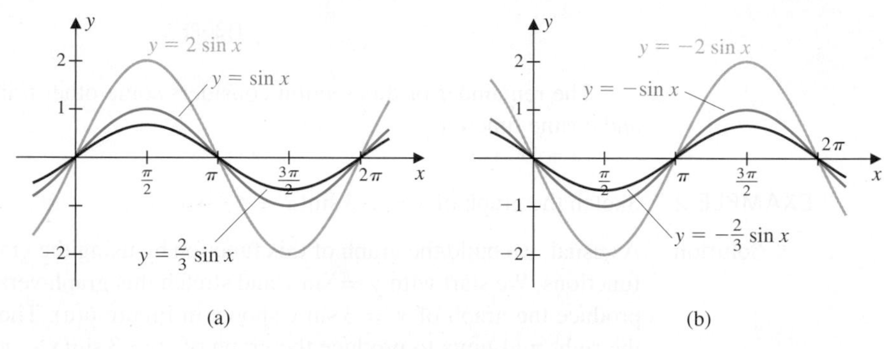

FIGURE 7

The transition from $y = A\sin x$ to $y = A\sin Bx$ requires a change in the period of the function. The sine function has period 2π, as does the graph of $y = A\sin x$. To find the period of $y = A\sin Bx$, note that

$$0 \le Bx \le 2\pi, \quad \text{implies that} \quad 0 \le x \le \frac{2\pi}{B}.$$

So the period is $2\pi/B$.

EXAMPLE 3 Sketch the graph of each trigonometric function.

a. $f(x) = 3\sin 2x$ **b.** $f(x) = \frac{1}{3}\sin\frac{x}{2}$

Solution **a.** The period of $f(x) = 3\sin 2x$ is $(2\pi)/2 = \pi$, so the graph is a horizontal compression of the graph of $y = 3\sin x$, whose maximum value is 3 and whose minimum value is -3. Four complete periods of the graph of f are shown in blue in Figure 8.

b. The period of $f(x) = \frac{1}{3}\sin\frac{x}{2}$ is $(2\pi)/(1/2) = 4\pi$, and the amplitude of f is $1/3$, so the graph is the horizontal elongation and vertical compression of the graph of $y = \sin x$ shown in red in Figure 8. ■

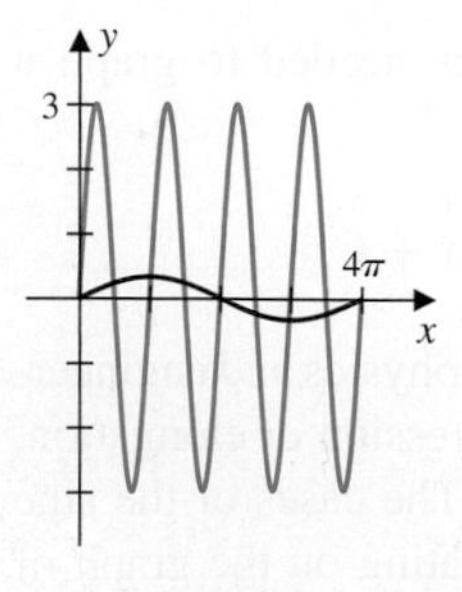

FIGURE 8

Sketching the graph of

$$f(x) = A\sin(Bx + C) = A\sin B\left(x + \frac{C}{B}\right)$$

requires a horizontal shift of the graph of $y = A\sin Bx$ by the amount $|C|/B$. As usual, this *phase shift* is to the left if C/B is positive and to the right if C/B is negative. A sample situation is shown in Figure 9.

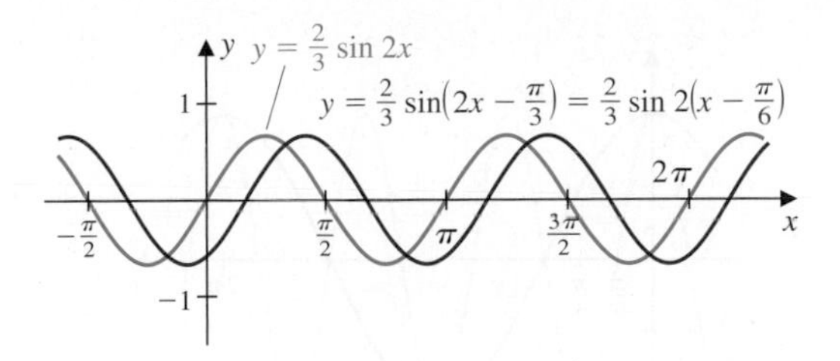

FIGURE 9

Sketching the graph of $y = A\cos(Bx + C)$ is similar, as illustrated in Example 4.

Graphs of $y = A\sin(Bx + C)$ and $y = A\cos(Bx + C)$ for $B > 0$

- The amplitude is $|A|$.
- The period is $2\pi/B$.
- The graph is horizontally shifted by $|C|/B$. The shift is left when $C/B > 0$ and right when $C/B < 0$.

EXAMPLE 4 Sketch the graph of $f(x) = 2\cos(3x - \pi/2)$.

Solution The first step is to factor the constant 3 from the terms in $3x - \pi/2$ to produce

$$y = 2\cos\left(3x - \frac{\pi}{2}\right) = 2\cos 3\left(x - \frac{\pi}{6}\right).$$

The graph of $y = 2\cos x$, shown in red in Figure 10, is a vertical stretching by a factor of 2 of the graph of $y = \cos x$.

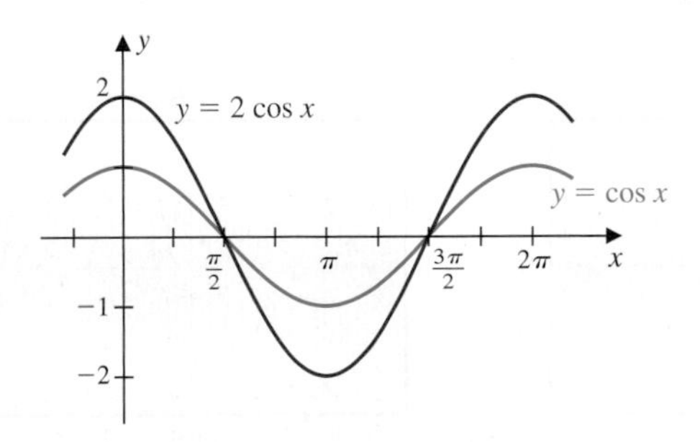

FIGURE 10

The period of $y = 2\cos x$ is 2π, so the period of $y = 2\cos 3x$ is $2\pi/3$. As a consequence, the graph of $y = 2\cos 3x$ is a horizontal compression of the graph of $y = 2\cos x$. Three periods of the graph of $y = 2\cos 3x$ are shown in yellow in both Figure 11(a) and Figure 11(b).

Finally, the graph of

$$y = 2\cos\left(3x - \frac{\pi}{2}\right) = 2\cos 3\left(x - \frac{\pi}{6}\right)$$

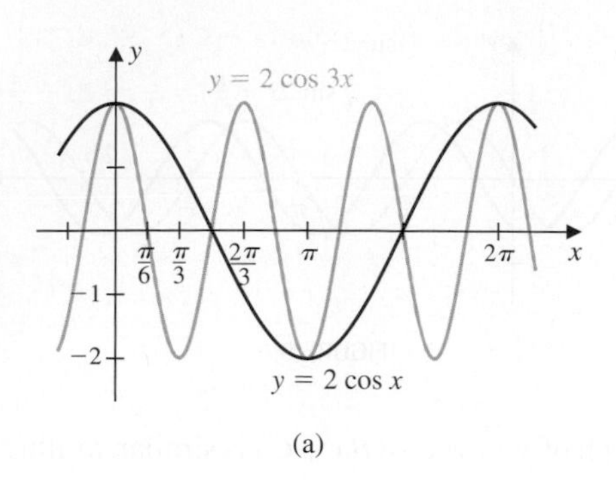

(a)

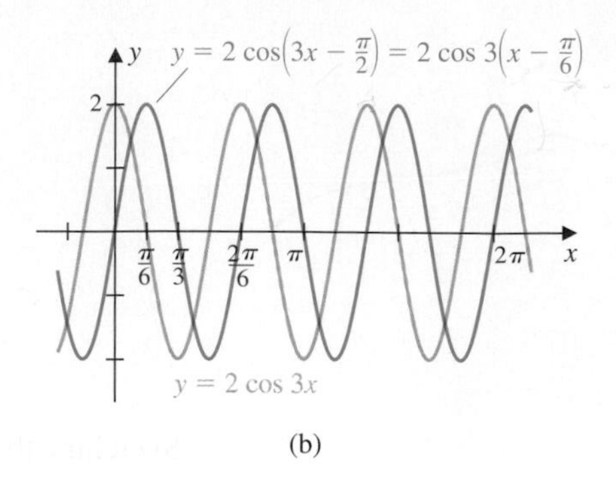

(b)

FIGURE 11

involves a translation of the graph of $y = 2\cos 3x$ to the right $\pi/6$ units, as shown in green in Figure 11(b). ■

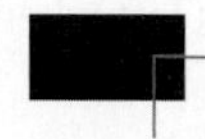

Using Graphing Devices

EXAMPLE 5 Use a graphing device to sketch the graph of $f(x) = \cos 75x$.

Solution Figures 12(a) and 12(b) show the graph of f using the viewing rectangles $[-10, 10] \times [-1.5, 1.5]$ and $[-8, 8] \times [-1.5, 1.5]$, respectively. The only similarity between these graphs and the graph of the cosine function seems to be that both have range $[-1, 1]$. The problem with using a graphing device is that there is too much oscillation for the point-plotting technique to work effectively. To produce a better representation for the graph, we first need the period, $2\pi/75 \approx 0.08$, so that we can prescribe a reasonable viewing rectangle.

Graphing devices are very useful in sketching the graphs of the trigonometric functions, but be careful to select a proper viewing rectangle.

$f(x) = \cos(75x)$

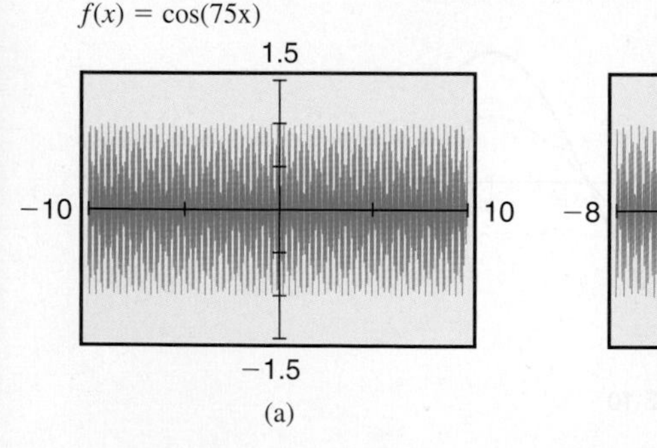

(a)

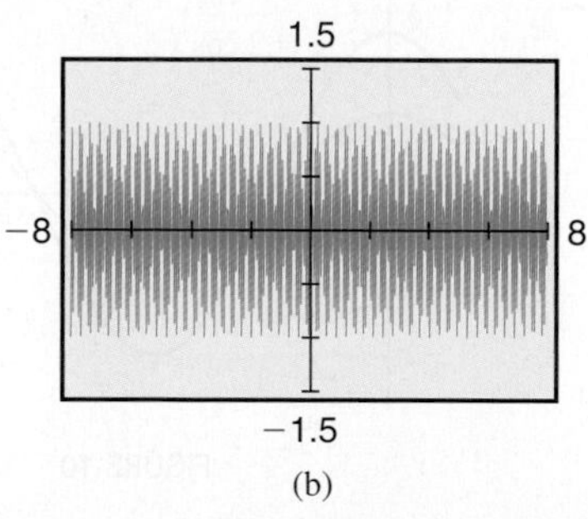

(b)

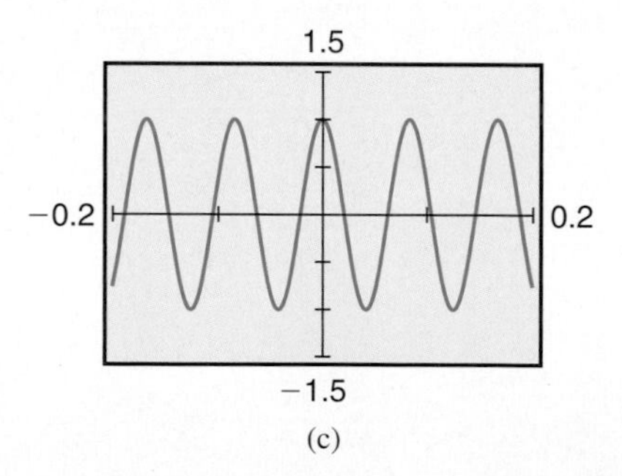

(c)

FIGURE 12

In Figure 12(c), the graph is shown in the viewing rectangle $[-0.2, 0.2] \times [-1.5, 1.5]$, which includes approximately

$$\frac{0.2 - (-0.2)}{0.08} = 5$$

periods of the function. ■

EXAMPLE 6 Use a graphing device to sketch a representation of the graph of

$$f(x) = \frac{\sin x}{x},$$

and describe the important features of the graph.

Solution In Example 5 a very narrow x-interval is required to obtain a good view of the graph. As Figure 13(a) and Figure 13(b) show, more information is found for this graph when a wider interval is used.

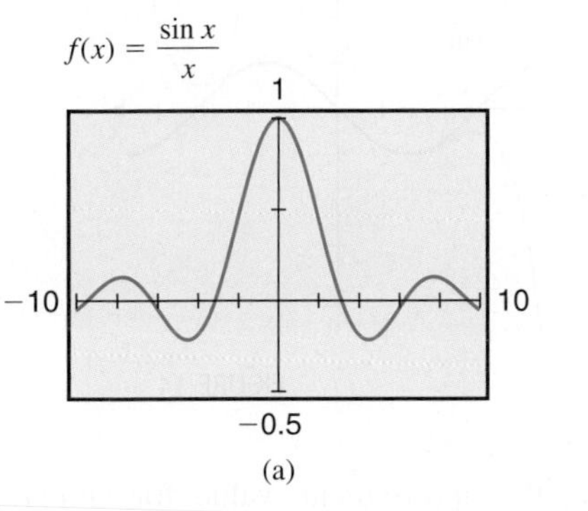

(a)

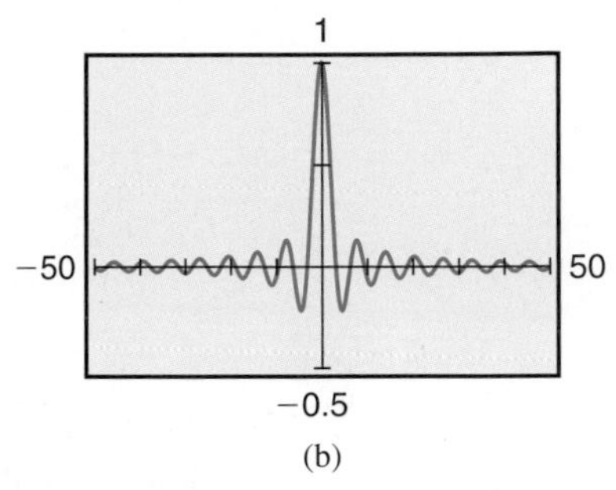

(b)

FIGURE 13

The function $f(x) = \sin x/x$ is important in calculus, where it is shown that

$$\frac{\sin x}{x} \to 1 \quad \text{as} \quad x \to 0.$$

This is why on the computer-generated graph the point (0, 1) appears to be included even though 0 is not in the domain.

Another interesting feature of the graph of $f(x) = \sin x/x$, shown more clearly in Figure 13(b), is that $y = f(x)$ has the horizontal asymptote $y = 0$ because

$$f(x) \to 0 \text{ as } x \to \infty \quad \text{and} \quad f(x) \to 0 \text{ as } x \to -\infty.$$

Also, the graph crosses its horizontal asymptote infinitely often, at every nonzero integer multiple of π. Finally, notice that both the numerator and denominator of f describe odd functions. Hence, f is an even function because

$$f(-x) = \frac{\sin(-x)}{(-x)} = \frac{-\sin x}{-x} = \frac{\sin x}{x} = f(x).$$

This explains the y-axis symmetry of the graph. ■

Applications

In calculus you will learn how to determine Taylor polynomials in order to approximate functions like $f(x) = \sin x$.

Polynomials are often used to approximate other functions over limited regions in their domains. In Example 7 we use a Taylor polynomial of degree 5 to approximate the sine function near the origin. The Taylor polynomials have frequent applications in calculus and other mathematical areas.

EXAMPLE 7 Use a graphing device to plot graphs of $y = \sin x$ and $y = x - \frac{1}{6}x^3 + \frac{1}{120}x^5$.

Solution Figure 14 shows the computer-generated graphs in the viewing rectangle $[-2\pi, 2\pi] \times [-5, 5]$. We can see that near the origin the polynomial is a good approximation to the sine curve.

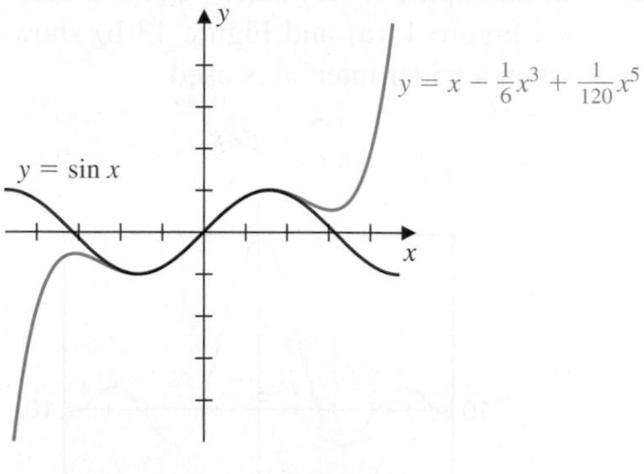

FIGURE 14

For example, the approximate value for sin(1) given from a calculator is 0.84147098, and the polynomial evaluated at 1 is approximately 0.84166666. ■

Periodic data can often be given an approximate fit using an appropriate sine or cosine function. In Example 8 we consider an application of this type.

EXAMPLE 8 Fit a sine wave of the form

$$y = A\sin(Bx + C) + D$$

to the data points shown in Figure 15(a).

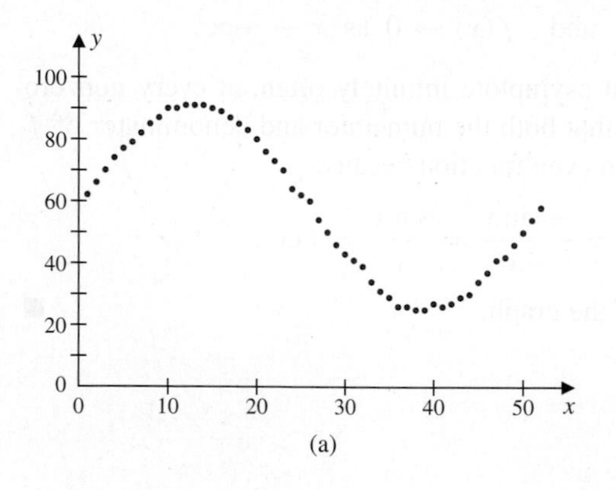

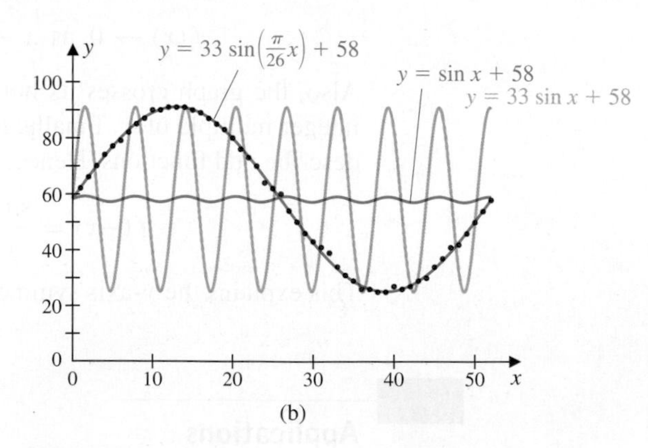

FIGURE 15

Solution The data points displayed in Figure 15(a) exhibit an approximate sine wave pattern centered vertically on the value 58. Since the graph of $y = \sin x$ passes through (0, 0), our first attempt at fitting the data might be to try

$$y = \sin x + 58.$$

The blue graph in Figure 15(b) indicates the curve is far too flat.

The maximum and minimum data values appear to be about 91 and 25, respectively, so the amplitude is approximately

$$\tfrac{1}{2}(91 - 25) = \tfrac{1}{2}(66) = 33.$$

The yellow graph in Figure 15(a) shows that to fit the original data points, the graph of $y = 33 \sin x + 58$ must also be stretched horizontally.

To stretch the curve $y = 33 \sin x + 58$ horizontally we increase its period. The original data appear to have a period of 52, so we set

$$\frac{2\pi}{B} = 52, \quad \text{which gives} \quad B = \frac{2\pi}{52} = \frac{\pi}{26}.$$

The green graph in Figure 15(b) shows that the curve

$$y = 33 \sin\left(\frac{\pi}{26}x\right) + 58,$$

fits the data quite well. ■

EXERCISE SET 4.5

In Exercises 1–4, use the graphs of the sine and cosine to sketch one period of the graph, specify the amplitude, and specify the period of the function.

1. a. $y = 3 \cos x$ **b.** $y = -2 \cos x$ **c.** $y = \frac{1}{4} \cos 4x$

2. a. $y = \sin 2x$ **b.** $y = \sin \frac{1}{2}x$ **c.** $y = -2 \sin 3x$

3. a. $y = 2 \cos \pi x$ **b.** $y = \cos 2\pi x$ **c.** $y = -2 \cos \frac{\pi}{2}x$

4. a. $y = -2 \sin \pi x$ **b.** $y = \sin \frac{\pi}{2}x$ **c.** $y = -3 \sin 2\pi x$

In Exercises 5–18, use the graphs of the sine and cosine to sketch one period of the graph of the function.

5. $y = 2 + \cos x$ **6.** $y = -2 + \sin x$

7. $y = \cos\left(x - \frac{\pi}{2}\right)$ **8.** $y = \sin(x - \pi)$

9. $y = -3 + \cos \pi x$ **10.** $y = 2 + \sin \pi x$

11. $y = 1 + \cos(2x + \pi)$ **12.** $y = -1 + \sin(3x + \pi)$

13. $y = -1 + 2 \sin\left(2x - \frac{\pi}{2}\right)$

14. $y = \frac{1}{3} \cos\left(\frac{\pi}{2} - 3x\right)$

15. $y = -2 \sin(x - 1) + 3$

16. $y = 2 - \cos(x - 1)$

17. $y = |2 \cos x|$ **18.** $y = |\sin x|$

*In Exercises 19–22, find **(a)** a cosine function and **(b)** a sine function whose graph matches the given curve.*

19.

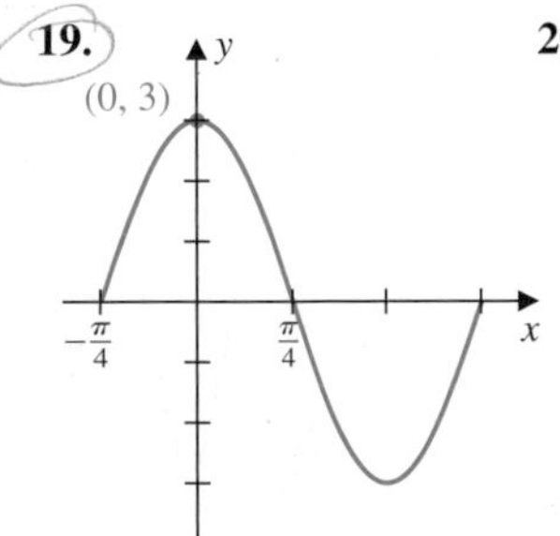

20.

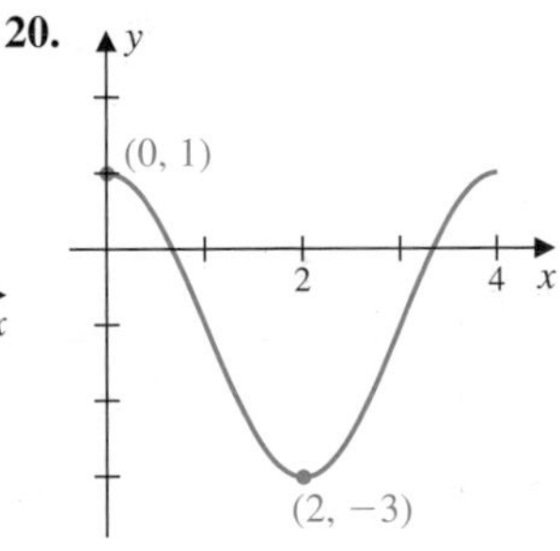

21. **22.**

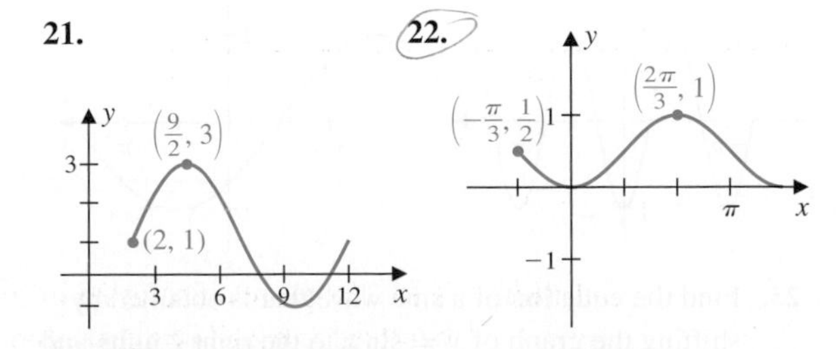

23. Match the equation with the graph.

a. $y = 2\sin\left(x - \frac{\pi}{2}\right)$ **b.** $y = \frac{1}{2}\sin\left(x + \frac{\pi}{2}\right)$

c. $y = 2\cos\left(x - \frac{\pi}{2}\right)$ **d.** $y = \frac{1}{2}\cos\left(x + \frac{\pi}{2}\right)$

i.

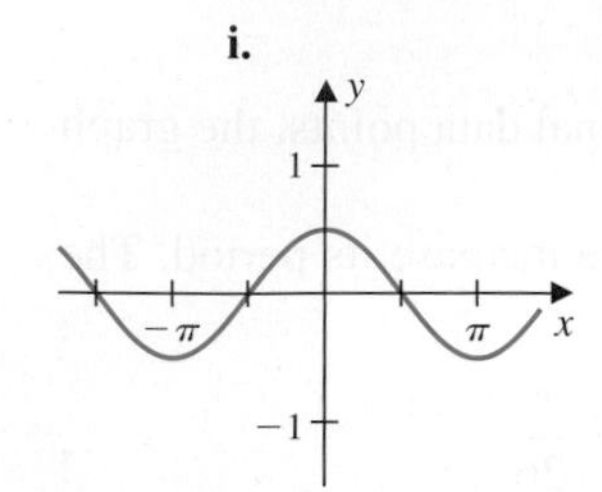

ii.

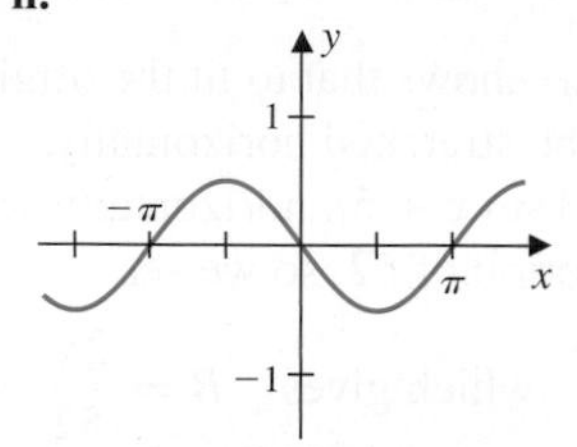

iii.

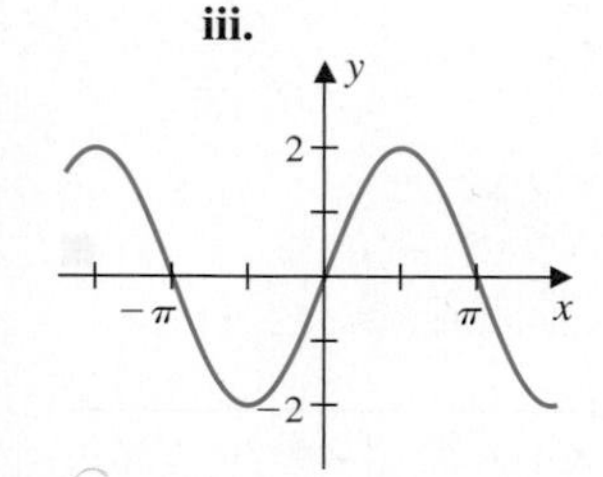

iv.

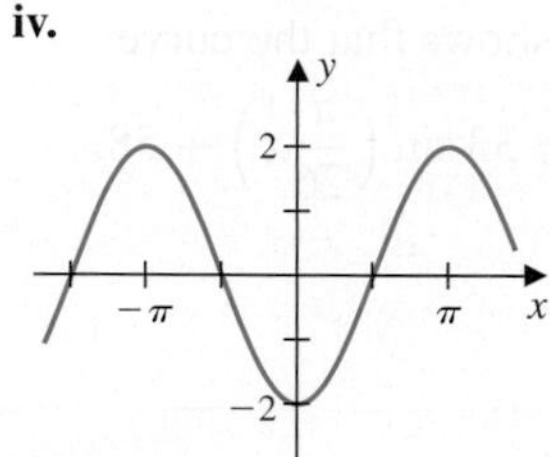

24. Match the equation with the graph.

a. $y = -\sin 2\left(x - \frac{\pi}{2}\right)$

b. $y = \sin\frac{1}{2}\left(x + \frac{\pi}{2}\right)$

c. $y = -\cos\frac{1}{2}\left(x - \frac{\pi}{2}\right)$

d. $y = \cos 2\left(x + \frac{\pi}{2}\right)$

i.

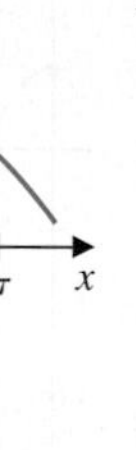

ii.

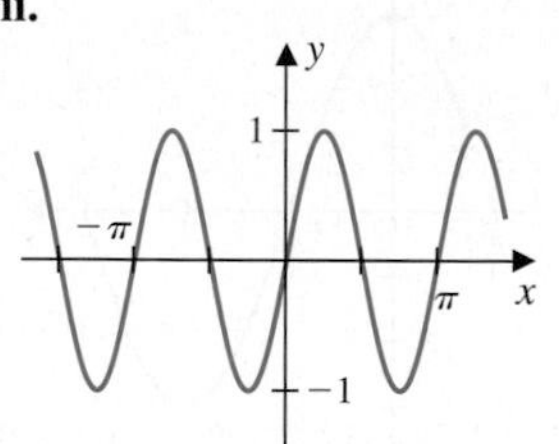

iii.

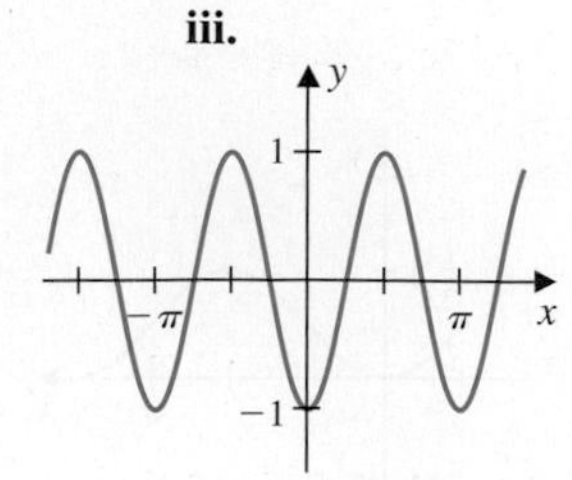

iv.

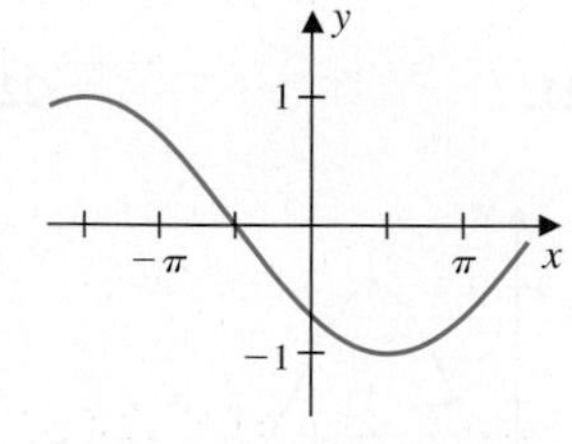

25. Find the equation of a sine wave that is obtained by shifting the graph of $y = \sin x$ to the right 2 units and downward 1 unit, and that is vertically compressed by a factor of 2 when compared to $y = \sin x$.

26. Find the equation of a cosine wave that is obtained by shifting the graph of $y = \cos x$ to the left 2 units and upward 1 unit, and that is horizontally compressed by a factor of 3 when compared to $y = \cos x$.

In Exercises 27–30, determine an appropriate viewing rectangle for the function and use it to sketch the graph.

27. $f(x) = \cos(100x)$

28. $f(x) = -5\sin(20x)$

29. $f(x) = \sin\left(\frac{x}{50}\right)$

30. $f(x) = x^2 + 5\sin(10x)$

In Exercises 31–34, use a graphing device to approximate, to one decimal place, the solutions to the equation.

31. $\cos x = x$

32. $\cos x = x^2$

33. $\sin x + \cos x = x$

34. $\sin x + x = x^3$

35. Find the smallest positive values of t for which

$$f(t) = 2\sin\left(3t - \frac{\pi}{4}\right)$$

a. is zero;

b. attains a maximum value;

c. attains a minimum value.

36. Find the smallest positive values of t for which

$$f(t) = -3\cos\left(2t + \frac{\pi}{6}\right)$$

a. is zero;

b. attains a maximum value;

c. attains a minimum value.

37. Fit a function of the form $f(x) = A + B\sin Cx$ to approximate the given data.

x	0	0.5	1	1.5	2	−0.5	−1	−1.5	−2
y	1	2.4	3	2.4	1	−0.4	−1	−0.4	1

38. Fit a function of the form $f(x) = A + B\cos C(x + D)$ to approximate the given data.

x	0	0.5	1	1.5	2	2.5	3	3.5	4
y	1.5	1.13	1	1.13	1.5	2	2.5	2.9	3

39. Sketch the graphs of $y = f(x)$, $y = g(x)$, and $y = f(x) + g(x)$ on the same set of axes. Discuss the effect on the graph of $y = f(x)$ caused by adding the graph of $y = g(x)$.

a. $f(x) = \sin x, \quad g(x) = x$

b. $f(x) = \sin x, \quad g(x) = -x$

c. $f(x) = \cos x, \quad g(x) = x$

d. $f(x) = \cos x, \quad g(x) = -x$

40. Use a graphing device to sketch each of the following graphs:

a. $y = \sin x^2$ **b.** $y = (\sin x)^2$

c. $y = \sin\left(\frac{1}{x}\right), x \neq 0$

d. $y = x \sin\left(\frac{1}{x}\right), x \neq 0$

e. $y = \frac{(\sin x)^2}{x}, x \neq 0$ **f.** $y = \sqrt{|\sin x|}$

41. Let A, B, and C be positive constants. Describe the effect on the graph of $y = A \cos B(x + C)$.

a. If B and C are fixed and A is doubled;

b. If A and C are fixed and B is doubled;

c. If A and B are fixed and C is doubled.

In Exercises 42 and 43, use the table of average monthly temperatures (in F°) for New York City and Portland, Maine.

Month (x)	New York	Portland
January (1)	41.4	30.2
February (2)	39.8	25.7
March (3)	45.1	34.7
April (4)	53.5	44.6
May (5)	60.2	51.8
June (6)	72.1	62.4
July (7)	76.8	70.3
August (8)	75.3	69.8
September (9)	67.5	59.8
October (10)	61.9	52.4
November (11)	50.4	41.8
December (12)	42.6	33.7

42. a. Fit a function $f(x) = A + B\sin(Cx + D)$ to approximate the data for New York City.

b. Graph the function found in part (a) along with the data points.

43. a. Fit a function $f(x) = A + B\sin(Cx + D)$ to approximate the data for Portland.

b. Graph the function found in part (a) along with the data points.

4.6 OTHER TRIGONOMETRIC FUNCTIONS

The most basic trigonometric functions, the sine and cosine functions, were defined for all real numbers in Section 4.4. In Section 4.3 we saw that for acute angles in right triangles there are four additional trigonometric functions. These functions involve the quotients and reciprocals of the sine and cosine functions. They can be defined for real numbers in the same way, but care must be taken to ensure that you do not divide by zero.

The Tangent, Cotangent, Secant, and Cosecant Functions

The **tangent**, **cotangent**, **secant**, and **cosecant** functions, written respectively as $\tan x$, $\cot x$, $\sec x$, and $\csc x$, are defined by the quotients

$$\tan x = \frac{\sin x}{\cos x}, \quad \sec x = \frac{1}{\cos x}, \quad \cot x = \frac{\cos x}{\sin x}, \quad \csc x = \frac{1}{\sin x}.$$

- The tangent and secant functions are defined whenever $x \neq n\pi + \pi/2$ for an integer n.
- The cotangent and cosecant functions are defined whenever $x \neq n\pi$ for an integer n.

Table 1 shows the signs of the trigonometric functions in the various quadrants of the plane. The entries for the sine and cosine were determined in Section 4.4. Those for the other trigonometric functions follow from the fact that a quotient is positive precisely when the signs of its numerator and denominator agree.

TABLE 1

Quadrant	I	II	III	IV
$\sin x$	+	+	−	−
$\cos x$	+	−	−	+
$\tan x = \frac{\sin x}{\cos x}$	+	−	+	−
$\cot x = \frac{\cos x}{\sin x}$	+	−	+	−
$\sec x = \frac{1}{\cos x}$	+	−	−	+
$\csc x = \frac{1}{\sin x}$	+	+	−	−

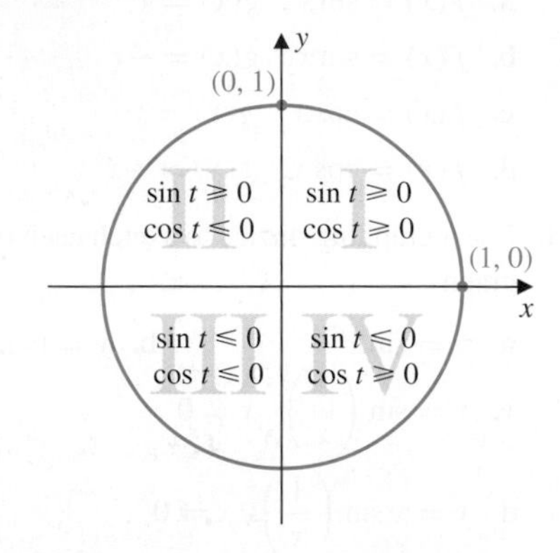

Because they involve quotients, the domains of the tangent, cotangent, secant, and cosecant functions are restricted to those numbers for which the denominator is nonzero. The denominator in the case of the tangent and the secant is the cosine function, so their domains do not contain odd multiples of $\pi/2$. In the case of the cotangent and cosecant, the denominator is the sine function, so their domains do not contain multiples of π. The values of the sine and cosine functions that we found in the previous sections give the entries in Table 2.

TABLE 2

x	0	$\frac{\pi}{6}$	$\frac{\pi}{4}$	$\frac{\pi}{3}$	$\frac{\pi}{2}$	$\frac{2\pi}{3}$	$\frac{3\pi}{4}$	$\frac{5\pi}{6}$	π
$\sin x$	0	$\frac{1}{2}$	$\frac{\sqrt{2}}{2}$	$\frac{\sqrt{3}}{2}$	1	$\frac{\sqrt{3}}{2}$	$\frac{\sqrt{2}}{2}$	$\frac{1}{2}$	0
$\cos x$	1	$\frac{\sqrt{3}}{2}$	$\frac{\sqrt{2}}{2}$	$\frac{1}{2}$	0	$-\frac{1}{2}$	$-\frac{\sqrt{2}}{2}$	$-\frac{\sqrt{3}}{2}$	−1
$\tan x$	0	$\frac{\sqrt{3}}{3}$	1	$\sqrt{3}$	—	$-\sqrt{3}$	−1	$-\frac{\sqrt{3}}{3}$	0
$\cot x$	—	$\sqrt{3}$	1	$\frac{\sqrt{3}}{3}$	0	$-\frac{\sqrt{3}}{3}$	−1	$-\sqrt{3}$	—
$\sec x$	1	$\frac{2\sqrt{3}}{3}$	$\sqrt{2}$	2	—	−2	$-\sqrt{2}$	$-\frac{2\sqrt{3}}{3}$	−1
$\csc x$	—	2	$\sqrt{2}$	$\frac{2\sqrt{3}}{3}$	1	$\frac{2\sqrt{3}}{3}$	$\sqrt{2}$	2	—

For example,

$$\tan\frac{\pi}{6} = \frac{\sin\frac{\pi}{6}}{\cos\frac{\pi}{6}} = \frac{\frac{1}{2}}{\frac{\sqrt{3}}{2}} = \frac{1}{\sqrt{3}} = \frac{\sqrt{3}}{3}$$

and

$$\sec\frac{3\pi}{4} = \frac{1}{\cos\frac{3\pi}{4}} = \frac{1}{-\frac{\sqrt{2}}{2}} = -\frac{2}{\sqrt{2}} = -\sqrt{2}.$$

A dash (—) in Table 2 is used to indicate a number that is not in the domain of the function. For example, there is a dash for the tangent function at $\pi/2$ since $\cos \pi/2 = 0$.

The trigonometric functions are all arithmetically related, so a complete knowledge of one of them and the quadrant location of x will give the values of the others.

EXAMPLE 1 Suppose that $\sin x = 2/3$ and that $\pi/2 < x < \pi$. Determine the values of the other trigonometric functions.

Solution We first find the value of the cosine function using the Pythagorean identity $(\sin x)^2 + (\cos x)^2 = 1$. This gives

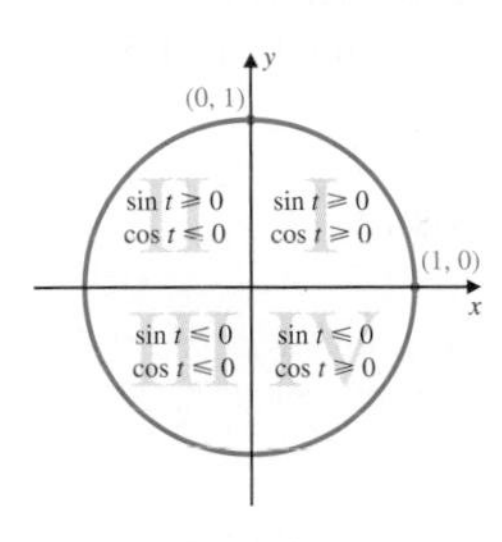

FIGURE 1

$$\cos x = \pm\sqrt{1 - (\sin x)^2} = \pm\sqrt{1 - \left(\frac{2}{3}\right)^2} = \pm\frac{\sqrt{5}}{3}.$$

Since x is in the interval $(\pi/2, \pi)$ (quadrant II), Figure 1 indicates that the cosine is negative, and $\cos x = -\sqrt{5}/3$. The remaining trigonometric functions are easily determined now that we have the values of both the sine and the cosine. They are

$$\tan x = \frac{\sin x}{\cos x} = -\frac{\frac{2}{3}}{\frac{\sqrt{5}}{3}} = -\frac{2\sqrt{5}}{5}, \qquad \cot x = \frac{\cos x}{\sin x} = -\frac{\frac{\sqrt{5}}{3}}{\frac{2}{3}} = -\frac{\sqrt{5}}{2},$$

$$\sec x = \frac{1}{\cos x} = -\frac{3}{\sqrt{5}} = -\frac{3\sqrt{5}}{5}, \quad \text{and} \quad \csc x = \frac{1}{\sin x} = \frac{3}{2}.$$

■

EXAMPLE 2 Find all values of x in the interval $[0, 2\pi]$ that satisfy $2\cos x - 3\tan x = 0$.

Solution First convert the equation to one involving only sines and cosines:

$$2\cos x - 3\frac{\sin x}{\cos x} = 0, \quad \text{so} \quad 2(\cos x)^2 - 3\sin x = 0.$$

Next convert the equation to involve only sines using the Pythagorean identity $(\cos x)^2 + (\sin x)^2 = 1$:

$$2\left(1 - (\sin x)^2\right) - 3\sin x = 0 \quad \text{and} \quad 2(\sin x)^2 + 3\sin x - 2 = 0.$$

This last equation factors as

$$(2\sin x - 1)(\sin x + 2) = 0, \quad \text{so} \quad \sin x = \frac{1}{2} \quad \text{or} \quad \sin x = -2.$$

The sine always lies between -1 and 1, so the only solutions are when

$$\sin x = \frac{1}{2}, \quad \text{which implies that} \quad x = \frac{\pi}{6} \quad \text{or} \quad x = \frac{5\pi}{6}.$$

■

The Graph of the Tangent Function

The tangent function is 0 precisely when the sine function is 0 and is undefined when the cosine function is 0. It is positive on $(0, \pi/2)$ since the sine and cosine are both positive there, but it is negative on $(\pi/2, \pi)$ since on this interval the sine is positive and the cosine is negative.

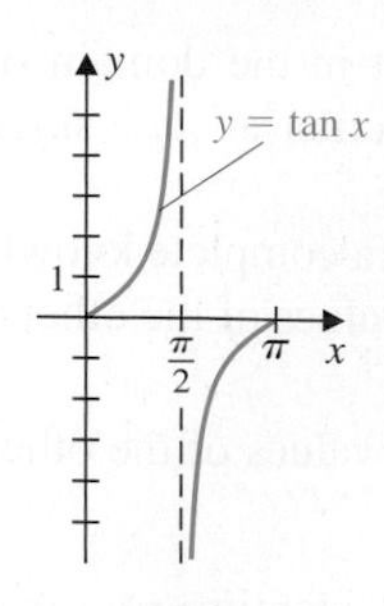

FIGURE 2

The portion of the graph of the tangent function for values in its domain that lie in the interval $[0, \pi]$ is shown in Figure 2. Notice the vertical asymptote at $x = \pi/2$. Since $\sin x > 0$ and $\cos x > 0$ on the interval $[0, \pi/2)$ and $\cos(\pi/2) = 0$, we have

$$\tan x \to \infty \text{ as } x \to \frac{\pi}{2}^{-},$$

and since $\sin x > 0$ and $\cos x < 0$ on the interval $(\pi/2, \pi]$, we have

$$\tan x \to -\infty \text{ as } x \to \frac{\pi}{2}^{+}.$$

The sine and cosine functions have period 2π, so the tangent function must also repeat every 2π units. Its period, however, is smaller than 2π. In Section 4.5 we showed that

$$\sin(x + \pi) = -\sin x \quad \text{and} \quad \cos(x + \pi) = -\cos x.$$

These results imply that

$$\tan(x + \pi) = \frac{\sin(x + \pi)}{\cos(x + \pi)} = \frac{-\sin x}{-\cos x} = \tan x.$$

Hence the tangent function repeats every π units. It is clear from the graph in Figure 2 that it repeats on no smaller interval, so the period of the tangent function is π. This means that we can sketch the graph of the entire tangent function from the portion shown in Figure 2. The graph of $y = \tan x$ is shown in Figure 3.

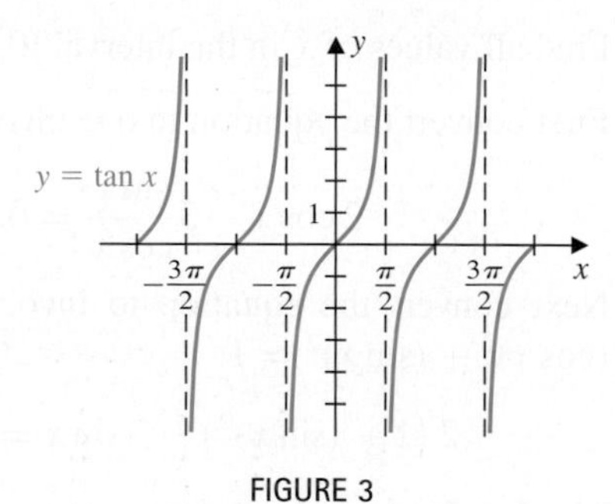

FIGURE 3

The graph given in Figure 3 shows that the tangent function is increasing on every interval that is completely contained in its domain, and that the graph has vertical asymptotes at the points that are not in its domain.

EXAMPLE 3 Sketch each of the graphs.

a. $y = \tan 2x$ **b.** $y = \tan\left(2x - \frac{\pi}{3}\right)$ **c.** $y = -\tan\left(2x - \frac{\pi}{3}\right) + 1$

Solution **a.** The graph of $y = \tan 2x$ is a horizontal compression, by a factor of 2, of the graph of the tangent function, so the resulting graph is as shown in Figure 4(a). The graph has been horizontally compressed by a factor of 2, so the period of the function has been reduced by the same factor, and $y = \tan 2x$ has period $\pi/2$. Notice that the first positive vertical asymptote is $x = \pi/4$, and the first positive zero is $x = \pi/2$.

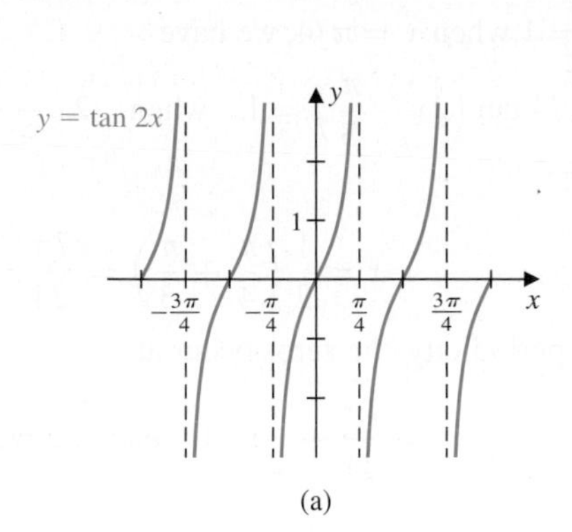

(a)

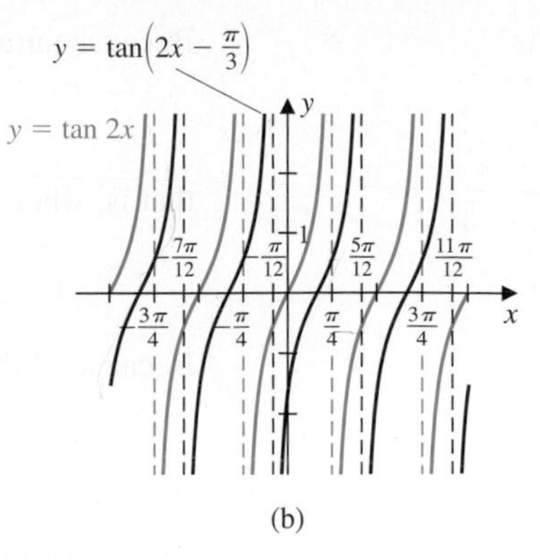

(b)

FIGURE 4

b. To sketch the graph of $y = \tan(2x - \pi/3)$, first factor the 2 from both the terms of $2x - \pi/3$ to produce

$$y = \tan\left(2x - \frac{\pi}{3}\right) = \tan 2\left(x - \frac{\pi}{6}\right).$$

This graph has the same shape as the graph of $y = \tan 2x$, but it is shifted to the right $\pi/6$ units, as shown in Figure 4(b). The first positive vertical asymptote is now at $x = \pi/4 + \pi/6 = 5\pi/12$, and the first two positive zeros are at $x = \pi/6$ and $x = \pi/2 + \pi/6 = 2\pi/3$.

c. The graph of $y = -\tan(2x - \pi/3)$ is the reflection about the x-axis of the graph of $y = \tan(2x - \pi/3)$, as shown in Figure 4(a). To obtain the graph of $y = -\tan(2x - \pi/3) + 1$, shift the graph of $y = -\tan(2x - \pi/3)$ upward 1 unit, as shown in green in Figure 5(b). Notice that the vertical asymptotes for this graph are the same as for the graph in Figure 5(a).

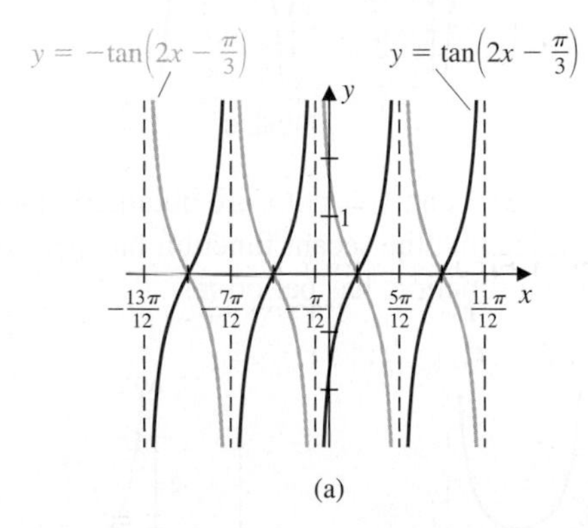

(a)

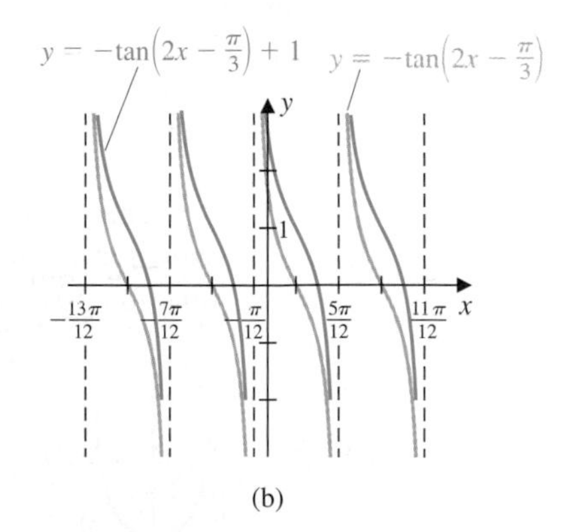

(b)

FIGURE 5

The locations of the zeros of $y = -\tan(2x - \pi/3) + 1$ are not obvious from the graph. However, the period of the function is $\pi/2$, so all the zeros can be found if one zero is found. First note that

$$0 = -\tan\left(2x - \frac{\pi}{3}\right) + 1 \quad \text{implies that} \quad \tan\left(2x - \frac{\pi}{3}\right) = 1.$$

Because $\tan x = 1$ when $x = \pi/4$, we have

$$\tan\left(2x - \frac{\pi}{3}\right) = 1 \quad \text{when} \quad 2x - \frac{\pi}{3} = \frac{\pi}{4},$$

that is, when

$$x = \frac{1}{2}\left(\frac{\pi}{4} + \frac{\pi}{3}\right) = \frac{7\pi}{24}.$$

Because of the periodicity, the zeros occur at

$$x = \frac{7\pi}{24} + \frac{\pi}{2}n, \text{ for each integer } n.$$

■

Now that we have the graphs of the sine, cosine, and tangent functions, we can determine the graphs of the remaining trigonometric functions by using the reciprocal graphing technique. Figure 6 shows the graph of $y = \csc x$, the reciprocal of the sine function. Notice that $y = \csc x$ decreases as $y = \sin x$ increases, and $y = \csc x$ increases as $y = \sin x$ decreases. The local maxima and local minima of $y = \csc x$ occur at the same points as $y = \sin x$, but with the roles reversed. The graph of $y = \csc x$ has vertical asymptotes at those values of x for which $\sin x = 0$, that is, at integer multiples of π. The period of the cosecant function is 2π, the same as the period of the sine function.

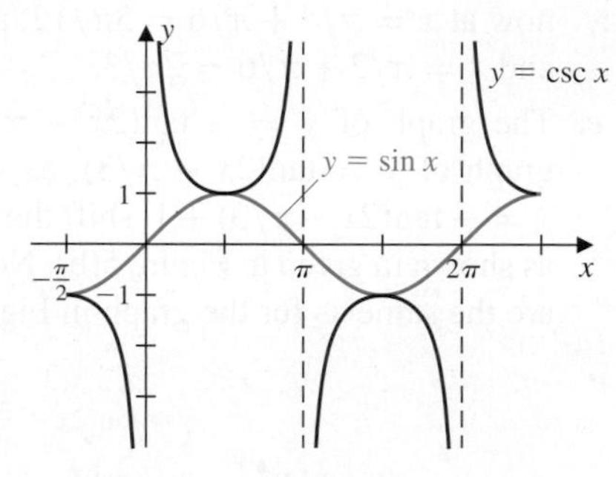

FIGURE 6

The graphs of $y = \sec x$ and $y = \cot x$ are obtained in a similar manner and are shown in Figure 7. Notice that the secant function has period 2π, but the cotangent function, like the tangent function, has period π.

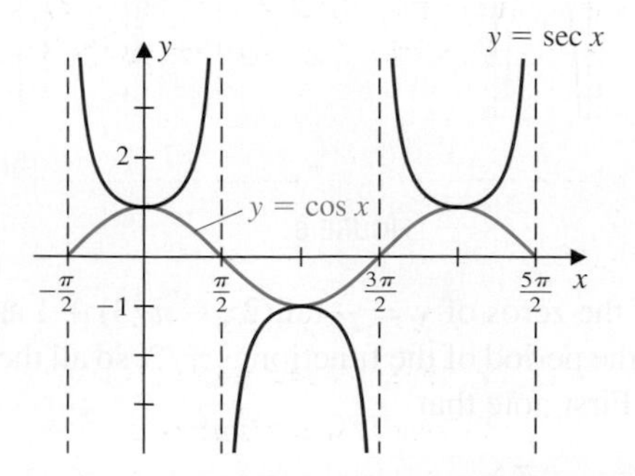

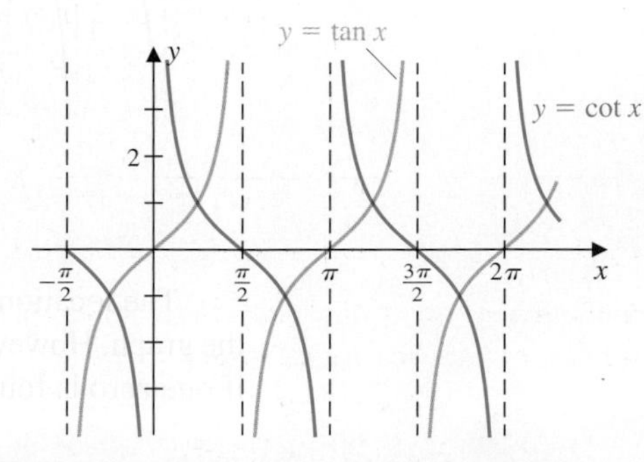

FIGURE 7

In Example 3, the graph in part (c) was systematically constructed from the graphs of $y = \tan 2x$ and $y = \tan(2x - \pi/3)$ needed in parts (a) and (b). Try to determine the strategy in Example 4 before reading the solution.

EXAMPLE 4 Sketch the graph of $y = \frac{1}{2}\csc(3x + \pi/2)$.

Solution The strategy we have chosen is to start with the graph of $y = \csc x$, which was given in Figure 6 and is reproduced in Figure 8(a). Then we modify this graph to produce the graphs of $y = \csc 3x$ and $y = \frac{1}{2}\csc 3x$. The graph that we want as our final result will be a horizontal shift of the graph of $y = \frac{1}{2}\csc 3x$.

The graph of $y = \csc 3x$ is a horizontal compression of the graph of $y = \csc x$ by a factor of 3, as shown in Figure 8(a). Notice that the period has been reduced from 2π to $2\pi/3$.

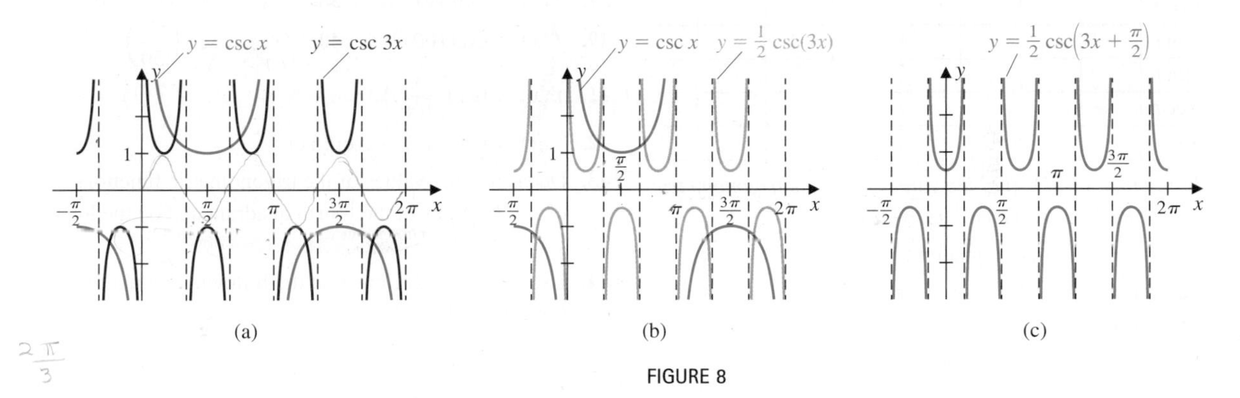

FIGURE 8

The graph of $y = \frac{1}{2}\csc 3x$, shown in yellow in Figure 8(b), is a vertical compression by a factor of 2 of the graph of $y = \csc 3x$. We now need to determine and perform the final horizontal shift on the graph of $y = \frac{1}{2}\csc 3x$. To do this, we write

$$y = \frac{1}{2}\csc\left(3x + \frac{\pi}{2}\right) = \frac{1}{2}\csc 3\left(x + \frac{\pi}{6}\right)$$

and find that the shift is to the left $\pi/6$ units, as shown in Figure 8(c). ■

The graph in Figure 8(c) looks suspiciously like the graph of a secant function that has been compressed horizontally and reduced vertically. Try to determine what you think the equation of this equivalent modified secant function is. In the next section you will see some trigonometric identities that you can use to verify your conjecture.

EXERCISE SET 4.6

In Exercises 1–6, find the quadrant in which $P(t)$ lies if the given conditions are satisfied.

1. $\sin t < 0$ and $\cos t < 0$

2. $\tan t < 0$ and $\cos t < 0$

3. $\sin t < 0$ and $\cot t > 0$

4. $\sec t > 0$ and $\tan t < 0$

5. $\sin t > 0$ and $\tan t > 0$

6. $\sin t < 0$ and $\sec t < 0$

7. Complete the table.

	$\frac{7\pi}{6}$	$\frac{5\pi}{4}$	$\frac{4\pi}{3}$	$\frac{3\pi}{2}$	$\frac{5\pi}{3}$	$\frac{7\pi}{4}$	$\frac{11\pi}{6}$	2π
$\sin t$								
$\cos t$								
$\tan t$								
$\cot t$								
$\sec t$								
$\csc t$								

8. Complete the table.

	$-\frac{\pi}{6}$	$-\frac{\pi}{4}$	$-\frac{\pi}{3}$	$-\frac{\pi}{2}$	$-\frac{2\pi}{3}$	$-\frac{3\pi}{4}$	$-\frac{5\pi}{6}$	$-\pi$
$\sin t$								
$\cos t$								
$\tan t$								
$\cot t$								
$\sec t$								
$\csc t$								

In Exercises 9–16, find the values of all the trigonometric functions from the given information.

9. $\cos t = \frac{3}{5}$, t is in quadrant I

10. $\cos t = -\frac{1}{2}$, $\pi \le t \le 2\pi$

11. $\cos t = -\frac{4}{5}$, $\pi \le t \le 3\pi/2$

12. $\sin t = \frac{1}{3}$, t is in quadrant II

13. $\tan t = 2$, $0 < t < \pi/2$

14. $\csc t = -\frac{3}{2}$, $3\pi/2 < t < 2\pi$

15. $\sec t = 3$, t is in quadrant IV

16. $\cot t = 3$, t is in quadrant III

In Exercises 17–28, sketch one period of the given curve.

17. $y = 3\tan x$

18. $y = \frac{1}{2}\cot x$

19. $y = -2\sec x$

20. $y = -\frac{1}{2}\csc x$

21. $y = \tan\left(x + \frac{\pi}{2}\right)$

22. $y = \tan\left(x - \frac{\pi}{2}\right)$

23. $y = \frac{1}{2}\sec\left(x + \frac{\pi}{4}\right)$

24. $y = 2\csc\left(x - \frac{\pi}{2}\right)$

25. $y = \tan\left(\frac{\pi x}{2}\right)$

26. $y = \cot\left(\frac{\pi x}{2}\right)$

27. $y = \tan\left(2x - \frac{\pi}{2}\right)$

28. $y = \sec(2x + \pi)$

In Exercises 29–36, find all values of t in the interval $[0, 2\pi]$ that satisfy the given equation.

29. $\tan t + 1 = 0$

30. $\cot t + \sqrt{3} = 0$

31. $(\tan t)^2 = \frac{1}{3}$

32. $(\cot t)^2 = 3$

33. $|\tan t| = 1$

34. $|\sec t| = 1$

35. $2\sin 2t - \sqrt{2}\tan 2t = 0$

36. $\tan t - 3\cot t = 0$

In Exercises 37–42, determine an appropriate viewing rectangle for the function and sketch the graph.

37. $f(x) = \tan(5x)$

38. $f(x) = \tan(8x - 10)$

39. $f(x) = \csc(100x)$

40. $f(x) = \sec\left(\frac{x}{50}\right)$

41. $f(x) = \tan\left(\frac{x}{100}\right)$

42. $f(x) = \tan(25x) - \csc(25x)$

43. Determine the values of the trigonometric functions of t if $P(t)$ lies in the fourth quadrant and on the line $y = -2x$.

44. Use the figure to show that the following reduction formulas hold.

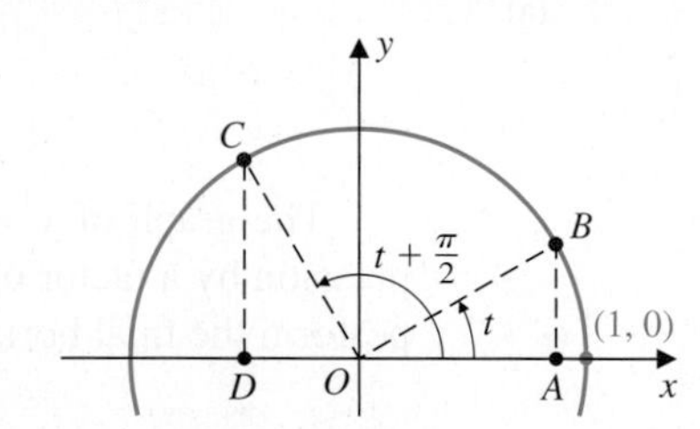

a. $\sin\left(t + \frac{\pi}{2}\right) = \cos t$ **b.** $\cos\left(t + \frac{\pi}{2}\right) = -\sin t$

c. $\tan\left(t + \frac{\pi}{2}\right) = -\cot t$

(*Hint*: First show that $\triangle OAB$ and $\triangle CDO$ are congruent.)

4.7 TRIGONOMETRIC IDENTITIES

In calculus you will want to know the *core identities* and be able to derive the others from this core.

Trigonometric identities are used to simplify the work involving the trigonometric functions. This dramatically increases their usefulness in applications. Our emphasis is on developing a small core of identities that are continually needed and can be used to determine a much larger collection. Although the number of core identities is quite small, an understanding of these and how to derive others from the core is essential for success when you study the calculus of trigonometry.

Pythagorean Identities

The first and most basic identity, one we have encountered previously, is the basic Pythagorean identity

$$(\sin x)^2 + (\cos x)^2 = 1.$$

By dividing this Pythagorean identity by $(\cos x)^2$, assuming it is nonzero, we obtain

$$\frac{(\sin x)^2}{(\cos x)^2} + 1 = \frac{1}{(\cos x)^2}, \quad \text{or} \quad (\tan x)^2 + 1 = (\sec x)^2.$$

In a similar manner, dividing the basic Pythagorean identity by $(\sin x)^2$ produces

$$(\cot x)^2 + 1 = (\csc x)^2.$$

These two identities are just modified forms of the same basic identity, so we call the collection the Pythagorean identities.

The Pythagorean Identities

At each real number x for which the functions are defined, we have

- $(\sin x)^2 + (\cos x)^2 = 1$
- $(\tan x)^2 + 1 = (\sec x)^2$
- $(\cot x)^2 + 1 = (\csc x)^2$

We recommend that you commit the original Pythagorean identity to memory, and derive the other two as needed.

Sum and Difference Formulas

The next identities we need are not as easy to derive. They involve the sine and cosine of the sum and difference of two numbers. These identities should be committed to memory, since they are frequently needed.

The Sum and Difference Formulas for Sine and Cosine

For every pair of real numbers x_1 and x_2, we have

- $\sin(x_1 \pm x_2) = \sin x_1 \cos x_2 \pm \cos x_1 \sin x_2$
- $\cos(x_1 \pm x_2) = \cos x_1 \cos x_2 \mp \sin x_1 \sin x_2$

In the sum and difference formulas, when the top (or bottom) sign is chosen on the left, the top (or bottom) sign is chosen on the right.

We first use the definition of the sine and cosine functions to show that the identity

$$\cos(x_1 - x_2) = \cos x_1 \cos x_2 + \sin x_1 \sin x_2$$

holds for the case when $0 < x_1 - x_2 < \pi/2$. This restriction on x_1 and x_2 is for convenience of illustration; the proof holds in other cases as well.

Figure 1 gives a typical illustration of the situation that we are considering.

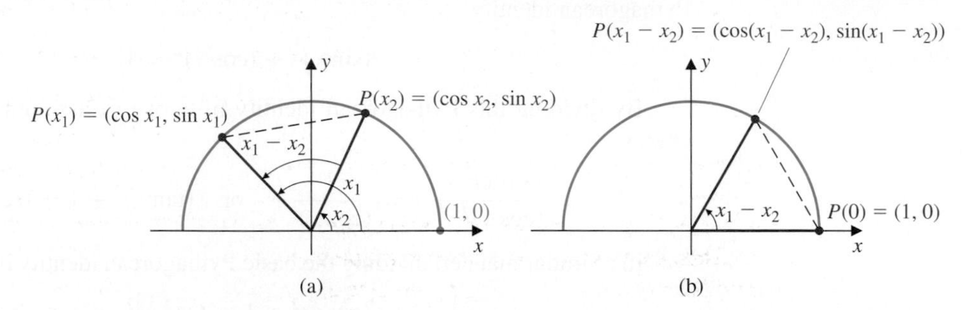

FIGURE 1

The triangle in Figure 1(b) is simply a clockwise rotation by x_2 radians of the triangle shown in Figure 1(a). So the distance from $P(x_1 - x_2)$ to $P(0)$ in Figure 1(b) is the same as the distance from $P(x_1)$ to $P(x_2)$ in Figure 1(a). Using the formula for the distance between two points gives

$$(\cos(x_1 - x_2) - 1)^2 + (\sin(x_1 - x_2) - 0)^2 = (\cos x_1 - \cos x_2)^2 + (\sin x_1 - \sin x_2)^2 .$$

Expanding the left side of this equation gives

$$(\cos(x_1 - x_2))^2 - 2\cos(x_1 - x_2) + 1 + (\sin(x_1 - x_2))^2$$

and expanding the right side gives

$$(\cos x_1)^2 - 2\cos x_1 \cos x_2 + (\cos x_2)^2 + (\sin x_1)^2 - 2\sin x_1 \sin x_2 + (\sin x_2)^2 .$$

But from the Pythagorean identity we know that

$$1 = (\sin(x_1 - x_2))^2 + (\cos(x_1 - x_2))^2 ,$$

$$1 = (\sin x_1)^2 + (\cos x_1)^2, \quad \text{and} \quad 1 = (\sin x_2)^2 + (\cos x_2)^2.$$

So the equation simplifies to

$$2 - 2\cos(x_1 - x_2) = 2 - 2\cos x_1 \cos x_2 - 2\sin x_1 \sin x_2.$$

Subtracting 2 from each side and then dividing by -2 gives the final identity

$$\cos(x_1 - x_2) = \cos x_1 \cos x_2 + \sin x_1 \sin x_2.$$

The derivation of this identity is somewhat involved, but, as so often happens in mathematics, the other identities now follow rather easily.

To show the identity involving the cosine of the sum, we use the fact that the cosine function is even and the sine function is odd, that is,

$$\cos(-x) = \cos x \quad \text{and} \quad \sin(-x) = -\sin x.$$

Converting the cosine of a sum into a cosine of a difference and using the difference identity gives

$$\begin{aligned}\cos(x_1 + x_2) &= \cos(x_1 - (-x_2)) \\ &= \cos x_1 \cos(-x_2) + \sin x_1 \sin(-x_2) \\ &= \cos x_1 \cos x_2 - \sin x_1 \sin x_2.\end{aligned}$$

Combining this with the difference identity gives

$$\cos(x_1 \pm x_2) = \cos x_1 \cos x_2 \mp \sin x_1 \sin x_2.$$

To show the identities that involve the sine function, we use a result that we discovered in Example 1 of Section 4.5. By shifting the graphs of the sine and cosine functions to the right $\pi/2$ units, as shown in Figure 2, we see that

$$\cos\left(x - \frac{\pi}{2}\right) = \sin x \quad \text{and} \quad \sin\left(x - \frac{\pi}{2}\right) = -\cos x.$$

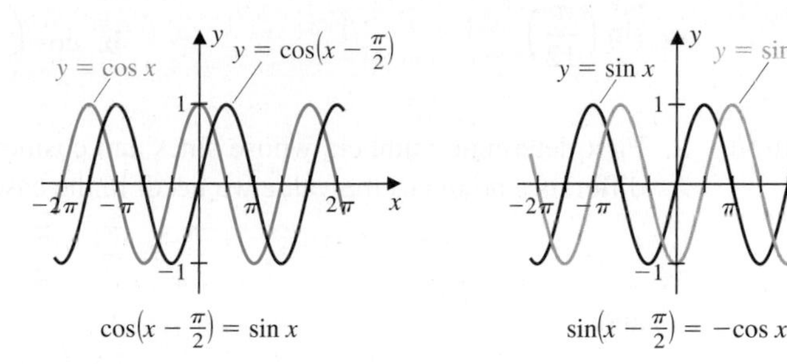

FIGURE 2

So

$$\begin{aligned}\sin(x_1 + x_2) &= \cos\left((x_1 + x_2) - \frac{\pi}{2}\right) \\ &= \cos\left(x_1 + \left(x_2 - \frac{\pi}{2}\right)\right) \\ &= \cos x_1 \cos\left(x_2 - \frac{\pi}{2}\right) - \sin x_1 \sin\left(x_2 - \frac{\pi}{2}\right) \\ &= \cos x_1 \sin x_2 - \sin x_1(-\cos x_2) \\ &= \sin x_1 \cos x_2 + \cos x_1 \sin x_2.\end{aligned}$$

Using this identity and the fact that

$$\cos(-x) = \cos x \quad \text{and} \quad \sin(-x) = -\sin x$$

gives

$$\begin{aligned}\sin(x_1 - x_2) &= \sin(x_1 + (-x_2)) \\ &= \sin x_1 \cos(-x_2) + \cos x_1 \sin(-x_2) \\ &= \sin x_1 \cos x_2 - \cos x_1 \sin x_2.\end{aligned}$$

Putting these identities together produces

$$\sin(x_1 \pm x_2) = \sin x_1 \cos x_2 \pm \cos x_1 \sin x_2,$$

so all four identities have been established.

We now have the three core identities for trigonometry.

Core Identities for Trigonometry

- $(\sin x)^2 + (\cos x)^2 = 1$
- $\sin(x_1 \pm x_2) = \sin x_1 \cos x_2 \pm \cos x_1 \sin x_2$
- $\cos(x_1 \pm x_2) = \cos x_1 \cos x_2 \mp \sin x_1 \sin x_2$

Example 1 shows how the sum and difference formulas for the sine and cosine can be used to extend our knowledge of these functions by allowing us to determine exact values for the sine and cosine.

EXAMPLE 1 Determine the exact value.

a. $\sin\left(\frac{\pi}{12}\right)$ **b.** $\cos\left(\frac{7\pi}{12}\right)$

Solution **a.** First determine numbers whose sines and cosines are known and whose sum or difference produces the value we need. In the case of $\pi/12$, we have

$$\frac{\pi}{12} = \frac{\pi}{3} - \frac{\pi}{4}$$

and

$$\sin\frac{\pi}{12} = \sin\left(\frac{\pi}{3} - \frac{\pi}{4}\right) = \sin\left(\frac{\pi}{3}\right)\cos\left(\frac{\pi}{4}\right) - \cos\left(\frac{\pi}{3}\right)\sin\left(\frac{\pi}{4}\right)$$
$$= \frac{\sqrt{3}}{2}\frac{\sqrt{2}}{2} - \frac{1}{2}\frac{\sqrt{2}}{2} = \frac{\sqrt{2}}{4}(\sqrt{3} - 1).$$

b. To find the cosine of $7\pi/12$, we can write

$$\frac{7\pi}{12} = \frac{\pi}{2} + \frac{\pi}{12},$$

but this requires finding the cosine of $\pi/12$ as well as the sine that we determined in part (a). Instead, notice that we can also write

$$\frac{7\pi}{12} = \frac{\pi}{3} + \frac{\pi}{4}.$$

So

$$\cos\frac{7\pi}{12} = \cos\left(\frac{\pi}{3} + \frac{\pi}{4}\right) = \cos\left(\frac{\pi}{3}\right)\cos\left(\frac{\pi}{4}\right) - \sin\left(\frac{\pi}{3}\right)\sin\left(\frac{\pi}{4}\right)$$
$$= \frac{1}{2}\frac{\sqrt{2}}{2} - \frac{\sqrt{3}}{2}\frac{\sqrt{2}}{2} = \frac{\sqrt{2}}{4}(1 - \sqrt{3}).$$ ■

All the other identities we will need are obtained from the core identities. For example, consider the formula for the tangent of the sum and difference of two values. The definition of the tangent function gives

$$\tan(x_1 \pm x_2) = \frac{\sin(x_1 \pm x_2)}{\cos(x_1 \pm x_2)}$$

so

$$\tan(x_1 \pm x_2) = \frac{\sin x_1 \cos x_2 \pm \cos x_1 \sin x_2}{\cos x_1 \cos x_2 \mp \sin x_1 \sin x_2}.$$

Dividing the numerator and denominator by $\cos x_1 \cos x_2$ gives

$$\tan(x_1 \pm x_2) = \frac{\dfrac{\sin x_1 \cos x_2}{\cos x_1 \cos x_2} \pm \dfrac{\cos x_1 \sin x_2}{\cos x_1 \cos x_2}}{\dfrac{\cos x_1 \cos x_2}{\cos x_1 \cos x_2} \mp \dfrac{\sin x_1 \sin x_2}{\cos x_1 \cos x_2}},$$

which simplifies to

$$\tan(x_1 \pm x_2) = \frac{\tan x_1 \pm \tan x_2}{1 \mp \tan x_1 \tan x_2}.$$

This is an interesting identity, but one that is easily derived from the sine and cosine formulas. It also has the weakness that it cannot be applied when either $\cos x_1 = 0$ or $\cos x_2 = 0$. There is a similar sum and difference identity for the cotangent function listed on the inside back cover.

EXAMPLE 2 Determine $\tan(\pi/12)$.

Solution Taking the same approach as in Example 1, write

$$\frac{\pi}{12} = \frac{\pi}{3} - \frac{\pi}{4},$$

and use the tangent identity we just derived to obtain

$$\tan\frac{\pi}{12} = \frac{\tan(\frac{\pi}{3}) - \tan(\frac{\pi}{4})}{1 + \tan(\frac{\pi}{3})\tan(\frac{\pi}{4})} = \frac{\sqrt{3} - 1}{1 + \sqrt{3} \cdot 1} = \frac{\sqrt{3} - 1}{\sqrt{3} + 1}.$$

Alternatively, we can use the cosine difference formula to compute

$$\cos\frac{\pi}{12} = \cos\frac{\pi}{3}\cos\frac{\pi}{4} + \sin\frac{\pi}{3}\sin\frac{\pi}{4} = \frac{1}{2}\frac{\sqrt{2}}{2} + \frac{\sqrt{3}}{2}\frac{\sqrt{2}}{2} = \frac{\sqrt{2}}{4}(1 + \sqrt{3}).$$

Then, using the value of $\sin(\pi/12)$ determined in Example 1, we have

$$\tan\frac{\pi}{12} = \frac{\sin\frac{\pi}{12}}{\cos\frac{\pi}{12}} = \frac{\frac{\sqrt{2}}{4}(\sqrt{3} - 1)}{\frac{\sqrt{2}}{4}(1 + \sqrt{3})} = \frac{\sqrt{3} - 1}{\sqrt{3} + 1}.$$

This second method might seem longer, but it has the advantage of simplicity. We only need to know the definition of the tangent function and the identities for the sum and difference of the sine and cosine functions. ■

Double-Angle Formulas

The *double-angle* formulas are also consequences of the sum formulas for the sine and cosine functions. For the sine function, we have the identity

$$\sin 2x = \sin(x + x) = \sin x \cos x + \sin x \cos x = 2 \sin x \cos x.$$

For the cosine, we have

$$\cos 2x = \cos x \cos x - \sin x \sin x,$$

which can be rewritten using the Pythagorean identity in a variety of ways:

$$\cos 2x = (\cos x)^2 - (\sin x)^2 = (\cos x)^2 - (1 - (\cos x)^2) = 2(\cos x)^2 - 1$$

or as

$$\cos 2x = (\cos x)^2 - (\sin x)^2 = 1 - (\sin x)^2 - (\sin x)^2 = 1 - 2(\sin x)^2.$$

These are called the *double-angle* formulas for the sine and cosine.

Double-Angle Formulas

For any real number x, we have

- $\sin 2x = 2 \sin x \cos x$
- $\cos 2x = (\cos x)^2 - (\sin x)^2 = 2(\cos x)^2 - 1 = 1 - 2(\sin x)^2$

EXAMPLE 3 Determine all the values of x in $[0, 2\pi]$ that satisfy the equation.

a. $\cos x = \sin 2x$ **b.** $\sin x = \cos 2x$ **c.** $1 = \sin x + \cos x$

Solution **a.** We can use the double-angle formula for the sine to write

$$\cos x = \sin 2x = 2 \sin x \cos x,$$

which implies that

$$0 = 2 \sin x \cos x - \cos x = (2 \sin x - 1) \cos x.$$

The solutions to this equation in $[0, 2\pi]$ occur when

$$\cos x = 0, \quad \text{that is,} \quad x = \frac{\pi}{2} \text{ or } x = \frac{3\pi}{2},$$

or when

$$\sin x = \frac{1}{2}, \quad \text{that is,} \quad x = \frac{\pi}{6} \text{ or } x = \frac{5\pi}{6}.$$

b. We can use the double-angle formula for the cosine in the form that involves only the sine function to write

$$\sin x = \cos 2x = 1 - 2(\sin x)^2,$$

which implies that

$$0 = 2(\sin x)^2 + \sin x - 1 = (\sin x + 1)(2 \sin x - 1).$$

The solutions to this equation in $[0, 2\pi]$ occur when

$$\sin x = -1, \quad \text{that is,} \quad x = \frac{3\pi}{2},$$

or when

$$\sin x = \frac{1}{2}, \quad \text{that is,} \quad x = \frac{\pi}{6} \text{ or } x = \frac{5\pi}{6}.$$

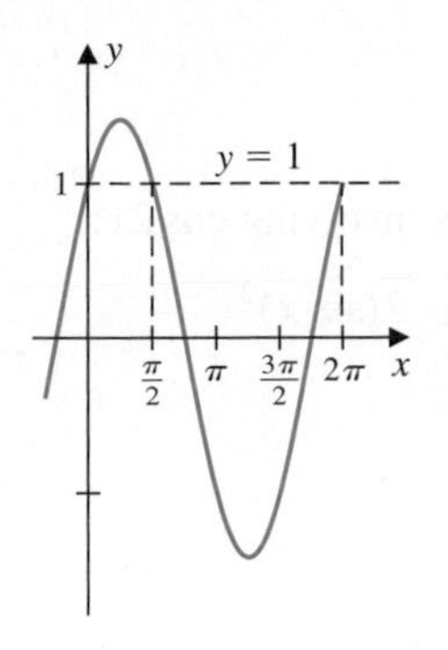

FIGURE 3

c. The equation $1 = \sin x + \cos x$ doesn't initially seem to belong with the others, but the graph of $f(x) = \sin x + \cos x$ shown in Figure 3 indicates that we have the graph of an expanded and translated sine function. The graph also indicates that $f(x) = 1$ on $[0, 2\pi]$ precisely when $x = 0$, $x = \pi/2$, or $x = 2\pi$. To verify these results algebraically, we first square both sides of the equation to give

$$1 = (\sin x + \cos x)^2 = (\sin x)^2 + (\cos x)^2 + 2 \sin x \cos x.$$

The Pythagorean and double-angle identities are used to simplify this to

$$1 = 1 + 2 \sin x \cos x = 1 + \sin 2x, \quad \text{so} \quad 0 = \sin 2x.$$

To have $\sin 2x = 0$, the argument $2x$ must be a multiple of π, that is, x must be a multiple of $\pi/2$. At first glance, then, we might conclude that the solutions are

$$x = 0, \quad x = \frac{\pi}{2}, \quad x = \pi, \quad x = \frac{3\pi}{2}, \quad \text{and} \quad x = 2\pi.$$

Often a graphing device can give visual information that leads to an algebraic solution.

Remember, though, that we squared the equation in the first step and that this might introduce *extraneous* solutions. In fact, this is what happened, since the only values of x in $[0, 2\pi]$ that satisfy the original equation, $1 = \sin x + \cos x$, are

$$x = 0, \quad x = \frac{\pi}{2}, \quad \text{and} \quad x = 2\pi.$$

This agrees with our graphical solution. ■

EXAMPLE 4 Determine the intervals contained in $[0, 2\pi]$, where the function $f(x) = \cos 2x + \cos x$ is positive and intervals where the function is negative.

Solution The double-angle formula for the cosine involving only cosines gives

$$\begin{aligned} f(x) = \cos 2x + \cos x &= (2(\cos x)^2 - 1) + \cos x \\ &= 2(\cos x)^2 + \cos x - 1 \\ &= (2 \cos x - 1)(\cos x + 1). \end{aligned}$$

So

$$f(x) = 0 \quad \text{when} \quad (2 \cos x - 1)(\cos x + 1) = 0,$$

which occurs when

$$\cos x = \frac{1}{2} \quad \text{or} \quad \cos x = -1.$$

The solutions to these last two equations for x in $[0, 2\pi]$ are

$$x = \frac{\pi}{3} \quad \text{and} \quad x = \frac{5\pi}{3} \quad \text{or } x = \pi.$$

The sign chart in Figure 4 shows that the function is positive on $[0, \pi/3) \cup (5\pi/3, 2\pi]$ and the function is negative on $(\pi/3, \pi) \cup (\pi, 5\pi/3)$. ■

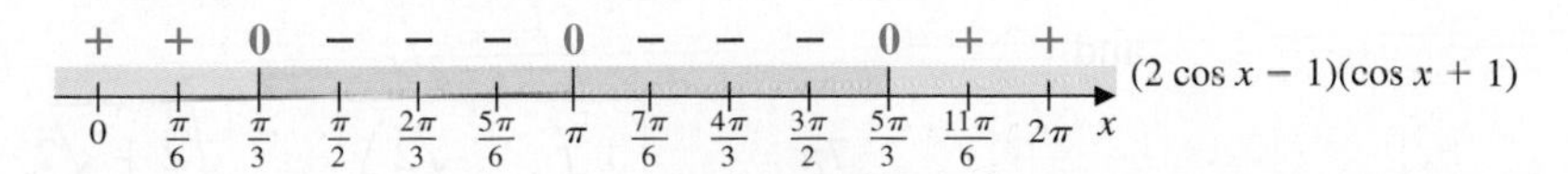

FIGURE 4

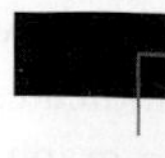

Half-Angle Formulas

The *half-angle* formulas come from the double-angle formulas involving $\cos 2x$:

$$\cos 2x = (\cos x)^2 - (\sin x)^2 = 2(\cos x)^2 - 1 = 1 - 2(\sin x)^2$$

when x is replaced by $x/2$. For the sine formula, we have

$$\cos x = 1 - 2\left(\sin \frac{x}{2}\right)^2,$$

which implies that

$$\left(\sin \frac{x}{2}\right)^2 = \frac{1 - \cos x}{2}.$$

For the cosine formula, we have

$$\cos x = 2\left(\cos \frac{x}{2}\right)^2 - 1 \quad \text{and} \quad \left(\cos \frac{x}{2}\right)^2 = \frac{1 + \cos x}{2}.$$

The second form of the half-angle formulas will be used in integral calculus.

Half-Angle Formulas

For any real number x, we have

- $\left(\sin \frac{x}{2}\right)^2 = \frac{1 - \cos x}{2}$ and $\left(\cos \frac{x}{2}\right)^2 = \frac{1 + \cos x}{2}$,

or, equivalently, by replacing x with $2x$,

- $(\sin x)^2 = \frac{1 - \cos 2x}{2}$ and $(\cos x)^2 = \frac{1 + \cos 2x}{2}$.

EXAMPLE 5 Determine $\sin(7\pi/8)$, $\cos(7\pi/8)$, and $\tan(7\pi/8)$.

Solution The half-angle formulas give

$$\left(\sin \frac{7\pi}{8}\right)^2 = \frac{1}{2}\left(1 - \cos \frac{7\pi}{4}\right) = \frac{1}{2}\left(1 - \frac{\sqrt{2}}{2}\right)$$

and

$$\left(\cos \frac{7\pi}{8}\right)^2 = \frac{1}{2}\left(1 + \cos \frac{7\pi}{4}\right) = \frac{1}{2}\left(1 + \frac{\sqrt{2}}{2}\right).$$

Since $7\pi/8$ is in the second quadrant, the sine is positive and the cosine is negative, so

$$\sin \frac{7\pi}{8} = \sqrt{\frac{1}{2}\left(1 - \frac{\sqrt{2}}{2}\right)} = \sqrt{\frac{2 - \sqrt{2}}{4}} = \frac{\sqrt{2 - \sqrt{2}}}{2}$$

and

$$\cos \frac{7\pi}{8} = -\sqrt{\frac{1}{2}\left(1 + \frac{\sqrt{2}}{2}\right)} = -\sqrt{\frac{2 + \sqrt{2}}{4}} = -\frac{\sqrt{2 + \sqrt{2}}}{2}.$$

The tangent is now given as

$$\tan\frac{7\pi}{8} = \frac{\sin\frac{7\pi}{8}}{\cos\frac{7\pi}{8}} = \frac{\sqrt{2-\sqrt{2}}}{-\sqrt{2+\sqrt{2}}} = -\sqrt{\frac{2-\sqrt{2}}{2+\sqrt{2}}}.$$ ■

Example 6 illustrates one of the many identities that can be derived using the half-angle formulas.

EXAMPLE 6 Show that when $0 < x < \pi$, we have $\tan\dfrac{x}{2} = \dfrac{1-\cos x}{\sin x}$.

Solution For $0 < x < \pi$, we have

$$0 < \frac{x}{2} < \frac{\pi}{2}, \quad \text{so} \quad \sin\frac{x}{2} > 0 \quad \text{and} \quad \cos\frac{x}{2} > 0.$$

Hence

$$\tan\frac{x}{2} = \frac{\sin\frac{x}{2}}{\cos\frac{x}{2}} = \frac{\sqrt{\dfrac{1-\cos x}{2}}}{\sqrt{\dfrac{1+\cos x}{2}}} = \sqrt{\frac{1-\cos x}{1+\cos x}}.$$

We want the numerator to be $1-\cos x$, so multiply the numerator and denominator within the square root by $1-\cos x$, which in effect "rationalizes the numerator." This gives

$$\tan\frac{x}{2} = \sqrt{\frac{(1-\cos x)^2}{(1+\cos x)(1-\cos x)}} = \sqrt{\frac{(1-\cos x)^2}{1-(\cos x)^2}}$$
$$= \sqrt{\frac{(1-\cos x)^2}{(\sin x)^2}} = \frac{|1-\cos x|}{|\sin x|}.$$

Since $\cos x \le 1$, the expression $1-\cos x$ is nonnegative for all x, and since $0 < x < \pi$, the sine is also nonnegative, so

$$\tan\frac{x}{2} = \frac{1-\cos x}{\sin x}.$$ ■

The half-angle formulas are needed in calculus to change powers of the sine or cosine functions into sums and differences in a manner illustrated in Example 7.

EXAMPLE 7 Express $(\sin x)^4$ as a sum that involves only constants and sine and cosine functions to the first power.

Solution The object is to reduce the power on the sine function, so we first use the half-angle formula for the sine to write

$$(\sin x)^4 = \left((\sin x)^2\right)^2 = \left(\frac{1-\cos 2x}{2}\right)^2 = \frac{1}{4}\left(1 - 2\cos 2x + (\cos 2x)^2\right).$$

Now we use the half-angle formula for the cosine to rewrite $(\cos 2x)^2$ in terms of $\cos 4x$. This gives

$$(\sin x)^4 = \frac{1}{4}\left(1 - 2\cos 2x + \left(\frac{1+\cos 4x}{2}\right)\right),$$

which simplifies to

$$(\sin x)^4 = \frac{3}{8} - \frac{1}{2}\cos 2x + \frac{1}{8}\cos 4x. \qquad ■$$

You might argue that the "simplified" expression for $(\sin x)^4$ in Example 7 looks more complicated than the original. In a sense you would be correct; it might be better to describe the process as changing $(\sin x)^4$ into a "standard" form. As we will now see, simple products of sines and cosines can also be expressed in this form.

It is the "standard" forms of the sines and cosines that are used in integral calculus.

Products of Sines and Cosines

Suppose we add the formulas for the sine of a sum and a difference,

$$\sin(x_1 + x_2) = \sin x_1 \cos x_2 + \cos x_1 \sin x_2,$$
$$\sin(x_1 - x_2) = \sin x_1 \cos x_2 - \cos x_1 \sin x_2.$$

Then the $\cos x_1 \sin x_2$ term cancels to give

$$\sin(x_1 + x_2) + \sin(x_1 - x_2) = 2\sin x_1 \cos x_2.$$

Dividing the result by 2 gives the identity

$$\sin x_1 \cos x_2 = \frac{1}{2}(\sin(x_1 - x_2) + \sin(x_1 + x_2)).$$

In a similar manner, the basic cosine identities,

$$\cos(x_1 + x_2) = \cos x_1 \cos x_2 - \sin x_1 \sin x_2$$
$$\cos(x_1 - x_2) = \cos x_1 \cos x_2 + \sin x_1 \sin x_2,$$

can be added to produce

$$\cos x_1 \cos x_2 = \frac{1}{2}(\cos(x_1 - x_2) + \cos(x_1 + x_2)).$$

They can also be subtracted to produce

$$\sin x_1 \sin x_2 = \frac{1}{2}(\cos(x_1 - x_2) - \cos(x_1 + x_2)).$$

Taken together, we have formulas for the products of sines and cosines.

Formulas for Products of Sines and Cosines

- $\sin x_1 \cos x_2 = \frac{1}{2}(\sin(x_1 - x_2) + \sin(x_1 + x_2))$
- $\cos x_1 \cos x_2 = \frac{1}{2}(\cos(x_1 - x_2) + \cos(x_1 + x_2))$
- $\sin x_1 \sin x_2 = \frac{1}{2}(\cos(x_1 - x_2) - \cos(x_1 + x_2))$

These identities are needed in differential equations, a course that follows calculus.

EXAMPLE 8 Write $(\sin 2x)^2(\cos 3x)^2$ as a sum of sine and cosine functions.

Solution We first use the half-angle formulas to rewrite the product as

$$(\sin 2x)^2(\cos 3x)^2 = \left(\frac{1 - \cos 4x}{2}\right)\left(\frac{1 + \cos 6x}{2}\right)$$
$$= \frac{1}{4}(1 - \cos 4x + \cos 6x - \cos 4x \cos 6x).$$

You can expect to see problems like Example 8 as the first, and most difficult, step in an integral calculus problem.

Then we use the cosine product formula to rewrite this as

$$(\sin 2x)^2(\cos 3x)^2 = \frac{1}{4}\left(1 - \cos 4x + \cos 6x - \frac{1}{2}\left(\cos(4x - 6x) + \cos(4x + 6x)\right)\right)$$

$$= \frac{1}{4} - \frac{1}{4}\cos 4x + \frac{1}{4}\cos 6x - \frac{1}{8}\cos(-2x) - \frac{1}{8}\cos 10x.$$

The cosine is an even function, so $\cos(-2x) = \cos 2x$. We can simplify this to

$$(\sin 2x)^2(\cos 3x)^2 = \frac{1}{4} - \frac{1}{8}\cos 2x - \frac{1}{4}\cos 4x + \frac{1}{4}\cos 6x - \frac{1}{8}\cos 10x. \quad ■$$

Applications

Example 9 illustrates how a double-angle formula can be used to simplify an equation.

EXAMPLE 9 In Example 5 of Section 4.4 we found that if a projectile is fired at an angle of elevation θ with an initial velocity of v_0 ft/sec, then the points $(x(t), y(t))$ on the path of the projectile at time t are given by

$$x(t) = (v_0 \cos\theta)t \quad \text{and} \quad y(t) = (v_0 \sin\theta)t - \frac{1}{2}gt^2,$$

where $g \approx 32\text{ft/sec}^2$ is the acceleration due to the earth's gravity. What angle θ will give the maximum horizontal distance for the projectile?

Solution Solving for t in terms of x in the horizontal equation gives

$$t = \frac{x}{v_0 \cos\theta}.$$

Then substituting this value of t into the vertical equation gives y as the quadratic function of x,

$$y = v_0 \sin\theta \frac{x}{v_0 \cos\theta} - \frac{1}{2}g\frac{x^2}{(v_0\cos\theta)^2} = \frac{gx}{(v_0\cos\theta)^2}\left(2\frac{v_0^2}{g}\sin\theta\cos\theta - x\right).$$

The maximum value for θ occurs when $y = 0$. This occurs when $x = 0$, that is, when the projectile is fired, and when it hits the ground, at

$$x = \frac{v_0^2}{g}(2\sin\theta\cos\theta) = \frac{v_0^2}{g}\sin 2\theta.$$

The maximum value for the sine function is 1 and this occurs when $2\theta = \pi/2$. So the maximum horizontal distance is obtained when $\theta = \pi/4$.

In summary, the maximum distance traveled by a projectile fired with initial velocity v_0 ft/sec occurs when the angle of elevation is $\pi/4 = 45°$. In this case the place where the projectile will land is approximately v_0^2/g feet from where it was fired. ■

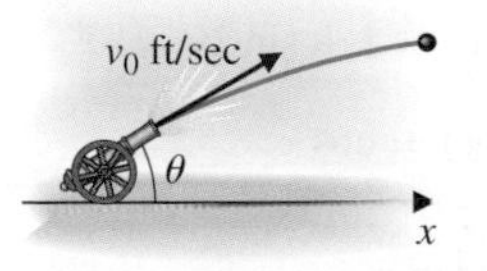

FIGURE 5

EXERCISE SET 4.7

In Exercises 1–6, use a sum or difference formula to determine the value of the trigonometric function.

1. $\sin\left(\frac{\pi}{2} - \frac{5\pi}{3}\right)$ **2.** $\cos\left(\frac{7\pi}{4} + \frac{\pi}{6}\right)$

3. $\cos\left(\frac{7\pi}{12}\right)$ **4.** $\sin\left(-\frac{\pi}{12}\right)$

5. $\tan\left(\frac{\pi}{12}\right)$ **6.** $\cot\left(-\frac{5\pi}{12}\right)$

In Exercises 7–12, use a double- or half-angle formula to determine the value of the trigonometric function.

7. $\sin\left(\frac{7\pi}{12}\right)$ **8.** $\cos\left(\frac{\pi}{8}\right)$

9. $\cos\left(\frac{5\pi}{8}\right)$ **10.** $\sin\left(\frac{5\pi}{12}\right)$

11. $\sin\left(\frac{13\pi}{12}\right)$ **12.** $\cos\left(\frac{11\pi}{8}\right)$

In Exercises 13–16, use the given information to find each value.

a. $\cos 2t$ **b.** $\sin 2t$ **c.** $\cos\left(\frac{t}{2}\right)$ **d.** $\sin\left(\frac{t}{2}\right)$

13. $\cos t = \frac{4}{5}, 0 < t < \pi/2$

14. $\sin t = -\frac{4}{5}, \pi < t < 3\pi/2$

15. $\tan t = \frac{5}{12}, \sin t < 0$

16. $\cot t = -\frac{24}{7}, \cos t > 0$

In Exercises 17–22, use the sum and difference formulas to verify the identity.

17. $\sin\left(t + \frac{3\pi}{2}\right) = -\cos t$

18. $\cos(t + \pi) = -\cos t$

19. $\sin\left(t + \frac{\pi}{2}\right) = \cos t$

20. $\cos\left(t + \frac{\pi}{2}\right) = -\sin t$

21. $\sin(\pi - t) = \sin t$

22. $\cos(\pi - t) = -\cos t$

In Exercises 23–26, use a half-angle formula to rewrite the given expression so that it involves the sum or difference of only constants and sine and cosine functions to the first power.

23. $(\cos 2x)^2$ **24.** $(\sin 3x)^2$

25. $(\sin x)^4$ **26.** $(\sin 4x)^2(\cos 4x)^2$

In Exercises 27–30, rewrite each product as a sum or difference.

27. $\sin 6t \cos 5t$ **28.** $\cos 5t \sin 8t$

29. $\cos 2t \cos 3t$ **30.** $\sin 3t \sin 5t$

In Exercises 31–38, verify the identities.

31. $(1 - (\cos x)^2)(\sec x)^2 = (\tan x)^2$

32. $\cot x + \tan x = \sec x \csc x$

33. $\tan x - \cot x = -2\cot 2x$

34. $(\sin x + \cos x)^2 = 1 + \sin 2x$

35. $\cos x = \sin x \sin 2x + \cos x \cos 2x$

36. $(\tan x)^2 - (\sin x)^2 = (\tan x)^2(\sin x)^2$

37. $\sec x - \cos x = \sin x \tan x$

38. $\cos x(\cot x + \tan x) = \csc x$

In Exercises 39–44, find all values of x in the interval $[0, 2\pi]$ that satisfy the given equation.

39. $\sin 2x = \sin x$

40. $\sin 2x = \cos x$

41. $2(\sin x)^2 + \cos x - 1 = 0$

42. $(\cos x)^2 - 3\sin x - 3 = 0$

43. $\tan x + \cot x = \frac{2}{\sin 2x}$

44. $2(\cot x)^2 + (\csc x)^2 - 2 = 0$

In Exercises 45–48, use the product-to-sum formulas on page 248 to verify the sum-to-product formula.

45. $\sin x + \sin y = 2\sin\frac{x+y}{2}\cos\frac{x-y}{2}$

46. $\sin x - \sin y = 2\cos\frac{x+y}{2}\sin\frac{x-y}{2}$

47. $\cos x + \cos y = 2\cos\frac{x+y}{2}\cos\frac{x-y}{2}$

48. $\cos x - \cos y = -2\sin\frac{x+y}{2}\sin\frac{x-y}{2}$

In Exercises 49–54, use a graphing device to sketch the graphs of f and g, and use the graphs to determine if $f(x) = g(x)$ is an identity.

49. $f(x) = (\sin x - \cos x)^2;\ g(x) = 1 - \sin 2x$

50. $f(x) = 2\left(\cos\frac{x}{2}\right)^2 - 1;\ g(x) = \cos x$

51. $f(x) = \frac{\sin 2x}{1 + \cos 2x};\ g(x) = \tan x$

52. $f(x) = \dfrac{2 \cot x}{1 + (\cot x)^2}$; $g(x) = \sin 2x$

53. $f(x) = (\sin x - \cos x)^2$; $g(x) = 1$

54. $f(x) = \tan\left(\dfrac{x}{2}\right)$; $g(x) = \dfrac{1 + \cos x}{\sin x}$

55. A center fielder throws a baseball to home plate from the right field 330 feet away. What is the minimal velocity that the ball can be thrown to reach the plate?

56. A 105-mm howitzer of type M102 firing an ME(M1) projectile has a muzzle velocity of 494 meters per second. What is the maximum distance that this projectile can theoretically travel? (Note: The acceleration due to gravity, in metric units, is approximately $g = 9.8$ m/sec.)

4.8 INVERSE TRIGONOMETRIC FUNCTIONS

In Example 4 of Section 4.3 we needed to recover the value of θ from the fact that $\tan\theta = \frac{1}{2}$. This calls for an inverse function. But in Section 2.5, we determined that a function has an inverse precisely when it is one-to-one. The trigonometric functions are all periodic, so they do not pass the Horizontal Line Test and cannot be one-to-one.

Like other functions we have studied, the inverse trigonometric functions are used in several different contexts in calculus.

The functions we consider in this section are not the inverses of the ordinary trigonometric functions. Instead, they are inverses of trigonometric functions for which the domains have been restricted to intervals on which they

- are one-to-one, and
- have the same range as the ordinary trigonometric functions.

The following review from Section 2.5 lists a few of the important facts about inverse functions.

Properties of Inverse Functions

Suppose that f is a one-to-one function.

- The inverse function f^{-1} is unique.
- The domain of f^{-1} is the range of f.
- The range of f^{-1} is the domain of f.
- If x is in the domain of f^{-1} and y is in the domain of f, then

$$f^{-1}(x) = y \quad \text{if and only if} \quad f(y) = x.$$

- If x is in the domain of f^{-1}, then $f(f^{-1}(x)) = x$.
- If x is in the domain of f, then $f^{-1}(f(x)) = x$.
- The graph of $y = f^{-1}(x)$ is the reflection of the graph of $y = f(x)$ about the line $y = x$.

Although we generally denote the inverse function by f^{-1}, the inverse functions that are particularly important for applications are given special notation. The inverse trigonometric functions fall into this category.

We begin, as usual, with the sine function. It certainly is not one-to-one, since its graph, shown in Figure 1(a), crosses every horizontal line between -1 and 1 an infinite number of times.

Suppose, however, that we consider the sine function with its domain restricted to the interval $[-\pi/2, \pi/2]$, as shown in Figure 1(b). On this interval the sine function is one-to-one and assumes all the values in its range, $[-1, 1]$. This restricted function has an inverse, which we call the inverse sine, or *arcsine*, function.

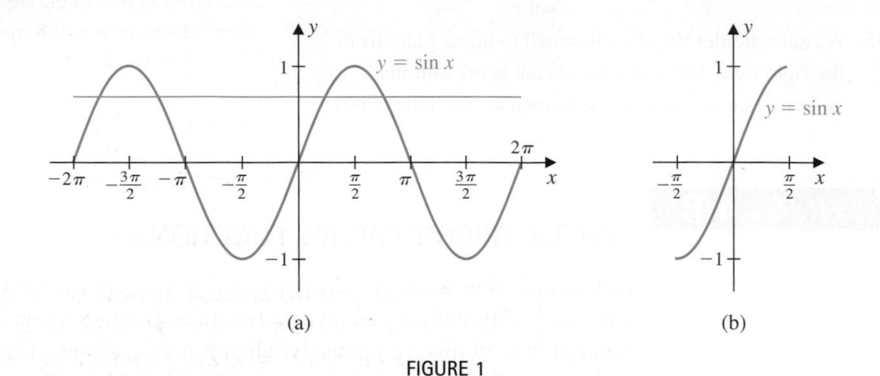

FIGURE 1

The arcsine function is denoted either as arcsin or as $\sin^{-1}$. We use the arcsin notation to avoid confusion between the inverse sine function and the reciprocal of the sine function, $\csc x = (\sin x)^{-1}$.

The Arcsine Function

The **arcsine** function, denoted **arcsin**, has domain $[-1, 1]$ and range $[-\pi/2, \pi/2]$, and is defined by

$$\arcsin x = y \quad \text{if and only if} \quad \sin y = x.$$

- For x in $[-1, 1]$, $\sin(\arcsin x) = x$.
- For x in $[-\pi/2, \pi/2]$, $\arcsin(\sin x) = x$.

The graph of $y = \arcsin x$ is the reflection about the line $y = x$ of the graph of the restricted sine function, as shown in Figure 2. Notice how steep the graph of $y = \arcsin x$ is at the ends of its domain. This corresponds to the flatness on the corresponding portions of the restricted sine function.

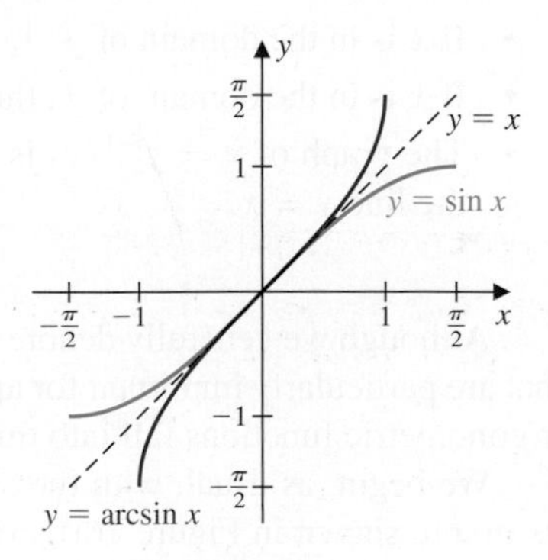

FIGURE 2

EXAMPLE 1 Evaluate each expression.

a. $\arcsin \dfrac{1}{2}$ **b.** $\arcsin\left(\sin \dfrac{\pi}{3}\right)$ **c.** $\arcsin\left(\sin \dfrac{3\pi}{4}\right)$

Solution

a. We need to find a number in the interval $[-\pi/2, \pi/2]$ whose sine is $\frac{1}{2}$. Since $\sin(\pi/6) = \frac{1}{2}$ and $\pi/6$ is in $[-\pi/2, \pi/2]$, we have

$$\arcsin \frac{1}{2} = \frac{\pi}{6}.$$

b. Since $\pi/3$ is in $[-\pi/2, \pi/2]$, the domain of the arcsine function, we have

$$\arcsin\left(\sin \frac{\pi}{3}\right) = \frac{\pi}{3}.$$

c. This part differs from part (b) because $3\pi/4$ is not in the domain of the arcsine function. In this case we need a number in $[-\pi/2, \pi/2]$ whose sine is the same as that of $3\pi/4$. Since $\sin(3\pi/4) = \sqrt{2}/2 = \sin(\pi/4)$, and $\pi/4$ is in $[-\pi/2, \pi/2]$, we have

$$\arcsin\left(\sin \frac{3\pi}{4}\right) = \arcsin\left(\sin \frac{\pi}{4}\right) = \frac{\pi}{4}.$$

■

The inverses for the other trigonometric functions are defined by making domain restrictions similar to those made for the sine function. In the case of the cosine function, whose graph is shown in Figure 3(a), we restrict the domain to the interval $[0, \pi]$, as shown in blue in Figure 3(b). The function on this interval is one-to-one and assumes all the values in the range of the ordinary cosine function. For this restricted function we have an inverse function, called the inverse cosine, or *arccosine*.

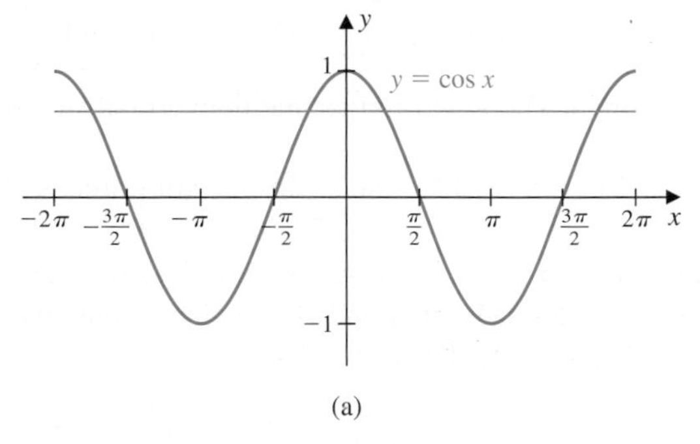

(a)

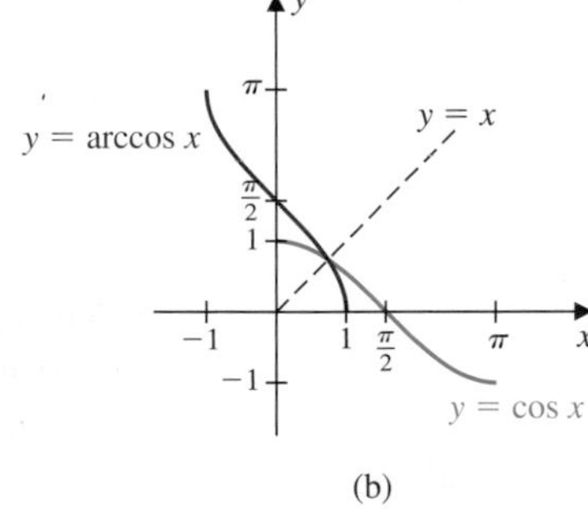

(b)

FIGURE 3

The Arccosine Function

The **arccosine** function, denoted **arccos**, has domain $[-1, 1]$ and range $[0, \pi]$ and is defined by

$$\arccos x = y \quad \text{if and only if} \quad \cos y = x.$$

- For x in $[-1, 1]$, $\cos(\arccos x) = x$.
- For x in $[0, \pi]$, $\arccos(\cos x) = x$.

The graph of $y = \arccos x$ is the reflection about the line $y = x$ of the graph of the restricted cosine function, as shown in red in Figure 3(b).

EXAMPLE 2 Evaluate each expression.

a. $\cos\left[\arccos\left(-\dfrac{1}{2}\right)\right]$ **b.** $\arccos\left(\cos\dfrac{\pi}{3}\right)$

c. $\arccos\left[\cos\left(-\dfrac{\pi}{4}\right)\right]$ **d.** $\sin\left[\arccos\left(-\dfrac{5}{13}\right)\right]$

Solution **a.** The problem is clearer when expressed in words. The $\arccos\left(-\frac{1}{2}\right)$ is the number in $[0, \pi]$ whose cosine is $-\frac{1}{2}$, so the cosine of this number must be $-\frac{1}{2}$. That is,

$$\cos\left[\arccos\left(-\frac{1}{2}\right)\right] = -\frac{1}{2}.$$

b. We need to find the number in the interval $[0, \pi]$ whose cosine is the same as the cosine of $\pi/3$. Since $\pi/3$ itself is in the interval $[0, \pi]$, we have

$$\arccos\left(\cos\frac{\pi}{3}\right) = \frac{\pi}{3}.$$

c. This problem is slightly more complicated than part (b) since $-\pi/4$ is not in the interval $[0, \pi]$. However, the cosine function is even, and $\pi/4$ is in $[0, \pi]$, so

$$\arccos\left(\cos\left(-\frac{\pi}{4}\right)\right) = \arccos\left(\cos\left(\frac{\pi}{4}\right)\right) = \frac{\pi}{4}.$$

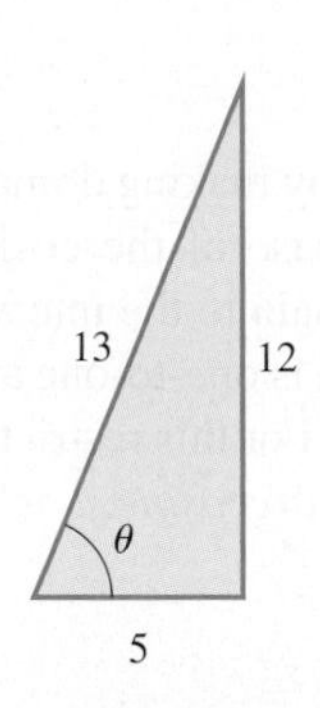

FIGURE 4

d. The triangle in Figure 4 shows the angle θ in $(0, \pi/2)$ with

$$\cos\theta = \frac{5}{13}, \quad \text{so} \quad \theta = \arccos\left(\frac{5}{13}\right).$$

The Pythagorean Theorem implies that the vertical side has length

$$\sqrt{13^2 - 5^2} = 12, \quad \text{so} \quad \sin\theta = \sin\left(\arccos\left(\frac{5}{13}\right)\right) = \frac{12}{13}.$$

The angle between 0 and π whose cosine is $-\frac{5}{13}$ lies in quadrant II. Since the sine is positive in both quadrants I and II, the sine of the $\arccos\left(-\frac{5}{13}\right)$ is the same as the sine of $\arccos\left(\frac{5}{13}\right)$. Hence

$$\sin\left(\arccos\left(-\frac{5}{13}\right)\right) = \frac{12}{13}.$$ ■

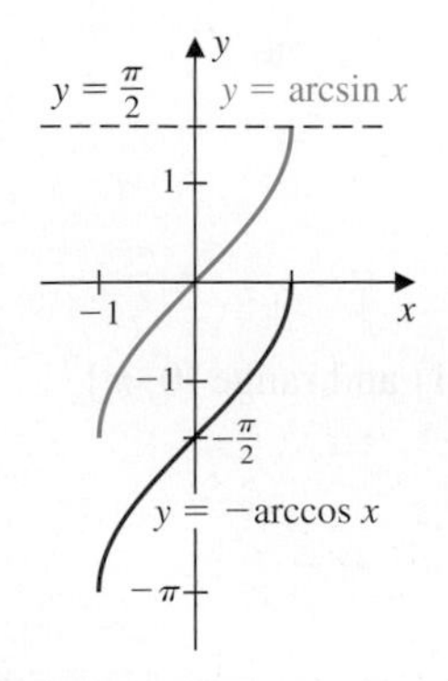

FIGURE 5

Just as there is a close relationship between the graphs of the sine and cosine functions, there is a similar connection between the graphs of their inverse functions. The graphs of $y = -\arccos x$ and $y = \arcsin x$ are shown in Figure 5. Notice that the shapes of the graphs are the same and that the graphs differ only by a vertical shift of $\pi/2$ units. Hence for all x in $[-1, 1]$, we have

$$\arcsin x = \frac{\pi}{2} - \arccos x \quad \text{and} \quad \arcsin x + \arccos x = \frac{\pi}{2}.$$

The arccosine function is not used extensively in calculus, but the arctangent function has many applications. The graph of the tangent function is shown in

Figure 6(a), and its restriction to a function that is one-to-one is shown in blue in Figure 6(b).

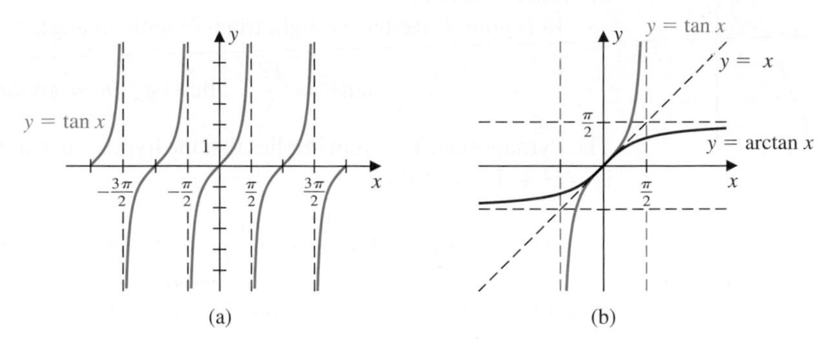

FIGURE 6

The Arctangent Function

The **arctangent** function, denoted **arctan**, has domain $(-\infty, \infty)$ and range $(-\pi/2, \pi/2)$, and is defined by

$$\arctan x = y \quad \text{if and only if} \quad \tan y = x.$$

- For x in $(-\infty, \infty)$, $\tan(\arctan x) = x$.
- For x in $(-\pi/2, \pi/2)$, $\arctan(\tan x) = x$.

The graph of $y = \arctan x$ is the reflection about the line $y = x$ of the graph of the restricted tangent function, as shown in Figure 6(b). The graph of $y = \tan x$ has vertical asymptotes at $x = -\pi/2$ and $x = \pi/2$. So the graph of $y = \arctan x$ has horizontal asymptotes at $y = -\pi/2$ and $y = \pi/2$, as shown in red in Figure 6(b).

In Example 4 of Section 4.3 we had to be satisfied with an answer for the billiard ball problem that was given in the form $\tan\theta = 1/2$. Having inverse trigonometric functions implies that

$$\theta = \arctan\frac{1}{2} \approx 0.464 \text{ radians} \approx 26.6^\circ.$$

EXAMPLE 3 Find $\cos\left(\arctan\frac{12}{5} + \arcsin\frac{3}{5}\right)$.

Solution This problem requires the sum formula for the cosine,

$$\cos(a + b) = \cos a\cos b - \sin a\sin b,$$

with $a = \arctan\frac{12}{5}$ and $b = \arcsin\frac{3}{5}$. Then we can rewrite the expression as

$$\cos\left(\arctan\frac{12}{5} + \arcsin\frac{3}{5}\right) = \cos\left(\arctan\frac{12}{5}\right)\cos\left(\arcsin\frac{3}{5}\right) - \sin\left(\arctan\frac{12}{5}\right)\sin\left(\arcsin\frac{3}{5}\right).$$

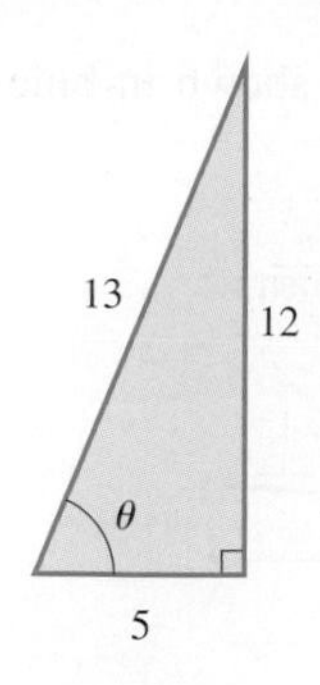

FIGURE 7

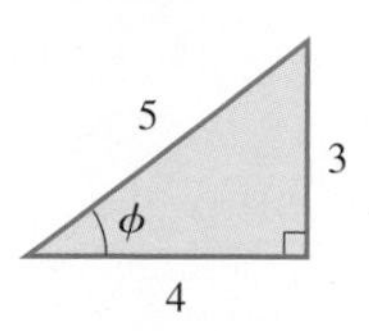

FIGURE 8

We first note that $\sin\left(\arcsin \frac{3}{5}\right) = \frac{3}{5}$. The other values can be obtained from trigonometric identities, but we have chosen to find them by considering the triangles shown in Figures 7 and 8.

In Figure 7, we have a right triangle with an angle θ satisfying

$$\tan\theta = \frac{12}{5}, \quad \text{that is,} \quad \theta = \arctan\frac{12}{5}.$$

The Pythagorean Theorem implies that the hypotenuse of the right triangle in Figure 7 is $\sqrt{5^2 + 12^2} = 13$, so

$$\sin\theta = \sin\left(\arctan\frac{12}{5}\right) = \frac{12}{13} \quad \text{and} \quad \cos\theta = \cos\left(\arctan\frac{12}{5}\right) = \frac{5}{13}.$$

In a similar manner, the right triangle in Figure 8 shows an angle ϕ with

$$\sin\phi = \frac{3}{5}, \quad \text{that is,} \quad \phi = \arcsin\frac{3}{5}.$$

The Pythagorean Theorem implies that the base of the right triangle in Figure 8 is $\sqrt{5^2 - 3^2} = 4$ and $\cos\phi = \cos\left(\arcsin\frac{3}{5}\right) = \frac{4}{5}$.

Combining our facts gives

$$\cos\left(\arctan\frac{12}{5} + \arcsin\frac{3}{5}\right) = \frac{5}{13}\cdot\frac{4}{5} - \frac{12}{13}\cdot\frac{3}{5} = -\frac{16}{65}.$$

■

The inverse trigonometric functions occur frequently in calculus because the composition of an inverse trigonometric function followed by a trigonometric function produces some useful algebraic relationships. Example 4 shows two of these.

EXAMPLE 4 Verify the identity.

a. $\sin(\arccos x) = \sqrt{1 - x^2}$ **b.** $\cos(2\arctan x) = \dfrac{1 - x^2}{1 + x^2}$

Solution **a.** The Pythagorean identity implies that

$$(\sin(\arccos x))^2 + (\cos(\arccos x))^2 = 1,$$

and $\cos(\arccos x) = x$. So

$$1 = (\sin(\arccos x))^2 + x^2, \quad \text{and} \quad \sin(\arccos x) = \pm\sqrt{1 - x^2}.$$

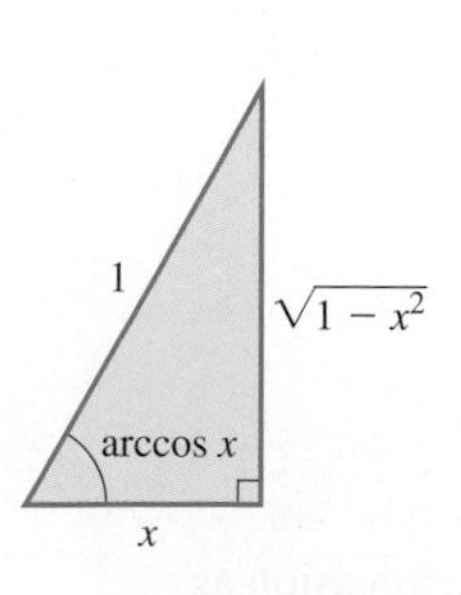

FIGURE 9

The range of the arccosine function is $[0, \pi]$, so the sine function is nonnegative, and

$$\sin(\arccos x) = \sqrt{1 - x^2}.$$

Alternatively, we can establish this fact by considering the right triangle in Figure 9, where for illustration purposes we have assumed that $x > 0$.

b. This identity is easiest to establish if we first introduce a variable, or parameter t, to represent the argument of the cosine. Then

$$t = 2\arctan x \quad \text{implies that} \quad x = \tan\frac{t}{2}.$$

We can reexpress the latter equation using the half-angle formulas for the sine and cosine as

$$x = \tan\frac{t}{2} = \frac{\sin(t/2)}{\cos(t/2)} = \frac{\sqrt{(1-\cos t)/2}}{\sqrt{(1+\cos t)/2}} = \frac{\sqrt{1-\cos t}}{\sqrt{1+\cos t}}.$$

Squaring the right and left sides of the last equation gives

$$x^2 = \frac{1-\cos t}{1+\cos t}, \quad \text{so} \quad x^2(1+\cos t) = 1-\cos t.$$

To solve for $\cos t$, we write

$$x^2 + x^2\cos t = 1-\cos t \quad \text{and} \quad x^2\cos t + \cos t = 1-x^2.$$

Factoring $\cos t$ from the left side gives

$$(x^2+1)\cos t = 1-x^2, \quad \text{so} \quad \cos t = \frac{1-x^2}{1+x^2}.$$

Making the substitution $t = 2\arctan x$ gives

$$\cos(2\arctan x) = \frac{1-x^2}{1+x^2}.$$ ■

Some calculus books restrict the secant function to $[0, \pi/2) \cup [\pi, 3\pi/2)$. Either definition is valid—we simply need to restrict the secant function to a set of real numbers where the function is one-to-one and where it assumes its entire range.

The arcsecant function is also used in calculus and is defined by restricting the secant function to the same basic interval as its reciprocal, the cosine function. Then features of the arccosine function can be used to generate facts about the arcsecant.

The graph of the secant function is shown in Figure 10(a), and a one-to-one restriction is shown in Figure 10(b).

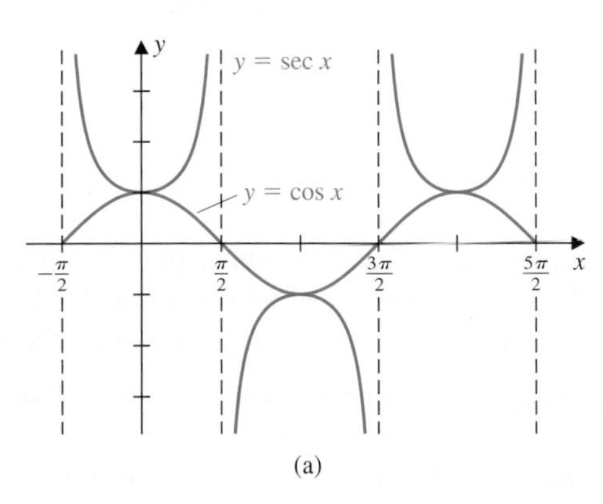

(a)

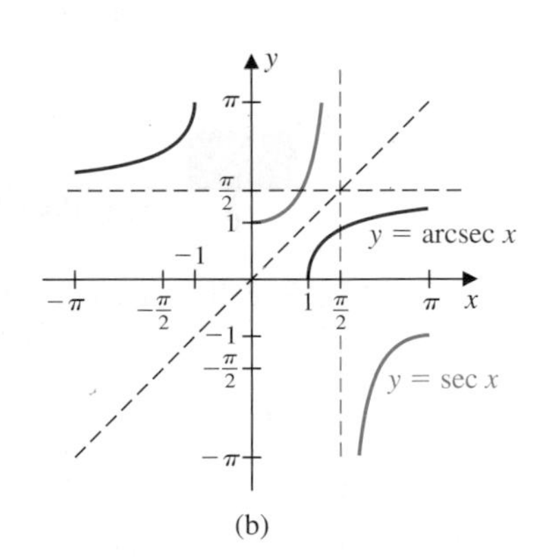

(b)

FIGURE 10

The Arcsecant Function

The **arcsecant** function, denoted **arcsec**, has domain $(-\infty, -1] \cup [1, \infty)$ and range $[0, \pi/2) \cup (\pi/2, \pi]$ and is defined by

This is a natural choice of domain restriction since it is as close as possible to that of its reciprocal, the cosine function.

$$\operatorname{arcsec} x = y \quad \text{if and only if} \quad \sec y = x.$$

- For x in $(-\infty, -1] \cup [1, \infty)$, $\sec(\operatorname{arcsec} x) = x$.
- For x in $[0, \pi/2) \cup (\pi/2, \pi]$, $\operatorname{arcsec}(\sec x) = x$.

The graph of $y = \operatorname{arcsec} x$ is the reflection about the line $y = x$ of the graph of the restricted secant function, as shown in red in Figure 10(b). Since the graph of $y = \sec x$ has a vertical asymptote at $x = \pi/2$, the graph of $y = \operatorname{arcsec} x$ has a horizontal asymptote at $y = \pi/2$.

To determine the relationship between the arcsecant and the arccosine, consider $\cos(\operatorname{arcsec} x)$. If we introduce a new variable $t = \operatorname{arcsec} x$, then $\sec t = x$. But $\sec t = 1/\cos t$, so $\cos t = 1/x$, and $t = \arccos(1/x)$. Hence we have

$$\operatorname{arcsec} x = \arccos \frac{1}{x}, \quad \text{whenever } |x| \geq 1.$$

The inverse cosecant function, denoted arccsc, and inverse cotangent, denoted arccot, are defined in a similar manner. Since they are not called on frequently in calculus, we do not consider them. Their graphs are shown in red in Figure 11.

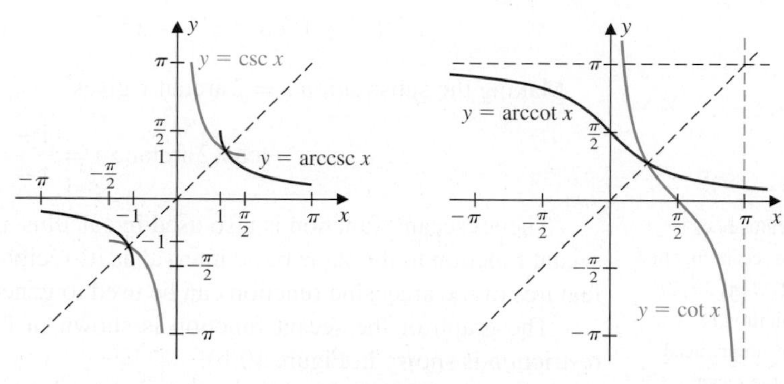

FIGURE 11

Applications

EXAMPLE 5 A satellite in orbit above the earth can see only a portion of the surface of the earth, as shown in Figure 12. Suppose that a satellite is in orbit about the equator a distance d miles above the earth. How many miles of the equator can be seen from the satellite?

Solution Let θ be the angle from the center of the earth to the most distant point P on the earth that the satellite can see, as shown in the Figure 12. The radius of the earth at the equator is approximately 3969 miles, so the distance from the center C of the earth to the satellite at S is approximately $3960 + d$ miles. The tangent PS to the circle is $90°$ to the radius at P, so

$$\cos\theta = \frac{3960}{3960 + d} \quad \text{and} \quad \theta = \arccos \frac{3960}{3960 + d}.$$

By the formula at the end of Section 4.2, the arc of the circle cut by this angle is

$$\arccos\left(\frac{3960}{3960 + d}\right) 3960 \text{ miles.}$$

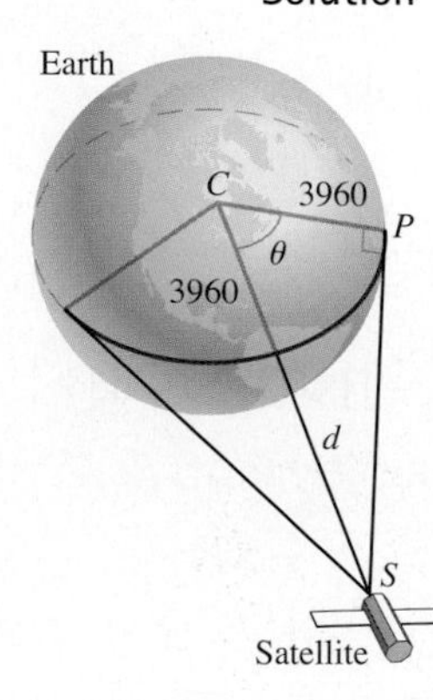

FIGURE 12

The amount of the equator that can be seen from the satellite is consequently

$$2 \arccos\left(\frac{3960}{3960 + d}\right) 3960 = 7920 \arccos\left(\frac{3960}{3960 + d}\right) \text{ miles.}$$

■

EXERCISE SET 4.8

In Exercises 1–14, find the exact value of the quantity, or explain why it is undefined.

1. $\arccos\left(\frac{\sqrt{3}}{2}\right)$ **2.** $\arcsin\left(-\frac{1}{2}\right)$

3. $\arcsin(1)$ **4.** $\arccos(-1)$

5. $\arccos\left(-\frac{\sqrt{2}}{2}\right)$ **6.** $\arcsin\left(\frac{\sqrt{3}}{2}\right)$

7. $\arctan(\sqrt{3})$ **8.** $\arctan\left(\frac{\sqrt{3}}{3}\right)$

9. $\arctan(-1)$ **10.** $\arctan(0)$

11. $\operatorname{arcsec}\left(\sqrt{2}\right)$ **12.** $\operatorname{arcsec}\left(-\frac{2\sqrt{3}}{3}\right)$

13. $\arccos(2)$ **14.** $\operatorname{arcsec}(0)$

In Exercises 15–36, find the exact value of each expression.

15. $\cos\left(\arccos\left(\frac{1}{2}\right)\right)$ **16.** $\sin\left(\arcsin\left(\frac{\sqrt{3}}{2}\right)\right)$

17. $\sin\left(\arccos\left(\frac{\sqrt{2}}{2}\right)\right)$ **18.** $\cos\left(\arcsin\left(\frac{\sqrt{2}}{2}\right)\right)$

19. $\tan\left(\arcsin\left(\frac{\sqrt{3}}{2}\right)\right)$ **20.** $\tan\left(\arcsin\left(-\frac{1}{2}\right)\right)$

21. $\arcsin\left(\sin\left(-\frac{\pi}{4}\right)\right)$ **22.** $\arccos\left(\cos\left(\frac{2\pi}{3}\right)\right)$

23. $\arccos\left(\cos\left(-\frac{\pi}{4}\right)\right)$ **24.** $\arcsin\left(\sin\left(\frac{3\pi}{4}\right)\right)$

25. $\arctan\left(\tan\left(\frac{\pi}{3}\right)\right)$ **26.** $\arctan\left(\tan\left(-\frac{\pi}{3}\right)\right)$

27. $\arctan\left(\tan\left(\frac{7\pi}{6}\right)\right)$ **28.** $\arctan\left(\tan\left(\frac{2\pi}{3}\right)\right)$

29. $\cos\left(\arcsin\left(\frac{4}{5}\right)\right)$ **30.** $\sin\left(\arccos\left(\frac{5}{13}\right)\right)$

31. $\tan\left(\arccos\left(\frac{4}{5}\right)\right)$ **32.** $\tan\left(\arcsin\left(\frac{12}{13}\right)\right)$

33. $\cos(\arctan(4))$ **34.** $\sin(\arctan(2))$

35. $\cos\left(\arcsin\left(\frac{3}{5}\right) + \arccos\left(\frac{4}{5}\right)\right)$

36. $\tan\left(\arcsin\left(\frac{1}{3}\right) + \arccos\left(\frac{1}{2}\right)\right)$

In Exercises 37–42, verify the given identity.

37. $\arcsin(-x) = -\arcsin(x)$, when $|x| \leq 1$

38. $\arccos(-x) = \pi - \arccos(x)$, when $|x| \leq 1$

39. $\arccos x = \arcsin(\sqrt{1-x^2})$, when $0 \leq x \leq 1$

40. $\arctan x = \arcsin\left(\dfrac{x}{\sqrt{1+x^2}}\right)$

41. $\tan(\arcsin x) = \dfrac{x}{\sqrt{1-x^2}}$, when $|x| < 1$ and $x \neq 0$

42. $\cos(\arcsin x) = \sqrt{1-x^2}$, when $|x| \leq 1$

In Exercises 43 and 44, (**a**) *solve each equation on the given interval, expressing the solution for x in terms of inverse trigonometric functions, and* (**b**) *use a calculator to approximate the solutions in part (a) to three decimal places.*

43. $(\tan x)^2 - \tan x - 2 = 0$ on $(-\pi/2, \pi/2)$

44. $6(\cos x)^2 - \cos x - 5 = 0$ on $[\pi/2, \pi]$

45. A lighthouse is 4 miles from a straight shoreline, as shown in the figure.

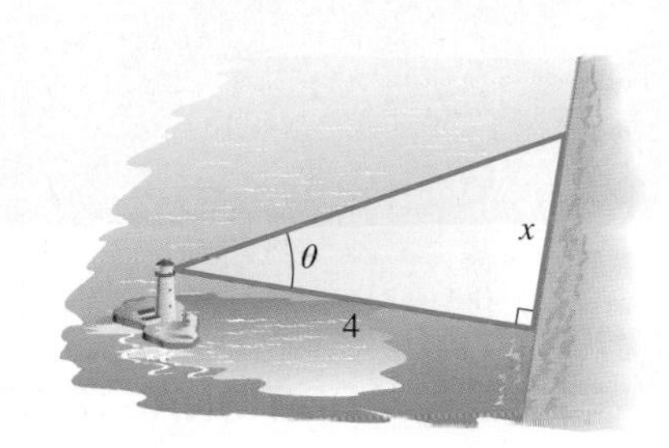

If the light from the lighthouse is moving along the shoreline, express the angle θ formed by the beam of light and the shoreline in terms of the distance x.

46. A large picture measures a feet from top to bottom. The bottom of the picture is b feet above the eye level of an observer who is standing x feet from the wall on which the picture is hung. Show that the angle θ subtended by the picture at the eye of the observer is given by

$$\theta = \arctan\left(\frac{a+b}{x}\right) - \arctan\left(\frac{b}{x}\right).$$

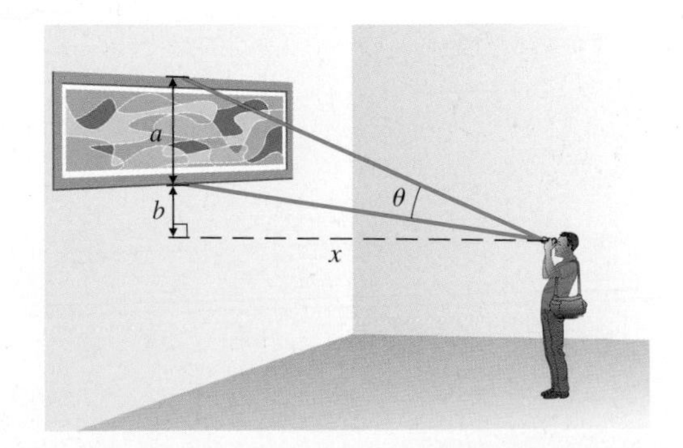

47. Commercial television satellites are commonly located in geosynchronous orbit about 22,300 miles above the surface of the Earth. At this distance they orbit the Earth at the same rate as the Earth rotates on its axis, so the satellites effectively remain stationary relative to the Earth. How much of the Earth's equator would such a satellite see if it were orbiting about the equator?

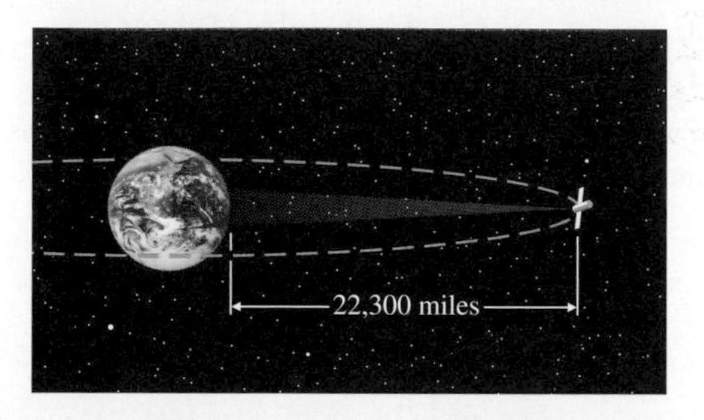

48. On April 12, 1961, Yuri Gagarin was the first person to orbit the Earth. His Russian Vostok capsule reached its apogee (the highest point above the earth) at a height of 203 miles as he crossed the equator in Africa. How much of the equator would it have been possible for him to see at this time?

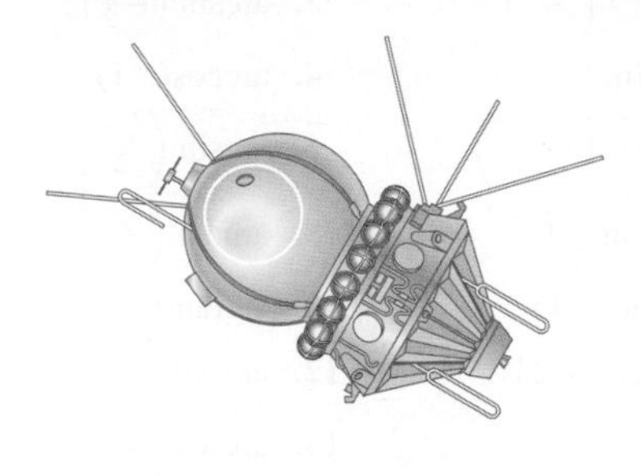

4.9 ADDITIONAL TRIGONOMETRIC APPLICATIONS

A Cessna Citation III business jet flying at 520 miles per hour is directly over Logan, Utah, and heading due south toward Phoenix, Arizona. Fifteen minutes later an F-15 Fighting Eagle passes over Logan traveling westward toward San Francisco at 1535 miles per hour. We want to determine a function that describes the distance between the planes in terms of the time after the F-15 passes over Logan until it reaches the California border 20 minutes later.

Let t be the time in hours after the time the F-15 passes Logan, $x(t)$ be the distance in miles that the F-15 has traveled, and $y(t)$ be the distance in miles the Cessna has traveled at time t. Converting the 20 minutes to $\frac{1}{3}$ hour and taking into consideration the $\frac{1}{4}$-hour lead time of the Cessna, we have

$$x(t) = 1535t \quad \text{and} \quad y(t) = \frac{1}{4}(520) + 520t, \quad \text{for } 0 \le t \le \frac{1}{3}.$$

The situation is illustrated in Figure 1.

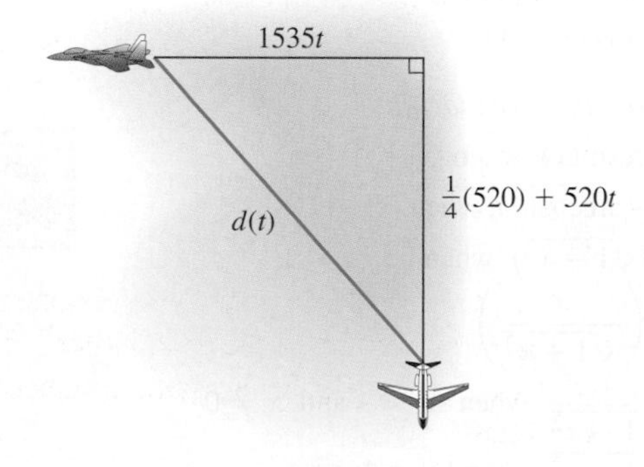

FIGURE 1

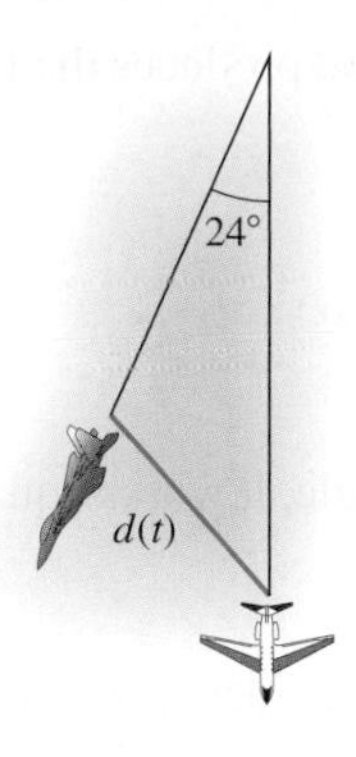

FIGURE 2

Since the triangle in the figure is a right triangle, the distance between the planes is

$$d(t) = \sqrt{[x(t)]^2 + [y(t)]^2} = \sqrt{(1535t)^2 + \left[520\left(t + \frac{1}{4}\right)\right]^2},$$

and they are $d\left(\frac{1}{3}\right) \approx 595$ miles apart when the F-15 reaches the California line.

This example is used to introduce this section because we want to show how dramatically the problem changes in the more likely situation where the paths of the planes are not perpendicular. Suppose, for example, that all the facts of the problem are the same except that the F-15 is heading over Logan on a course 24° west of south toward Nellis Air Force Base, near Las Vegas, Nevada, 395 miles away. Figure 2 illustrates this new situation, from which we can see that the triangle generally has no right angle. We will reconsider this problem in Example 3, after we have determined the Law of Cosines.

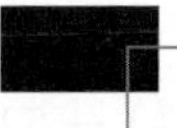

Law of Cosines

To handle problems involving nonright triangles we need the Law of Cosines, of which the Pythagorean Theorem is a special case. The Law of Cosines is used to find the missing part of a nonright triangle provided that two sides and an angle are known, although it is easier to apply if the known angle is formed by the known sides.

Law of Cosines

Suppose that a triangle has sides of lengths a, b, and c and has corresponding opposite angles α, β, and γ, as shown in Figure 3(a). Then

$$a^2 = b^2 + c^2 - 2bc\cos\alpha,$$

and, similarly,

$$b^2 = a^2 + c^2 - 2ac\cos\beta \quad \text{and} \quad c^2 = a^2 + b^2 - 2ab\cos\gamma.$$

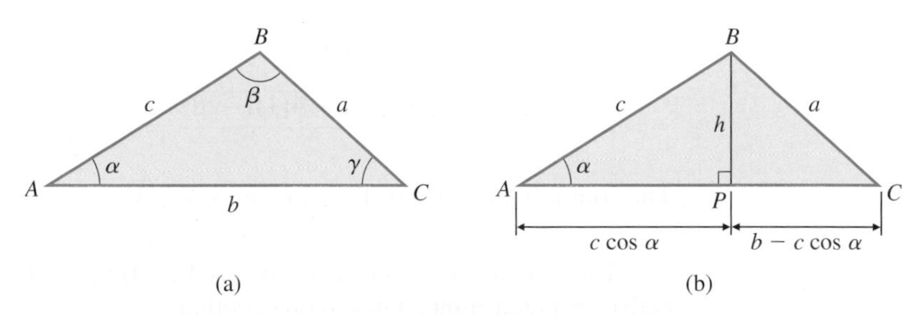

FIGURE 3

To see why the Law of Cosines holds, consider the triangles shown in Figure 3(b). We have labeled P as the point where the vertical from B intersects the base of the triangle and called this vertical distance h. Since $\triangle APB$ and $\triangle BPC$ are both right triangles,

$$c^2 = h^2 + (c\cos\alpha)^2 \quad \text{and} \quad a^2 = h^2 + (b - c\cos\alpha)^2.$$

Solving for h^2 in the first equation and substituting into the second produces the Law of Cosines:

$$\begin{aligned} a^2 &= [c^2 - (c\cos\alpha)^2] + (b - c\cos\alpha)^2 \\ &= c^2 - (c\cos\alpha)^2 + b^2 - 2bc\cos\alpha + (c\cos\alpha)^2 \\ &= b^2 + c^2 - 2bc\cos\alpha. \end{aligned}$$

Although we have illustrated the situation when α is an acute angle, it works equally well for obtuse angles.

EXAMPLE 1 A triangle has sides of length 6 and 8, and the angle between these sides is 60°. What is the length of the third side?

Solution Note that degree measure for angles is generally used for applications we will consider in this section. The triangle is shown in Figure 4. The Law of Cosines implies that the length x of the third side satisfies

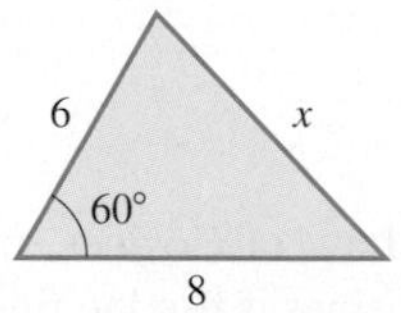

$$x^2 = 6^2 + 8^2 - 2(6)(8)\cos 60^\circ = 100 - 96\left(\tfrac{1}{2}\right) = 52$$

and

$$x = \sqrt{52} = 2\sqrt{13}.$$

FIGURE 4

EXAMPLE 2 A triangle has sides of length 6 and 8, and a 60° angle opposite the side of length 8, as shown in Figure 5. What is the length of the third side?

Solution In this example the given angle is not formed by the two given sides, but with a little extra work the Law of Cosines can be used to find the length x. The Law of Cosines implies that

$$8^2 = x^2 + 6^2 - 2(6)(x)\cos 60^\circ = x^2 - 6x + 36.$$

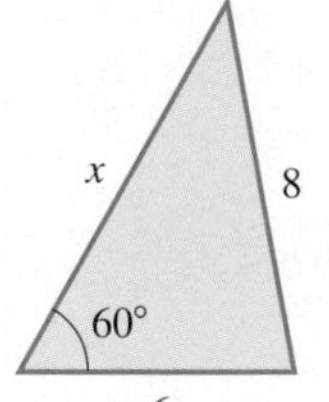

FIGURE 5

This equation reduces to

$$x^2 - 6x - 28 = 0,$$

and the Quadratic Formula gives

$$x = \frac{6 \pm \sqrt{36 - 4(1)(-28)}}{2} = \frac{6 \pm \sqrt{148}}{2} = 3 \pm \sqrt{37}.$$

The solution must be positive, so $x = 3 + \sqrt{37} \approx 9.08$.

The Law of Cosines can also be used to solve the problem of the aircraft whose paths are not at right angles to one another.

EXAMPLE 3 A Cessna Citation III business jet flying at 520 miles per hour is directly over Logan, Utah, and heading due south to Phoenix. Fifteen minutes later an F-15 Eagle passes over Logan traveling toward Nellis Air Force Base, 395 miles away. It travels at 1535 miles per hour on a course of 24° west of south. Determine a function that describes the distance between the planes after the F-15 passes over Logan until it reaches Nellis Air Force Base approximately $t = 0.258$ hour later.

Solution Figure 6 illustrates the situation. For $t \geq 0$, let $x(t)$ be the distance in miles that the F-15 has traveled and let $y(t)$ be the distance in miles that the Cessna has traveled. The values of $x(t)$ and $y(t)$ are the same as in the opening example, namely,

$$x(t) = 1535t \quad \text{and} \quad y(t) = \tfrac{1}{4}(520) + 520t = 520\left(t + \tfrac{1}{4}\right),$$

with

$$x(0.258) = 1535(0.258) \approx 396 \text{ miles},$$

and

$$y(0.258) = 520(0.258 + 0.25) \approx 264 \text{ miles}.$$

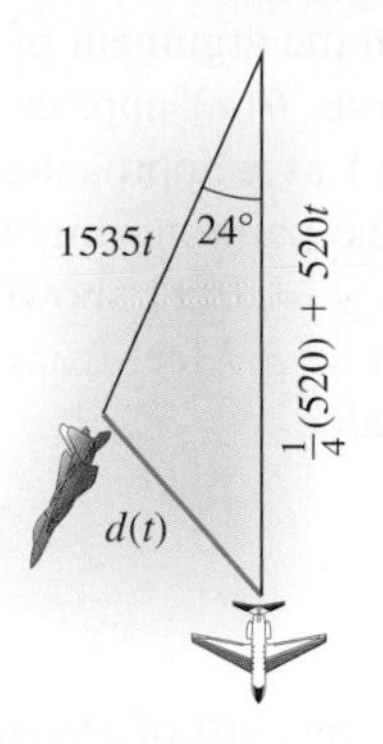

FIGURE 6

In calculus, you will be asked to determine the rate at which the distance $d(t)$ is changing.

The Law of Cosines implies that in this situation

$$d(t) = \sqrt{[x(t)]^2 + [y(t)]^2 - 2x(t)y(t)\cos 24^\circ},$$

where $d(t)$ is in miles and t is in hours. When the F-15 reaches Nellis Air Force Base, the distance separating the planes is

$$d(0.258) \approx \sqrt{(396)^2 + (264)^2 - 2(396)(264)\cos 24^\circ} \approx 188 \text{ miles}. \quad \blacksquare$$

EXAMPLE 4 A picture in an art museum is 5 feet high and hung so that its base is 8 feet above the ground. Find the viewing angle $\theta(x)$ of a 6-foot-tall viewer who is standing x feet from the wall.

Solution The situation is illustrated in Figure 7. Since $\triangle APB$ and $\triangle APC$ are both right triangles, we have $AC = \sqrt{49 + x^2}$ and $AB = \sqrt{4 + x^2}$.

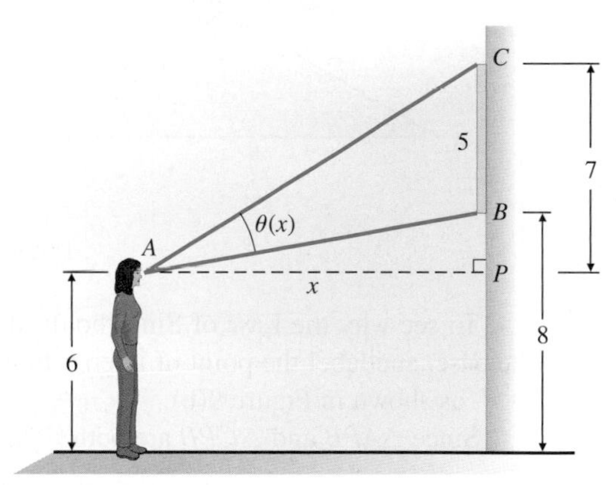

FIGURE 7

You can expect to see problems like this as the first, and most difficult, step in a calculus problem.

The Law of Cosines applied to $\triangle ABC$ implies that

$$25 = (4 + x^2) + (49 + x^2) - 2\sqrt{4 + x^2}\sqrt{49 + x^2}\cos\theta(x),$$

which gives

$$\cos\theta(x) = \frac{53 + 2x^2 - 25}{2\sqrt{4 + x^2}\sqrt{49 + x^2}}.$$

So

$$\theta(x) = \arccos\frac{14 + x^2}{\sqrt{(4 + x^2)(49 + x^2)}}. \quad \blacksquare$$

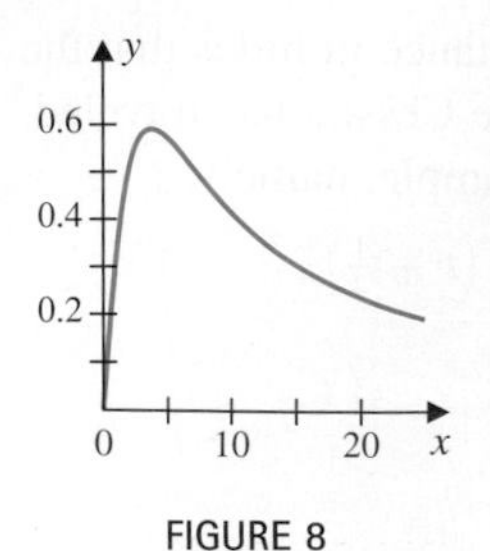

FIGURE 8

Notice in Example 4 that as x becomes large, the constants in both the numerator and the denominator become less significant and the fraction in the argument of the arccosine approaches 1. As a consequence, as x becomes large, $\theta(x)$ approaches $\arccos(1) = 0$. The argument of the arccosine also approaches 1 as x approaches 0, so $\theta(x)$ approaches 0 in this case as well. The best view of the painting might be defined as occurring when $\theta(x)$ is a maximum. The graph of $y = \theta(x)$ shown in Figure 8 indicates that the maximum occurs at approximately $x = 3.7$ feet from the wall. Methods of calculus can be used to determine the exact value.

Law of Sines

We have seen that the Law of Cosines can be used to find the missing parts of a triangle provided that two sides and an angle are known. In other situations, for example, when the angles and one side are known, we use the Law of Sines.

The Law of Sines

Suppose that a triangle has sides of length a, b, and c with corresponding opposite angles α, β, and γ, as shown in Figure 9(a). Then

$$\frac{\sin\alpha}{a} = \frac{\sin\beta}{b} = \frac{\sin\gamma}{c}.$$

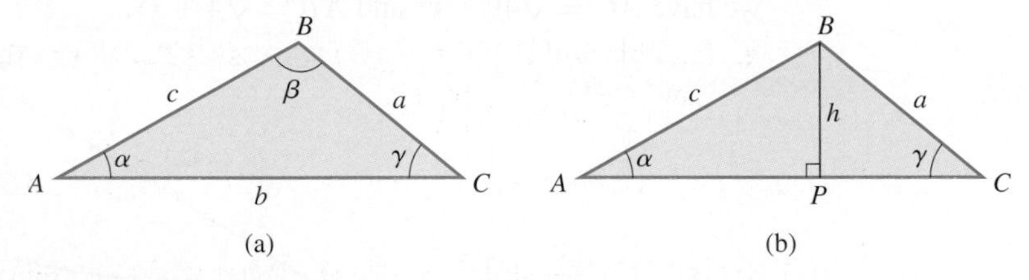

FIGURE 9

To see why the Law of Sines holds, drop a perpendicular from the vertex B to the base, and label the point of intersection P. Let h be the vertical distance from B to P, as shown in Figure 9(b).

Since $\triangle APB$ and $\triangle CPB$ are both right triangles, we have

$$h = c\sin\alpha \quad \text{and} \quad h = a\sin\gamma.$$

Eliminating h in the last two equations gives

$$c\sin\alpha = a\sin\gamma, \quad \text{or} \quad \frac{\sin\alpha}{a} = \frac{\sin\gamma}{c}.$$

In a similar manner, we can show that

$$\frac{\sin\beta}{b} = \frac{\sin\alpha}{a} = \frac{\sin\gamma}{c}.$$

EXAMPLE 5 In Example 1 we were given a triangle having sides 6 and 8, with an angle of 60° between them. We used the Law of Cosines to find that the third side had length $2\sqrt{13}$. Determine the angles α and β in Figure 10.

Solution If we know the angle α, then we can easily find β since

$$\beta = 180 - 60 - \alpha = 120 - \alpha \text{ degrees.}$$

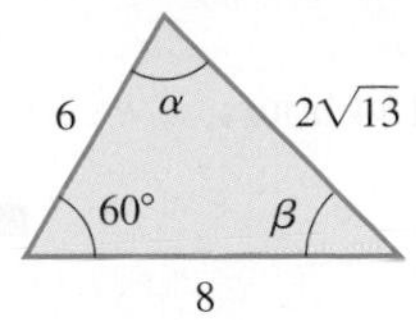

6 α $2\sqrt{13}$ 60° β 8

FIGURE 10

To find α we could use the Law of Cosines to obtain

$$8^2 = 6^2 + (2\sqrt{13})^2 - 2(6)(2\sqrt{13})\cos\alpha.$$

But it is easier to use the Law of Sines, which gives

$$\frac{\sin\alpha}{8} = \frac{\sin 60^\circ}{2\sqrt{13}}, \quad \text{so} \quad \sin\alpha = \frac{8}{2\sqrt{13}} \cdot \frac{\sqrt{3}}{2} = \frac{2\sqrt{39}}{13}.$$

Hence,

$$\alpha = \arcsin\left(\frac{2\sqrt{39}}{13}\right) \approx 1.2898 \approx 73.9^\circ \quad \text{and} \quad \beta = 120 - \alpha \approx 46.1^\circ.$$ ■

EXAMPLE 6 The aircraft carrier *Carl Vinson* leaves the Pearl Harbor naval shipyard in Hawaii and heads due west at 28 knots. A helicopter is 175 nautical miles from the carrier at an angle of 35° south of west.

a. What course should the helicopter travel at its cruising speed of 130 knots to intercept the aircraft carrier?

b. How long will it take the helicopter to reach the carrier?

Solution Figure 11 gives an illustration of the situation, assuming that the intersection point occurs at time t, which at present is unknown.

a. First we will find the angle θ that gives the course the helicopter should fly. The Law of Sines implies that

$$\frac{\sin\theta}{28t} = \frac{\sin 35^\circ}{130t},$$

so

$$\sin\theta = \frac{28t}{130t}\sin 35^\circ = \frac{28}{130}\sin 35^\circ \approx 0.1235, \quad \text{and} \quad \theta = \arcsin(0.1235) \approx 7^\circ.$$

The helicopter should consequently fly a course that is approximately $35+7 = 42^\circ$ to the north of east, as shown in Figure 11(b).

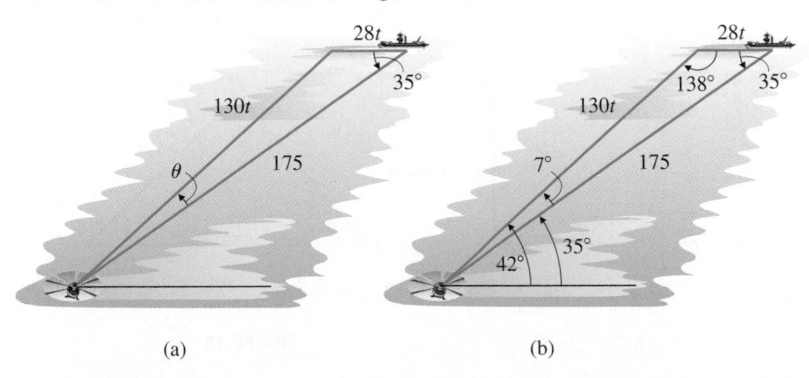

FIGURE 11

b. To determine the time required to reach the carrier, we again use the Law of Sines, but now we involve the side that gives the original distance between the carrier and the helicopter, as shown in Figure 11(b). Because the remaining angle of the triangle has measure $180 - 35 - 7 = 138°$,

$$\frac{\sin 138°}{175} = \frac{\sin 35°}{130t} \quad \text{and} \quad t = \frac{175 \sin 35°}{130 \sin 138°} \approx 1.154 \text{ hours},$$

or approximately 1 hour and 9 minutes. ■

The Law of Sines is easier to apply than the Law of Cosines, but it can lead to some interesting situations when the given information consists of an angle and two sides, one of which is opposite the given angle. In Figure 12(a) we present a typical situation, where the parts of the triangle are shown in red, that is, the angle α, the length a of the side opposite of α, and the length c of a side adjacent to α. What happens depends on the value a.

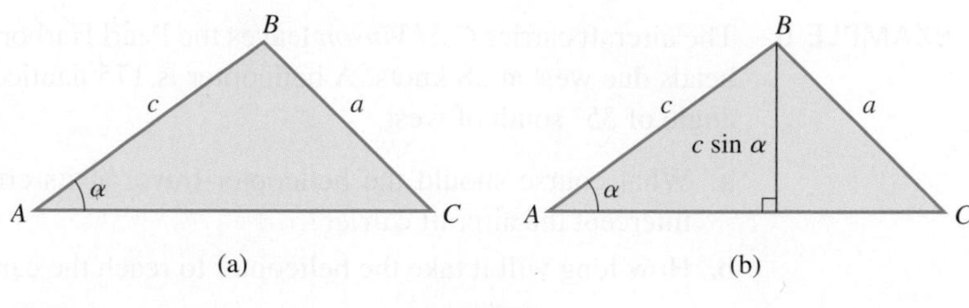

FIGURE 12

Drop a perpendicular line from B to the base, as in Figure 12(b), and notice that the length of this vertical line segment is $c \sin \alpha$.

- If we were given $a < c \sin \alpha$, there would be no triangle formed, as shown in Figure 13(a).
- If $a = c \sin \alpha$, there is precisely one triangle that meets the requirement, the right triangle shown in Figure 13(b).
- If $a \geq c$, there is also only one possibility, as shown in Figure 13(c).
- But if $c \sin \alpha < a < c$, there are two possibilities, $\triangle ABC$ and $\triangle ABC'$ in Figure 13(d). Since $\triangle C'BC$ in this figure is isosceles, $\gamma' = \pi - \gamma$.

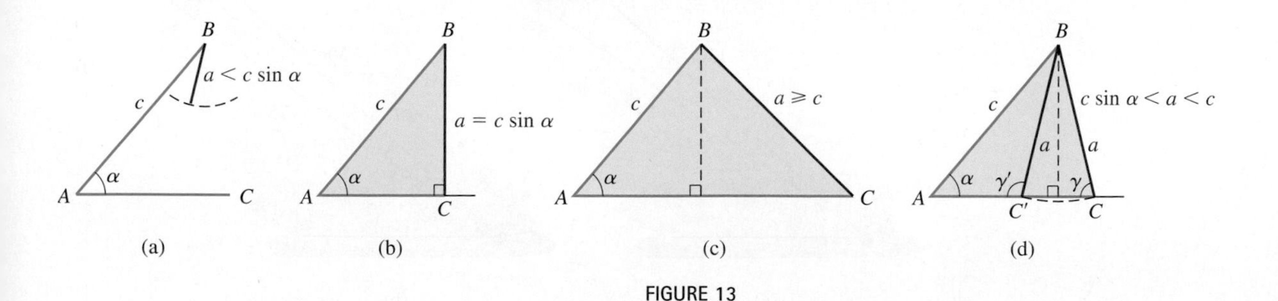

FIGURE 13

Generally the physical circumstances of the problem will dictate the required situation, but it is important to check the reasonableness of the solution to the problem involving this application of the Law of Sines.

EXAMPLE 7 A campground lies at the west end of an east–west road in a relatively flat, but dense, forest. The starting point for a hike lies 30 kilometers to the northeast of the campground. A hiker begins at the starting point and travels in the general direction of the campground, reaching the road after 25 kilometers. Approximately how far is the hiker from the campground after reaching the road?

Solution The situation facing the hiker is illustrated in Figure 14, where the campground is located at A and the hiker begins at B. We assume that $\triangle ABC$ gives the correct solution to the problem, since traveling along the line BC' would put the hiker badly off course.

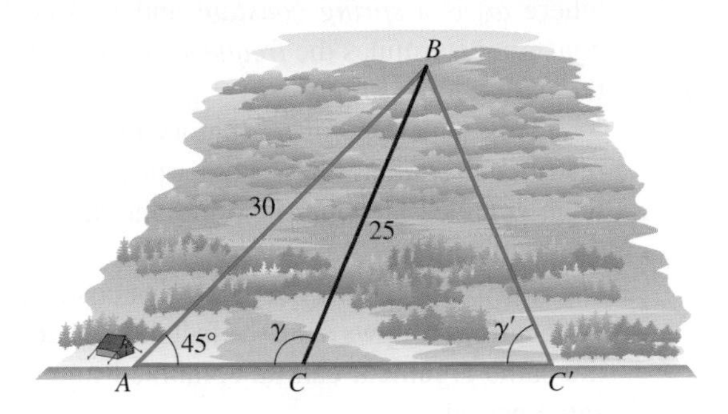

FIGURE 14

Our first step is to use the Law of Sines to determine the angle γ. Since

$$\frac{\sin 45^\circ}{25} = \frac{\sin \gamma}{30}, \quad \text{we have} \quad \sin \gamma = \frac{30 \sin 45^\circ}{25} = \frac{3\sqrt{2}}{5}.$$

So

$$\gamma = \arcsin \frac{3\sqrt{2}}{5} \approx 58^\circ \quad \text{or} \quad \gamma \approx 180^\circ - 58^\circ = 122^\circ.$$

The larger angle is correct, unless the hiker is lost, so the angle at B is $180^\circ - 122^\circ - 45^\circ = 13^\circ$. Using the Law of Sines again we can find AC from

$$\frac{\sin 45^\circ}{25} = \frac{\sin 13^\circ}{AC}, \quad \text{so} \quad AC = \frac{25 \sin 13^\circ}{\sin 45^\circ} \approx 7.95 \text{ kilometers}.$$

If the hiker is lost, but measured the distance correctly, the angle at B is $180^\circ - 58^\circ - 45^\circ = 77^\circ$, and the distance to the camp is

$$AC' = \frac{25 \sin 77^\circ}{\sin 45^\circ} \approx 34.5 \text{ kilometers}.$$

■

EXAMPLE 8 Is there a triangle that satisfies $a = 2$, $c = 3$, and $\alpha = 55^\circ$?

Solution If there is such a triangle, then the Law of Sines implies that

$$\frac{\sin \alpha}{a} = \frac{\sin \gamma}{c}.$$

Therefore

$$\frac{\sin 55^\circ}{2} = \frac{\sin \gamma}{3} \quad \text{and} \quad \sin \gamma = \frac{3}{2} \sin 55^\circ \approx 1.23.$$

The sine of an angle can never exceed 1, so a triangle with these properties cannot be constructed, as illustrated in Figure 15. ■

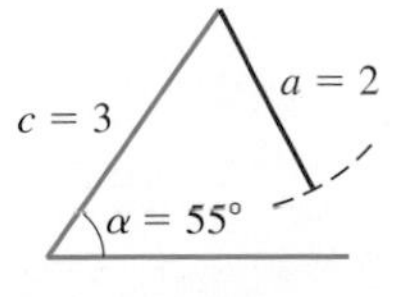

FIGURE 15

Simple Harmonic Motion

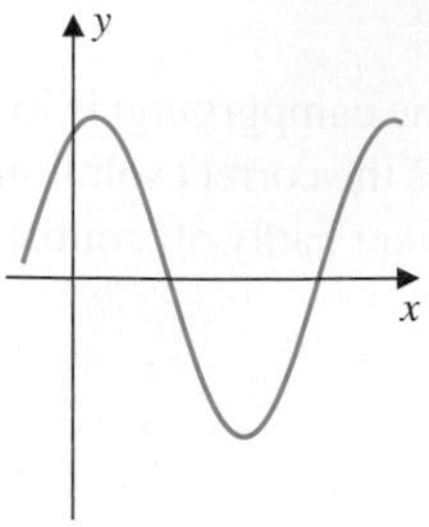

Oscillating motion is common in physical problems. For example, if we ignore friction, a weight attached to a spring that is displaced from its equilibrium position will vibrate in a oscillatory manner described by

$$y(t) = A \sin \omega_0 t + B \cos \omega_0 t,$$

where ω_0 is a *spring constant* and $y(t)$ is the displacement at time t. The spring constant determines the *frequency* of the motion, which is $\omega_0/(2\pi)$, the reciprocal of the period. (See Figure 16.)

FIGURE 16

This same type of oscillatory motion is common in electrical circuits, musical instruments, and numerous other physical applications. All these phenomena exhibit what is called *harmonic* motion and are described using combinations of sine and cosine functions.

One of the features of the sine or cosine function that finds regular use in physical applications follows from the fact that linear combinations of these functions that have the same argument can be combined into a single sine or cosine function with the same period.

Musical tones are described mathematically using equations for harmonic motion. This connection between mathematics and music has been known since the time of Pythagoras, over 2500 years ago.

Sine Combination Formula

For every real number x and nonzero constants A and B, an expression of the form $A \sin x + B \cos x$ can be written as

$$A \sin x + B \cos x = C \sin(x + \delta),$$

where

$$C^2 = A^2 + B^2 \quad \text{and} \quad \delta = \arctan \frac{B}{A}.$$

This relationship can be seen by examining the right triangle shown in Figure 17. In this triangle we have

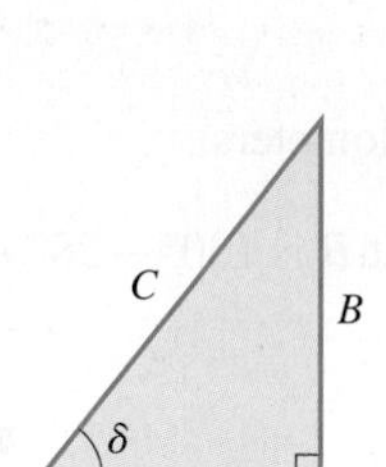

FIGURE 17

$$A = C \cos \delta \quad \text{and} \quad B = C \sin \delta,$$

which gives

$$\begin{aligned} A \sin x + B \cos x &= C \cos \delta \sin x + C \sin \delta \cos x \\ &= C(\sin x \cos \delta + \sin \delta \cos x) = C \sin(x + \delta). \end{aligned}$$

In addition,

$$A^2 + B^2 = C^2,$$

and

$$\frac{B}{A} = \frac{C \sin \delta}{C \cos \delta} = \tan \delta, \quad \text{so} \quad \delta = \arctan \frac{B}{A}.$$

There is a similar relationship involving the cosine function.

Cosine Combination Formula

For every real number x and nonzero constants A and B, an expression of the form $A \sin x + B \cos x$ can be written as

$$A \sin x + B \cos x = C \cos(x - \delta),$$

where

$$C^2 = A^2 + B^2 \quad \text{and} \quad \delta = \arctan \frac{A}{B}.$$

EXAMPLE 9 Suppose that we have a spring with the spring constant k and a system with mass m, an initial position y_0, and a velocity v_0, as shown in Figure 18. Some basic rules of physics tell us that the motion of the mass at the end of the spring is given by

$$y(t) = v_0\sqrt{\frac{m}{k}} \sin \sqrt{\frac{k}{m}}t + y_0 \cos \sqrt{\frac{k}{m}}t.$$

Sketch the graph of the motion of a spring–mass system that has a mass $m = 16$ kilograms, a spring constant $k = 1$ newtons per meter, an initial position $y_0 = 3$ meters, and an initial velocity of 1 meter per second.

Solution Since $\sqrt{m/k} = \sqrt{16} = 4$, the motion of the system is given by the equation

$$y(t) = 4 \sin \frac{t}{4} + 3 \cos \frac{t}{4}.$$

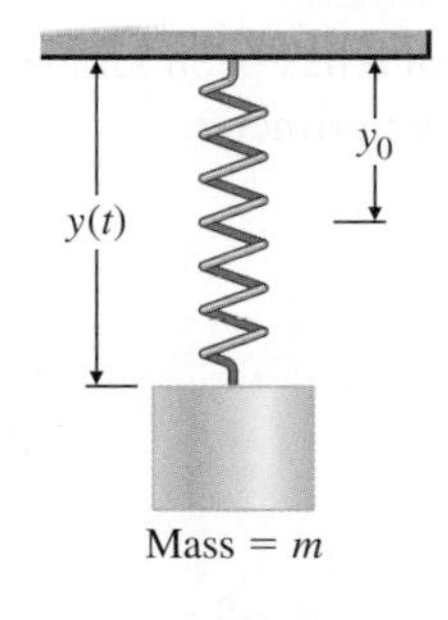

FIGURE 18

This can be reexpressed, using the sine combination formula, as

$$y(t) = \sqrt{4^2 + 3^2} \sin \left(\frac{t}{4} + \arctan \frac{3}{4}\right) = 5 \sin \frac{1}{4} \left(t + 4 \arctan \frac{3}{4}\right).$$

To sketch this graph we first sketch the graph of $y = 5 \sin t$, which is shown in Figure 19(a). We expand the graph of $y = 5 \sin t$ horizontally by a factor of 4 to produce the graph of $y = 5 \sin(t/4)$ shown in Figure 19(b). The graph of

$$y(t) = 4 \sin \frac{t}{4} + 3 \cos \frac{t}{4} = 5 \sin \frac{1}{4} \left(t + 4 \arctan \frac{3}{4}\right)$$

is the shift to the left $4 \arctan\left(\frac{3}{4}\right) \approx 2.57$ units, as shown in Figure 19(c). ■

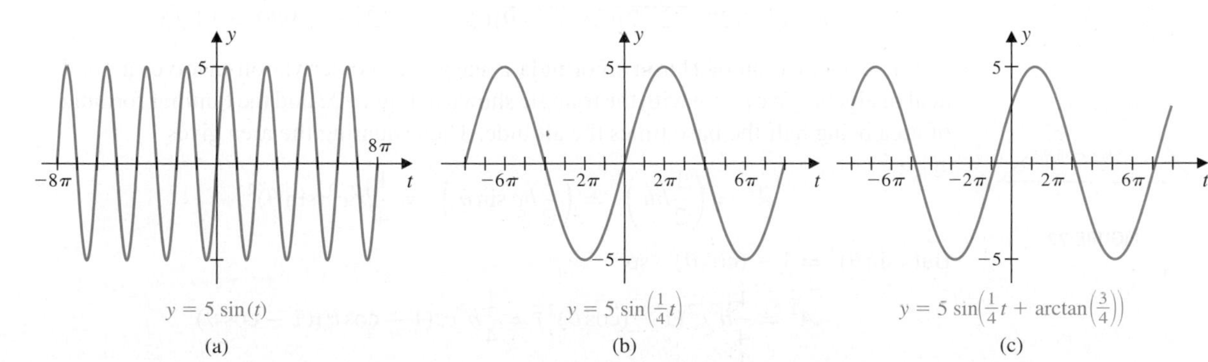

FIGURE 19

Heron's Formula

Heron's Formula provides an alternative method for finding the area of a triangle. The formula is very useful when only the lengths of the sides are known. The derivation of the formula requires the application of the Law of Cosines.

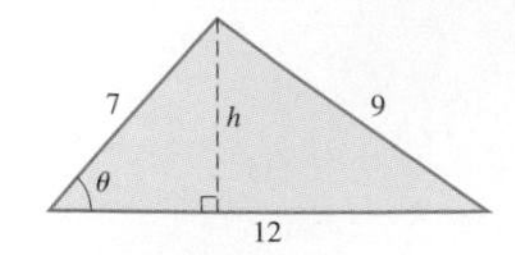

FIGURE 20

If you ask most people for a formula for the area of a triangle, a typical response is "take one-half the base times the altitude." But this is an awkward formula to use when only the lengths of the sides of a triangle are known. Consider, for example, the triangle shown in Figure 20, whose sides have lengths 7, 9, and 12.

The altitude of the triangle is

$$h = 7 \sin \theta.$$

To determine $\sin \theta$, we first use the Law of Cosines to determine $\cos \theta$. This gives

$$9^2 = 7^2 + 12^2 - 2(7 \cdot 12) \cos \theta, \quad \text{so} \quad \cos \theta = \frac{49 + 144 - 81}{2(84)} = \frac{2}{3}.$$

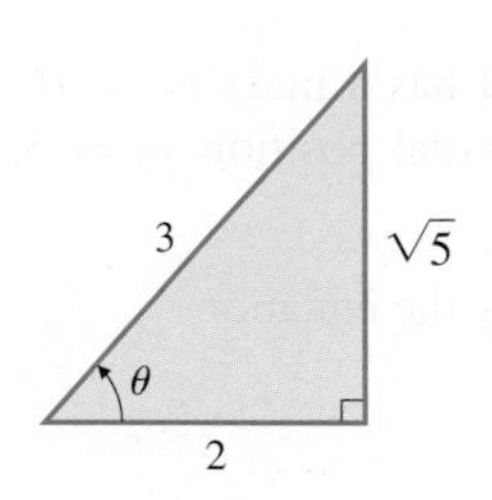

FIGURE 21

Consider the right triangle in Figure 21, which also has $\cos \theta = 2/3$. This triangle is similar to the right triangle to the left of the altitude in Figure 20, so in each case $\sin \theta = \sqrt{5}/3$. Hence the area of the triangle in Figure 20 is

$$\mathcal{A} = \frac{1}{2}(12)h = \frac{1}{2}(12)(7 \sin \theta) = 42 \cdot \frac{\sqrt{5}}{3} = 14\sqrt{5}.$$

There is an easier formula to apply in this situation, one which has been known for over 2000 years and uses only the lengths of the sides and the perimeter.

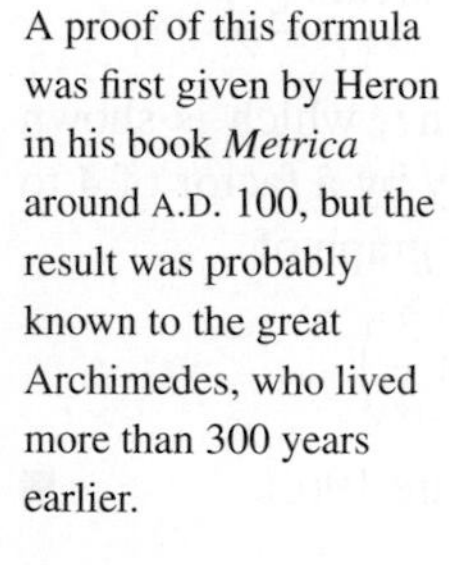
A proof of this formula was first given by Heron in his book *Metrica* around A.D. 100, but the result was probably known to the great Archimedes, who lived more than 300 years earlier.

Heron's Formula

A triangle with sides of length a, b, and c has area given by

$$\mathcal{A} = \tfrac{1}{4}\sqrt{P(P-2a)(P-2b)(P-2c)},$$

where $P = a + b + c$ is the perimeter of the triangle.

It is much easier to use Heron's Formula to find the area of the triangle in Figure 20. The triangle has a perimeter $P = 7 + 9 + 12 = 28$, so Heron's formula implies that the area is

$$\mathcal{A} = \tfrac{1}{4}\sqrt{28(28 - 2 \cdot 7)(28 - 2 \cdot 9)(28 - 2 \cdot 12)} = \sqrt{980} = 14\sqrt{5}.$$

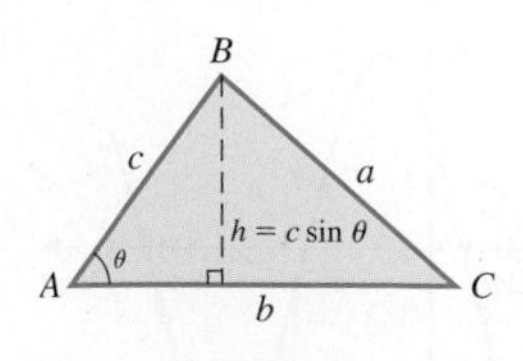

FIGURE 22

The application of Heron's Formula is easy, but the derivation involves a good deal of algebra. We begin with the triangle shown in Figure 22 and the familiar formula of area being half the base times the altitude. Then squaring the area gives

$$\mathcal{A}^2 = \left(\frac{1}{2}bh\right)^2 = \left(\frac{1}{2}bc \sin \theta\right)^2 = \frac{1}{4}b^2c^2(\sin \theta)^2.$$

But $(\sin \theta)^2 = 1 - (\cos \theta)^2$, so

$$\begin{aligned} \mathcal{A}^2 &= \frac{1}{4}b^2c^2(1 - (\cos \theta)^2) = \frac{1}{4}b^2c^2(1 + \cos \theta)(1 - \cos \theta) \\ &= \frac{1}{4}(bc + bc \cos \theta)(bc - bc \cos \theta). \end{aligned}$$

Now the Law of Cosines comes into play. Since

$$a^2 = b^2 + c^2 - 2bc\cos\theta \quad \text{we have} \quad bc\cos\theta = \frac{1}{2}(b^2 + c^2 - a^2).$$

Hence,

$$\begin{aligned} \mathcal{A}^2 &= \frac{1}{4}\left(bc + \frac{1}{2}(b^2 + c^2 - a^2)\right)\left(bc - \frac{1}{2}(b^2 + c^2 - a^2)\right) \\ &= \frac{1}{16}(2bc + b^2 + c^2 - a^2)(2bc - b^2 - c^2 + a^2). \end{aligned}$$

By regrouping and factoring, we can rewrite this as

$$\begin{aligned} \mathcal{A}^2 &= -\frac{1}{16}(a^2 - (b^2 + 2bc + c^2))(a^2 - (b^2 - 2bc + c^2)) \\ &= -\frac{1}{16}(a^2 - (b + c)^2)(a^2 - (b - c)^2). \end{aligned}$$

Factoring this last formula gives

$$\begin{aligned} \mathcal{A}^2 &= -\frac{1}{16}(a + (b + c))(a - (b + c))(a - (b - c))(a + (b - c)) \\ &= \frac{1}{16}(a + b + c)(b + c - a)(a + c - b)(a + b - c). \end{aligned}$$

Suppose we now add and subtract a, b, and c, respectively, in each of the last three factors. Then

$$\begin{aligned} \mathcal{A}^2 &= \frac{1}{16}(a + b + c)(a + b + c - 2a)(a + b + c - 2b)(a + b + c - 2c) \\ &= \frac{1}{16}(P)(P - 2a)(P - 2b)(P - 2c), \end{aligned}$$

so

$$\mathcal{A} = \frac{1}{4}\sqrt{P(P - 2a)(P - 2b)(P - 2c)}.$$

EXERCISE SET 4.9

In Exercises 1–4, let the angles of a triangle be α, β, and γ, with opposite sides of length a, b, and c, respectively. Use the Law of Cosines to find the remaining side and one of the other angles.

1. $\alpha = 55°;\ b = 12;\ c = 20$

2. $\gamma = 115°;\ a = 14;\ b = 18$

3. $\beta = 30°;\ a = 25;\ c = 32$

4. $\alpha = 60°;\ b = 50;\ c = 35$

In Exercises 5–8, let the angles of a triangle be α, β, and γ, with opposite sides of length a, b, and c, respectively. Use the Law of Sines to find the remaining sides.

5. $\alpha = 40°;\ \beta = 87°;\ c = 115$

6. $\alpha = 70°;\ \beta = 38°;\ a = 35$

7. $\beta = 100°;\ \gamma = 30°;\ c = 20$

8. $\alpha = 65°;\ \gamma = 50°;\ b = 10$

In Exercises 9–12, let the angles of a triangle be α, β, and γ, with opposite sides of length a, b, and c, respectively. Use the Law of Cosines and the Law of Sines to find the remaining parts of the triangle.

9. $\alpha = 130°;\ b = 5;\ c = 7$

10. $\beta = 53.5°;\ a = 9;\ c = 12.5$

11. $a = 8;\ b = 9;\ c = 13$

12. $a = 3;\ b = 5;\ c = 7$

13. Show that there is no triangle satisfying the conditions $a = 3$, $b = 10$, and $\alpha = 25.4°$.

14. Is there a triangle that satisfies $a = 2$, $b = 11$, and $\alpha = 63°$?

15. Solve for the missing parts of the triangle satisfying $a = 125$, $b = 150$, $\alpha = 55°$. (There are two solutions.)

16. Solve for the missing parts of the triangle satisfying $a = 4$, $b = 5$, and $\alpha = 53°$.

17. Find the area of the triangle with sides of lengths 12 centimeters, 14 centimeters, and 16 centimeters.

18. A triangular lot has sides of lengths 325 feet, 175 feet, and 200 feet. Find the area of the lot.

19. A gas pipeline is to be constructed between towns A and B. The engineers have two alternatives. They can connect A and B directly, but then they must build the pipeline through a swamp. Alternatively, they can build the pipeline from town A to town C, which is 3 miles directly west of A, and then to town B, which is 2 miles directly northwest of C. The cost of construction through the swamp from A to B is \$125,000 per mile and the cost to go through C is \$100,000 per mile.

a. Which alternative should the engineers select?

b. At what cost per mile for construction through C would there be no price difference in the two alternatives?

20. A surveyor needs to determine the distance across a pond and makes the measurements shown in the figure. What is the distance from A to B?

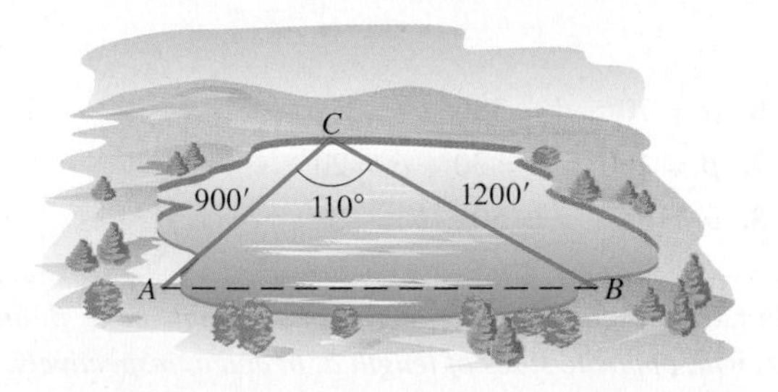

21. Points A and B are on opposite sides of a river. To find the distance between the points, a third point C is located on the same side of the river as point A. The distance between A and C is 45 feet, $\angle ACB$ is determined to be 42°, and $\angle BAC$ is 105°. Find the distance between A and B.

22. A boat leaves port and travels at a bearing S 48° E (48° south of east) at 15 miles per hour. A Coast Guard cutter is located 23 miles due east of the port. If the cutter can average 25 miles per hour, what bearing should it travel in order to intercept the boat? When will the cutter intercept the boat?

23. Two ships leave port at 10:00 A.M. Ship #1 travels at a bearing of N 62° E (62° north of east) at 20 miles per hour and Ship #2 travels at a bearing of S 75° E at 25 miles per hour. How far apart are the ships at noon?

24. In tracking the relative location of two aircraft, a controller determines that the distance from the control tower to the first aircraft is 150 miles and the distance to the second is 100 miles. The angle between the two aircraft is 50°. How far apart are they?

25. The National Forest Service maintains observation towers to check for the outbreak of forest fires. Suppose two towers are at the same elevation, one at point A and another 10 miles due west at a point B (see the figure).

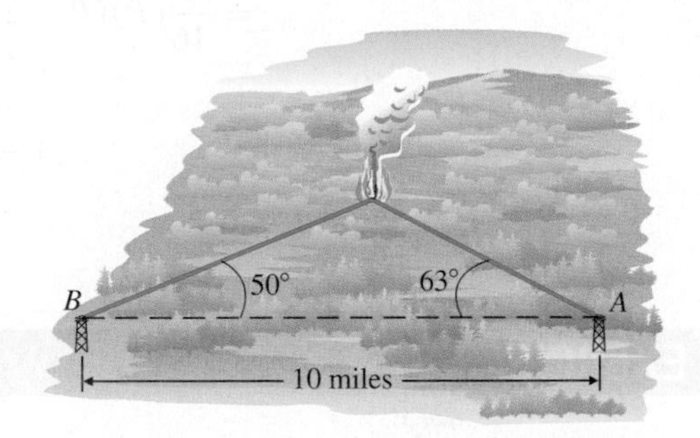

The ranger at A spots a fire in the northwest whose line of sight makes an angle of 63° with the line between the towers, and contacts the ranger at B. This ranger locates the fire along a line of sight that makes a 50° angle with the line between the towers. How far is the fire from the tower at B?

26. The lengths of the sides of a triangular parcel of land are approximately 200 feet, 300 feet, and 450 feet. If land is valued at \$2000 per acre (1 acre is 43,560 ft^2), what is the value of the parcel of land?

REVIEW EXERCISES FOR CHAPTER 4

In Exercises 1–6, find **(a)** $P(t)$, *the terminal point on the unit circle determined by* t; **(b)** *the reference number for* t; *and* **(c)** *the values of the six trigonometric functions of* t.

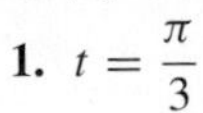

1. $t = \dfrac{\pi}{3}$ **2.** $t = \dfrac{5\pi}{3}$

3. $t = \dfrac{5\pi}{4}$ **4.** $t = \dfrac{8\pi}{3}$

5. $t = -\dfrac{19\pi}{6}$

6. $t = -\dfrac{23\pi}{3}$

In Exercises 7–10, find the values of all the trigonometric functions from the given information.

7. $\cos t = \dfrac{3}{5}, \ \dfrac{3\pi}{2} < t < 2\pi$

8. $\sin t = -\frac{1}{2}, \ \cos t < 0$

9. $\tan t = \dfrac{1}{4}, \ 0 < t < \dfrac{\pi}{2}$

10. $\sec t = -5, \ \dfrac{\pi}{2} < t < \pi$

In Exercises 11–18, find all values of x in $[0, \pi]$ that satisfy the given equation.

11. $\cos \dfrac{x}{3} = \dfrac{1}{2}$ **12.** $\sin 4x = -\dfrac{\sqrt{3}}{2}$

13. $2(\sin x)^2 - 3 \sin x + 1 = 0$

14. $2(\cos x)^2 - \sin x - 1 = 0$

15. $(\tan x)^3 - 4 \tan x = 0$

16. $2 \tan x - 3 \cot x = 0$

17. $\cot x - \csc x = 1$

18. $\sin 2x + \cos x = 0$

In Exercises 19–22, determine whether the function is even, odd, or neither.

19. $f(x) = (\sin x)^3$ **20.** $f(x) = \sin x \cos x$

21. $f(x) = x^2(\cos x)^2$ **22.** $f(x) = x^5 + \cos x$

23. Match the equation with the graph.

a. $y = \sin 2\left(x + \dfrac{\pi}{2}\right)$

b. $y = \sin \dfrac{1}{2}\left(x + \dfrac{\pi}{2}\right)$

c. $y = \cos 2\left(x + \dfrac{\pi}{2}\right)$

d. $y = \cos \dfrac{1}{2}\left(x + \dfrac{\pi}{2}\right)$

i.

ii.

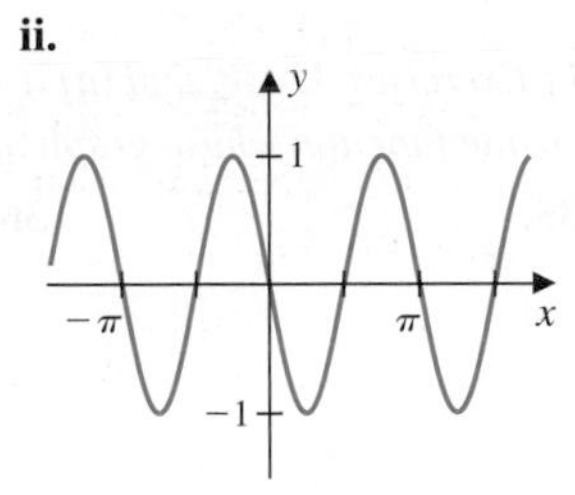

iii.

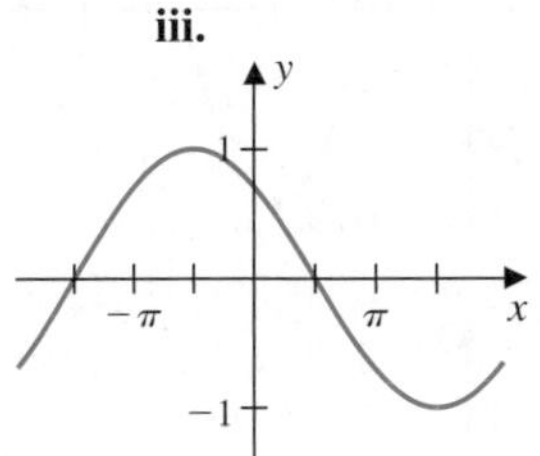

iv.

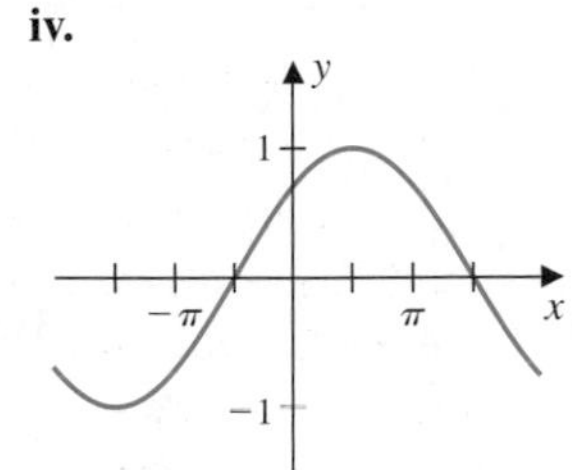

24. Match the equation with the graph.

a. $y = \sin(x + \pi)$

b. $y = 1 - \sin(x - \pi)$

c. $y = 1 - \cos(x - \pi)$

d. $y = -\cos(x + \pi)$

i.

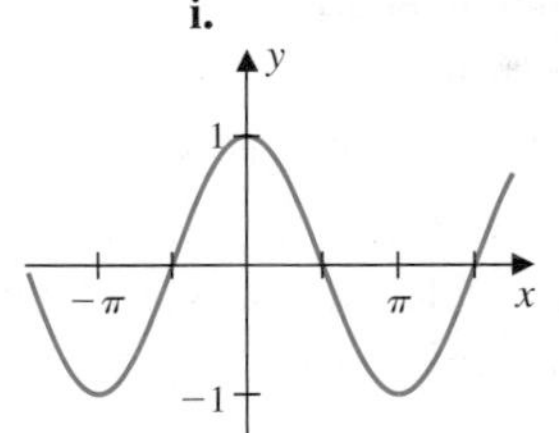

ii.

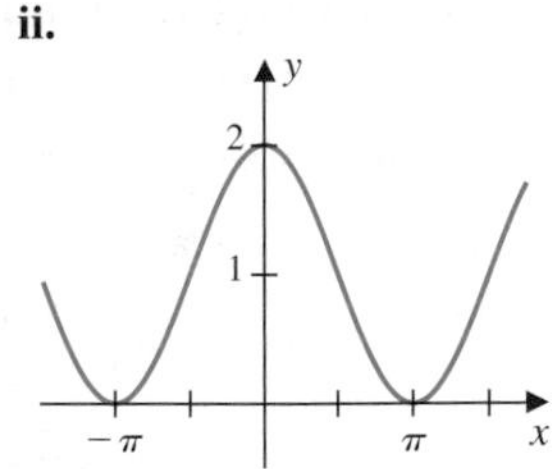

iii.

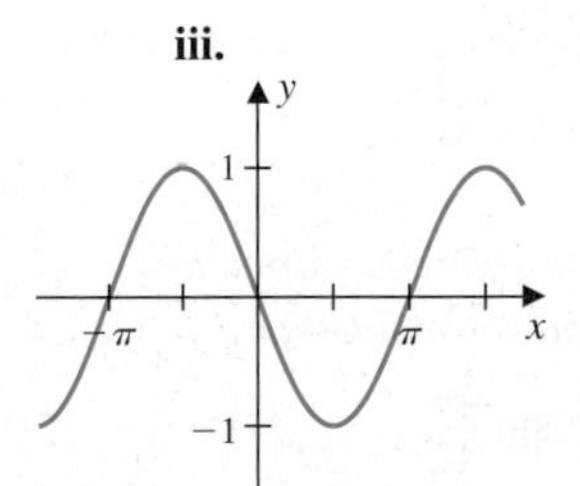

iv.

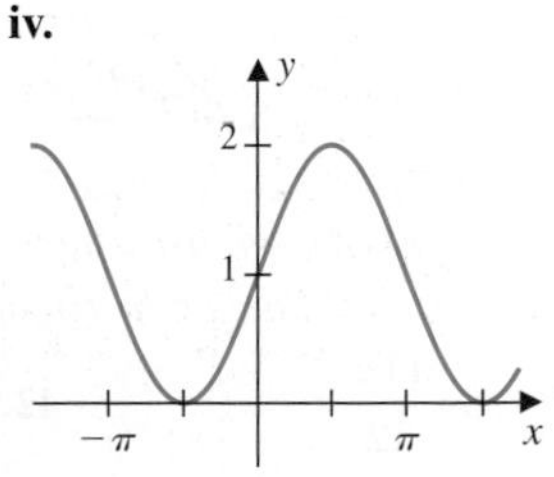

In Exercises 25–34, sketch one period of the graph of the function.

25. $y = 3 \sin \frac{1}{3}x$ **26.** $y = 4 \cos 4\pi x$

27. $y = -3 \cos 2x$ **28.** $y = 2 + 4 \sin 4x$

29. $y = \cos(3x - \pi)$ **30.** $y = -3 \cos\left(2x - \dfrac{\pi}{3}\right)$

31. $y = \cot\left(x + \frac{\pi}{6}\right)$

32. $y = -\tan\left(x - \frac{\pi}{2}\right)$

33. $y = \sec(3\pi x)$

34. $y = -3\csc\left(x - \frac{\pi}{2}\right)$

In Exercises 35–38, find ***(a)*** *a sine function and* ***(b)*** *a cosine function whose graph matches the given curve.*

35.

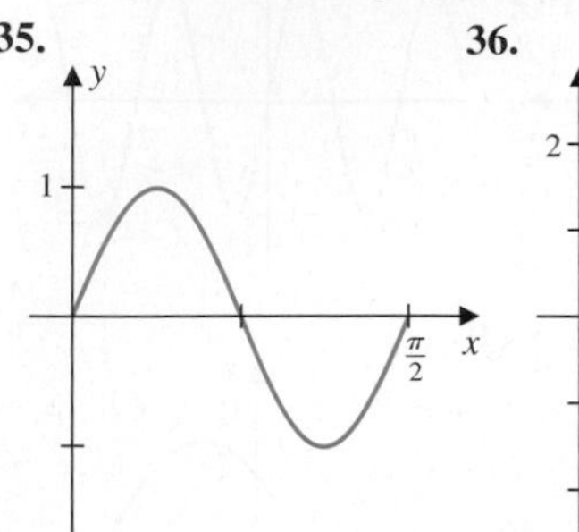

36.

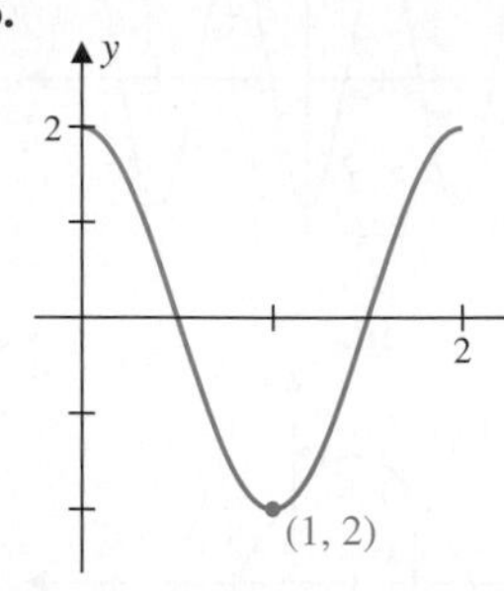

37.

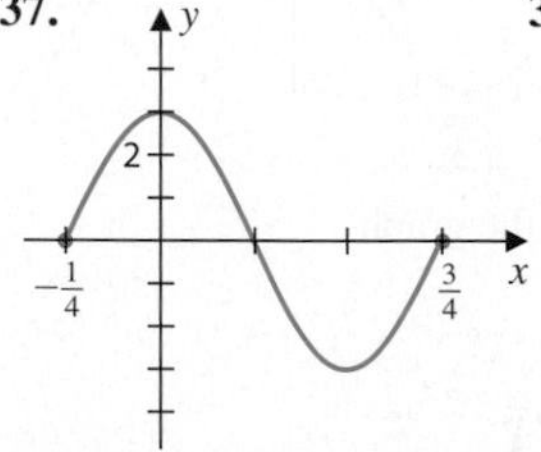

38.

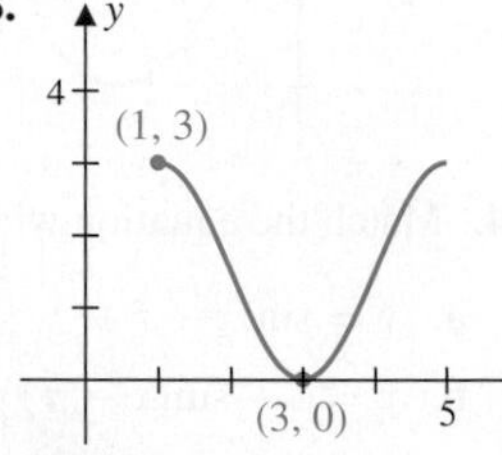

In Exercises 39 and 40, refer to the right triangle in the figure. ***(a)*** *From the information given, find any missing angles or sides.* ***(b)*** *Find the values of the six trigonometric functions of the angle α.*

39. $\alpha = 60°; AC = 13$

40. $\alpha = 45°; BC = 12$

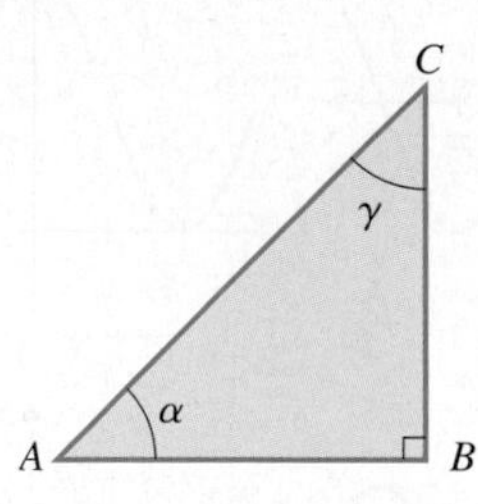

In Exercises 41–44, use a sum or difference formula to find the exact value of the trigonometric function.

41. $\cos\frac{11\pi}{12}$

42. $\sin\frac{5\pi}{12}$

43. $\cos\left(-\frac{13\pi}{12}\right)$

44. $\tan\left(-\frac{\pi}{12}\right)$

In Exercises 45–48, use a double-angle formula to find the exact value of the trigonometric function.

45. $\sin\frac{5\pi}{8}$

46. $\cos\frac{11\pi}{12}$

47. $\tan\frac{7\pi}{12}$

48. $\cot\frac{3\pi}{8}$

In Exercises 49–52, use the addition and subtraction formulas to verify the given formula.

49. $\sin\left(t - \frac{\pi}{2}\right) = -\cos t$

50. $\cos\left(t - \frac{\pi}{2}\right) = \sin t$

51. $\sin\left(\frac{3\pi}{2} - t\right) = -\cos t$

52. $\cos\left(\frac{3\pi}{2} - t\right) = -\sin t$

In Exercises 53–56, rewrite the expression so that it involves the sum or difference of only constants and sines and cosines to the first power.

53. $(\cos 5x)^2$

54. $(\sin 6x)^2$

55. $(\sin x)^4$

56. $(\sin 2x)^2(\cos 2x)^2$

In Exercises 57–60, rewrite each product as a sum or difference.

57. $\sin 3t \cos 4t$

58. $\cos 8t \sin 4t$

59. $\cos 2t \cos 4t$

60. $\sin 3t \sin 5t$

In Exercises 61–64, rewrite each sum or difference as a product.

61. $\sin 2t + \sin 6t$

62. $\sin 3t + \sin 7t$

63. $\cos 4t + \cos 2t$

64. $\cos 5t - \cos 3t$

In Exercises 65–68, verify the identities.

65. $(\cos x)^4 - (\sin x)^4 = \cos 2x$

66. $(\sin x - \cos x)^2 = 1 - \sin 2x$

67. $\frac{\sin x}{1 - \cos x} = \cot x + \csc x$

68. $\frac{\cos x}{1 - \tan x} + \frac{\sin x}{1 - \cot x} = \sin x + \cos x$

In Exercises 69–72, find the exact value of the quantity.

69. $\sin(\arctan(\sqrt{3}))$

70. $\cos\left(\arcsin\left(\frac{4}{5}\right)\right)$

71. $\sin\left(\arcsin\left(\frac{3}{5}\right) - \arcsin\left(\frac{5}{13}\right)\right)$

72. $\tan\left(\arccos\left(\frac{1}{4}\right) + \arcsin\left(\frac{1}{2}\right)\right)$

In Exercises 73–78, assume the angles of the triangle are α, β, and γ with opposite sides of length a, b, and c, respectively. Use either the Law of Cosines or the Law of Sines to find the missing parts of the triangle.

73. $\alpha = 25°;\ b = 12;\ c = 20$

74. $\alpha = 30°;\ \beta = 100°;\ c = 25$

75. $a = 6;\ b = 8;\ c = 10$

76. $a = 65;\ \beta = 30°;\ c = 35$

77. $\beta = 76°;\ \gamma = 50°;\ b = 10.5$

78. $a = 8;\ c = \sqrt{3};\ \gamma = 45°$

In Exercises 79 and 80, use a graphing device to approximate solutions to the equation.

79. $\sin x = x^2$ **80.** $\cos x = x^3$

In Exercises 81 and 82, determine an appropriate viewing rectangle for the function and use it to sketch the graph.

81. $f(x) = 4\cos(125x)$

82. $f(x) = x - 10\sin(25x)$

83. A gutter is to be made from a long strip of tin by bending up the sides so that the base and the sides have the same length b. Express the area of a cross section of the gutter as a function of the angle θ that the outsides make with the base.

84. Two ships A and B leave port at the same time, ship A traveling 45° NE and ship B traveling 45° SE. If ship A is moving at an average speed of 20 miles per hour and ship B at an average speed of 35 miles per hour, how far apart are the ships after 2 hours? What is the bearing of ship B from ship A?

85. An airplane takes off from an airport and travels at a heading of 150°, measured clockwise from north. If the average speed is 380 miles per hour, how far south and east is the plane from the airport after 2.5 hours?

86. From the top of a 200-foot lighthouse, the angle of depression to a ship in the water is 28°. How far is the ship from the lighthouse?

CHAPTER 4: EXERCISES FOR CALCULUS

In Exercises 1 and 2, use a graphing device to sketch the graph of the function, and approximate the absolute maximum and minimum of the function on the specified interval.

1. $f(x) = x - \sin x;\ [-2, 3]$

2. $f(x) = \sin x + \cos x;\ [-\pi, \pi]$

3. Make the indicated trigonometric substitution and simplify the expression, assuming that $a > 0$.

a. $\sqrt{a^2 - u^2},\ u = a\sin t,\quad -\frac{\pi}{2} \le t \le \frac{\pi}{2}$

b. $\sqrt{u^2 + a^2},\ u = a\tan t,\quad -\frac{\pi}{2} < t < \frac{\pi}{2}$

c. $\sqrt{u^2 - a^2},\ u = a\sec t,\quad 0 < t < \frac{\pi}{2}$

d. $\frac{\sqrt{u^2 - a^2}}{u},\ u = a\sec t,\quad 0 < t < \frac{\pi}{2}$

4. Express the area of the rectangle inscribed in a semicircle of radius r in terms of the angle θ shown in the figure.

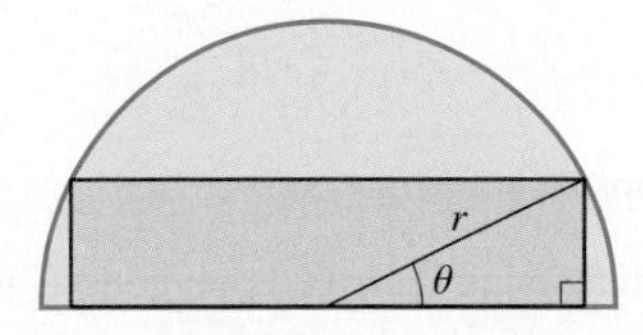

5. Approximate the area of the quadrilateral shown in the figure.

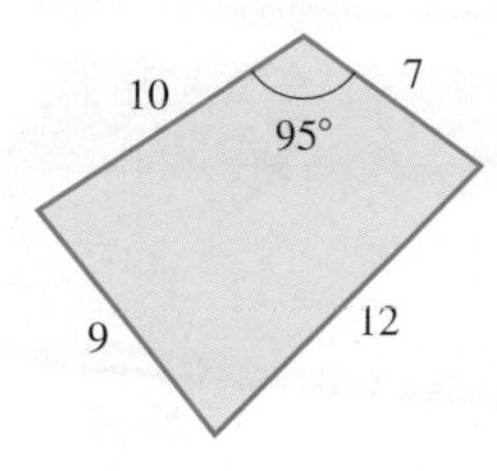

6. Suppose that points A and B are a distance d meters apart and that the angles formed with the horizontal and the top of a hill are α and β, respectively, as shown in the figure. Show that

$$h = \frac{d}{\cot\alpha - \cot\beta}.$$

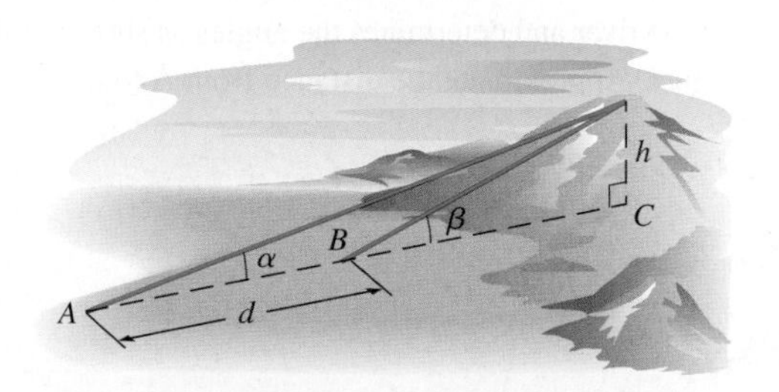

7. A reconnaissance plane is flying at an altitude of 33,000 feet above two ships A and B, as shown in the figure. The angle of depression from the plane to ship A is 32°, and the angle of depression to ship B is 47°.

Approximate the distance between the two ships, rounding your answer to the nearest 100 feet.

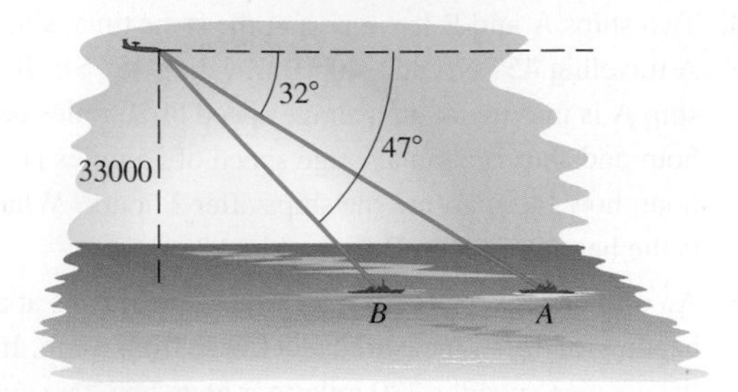

8. A pilot needs to determine the bearing of city B from city A in order to deliver cargo.

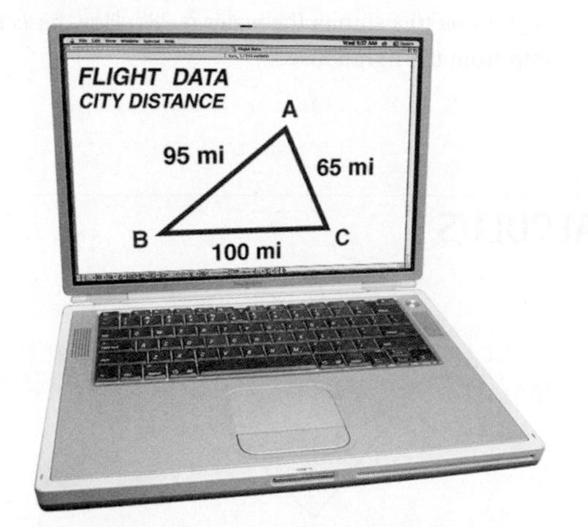

The pilot measures the distances between cities A and B, and a third city, C, that is due east of B, as shown in the figure. What bearing should the pilot fly?

9. To find the distance across a river, a surveyor selects two reference points, A and B, which are 300 feet apart on the same side of the river. The surveyor then selects a third reference point, C, on the opposite side of the river and determines the angles as shown in the figure. Approximate the distance from A to C.

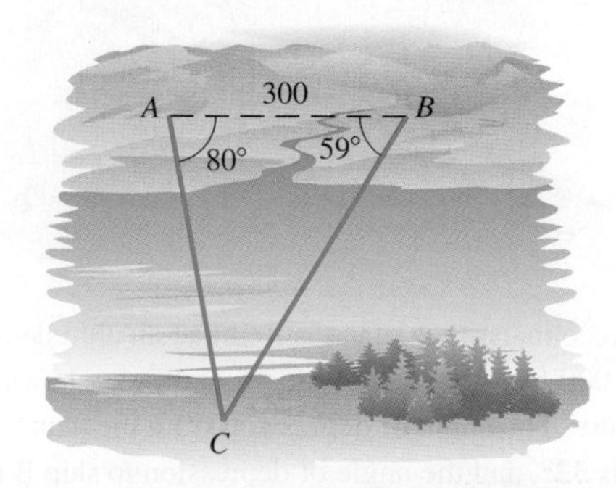

10. a. Use a graphing device to sketch the graphs of

$$y = \frac{\sin x}{x} \quad \text{and} \quad y = \frac{\cos x - 1}{x},$$

and make a conjecture about the values of $\sin x/x$ and $(\cos x - 1)/x$ as $x \to 0$.

b. Let $f(x) = \sin x$ and $g(x) = \cos x$, and use the sum formula for the sine and cosine functions to simplify

$$\frac{f(x+h) - f(x)}{h} = \frac{\sin(x+h) - \sin x}{h}$$

and

$$\frac{g(x+h) - g(x)}{h} = \frac{\cos(x+h) - \cos x}{h}.$$

c. Make a conjecture concerning the difference quotients in part (b) as $x \to 0$.

11. An n-sided regular polygon (all of whose sides are of equal length) inscribed in a circle of radius 1 has been divided into n congruent triangles, each with central angle $2\pi/n$ radians.

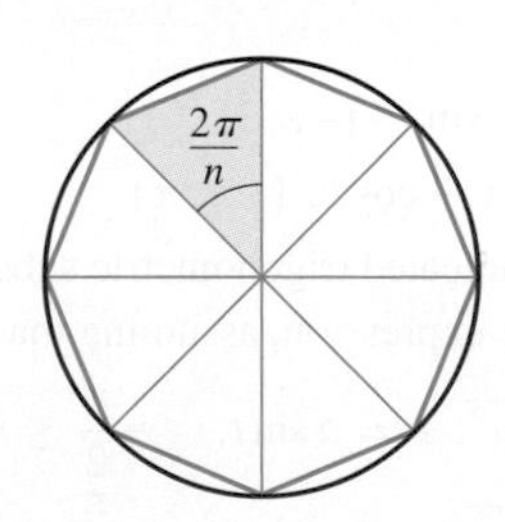

a. Show that the area of each triangle is

$$\frac{1}{2}\sin\frac{2\pi}{n}.$$

b. What value would you expect

$$\frac{n}{2}\sin\frac{2\pi}{n}$$

to approach as $n \to \infty$?

12. Rather than using inscribed polygons as in Exercise 11, circumscribed polygons can be used, as in the accompanying figure, where again the polygon has been divided into n congruent triangles.

a. Show that the area of each triangle is

$$\tan\frac{\pi}{n}.$$

b. What value would you expect

$$n\tan\frac{\pi}{n}$$

to approach as $n \to \infty$?

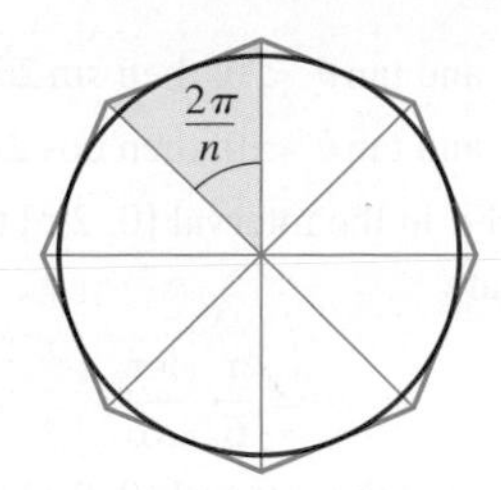

CHAPTER 4: CHAPTER TEST

Determine whether the statement is true or false. If false, describe how the statement might be changed to make it true.

1. The number of degrees in $11\pi/4$ radians is 450°.

2. The number of degrees in $-5\pi/6$ radians is $-210°$.

3. The number of radians in 240° is $\frac{4\pi}{3}$.

4. The number of radians in $-210°$ is $-7\pi/6$.

5. The reference number for $7\pi/4$ is $\pi/4$.

6. The reference angle for 480° is 80°.

7. If $\cos\theta > 0$ and the $\tan\theta < 0$, then θ is in quadrant III.

8. If $\sin\theta > 0$ and the $\sec\theta > 0$, then θ is in quadrant II.

9. If $\sin\theta < 0$ and the $\cot\theta > 0$, then θ is in quadrant III.

10. If $\csc\theta > 0$ and the $\cos\theta > 0$, then θ is in quadrant I.

11. If $\tan\theta > 0$, then θ is in quadrant I or in quadrant II.

12. If $\csc\theta < 0$, then θ is in quadrant II or in quadrant III.

13. If $\theta = 5\pi/6$, then $\sin\theta = \sqrt{3}/2$.

14. If $\theta = -11\pi/4$, then $\sin\theta = \sqrt{2}/2$.

15. If $\theta = -\frac{2\pi}{3}$, then $\cos\theta = -\frac{1}{2}$.

16. If $\theta = 17\pi/6$, then $\cos\theta = -\sqrt{3}/2$.

17. If $\theta = 7\pi/6$, then $\tan\theta = \sqrt{3}/3$.

18. If $\theta = -5\pi/3$, then $\cot\theta = -\sqrt{3}/3$.

19. If $\theta = \pi$, then $\sec\theta = 1$.

20. If $\theta = -5\pi/2$, then $\csc\theta = -1$.

In Exercises 21–24, use the right triangle in the figure.

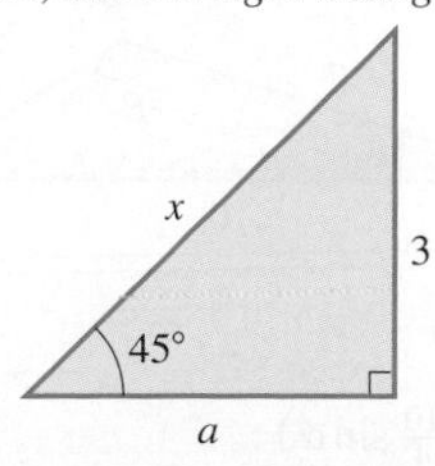

21. The hypotenuse x of the triangle has length $3\sqrt{2}$.

22. The missing angle of the triangle is 30°.

23. The side a of the triangle has length $3\sqrt{2}$.

24. The hypotenuse x of the triangle has length

$$\frac{3}{\cos 45°}.$$

In Exercises 25–28, use the right triangle in the figure.

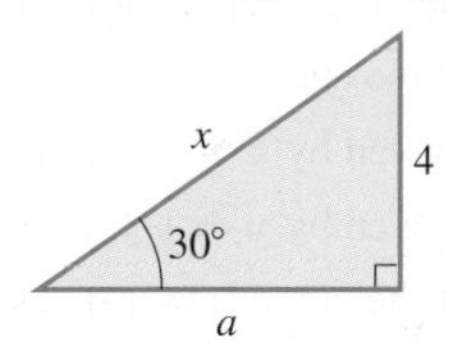

25. The hypotenuse of the triangle is 8.

26. The missing side of the triangle has length $a = 8\cos 30°$.

27. The side a of the triangle has length $4\tan 30°$.

28. The hypotenuse x of the triangle has length $4\csc 60°$.

In Exercises 29–32, use the right triangle in the figure.

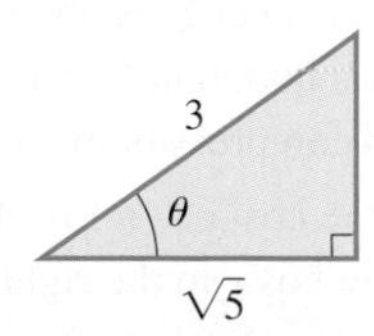

29. $\sin\theta = \frac{2}{3}$

30. $\cos\theta = \frac{2}{3}$

31. $\tan\theta = \frac{\sqrt{5}}{5}$

32. $\sec\theta = \frac{3}{\sqrt{5}}$

33. If $\cos\theta = \frac{4}{5}$ and θ lies in quadrant IV, then $\sin\theta = \frac{3}{5}$.

34. If $\tan\theta = -\frac{\sqrt{3}}{2}$ and θ lies in quadrant II, then $\cos\theta = -\frac{2\sqrt{3}}{3}$.

35. If $\sec\theta = \frac{3}{2}$ and $\tan\theta < 0$, then $\sin 2\theta = -\frac{4\sqrt{5}}{9}$.

36. If $\sec\theta = \frac{3}{2}$ and $\tan\theta < 0$, then $\cos 2\theta = -\frac{1}{9}$.

37. All values of θ in the interval $[0, 2\pi]$ that satisfy $\cos 2\theta = \frac{1}{2}$ are

$$\theta = \frac{\pi}{6}, \frac{5\pi}{6}.$$

38. All values of θ in the interval $[0, 2\pi]$ that satisfy $\sin 3\theta = \frac{1}{2}$ are

$$\theta = \frac{\pi}{18}, \frac{5\pi}{18}, \frac{13\pi}{18}, \frac{17\pi}{18}, \frac{25\pi}{18}, \frac{29\pi}{18}.$$

In Exercises 39–42, use the figure.

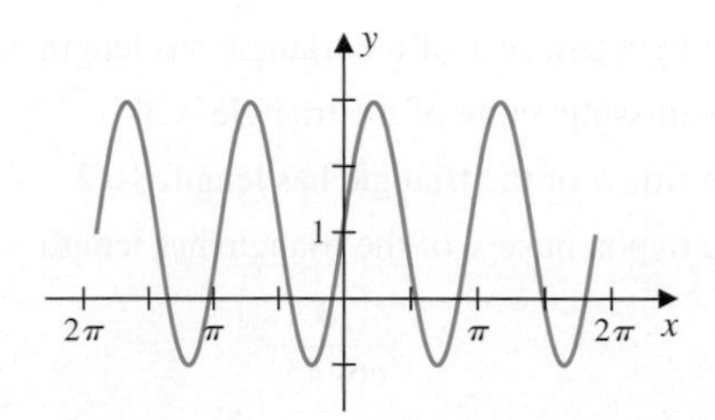

39. The amplitude of the curve shown in the figure is 2.

40. The period of the curve is π.

41. The curve is given by $y = 1 + 2\sin 2x$.

42. The curve is given by $y = 1 + 2\cos(2x - \pi)$.

43. The period of the curve $y = 1 + 2\sin 2x$ is π.

44. The period of the curve $y = \cos\frac{1}{2}x$ is 4π.

45. The amplitude of the curve $y = -2 + 3\sin 2x$ is 3.

46. The amplitude of the curve $y = -2\cos\frac{1}{2}x$ is 2.

47. The graph of $y = 2\cos\frac{1}{2}x$ is obtained from the graph of $y = \cos x$ through a horizontal stretching by a factor of $\frac{1}{2}$ and a vertical stretching by a factor of 2.

48. The graph of $y = \frac{1}{2}\sin 2x$ is obtained from the graph of $y = \sin x$ by a horizontal stretching by a factor of 2 and a vertical compression by a factor of 2.

49. The graph of $y = \cos(x - \pi)$ is obtained by shifting the graph of $y = \cos x$ to the right π units.

50. The graph of $y = -\sin(x + \pi)$ is obtained by shifting the graph of $y = \sin x$ to the left π units and reflecting the result about the x-axis.

51. The graph of $y = \frac{1}{2}\sin\left(x - \frac{\pi}{4}\right)$ is obtained from the graph of $y = \sin x$ from a vertical compression by a factor of 2 followed by a shift to the right of $\frac{\pi}{4}$ units.

52. The graph of $y = \sin\left(2x + \frac{\pi}{2}\right)$ is obtained by shifting the graph of $y = \sin x$ to the left $\frac{\pi}{2}$ units.

53. The graph of $y = 1 + \sin(2x - 3)$ is obtained from the graph of $y = \sin x$ from a horizontal compression by a factor of 2, followed by a shift to the right 3 units and upward 1 unit.

54. The graph of $y = -1 + 2\cos\left(2x - \frac{\pi}{3}\right)$ is obtained from the graph of $y = \cos x$ from a horizontal compression by a factor of 2 and a vertical stretching by a factor of 2, followed by a shift to the right $\frac{\pi}{3}$ units and downward 1 unit.

55. The values of θ in the interval $[0, 2\pi]$ that satisfy $2\cos^2\theta - \cos\theta - 1 = 0$ are

$$\theta = \frac{2\pi}{3}, \frac{4\pi}{3}, 0, \text{ and } 2\pi.$$

56. The values of θ in the interval $[0, 2\pi]$ that satisfy $2\sin 2\theta = 4\sin\theta$ are $\theta = 0, 2\pi$.

57. $\sin\left(\dfrac{7\pi}{12}\right) = \dfrac{\sqrt{2}}{4}\left(\sqrt{3} + 1\right)$

58. $\cos\left(\dfrac{\pi}{12}\right) = \dfrac{\sqrt{2}}{4}\left(\sqrt{3} - 1\right)$

59. For all θ, $\cos\left(\theta - \dfrac{\pi}{2}\right) = \sin\theta$.

60. For all θ, $\sin\left(\theta - \dfrac{\pi}{2}\right) = \sin\theta$.

61. If x is a real number in the interval $[-1, 1]$, then $\sin(\arcsin(x)) = x$.

62. If x is a real number in the interval $[-\pi/2, \pi/2]$, then $\arcsin(\sin(x)) = x$.

63. $\arccos\left(\frac{\sqrt{3}}{2}\right) = \frac{\pi}{3}$.

64. $\arcsin\left(-\frac{1}{2}\right) = \frac{11\pi}{6}$.

65. If $\theta = \arcsin\left(\frac{3}{5}\right)$, then $\cos\theta = \frac{4}{5}$.

66. If $\theta = \arctan\left(\frac{1}{3}\right)$, then $\sin\theta = \frac{\sqrt{3}}{3}$.

67. The domain of the function $f(x) = \arcsin(2x)$ is $[-1/2, 1/2]$.

68. The domain of the function $f(x) = \arctan(2x)$ is $(-\infty, \infty)$.

In Exercises 69 and 70, use the figure.

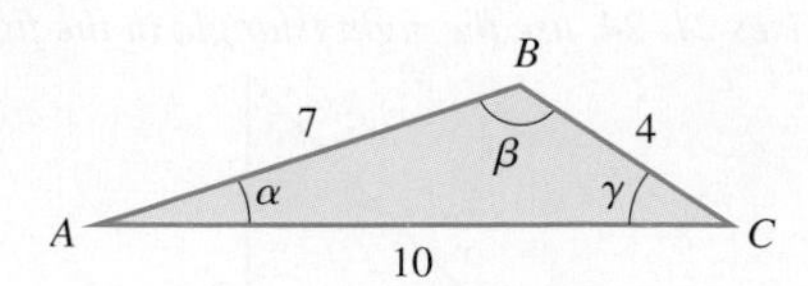

69. $\cos\alpha = \frac{133}{140}$.

70. $\beta = \arcsin\left(\frac{10}{4}\sin\alpha\right)$.

Exponential and Logarithm Functions

5

Calculus Connections

On September 19, 1991, the remains of a prehistoric man were found encased in ice near the border of Italy and Switzerland. The remains were clearly old, but how old were they?

Police Austria

A common way to date objects of the likely age of the prehistoric man is to use *radioactive carbon dating.* All living things contain carbon, which has a number of isotopes, predominant among which is the stable isotope carbon 12. Carbon in living beings also contains minute, but accurately measurable, amounts of the radioactive isotope carbon 14. The relative proportion of these two isotopes in living beings is well known to scientists.

It is assumed that the carbon is replenished from the atmosphere, and that the relative amounts of the isotopes in living beings have remained constant throughout time. When the organism dies, the replenishment of carbon ceases. The radioactive isotope carbon 14 decays with time, so the proportion of that isotope decreases. The decay rate of carbon 14 is such that half of a given amount will be lost after approximately 5730 years. Based on this rate of decay, scientists can estimate the date at which an organism died. This valuable technique has been used since the early 1950s, when it was proposed by the chemist Willard Libby, who won a Nobel Prize for his discovery.

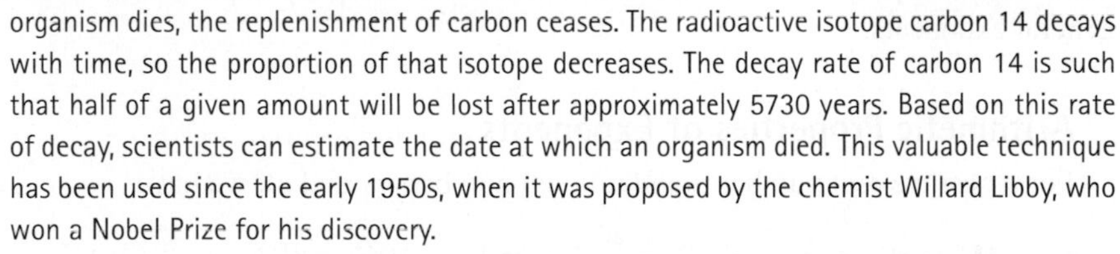

Radioactive carbon dating is only one of the many applications of calculus that requires *exponential* functions for its solution. In the example described here, the quantity $Q(t)$ of carbon 14 present at time t takes the form

$$Q(t) = Q_0 e^{-0.000121t}$$

where Q_0 is the estimated amount of the isotope in the sample at the time the man died and $e \approx 2.72$. The function in this solution is the *natural exponential function.* Researchers estimated that 47.590% of the amount of the carbon 14 had decayed by the time the remains of the prehistoric man were found, indicating that he was about 5300 years old.

The radioactive carbon dating technique has limitations and is not without its detractors, but scientists have found it to be a valuable tool for dating archaeological and anthropological materials.

5.1 INTRODUCTION

In Chapter 2 we saw that one-to-one functions have inverses, which reverse the function process. The function described by $f(x) = x^3$, for example, has the inverse $f^{-1}(x) = x^{1/3}$, and for every value of x the inverse relationship gives

$$f\left(f^{-1}(x)\right) = f\left(x^{1/3}\right) = \left(x^{1/3}\right)^3 = x$$

and

$$f^{-1}(f(x)) = f^{-1}\left(x^3\right) = \left(x^3\right)^{1/3} = x.$$

In Chapter 4 we saw that by restricting the domain of the trigonometric functions to a portion on which they are one-to-one, we can define inverse trigonometric functions, known as the *arc functions*. For example, after restricting the domain of the sine function to the interval $[-\pi/2, \pi/2]$, the arcsine function is defined by

$$\arcsin(\sin x) = x \quad \text{for each } x \text{ in } [-\pi/2, \pi/2]$$

and

$$\sin(\arcsin x) = x \quad \text{for each } x \text{ in } [-1, 1].$$

The most important function–inverse function pair in calculus is the natural exponential function with its inverse, the natural logarithm function.

In this chapter we introduce another class of functions, the exponentials, and their inverses, the logarithms.

One of the first topics discussed in algebra courses concerns the laws of exponents. These tell us that for every positive real number a and rational number $r = p/q$, where p and q have no common factors and $q > 0$, we have the root and power definition

$$a^r = a^{p/q} = \sqrt[q]{a^p} = \left(\sqrt[q]{a}\right)^p.$$

For example,

$$27^{2/3} = \left(\sqrt[3]{27}\right)^2 = 3^2 = 9 \quad \text{and} \quad 9 = \sqrt[3]{9^3} = \sqrt[3]{(3^2)^3} = \sqrt[3]{(3^3)^2} = \sqrt[3]{(27)^2}.$$

The *arithmetic properties of exponents* are used to simplify expressions involving general exponents.

Arithmetic Properties of Exponents

For a positive real number a and rational numbers r_1 and r_2, we have

- $a^{r_1+r_2} = a^{r_1}a^{r_2}, \qquad a^{r_1-r_2} = \dfrac{a^{r_1}}{a^{r_2}}, \qquad$ and $\qquad (a^{r_1})^{r_2} = a^{r_1 r_2}.$

For a pair of positive real numbers a_1 and a_2 and a single rational number r, we have

- $a_1^r a_2^r = (a_1 a_2)^r \qquad$ and $\qquad \dfrac{a_1^r}{a_2^r} = \left(\dfrac{a_1}{a_2}\right)^r.$

Exponents can be extended to include all real numbers, not just rational numbers. We will define, for every positive number $a \neq 1$, an exponential function $f(x) = a^x$ that is valid for all real numbers x. The definition must ensure that the arithmetic properties of exponents continue to hold, and the definition must reduce to the root and power definition when the exponent is rational.

There are various equivalent ways to define exponential functions; one is to take the limit of rational approximations. In Figure 1 we see a representation of $y = 3^x$ for rational numbers x. We have indicated that the irrational numbers are missing from the domain by using a dotted line for the graph. We extend the domain of this function to include all the real numbers, essentially filling the holes in the graph.

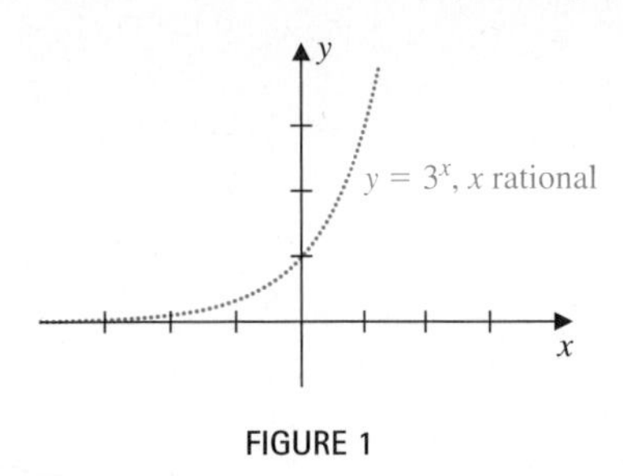

FIGURE 1

To illustrate, suppose we want to define $3^{\sqrt{2}}$. Since $\sqrt{2}$ is irrational, $3^{\sqrt{2}}$ cannot be defined in terms of roots and powers of 3. But we can use roots and powers of 3 to get arbitrarily accurate approximations to $3^{\sqrt{2}}$. The decimal expansion of $\sqrt{2}$ is

$$\sqrt{2} = 1.414213562\ldots,$$

so we define $3^{\sqrt{2}}$ as the value that approximations of the form

The intuitive notion of "approaching" is made precise in calculus with the concept of the limit.

$$\begin{aligned}
3^{1.4} &= 3^{14/10} = 4.65553672\ldots, \\
3^{1.41} &= 3^{141/100} = 4.70696500\ldots, \\
3^{1.414} &= 3^{1414/1000} = 4.72769503\ldots, \\
3^{1.4142} &= 3^{14142/10000} = 4.72873393\ldots, \\
3^{1.41421} &= 3^{141421/100000} = 4.72878588\ldots,
\end{aligned}$$

approach as more digits of $\sqrt{2}$ are used. To complete this process rigorously requires more mathematical analysis than we have available, but it certainly appears that this limiting process is converging to a value that we can use to define $3^{\sqrt{2}}$, a number that is close to 4.729.

In like manner, we can define a^x whenever $a > 0$ and x is a real number. The specific value for various choices of a and x need not concern us at this time; what we want is the class of functions defined in this manner.

One exception to this process is the case $a = 1$. Since $1^r = 1$ for every rational number r, we also have $1^x = 1$ for every real number x, which gives the constant function $f(x) = 1^x \equiv 1$. We have already considered constant functions, so the base $a = 1$ will be excluded from further discussion in this chapter.

- When we discuss exponential functions, we will mean that we have a function of the form $f(x) = a^x$, where a is a positive real number and $a \neq 1$.

5.2 THE NATURAL EXPONENTIAL FUNCTION

In Section 5.1 we saw that for every positive number $a \neq 1$, called the *base*, we can define an *exponential function* of the form

$$f(x) = a^x,$$

whose domain is the set of all real numbers. The graphs of some exponential functions with base $a > 1$ are shown in Figure 1. Notice that each of these functions has $y = 0$ as a horizontal asymptote, since for $a > 1$,

$$a^x \to 0 \quad \text{as} \quad x \to -\infty.$$

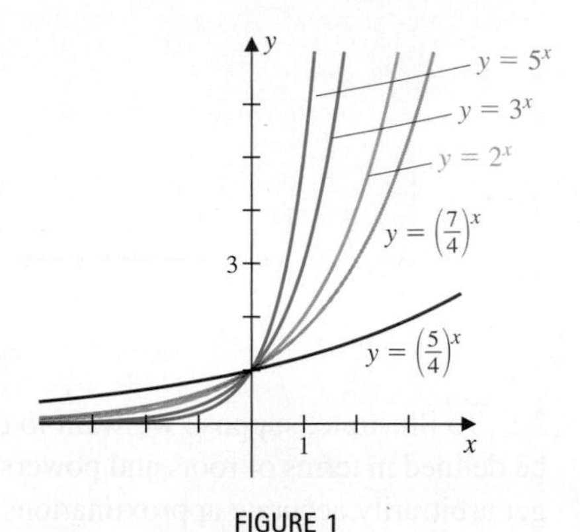

FIGURE 1

In addition, the graphs all pass through the point (0, 1). These exponential functions are always increasing, and the larger the base a, the faster the rate of increase. The range of $f(x) = a^x$ is $(0, \infty)$ because

$$a^x \to \infty \quad \text{as} \quad x \to \infty.$$

When $0 < a < 1$ we have $1 < 1/a$, so the graphs of the exponential functions with base less than 1 are found using the reciprocal graphing technique and the fact that

$$y = \left(\frac{1}{a}\right)^x = \frac{1}{a^x}.$$

This is illustrated in Figure 2. These graphs pass through (0, 1) and have $y = 0$ as a horizontal asymptote, because

$$a^x \to 0 \quad \text{as} \quad x \to \infty.$$

They are always decreasing, and the smaller the base a, the faster the rate of decrease, as shown in Figure 3. The range is again $(0, \infty)$.

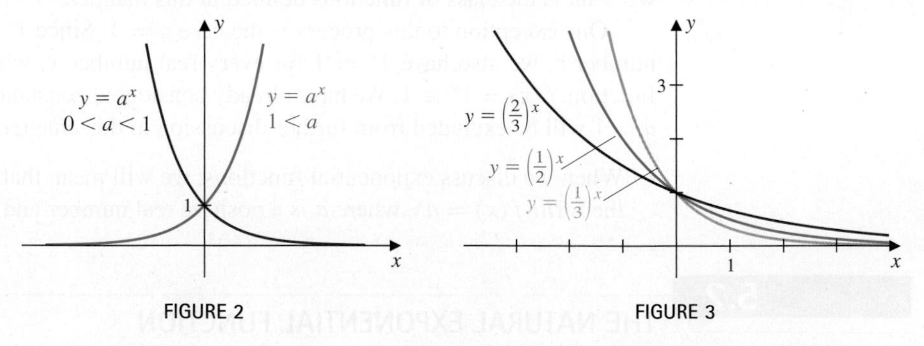

FIGURE 2 FIGURE 3

For any positive number $a \neq 1$ the exponential function $f(x) = a^x$ is one-to-one, because it is either always increasing (when $a > 1$) or always decreasing (when $0 < a < 1$). This fact is used in the solution to the problems in Example 1.

EXAMPLE 1 Determine all values of x that satisfy the equation.

a. $2^{2x-2} = 64$ **b.** $3^{x-2} = 27^{x+5}$

c. $x^2 4^{x/2} - 2x2^{x+1} - 32^x = 0$

Solution **a.** Since $64 = 2^6$ and the exponential functions are one-to-one, $2^{2x-2} = 64 = 3^6$ implies that

$$2x - 2 = 6 \quad \text{so} \quad x = 4.$$

b. Since $27 = 3^3$ we have $27^{2x+5} = \left(3^3\right)^{x+5} = 3^{3\cdot(x+5)}$, so

$$x - 3 = 3 \cdot (x + 5), \quad \text{which implies that} \quad x - 3 = 3x + 15 \quad \text{and} \quad x = -9.$$

c. First note that

$$4^{x/2} = \left(2^2\right)^{x/2} = 2^{2\cdot x/2} = 2^x \quad \text{and} \quad 2^{x+1} = 2 \cdot 2^x.$$

So the original equation can be rewritten and factored as

$$0 = x^2 4^{x/2} - 2x2^{x+1} - 32^x = x^2 2^x - 2x2^x - 32^x = (x^2 - 2x - 3)2^x = (x+1)(x-3)2^x.$$

Since $2^x > 0$ for all values of x, the only solutions are $x = -1$ and $x = 3$. ■

In Example 2, the graph of $y = 2^x$ and the graphing techniques from Chapters 1 and 2 are used to sketch the graphs of some modified exponential functions.

EXAMPLE 2 Use the graph of $y = 2^x$ to sketch the graph.

a. $f(x) = 2^x - 3$ **b.** $g(x) = -3 \cdot 2^{(x-1)} + 1$

Solution **a.** The graph of $f(x) = 2^x - 3$ is obtained by shifting the graph of $y = 2^x$ downward 3 units, as shown in Figure 4. The domain of f is $(-\infty, \infty)$, the range is $(-3, \infty)$, and $y = -3$ is the horizontal asymptote.

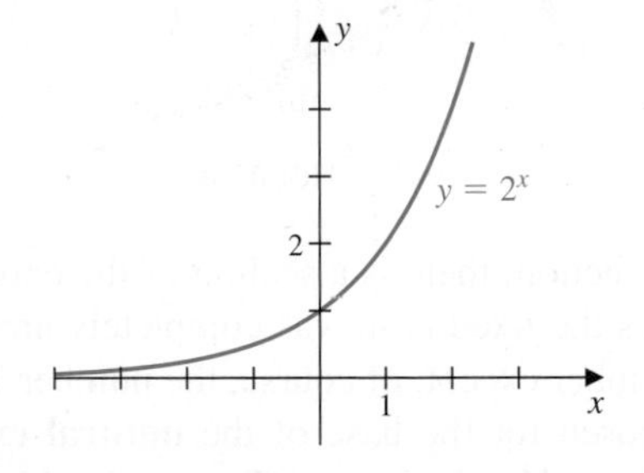

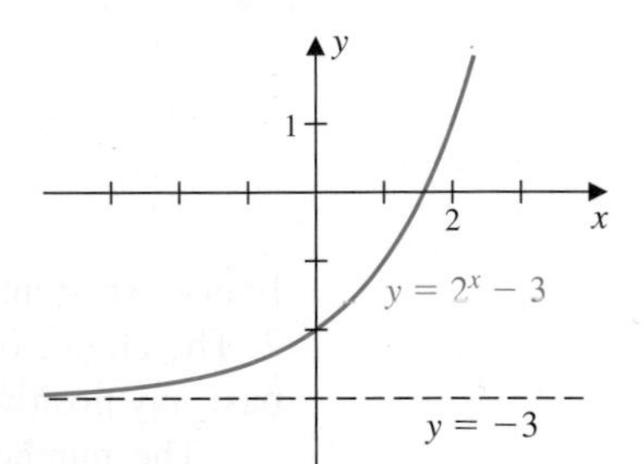

FIGURE 4

b. To sketch the graph of $g(x) = -3 \cdot 2^{(x-1)} + 1$, first stretch the graph of $y = 2^x$ vertically by a factor of 3 to obtain the graph of $y = 3 \cdot 2^x$, shown in Figure 5(a). Then shift this graph to the right 1 unit to produce the graph of $y = 3 \cdot 2^{(x-1)}$, shown in Figure 5(b).

The final graph of $g(x) = -3 \cdot 2^{(x-1)} + 1$ is obtained by reflecting the graph of $y = 3 \cdot 2^{(x-1)}$ about the x-axis and then shifting this graph upward 1 unit, as shown in Figure 5(c). The line $y = 1$ is the horizontal asymptote. ■

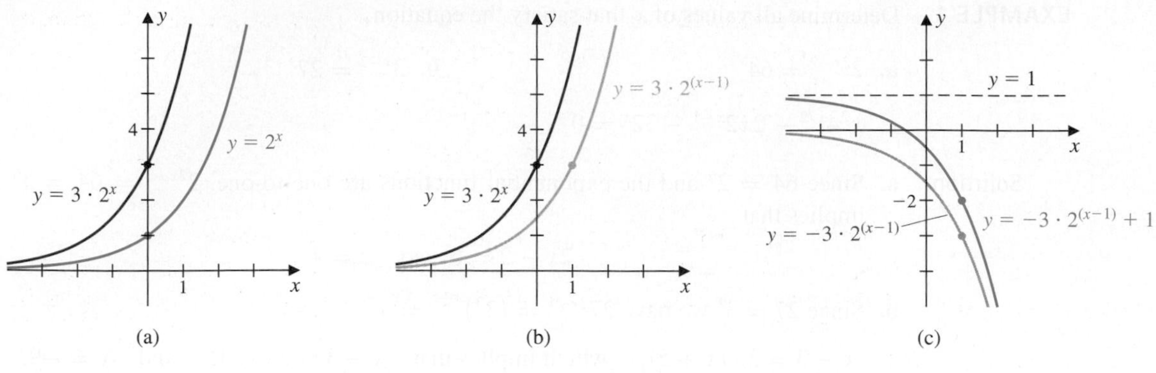

FIGURE 5

The exponential functions are even more closely related than their graphs might lead one to believe. Each exponential function can be converted into a scaled function of any single exponential function. Suppose, for example, that we want to write an arbitrary exponential function $f(x) = a^x$ in terms of an exponential function with base 2. The graph of $y = a^x$ satisfies the Horizontal Line Test, so each of these exponential functions is one-to-one, as shown in Figure 6(a). The range of $f(x) = a^x$ is $(0, \infty)$, so there is a unique real number k with $a = 2^k$. (See Figure 6(b).) For this value of k we have, as shown in Figure 6(c),

$$f(x) = a^x = (2^k)^x = 2^{kx}.$$

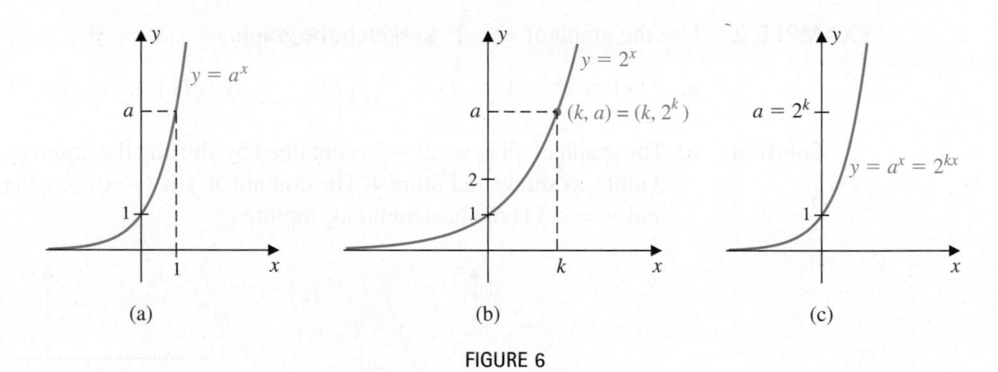

FIGURE 6

Every exponential function, then, is a scaling of the exponential function with base 2. The choice of 2 as the fixed base was completely arbitrary. We could use as our base any positive number except, of course, the number 1.

The number chosen for the base of the **natural exponential function** is an irrational number denoted by the letter e. To see why this is the *natural* base requires some background. Throughout the book we have remarked that much of the study of calculus involves determining the limiting values of *difference quotients* of the form

The limit of this difference quotient is the most basic concept in differential calculus.

$$\frac{f(x+h) - f(x)}{h} \quad \text{as} \quad h \to 0.$$

This limit describes the *slope* of the curve at the point $(x, f(x))$. For certain functions this limiting value is easy to determine. For example, if $f(x) = 3x$, then

$$\frac{f(x+h) - f(x)}{h} = \frac{3(x+h) - 3x}{h} = \frac{3x + 3h - 3x}{h} = \frac{3h}{h} = 3,$$

regardless of the values of x and h. This is certainly reasonable, because the graph of $f(x) = 3x$ is a line with slope 3. If $f(x) = x^2$, the difference quotient is

$$\frac{f(x+h) - f(x)}{h} = \frac{(x+h)^2 - x^2}{h} = \frac{x^2 + 2xh + h^2 - x^2}{h} = \frac{(2x+h)h}{h} = 2x + h.$$

The *derivative* in calculus is based on the difference quotient and describes the slope of the graph. It is the key to solving many diverse problems in mathematics.

The difference quotient for this function depends on the values of both x and h. As h approaches zero, the difference quotient approaches $2x$ and the slope of the graph at any point is twice the x-coordinate. Some illustrations of this result are shown in Figure 7.

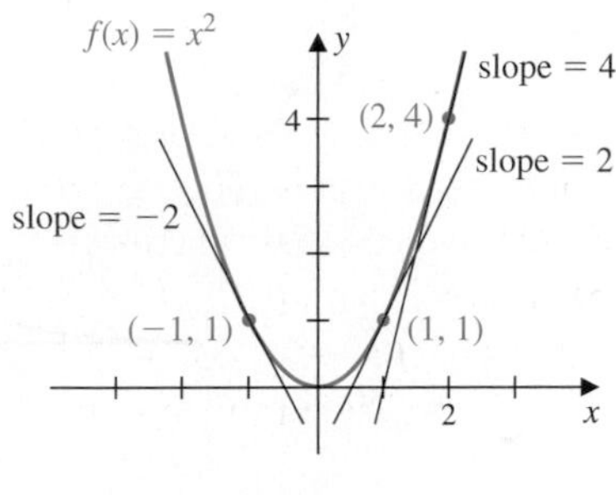

FIGURE 7

For an exponential function of the form $f(x) = a^x$, the difference quotient is

$$\frac{f(x+h) - f(x)}{h} = \frac{a^{x+h} - a^x}{h} = \frac{a^x a^h - a^x}{h} = a^x \left(\frac{a^h - 1}{h}\right).$$

The value of a that gives us the *natural* exponential function is the value for which

$$\frac{a^h - 1}{h} \to 1 \quad \text{as} \quad h \to 0.$$

A fundamental property of the exponential function that is used repeatedly in calculus is that the slope of the graph at a point is the same as the value of the function.

- The natural exponential function has the property that at each point on the graph, the slope of the graph is the same as the value of the function at that point.

Table 1 contains values, up to three decimal places, of the quotient

$$\frac{a^h - 1}{h}$$

for various values of a and h that give limiting values close to 1.

TABLE 1

h	$a = 2$	$a = 3$	$a = 2.5$	$a = 2.75$	$a = 2.675$	$a = 2.7125$
0.1	0.718	1.161	0.960	1.064	1.034	1.049
0.01	0.696	1.105	0.921	1.017	0.989	1.003
0.001	0.693	1.099	0.917	1.012	0.984	0.998
0.0001	0.693	1.099	0.916	1.012	0.984	0.998

From Table 1 it appears that the base a with

$$\frac{a^h - 1}{h} \to 1, \quad \text{as} \quad h \to 0$$

is a little greater than the last entry, 2.7125. In fact, the unique value that satisfies our condition is the irrational number $e \approx 2.718281828459045\ldots$. The graph of $f(x) = e^x$, shown in Figure 8, has the property that

In calculus, the exponential function of primary interest is the natural exponential function, which has base e.

$$\frac{e^{x+h} - e^x}{h} = e^x \left(\frac{e^h - 1}{h} \right) \to e^x \cdot 1 = e^x \quad \text{as} \quad h \to 0.$$

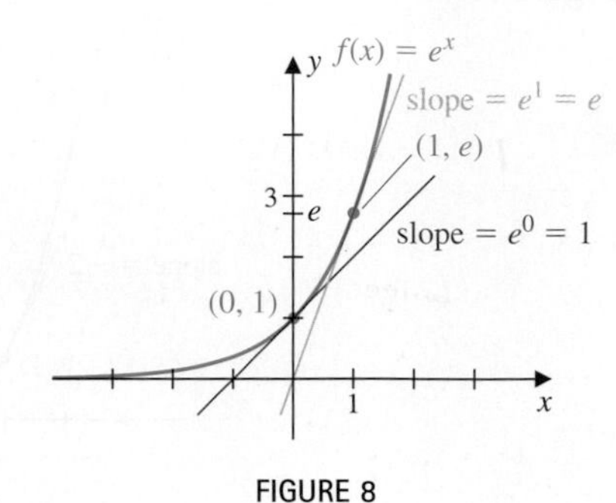

FIGURE 8

In addition, since

$$\frac{e^h - 1}{h} \approx 1 \quad \text{for } h \text{ close to } 0,$$

we have

$$e^h - 1 \approx h \quad \text{and} \quad e^h \approx 1 + h \quad \text{for } h \text{ close to } 0.$$

These results imply that

$$(1 + h)^{1/h} \to (e^h)^{1/h} = e \quad \text{as} \quad h \to 0.$$

In Table 2 you can see that $(1+h)^{1/h}$ does indeed appear to approach the approximation we have given for e.

TABLE 2

h	0.1	0.01	0.001	0.0001	0.00001	0.000001
$(1+h)^{1/h}$	2.593742	2.704814	2.716924	2.718146	2.718268	2.718281

Since e is between 2 and 3, the graph of $y = e^x$ shown in yellow in Figure 9 lies between the graphs of $y = 2^x$, shown in blue, and $y = 3^x$, shown in red, and is closer to 3^x. The graph of $y = e^x$ also passes through the points $(0, 1)$, $(1, e)$, and $(-1, 1/e)$.

The following list contains important properties of the natural exponential function $f(x) = e^x$ and its reciprocal $g(x) = e^{-x}$.

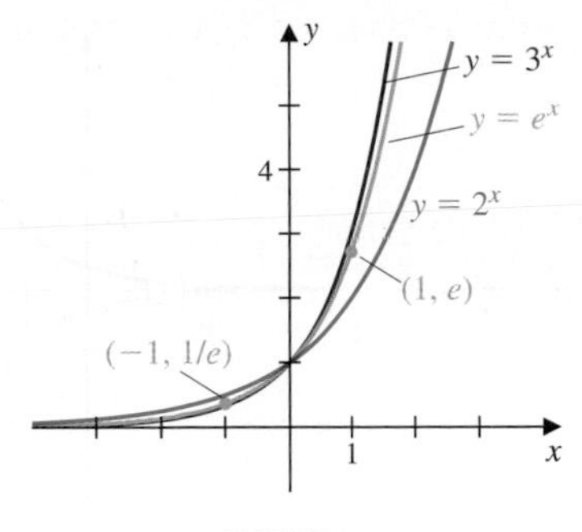

FIGURE 9

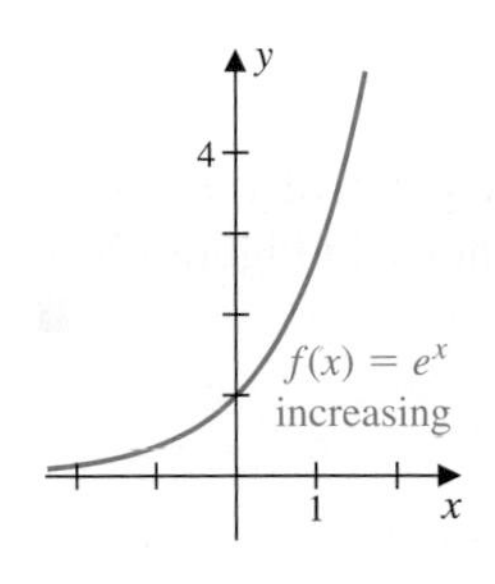

FIGURE 10

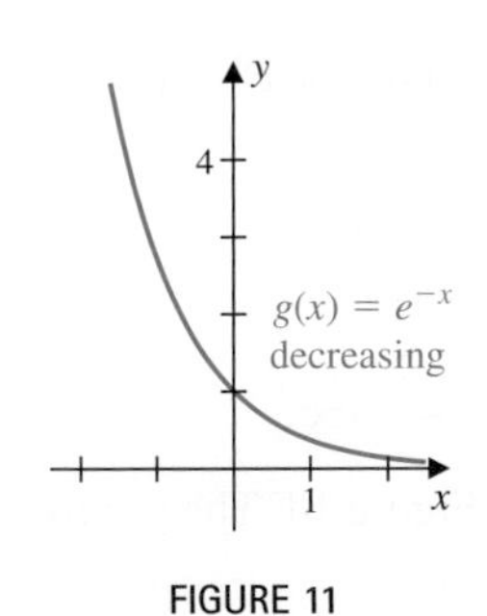

FIGURE 11

Properties of $f(x) = e^x$ and $g(x) = e^{-x}$

- The domain of the functions f and g is the set of all real numbers.
- The range of the functions f and g is the set of all positive real numbers.
- The y-intercept for the graphs of $y = e^x$ and $y = e^{-x}$ is the point $(0, 1)$.
- The function $f(x) = e^x$ is always increasing, and $g(x) = e^{-x}$ is always decreasing.
- The graphs of $y = e^x$ and $y = e^{-x}$ have the x-axis as horizontal asymptote.
- The graph of $y = e^{-x}$ is the reflection about the y-axis of the graph of $y = e^x$.
- $e^x \to \infty$ as $x \to \infty$ and $e^x \to 0$ as $x \to -\infty$
- $e^{-x} \to 0$ as $x \to \infty$ and $e^{-x} \to \infty$ as $x \to -\infty$

(See Figures 10 and 11.)

The general arithmetic rules for combining exponentials also apply to the natural exponential function, and are stated as follows.

The natural exponential function is used repeatedly in calculus. It is exceptionally important for applications.

Arithmetic Properties of the Natural Exponential Function

For each pair of real numbers x_1, x_2 we have

- $e^{x_1} \cdot e^{x_2} = e^{x_1+x_2}$,
- $(e^{x_1})^{x_2} = e^{x_1x_2}$, and
- $\dfrac{e^{x_1}}{e^{x_2}} = e^{x_1-x_2}$.

These arithmetics properties are used in the next examples.

EXAMPLE 3 Sketch the graph of $f(x) = 2 - e^{x-1}$.

Solution The graph of $y = e^{x-1}$ shown in Figure 12(a) is obtained by translating the graph of $y = e^x$ to the right 1 unit. The graph of $y = -e^{x-1}$ is the reflection about the x-axis of

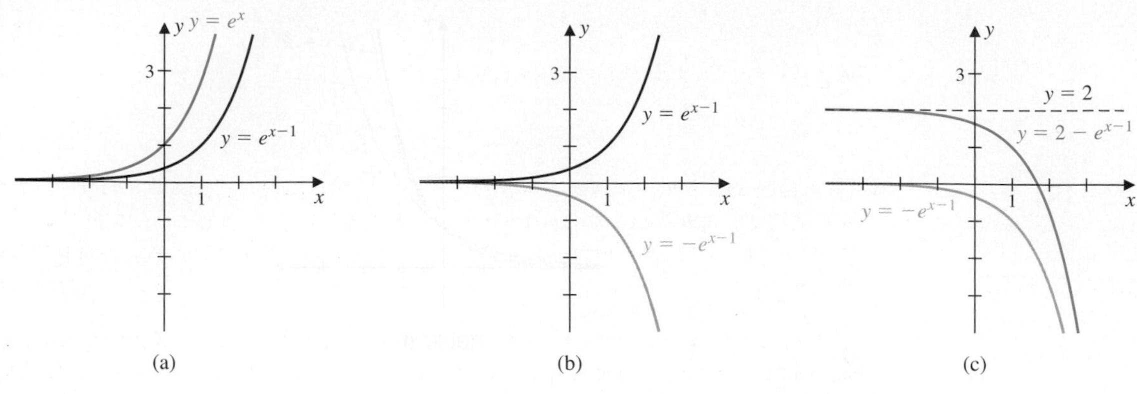

FIGURE 12

the graph of $y = e^{x-1}$, as shown in Figure 12(b). Translating the graph of $y = -e^{x-1}$ vertically upward 2 units gives the graph of $f(x) = 2 - e^{x-1}$, shown in Figure 12(c). The graph has the horizontal asymptote $y = 2$. ■

EXAMPLE 4 Use the graph of $y = e^x$ to determine the graph of $f(x) = e^x + e^{-x}$.

Solution First recall that the graph of $y = e^{-x} = 1/e^x$ is the reflection about the y-axis of the graph of $y = e^x$, as shown in Figure 13.

The graph of $f(x) = e^x + e^{-x}$ passes through (0, 2) since $f(0) = e^0 + e^{-0} = 1 + 1 = 2$. Also,

$$e^{-x} \to 0 \text{ as } x \to \infty, \quad \text{so} \quad e^x + e^{-x} \to e^x \text{ as } x \to \infty$$

and

$$e^x \to 0 \text{ as } x \to -\infty, \quad \text{so} \quad e^x + e^{-x} \to e^{-x} \text{ as } x \to -\infty.$$

The graph of $y = e^x + e^{-x}$ is similar to that shown in yellow in Figure 14. The y-axis symmetry follows from the fact that

$$f(-x) = e^{(-x)} + e^{-(-x)} = e^{-x} + e^x = f(x).$$ ■

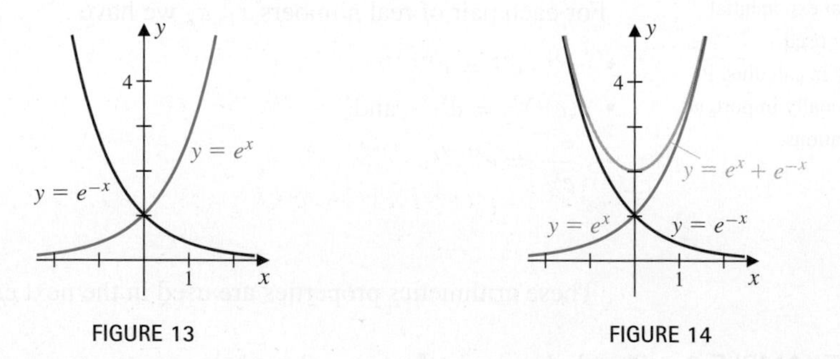

FIGURE 13 FIGURE 14

The natural exponential function has applications in many circumstances, and modifications of this function are particularly important in statistics.

EXAMPLE 5 Sketch the graph of $f(x) = e^{-x^2}$.

Solution The graph passes through (0, 1) because $f(0) = e^{-0^2} = 1$. Writing

$$f(x) = e^{-x^2} = \frac{1}{e^{x^2}},$$

we see that as x becomes large, $f(x)$ rapidly approaches 0. In addition, $f(-x) = f(x)$, so the graph has y-axis symmetry. The graph has the appearance of the *bell-shaped* curve, shown in Figure 15. This curve is associated with the normal distribution in statistics. ■

FIGURE 15

We have seen that the natural exponential function grows rapidly for positive values of x. Example 6 gives a quantitative illustration of just how rapid this growth is.

EXAMPLE 6 Use a graphing device to compare the growth of $f(x) = e^x$ and $g(x) = x^4$ for large values of x.

Solution In the viewing rectangle $[-5, 5] \times [-5, 5]$, we have the graphs shown in Figure 16(a), where it appears that x^4 grows faster than e^x. The viewing rectangle $[0, 20] \times [0, 6800]$ shown in Figure 16(b) indicates that there is another intersection point of the graphs, and it appears that e^x exceeds x^4 after this third intersection point. The viewing rectangle $[0, 20] \times [0, 100000]$ shown in Figure 16(c) shows that e^x greatly exceeds x^4 for values of x greater than 9. By zooming in on the curves, we find that the three intersection points occur when x is approximately -0.82, 1.43, and 8.61. ■

The natural exponential function eventually grows faster than x^n for any $n \geq 0$.

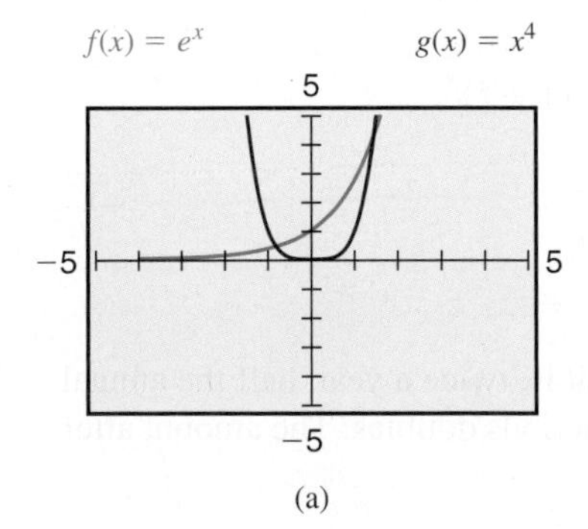

(a)

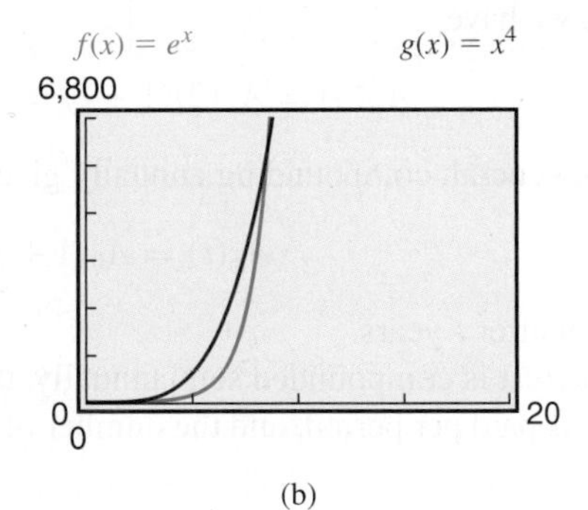

(b)

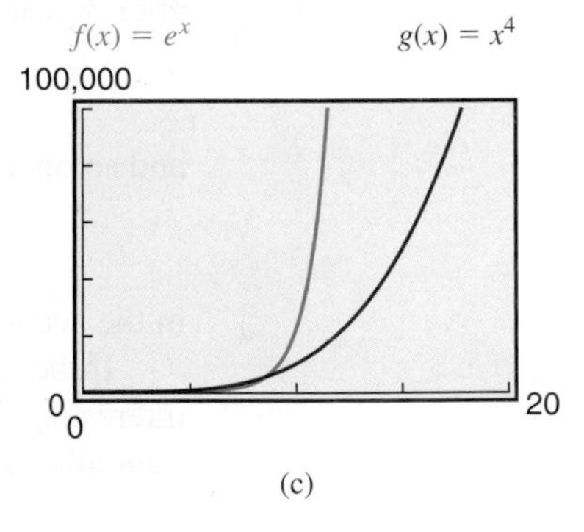

(c)

FIGURE 16

EXAMPLE 7 Determine the domain of $f(x) = \sqrt{x^2e^x - 2xe^x - 8e^x}$.

Solution The domain of f is the set of all real numbers x with

$$0 \leq x^2e^x - 2xe^x - 8e^x = e^x(x^2 - 2x - 8) = e^x(x + 2)(x - 4).$$

Since $e^x > 0$ for all real numbers x, the inequality is equivalent to solving

$$(x + 2)(x - 4) \geq 0.$$

The sign chart in Figure 17 gives the solution $(-\infty, -2] \cup [4, \infty)$. ■

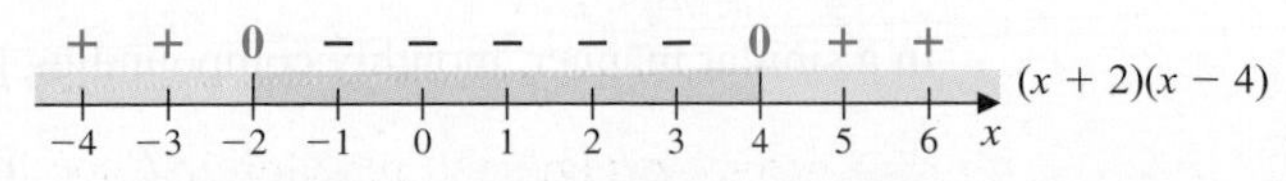

FIGURE 17

Applications

The natural exponential function is regularly used to model physical situations, but calculus is needed for a full appreciation of its value. We introduce this topic using **compound interest** as our model, but the applications of this behavior are much more far reaching than this example suggests. In Section 5.4 we consider other applications.

Suppose that we invest a sum A_0 in a savings account. The amount in the account at the end of t years depends on the interest rate i and on the number of times per year that the interest is compounded. In the past it was common to compound the interest only a few times a year, but most financial institutions now compound interest daily. The more frequently the interest is compounded, the faster the growth in the account. The best of all possible situations for the investor occurs when the interest is compounded *continuously*. To examine this statement, let us consider the consequences of various methods for compounding interest.

If the interest is compounded yearly, with annual interest rate i and an initial amount A_0, the amount in the account after 1 year is

$$A_y(1) = A_0 + iA_0 = A_0(1+i).$$

After 2 years we have the amount

$$A_y(2) = A_y(1)(1+i) = A_0(1+i)^2,$$

after 3 years we have

$$A_y(3) = A_y(2)(1+i) = A_0(1+i)^3,$$

and so on. In general, compounding annually gives

$$A_y(t) = A_0(1+i)^t$$

in the account after t years.

If the interest is compounded semiannually, that is, twice a year, half the annual interest, $i/2$, is paid per period, and the number of periods doubles. The amount after 6 months is

$$A_s\left(\frac{1}{2}\right) = A_0\left(1+\frac{i}{2}\right),$$

so the amount after 1 year is

$$A_s(1) = A_0\left(1+\frac{i}{2}\right)^2,$$

and the amount after t years is

$$A_s(t) = A_0\left(1+\frac{i}{2}\right)^{2t}.$$

In a similar manner, monthly compounding produces the amount

$$A_m(t) = A_0\left(1+\frac{i}{12}\right)^{12t},$$

and daily compounding gives

$$A_d(t) = A_0\left(1 + \frac{i}{365}\right)^{365t}.$$

In general, the interest i compounded n times a year produces the amount

$$A_n(t) = A_0\left(1 + \frac{i}{n}\right)^{nt}.$$

Continuous compounding is the limiting case that occurs when the number of compounding periods per year increases without bound. To determine a function describing this situation, introduce the variable change $h = i/n$ into the formula for $A_n(t)$. Then $n = i/h$, and using the arithmetic properties of exponents gives

$$A_n(t) = A_0\left(1 + \frac{i}{n}\right)^{nt} = A_0(1+h)^{(i/h)t} = A_0\left((1+h)^{1/h}\right)^{it}.$$

At the beginning of this section we saw that

$$(1+h)^{1/h} \to e \quad \text{as } h \to 0.$$

As a result,

$$A_n(t) \to A_0e^{it} \quad \text{as } h \to 0.$$

Hence, $A_n(t)$ approaches A_0e^{it} as the number $n = i/h$ of compounding periods increases indefinitely. This limiting term gives the amount when the interest is *compounded continuously*, and it is denoted by

$$A_c(t) = A_0e^{it}.$$

The value of $A_c(t)$ is larger than the amount produced by any compounding method, and is quite close to that given by daily compounding, as Example 8 demonstrates.

EXAMPLE 8 Determine the value of a CD (certificate of deposit) in the amount of \$1000 that matures in 6 years and pays 5% per year compounded annually, monthly, daily, and continuously.

Solution The decimal equivalent of 5% is 0.05, so the various compounding methods give

$$A_y(6) = 1000(1 + 0.05)^6 \approx \$1340.10,$$

$$A_m(6) = 1000\left(1 + \frac{0.05}{12}\right)^{12(6)} \approx \$1349.02,$$

$$A_d(6) = 1000\left(1 + \frac{0.05}{365}\right)^{365(6)} \approx \$1349.83,$$

and

$$A_c(6) = 1000e^{0.05(6)} \approx \$1349.86.$$

■

Notice that in Example 8 there is very little difference between the amount produced by daily compounding and that given by continuous compounding. In addition, the exact compounding formulas are valid only for values of t that are integral multiples or fractions of the compounding period, whereas $A_c(t)$ is defined for all positive values of t. Example 9 shows how this can be useful for accurately approximating the future value of an account.

EXAMPLE 9 Determine the approximate length of time it takes an amount to double in value if it earns 9% compounded daily.

Solution First, assume that instead of compounding daily, the account compounds continuously. Then the time, t, that it takes for the amount A_0 to double is obtained from the equation

$$2A_0 = A_c(t) = A_0 e^{0.09t}, \quad \text{and dividing by } A_0 \text{ gives} \quad 2 = e^{0.09t}.$$

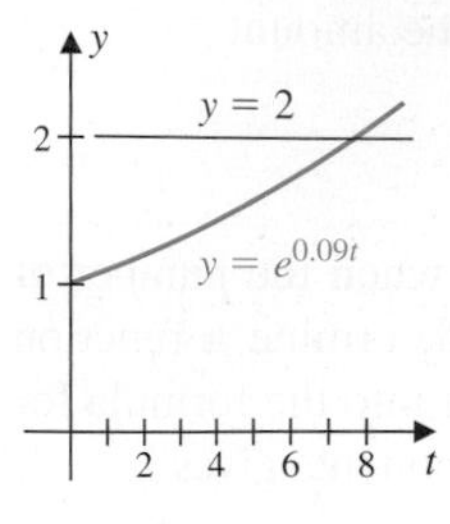

FIGURE 18

We cannot yet find the exact solution to this problem, but the graphs of $y = e^{0.09t}$ and $y = 2$ shown in Figure 18 indicate that $t \approx 7.7$.

If we estimate conservatively, a reasonable estimate is that it will take about 7.75 years, or 7 years and 9 months, to double our investment at this rate. The exact value after this time is

$$A_d(7.75) = A_0\left(1 + \frac{0.09}{365}\right)^{365(7.75)} \approx 2.0086A_0. \quad \blacksquare$$

EXAMPLE 10 A biologist observes that in the early stages of development each cell of a frog embryo divides approximately every half hour. Use an exponential function to predict the result of growth of a single cell after the first 8 hours.

Solution The numbers of cells after the first 5 half hours are given in Table 3. If t denotes time in hours, then the number of cells at time t is given by $Q(t) = 2^{2t}$. This exponential model predicts that the cell will produce $2^{16} = 65{,}536$ cells at the end of 8 hours. ■

TABLE 3

Half Hour	Cells
1	$2 = 2^1$
2	$4 = 2^2$
3	$8 = 2^3$
4	$16 = 2^4$
5	$32 = 2^5$

EXERCISE SET 5.2

In Exercises 1–14, sketch the graphs, showing any horizontal asymptotes.

1. $f(x) = 2^x + 1$
2. $f(x) = 2^{x-1} - 3$
3. $f(x) = -4^x$
4. $f(x) = 10^{-x}$
5. $f(x) = 3 \cdot \left(\frac{1}{4}\right)^{x-1} + 2$
6. $f(x) = -4 \cdot \left(\frac{1}{3}\right)^{x+1} - 1$
7. $f(x) = -e^{x-1}$
8. $f(x) = e^{x-2} + 1$
9. $f(x) = 2 - e^{-(x-3)}$
10. $f(x) = e^{-x} + 2$
11. $f(x) = e^{2x}$
12. $f(x) = e^{2(x-1)} - 1$
13. $f(x) = -e^{|x|}$
14. $f(x) = -e^{-|x|}$

In Exercises 15–18, use the graph of the exponential function to solve the inequality.

15. $e^x > 1$
16. $2^x \le 8$

17. $\left(\frac{1}{4}\right)^x \geq 2$ **18.** $\left(\frac{1}{9}\right)^x < 3$

19. Match the equation with the graph.

a. $y = e^{x+1} - 2$

b. $y = -e^x + 1$

c. $y = 2e^x$

d. $y = e^{-x}$

i.

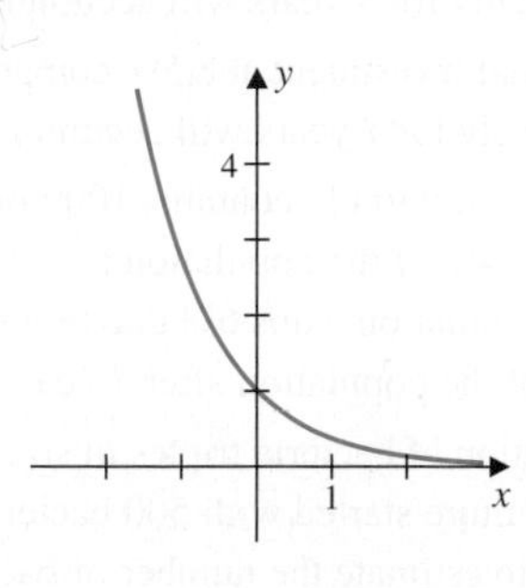

ii.

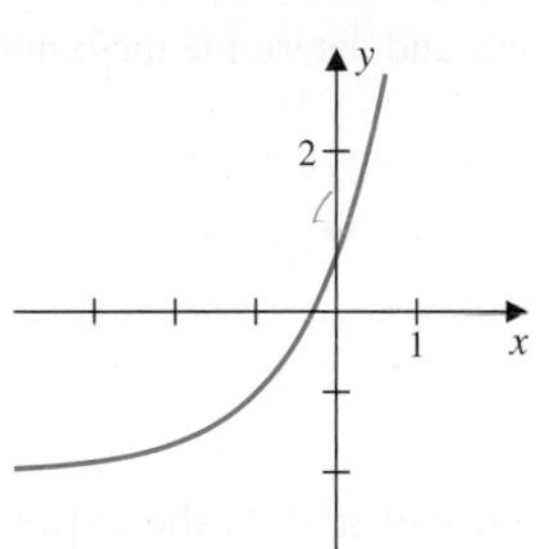

iii.

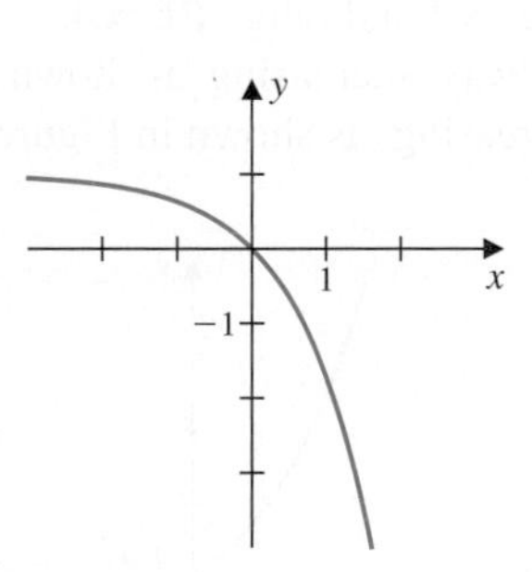

iv.

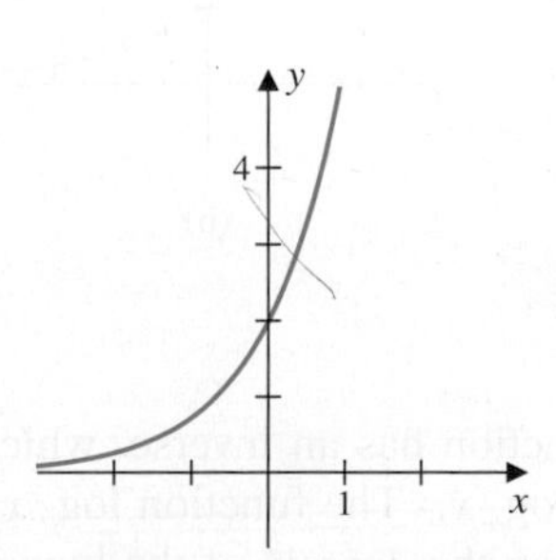

20. Sketch the graphs of $f(x) = ce^{x-1}$, for $c = 1, 2, -1, -2$, on the same set of axes.

In Exercises 21–24, use a graphing device to approximate, to one decimal place, all solutions to the equation.

21. $e^{x-2} = x$

22. $e^x = x^2$

23. $e^{-x} = (x-2)^2$

24. $xe^x = x^2 + 4x + 2$

In Exercises 25–30, ***(a)*** *use a graphing device to sketch the graphs of the functions and* ***(b)*** *approximate, to one decimal place, the intervals on which the function is increasing and on which it is decreasing.*

25. $f(x) = xe^x$

26. $f(x) = \dfrac{e^x}{x}$

27. $f(x) = e^{-x^2-x}$

28. $f(x) = e^{x^3-x}$

29. $f(x) = e^x - e^{-x}$

30. $f(x) = x + e^x$

31. Approximate the value of k for which $3^x = e^{kx}$.

32. a. Use a calculator to approximate the value of $(1 + 1/n)^n$ for $n = 1, 5, 10, 10^2, 10^3, 10^4$, and 10^5.

b. Use a graphing device to plot $f(x) = (1 + 1/x)^x$ and $y = e$ in the viewing rectangle $[0, 30] \times [0, 3]$.

33. a. Use a graphing device to compare the rates of growth of the functions $f(x) = 2^x$ and $g(x) = x^5$ by graphing the two functions in the following viewing rectangles:

i. $[-5, 5] \times [-5, 5]$

ii. $[0, 10] \times [0, 10^3]$

iii. $[0, 30] \times [0, 10^7]$

b. Approximate the solutions to $2^x = x^5$.

34. Use a graphing device to compare the rates of growth of $f(x) = e^x$ and $g(x) = x^{10}$ by graphing the functions together in several appropriate viewing rectangles. Approximate the solutions to $e^x = x^{10}$.

35. Determine the value of an investment in the amount of \$5000 that matures in 5 years and pays 6.5% per year compounded as indicated.

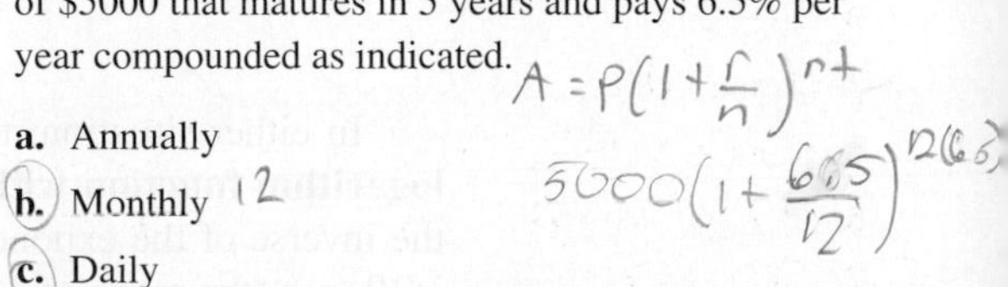

a. Annually

b. Monthly

c. Daily

d. Continuously

36. Suppose that \$1000 is invested at 10% interest and the interest rate remains fixed for 8 years. Complete the following table:

Interest Compounded	Value after 8 Years
Annually	
Semiannually	
Quarterly	
Monthly	
Weekly	
Daily	
Hourly	
Continuously	

37. If \$10,000 is invested and interest is compounded quarterly, determine the amount of the investment after 5 years for the given interest rates.

a. 8% **b.** 6.5% **c.** 6% **d.** 5.5%

38. Which interest rate and compounding period gives the best return?

a. 8% compounded annually

b. 7.5% compounded semiannually

c. 7% compounded continuously

39. What initial investment at 8% compounded semiannually for 5 years will accumulate to \$10,000?

40. What initial investment at 8.5% compounded continuously for 7 years will accumulate to \$50,000?

41. A population initially contains 1000 individuals. Each year 4% of the population is lost due to deaths and the population gains 6% due to births. What is the size of the population after 7 years?

42. A population of bacteria triples in size every 6 hours, and the culture started with 500 bacteria. Find a function to estimate the number of bacteria present at any time t, and determine the number present after 3 days.

5.3 LOGARITHM FUNCTIONS

In Section 5.2 we saw that for each positive number $a \neq 1$, the exponential function $f(x) = a^x$ is one-to-one with domain $(-\infty, \infty)$ and range $(0, \infty)$.

When $a > 1$ the function $f(x) = a^x$ is always increasing, as shown in Figure 1(a). When $0 < a < 1$, $f(x) = a^x$ is always decreasing, as shown in Figure 1(b).

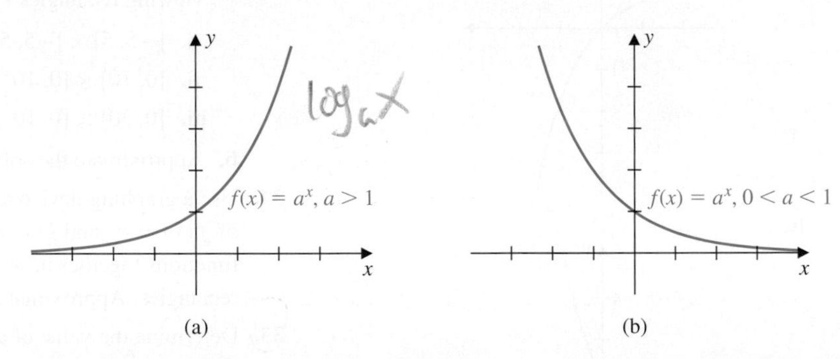

FIGURE 1

In either situation, the exponential function has an inverse, which we call the **logarithm function with base a**, written $\log_a x$. The function $\log_a x$ is defined as the inverse of the exponential function a^x, so the domain of the logarithm function is $(0, \infty)$ (the range of a^x) and the range of the logarithm function is $(-\infty, \infty)$ (the domain of a^x).

The Logarithm Function with Base a

For each positive number $a \neq 1$ and each x in $(0, \infty)$,

- $y = \log_a x$ if and only if $x = a^y$.

For each x in $(0, \infty)$ and each real number y,

- $\log_a a^y = y$ and $a^{\log_a x} = x$.

EXAMPLE 1 Evaluate the expression.

a. $\log_{10} 10000$ **b.** $\log_{10} 0.1$ **c.** $\log_2 32$

d. $\log_4 2$ **e.** $\log_5 5^2$ **f.** $3^{\log_3 8}$

Solution The inverse relationship between logarithms and exponentials is used to evaluate the expression. That is,

$$\log_a x = y \text{ is equivalent to } a^y = x.$$

a. $10^3 = 1000$, so $\log_{10} 1000 = 3$.
b. $10^{-1} = 0.1$, so $\log_{10} 0.1 = -1$.
c. $2^5 = 32$, so $\log_2 32 = 5$.
d. $4^{1/2} = \sqrt{4} = 2$, so $\log_4 2 = \frac{1}{2}$.
e. $y = \log_5 x$ and $y = 5^x$ are inverses, so $\log_5 5^2 = 2$.
f. $y = \log_3 x$ and $y = 3^x$ are inverses, so $3^{\log_3 8} = 8$. ■

EXAMPLE 2 Use the inverse relationship with the exponential functions to determine x.

a. $x = \log_3 81$ **b.** $\log_5 x = 3$ **c.** $\log_2(x^2 - 2x) = 3$

Solution **a.** The exponential-logarithm conversion implies that

$$x = \log_3 81 \quad \text{is equivalent to} \quad 3^x = 81.$$

Since $81 = 3^4$, we have $x = 4$.

b. The exponential-logarithm conversion gives

$$\log_5 x = 3 \quad \text{is equivalent to} \quad x = 5^3 = 125.$$

c. The conversion in this instance implies that

$$\log_2(x^2 - 2x) = 3 \quad \text{is equivalent to} \quad x^2 - 2x = 2^3 = 8.$$

Solving this quadratic equation gives

$$0 = x^2 - 2x - 8 = (x - 4)(x + 2), \quad \text{so } x = 4 \quad \text{or} \quad x = -2.$$ ■

Notice how frequently the reflection property is used to find graphs of inverse functions.

The graph of an inverse function is found by reflecting the graph of the function about the line $y = x$. So the graph of $y = \log_a x$ when $a > 1$ has the form shown in Figure 2(a). When $0 < a < 1$, the graph $y = \log_a x$ has the form shown in Figure 2(b).

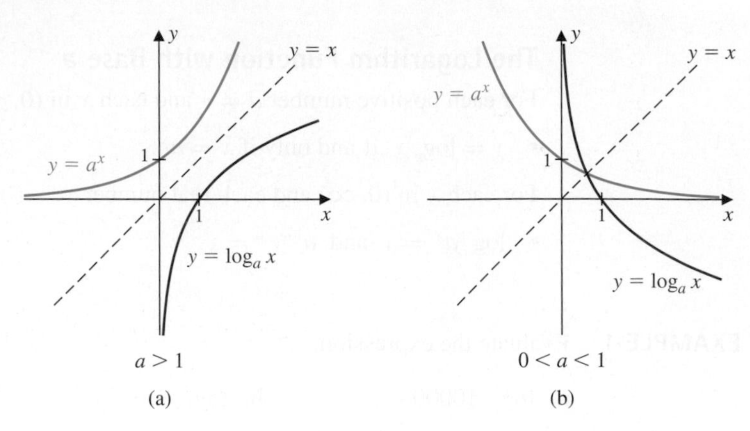

FIGURE 2

Properties of the Logarithm Function with Base a

- For $a > 0$ and $a \neq 1$,

$$\log_a 1 = 0 \quad \text{and} \quad \log_a a = 1.$$

- The graph of $y = \log_a x$ has the vertical asymptote $x = 0$.
  - For $a > 1$: $\log_a x \to -\infty$ as $x \to 0^+$.
  - For $0 < a < 1$: $\log_a x \to \infty$ as $x \to 0^+$.
- The graph of $y = \log_a x$ has no horizontal asymptote.
  - For $a > 1$: $\log_a x \to \infty$ as $x \to \infty$.
  - For $0 < a < 1$: $\log_a x \to -\infty$ as $x \to \infty$.

EXAMPLE 3 Sketch the graph of the function.

a. $f(x) = \log_2 x$ **b.** $g(x) = \log_2(x - 1) + 3$

Solution **a.** The general shape of the graph is as shown in red in Figure 3. Since

$$2 = 2^1, \quad 4 = 2^2, \quad 8 = 2^3, \quad \text{and} \quad \frac{1}{2} = 2^{-1},$$

the points $(1, 2)$, $(2, 4)$, $(3, 8)$, and $\left(-1, \frac{1}{2}\right)$ lie on the graph of $y = 2^x$. Since $y = 2^x$ precisely when $x = \log_2 y$, we have

$$\log_2 2 = 1, \quad \log_2 4 = 2, \quad \log_2 8 = 3, \quad \text{and} \quad \log_2 \frac{1}{2} = -1.$$

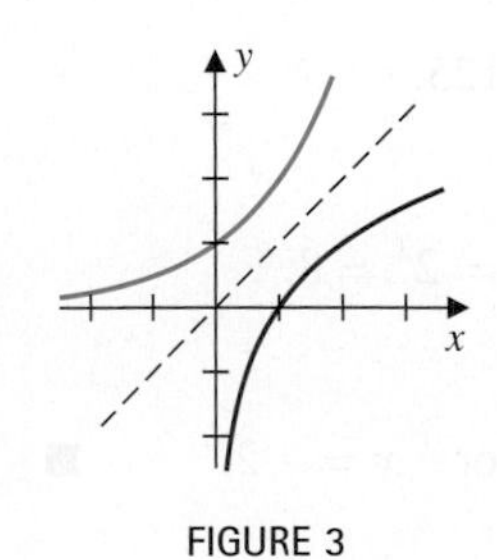

FIGURE 3

So the points $(2, 1)$, $(4, 2)$, $(8, 3)$, and $\left(\frac{1}{2}, -1\right)$ lie on the graph of $f(x) = \log_2 x$, as shown in Figure 4. The graph of $y = \log_2 x$ is the reflection about $y = x$ of the graph of $y = 2^x$.

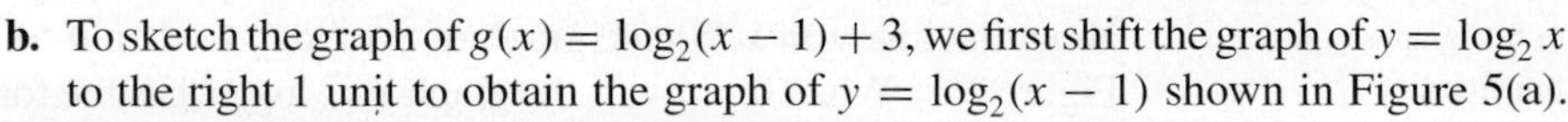

b. To sketch the graph of $g(x) = \log_2(x - 1) + 3$, we first shift the graph of $y = \log_2 x$ to the right 1 unit to obtain the graph of $y = \log_2(x - 1)$ shown in Figure 5(a).

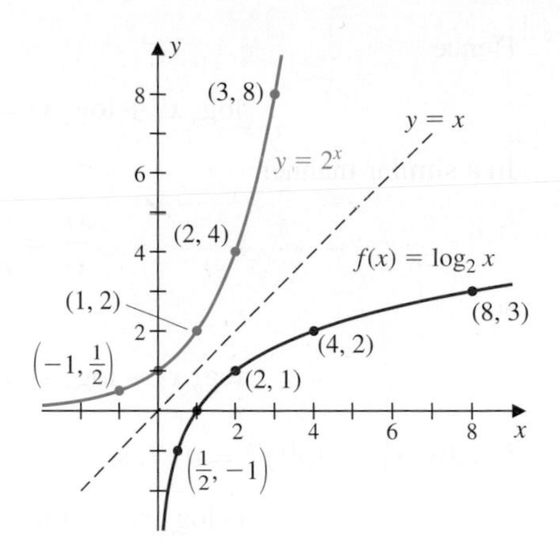

FIGURE 4

Shifting the graph of $y = \log_2(x - 1)$ upward 3 units gives the graph of $g(x) = \log_2(x - 1) + 3$ shown in Figure 5(b). The domain of g is the interval $(1, \infty)$, the range is the set of all real numbers, and the graph has a vertical asymptote at $x = 1$. ■

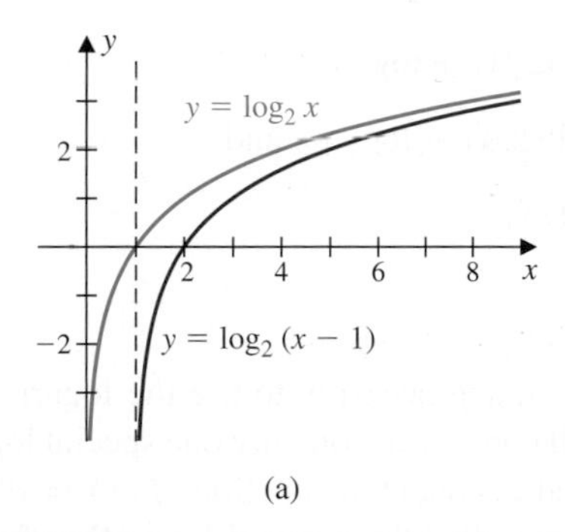

(a)

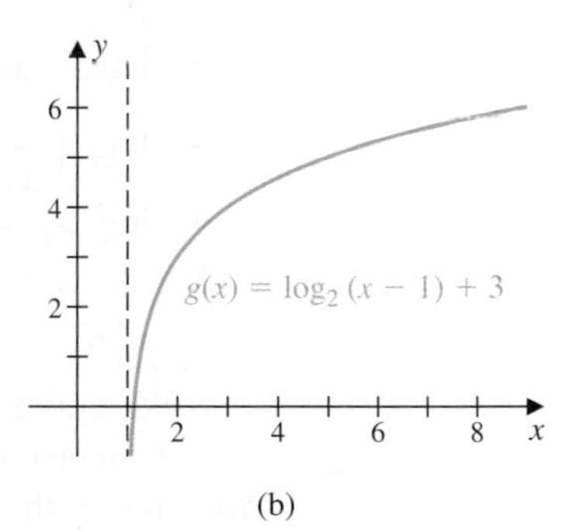

(b)

FIGURE 5

At the beginning of Section 5.1 we listed the arithmetic properties of exponents that hold for all positive numbers a and all real numbers r_1 and r_2:

$$a^{r_1+r_2} = a^{r_1}a^{r_2}, \qquad a^{r_1-r_2} = \frac{a^{r_1}}{a^{r_2}}, \qquad \text{and} \qquad (a^{r_1})^{r_2} = a^{r_1 r_2}.$$

Each of these rules has a logarithm equivalent that follows from the inverse relationship with the exponential function. For example, if we let

$$\log_a x_1 = r_1 \quad \text{and} \quad \log_a x_2 = r_2,$$

then the exponential-logarithm conversion tells us that

$$x_1 = a^{r_1} \quad \text{and} \quad x_2 = a^{r_2},$$

so

$$x_1x_2 = a^{r_1}a^{r_2} = a^{r_1+r_2}.$$

The inverse relationship between the logarithm and exponential function implies that

$$r_1 + r_2 = \log_a (x_1x_2).$$

Hence

$$\log_a x_1 + \log_a x_2 = r_1 + r_2 = \log_a (x_1 x_2).$$

In a similar manner,

$$\frac{x_1}{x_2} = \frac{a^{r_1}}{a^{r_2}} = a^{r_1 - r_2},$$

so

$$\log_a x_1 - \log_a x_2 = r_1 - r_2 = \log_a \left(\frac{x_1}{x_2}\right).$$

Finally, $x_1^{r_2} = (a^{r_1})^{r_2} = a^{r_1 r_2}$, so

$$r_2 \log_a x_1 = \left(\log_a x_1\right) r_2 = r_1 r_2 = \log_a x_1^{r_2}.$$

In summary, the following arithmetic properties are true for every logarithm function.

Arithmetic Properties of Logarithms

Suppose that x_1 and x_2 are positive real numbers, and that r is a real number. Then

- $\log_a (x_1 x_2) = \log_a x_1 + \log_a x_2$,
- $\log_a \left(\dfrac{x_1}{x_2}\right) = \log_a x_1 - \log_a x_2$, and
- $\log_a x_1^r = r \log_a x_1$.

We will not have much occasion to use the logarithm functions with base $0 < a < 1$. In fact, we will concentrate on only one special logarithm function, the inverse function to the natural exponential function, $f(x) = e^x$. The inverse to the natural exponential function is called the **natural logarithm function** and is written using the special notation $\ln x$.

- $\ln x$ is equivalent to $\log_e x$, the inverse function of $f(x) = e^x$.

The Natural Logarithm Function

For each x in $(0, \infty)$,

- $y = \ln x$ if and only if $x = e^y$.

So for each x in $(0, \infty)$ and each real number y,

- $\ln e^y = y$ and $e^{\ln x} = x$.

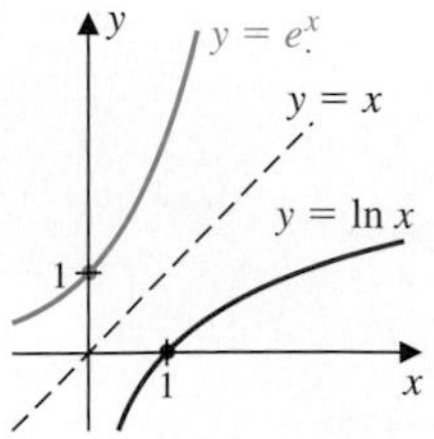

FIGURE 6

The graph of the natural logarithm function is shown in Figure 6. The properties of the natural logarithm function $f(x) = \ln x$ listed in the box follow from the corresponding properties of the natural exponential function $g(x) = e^x$.

Properties of $f(x) = \ln x$ and $g(x) = e^x$

- The domain of f and the range of g is $(0, \infty)$.
- The range of f and the domain of g is $(-\infty, \infty)$.
- The x-intercept for the graph of $y = \ln x$ is $(1, 0)$.
- The function $f(x) = \ln x$ is always increasing.
- The graph of $y = \ln x$ has the y-axis as a vertical asymptote.
- $\ln x \to \infty \quad \text{as} \quad x \to \infty \quad \text{and} \quad \ln x \to -\infty \quad \text{as} \quad x \to 0^+$
- Since exponential growth is very fast, logarithmic growth is very slow.

The arithmetic rules for logarithms are restated for the natural logarithm function, because they will be used frequently.

Arithmetic Properties of Natural Logarithms

The arithmetic properties of logarithms are used to simplify complicated logarithmic expressions.

Suppose that x_1 and x_2 are positive real numbers, and that r is a real number. Then

- $\ln(x_1 x_2) = \ln x_1 + \ln x_2$,
- $\ln\left(\dfrac{x_1}{x_2}\right) = \ln x_1 - \ln x_2$, and
- $\ln x_1^r = r \ln x_1$.

EXAMPLE 4 Determine the values of x that satisfy the expression.

a. $\ln(x^2 + 1) = \ln(x - 1) + \ln(x + 2)$ **b.** $\ln 3 + \ln(2x - 1) = 2\ln(x + 1)$

c. $\ln(x - 1) + \ln(x - 3) = 3\ln 2$ **d.** $\frac{1}{2} = \ln\sqrt{(1 - x)/(1 + x)}$

Solution **a.** The product rule for logarithms implies that

$$\ln(x^2 + 1) = \ln(x - 1) + \ln(x + 2) = \ln(x - 1)(x + 2).$$

The natural logarithm is a one-to-one function, so we have

$$x^2 + 1 = (x - 1)(x + 2) = x^2 + x - 2.$$

This simplifies to $1 = x - 2$, so $x = 3$.

b. In this case, we use both the product and exponent rules to produce

$$\ln 3 + \ln(2x - 1) = \ln 3(2x - 1) = \ln(6x - 3)$$

and

$$2\ln(x + 1) = \ln(x + 1)^2 = \ln(x^2 + 2x + 1).$$

The two expressions are equal provided that

$$x^2 + 2x + 1 = 6x - 3.$$

This equation can be simplified to

$$0 = x^2 - 4x + 4 = (x-2)^2, \quad \text{so} \quad x = 2.$$

c. The equation can be written in the form

$$0 = \ln(x-1) + \ln(x-3) - 3\ln 2,$$

so we can use the arithmetic properties to rewrite the equation as

$$0 = \ln(x-1)(x-3) - \ln 2^3 = \ln \frac{(x-1)(x-3)}{8}.$$

Applying the exponential-logarithm conversion and using the fact that $e^0 = 1$ gives

$$1 = \frac{(x-1)(x-3)}{8} \quad \text{so} \quad 8 = (x-1)(x-3) = x^2 - 4x + 3.$$

As a consequence,

$$0 = x^2 - 4x - 5 = (x-5)(x+1), \quad \text{so} \quad x = 5 \text{ or } x = -1.$$

However, the value $x = -1$ cannot be substituted into the original equation because the natural logarithm function is not defined for negative numbers. So $x = -1$ is an *extraneous* solution.

We do have a solution at $x = 5$, because substituting this value in the equation gives

$$\ln(5-1) + \ln(5-3) = \ln 4 + \ln 2 = \ln 2^2 + \ln 2 = 2\ln 2 + \ln 2 = 3\ln 2.$$

d. We can rewrite this equation as

$$\frac{1}{2} = \ln \sqrt{\frac{1-x}{1+x}} = \ln \left(\frac{1-x}{1+x}\right)^{1/2} = \frac{1}{2} \ln \frac{1-x}{1+x},$$

so

$$1 = \ln \frac{1-x}{1+x} \quad \text{and} \quad \frac{1-x}{1+x} = e.$$

We now solve this equation for x. Since

$$1 - x = (1+x)e = e + ex, \quad \text{we have} \quad 1 - e = x + ex = (1+e)x,$$

and

$$x = \frac{1-e}{1+e}.$$

■

EXAMPLE 5 Use the properties of logarithms to determine the values of x that satisfy

$$e^{1-2x} = 2 \cdot 3^{x+1}.$$

Solution We first take the natural logarithm of both sides of the equation. This gives

$$\ln e^{1-2x} = \ln(2 \cdot 3^{x+1}), \quad \text{so} \quad 1 - 2x = \ln 2 + \ln 3^{x+1} = \ln 2 + (x+1)\ln 3.$$

Isolating x in the last equation gives

$$-2x - x\ln 3 = \ln 2 + \ln 3 - 1 \quad \text{so} \quad x(-2 - \ln 3) = \ln 2 + \ln 3 - 1$$

and

$$x = \frac{1 - \ln 2 - \ln 3}{2 + \ln 3}.$$

■

EXAMPLE 6 Find the domain of the function

$$f(x) = \ln\left(\frac{x^2 - 1}{x - 3}\right).$$

Solution The natural logarithm is defined when

$$\frac{x^2 - 1}{x - 3} > 0, \quad \text{that is, when} \quad \frac{(x+1)(x-1)}{x - 3} > 0.$$

The sign chart in Figure 7 shows that the domain of the function is $(-1, 1) \cup (3, \infty)$.

■

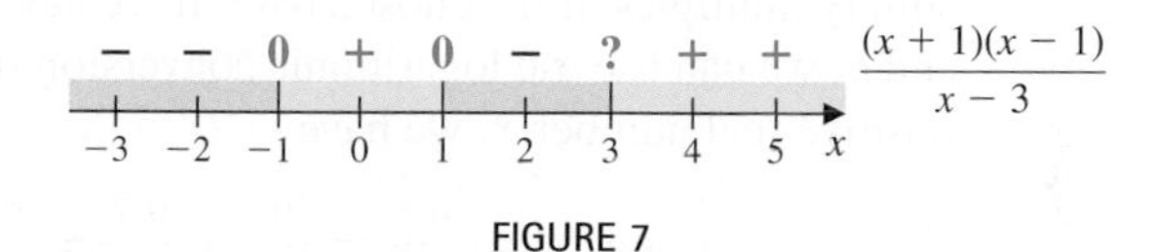

FIGURE 7

In Section 5.2 we found that any exponential functions is a scaling of any other exponential function. In particular, then, all exponential functions are a scaling of the natural exponential function. For any positive $a \neq 1$ there is a a number k with

$$a^x = e^{kx}.$$

In Section 5.2 we could not find the exact value for k, but the arithmetic properties of the logarithm permit us to determine k.

Applying the natural logarithm to both sides of the equation $a^x = e^{kx}$ gives

$$\ln a^x = \ln e^{kx} \text{ is equivalent to } x \ln a = kx \ln e = kx \cdot 1 = kx, \quad \text{so } k = \ln a.$$

So for any exponential function $f(x) = a^x$ we have

$$a^x = e^{(\ln a)x}.$$

This relationship between a general exponential function of the form $f(x) = a^x$ and the natural exponential function $f(x) = e^x$ is critical for working with exponential functions in calculus. In fact, if you remember only one fact about general exponential functions, this is the one you want.

This conversion formula gives the critical relationship between a^x and e^x.

General Exponential to Natural Exponential Conversion

For any positive real number a and every real number x,

- $a^x = e^{(\ln a)x}$.

We can gain even more from this conversion formula. Let $y = \log_a x$. Since

$$y = \log_a x \quad \text{is equivalent to} \quad x = a^y = e^{(\ln a)y},$$

we have

$$\ln x = \ln\left(e^{(\ln a)y}\right) = (\ln a) y \ln e = (\ln a) y \cdot 1 = (\ln a)\log_a x.$$

As a consequence, every logarithm function is simply a multiple of the natural logarithm function.

This conversion formula gives the critical relationship between logarithms.

General Logarithm to Natural Logarithm Conversion

For every positive real number $a \neq 1$ and every $x > 0$,

- $\log_a x = \dfrac{\ln x}{\ln a}$.

This result implies that all logarithm functions are essentially equivalent, and we can choose any one of them for our basic function; the other logarithm functions are simply multiples of the chosen base. If we have two logarithm functions with bases a and b, we can use the logarithmic conversion property twice to deduce that for every positive real number x, we have

$$\log_b x = \frac{\ln x}{\ln b} = \frac{\ln x}{\ln b} \cdot \frac{\ln a}{\ln a} = \frac{\ln a}{\ln b} \cdot \frac{\ln x}{\ln a} = \frac{\ln a}{\ln b} \log_a x.$$

This shows that every logarithm function is a constant multiple of any other logarithm function.

Notice that when we use $x = a$ in this result we obtain

$$\log_b a = \frac{\ln a}{\ln b} \cdot \log_a a = \frac{\ln a}{\ln b}.$$

In a similar manner, we have

$$\log_a b = \frac{\ln b}{\ln a}.$$

Putting these results together tells us that for every pair of positive real numbers $a \neq 1$ and $b \neq 1$, we have

- $\log_b a = \dfrac{1}{\log_b a}$.

Applications

EXAMPLE 7 In Example 10 of Section 5.2 we found that the function $Q(t) = 2^{2t}$ predicts the number of cells produced by a single cell in a frog embryo if each cell divides approximately every half hour. How many hours will it take for 1 cell to produce 50,000 cells?

Solution We need to determine t so that

$$50000 = Q(t) = 2^{2t} = 2^{2t} = e^{2t \ln 2}.$$

Converting to logarithms gives

$$2t \ln 2 = \ln 50000, \quad \text{so} \quad t = \frac{\ln 50000}{2 \ln 2} \approx 7.8 \text{ hours.}$$

■

For calculus applications the natural logarithm function is basic, but in some instances it is more useful to choose the logarithm with base 10. This base-10 logarithm is called the *common logarithm.* Because computers represent numbers using the binary system, which uses the digits 0 and 1 only, it is also common in computer science to use the logarithm with base 2.

EXERCISE SET 5.3

In Exercises 1–16, evaluate the expression.

1. $\log_4 4^3$

2. $\log_2 32$

3. $\log_4 64$

4. $\log_8 2^{12}$

5. $\log_4 2$

6. $\log_{25} 5$

7. $\log_{10} 0.001$

8. $\log_{10} 10000$

9. $\log_2 \frac{1}{8}$

10. $\log_4 \frac{1}{2}$

11. $e^{\ln 5}$

12. $5^{\log_5 6}$

13. $\ln e^{1/3}$

14. $\log_2 2^{\sqrt{3}}$

15. $e^{2 \ln \pi}$

16. $e^{-1/2 \ln 16}$

In Exercises 17–26, use the properties of logarithms to simplify the expression so that the result does not contain logarithms of products, quotients, or powers.

17. $\ln x(x+1)$

18. $\ln \dfrac{1}{x}$

19. $\log_3 \dfrac{x^4}{x+1}$

20. $\log_2 (2x-1)^5$

21. $\ln \dfrac{2x^3}{(x+4)^2}$

22. $\ln \dfrac{x\sqrt[3]{x^2}}{(x+2)^3}$

23. $\log_3 \dfrac{(3x+2)^{3/2}(x-1)^3}{x\sqrt{x+1}}$

24. $\ln \dfrac{x^2\sqrt{x+1}}{\sqrt[3]{x^2+2x+1}}$

25. $\ln \sqrt{x\sqrt{x+1}}$

26. $\ln \sqrt{x^2 \sqrt{\dfrac{x-2}{x+3}}}$

In Exercises 27–32, rewrite the expression as a single logarithm.

27. $\ln x + 2\ln(x+1)$

28. $2\ln(x+2) + 3\ln(x-1)$

29. $\frac{1}{2}\ln x - 2\ln(x-1)$

30. $2\ln x - \frac{1}{3}\ln(x+1)$

31. $\ln(x-1) + \frac{1}{2}\ln x - 2\ln x$

32. $\ln(x^2+x+1) - 3\ln(x+2) + \ln x$

In Exercises 33–54, use the properties of logarithms to solve the equation for x.

33. $\log_3 x = 4$

34. $\log_2 x = 5$

35. $\log_2(3x-4) = 3$

36. $\log_3(2-x) = 2$

37. $\log_x 4 = 2$

38. $\log_x 3 = \frac{1}{3}$

39. $\ln(2-x) = 4$

40. $1 - \ln(3x+2) = 0$

41. $\ln 2 + \ln(x+1) = \ln(4x-7)$

42. $2\ln x = \ln 4 + \ln(x+3)$

43. $2\ln x = \ln(4x+6) - \ln 2$

44. $\ln x + \ln(x-1) = \ln 2$

45. $\ln(2x-1) - \ln(x-1) = \ln 5$

46. $2\ln(x+2) - \ln x = \ln 8$

47. $\log_3(2x^2+17x) = 2$

48. $\log_2(5x^2-8x) = 2$

49. $4^x = 3$

50. $5^{2x-1} = 2$

51. $e^{2x} = 3^{x-4}$

52. $2^{x-2} = e^{x/2}$

53. $2 \cdot 3^{-x} = 2^{3x}$

54. $3e^{-x} = 4^{3x-1}$

In Exercises 55–64, sketch the graph of the function.

55. $y = \log_2(x-3)$

56. $y = -\log_2(x+4)$

57. $y = 2 - \log_2(x-1)$

58. $y = \log_3(x-2) + 1$

59. $y = 2\ln(x+1) - 3$

60. $y = -2\ln(x-1) + 1$

61. $y = \ln(-x)$

62. $y = \ln(3-x)$

63. $y = |\ln x|$

64. $y = \ln|x|$

65. Match the equation with the graph.

a. $y = -\ln x + 1$

b. $y = \ln(x - 1) - 1$

c. $y = \ln(-x)$

d. $y = 2\ln(x - 1) - 1$

i.

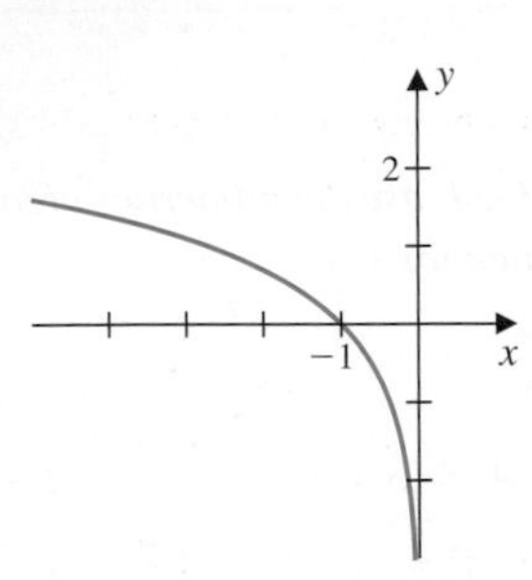

ii.

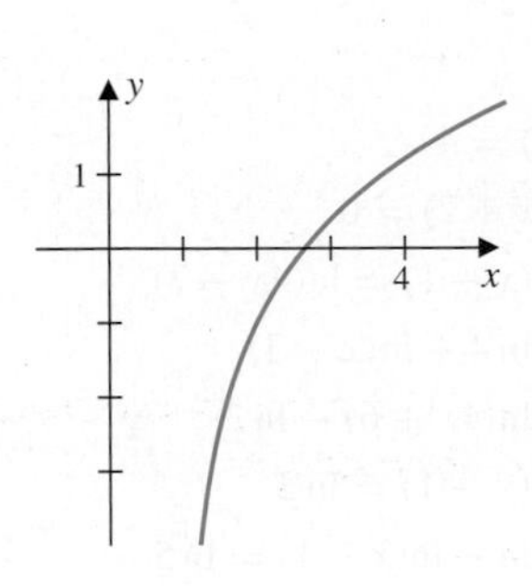

iii.

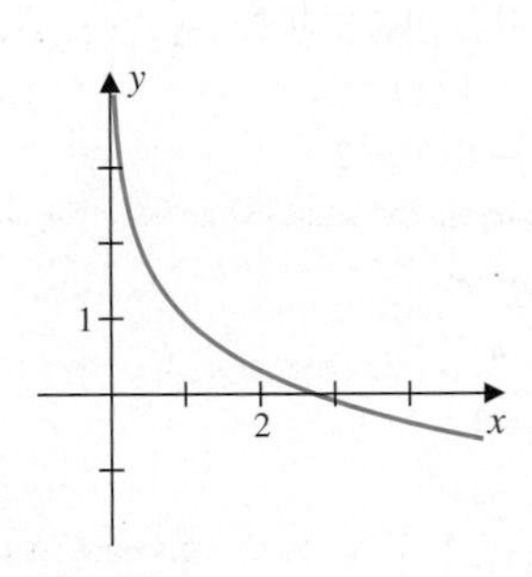

iv.

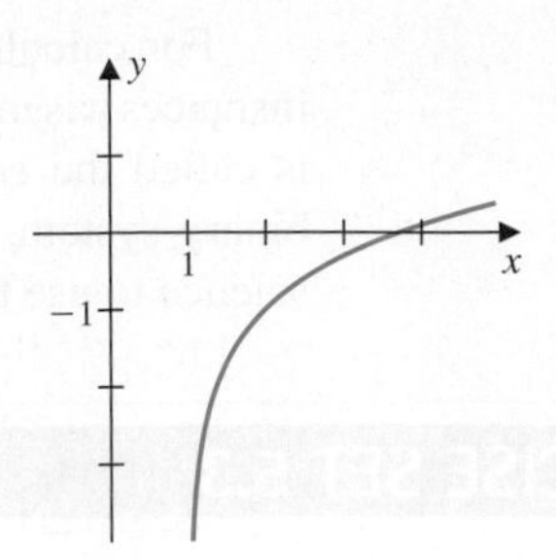

66. Sketch the graphs of $f(x) = c\ln(x - 1) + 1$, for $c = 1, 2, -1$, and -2, on the same set of axes.

In Exercises 67–70, use a graphing device to sketch the graph of the function.

67. $f(x) = \ln(4 - x^2)$

68. $f(x) = x^2 - \ln x$

69. $f(x) = \dfrac{\ln x}{x}$

70. $f(x) = \ln|x^2 - 1|$

71. Let $f(x) = a + \ln x$ and $g(x) = \sqrt[n]{x}$ for a positive integer n and a real number a. Use a graphing device to sketch the graphs of f and g for different values of a and n, and determine which functions grow more rapidly as $x \to \infty$.

72. Determine the value of k for which $3^x = e^{kx}$, and compare this result to the approximation found in Exercise 31 of Section 5.2.

73. A population initially contains 1000 individuals. Each year 4% of the population is lost due to deaths and the population gains 6% due to births. Approximately how much time will it take for this population to double?

74. A population of bacteria triples in size every 6 hours, and the culture started with 500 bacteria. Determine the number of hours it will take for this population to grow to 5000.

5.4 EXPONENTIAL GROWTH AND DECAY

In many applications calculus is needed to get started, but the final solution requires only an understanding of the natural exponential and logarithm functions.

In many natural settings, quantities grow or decay at a rate that is approximately proportional to the amount of the quantity present at the time. For example, the rate at which some cultures of bacteria and animal populations increase is proportional to the size of the population. The mass of a radioactive substance *decays*, or decreases with time, at a rate proportional to the mass. Suppose that the rate of growth or decay of some quantity at a given instant is proportional to the amount present at that instant. Then the quantity present at any time can be described using the natural exponential function.

Suppose that Q_0 denotes the initial amount of a quantity and $Q(t)$ is the amount at any time t. In addition, assume that the change in $Q(t)$ is proportional to $Q(t)$. Then $Q(t)$ is described by

$$Q(t) = Q_0 e^{kt}.$$

- If $k > 0$, then Q **grows exponentially** with time.
- If $k < 0$, then Q **decays exponentially** with time.

The constant k is called the **constant of proportionality**. A typical instance of these concepts is illustrated in Figure 1.

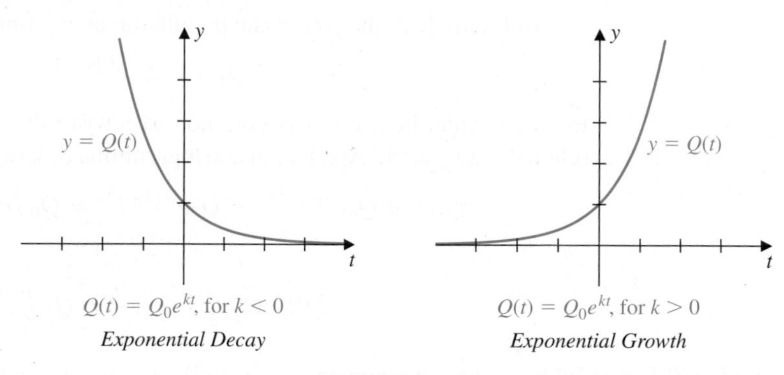

$Q(t) = Q_0 e^{kt}$, for $k < 0$
Exponential Decay

$Q(t) = Q_0 e^{kt}$, for $k > 0$
Exponential Growth

FIGURE 1

In Section 5.2 we saw that if interest is compounded continuously at an interest rate i, then an initial investment of A_0 dollars is worth

$$A(t) = A_0 e^{it}$$

dollars after t years. This is an example of exponential growth. The next examples illustrate this behavior.

EXAMPLE 1 What initial investment compounded continuously at 8% will grow to \$150,000 in 20 years?

Solution Let A_0 denote the initial investment. The amount at any time t is

$$A(t) = A_0 e^{0.08t}.$$

The investment is to grow to \$150,000 in 20 years, so the required initial investment is found from the equation

$$150000 = A_0 e^{0.08 \cdot 20}, \quad \text{and} \quad A_0 = \frac{150000}{e^{1.6}} \approx \$30{,}285.$$ ■

EXAMPLE 2 A laboratory researcher observes that a culture of cells triples in size in 2 days. How large will the culture be in 5 days if the population grows exponentially?

Solution Let the initial size be Q_0, and assume that

$$Q(t) = Q_0 e^{kt}.$$

To determine k, we use the fact that the population triples in 2 days. This implies that

$$Q(2) = 3Q_0 = Q_0 e^{2k} \quad \text{and} \quad e^{2k} = 3.$$

Converting the last equation using natural logarithms gives

$$2k = \ln 3 \quad \text{and} \quad k = \frac{\ln 3}{2}.$$

As a consequence, the size at time t is

$$Q(t) = Q_0 e^{((\ln 3)/2)t}.$$

After 5 days the size of the culture is

$$Q(5) = Q_0 e^{5(\ln 3)/2} \approx Q_0 e^{2.747} \approx 15.6 Q_0.$$

■

In Example 2, the size of the population at any time t is found to be

$$Q(t) = Q_0 e^{((\ln 3)/2)t}.$$

In some scientific areas it is common to rewrite this expression using the inverse relation between the exponential and logarithmic functions. For example, we can write

$$Q(t) = Q_0 e^{((\ln 3)/2)t} = Q_0 e^{(\ln 3)(t/2)} = Q_0 \left(e^{\ln 3}\right)^{t/2} = Q_0 (3)^{t/2}$$

so

$$Q(t) = Q_0 ((3)^{1/2})^t = Q_0 \left(\sqrt{3}\right)^t.$$

EXAMPLE 3 Table 1 gives the population, in millions, of the United States for the years 1940 through 2010. Assume that the population changes by an amount proportional to the amount of population present. Predict the population in the years 2000 and 2010 using the population figures given.

a. 1940 and 1950 **b.** 1980 and 1990

TABLE 1

Year	Population	Year	Population
1940	131	1980	226
1950	150	1990	250
1960	179	2000	281
1970	203	2010	309

Solution **a.** Suppose that $Q(t)$ represents the population t years after 1930, and that

$$Q(t) = Q_0 e^{kt}$$

The initial population is the 1940 population, so $Q(0) = Q_0 = 131$. To find the proportionality constant k, we will use the population 10 years later, in 1950. Then

$$150 = Q(10) = 131 e^{10k}, \quad \text{so} \quad \frac{150}{131} = e^{10k} \quad \text{and} \quad k = \frac{1}{10} \ln\left(\frac{150}{131}\right).$$

These models of population growth are quite crude. You will see better models in calculus and differential equations.

Hence

$$Q(t) = 131 e^{[(1/10)\ln(150/131)]t} \approx 131 e^{0.0135t}.$$

The years 2000 and 2010 correspond to $t = 60$ and $t = 70$ years after 1940, respectively, so the populations are predicted to be approximately

$$Q(60) = 131e^{(0.0135)(60)} \approx 295 \text{ million}$$

and

$$Q(80) = 131e^{(0.0135)(70)} \approx 338 \text{ million.}$$

Both predictions are too large, as Figure 2 indicates.

To plot the original data points and this exponential approximation on the same set of axes, we need to shift the exponential function to the right by 1940 so that the starting time is $t = 0$. That is, we plot

$$y = 131e^{[(1/10)\ln(150/131)](t-1940)}.$$

The data points are shown in Figure 2 together with the graph (in blue).

A graphing device can quickly plot the original data points and the function used to fit the data.

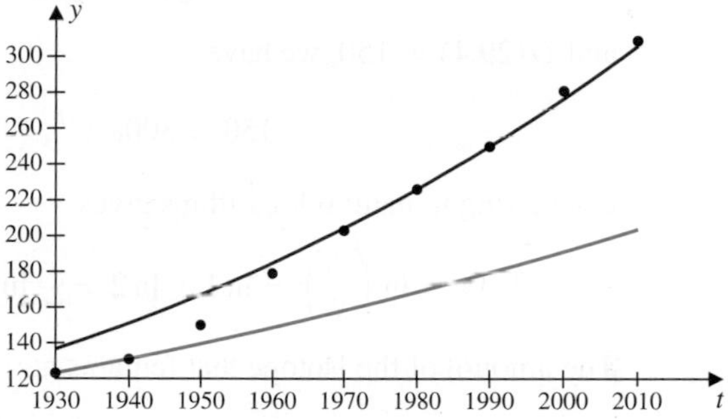

FIGURE 2

b. Suppose we instead use the 1980 and 1990 figures to generate the exponential function. We now let $Q(t)$ represent the population t years after 1980. Then the same type of calculations as in part (a) gives

$$Q(t) = 226e^{[(1/10)\ln(250/226)]t} \approx 226e^{0.0101t}.$$

The new estimates for the populations in 2000 and 2010 are approximately

$$Q(20) = 226e^{(0.0101)(20)} \approx 277 \text{ million}$$

and

$$Q(30) = 226e^{(0.0101)(30)} \approx 306 \text{ million,}$$

which are quite accurate. The red graph in Figure 2 shows

$$y = 226e^{[(1/10)\ln(250/226)](t-1980)}.$$

This function gives a much better approximation to the later data points, one that is more in line with the U.S. Census data. However, it is a poor model of the population before 1960. ■

Population growth is only approximated by an exponential function over small intervals of time, because the assumption of a constant growth rate is too restrictive. The growth rate is usually dependent on additional conditions such as the availability of a food supply and pressures due to overcrowding. Example 4 concerns the decay of a radioactive substance. In this case the exponential function gives a true quantitative picture.

EXAMPLE 4 The radioactive isotope strontium 90 has a *half-life* of 29.1 years. This half-life is the time it takes for one half of the original amount to decay into another substance.

a. How much strontium 90 will remain after 20 years from an initial amount of 300 kilograms?

b. How long will it take for 80% of the original amount to decay?

Solution Let $Q(t)$ be the amount that remains after t years. Then $Q(0) = 300$ and, because the half-life is 29.1 years, $Q(29.1) = 150$. Using this information, we can find the proportionality constant k. Since

$$Q(t) = Q_0 e^{kt} = 300e^{kt},$$

and $Q(29.1) = 150$, we have

$$150 = 300e^{29.1k}, \quad \text{so} \quad e^{29.1k} = \frac{1}{2}.$$

Converting to natural logarithms gives

$$29.1k = \ln\left(\frac{1}{2}\right) = \ln 1 - \ln 2 = -\ln 2, \quad \text{so} \quad k = -\frac{\ln 2}{29.1} \approx -0.0238.$$

The amount of the isotope that remains at time t is

$$Q(t) = 300e^{(-\ln 2/29.1)t} \approx 300e^{-0.0238t}.$$

a. With $t = 20$ the amount after 20 years is

$$Q(20) \approx 300e^{-0.0238(20)} = 300e^{-0.476} \approx 300(0.621) \approx 186 \text{ kilograms}.$$

b. When 80% of the original amount has decayed, there will be 20%, or 60 kilograms, remaining. To find the time t, we solve $Q(t) = 60$ for t, that is,

$$300e^{(-\ln 2/29.1)t} = Q(t) = 60, \quad \text{so} \quad e^{(-\ln 2/29.1)t} = 0.2.$$

Hence 80% will have decayed when

$$-\frac{\ln 2}{29.1}t = \ln 0.2, \quad \text{and} \quad t = -\frac{29.1 \ln 0.2}{\ln 2} \approx 67.6 \text{ years}.$$ ■

At the beginning of the chapter we discussed how scientists can determine the approximate age of ancient objects by a technique called radioactive carbon dating. All living tissue contains carbon isotopes. Approximately 98.89% of this carbon consists of the stable isotope carbon 12, and most of the remainder is the stable isotope carbon 13. But a small amount, about one part in one trillion (1 part in 10^{12}), is the radioactive isotope carbon 14. The carbon 14 isotope decays over time to produce nitrogen 14.

Radioactive carbon dating makes the assumption that the percentage of various isotopes of carbon in all living things remains constant throughout history. The date at which an ancient organism died can be estimated by comparing the current carbon 14 to carbon 12 proportion to the original proportion of these isotopes. The half-life of carbon 14 is about 5730 years.

EXAMPLE 5 Estimate the age of the Ice Man discussed in the opening of this chapter, assuming that 52.4% of the original amount of carbon 14 remained at the time of the discovery.

Solution Let $Q(t)$ represent the amount of carbon 14 present t years after the man died. Then

$$Q(t) = Q(0)e^{kt} = Q_0 e^{kt}.$$

The half-life of carbon 14 is 5730 years, so

$$\frac{1}{2}Q_0 = Q(5730) = Q_0 e^{5730k}.$$

Thus

$$5730k = \ln\left(\frac{1}{2}\right) = \ln 1 - \ln 2 = -\ln 2, \quad \text{so} \quad k = -\frac{\ln 2}{5730} \approx -0.000121.$$

This gives

$$Q(t) = Q_0 e^{(-\ln 2/5730)t} \approx Q_0 e^{-0.000121t}.$$

Since 52.4% of the carbon 14 was actually present, we need to find t with

$$0.524Q_0 = Q_0 e^{(-\ln 2/5730)t}, \quad \text{that is,} \quad t = -\frac{5730 \ln 0.524}{\ln 2} \approx 5342.$$

Consequently, the Ice Man lived about 5300 years ago, around 3000 B.C.E. ■

Many physical problems involve rates and proportions that can be solved by rewriting the problem in the form of a growth or decay problem. One example is Newton's Law of Cooling, which is used to predict the temperature $T(t)$ of an object.

Suppose an object with initial temperature $T(0)$ is brought into an area of constant temperature T_m. The rate at which the temperature changes is proportional to $T(t) - T_m$. To see how this problem is changed into the growth-decay format, define

$$S(t) = T(t) - T_m.$$

Since T_m is constant, the rate at which $S(t)$ changes is the same as the rate at which $T(t)$ changes.

Newton's Law of Cooling states that $T(t)$ changes at a rate proportional to $T(t) - T_m = S(t)$, so the rate at which $S(t)$ changes is proportional to $S(t)$ and, for some constant k,

$$S(t) = S(0)e^{kt}.$$

Changing back to the function T gives us

$$T(t) - T_m = (T(0) - T_m)\, e^{kt} \quad \text{and} \quad T(t) = T_m + (T(0) - T_m)\, e^{kt}.$$

EXAMPLE 6 A well-cooked turkey with uniform temperature 170° is taken out of an oven and placed in a 70° room. After 5 minutes, the turkey's temperature has decreased to 160°. What would be the expected temperature of the bird after another 15 minutes?

Solution If $T(t)$ is the temperature of the turkey t minutes after being brought out of the oven, then $T(0) = 170$, $T_m = 70$, and

$$T(t) = T_m + (T(0) - T_m)\, e^{kt} = 70 + 100e^{kt} \quad \text{for some constant } k.$$

Since $T(5) = 160$, we can determine k from the equation

$$160 = 70 + 100e^{5k}, \quad \text{so} \quad 90 = 100e^{5k}$$

and

$$k = \frac{1}{5}\ln\frac{9}{10} \approx -0.0211.$$

The temperature 20 minutes after the turkey is taken from the oven is

$$T(20) = 70 + 100e^{(1/5)\ln(9/10)(20)} \approx 70 + 100e^{-0.0211(20)} \approx 136°.$$ ■

EXERCISE SET 5.4

1. A bacteria culture starts with 2000 bacteria and the population doubles every 3 hours.

a. Find an expression for the number of bacteria after t hours.

b. Find the number of bacteria that will be present after 6 hours.

c. When will the population reach 22,000?

2. Under ideal conditions, a cell of the bacteria *Escherichia coli*, commonly found in the human intestine, divides to create two cells in approximately 22 minutes. Assume the initial population is 200 cells.

a. Find an expression for the number of cells after t hours.

b. Find the number of cells that will be present after 10 hours.

c. When will the population reach 10,000 cells?

3. The bacteria in a culture increase from 500 at 1:00 P.M. to 4000 at 6:00 P.M.

a. Find an expression for the number of bacteria t hours after 1:00 P.M.

b. Find the number of bacteria that will be present at 7:00 P.M.

c. When will the population reach 15,000?

d. How long does it take the population to double in size?

4. A bacteria culture started with 8500 bacteria and increased by 15% in the first 2 hours.

a. Find an expression for the number of bacteria t hours after the culture was first observed.

b. Find the number of bacteria that will be present after 8 hours.

c. When will the population reach 35,000?

d. How long does it take the population to triple in size?

5. The radioactive isotope thorium 234 has a half-life of approximately 578 hours.

a. If a sample has a mass of 64 milligrams, find an expression for the mass after t hours.

b. How much will remain after 75 hours?

c. When will the initial mass decay to 12 milligrams?

6. A certain radioactive substance has a half-life of 8 years.

a. If a sample has mass 200 grams, find an expression for the mass after t years.

b. How much of a 200-gram sample will remain after 15 years?

c. How long will it take for 90% of the sample to decay?

7. A culture of bacteria doubles in size every 2 hours. How long will it take to triple in size?

8. A culture of bacteria doubles after 4 hours. What proportion of the original number of bacteria will be present **(a)** after 8 hours and **(b)** after 16 hours?

9. Find the half-life of a radioactive substance if 220 grams of the substance decays to 200 grams in 4 years.

10. Find the half-life of a radioactive substance that decays by 5% in 9 years.

11. The table gives estimates of the world population, in millions, from 1950 to 2000, taken from the U.S. Census Bureau.

Year	Population
1950	2555
1960	3040
1970	3708
1980	4455
1990	5275
2000	6079

a. Use the exponential model and the population figures from 1950 and 1960 to predict the world population in the year 2050.

b. Use the exponential model and the population figures from 1990 and 2000 to predict the world population in the year 2050.

12. The population of the world was approximately 6.3 billion in July 2003. Assuming that the population will grow by 2% in 1 year, estimate the following.

a. When the population of the world will double.

b. When the population of the world will triple.

13. If $10,000 is invested in an account that returns 8% per year compounded continuously, how long will it take for the investment to **(a)** double and **(b)** triple?

14. If $25,000 is invested in an account that returns 7% per year compounded continuously, how long will it take for the investment to **(a)** double and **(b)** triple?

15. An initial investment of $10,000 is made in an account where interest is compounded continuously. If the investment is to grow to $25,000 in 5 years, what is the required interest rate?

16. The parents of a 3-year-old child put $20,000 into an account with the hope that the amount will grow to $120,000 when the child starts college in 15 years. What rate of continuously compounded interest is necessary for this goal to be met?

17. An outdoor thermometer reading $-3°C$ is brought into a room at 20°C. One minute later the thermometer reads 5°C. How long will it take to reach 19.5°C?

18. A potato at room temperature of 20°C is placed in an oven whose temperature is 200°C. An hour later the temperature of the potato is 150°C. Under the (unreasonable) assumption that the potato's temperature is always uniform, how long did it take for the potato to reach 50°C?

19. A body is found floating face down in a lake. When the body was taken from the water at noon its temperature was 66°F. The temperature of the body when first found at 11:00 A.M. was 67°F. The lake has a constant temperature of 62°F and the body of the victim was a normal 98.6°F before going into the water. When did the victim drown?

20. The rate of change of air pressure P with respect to altitude h is proportional to P when the temperature is constant. Suppose at sea level the pressure is 1.01×10^5 pascals (Pa), and at altitude $h = 2$ kilometers the pressure is 8.08×10^4 Pa. Find the atmospheric pressure at 5 kilometers.

21. Archaeologists call piles of empty shells *middens*, and these are often used to help date the periods of ancient civilizations. The oldest known oyster shell midden is located in Dobbs Ferry, New York, and shells in this pile have been determined to have only about 33.97% of the amount of carbon 14 that they had when the oysters were alive. In about what year were the oysters consumed?

22. Prehistoric cave drawings have been discovered in a number of locations in south-eastern France. The amount of carbon 14 remaining in the artifacts is estimated at 21%. What is the approximate age of the drawings?

REVIEW EXERCISES FOR CHAPTER 5

1. Match the equation with the graph.

a. $y = -e^{1-x}$

b. $y = e^{x-2} - 1$

c. $y = e^{x+2} - 1$

d. $y = -e^{x+1}$

i.

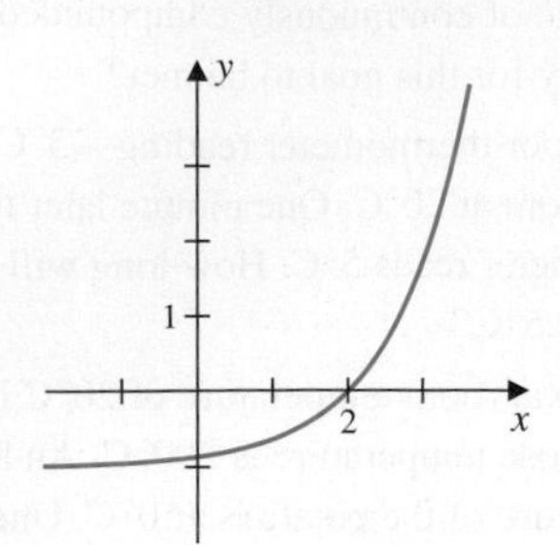

ii.

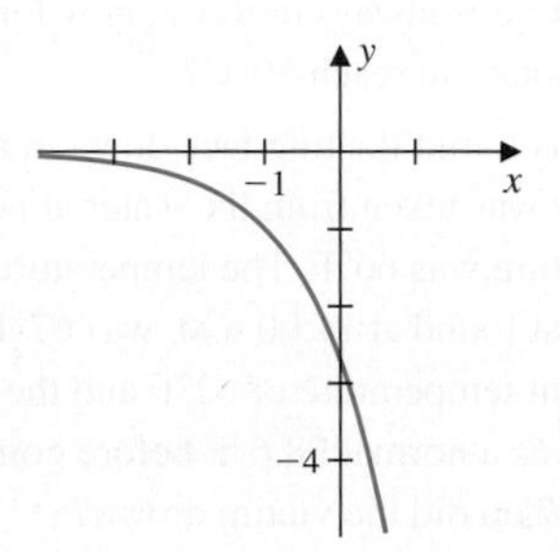

iii.

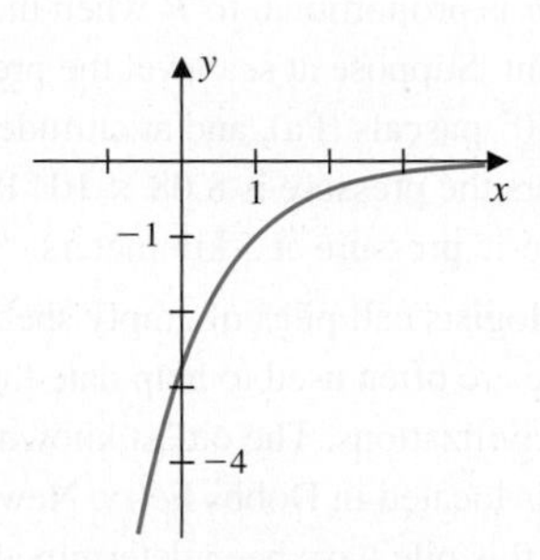

iv.

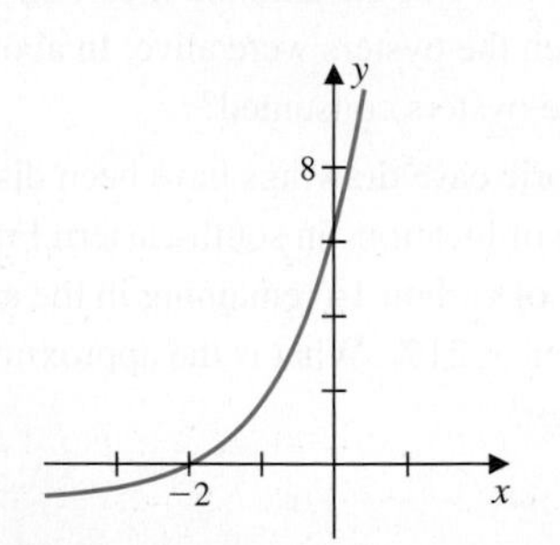

i.

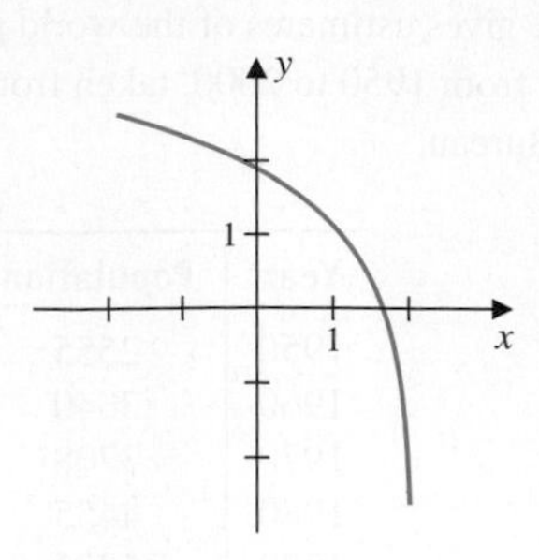

ii.

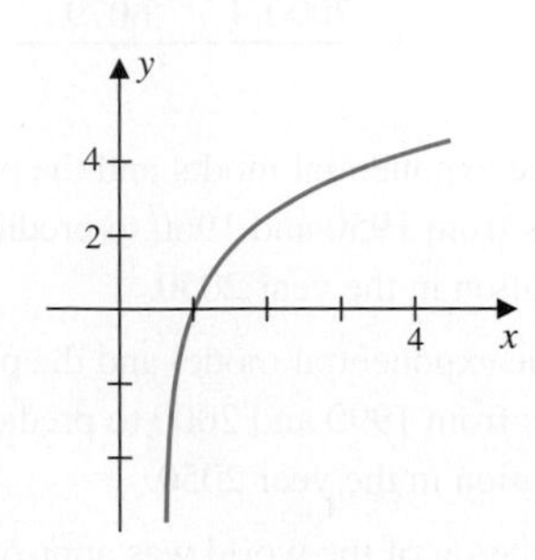

iii.

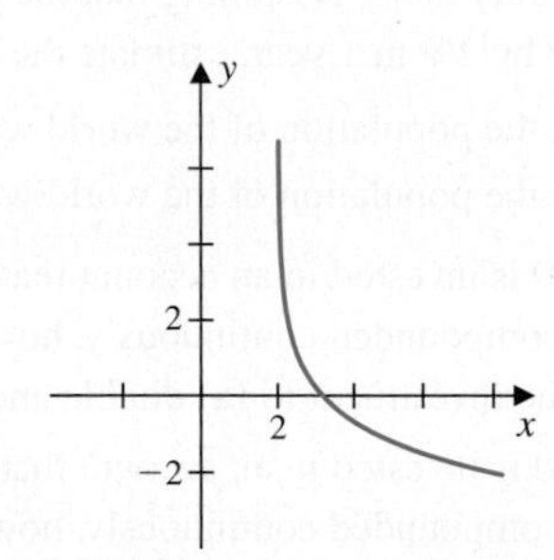

iv.

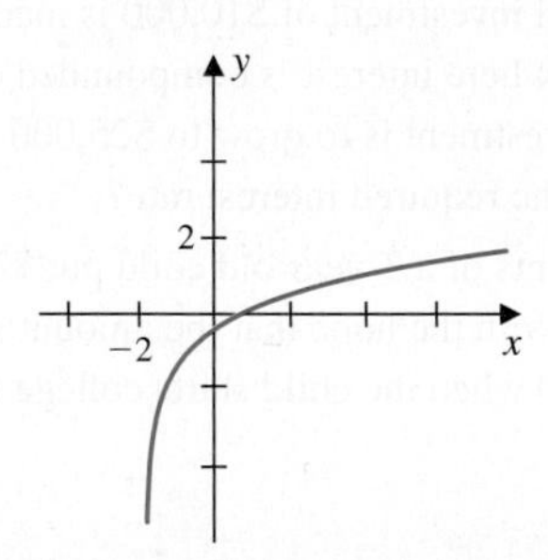

2. Match the equation with the graph.

a. $y = \ln(2 - x) + 1$

b. $y = -\ln(x - 2)$

c. $y = \ln(x + 2) - 1$

d. $y = \ln(x - 1) + 2$

In Exercises 3–14, sketch the graph of the function.

3. $f(x) = 2^{x-1} - 3$ **4.** $f(x) = e^{x-2}$

5. $f(x) = e^{-x} - 2$ **6.** $f(x) = 1 - 3^{2-x}$

7. $f(x) = 3e^{1-x}$ **8.** $f(x) = -2e^{x+1} + 1$

9. $f(x) = 2\ln x$ **10.** $f(x) = \ln(x - 3)$

11. $f(x) = 3 - \log_2(x+1)$

12. $f(x) = \log_{10}(3-x) + 2$

13. $f(x) = e^{-x^2+5x-6}$ 14. $f(x) = \ln x^{-2}$

In Exercises 15–22, evaluate the expression without using a calculator.

15. $\log_5 1$ 16. $\log_{10} 0.000001$

17. $2^{\log_2 15}$ 18. $\log_2 \frac{1}{32}$

19. $\log_4 2$ 20. $\log_2 256$

21. $e^{3\ln 4}$ 22. $\log_5 e^{-2\ln 5}$

In Exercises 23–26, rewrite the expression so that the result does not contain logarithms of products, quotients, or powers.

23. $\ln \dfrac{3x^2}{\sqrt{x-1}}$ 24. $\log_2 \left(\dfrac{x^2-1}{x^2-4}\right)$

25. $\log_{10} \dfrac{\sqrt{x+1}\sqrt[3]{x-1}}{x\,(x+3)^{5/2}}$

26. $\ln \sqrt{\dfrac{x\sqrt{x+1}}{x+2}}$

In Exercises 27–30, rewrite the expression as a single logarithm.

27. $\ln x + \frac{1}{3}\ln x(x+1) + 2\ln(x-1)$

28. $\frac{1}{2}\ln(2x+1) + \ln(x-1) - \ln\left(x^2+1\right)$

29. $3\ln\left(x^3+2\right) + \ln 5 - \frac{1}{2}\ln\left(x^5-1\right)$

30. $\frac{3}{2}\ln\left(x^2-2\right) - 2\ln(x+1)$

In Exercises 31–38, determine the value of x without using a calculator.

31. $\ln(2x-3) = 4$ 32. $e^{3x-4} = 5$

33. $\ln(2x-1) + \ln(3x-2) = \ln 7$

34. $\ln(x-1) - \ln(x-3) = 1$

35. $3^x \cdot 5^{x-2} = 3^{4x}$ 36. $3 \cdot 4^x = 2^{2x+1}$

37. $2e^x x^2 - e^x x = e^x$ 38. $x\ln x - x = 0$

In Exercises 39–44, use a graphing device to approximate the solution.

39. $e^{x^2} = x - 2$

40. $\ln(x+1) = x^3 - 2$

41. $e^x > x^4$

42. $\ln x < 2x - 3$

43. $e^{x-1} - 3 < x^5$

44. $\ln x^2 > x^3 - 2x^2 - x - 2$

45. Use a graphing device to determine the intervals where $f(x) = x^2 e^{1-x^2}$ is increasing, the intervals where it is decreasing, and any local maximums and minimums.

46. Use a graphing device to determine the intervals where $f(x) = x + \ln(1-x^3)$ is increasing, where it is decreasing, and any local maximums and minimums.

47. How long does it take an amount of money to double, if it is deposited at 6% compounded continuously?

48. The radioactive isotope uranium 235 has a half-life of 8.8×10^8 years.

a. How much of a 1-gram sample will decay after 1000 years?

b. How long will it take for 90% of the mass to decay?

49. A bacteria culture is known to grow at a rate proportional to the amount present. After 1 hour, 1000 bacteria are present and after 4 hours, 3000 bacteria are present.

a. Find an expression for the number of bacteria at any time t.

b. Find the number of bacteria that will be present after 5 hours.

c. When will the population reach 20,000?

d. How long does it take for the population to triple?

50. Determine the value of a CD (certificate of deposit) in the amount of $10,000 that matures in 8 years and pays 10% per year compounded as indicated.

a. Annually

b. Monthly

c. Daily

d. Continuously

51. Determine the approximate length of time it takes an initial investment to double in value if it earns 10% compounded quarterly.

52. Determine the approximate length of time it takes an initial investment to triple in value if it earns 10% compounded daily.

53. Determine the length of time it takes an initial investment to double in value if it earns 9% compounded continuously.

54. Determine the length of time it takes an initial investment to triple in value if it earns 10% compounded continuously.

CHAPTER 5: EXERCISES FOR CALCULUS

1. Sketch the graph of each pair of functions and determine when $f = g^{-1}$.

a. $f(x) = 2\ln x;\ g(x) = e^{x/2}$

b. $f(x) = \ln \frac{x}{2};\ g(x) = e^{2x}$

c. $f(x) = \ln |x|;\ g(x) = e^{|x|}$

d. $f(x) = -\ln x;\ g(x) = e^{-x}$

e. $f(x) = 1 + \ln x;\ g(x) = e^{x-1}$

f. $f(x) = 2\ln x;\ g(x) = \frac{1}{2}e^{x}$

2. Use a graphing device to compare the growth rates of $f(x) = x^n$ and $g(x) = a^x$ as $x \to \infty$. Use various values of n and a.

3. Use a graphing device to compare the long-term growth rates of the following functions as $x \to \infty$. Arrange them in order according to increasing long-term growth rates.

$$\ln x,\ x^x,\ e^{3x},\ x^{20},\ x^{-4},\ x^{10}e^{-x},\ x^{1/20},\ \frac{e^{6x}}{x^8}$$

4. Explain how the graph of $y = 3e^{x-2}$ can be obtained from $y = e^x$ using only a horizontal translation.

5. Explain how the graph of $y = 3e^{x-2}$ can be obtained from $y = e^x$ using only vertical scaling.

6. Explain how the graph of $y = 3 + \ln 2x$ can be obtained from $y = \ln x$ using only horizontal scaling.

7. Explain how the graph of $y = 3 + \ln 2x$ can be obtained from the graph of $y = \ln x$ using only vertical translation.

8. Archaeologists have found an animal bone and estimate that 78% of the original radioactive carbon 14 remains. Determine the age of the bone.

9. Bankers approximate the time it takes to double the amount of an investment made at a fixed interest rate by dividing the percent of the annual interest rate into 70. For example, \$10,000 invested at 8.75% per year will become \$20,000 in approximately $70/8.75 = 8$ years. Explain why this is a reasonable estimate.

10. Populations whose growth is limited can often be modeled using the *logistic equation*, which has the form

$$P(t) = \frac{A}{1 + Be^{-Ct}},$$

where A, B, and C are positive constants.

a. What is the initial population in terms of the constants?

b. What is the limiting population in terms of the constants?

c. Sketch the graph of $P(t)$ when $A = 5000$, $B = 300$, and $C = -0.6$.

11. The concentration of a drug in the bloodstream after a single injection decreases in time as the drug is absorbed by the bloodstream, and the rate of decrease of the concentration at time t is proportional to the concentration at time t.

a. The initial concentration of a drug in the bloodstream is 20 milligrams per liter, and 3 hours later it is 12 milligrams per liter. Determine an expression for the concentration at time t.

b. What is the half-life of the drug described in part (a)?

c. Sodium phentobarbital, which has a half-life of 5 hours, is used to anesthetize a dog for an operation. The dog is anesthetized when its bloodstream contains at least 30 milligrams of the drug for each kilogram of body weight. What dose of sodium phentobarbital should be administered if a 25-kilogram dog is to be anesthetized for at least 1 hour?

12. a. The *hyperbolic cosine* function is defined by

$$\cosh x = \frac{e^x + e^{-x}}{2}.$$

Plot $f(x) = \cosh x$, $y = e^x/2$, and $y = e^{-x}/2$ on the same set of axes.

b. The *hyperbolic sine* function is defined by

$$\sinh x = \frac{e^x - e^{-x}}{2}.$$

Plot $f(x) = \sinh x$, $y = e^x/2$, and $y = -e^{-x}/2$ on the same set of axes.

c. Verify the identities

$$(\cosh x)^2 - (\sinh x)^2 = 1$$

and

$$\sinh(a + b) = \sinh a \cosh b + \cosh a \sinh b.$$

d. The graph of a cable hanging between two supports has the equation $y = a\cosh(x/a)$, where $a > 0$. This curve is called a *catenary*. Describe the effect of a on the shape of the catenary.

CHAPTER 5: CHAPTER TEST

Determine whether the statement is true or false. If false, describe how the statement might be changed to make it true.

1. When evaluated, $e^{2\ln 3}$ is 9.

2. When evaluated, $2\log_a a^{1/2}$ is $\frac{1}{4}$.

3. The solution to the equation $\log_2 x = 5$ is 25.

4. The solution to the equation $\log_3 x = 2$ is 8.

5. For all real numbers x, we have $\ln e^x = x$.

6. For all real numbers $x > 0$, we have $e^{\ln x} = x$.

7. The domain of the function $f(x) = \ln(x-2)$ is $(2, \infty)$.

8. The range of the function $f(x) = 1 + e^{x-2}$ is $[2, \infty)$.

9. The range of $f(x) = 2 - \ln(x-1)$ is $(-\infty, \infty)$.

10. The domain of $f(x) = 3 + e^{2-x}$ is $(-\infty, \infty)$.

11. The only solution to the equation $3^{2x+5} = 27$ is $x = 11$.

12. The only solution to the equation $2^{x+3} = 4$ is

$$x = \frac{\ln 4 - 3\ln 2}{\ln 2}.$$

13. The only solution to the equation $2^{2x+1} = 3^{x-2}$ is

$$x = -\frac{2\ln 3 + \ln 2}{\ln 4 - \ln 3}.$$

14. The only solution to the equation $4^x = 5^{2x-1}$ is

$$x = \frac{\ln 5}{2(\ln 5 - \ln 2)}.$$

15. The expression $\ln xy^2$ is equivalent to $\ln x + 2\ln y$.

16. The expression $2\ln x - \ln y + \ln(x+y)$ is equivalent to

$$\ln\frac{x^3 + x^2 y}{y}.$$

17. The only solutions to the equation $\ln x + \ln(x-1) = \ln(4x+6)$ are $x = -1$ and $x = 6$.

18. The only solutions to the equation $\ln(x+6) - \ln(x+1) = \ln(x-2)$ are $x = -4$ and $x = 4$.

19. The graph of $y = e^x$ has a vertical asymptote $y = 0$.

20. The graph of $y = 2 + e^{x-1}$ has a horizontal asymptote $y = 2$.

21. The graph of $y = -1 + e^{x-2}$ is obtained by shifting the graph of $y = e^x$ to the left 2 units and upward 1 unit.

22. The graph of $y = e^{x+3}$ can be obtained by shifting the graph of $y = e^x$ to the left 1 unit and vertically stretching the result by a factor of e^3.

23. The graph of $y = \ln x$ has a vertical asymptote $x = 2$.

24. The graph of $y = 3 + \ln(x-1)$ has a vertical asymptote $x = -3$.

25. The graph of $y = 2 + \ln(x-1)$ is obtained by shifting the graph of $y = \ln x$ to the right 2 units and downward 1 unit.

26. The graph of $y = \ln(2x-1)$ is obtained by shifting the graph of $y = \ln x$ to the left $\frac{1}{2}$ units and upward $\ln 2$ units.

In Exercises 27–30, use the figure and assume the graphs are all transformations of $y = e^x$.

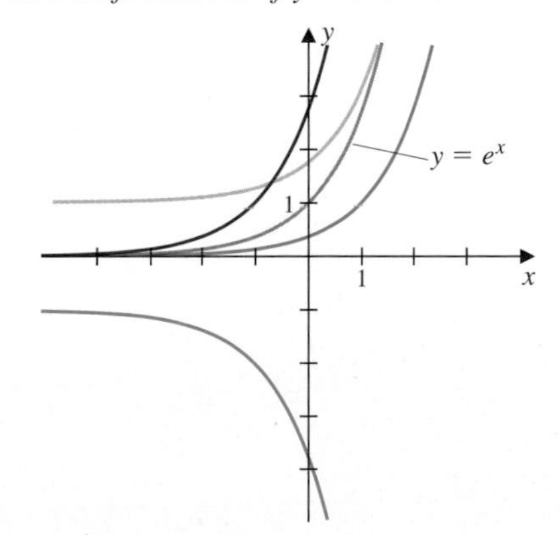

27. The blue graph is $y = e^{x-1}$.

28. The red graph is $y = e^{x-1}$.

29. The green graph is $y = -1 - e^{x+1}$.

30. The yellow graph is $y = 1 + e^{2x-1}$.

In Exercises 31–34, use the figure and assume the graphs are all transformations of $y = \ln x$.

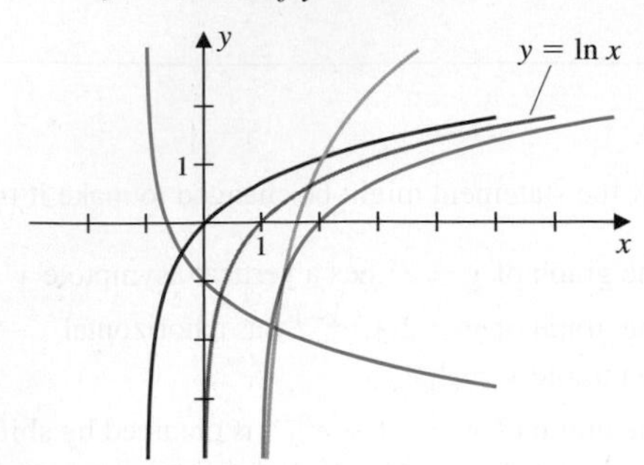

31. The blue graph is $y = \ln(x + 1)$.

32. The red graph is $y = \ln(x + 1)$.

33. The green graph is $y = -1 - \ln(x - 1)$.

34. The yellow graph is $y = 1 - 2\ln(x - 1)$.

35. The inverse of $f(x) = 2^{x+1}$ is $f^{-1}(x) = -1 + \log_2 x$.

36. The inverse of $f(x) = 1 + e^{x-1}$ is $f^{-1}(x) = 1 + \ln(x - 1)$.

37. If a bacteria culture starts with 1000 bacteria and after 3 hours there are 2500 bacteria, the number of bacteria at any time t is given by $Q(t) = 1000e^{(t \ln 2500)/3}$.

38. If a radioactive substance decays 5% in 10 years, half of the initial mass will decay in

$$t = \frac{10 \ln 2}{\ln(0.95)} \text{ years.}$$

39. An initial investment deposited in an account returning 7% interest compounded continuously will double in approximately 9 years (rounded to the nearest year).

40. The amount of an initial investment returning 10% compounded continuously that will accumulate to \$150,000 in 18 years is approximately \$25,000 (rounded to the nearest thousand).

Conic Sections, Polar Coordinates, and Parametric Equations

6

Calculus Connections

Much of calculus involves the study of curves that occur in our physical world. By determining equations and functions that describe the curves, we can better examine their properties and those of the physical objects they represent. Take, as an example, some of the physical problems associated with the orbiting satellites that are part of the global positioning system (GPS). These satellites transmit signals that permit objects to be located to a degree of accuracy only dreamed of a few years ago. The present applications of the system range from the navigational displays in our cars to the scheduling of trucks on our highways to the allocation of personnel at some of the world's major airports. These applications have only scratched the surface; the effect of this precise locating system will probably not be fully realized for a number of years.

© Image copyright Neo Edmund, 2009. Used under license from Shutterstock.com

Parabolic dish

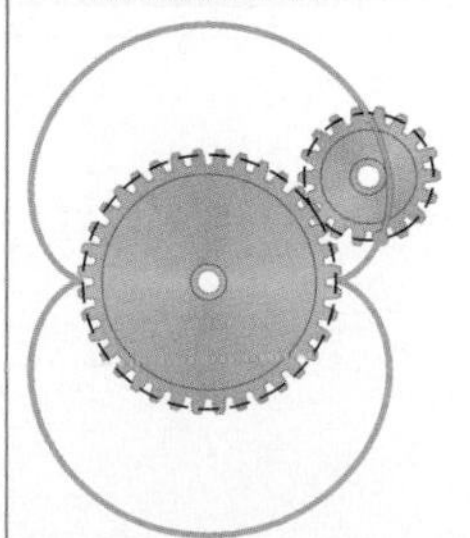

Satellites move in an elliptical orbit about Earth, just as Earth moves in an elliptical orbit about the sun. Signals that satellites relay to Earth are often collected in dishes that have a parabolic shape, and some satellites use a location scheme based on properties of hyperbolas. These three types of curves, the parabola, the ellipse, and the hyperbola, are the conic sections—curves we study in the first part of this chapter.

Curves that arise from rotating a circle about an object occur frequently in physical situations. The path of a tooth on a gear that is turning on another gear is one example; a robot that rotates to paint or weld a car on an assembly line is another.

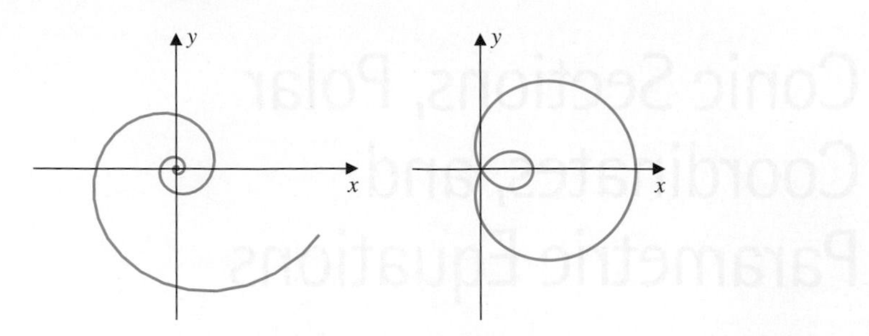

Curves described in this manner are often difficult to represent using the familiar rectangular (Cartesian) coordinate system. For that matter, even common circles cannot be represented by a function in the rectangular coordinate system. Patching together rectangular equations to represent curves that spiral or overlap, such as the ones shown in the figure, would be incredibly complicated.

To better describe these curves we introduce the polar coordinate system, which uses a circular rather than a rectangular method for describing points in the plane. We examine this system later in the chapter.

The Gateway Arch in St. Louis has the form of an inverted cycloid.
© David R. Frazier Photolibrary, Inc. / Alamy

Some other commonly occurring curves cannot easily be represented in a direct manner using either the rectangular or the polar coordinate system, but can be described if an additional variable called a *parameter* is introduced. Parametric representation is particularly important in physics, where the position of an object is often described using a time parameter. A problem as simple as describing a specific point on a car tire as the car moves down a road has this form; it produces a curve called a *cycloid.* Paths of objects as they travel through space are almost always of this form, but the discussion of curves in space will be postponed for a year or so, until you are in the final term of calculus. In this chapter we provide only an introduction to the notion of parametric representation of curves, but this introduction will give you a solid basis of experience for a later study of general curves in space.

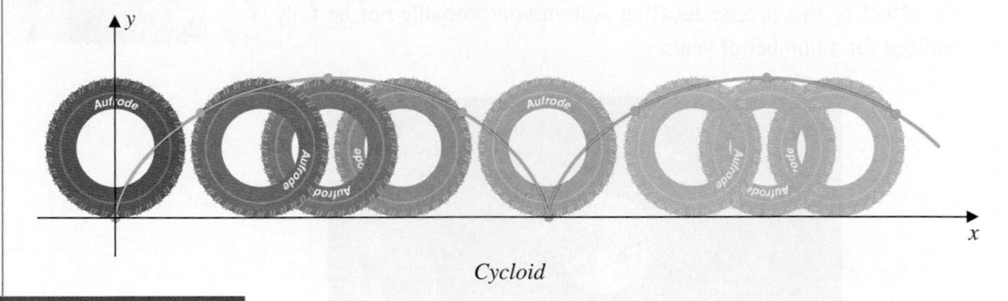

Cycloid

6.1 INTRODUCTION

The general **quadratic equation** in x and y has the form

$$Ax^2 + Bxy + Cy^2 + Dx + Ey + F = 0,$$

where A, B, C, D, E, and F are constants. The graphs of these equations are called **conic sections**, or simply **conics**, since they were historically developed by considering the various curves generated when a double-napped cone (see Figure 1) is cut by planes. These are the first natural extensions of the general equation of a line, which has the form $Ax + By + C = 0$ (as we saw at the end of Section 1.7).

FIGURE 1

The conic sections are basic graphs, some of whose special cases we have looked at often in this book. For example, the graphs of the parabola with equation $y = x^2$, and the circle with equation $x^2 + y^2 = 1$, are conic sections.

Three basic distinct curves result from the intersection of a plane with the double-napped cone, as shown in Figure 2.

- A **parabola**, if the plane intersects only one nappe of the cone and is parallel to one of its *generators* (the lines forming the surface of the cone).
- An **ellipse**, if the plane intersects only one nappe of the cone but is not parallel to one of its generators.
- A **hyperbola**, if the plane intersects both nappes of the cone.

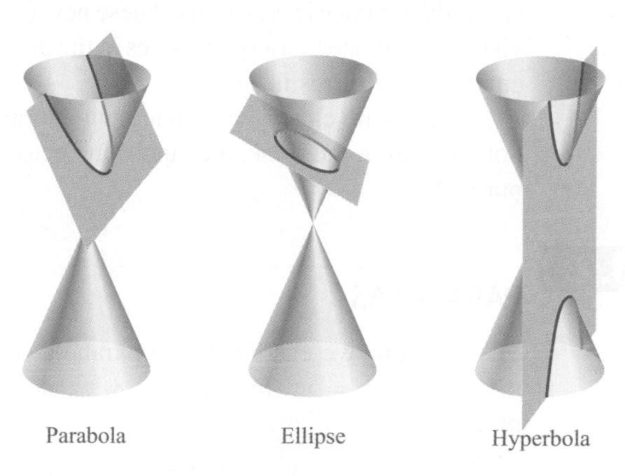

FIGURE 2

In calculus, you will be expected to recognize equations describing parabolas, circles, ellipses, and hyperbolas.

Curves such as a point, circle, line, or pair of intersecting lines can also be produced by intersecting a plane with a double-napped cone, but these are only special, or *degenerate*, cases of either the parabola, the ellipse, or the hyperbola.

This geometric approach to the study of conics leads to a number of important applications of these curves, which were first studied extensively by the ancient Greeks. The mathematician Apollonius, who lived in the 3rd century B.C.E., wrote eight volumes on the subject.

Curves considered previously have been graphs of functions, but curves are not always presented this way.

In the 16th century, Galileo found that the path of a projectile fired upward at an angle is a parabola. In 1609, the German mathematician and astronomer Johannes Kepler deduced from observed data that the planets and other objects in our solar system orbit the sun in elliptical orbits. Later in the 17th century Isaac Newton proved that Kepler's observations were correct. Parabolas and hyperbolas are used in the construction of reflecting telescopes. The applications of conic sections are so extensive that they encompass many areas, including engineering, physics, astronomy, architecture, and optics.

Our study will concentrate on the different types of curves that are generated by quadratic equations rather than on the intersection properties with the cone. In the first three sections we consider conics that occur as graphs of the quadratic equation

$$Ax^2 + Cy^2 + Dx + Ey + F = 0,$$

obtained by setting $B = 0$ in the general quadratic equation. The general form of the quadratic equation with $B \neq 0$ involves a rotation of the graph of a conic section, and will not be considered. A discussion of this situation, however, can be downloaded as a pdf file from our website at

http://www.math.ysu.edu/~faires/PreCalculus/Rotated_Conics

We use the polar coordinate system to plot complicated curves in a plane.

In Section 6.2 we look at parabolas, curves produced when either A or C (but not both) is zero. Section 6.3 concerns ellipses, which occur when A and C are both nonzero and of the same sign. Section 6.4 considers the situation when A and C are nonzero, but of different signs, which produces hyperbolas.

Representing curves parametrically permits us to graph many complicated curves.

In this chapter we also study two additional methods for describing curves in the plane. The polar coordinate system is discussed in Section 6.5, and in Section 6.6 we consider equations for the conic sections written in polar coordinates. In Section 6.7 we study parametric equations. These new methods of representing curves allow us to visualize a greater variety of curves in the plane and lay the groundwork for describing curves and surfaces in space.

Vectors are closely associated with parametric equations. Although this topic is not covered in the book, a chapter on vectors can be downloaded as a pdf file from our website.

6.2 PARABOLAS

The first time we encountered a parabola was in Section 1.4, where we considered the graph of $y = x^2$. In Section 1.8, we described more general quadratic functions of the form

$$y = f(x) = ax^2 + bx + c, \quad \text{where} \quad a \neq 0.$$

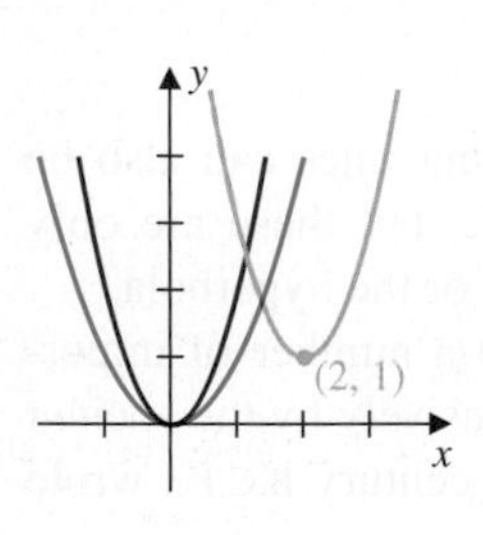

FIGURE 1

The graphs of the general quadratic functions are similar to those of $y = x^2$ (see the blue graph in Figure 1), but might involve a stretch or compression (if $|a| \neq 1$), shown in red for the graph of $y = 2x^2$, as well as horizontal and vertical shifts. When $a < 0$, a reflection about the x-axis is also involved.

Completing the square on a quadratic reveals its secrets. For example, the graph of

$$y = f(x) = 2x^2 - 8x + 9 = 2(x^2 - 4x + 4) - 2 \cdot 4 + 9 = 2(x - 2)^2 + 1,$$

shown in yellow in Figure 1, is a translation of the graph of $y = x^2$ to the right 2 units and upward 1 unit.

The graphs of parabolas, then, have been studied quite thoroughly. In this section we show some of the geometric properties of parabolas that are important in certain applications. To do this we introduce an alternative, geometric, definition for the parabola.

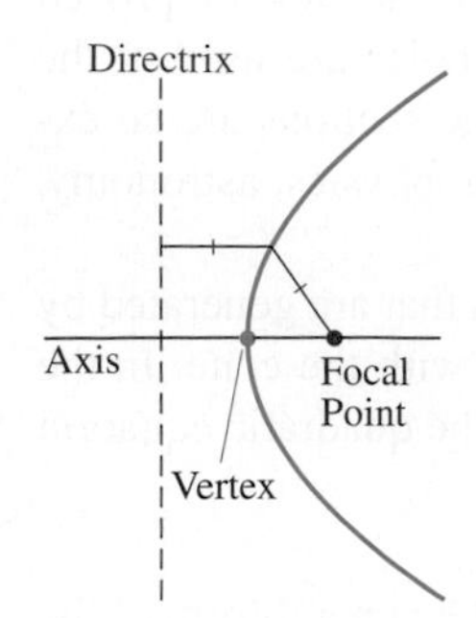

FIGURE 2

Parabola

A **parabola** is the set of points in a plane that are equidistant from a given point, called the **focal point**, and a given line, called the **directrix**, that does not contain the focal point.

The **axis** of the parabola is the line that goes through the focal point and is perpendicular to the directrix. The point of intersection of the axis and the parabola is the **vertex**. (See Figure 2.)

A *standard form* for the equation of a parabola is derived by superimposing an xy-coordinate system so that the axis of the parabola corresponds with the y-axis, and the vertex of the parabola, which is located midway between the focal point and the directrix, is at the origin. If the focal point is located at $(0, c)$, then the directrix has equation $y = -c$. The situation when $c > 0$ is shown in Figure 3(a), and that when $c < 0$ is shown in Figure 3(b).

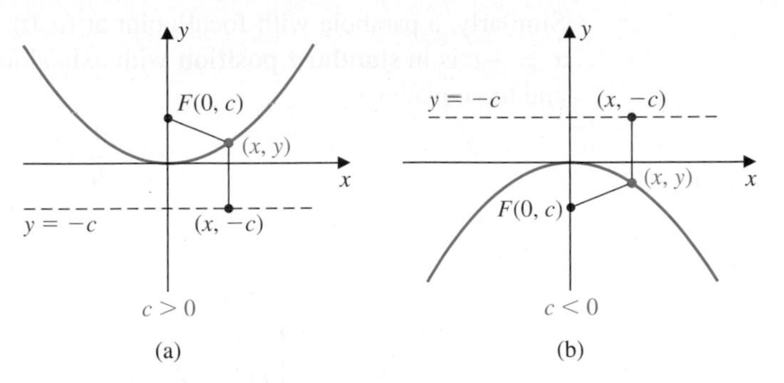

FIGURE 3

By definition, the distance from an arbitrary point $P(x, y)$ on the curve to the focal point $F(0, c)$ is the same as the distance from $P(x, y)$ to the directrix. These distances are

$$d((x, y), (0, c)) = \sqrt{(x - 0)^2 + (y - c)^2}$$

and

$$d((x, y), (x, -c)) = \sqrt{(x - x)^2 + (y + c)^2}.$$

We equate the radicals and square to obtain

$$x^2 + (y - c)^2 = (y + c)^2, \quad \text{or} \quad x^2 + y^2 - 2cy + c^2 = y^2 + 2cy + c^2.$$

The last equation simplifies to the equation $y = \frac{1}{4c}x^2$. Hence the parabola has the usual form $y = ax^2$, where $a = \frac{1}{4c}$.

From this derivation we can see that the basic parabola $y = x^2$, which we have seen so often, has its vertex at the origin and $c = \frac{1}{4}$. Its focal point is at $\left(0, \frac{1}{4}\right)$, and the equation of its directrix is $y = -\frac{1}{4}$, as shown in Figure 4.

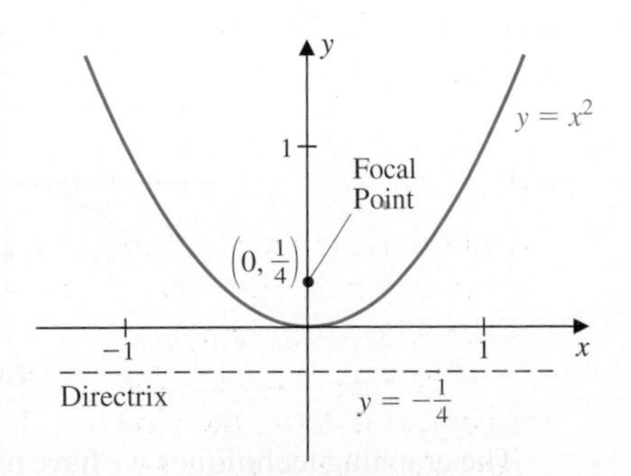

FIGURE 4

Standard-Position Parabolas

A parabola with focal point at $(0, c)$, vertex at $(0, 0)$, and directrix $y = -c$ is in **standard position** with axis along the y-axis (see Figure 5(a)) and has equation

$$y = \frac{1}{4c}x^2.$$

Similarly, a parabola with focal point at $(c, 0)$, vertex at $(0, 0)$, and directrix $x = -c$ is in **standard position** with axis along the x-axis (see Figure 5(b)) and has equation

$$x = \frac{1}{4c}y^2.$$

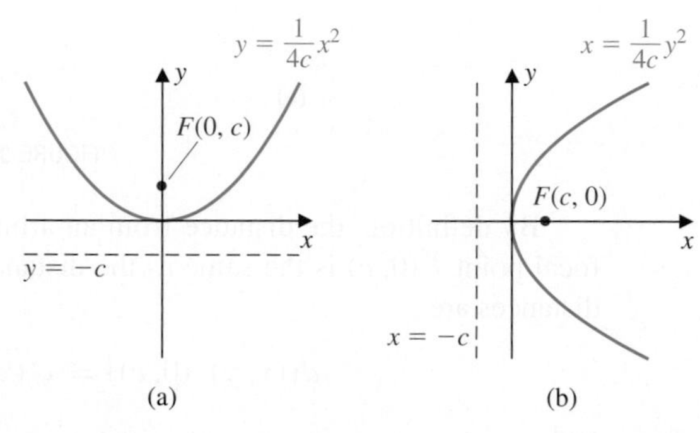

FIGURE 5

EXAMPLE 1 Find an equation of the parabola with focal point at $(0, -2)$ and directrix $y = 2$.

Solution The focal point lies along the y-axis. Since the vertex is the midpoint of the line segment from the focal point to the directrix, the vertex is at the origin. So the parabola is in standard position with axis along the y-axis (see Figure 6). The equation of the parabola is

$$y = \frac{1}{4(-2)}x^2 = -\frac{1}{8}x^2.$$

■

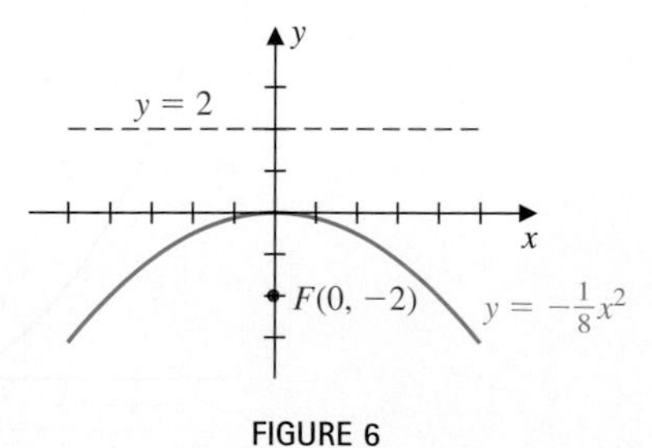

FIGURE 6

The graphing techniques we have used so frequently permit us to sketch the graph of any parabola whose directrix is parallel to one of the coordinate axes.

EXAMPLE 2 Find the focal point and directrix of the parabola with equation $2y^2+8y-x+7=0$.

Solution We first complete the square on the y-variable so that we can compare this equation with the equation of a parabola in standard position. We have

$$2(y^2+4y+4)-2(4)-x+7=0,$$

which gives

$$2(y+2)^2-x-1=0, \quad \text{or} \quad x+1=2(y+2)^2.$$

This parabola will be compared to the parabola in standard position that has the equation

$$x=2y^2=\frac{1}{4c}y^2.$$

Since

$$\frac{1}{4c}=2, \quad \text{we have} \quad c=\frac{1}{8}.$$

The standard position parabola has focal point at $\left(\frac{1}{8},0\right)$ and directrix $x=-\frac{1}{8}$. (See Figure 7.)

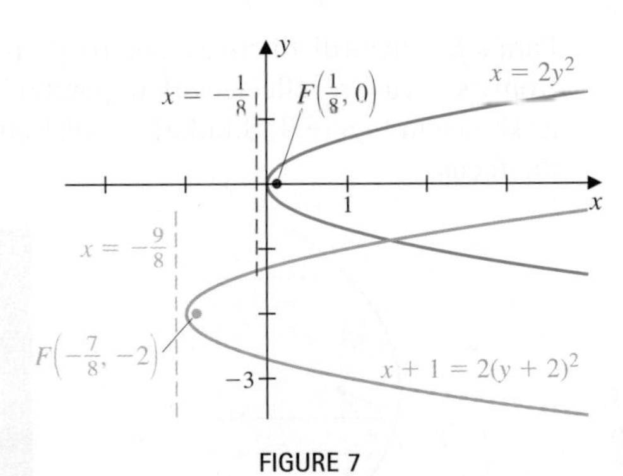

FIGURE 7

The graph of the parabola with equation $x+1=2(y+2)^2$ is the translation of the graph of $x=2y^2$, to the left 1 unit and downward 2 units. Consequently, the focal point of the new parabola is at $\left(\frac{1}{8}-1,0-2\right)=\left(-\frac{7}{8},-2\right)$, and the directrix is $x=-\frac{1}{8}-1=-\frac{9}{8}$, also shown in Figure 7. The vertex of the parabola is at $(-1,-2)$. ■

EXAMPLE 3 Determine the equation of the parabola with focal point at $(1,5)$ and directrix $y=-1$.

Solution The vertex of the parabola lies midway between the focal point and the directrix, so the distance from the vertex to the focal point is

$$c=\frac{1}{2}(5-(-1))=3.$$

Hence the vertex is at $(1,2)$. Figure 8(a) shows the standard-position parabola whose focal point is 3 vertical units above its vertex, and this parabola has equation

$$y=\frac{1}{4(3)}x^2=\frac{1}{12}x^2.$$

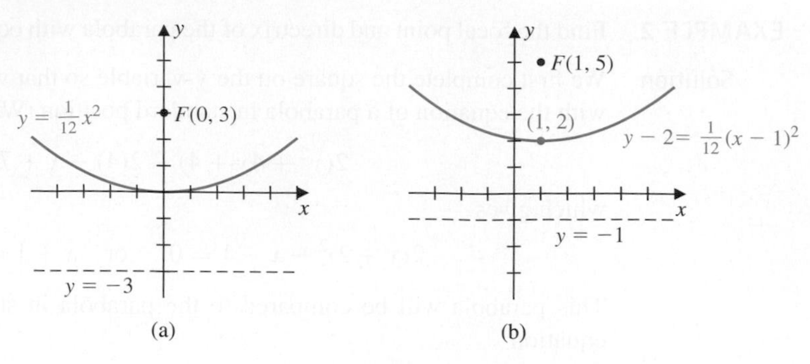

FIGURE 8

The parabola whose vertex is translated to (1, 2) is shown in Figure 8(b) and has equation

$$y - 2 = \frac{1}{12}(x - 1)^2, \quad \text{or} \quad y = \frac{1}{12}(x - 1)^2 + 2. \qquad \blacksquare$$

Reflection Property

Parabolas have a distinctive reflective property. Sound, light, or other waves emanating from the focus are reflected off the parabolic surface parallel to the axis of the parabola, as shown in Figure 9. Similarly, waves collected by a parabolic dish are reflected to the focus.

FIGURE 9 FIGURE 10

If a parabola is rotated about its axis to construct a surface, called a *paraboloid*, then cross sections containing the axis are parabolic and have a common focal point, as shown in Figure 10. In the case of a flashlight or searchlight, light emitted from the focus is reflected in parallel rays, creating a concentrated beam of light. Similarly, reflecting telescopes and satellite receivers have parabolic shapes, since light or radio waves bouncing off the surface are reflected to the focus, where they are collected and amplified.

Applications

EXAMPLE 4 A satellite dish receiver has its amplifier in line with the edge of the dish, as shown in Figure 11(a). The diameter of the dish at the edge is 1 meter. How deep is the dish?

This type of problem is commonly assigned in a calculus course, even though no calculus is required in its solution.

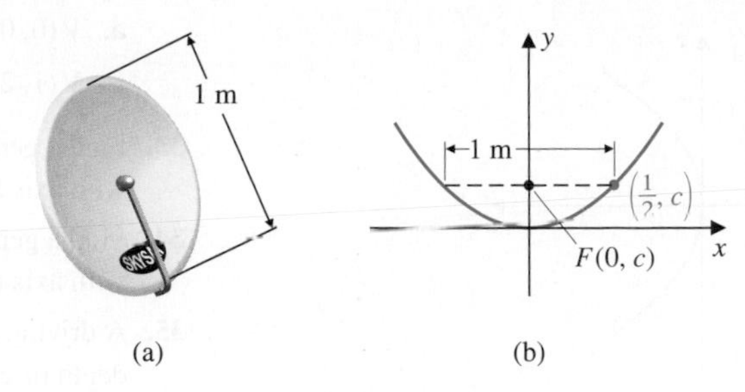

FIGURE 11

Solution A cross section of the dish is shown in Figure 11(b), where the focal point (the receiver) is at $(0, c)$. The equation of a parabola in standard position with axis along the y-axis is

$$y = \frac{1}{4c}x^2.$$

The point $\left(\frac{1}{2}, c\right)$ lies on the graph of the parabola, so

$$c = \frac{1}{4c}\left(\frac{1}{2}\right)^2 \quad \text{gives} \quad c^2 = \frac{1}{16}, \quad \text{and} \quad c = \frac{1}{4}.$$

The distance from the focal point to the vertex and, hence, the depth of the dish, is $c = 0.25$ meters $= 25$ centimeters. ■

EXERCISE SET 6.2

In Exercises 1–18, sketch the graph of the parabola showing the vertex, focal point, and directrix.

1. $y = 2x^2$
2. $9y = 16x^2$
3. $9y = -16x^2$
4. $y = -2x^2$
5. $y^2 = 2x$
6. $9y^2 = 16x$
7. $9y^2 = -16x$
8. $y^2 = -2x$
9. $x^2 - 6x + 9 = 2y$
10. $x^2 + 4x + 4 = 2y$
11. $x^2 - 4x - 2y + 2 = 0$
12. $x^2 - 4x + 2y + 6 = 0$
13. $y^2 - 8y + 12 = 2x$
14. $y^2 + 6y + 6 - 3x = 0$
15. $2x^2 + 4x - 9y + 20 = 0$
16. $2x^2 - 4x + 3y - 4 = 0$
17. $3x^2 - 12x + 4y + 8 = 0$
18. $3x^2 - 6x - 2y - 1 = 0$

In Exercises 19–22, find an equation for the given parabola.

19.

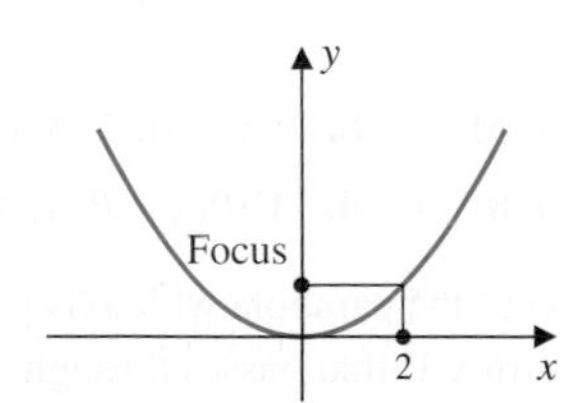

20.

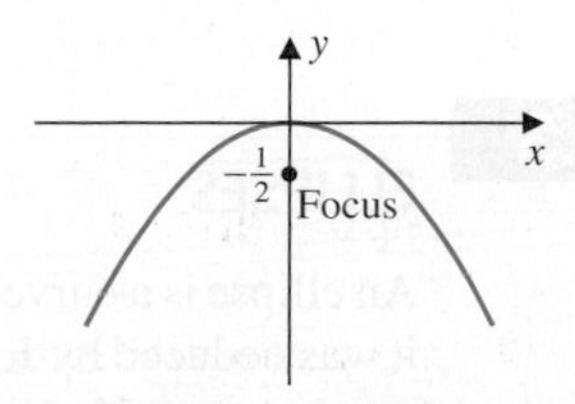

21.

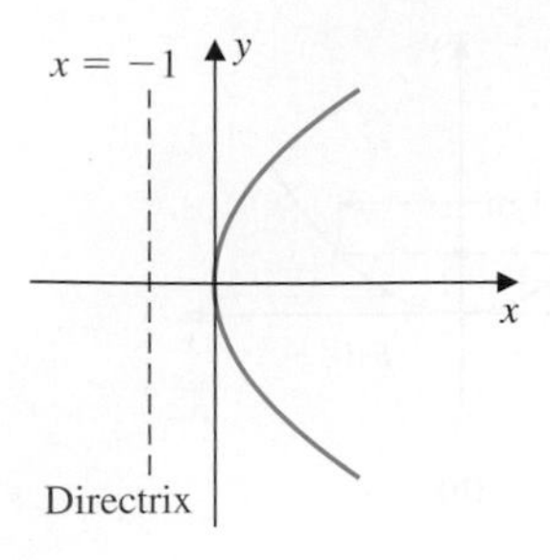

22.

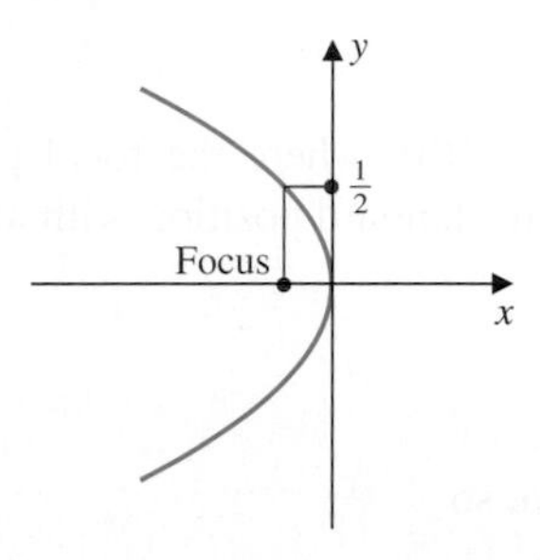

In Exercises 23–30, determine the equation of the parabola that satisfies the given conditions.

23. Focus at $(-2, 2)$, directrix $y = -2$

24. Focus at $(-2, 2)$, directrix $x = 2$

25. Vertex at $(-2, 2)$, directrix $x = 2$

26. Vertex at $(-2, 2)$, directrix $y = -2$

27. Vertex at $(-2, 2)$, focus at $(-2, 0)$

28. Vertex at $(-2, 2)$, focus at $(-2, 4)$

29. Vertex at $(-2, 2)$, focus at $(-4, 2)$

30. Vertex at $(-2, 2)$, focus at $(2, 2)$

31. Find an equation of the parabola with axis parallel to the y-axis and vertex V that passes through the point P.

a. $V(0, 0)$, $P(4, 6)$ **b.** $V(1, 0)$, $P(5, 6)$

c. $V(1, 2)$, $P(5, 8)$ **d.** $V(0, 2)$, $P(4, 8)$

32. Find an equation of the parabola with axis parallel to the x-axis and vertex V that passes through the point P.

a. $V(0, 0)$, $P(4, 6)$ **b.** $V(1, 0)$, $P(5, 6)$

c. $V(1, 2)$, $P(5, 8)$ **d.** $V(0, 2)$, $P(4, 8)$

33. Find a general form for the equation of a parabola with axis the y-axis and passing through $(1, 2)$.

34. Find a general form for the equation of a parabola with axis the x-axis and passing through $(1, 2)$.

35. A driving light has a parabolic cross section with a depth of 2 inches and a cross-section height of 4 inches. Where should the light source be placed to produce a parallel beam of light?

36. A reflector for a satellite dish has a parabolic cross section with the receiver at the focus. The dish is 4 inches deep and 20 inches wide at the edge. How far is the receiver from the vertex of the parabolic dish?

37. A ball thrown horizontally from the top edge of a building follows a parabolic curve with vertex at the top edge of the building and axis along the side of the building. The ball passes through a point 100 feet from the building when it is a vertical distance of 16 feet from the top.

a. How far from the building will the ball land if the building is 64 feet high?

b. Recompute the answer if instead the ball is thrown from the top of the Sears Tower in Chicago, which has a height of 1450 feet.

38. A projectile fired from the ground follows a parabolic path. The maximum height of the projectile is 200 feet and it strikes the ground 1000 feet from the firing point. When was the projectile 150 feet above the ground?

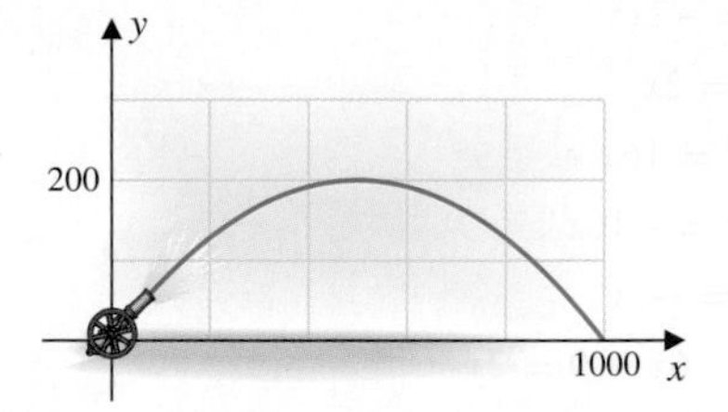

6.3 ELLIPSES

An ellipse is a curve that has the appearance of an elongated circle. In the 17th century it was deduced by Johannes Kepler, and later proved by Isaac Newton, that the planets revolve about the sun in elliptical orbits.

There are many equivalent definitions for the ellipse. The definition we use involves two fixed points and a given distance. In physics you are more likely to see a definition that uses a given point, a given line, and a number called the *eccentricity*. This eccentricity describes how close the ellipse is to being a circle.

Ellipse

An **ellipse** is the set of points in a plane for which the sum of the distances from two fixed points is a given constant. The two fixed points are the **focal points** of the ellipse, and the line passing through the focal points is the **axis**. The points of intersection of the axes and the ellipse are the **vertices**. (See Figure 1(a).)

Calculus is used to verify that Kepler's laws of planetary motion follow from Newton's laws of motion. Newton's laws revolutionized scientific thought in the 17th century.

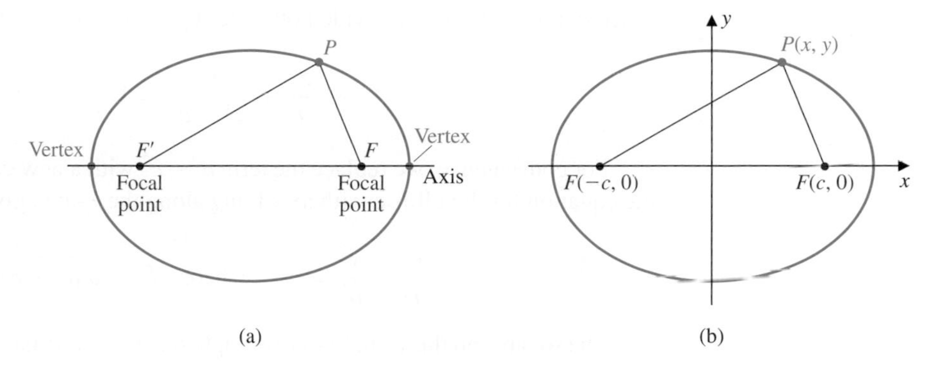

FIGURE 1

To determine standard equations for the ellipse, position an xy-coordinate system with the x-axis along the axis of the ellipse and the origin midway between the focal points, as shown in Figure 1(b). This derivation will give you an excellent chance to apply your algebraic skills. The focal points have been given the coordinates $F(c, 0)$ and $F'(-c, 0)$. Since the distance from $F(c, 0)$ to $F'(-c, 0)$ is $2c$, the fixed constant specified in the definition must be greater than this value.

Call the fixed distance $2a$, where $a > c > 0$. Then the vertices have the coordinates $V(a, 0)$ and $V'(-a, 0)$, because

$$d(V, F) + d(V, F') = (a - c) + (a + c) = 2a$$

and

$$d(V', F) + d(V', F') = (a + c) + (a - c) = 2a.$$

In general, a point (x, y) is on the ellipse precisely when

$$\begin{aligned} 2a &= d((x, y), (c, 0)) + d((x, y), (-c, 0)) \\ &= \sqrt{(x - c)^2 + (y - 0)^2} + \sqrt{(x + c)^2 + (y - 0)^2}, \end{aligned}$$

that is, when

$$\sqrt{(x - c)^2 + y^2} = 2a - \sqrt{(x + c)^2 + y^2}.$$

Squaring each side of this last equation gives

$$(x - c)^2 + y^2 = 4a^2 - 4a\sqrt{(x + c)^2 + y^2} + (x + c)^2 + y^2,$$

which, after completing the multiplication and dividing by 4, simplifies to

$$a\sqrt{(x+c)^2+y^2} = a^2 + cx.$$

Squaring again, then expanding and subtracting $2a^2cx$ from both sides of the equation gives

$$a^2x^2 + a^2c^2 + a^2y^2 = a^4 + c^2x^2,$$

which simplifies to

$$(a^2 - c^2)x^2 + a^2y^2 = a^4 - a^2c^2.$$

Since $a > c > 0$, we can divide both sides by $a^2 - c^2$ and a^2 to obtain

$$\frac{x^2}{a^2} + \frac{y^2}{a^2 - c^2} = 1.$$

For convenience, we replace the term $a^2 - c^2$ with a new constant, b^2, to produce the equation for the ellipse with axis lying along the x-axis given by

$$\frac{x^2}{a^2} + \frac{y^2}{b^2} = 1, \quad \text{where} \quad b = \sqrt{a^2 - c^2}.$$

The squares on the x- and y-terms imply that the graph has both y-axis and x-axis symmetry. The x-intercepts occur at $(a, 0)$ and $(-a, 0)$, and the y-intercepts occur at $(0, b)$ and $(0, -b)$, where $b = \sqrt{a^2 - c^2} < a$.

The *major axis* of an ellipse is the longest line segment joining two points on the ellipse, and the *minor axis* is the shortest. The major axis for this ellipse has length $2a$ and joins the points $(a, 0)$ and $(-a, 0)$. The minor axis has length $2b$ and joins $(0, b)$ and $(0, -b)$. An ellipse centered at the origin with axis along the x-axis is said to be in *standard position*. (See Figure 2(a).) A similar type of equation results when the roles of the axes are interchanged. (See Figure 2(b).)

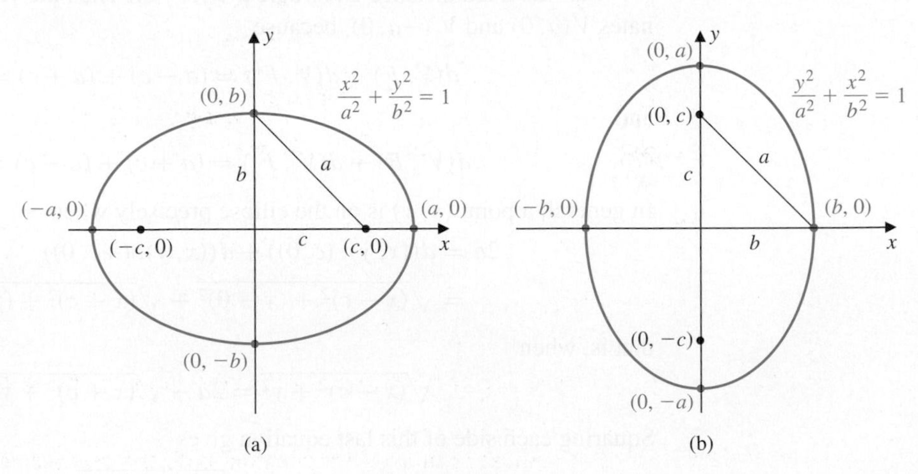

FIGURE 2

Standard-Position Ellipses

An ellipse with focal points at $(c, 0)$ and $(-c, 0)$ and vertices at $(a, 0)$ and $(-a, 0)$, where $a > c > 0$, is in **standard position** with axis along the x-axis (see Figure 2(a)) and has equation

$$\frac{x^2}{a^2} + \frac{y^2}{b^2} = 1, \quad \text{where} \quad b = \sqrt{a^2 - c^2}.$$

Similarly, an ellipse with focal points at $(0, c)$ and $(0, -c)$ and vertices at $(0, a)$ and $(0, -a)$, where $a > c > 0$, is in **standard position** with axis along the y-axis (see Figure 2(b)) and has equation

$$\frac{y^2}{a^2} + \frac{x^2}{b^2} = 1, \quad \text{where} \quad b = \sqrt{a^2 - c^2}.$$

Notice that since $a > c > 0$, we always have

$$b = \sqrt{a^2 - c^2} < \sqrt{a^2} = a.$$

EXAMPLE 1 Sketch the graph of the ellipse $9x^2 + 16y^2 = 144$, and find its focal points.

Solution Dividing both sides of the equation by 144 gives

$$\frac{x^2}{16} + \frac{y^2}{9} = 1.$$

The coefficient in the denominator of x^2 is larger than the coefficient in the denominator of y^2, so the ellipse is in standard position with axis along the x-axis. Since $a^2 = 16$ and $b^2 = 9$, we have $a = 4$ and $b = 3$. The axis intercepts are at $(4, 0)$, $(-4, 0)$, $(0, 3)$, and $(0, -3)$, as shown in Figure 3. Since $b^2 = a^2 - c^2$, we have $c^2 = a^2 - b^2 = 16 - 9 = 7$, and the focal points are at $(\sqrt{7}, 0)$ and $(-\sqrt{7}, 0)$. ■

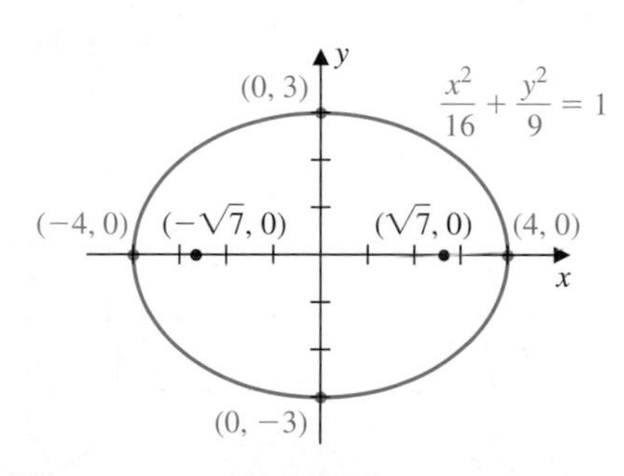

FIGURE 3

EXAMPLE 2 Find an equation of the ellipse in standard position with a vertex at $(5, 0)$ and a focal point at $(3, 0)$.

Solution Since the ellipse is in standard position, both the vertices and the focal points are centered about the origin. So the other vertex is at $(-5, 0)$, and the other focal point is at $(-3, 0)$. Since $a = 5$ and $c = 3$, we have

$$b = \sqrt{a^2 - c^2} = \sqrt{25 - 9} = 4,$$

and the ellipse, shown in Figure 4, has equation

$$\frac{x^2}{25} + \frac{y^2}{16} = 1, \quad \text{or} \quad 16x^2 + 25y^2 = 400.$$

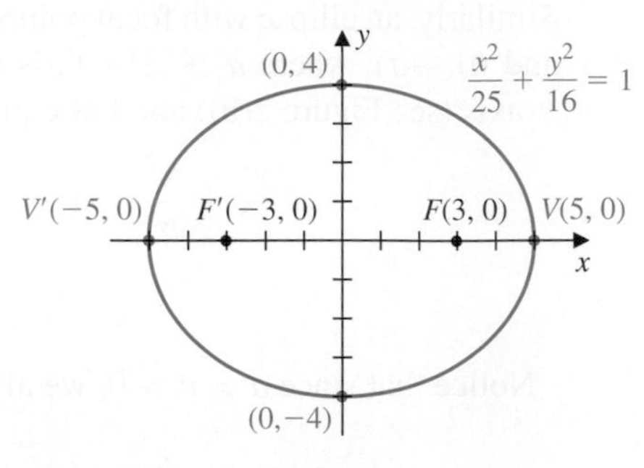

FIGURE 4

EXAMPLE 3 Find the equation of the ellipse with vertices at $(-2, -4)$ and $(-2, 2)$ and a focal point at $(-2, 1)$.

Solution The vertices lie on the vertical line $x = -2$, so the major axis of the ellipse is on the line $x = -2$. The distance between the vertices is 6, which is the length $2a$ of the major axis, so $a = 3$. The center lies midway between the vertices and is the point $(-2, -1)$, as shown in Figure 5. The distance between the center and the focal point $(-2, 1)$ is 2, so $c = 2$. The value b^2 can now be found from the equation

$$b^2 = a^2 - c^2 = 9 - 4 = 5.$$

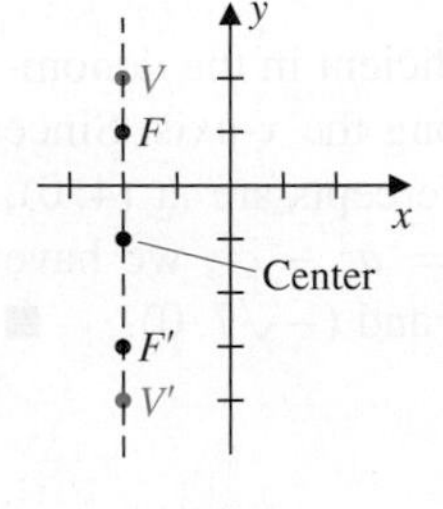

FIGURE 5

The equation of the ellipse, shown in Figure 6, is consequently

$$\frac{(y + 1)^2}{9} + \frac{(x + 2)^2}{5} = 1.$$

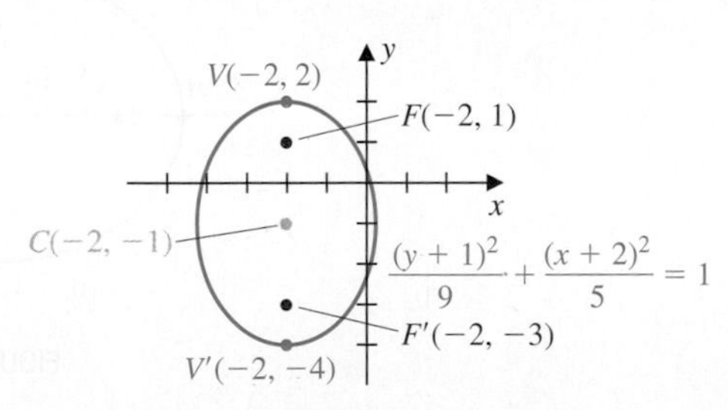

FIGURE 6

In the introduction to the chapter, we indicated that quadratic equations of the form

$$Ax^2 + Cy^2 + Dx + Ey + F = 0$$

produce ellipses when A and C have the same sign, that is, when $AC > 0$. In fact, the ellipses that are produced are just vertical and horizontal translations of ellipses in standard position. The usual technique of completing the square is used to determine how the curve is translated, as illustrated in Example 4.

EXAMPLE 4 Sketch the graph of the ellipse with equation $9x^2 - 72x + 4y^2 + 16y + 124 = 0$, and find its focal points and vertices.

Solution We first complete the square in both x and y so that this equation can be compared to the equation of an ellipse in standard position. The equation can be written as

$$9(x^2 - 8x) + 4(y^2 + 4y) = -124,$$

so

$$9(x^2 - 8x + 16) - 9(16) + 4(y^2 + 4y + 4) - 4(4) = -124$$

and

$$9(x - 4)^2 + 4(y + 2)^2 = 36.$$

Dividing by 36 gives

$$\frac{(x-4)^2}{4} + \frac{(y+2)^2}{9} = 1.$$

Since the denominator of the y^2-term in the standard position equation

$$\frac{y^2}{9} + \frac{x^2}{4} = 1$$

is larger than the denominator of the x^2-term, this ellipse is in standard position with axis along the y-axis as shown in Figure 7(a). The vertices of the standard-position ellipse are at $(0, 3)$ and $(0, -3)$. Also

$$c^2 = a^2 - b^2 = 9 - 4 = 5,$$

so the focal points of the ellipse in standard position occur at $(0, \sqrt{5})$ and $(0, -\sqrt{5})$.

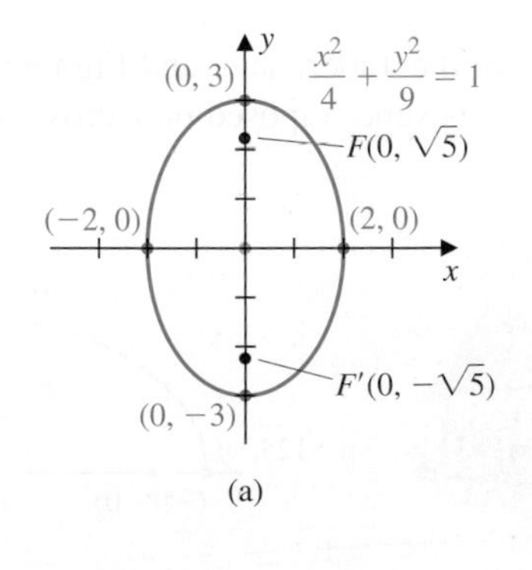

(a)

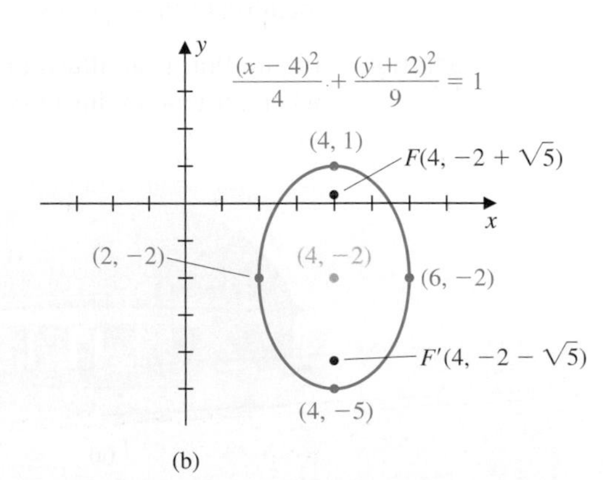

(b)

FIGURE 7

The graph of the original equation,

$$\frac{(y+2)^2}{9} + \frac{(x-4)^2}{4} = 1,$$

is simply a translation of the graph of the standard-position ellipse, downward 2 units and to the right 4 units. Hence, the focal points occur at $(4, -2+\sqrt{5})$ and $(4, -2-\sqrt{5})$, and the vertices are at (4, 1) and (4, −5), as shown in Figure 7(b). ■

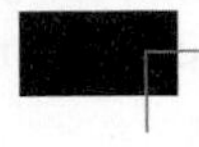

Reflection Property

Like the parabola, the ellipse has an interesting reflection property. If a light or sound source is placed at one focus of a reflecting surface with elliptical cross sections, then the wave emitted from the source is reflected off the surface to the other focus (see Figure 8).

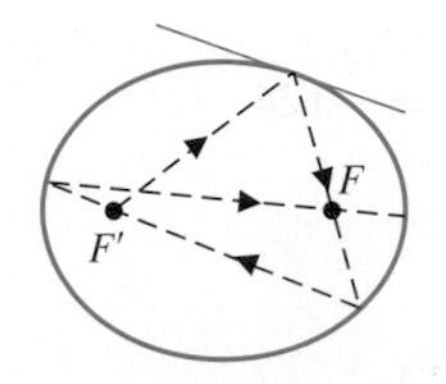

FIGURE 8

When a person stands at one of the focal points in a building with an elliptical ceiling, for example, sound made at that point is reflected from the ceiling to the other focal point. Rooms with this property are called "whispering galleries." These whispering galleries are no recent innovation. Statuary Hall in the Capitol Building in Washington, D.C. and the Mormon Tabernacle in Salt Lake City, Utah were built over a century ago and use this principle, as does St. Paul's Cathedral in London, which was designed by the outstanding mathematician and architect Sir Christopher Wren in the decade after the great fire of 1666. In fact, the principles of this design were known to the Greek geometers of Euclid's time, over 2300 years ago.

Applications

EXAMPLE 5 The dome of the Mormon Tabernacle in Salt Lake City is 250 feet long, 150 feet wide, and 80 feet high, and its longitudinal cross section has the shape of an ellipse. The conductor for musical performances stands near one of the focal points of the ellipse, and recording equipment can be placed at the other. In this way the sound heard by the conductor corresponds very closely to the sound being recorded. Determine the location of these points.

Solution Figure 9(a) is an illustration of the Tabernacle, and Figure 9(b) shows the situation when an xy-coordinate system is superimposed on a cross section of the dome. The

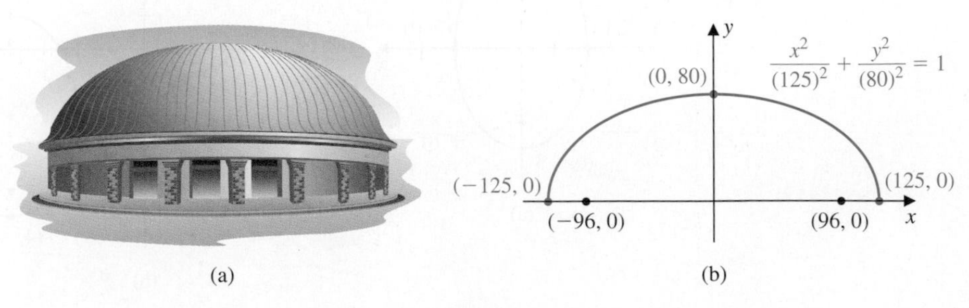

FIGURE 9

width of the building plays no part in the calculations, but the length and height are used to determine the lengths of the major and minor axes, 250 and 160 feet, respectively. The ellipse in standard position has equation

$$\frac{x^2}{125^2} + \frac{y^2}{80^2} = 1.$$

The focal points occur at a distance of

$$c = \sqrt{a^2 - b^2} = \sqrt{125^2 - 80^2} = \sqrt{9225} \approx 96 \text{ feet}$$

from the center of the building. So the conductor should be located approximately $125 - 96 = 29$ feet from one end of the building, with the recording equipment placed an equal distance from the other end. ■

The **eccentricity** of an ellipse tells how the ellipse differs from a circle. For an ellipse in standard position, the eccentricity is defined to be

$$e = \frac{c}{a} = \frac{\sqrt{a^2 - b^2}}{a},$$

with the usual definitions of a, b, and c. Since $a > b > 0$ and $a > c$, we have $0 < e < 1$. If we permitted c to be 0, then the ellipse would degenerate to a circle with $a = b$ and $e = 0$. Figure 10 shows various ellipses with the same vertices. Notice that as the focal point approaches the origin, the ellipse approaches a circle, and e approaches 0. The ellipse becomes increasingly elongated as the focal points approach the vertices, which occurs when e approaches 1 from the left.

Ellipses play an important role in astronomy. In the early 17th century, Johannes Kepler used an extensive collection of data to deduce that the orbits of the planets are ellipses with the sun at one focal point. The earth orbits the sun in a nearly circular orbit with eccentricity 0.0167. In fact, the orbits of all the planets are nearly circular, as you can see from the data in Table 1.

In 2006, Pluto, long considered one of the planets orbiting the sun in our solar system, was reclassified as a "dwarf planet". Its orbit has an eccentricity of 0.2481, which is larger than the eccentricity of any of the planets. Even so, its orbit is nearly circular because

$$0.2481 = \frac{\sqrt{a^2 - b^2}}{a} \quad \text{implies that} \quad b = 0.969a.$$

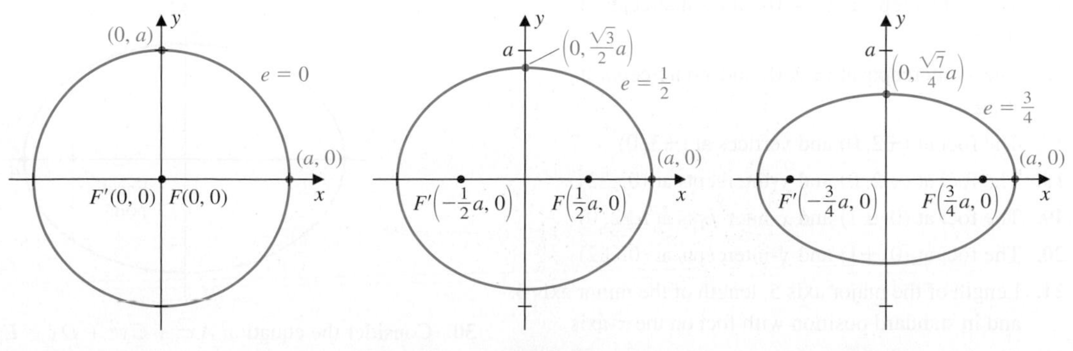

FIGURE 10

TABLE 1

Planet	Major Axis	Eccentricity	Planet	Major Axis	Eccentricity
Mercury	1.159×10^8 km	0.2056	Jupiter	1.556×10^9 km	0.0484
Venus	2.162×10^8 km	0.0068	Saturn	2.854×10^9 km	0.0543
Earth	2.991×10^8 km	0.0167	Uranus	5.741×10^9 km	0.0460
Mars	4.557×10^8 km	0.0934	Neptune	9.000×10^9 km	0.0082

On the other hand, the orbit of Halley's comet about the sun is an example of a very non-circular elliptic orbit. It has a major axis of 5.39×10^9 kilometers and eccentricity of $e = 0.967$. Its orbit takes 76 years and it was last close to the earth in 1986.

EXERCISE SET 6.3

In Exercises 1–14, sketch the graph of the ellipse showing the vertices and focal points.

1. $\dfrac{x^2}{4} + y^2 = 1$ **2.** $\dfrac{x^2}{16} + y^2 = 1$

3. $x^2 + \dfrac{y^2}{9} = 1$ **4.** $\dfrac{x^2}{4} + \dfrac{y^2}{16} = 1$

5. $16x^2 + 25y^2 = 400$ **6.** $25x^2 + 16y^2 = 400$

7. $3x^2 + 2y^2 = 6$ **8.** $3x^2 + 4y^2 = 12$

9. $4x^2 + y^2 + 16x + 12 = 0$

10. $2x^2 + 4y^2 - 4x - 14 = 0$

11. $x^2 + 4y^2 - 2x - 16y + 13 = 0$

12. $4x^2 + 9y^2 - 16x + 90y + 97 = 0$

13. $2x^2 + 4y^2 + 4x - 16y + 2 = 0$

14. $3x^2 + 4y^2 + 12x + 8y + 4 = 0$

In Exercises 15–28, find an equation of the ellipse that satisfies the stated conditions.

15. The x-intercepts at $(\pm 4, 0)$ and y-intercepts at $(0, \pm 3)$

16. The x-intercepts at $(\pm 2, 0)$ and y-intercepts at $(0, \pm 5)$

17. The foci at $(\pm 2, 0)$ and vertices at $(\pm 3, 0)$

18. The foci at $(\pm 2, 0)$ and y-intercepts at $(0, \pm 2)$

19. The foci at $(0, \pm 1)$ and x-intercepts at $(\pm 2, 0)$

20. The foci at $(0, \pm 1)$ and y-intercepts at $(0, \pm 2)$

21. Length of the major axis 5, length of the minor axis 3, and in standard position with foci on the x-axis

22. The foci at $(0, \pm 2)$ and length of the major axis 8

23. The foci at $(3, 0)$ and $(1, 0)$ and a vertex at $(0, 0)$

24. The foci at $(0, 4)$ and $(0, 8)$ and a vertex at $(0, 2)$

25. The vertices at $(-4, -2)$ and $(-4, 8)$ and a focal point at $(-4, 0)$

26. The vertices at $(2, 2)$ and $(6, 2)$ and a focal point at $(5, 2)$

27. The vertices at $(3, 3)$ and $(3, -1)$ and passing through $(2, 1)$

28. The vertices at $(-4, -2)$ and $(2, -2)$ and passing through $(-1, 1)$

29. The *latus rectum* of an ellipse is a line segment that passes through a focal point perpendicular to the major axis and joins the points on the ellipse (see the figure). Find the length of the latus rectum of the ellipse with equation

$$\frac{x^2}{a^2} + \frac{y^2}{b^2} = 1, \quad \text{when} \quad a > b > 0.$$

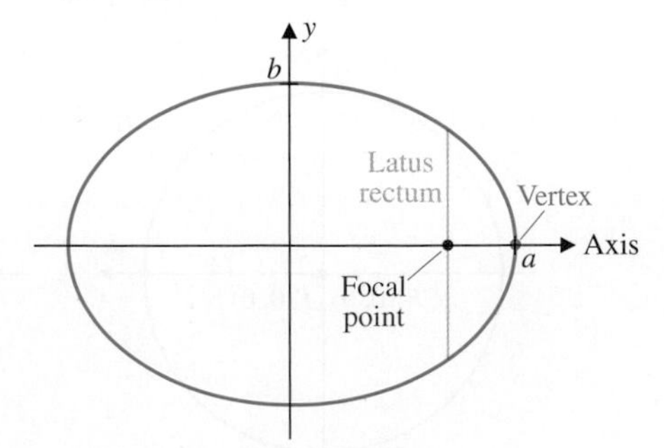

30. Consider the equation $Ax^2 + Cy^2 + Dx + Ey + F = 0$, where A and C are positive constants. Find conditions on the constants that ensure that this equation describes each of the following:

a. An ellipse

b. A circle

c. A single point

d. No points

31. Olympic Stadium in Montreal, shown in the figure below, is constructed in the shape of an ellipse with major and minor axes 480 and 280 meters, respectively. Find an equation for this ellipse.

32. Halley's comet is named to honor Edmund Halley (1656–1742). In 1682 Halley determined the orbit of this comet and predicted that it would return about 76 years later. The orbit is elliptical with a focal point at the sun, its major axis is approximately 5.39×10^9 kilometers, and its minor axis is approximately 1.36×10^9 kilometers. How close does this comet pass to the sun?

33. A satellite is placed in elliptical earth orbit with a minimum distance of 160 kilometers and a maximum distance of 16,000 kilometers above the earth. Find the eccentricity and equation of the orbit, assuming that the center of the earth is at one focal point and that the radius of the earth is 6380 kilometers.

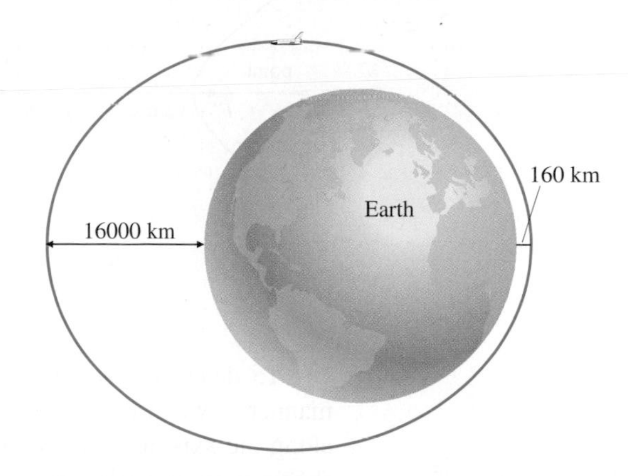

34. A doorway 40 inches wide is capped with a semi-ellipse that is 12 inches high at its center. What is the height of the doorway 16 inches from the center?

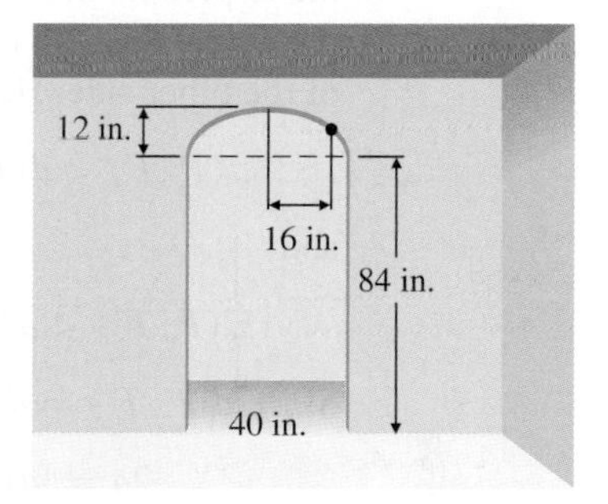

6.4 HYPERBOLAS

Although the shapes of ellipses and hyperbolas are very different, their definitions are quite similar. Both of these families of curves are defined using two fixed points and a fixed distance. In the case of an ellipse, the fixed distance is the sum of the distances between the two fixed points, whereas the hyperbola uses the difference of these distances.

Hyperbola

A **hyperbola** is the set of points in a plane for which the magnitude of the difference between the distances from two fixed points is a given constant. The two fixed points are the **focal points**, and the line passing through the focal points is the **axis** of the hyperbola. The points of intersection of the axes and the hyperbola are the **vertices**. (See Figure 1(a).)

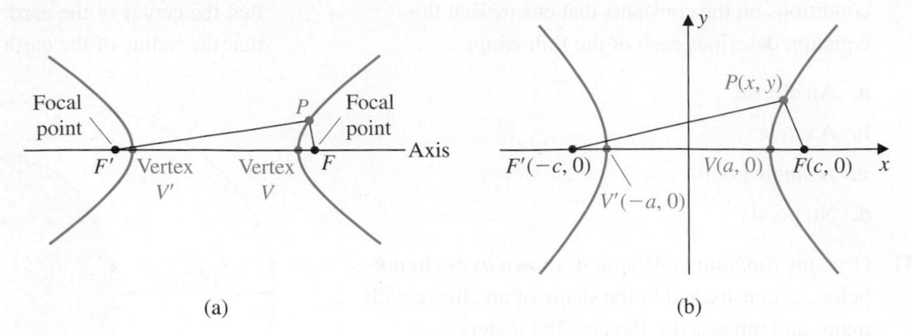

FIGURE 1

To determine standard equations for the hyperbola, we proceed in the same manner as we did for the ellipse. An xy-coordinate system is placed with the x-axis along the axis of the hyperbola and the origin midway between the focal points, as shown in Figure 1(b). As in the case of the ellipse, the focal points have been given the coordinates $F(c, 0)$ and $F'(-c, 0)$.

The vertices are located at $V(a, 0)$ and $V'(-a, 0)$, where $a \neq c$, and the fixed difference is $2a$. The vertices for the ellipse were located outside the focal points, but the hyperbola has vertices located between the focal points. To see why this is true, recall that the sum of the lengths of two sides of a triangle always exceeds the length of the other side. Hence,

$$d(P, F) < d(P, F') + d(F', F) \quad \text{so} \quad d(P, F) - d(P, F') < d(F', F),$$

and

$$d(P, F') < d(P, F) + d(F, F') \quad \text{so} \quad d(P, F') - d(P, F) < d(F, F').$$

Since $d(F, F') = d(F', F)$, when taken together, this implies that

$$2a = |d(P, F) - d(P, F')| < d(F, F') = 2c, \quad \text{and} \quad a < c.$$

To derive the equation of the hyperbola, consider a point (x, y) on the hyperbola with $x > 0$. The hyperbola has y-axis symmetry, so the equation derived also holds when $x < 0$.

The point (x, y) is on the graph precisely when

$$\begin{aligned} 2a &= |d((x, y), (-c, 0)) - d((x, y), (c, 0))| \\ &= \left|\sqrt{(x + c)^2 + (y - 0)^2} - \sqrt{(x - c)^2 + (y - 0)^2}\right|. \end{aligned}$$

Since $x > 0$ and $c > 0$, we have $(x + c)^2 > (x - c)^2$, and the absolute values are not needed. So

$$2a = d((x, y), (-c, 0)) - d((x, y), (c, 0)) = \sqrt{(x + c)^2 + y^2} - \sqrt{(x - c)^2 + y^2},$$

and

$$\sqrt{(x + c)^2 + y^2} = 2a + \sqrt{(x - c)^2 + y^2}.$$

Squaring both sides of the last equation gives

$$(x + c)^2 + y^2 = 4a^2 + 4a\sqrt{(x - c)^2 + y^2} + (x - c)^2 + y^2.$$

Completing the multiplication and dividing by 4 simplifies this to

$$-a\sqrt{(x-c)^2+y^2} = a^2 - cx.$$

Squaring again and adding $2a^2cx$ to both sides of the equation gives

$$a^2x^2 + a^2c^2 + a^2y^2 = a^4 + c^2x^2,$$

the same equation as we found for the ellipse. It simplifies in the same way as it did for the ellipse to

$$\frac{x^2}{a^2} + \frac{y^2}{a^2 - c^2} = 1,$$

or, more appropriately in this case, since $0 < a < c$,

$$\frac{x^2}{a^2} - \frac{y^2}{c^2 - a^2} = 1.$$

If we replace $c^2 - a^2$ with a new constant b^2, we produce the equation for a hyperbola with axis lying along the x-axis given by

$$\frac{x^2}{a^2} - \frac{y^2}{b^2} = 1, \quad \text{where} \quad b = \sqrt{c^2 - a^2}.$$

Notice that the squares on the x- and y-terms imply that the graph has both x-axis and y-axis symmetry. The x-intercepts occur at the vertices $(a, 0)$ and $(-a, 0)$. There are no y-intercepts. This follows from the fact that

$$y^2 = \frac{b^2}{a^2}(x^2 - a^2),$$

which implies that the equation is valid only for $x^2 \geq a^2$, that is, when $x \geq a$ or $x \leq -a$. A hyperbola with equation in this form is said to be in *standard position* with axis along the x-axis. A similar type of equation results when the roles of the axes are reversed. (See Figure 2.)

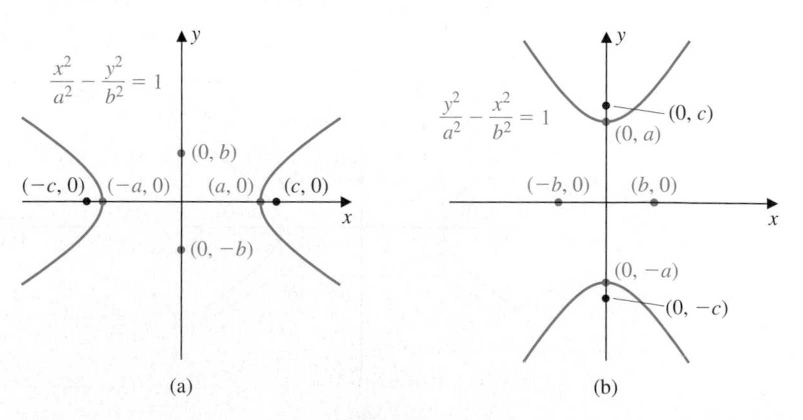

FIGURE 2

Standard-Position Hyperbolas

A hyperbola with focal points at $(c, 0)$ and $(-c, 0)$ and vertices at $(a, 0)$ and $(-a, 0)$, where $c > a > 0$, is in **standard position** with axis along the x-axis (see Figure 2(a)) and has equation

$$\frac{x^2}{a^2} - \frac{y^2}{b^2} = 1, \quad \text{where} \quad b = \sqrt{c^2 - a^2}.$$

Similarly, a hyperbola with focal points at $(0, c)$ and $(0, -c)$ and vertices at $(0, a)$ and $(0, -a)$, where $c > a > 0$, is in **standard position** with axis along the y-axis (see Figure 2(b)) and has equation

$$\frac{y^2}{a^2} - \frac{x^2}{b^2} = 1, \quad \text{where} \quad b = \sqrt{c^2 - a^2}.$$

In the case of the ellipse we found that we always have $0 < b < a$. This is no longer true for hyperbolas. All we can conclude is that $b = \sqrt{c^2 - a^2} < \sqrt{c^2} = c$.

The standard-position equations can be used to determine the end behavior of the hyperbola. Consider the situation in the first quadrant for a hyperbola in standard position with axis along the x-axis. Solving for y in terms of x in

$$\frac{x^2}{a^2} - \frac{y^2}{b^2} = 1 \quad \text{gives} \quad y = \sqrt{\frac{b^2}{a^2}(x^2 - a^2)} = \frac{b}{a}\sqrt{x^2 - a^2}.$$

As x becomes increasingly large, the constant term, $-a^2$, under the radical becomes less important, and

$$y = \frac{b}{a}\sqrt{x^2 - a^2} \approx \frac{b}{a}\sqrt{x^2} = \frac{b}{a}x.$$

Hence the graph of the hyperbola approaches the line $y = (b/a)x$ as x becomes large. This line is a *slant asymptote* to the graph of the hyperbola. The hyperbola has both x-axis and y-axis symmetry, so the graph also approaches $y = (b/a)x$ in the third quadrant, and approaches $y = -(b/a)x$ in quadrants II and IV. As a consequence, the hyperbola shown in Figure 3(a) with equation

$$\frac{x^2}{a^2} - \frac{y^2}{b^2} = 1 \quad \text{has slant asymptotes} \quad y = \pm\frac{b}{a}x.$$

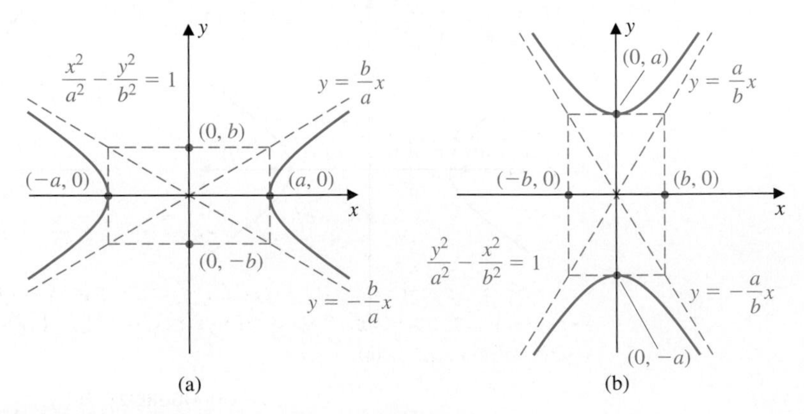

FIGURE 3

In a similar manner, as shown in Figure 3(b),

$$\frac{y^2}{a^2} - \frac{x^2}{b^2} = 1 \quad \text{has slant asymptotes} \quad y = \pm\frac{a}{b}x.$$

EXAMPLE 1 Sketch the graph of the hyperbola with equation $16x^2 - 9y^2 = 144$.

Solution The equation can be rewritten as

$$\frac{x^2}{9} - \frac{y^2}{16} = 1.$$

The hyperbola is in standard position with axis along the x-axis with $a = 3$ and $b = 4$, and the asymptotes are $y = \pm\frac{4}{3}x$. This is enough information to give the reasonable sketch shown in Figure 4. Since $c = \sqrt{a^2 + b^2} = \sqrt{9 + 16} = 5$, the focal points are at $(5, 0)$ and $(-5, 0)$. ■

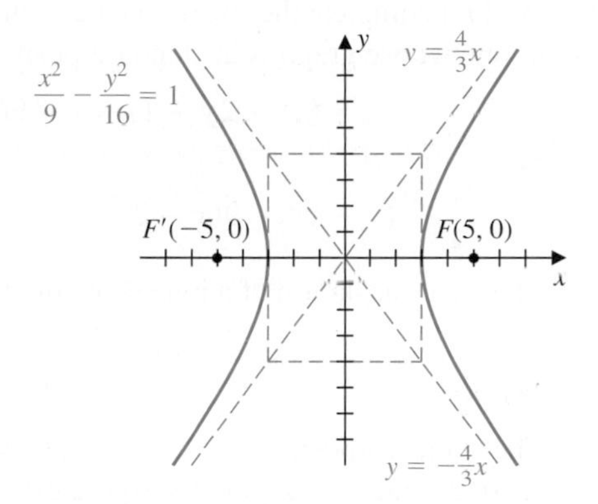

FIGURE 4

EXAMPLE 2 Find the equation of the hyperbola with a focal point at $(2, -1)$ and vertices at $(1, -1)$ and $(-3, -1)$.

Solution Since the vertices and the focal point lie on the horizontal line $y = -1$, the axis of the hyperbola is parallel to the x-axis. The vertices are centered about the point $(-1, -1)$ and the distance from this center point to the vertices is 2, so $a = 2$. The focus $(2, -1)$ is 3 units to the right of the center, so $c = 3$. The value b^2 can now be found from the equation $b = \sqrt{c^2 - a^2}$ as

$$b^2 = 9 - 4 = 5.$$

The standard-position hyperbola with $a^2 = 4$, $b^2 = 5$, $c^2 = 9$, and axis along the x-axis has equation

$$\frac{x^2}{4} - \frac{y^2}{5} = 1 \quad \text{and asymptotes } y = \pm\frac{\sqrt{5}}{2}x.$$

The hyperbola with axis parallel to the x-axis on the line $y = -1$ and center $(-1, -1)$ is the standard-position hyperbola shifted to the left 1 unit and downward 1 unit. It has equation

$$\frac{(x+1)^2}{4} - \frac{(y+1)^2}{5} = 1$$

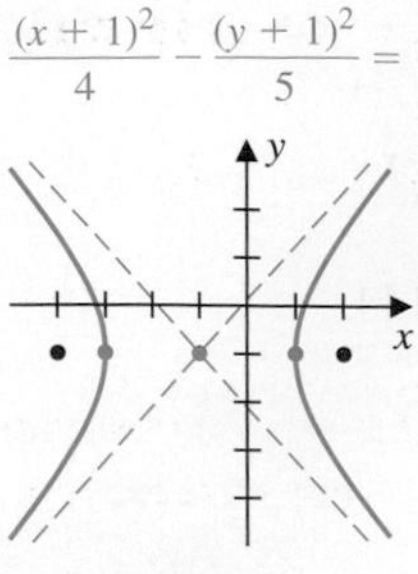

$$\frac{(x+1)^2}{4} - \frac{(y+1)^2}{5} = 1$$

FIGURE 5

and is shown in Figure 5. This hyperbola has asymptotes

$$y + 1 = \pm\frac{\sqrt{5}}{2}(x+1).$$

■

In Section 6.1, we indicated that quadratic equations of the form

$$Ax^2 + Cy^2 + Dx + Ey + F = 0$$

produce hyperbolas when A and C have opposite signs, that is, when $AC < 0$. These hyperbolas are just vertical and horizontal translations of hyperbolas in standard position. As in the case of the ellipse, we use completion of the square to determine how the curve is translated, which is illustrated in Example 3.

EXAMPLE 3 Sketch the graph of the hyperbola with equation

$$y^2 - 2y - 9x^2 + 36x = 39.$$

Solution We first complete the square on the x- and y-terms so that we can relate this equation to one whose graph is in standard position. This gives

$$(y^2 - 2y + 1) - 1 - 9(x^2 - 4x + 4) + 9(4) = 39,$$

so

$$(y-1)^2 - 9(x-2)^2 = 4, \quad \text{and} \quad \frac{(y-1)^2}{4} - \frac{(x-2)^2}{\frac{4}{9}} = 1.$$

This is a translation of a hyperbola in standard position with axis along the y-axis:

$$\frac{y^2}{4} - \frac{x^2}{\frac{4}{9}} = 1.$$

The standard-position hyperbola has $a = 2$ and $b = \sqrt{4/9} = 2/3$, so its vertices are at $(0, 2)$ and $(0, -2)$ and its asymptotes have equations

$$y = \pm\frac{a}{b}x = \pm\frac{2}{\frac{2}{3}}x = \pm 3x,$$

as shown in Figure 6(a).

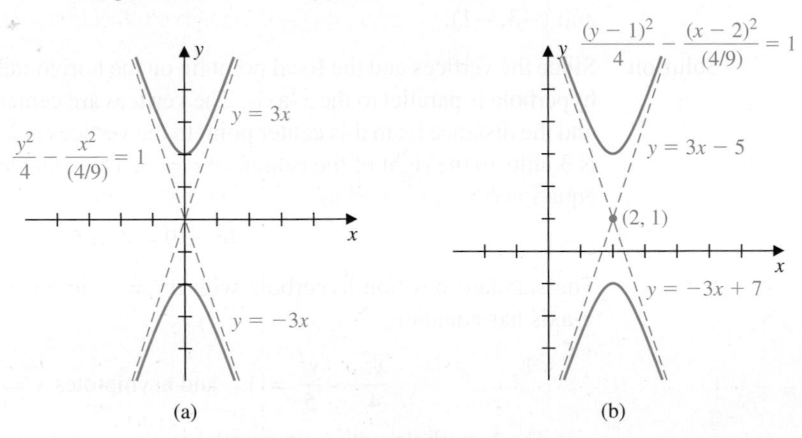

FIGURE 6

The graph of

$$\frac{(y-1)^2}{4} - \frac{(x-2)^2}{\frac{4}{9}} = 1$$

has the same shape but is shifted to the right 2 units and upward 1 unit, as shown in Figure 6(b). The asymptotes for the graph are also shifted, to become

$$y - 1 = \pm 3(x - 2), \quad \text{that is,} \quad y = 3x - 5 \quad \text{and} \quad y = -3x + 7. \qquad ■$$

Reflection Property

The hyperbola, like the parabola and ellipse, has a reflection property. Light or sound emitted from one of the focal points of the hyperbola is reflected off the surface along a line directly away from the other focal point. (See Figure 7.)

The hyperbola reflection property finds application in the construction of large telescopes, such as the Hale telescope at the Mount Palomar Observatory in California. Within this telescope there is a Cassegrain configuration that consists of a hyperbolic mirror inserted between the parabolic reflector and the parabola's focal point. This hyperbola has one focal point coinciding with the focal point of the parabola and reflects the image back through a hole in the center of the parabolic mirror, as shown in Figure 8. From there it goes to the other focal point of the hyperbolic mirror, located beyond the vertex of the parabolic mirror.

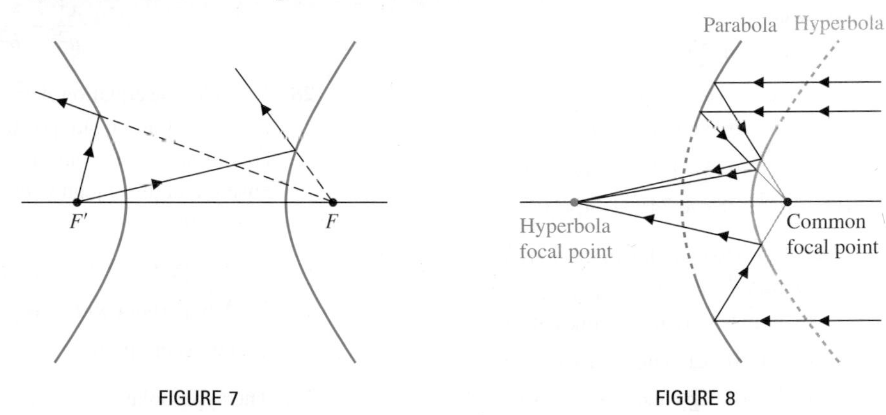

FIGURE 7

FIGURE 8

Eccentricity is also defined for hyperbolas. For a hyperbola in standard position, the eccentricity is

$$e = \frac{c}{a} = \frac{\sqrt{a^2 + b^2}}{a},$$

with our usual definitions of a, b, and c. Since $0 < a < c$ we have $e > 1$. Figure 9 shows various hyperbolas with the same vertices. Notice that as the focal point

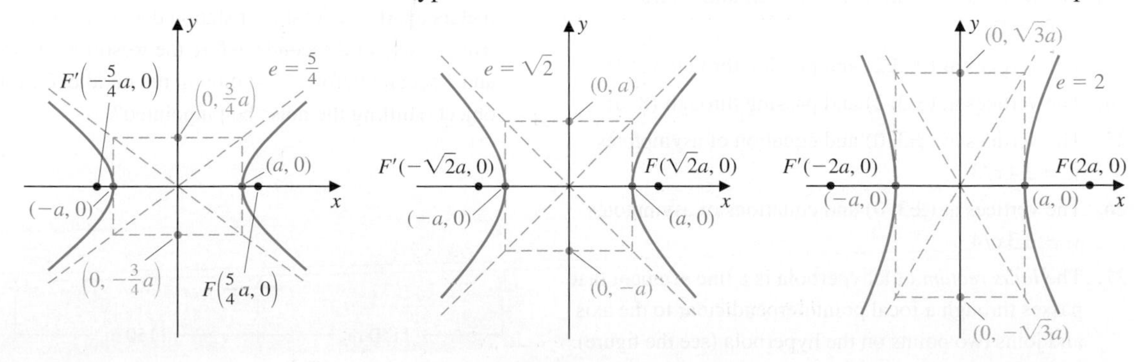

FIGURE 9

approaches the vertex, the hyperbola becomes increasingly narrow and e approaches 1 from the right. As the focal point moves away from the vertex, the hyperbola becomes wider, approaching the pair of vertical lines $x = \pm a$.

EXERCISE SET 6.4

In Exercises 1–14, sketch the graph of the hyperbola showing the vertices, focal points, and asymptotes.

1. $\frac{x^2}{4} - \frac{y^2}{9} = 1$
2. $\frac{x^2}{9} - \frac{y^2}{4} = 1$
3. $\frac{y^2}{4} - \frac{x^2}{9} = 1$
4. $\frac{y^2}{9} - \frac{x^2}{4} = 1$
5. $x^2 - y^2 = 1$
6. $y^2 - 4x^2 = 1$
7. $9y^2 - 18y - 4x^2 = 27$
8. $x^2 + 2x - 4y^2 = 3$
9. $3x^2 - y^2 = 6x$
10. $2y^2 + 8y = 9x^2$
11. $9x^2 - 4y^2 - 18x - 8y = 31$
12. $y^2 - 4x^2 - 2y - 16x = 19$
13. $9y^2 - 4x^2 - 36y + 16x - 16 = 0$
14. $4x^2 - 6y^2 - 8x + 24y - 44 = 0$

In Exercises 15–26, find an equation of the hyperbola that satisfies the stated conditions.

15. The foci at $(\pm 5, 0)$ and vertices at $(\pm 3, 0)$
16. The foci at $(0, \pm 13)$ and vertices at $(0, \pm 12)$
17. The foci at $(0, \pm 5)$ and vertices at $(0, \pm 4)$
18. The foci at $(\pm 13, 0)$ and vertices at $(\pm 5, 0)$
19. A focus at $(2, 2)$ and vertices at $(2, 1)$ and $(2, -3)$
20. A focus at $(-3, 3)$ and vertices at $(-3, 0)$ and $(-3, -6)$
21. The foci at $(-1, 4)$ and $(5, 4)$ and a vertex at $(0, 4)$
22. The foci at $(-1, -2)$ and $(-7, -2)$ and a vertex at $(-2, -2)$
23. The vertices at $(0, \pm 2)$ and passing through $(3, 4)$
24. The vertices at $(\pm 2, 2)$ and passing through $(8, 8)$
25. The vertices at $(\pm 3, 0)$ and equation of asymptotes $y = \pm 4x/3$
26. The vertices at $(\pm 3, 0)$ and equations of asymptotes $y = \pm 3x/4$
27. The *latus rectum* of a hyperbola is a line segment that passes through a focal point perpendicular to the axis and joins two points on the hyperbola (see the figure).

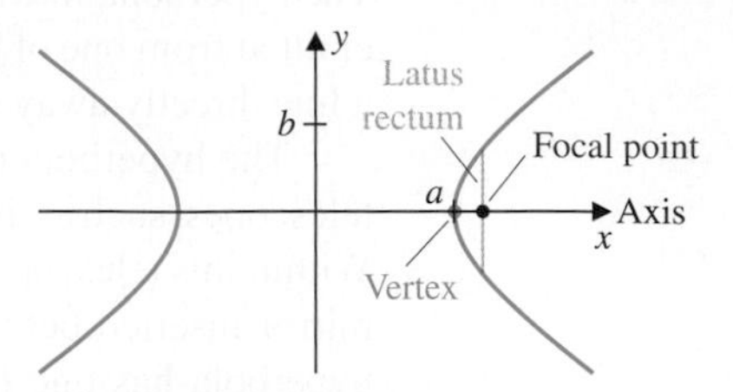

Find the length of a latus rectum of the hyperbola with equation

$$\frac{x^2}{a^2} - \frac{y^2}{b^2} = 1.$$

28. Consider the equation $Ax^2 - Cy^2 + Dx + Ey + F = 0$, where A and C are positive constants. Find conditions on the constants A, C, D, E, and F that ensure that this equation describes each of the following:

 a. A hyperbola with axis parallel to the x-axis

 b. A hyperbola with axis parallel to the y-axis

 c. Intersecting lines

29. The hyperbolas

$$\frac{x^2}{a^2} - \frac{y^2}{b^2} = 1 \quad \text{and} \quad \frac{y^2}{b^2} - \frac{x^2}{a^2} = 1$$

are *conjugates* of each other. How are their graphs related?

30. Three detection stations lie on an east–west line 1150 meters apart. The eastmost station detects a sound from an object 2 seconds before the westmost station and 1 second before the station in the middle. Can the object emitting the noise be pinpointed?

31. A company has two manufacturing plants that produce identical automobiles. Because of differing manufacturing and labor conditions in the plants, it costs \$130 more to produce a car in plant A than in plant B. The shipping costs from both plants are the same, \$1 per mile, as are the loading and unloading costs, \$25 per car. State the criteria for determining from which plant a car should be shipped.

6.5 POLAR COORDINATES

A problem can often be simplified by changing from rectangular to polar coordinates.

The rectangular (Cartesian) coordinate system has been used throughout the book to represent points and curves in the plane, but there is another common way, called the *polar coordinate system*, to represent points in the plane. In the rectangular coordinate system a point in the plane is specified by an ordered pair (x, y) that describes the distances of the point from the x- and y-axes, as shown in Figure 1(a). In the polar coordinate system a point in the plane is represented using an ordered pair (r, θ) that describes a distance and direction from a fixed reference point, as shown in Figure 1(b).

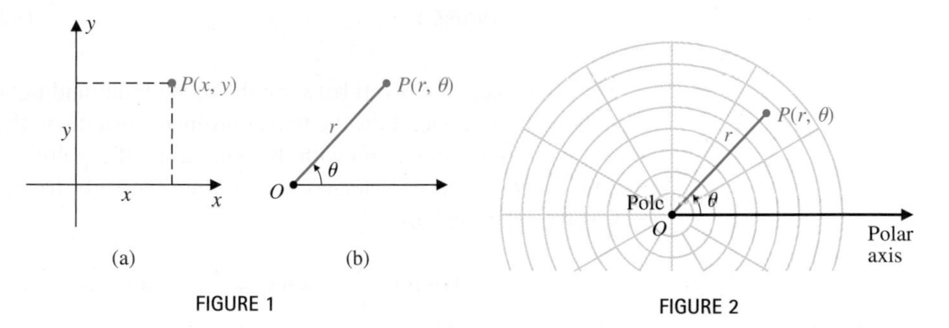

FIGURE 1

FIGURE 2

To describe the polar coordinate system, we first choose a fixed point O, called the origin, or **pole**. We then draw a half-line, or **ray**, called the **polar axis**, originating at the pole and extending to the right. To determine polar coordinates of a point P in the plane, we use an angle θ and a directed distance r. The angle θ is formed by the polar axis and the line segment $\overline{OP}$. The directed distance r is determined by the length of the line segment $\overline{OP}$, as shown in Figure 2. The ordered pair (r, θ) is a pair of **polar coordinates** for the point P.

In the rectangular coordinate system, each point in the plane has a unique representation, but this is not true in the polar coordinate system. The angles $\theta + 2n\pi$, for $n = 0, \pm 1, \pm 2, \ldots$, all have the same terminal side, so each point in the polar coordinate system has many representations. For example, $(1, \pi/6)$, $(1, 13\pi/6)$, and $(1, -11\pi/6)$ all represent the same point, as shown in Figure 3.

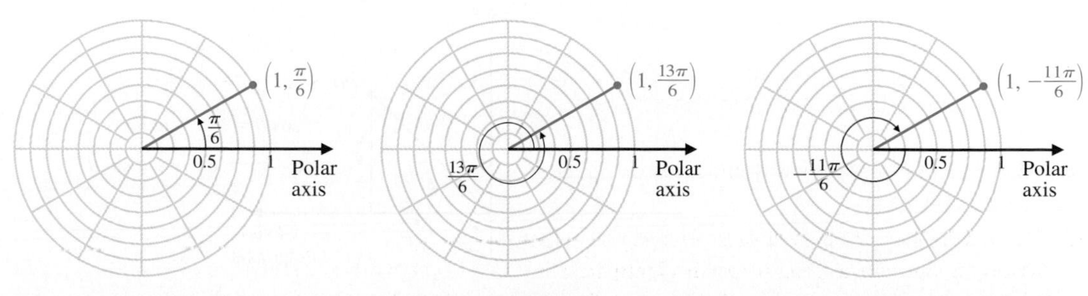

FIGURE 3

It is also convenient to allow r to be negative, with the understanding in this case that (r, θ) is the point $|r|$ units from the origin in the direction *opposite* to that given by θ. For example, the points $(-4, \pi/6)$ and $(-3, 3\pi/4)$ are as shown in Figure 4. This implies, for example, that the polar coordinates (r, θ) and $(-r, \theta + \pi)$ always represent the same point. (See Figure 5.)

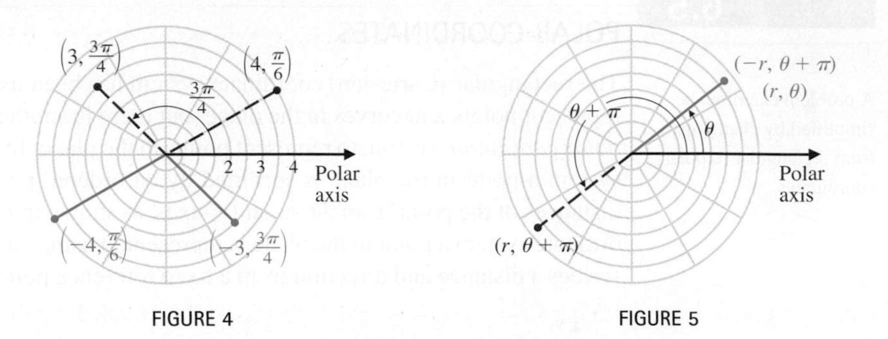

FIGURE 4 FIGURE 5

We often need to convert between the rectangular and polar coordinate systems. To see how this is done, draw the two coordinate systems with a common origin and the positive x-axis coinciding with the polar axis. If a point P has polar coordinates (r, θ) and rectangular coordinates (x, y), then the right triangle shown in Figure 6 gives us the relationships

$$\cos\theta = \frac{x}{r}, \quad \sin\theta = \frac{y}{r}, \quad \text{and} \quad \tan\theta = \frac{y}{x}.$$

Solving for x and y in these equations gives

$$x = r\cos\theta \quad \text{and} \quad y = r\sin\theta.$$

In addition, for a given set of rectangular coordinates $P(x, y)$, one set of polar coordinates for P, when $x \neq 0$, is

$$\theta = \arctan\frac{y}{x} \quad \text{and} \quad r = \pm\sqrt{x^2 + y^2},$$

where the sign for r is chosen to ensure that the point is in the correct quadrant. When $x = 0$, one set of polar coordinates is $\theta = \pi/2$ and $r = y$.

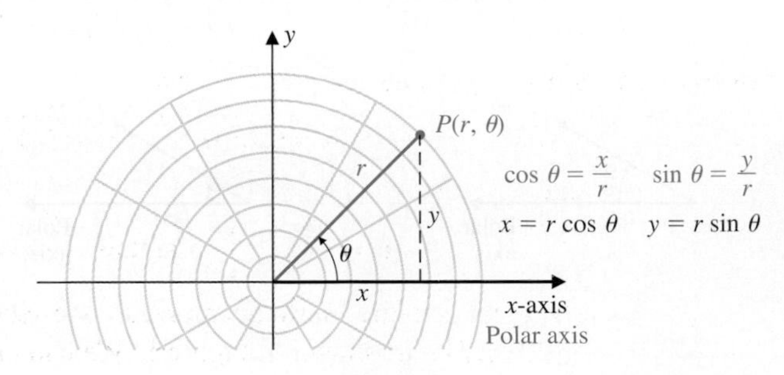

FIGURE 6

Relationship between Rectangular and Polar Coordinates

- A point with polar coordinates $P(r, \theta)$ has rectangular coordinates $P(x, y)$, where
$$x = r\cos\theta \quad \text{and} \quad y = r\sin\theta.$$
- One set of polar coordinates for the point with rectangular coordinates $P(x, y)$ is given, when $x \neq 0$, by
$$\theta = \arctan\frac{y}{x} \quad \text{and} \quad r = \pm\sqrt{x^2 + y^2},$$
where the sign for r is chosen so that the point is in the correct quadrant. When $x = 0$, one set is $\theta = \frac{\pi}{2}$ and $r = y$.

EXAMPLE 1 Find rectangular coordinates of the point that has polar coordinates $(2, 2\pi/3)$.

Solution Since $r = 2$ and $\theta = 2\pi/3$, we have, as shown in Figure 7,
$$x = r\cos\theta = 2\cos\left(\frac{2\pi}{3}\right) = 2\left(-\frac{1}{2}\right) = -1$$
and
$$y = r\sin\theta = 2\sin\left(\frac{2\pi}{3}\right) = 2\left(\frac{\sqrt{3}}{2}\right) = \sqrt{3}.$$ ■

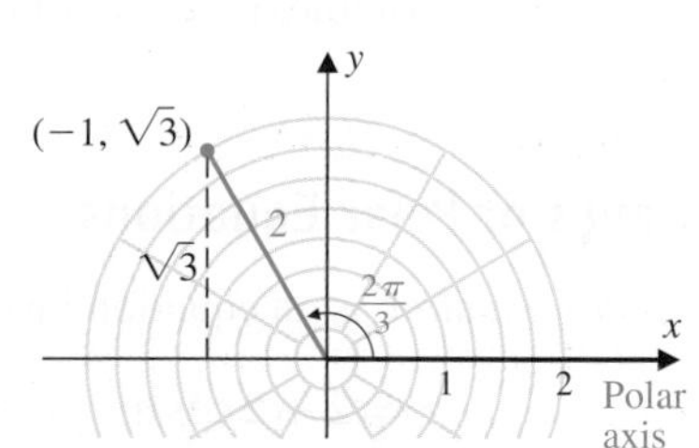

Rectangular coordinates $(-1, \sqrt{3})$
Polar coordinates $\left(2, \frac{2\pi}{3}\right)$

FIGURE 7

EXAMPLE 2 Find polar coordinates of the point given in rectangular coordinates.

a. $(1, -1)$ **b.** $(-\sqrt{3}/2, 1/2)$

Solution **a.** Since $x = 1$ and $y = -1$, we have
$$r = \pm\sqrt{x^2 + y^2} = \pm\sqrt{(1)^2 + (-1)^2} = \pm\sqrt{2},$$
and
$$\theta = \arctan\frac{y}{x} = \arctan\frac{-1}{1} = \arctan(-1) = -\frac{\pi}{4}.$$
The point is in the fourth quadrant, so one set of polar coordinates for the point is $(\sqrt{2}, -\pi/4)$, as shown in Figure 8. We can also use, for example, $(-\sqrt{2}, 3\pi/4)$ or $(\sqrt{2}, 7\pi/4)$.

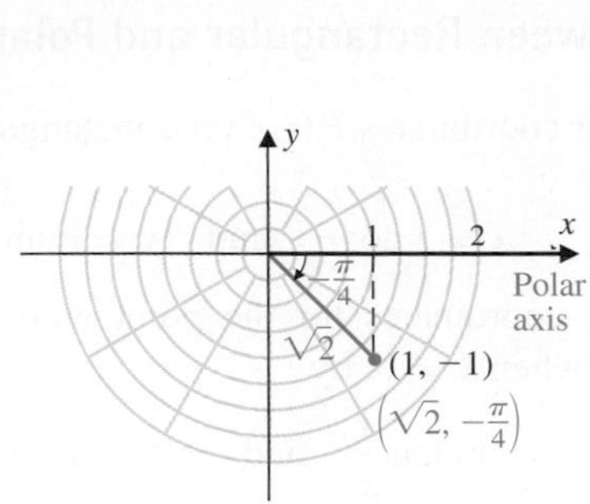

Rectangular coordinates $(1, -1)$
Polar coordinates $\left(\sqrt{2}, -\frac{\pi}{4}\right)$

FIGURE 8

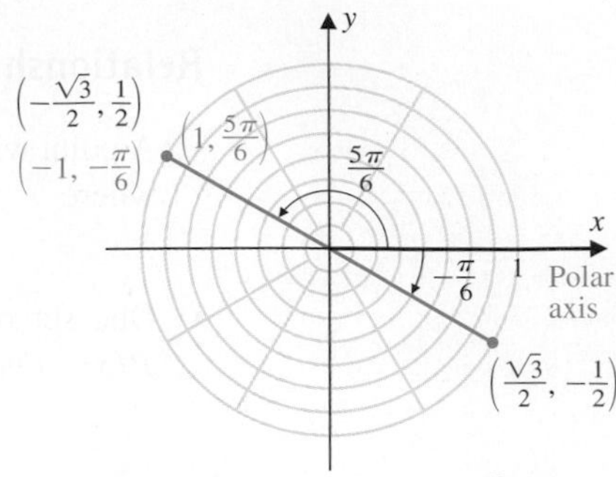

Rectangular coordinates $\left(-\frac{\sqrt{3}}{2}, \frac{1}{2}\right)$
Polar coordinates $\left(-1, -\frac{\pi}{6}\right)$

FIGURE 9

b. Since $x = -\sqrt{3}/2$ and $y = 1/2$, the choices for r are

$$r = \pm\sqrt{\left(-\frac{\sqrt{3}}{2}\right)^2 + \left(\frac{1}{2}\right)^2} = \pm 1.$$

The point $(-\sqrt{3}/2, 1/2)$ lies on the unit circle and is associated with the angle $5\pi/6$, as shown in Figure 9. So one set of polar coordinates for the point is $(1, 5\pi/6)$. Another polar coordinate representation for the point is $(-1, -\pi/6)$. ■

Graphs of Polar Equations

Many interesting curves have complicated rectangular representations, but changing to polar coordinates can simplify the problem.

A *polar equation* is an equation involving the polar variables r and θ. For example,

$$r = 2 + 2\cos\theta, \quad r = \sin 2\theta, \quad r = 3, \quad \text{and} \quad \theta = \frac{\pi}{2}$$

are all polar equations. The curve described by the equation is the collection of all points in the plane $P(r, \theta)$ that have *at least one* polar representation (r, θ) that satisfies the equation.

Quite often a polar equation is represented as $r = f(\theta)$, so r is a function of the independent variable θ. Computer algebra systems and graphing calculators can sketch curves in polar coordinates and permit us to see curves in the plane that might otherwise be very difficult to represent. Some of these examples are shown in Figure 10. In this brief introduction to polar curves, we look at just a few of those that are commonly seen in calculus.

EXAMPLE 3 Sketch the graph of the polar equation.

a. $r = 3$ **b.** $\theta = \pi/3$

Solution **a.** The graph consists of all points with r-coordinate 3, that is, all points that are 3 units from the origin. The graph is the circle with center at the origin and radius 3, shown in Figure 11(a). The rectangular equation of the circle is $x^2 + y^2 = 9$.

Graphing devices permit us to visualize many interesting and beautiful curves in the plane that we otherwise could not hope to sketch.

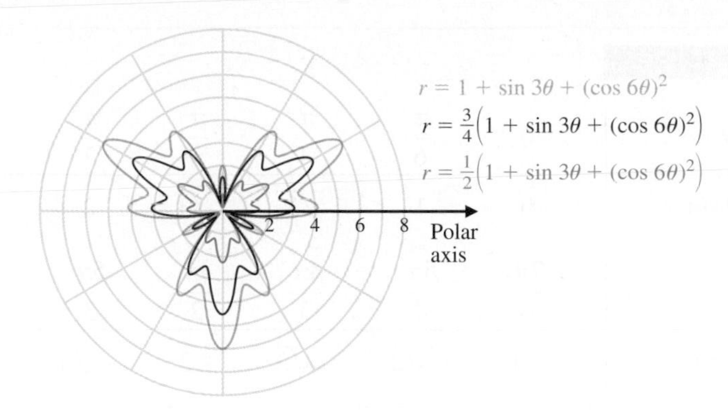

FIGURE 10

b. The graph consists of all points with θ-coordinate $\pi/3$. The graph is the line passing through the origin making an angle of $\pi/3$ radians with the polar axis, as shown in Figure 11(b). When $r > 0$, $(r, \pi/3)$ lies in the first quadrant and when $r < 0$, $(r, \pi/3)$ lies in the third quadrant.

The equation for a circle centered at the origin is much simpler in polar than in rectangular coordinates.

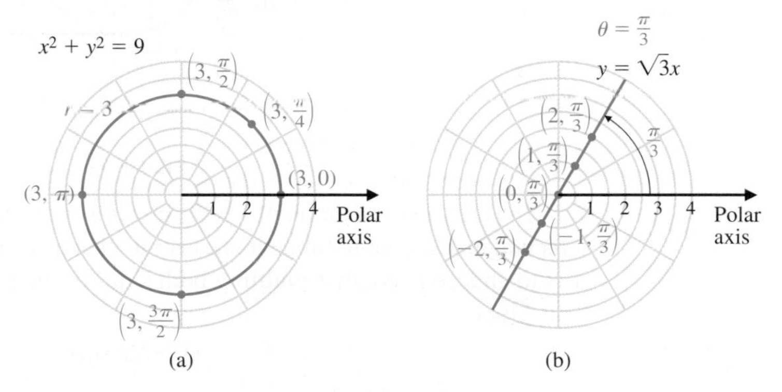

FIGURE 11

A point on this line has rectangular coordinates (x, y), where

$$\frac{y}{x} = \tan\theta = \tan\frac{\pi}{3} = \sqrt{3}.$$

So the line has the rectangular equation $y = \sqrt{3}x$. ■

EXAMPLE 4 Sketch the graph of the polar equation $r = 2\sin\theta$, and transform this polar equation into a rectangular equation.

Solution Table 1 lists sample values of θ and the corresponding values of r. Although θ can assume any real value, the sine function has period 2π, so we need only consider values of θ between 0 and 2π. Connecting the points in Table 1 with a smooth curve gives the graph shown in Figure 12. The graph appears to be a circle of radius 1 with center at the point with polar coordinates $(1, \pi/2)$.

Notice that the curve is traced, once for those points when θ is in $[0, \pi]$, and again when θ is in $[\pi, 2\pi]$.

TABLE 1

θ	0	$\frac{\pi}{6}$	$\frac{\pi}{4}$	$\frac{\pi}{3}$	$\frac{\pi}{2}$	$\frac{2\pi}{3}$	$\frac{3\pi}{4}$	$\frac{5\pi}{6}$	π
$r = 2\sin\theta$	0	1	$\sqrt{2}$	$\sqrt{3}$	2	$\sqrt{3}$	$\sqrt{2}$	1	0
θ	$\frac{7\pi}{6}$	$\frac{5\pi}{4}$	$\frac{4\pi}{3}$	$\frac{3\pi}{2}$	$\frac{5\pi}{3}$	$\frac{7\pi}{4}$	$\frac{11\pi}{6}$	2π	
$r = 2\sin\theta$	-1	$-\sqrt{2}$	$-\sqrt{3}$	-2	$-\sqrt{3}$	$-\sqrt{2}$	-1	0	

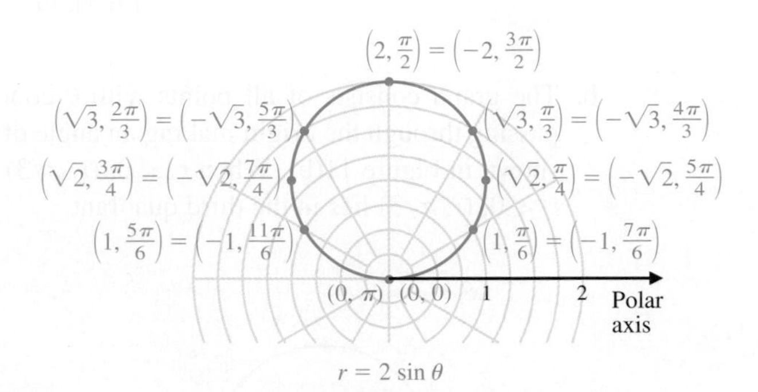

FIGURE 12

To verify that the graph of $r = 2\sin\theta$ is a circle, we change the polar equation to a rectangular equation. Since it is easier to change r^2 to rectangular coordinates than it is to change r, we first multiply both sides of the polar equation $r = 2\sin\theta$ by r. This gives

$$r^2 = 2r\sin\theta.$$

Since $r^2 = x^2 + y^2$ and $y = r\sin\theta$, we have

$$x^2 + y^2 = 2y, \quad \text{and} \quad x^2 + y^2 - 2y = 0.$$

Completing the square on the y-terms gives

$$x^2 + y^2 - 2y + 1 = 1, \quad \text{and} \quad x^2 + (y - 1)^2 = 1.$$

This equation describes the circle with radius 1 and center with rectangular coordinates (0, 1), as shown in Figure 12. ■

The equations in Examples 3(a) and 4 are special cases of the family of circles shown in Figure 13.

EXAMPLE 5 Sketch the graph of $r = 1 + \cos\theta$, and transform the equation to a rectangular equation.

Solution Since $\cos(-\theta) = \cos\theta$, the graph has x-axis symmetry, so we only need to consider values of θ between 0 and π. The values in Table 2, which decrease from 2 to 0 as θ increases from 0 to π, give the graph above the x- or polar axis shown in Figure 14(a).

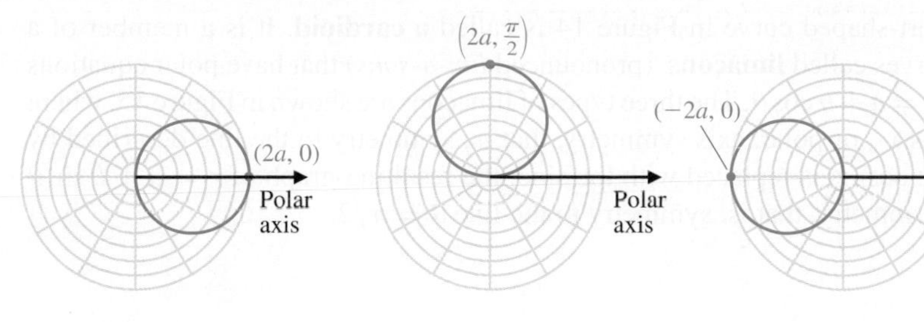

$r = 2a \cos\theta$
Center: $(a, 0)$
Radius: a

$r = 2a \sin\theta$
Center: $\left(a, \frac{\pi}{2}\right)$
Radius: a

$r = -2a \cos\theta$
Center: (a, π)
Radius: a

$r = -2a \sin\theta$
Center: $\left(a, \frac{3\pi}{2}\right)$
Radius: a

FIGURE 13

TABLE 2

θ	0	$\frac{\pi}{6}$	$\frac{\pi}{4}$	$\frac{\pi}{3}$	$\frac{\pi}{2}$	$\frac{2\pi}{3}$	$\frac{3\pi}{4}$	$\frac{5\pi}{6}$	π
$r = 1 + \cos\theta$	2	$1 + \frac{\sqrt{3}}{2}$	$1 + \frac{\sqrt{2}}{2}$	$\frac{3}{2}$	1	$\frac{1}{2}$	$1 - \frac{\sqrt{2}}{2}$	$1 - \frac{\sqrt{3}}{2}$	0

The graph below this axis is simply the reflection of the upper portion about the x- or polar axis, so the complete graph is as shown in Figure 14(b).

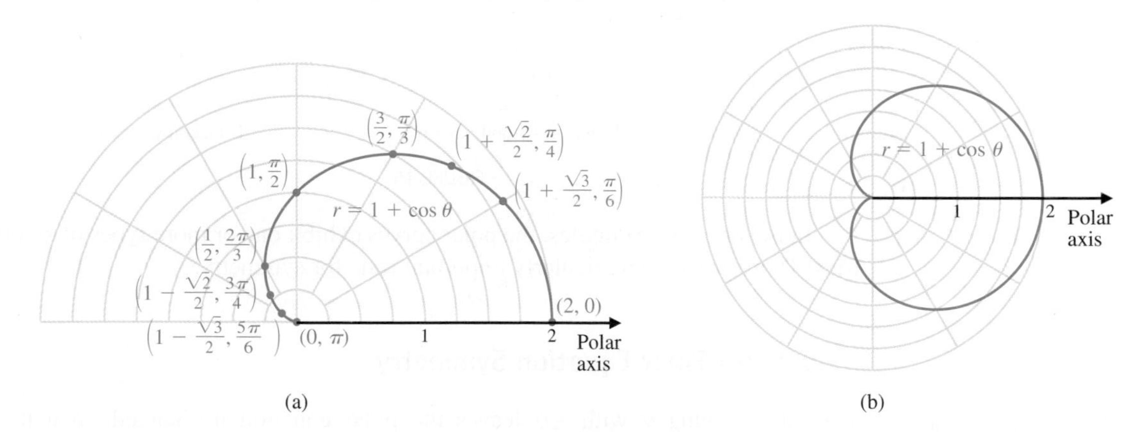

FIGURE 14

To transform the polar equation to a rectangular equation, we multiply both sides of $r = 1 + \cos\theta$ by r and use the relations $r^2 = x^2 + y^2$ and $x = r\cos\theta$. This gives

$$r^2 = r + r\cos\theta, \quad \text{so} \quad x^2 + y^2 = r + x \quad \text{and} \quad x^2 - x + y^2 = r.$$

Squaring both sides of the last equation and using the fact that $r^2 = x^2 + y^2$ gives

$$(x^2 - x + y^2)^2 = r^2 = x^2 + y^2, \quad \text{so} \quad (x^2 - x + y^2)^2 = x^2 + y^2.$$

The resulting rectangular equation is quite complicated and would be difficult to graph without its polar equivalent. ■

The heart-shaped curve in Figure 14 is called a **cardioid**. It is a member of a family of curves called **limaçons** (pronounced *lim-a-sons*) that have polar equations of the form $r = a + b\cos\theta$. The three types of limaçons are shown in Figure 15, where the curve has x-, or polar, axis symmetry, that is, symmetry to the line described by $\theta = 0$. If the cosine is replaced with the sine, the resulting graph of $r = a + b\sin\theta$ has y-axis symmetry, that is, symmetry to the line $\theta = \pi/2$.

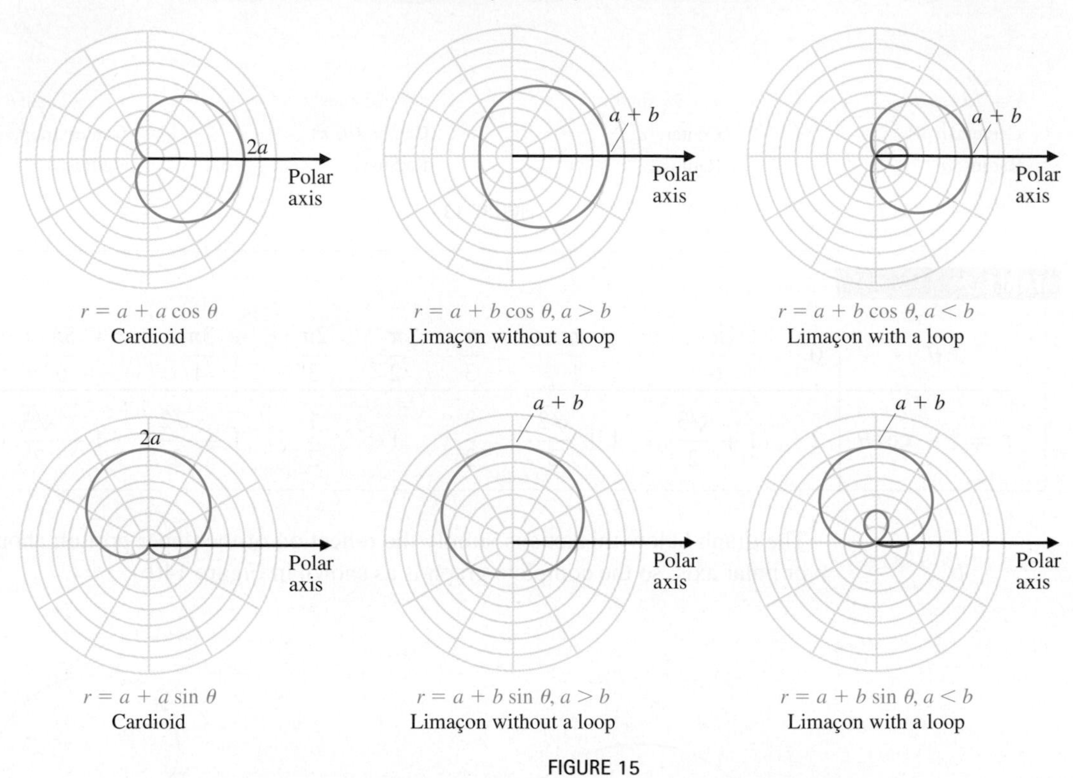

FIGURE 15

In the previous examples, the polar curves exhibit one or more types of symmetries. There are three particularly important tests for symmetry.

Tests for Polar Equation Symmetry

Recognizing symmetries of polar curves can greatly simplify the graphing process.

- If replacing θ with $-\theta$ leaves the polar equation unchanged, then the graph has symmetry with respect to $\theta = 0$, that is, x-axis symmetry. (See Figure 16(a).)
- If replacing θ with $\pi - \theta$ leaves the polar equation unchanged, then the graph has symmetry with respect to $\theta = \pi/2$, that is, y-axis symmetry. (See Figure 16(b).)
- If replacing r with $-r$ leaves the polar equation unchanged, then the graph has symmetry with respect to the pole, that is, origin symmetry. (See Figure 16(c).)

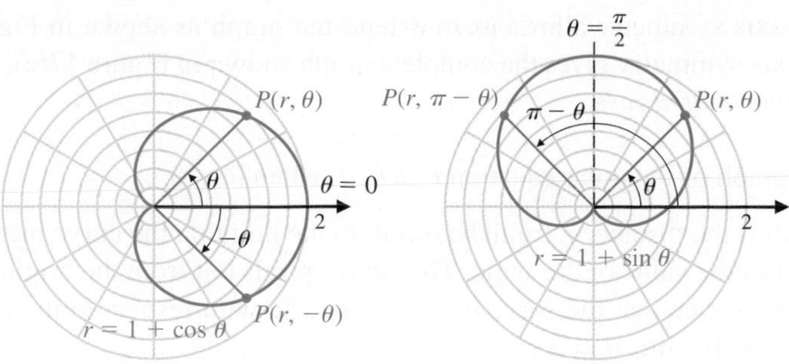

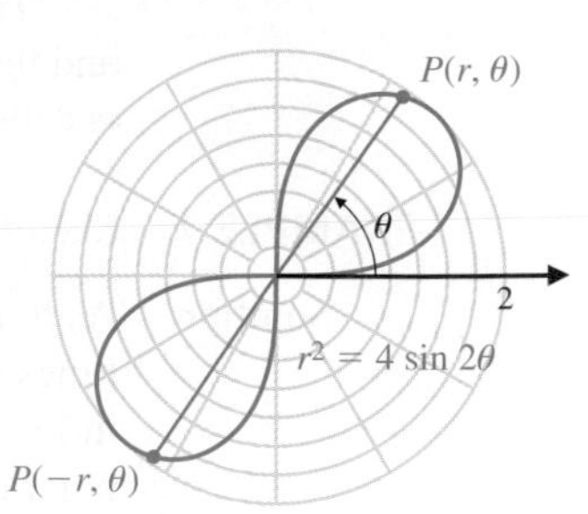

Symmetry with respect to $\theta = 0$
$\cos(-\theta) = \cos(\theta)$
(a)

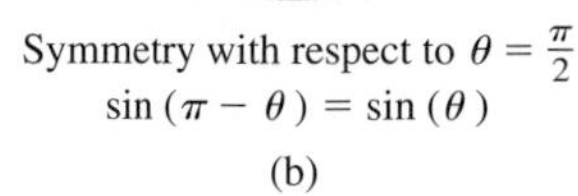

Symmetry with respect to $\theta = \frac{\pi}{2}$
$\sin(\pi - \theta) = \sin(\theta)$
(b)

Symmetry with respect to the pole
$(-r)^2 = r^2$
(c)

FIGURE 16

EXAMPLE 6 Sketch the graph of $r = \cos 2\theta$.

Solution Since the cosine function is even, we have $\cos(-\theta) = \cos\theta$ for all values of θ, and the graph has x-axis symmetry. The graph also has y-axis symmetry, since

$$\cos 2(\pi - \theta) = \cos(2\pi - 2\theta) = \cos(-2\theta) = \cos 2\theta.$$

Because of this symmetry, the graph is determined once we know the shape of the graph in the first quadrant. Table 3 lists values of θ in $[0, \pi/2]$ and the corresponding values of r.

TABLE 3

θ	0	$\frac{\pi}{12}$	$\frac{\pi}{8}$	$\frac{\pi}{6}$	$\frac{\pi}{4}$	$\frac{\pi}{3}$	$\frac{3\pi}{8}$	$\frac{5\pi}{12}$	$\frac{\pi}{2}$
$r = \cos 2\theta$	1	$\frac{\sqrt{3}}{2}$	$\frac{\sqrt{2}}{2}$	$\frac{1}{2}$	0	$-\frac{1}{2}$	$-\frac{\sqrt{2}}{2}$	$-\frac{\sqrt{3}}{2}$	-1

Since $r = 0$ when $\theta = \pi/4$, the graph approaches the pole along the line $\theta = \pi/4$, as shown in Figure 17(a). Notice that for θ between $\pi/3$ and $\pi/2$ the value of r is negative, so the points reflect through the pole and lie in quadrant III.

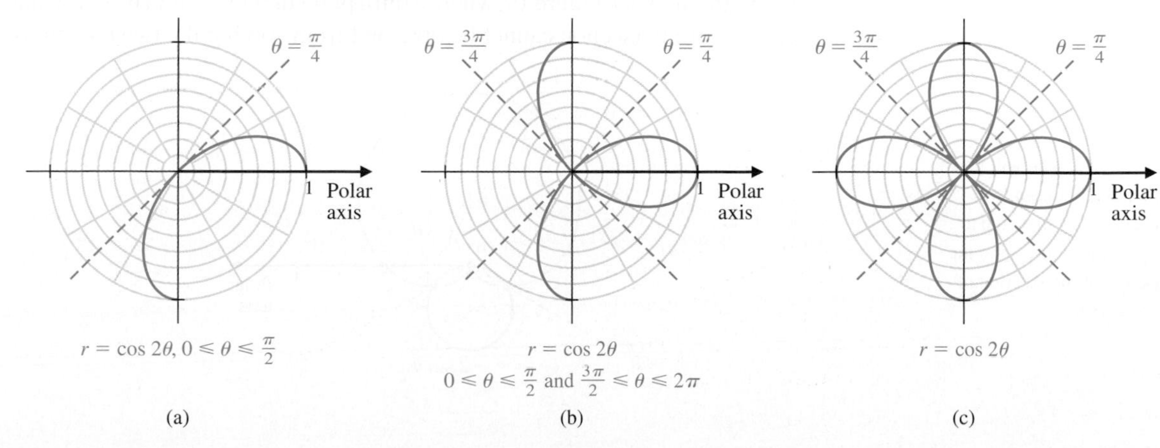

$r = \cos 2\theta, 0 \leqslant \theta \leqslant \frac{\pi}{2}$
(a)

$r = \cos 2\theta$
$0 \leqslant \theta \leqslant \frac{\pi}{2}$ and $\frac{3\pi}{2} \leqslant \theta \leqslant 2\pi$
(b)

$r = \cos 2\theta$
(c)

FIGURE 17

The x-axis symmetry allows us to extend the graph as shown in Figure 17(b), and the y-axis symmetry gives the complete graph shown in Figure 17(c). The graph is called a *four-leafed rose*. ■

EXAMPLE 7 Sketch the graph of the polar equation $r = \theta/2$, when $\theta \geq 0$.

Solution As θ increases, the distance r from the origin to the point on the curve increases four times as fast as the value of the angle. The curve spirals out from the origin, as shown in Figure 18. Notice that the curve is not symmetric with respect to the pole, to the polar axis, or to the line $\theta = \pi/2$. ■

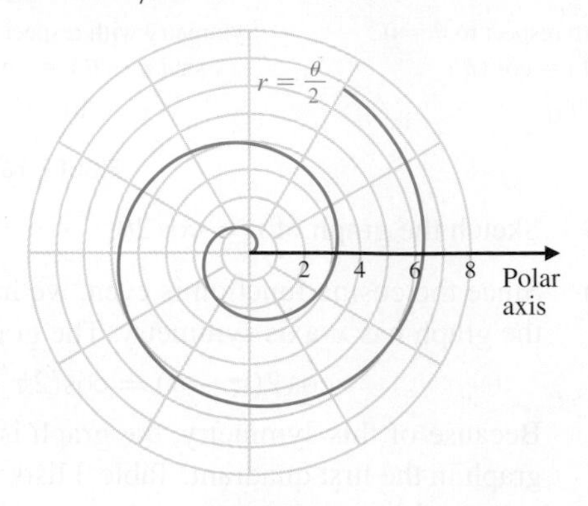

FIGURE 18

EXAMPLE 8 Sketch the graphs of $r = -2\sin\theta$ and $r = 2 + 2\sin\theta$, and find the points of intersection.

Solution Equating the two r values and solving for θ gives

$$-2\sin\theta = 2 + 2\sin\theta, \quad 4\sin\theta + 2 = 0, \quad \text{and} \quad \sin\theta = -\frac{1}{2}.$$

The sine is negative in quadrants II and III and $\sin\pi/6 = 1/2$, so

$$\theta = \frac{7\pi}{6}, \quad \text{or} \quad \theta = \frac{11\pi}{6}.$$

Substituting these two values into either of the original equations gives $r = 1$, so points of intersection have polar coordinates $(1, 7\pi/6)$ and $(1, 11\pi/6)$.

The graphs are shown in Figure 19, where a third point of intersection is seen at the origin. This point of intersection cannot be obtained from solving the two equations

In calculus you will find the area between two curves. The first step to the solution is to find the points of intersection of the curves.

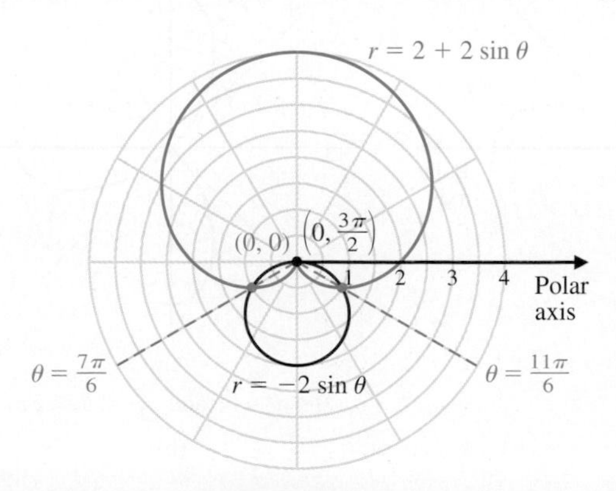

FIGURE 19

simultaneously. On $r = -2\sin\theta$ the origin is represented by the coordinates $(0, 0)$, but on $r = 2 + 2\sin\theta$ it is represented by the coordinates $(0, 3\pi/2)$. ■

EXERCISE SET 6.5

*In Exercises 1–12, **(a)** plot the point with the given polar coordinates, **(b)** convert the polar coordinates to rectangular coordinates, and **(c)** give two other pairs of polar coordinates that represent the point, one with $r > 0$ and one with $r < 0$.*

1. $(2, \pi/3)$ **2.** $(2, 3\pi/4)$
3. $(3, -\pi/4)$ **4.** $(2, -2\pi/3)$
5. $(5, -4\pi/3)$ **6.** $(1, -7\pi/4)$
7. $(8, 7\pi/4)$ **8.** $(2, 5\pi/3)$
9. $(-2, 5\pi/6)$ **10.** $(-4, \pi/3)$
11. $(-1, -2\pi/3)$ **12.** $(-3, -\pi/4)$

In Exercises 13–20, convert the rectangular coordinates to polar coordinates.

13. $(2, 0)$ **14.** $(-2, 0)$
15. $(0, -4)$ **16.** $(0, 4)$
17. $(1, -\sqrt{3})$ **18.** $(\sqrt{3}, -1)$
19. $(-4, 4)$ **20.** $(-3, -3\sqrt{3})$

In Exercises 21–26, convert the polar equation to a rectangular equation.

21. $r = 4$ **22.** $r = 2$
23. $\theta = 3\pi/4$ **24.** $\theta = \pi/6$
25. $r = 2\cos\theta$ **26.** $r = 4\sin\theta$

In Exercises 27–32, convert the rectangular equation to a polar equation.

27. $y = x$ **28.** $y = \sqrt{3}x$
29. $x^2 + y^2 = 9$ **30.** $x^2 + y^2 = 5$
31. $x^2 + y^2 = 2y$ **32.** $x^2 + y^2 = 3x$

In Exercises 33–56, sketch the graph of the polar equation.

33. $r = 3$ **34.** $r = 5\pi/3$
35. $\theta = 5\pi/3$ **36.** $\theta = 3$
37. $r = 3\cos\theta$ **38.** $r = 4\sin\theta$
39. $r = -4\sin\theta$ **40.** $r = -3\cos\theta$
41. $r = 2 + 2\cos\theta$ **42.** $r = 1 + \sin\theta$
43. $r = 2 + \sin\theta$ **44.** $r = 2 + \cos\theta$
45. $r = 1 - 2\cos\theta$ **46.** $r = 2 - 4\sin\theta$
47. $r = 3\sin 3\theta$ **48.** $r = 4\cos 3\theta$
49. $r = 3\cos 2\theta$ **50.** $r = 4\sin 2\theta$
51. $r^2 = 16\cos 2\theta$ **52.** $r^2 = 4\sin 2\theta$
53. $r = \theta$ **54.** $r = 2^{-\theta}$
55. $r = e^{\theta}$ **56.** $r = \ln\theta$

In Exercises 57–62, compare the graphs of the polar equations. Place a dot to show the location of $(r, 0)$ and add arrows to the graphs to indicate the direction of increasing values of θ.

57. a. $r = \sin\theta$ **b.** $r = \cos\theta$
c. $r = -\sin\theta$ **d.** $r = -\cos\theta$

58. a. $r = \cos\theta$ **b.** $r = 1 + \cos\theta$
c. $r = 2 + \cos\theta$ **d.** $r = 3 + \cos\theta$

59. a. $r = \sin\theta$ **b.** $r = 1 + \sin\theta$
c. $r = 2 + \sin\theta$ **d.** $r = 3 + \sin\theta$

60. a. $r = 1 + \cos\theta$ **b.** $r = 1 - \cos\theta$
c. $r = \cos\theta - 1$ **d.** $r = -\cos\theta - 1$

61. a. $r = 1 + \sin\theta$ **b.** $r = 1 - \sin\theta$
c. $r = \sin\theta - 1$ **d.** $r = -\sin\theta - 1$

62. a. $r = \sin\theta$ **b.** $r = \sin 2\theta$
c. $r = \sin 3\theta$ **d.** $r = \sin 4\theta$

63. Find the coordinates of the bases on a baseball field if a polar coordinate system is placed on the field with the pole at home plate and the polar axis aligned as indicated. (Assume that the infield is a square with sides of 90 feet.)

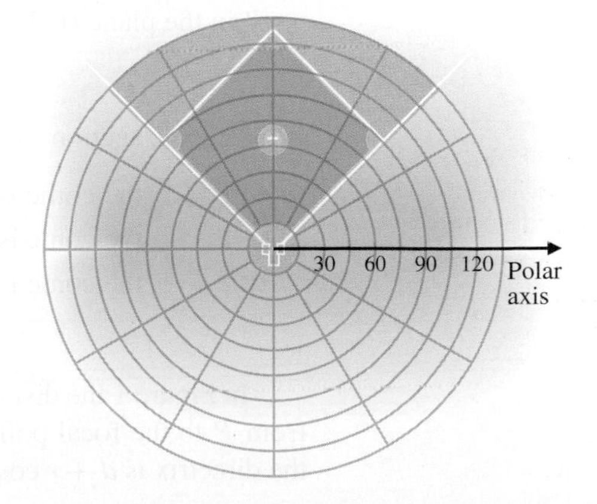

a. Parallel to the line between first and third base.

b. Along the first-base line.

64. The lengthwise cross section of an apple can be reasonably approximated using a polar curve that we have seen in this section. What would be the form of the equation if the polar axis were along the stem of the apple with the pole at the base of the stem?

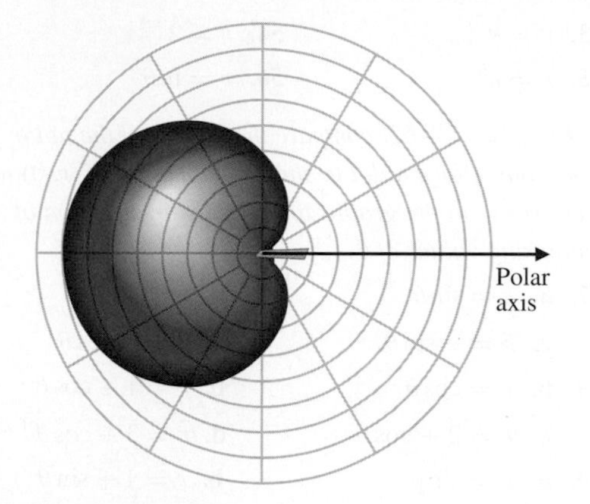

In Exercises 65–68, find the rectangular coordinates of all the points of intersection of the two curves.

65. $r = 1 + 2\cos\theta,\ r = 1$

66. $r = \sin 2\theta,\ r = 1$

67. $r = 2 - 2\sin\theta,\ r = 2\sin\theta$

68. $r = 2 + 2\cos\theta,\ r = -2\cos\theta$

69. Use a graphing device to sketch several curves from the family of curves

$$r = 1 + \sin(n\theta) + (\cos(2n\theta))^2, \quad \text{where } n \geq 1.$$

70. Use a graphing device to sketch several curves from the family of curves

$$r = (\sin m\theta)(\cos n\theta),$$

where m and n are positive integers. Discuss the effect of m and n on the curves.

71. Use a graphing device to compare the curves

$$r = \sin m\theta \quad \text{and} \quad r = |\sin m\theta|,$$

when $m = 1$, 2, and 3.

6.6 CONIC SECTIONS IN POLAR COORDINATES

The polar form of an ellipse is used to verify that the orbit of a planet around the sun is an ellipse.

Earlier in this chapter we found the rectangular equation for a parabola in terms of its focal point and directrix. We also found rectangular equations for ellipses and hyperbolas in terms of two foci. In many physical problems it is more useful to define all the conic sections in terms of a focus and a directrix and use these to determine *polar* equations.

The *eccentricity* as defined here agrees with the earlier definition for ellipses and hyperbolas.

Conics Defined by Eccentricity

Let F be a fixed point, called the **focus**, l be a fixed line, called the **directrix**, and e, called the **eccentricity**, be a fixed positive number. The set of all points P in the plane such that

$$\text{distance}\,(P, F) = e \cdot \text{distance}\,(P, l)$$

is a conic section. If

- $e = 1$, the conic is a parabola;
- $e < 1$, the conic is an ellipse;
- $e > 1$, the conic is a hyperbola.

In Figure 1 the distance from the focal point to the directrix is d, and the distance from P to the focal point is r. So the distance from a point $P(r, \theta)$ on the graph to the directrix is $d + r\cos\theta$.

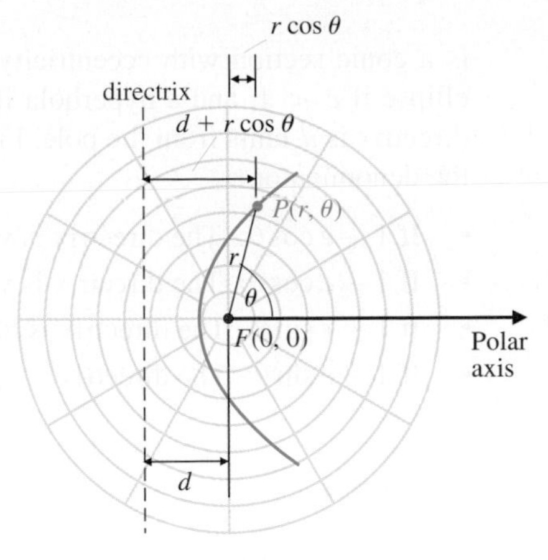

FIGURE 1

Hence we have

$$r = \text{distance}\,(P, F) = e \cdot \text{distance}\,(P, l) = e(d + r\cos\theta).$$

Squaring both sides of this equation and using the fact that $r^2 = x^2 + y^2$ and $x = r\cos\theta$ gives the rectangular coordinate equation

$$x^2 + y^2 = e^2(d + x)^2 = e^2d^2 + 2e^2dx + e^2x^2.$$

This equation is quadratic in x and y, so it does describe a conic section. To discover which conic section it represents, we rewrite the equation as

$$(1 - e^2)x^2 - 2e^2dx + y^2 = e^2d^2.$$

Since the coefficient of y^2 is 1,

- If $e = 1$, the coefficient of x^2 is zero and the curve is a parabola.
- If $0 < e < 1$, the coefficient of x^2 is positive and the curve is an ellipse.
- If $e > 1$, the coefficient of x^2 is negative and the curve is a hyperbola.

Solving the defining equation

$$r = e(d + r\cos\theta)$$

for r gives

$$r(1 - e\cos\theta) = ed, \quad \text{or} \quad r = \frac{ed}{1 - e\cos\theta},$$

when the focal point is at the pole and the directrix is vertical and to the left of the pole. The following result describes the general situation for conics with a focus at the pole and the directrix parallel to a coordinate axis.

Polar Equations of Conic Sections

The graph of a polar equation of the form

$$r = \frac{ed}{1 \pm e\cos\theta} \quad \text{or} \quad r = \frac{ed}{1 \pm e\sin\theta}$$

is a conic section with eccentricity e. The graph is a parabola if $e = 1$, an ellipse if $e < 1$, and a hyperbola if $e > 1$. The focus is at the pole, and the directrix is d units from the pole. Figure 2 shows that the position depends on the denominator.

- If $1 + e\cos\theta$: The directrix is vertical and to the right of the pole.
- If $1 - e\cos\theta$: The directrix is vertical and to the left of the pole.
- If $1 + e\sin\theta$: The directrix is horizontal and above the pole.
- If $1 - e\sin\theta$: The directrix is vertical and below the pole.

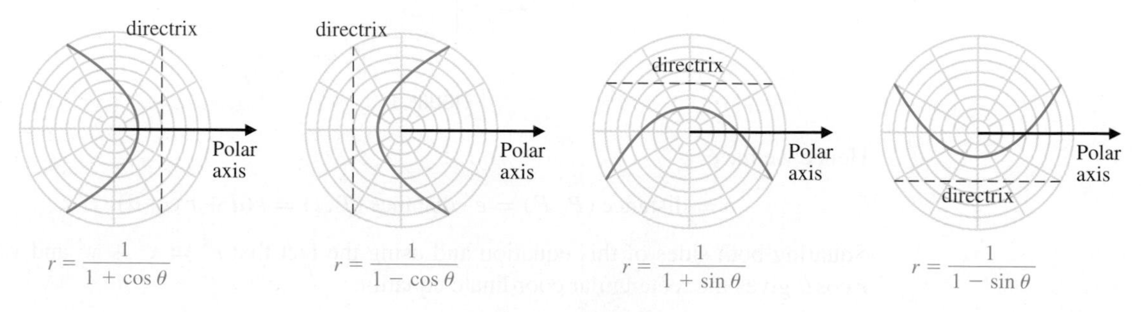

FIGURE 2

Note that a conic in standard polar position has a focus at the pole, that is, the origin in rectangular coordinates. However, in standard rectangular position, the focus will *not* be at the origin. So *standard* position depends on the coordinate system being used.

EXAMPLE 1 Sketch the graph of the parabola whose polar equation is

$$r = \frac{2}{1 + \sin\theta}.$$

Solution In this equation, $e = 1$ and $d = 2$. The focus of the parabola is at the origin, and the directrix is the horizontal line $y = 2$, which is 2 units above the pole. The vertex of the parabola is midway between the focus and the directrix, so it has polar coordinates $(1, \pi/2)$ and rectangular coordinates $(0, 1)$. The parabola opens downward and passes through the points with polar coordinates $(2, 0)$ and $(2, \pi)$, as shown in Figure 3. ■

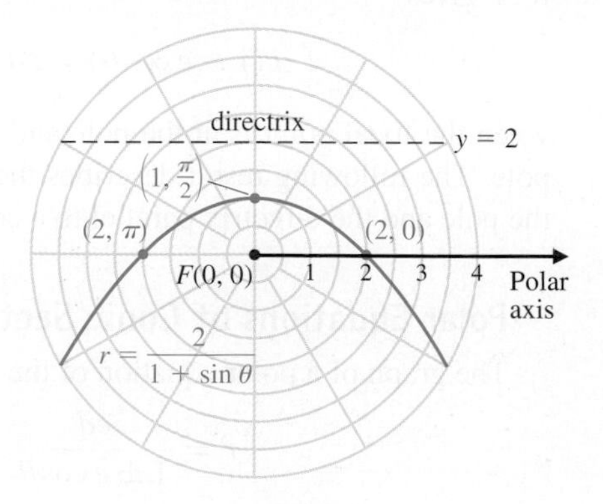

FIGURE 3

EXAMPLE 2 Sketch the graph of the conics.

a. $r = \dfrac{16}{5 + 3\cos\theta}$ **b.** $r = \dfrac{10}{2 + 5\sin\theta}$

Solution **a.** The standard form of the conic section with focal point at the origin and vertical directrix d units to the right of the focal point is

$$r = \frac{ed}{1 + e\cos\theta}.$$

Our equation is

$$r = \frac{16}{5 + 3\cos\theta} = \frac{16}{5(1 + \frac{3}{5}\cos\theta)} = \frac{\frac{16}{5}}{1 + \frac{3}{5}\cos\theta},$$

so $e = \frac{3}{5}$. Since $ed = \frac{16}{5}$, we have $d = \frac{16}{3}$. The eccentricity is less than 1, so the equation describes an ellipse. Its vertices, which occur when $\theta = 0$ and $\theta = \pi$, have polar coordinates $(2, 0)$ and $(8, \pi)$. The graph is shown in Figure 4.

b. Writing the equation in standard form gives

$$r = \frac{10}{2 + 5\sin\theta} = \frac{5}{1 + \frac{5}{2}\sin\theta}.$$

Since $e = \frac{5}{2} > 1$, the conic is a hyperbola with horizontal directrix above the x-axis a distance

$$d = \frac{5}{e} = \frac{5}{\frac{5}{2}} = 2.$$

The vertices of the hyperbola occur when $\theta = \pi/2$ and $3\pi/2$, at the points with polar coordinates $(10/7, \pi/2)$ and $(-10/3, 3\pi/2)$. The portion of the hyperbola lying below the directrix is obtained by plotting $(10/7, \pi/2)$ and the points of intersection with the x-axis. These occur at the points with polar coordinates $(5, \pi)$ and $(5, 0)$. The portion of the hyperbola lying above the directrix is the reflection of the lower part about the horizontal line equidistant from $y = \frac{10}{7}$ and $y = \frac{10}{3}$, as shown in Figure 5. ■

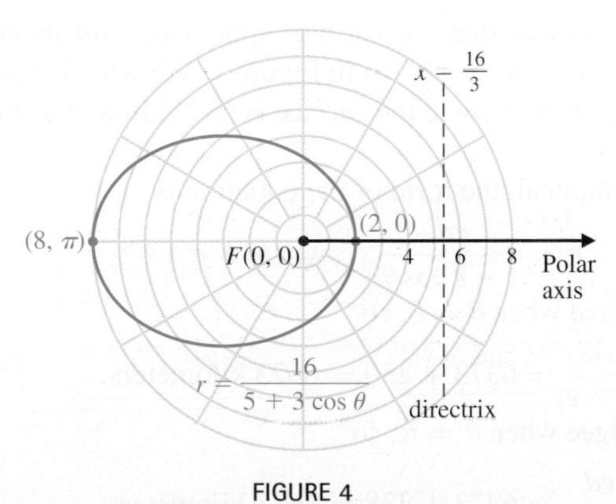

FIGURE 4

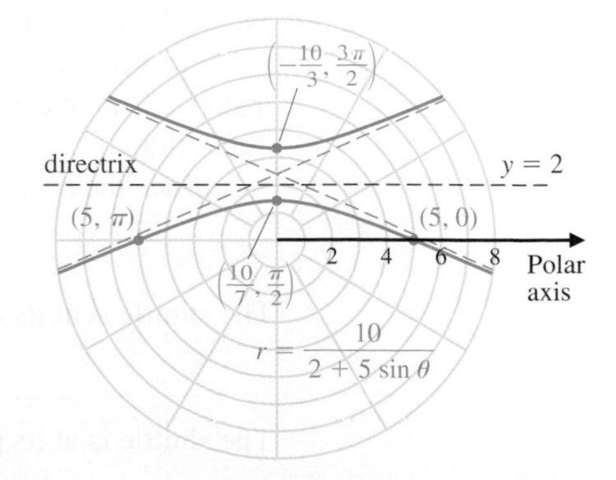

FIGURE 5

EXAMPLE 3 Find a polar equation of the conic with a focus at the origin, eccentricity $e = 1$, and directrix $y = \frac{1}{4}$.

Solution Since the directrix is horizontal and above the pole, the conic has the form

$$r = \frac{ed}{1 + e\sin\theta}.$$

The distance from the pole to the directrix is $\frac{1}{4}$, so $d = \frac{1}{4}$. Using this and the fact that the eccentricity $e = 1$ gives the equation of the conic as

$$r = \frac{\frac{1}{4}}{1 + \sin\theta} = \frac{1}{4 + 4\sin\theta}.$$

The graph is the parabola shown in Figure 6. ■

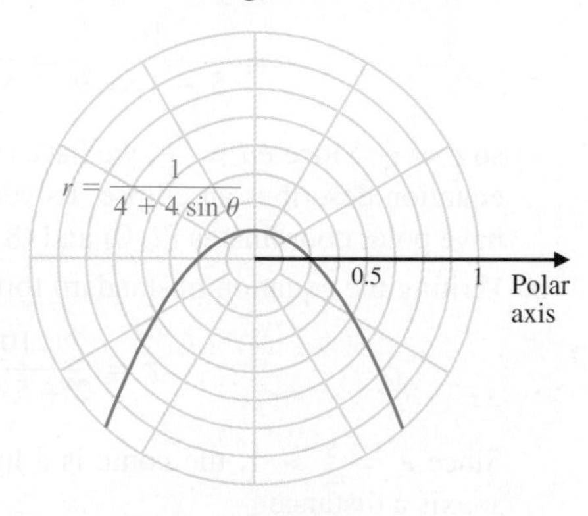

FIGURE 6

Applications

Example 4 illustrates that polar representation is more convenient than rectangular representation for problems involving orbits.

EXAMPLE 4 The initial flight of the NASA space shuttle occurred when the *Columbia* was placed in elliptical earth orbit on April 12, 1981. The *apogee* (the maximum height above the earth) was 250 kilometers and the *perigee* (the minimum height above the earth) was 238 kilometers, with a focal point of the orbit located at the center of mass of the earth. Determine the eccentricity of the orbit.

Solution We will assume for simplicity that the earth is spherical, with its center of mass located at the center of the earth, as shown in Figure 7. We will also assume that the distance from the center of the earth to the surface is 6373 kilometers (approximately 3960 miles).

Since the orbit is elliptical, the form of the equation is

$$r = \frac{ed}{1 - e\cos\theta}, \quad \text{where } e < 1.$$

The shuttle is at its apogee when $\theta = 0$, so

$$\frac{ed}{1 - e} = 6373 + 250 = 6623 \text{ kilometers}.$$

The shuttle is at its perigee when $\theta = \pi$, so

$$\frac{ed}{1 + e} = 6373 + 238 = 6611 \text{ kilometers}.$$

With results from calculus we can calculate the length of time it took this shuttle to orbit the earth.

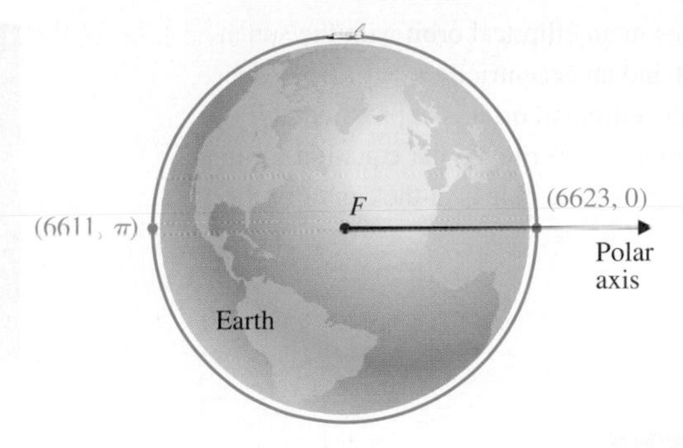

FIGURE 7

Solving these two equations for ed and equating them gives

$$6623(1 - e) = ed = 6611(1 + e),$$

so

$$e = \frac{6623 - 6611}{6623 + 6611} = \frac{12}{13234} \approx 9.1 \times 10^{-4}.$$

This very small value for the eccentricity reflects the fact that the orbit is nearly circular. At its perigee it is 6611 kilometers from the center of the earth, only 12 kilometers, about 0.18%, less than at its apogee. ■

EXERCISE SET 6.6

*In Exercises 1–8, **(a)** sketch the graph of the conic section, and **(b)** find a corresponding rectangular equation.*

1. $r = \dfrac{2}{1 + \cos\theta}$

2. $r = \dfrac{3}{4 + 4\sin\theta}$

3. $r = \dfrac{2}{2 - \sin\theta}$

4. $r = \dfrac{2}{4 + \cos\theta}$

5. $r = \dfrac{1}{1 + 2\cos\theta}$

6. $r = \dfrac{2}{1 - 2\sin\theta}$

7. $r = \dfrac{3}{1 - 2\sin\theta}$

8. $r = \dfrac{4}{2 + 3\cos\theta}$

In Exercises 9–16, find a polar equation of the conic with a focus at the origin that satisfies the given conditions.

9. $e = 2$; directrix $x = -4$

10. $e = \frac{1}{2}$; directrix $y = -2$

11. $e = 1$; directrix $y = -\frac{1}{4}$

12. $e = \frac{1}{3}$; directrix $x = 1$

13. $e = 3$; directrix $x = 2$

14. $e = \frac{2}{3}$; directrix $y = 2$

15. $e = \frac{1}{4}$; directrix $y = 1$

16. $e = 1$; directrix $x = -3$

17. Find a polar equation of the ellipse in standard polar position that has vertices at the points with polar coordinates $(0, 1)$ and $(3, \pi)$.

18. Find a polar equation of the parabola in standard polar position that has its vertex at the point with rectangular coordinates $(-6, 0)$.

19. The world's first orbiting satellite, *Sputnik I*, was launched in the Soviet Union on October 4, 1957. Its elliptical orbit reached a maximum height of 560 miles above the earth and a minimum height of 145 miles above the earth. Write a polar equation of the orbit, assuming that the pole is placed at the center of the earth, which is 3960 miles below the surface.

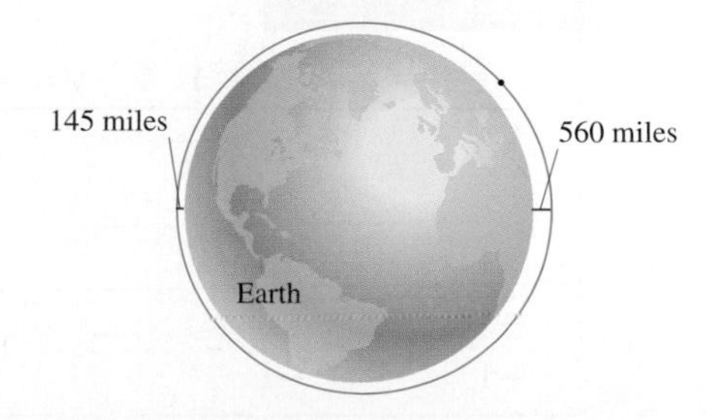

20. The earth moves in an elliptical orbit with the sun at one focal point and an eccentricity $e = 0.0167$. The major axis of the elliptical orbit is approximately 2.99×10^8 kilometers. Write a polar equation for this ellipse, assuming that the pole is at the center of the sun.

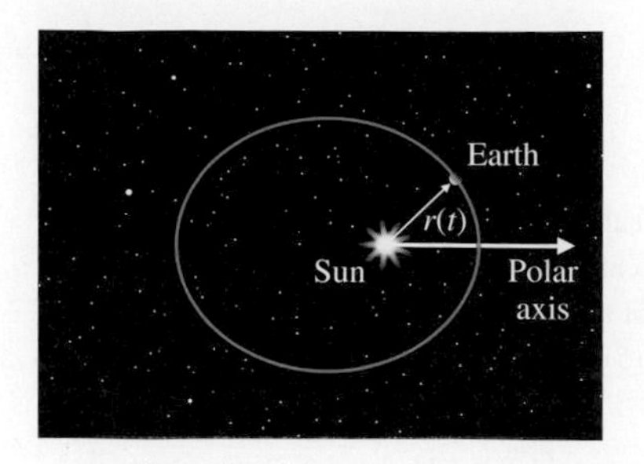

6.7 PARAMETRIC EQUATIONS

We have represented curves using rectangular coordinates and using polar coordinates. In each case the curve consists of all points in the plane that satisfy a single equation. An alternative method of describing a curve in the plane is to express the x- and y-coordinates of points on the curve separately as functions of a third variable, called a *parameter*. This permits a flexibility that is not available using strictly rectangular or polar equations. It is also the basis for the method used to describe curves in three-dimensional space.

You will study parametric surfaces in *multivarable calculus*. Parametric surfaces are used in three-dimensional computer graphics.

Parametric Equations

The equations

$$x = f(t) \quad \text{and} \quad y = g(t)$$

are **parametric** equations for the curve determined by the points $(x, y) = (f(t), g(t))$, when the **parameter** t assumes all the values in the common domain of f and g.

EXAMPLE 1 Describe and sketch the curve with parametric equations

$$x = 2t + 1 \quad \text{and} \quad y = 4t^2 - 1, \quad \text{for} \quad -1 \le t \le 1.$$

Solution By choosing representative values of t in the interval $[-1, 1]$, we obtain the points in Table 1 and on the graph shown in Figure 1. Instead of proceeding in this manner, however, we can eliminate the parameter to determine a direct relation between x and y.

TABLE 1

t	$x = 2t + 1$	$y = 4t^2 - 1$
0	1	−1
$\frac{1}{2}$	2	0
1	3	3
$-\frac{1}{2}$	0	0
−1	−1	3

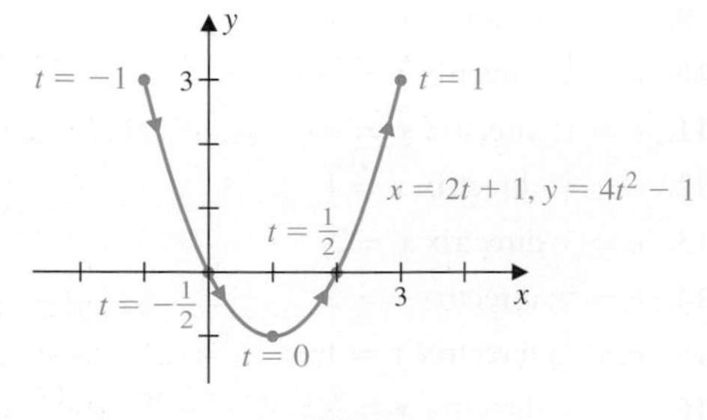

FIGURE 1

Solving for t in the first equation and substituting into the second gives

$$t = \frac{x-1}{2}, \quad \text{so} \quad y = 4\left(\frac{x-1}{2}\right)^2 - 1 = (x-1)^2 - 1.$$

This is the equation of a parabola with vertex at $(1, -1)$. As t varies from -1 to 1, points on the curve are traced from $(-1, 3)$ to $(3, 3)$. The x-intercepts occur at values of t making $y = 0$. Setting

$$0 = y = 4t^2 - 1 \quad \text{gives} \quad t^2 = \frac{1}{4}, \quad \text{so} \quad t = \pm\frac{1}{2}.$$

Substituting these values into the parametric equation for x gives the x-intercepts $(0, 0)$ and $(2, 0)$. The arrow heads on the curve in Figure 1 indicate the direction traced by increasing values of t. ■

EXAMPLE 2 Find parametric equations for the line through $(1, 2)$ with slope $\frac{1}{2}$.

Solution A rectangular equation for this line is

$$y - 2 = \frac{1}{2}(x-1).$$

One set of parametric equations is found by setting $t = x - 1$. Then

$$y - 2 = \frac{1}{2}t, \quad \text{so} \quad y = \frac{1}{2}t + 2.$$

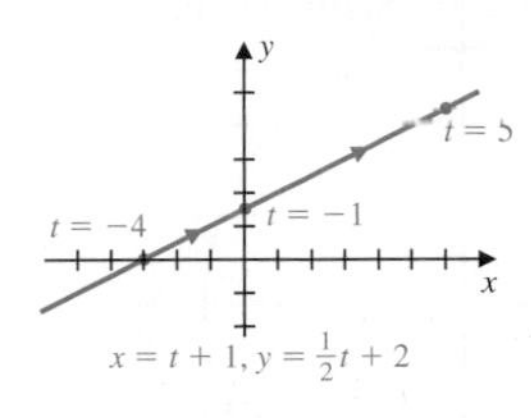

FIGURE 2

The line, shown in Figure 2, is described by the parametric equations

$$x = t + 1 \quad \text{and} \quad y = \frac{1}{2}t + 2, \quad \text{for } -\infty < t < \infty.$$

Parametric representations of curves are not unique. Another set of parametric equations for the line is

$$x = t \quad \text{and} \quad y = \frac{1}{2}(t-1) + 2, \quad \text{for } -\infty < t < \infty.$$

The only conditions that need to be fulfilled are that both x and y assume all real values and that

$$y = \frac{1}{2}(x-1) + 2.$$ ■

Notice that the parametric equations

$$x = t^2 \quad \text{and} \quad y = \frac{1}{2}(t^2 - 1) + 2 = \frac{1}{2}(t^2 + 3), \quad \text{for } -\infty < t < \infty$$

also satisfy the equation

$$y = \frac{1}{2}(x-1) + 2.$$

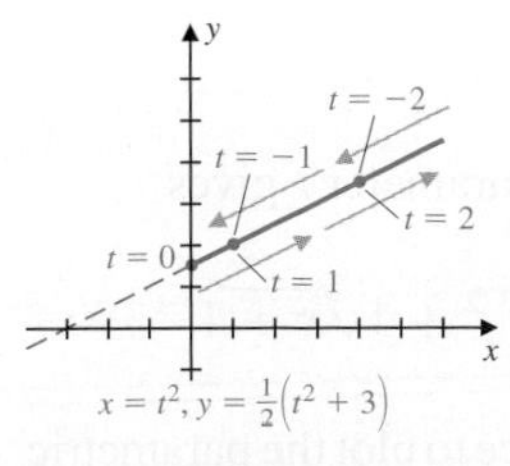

FIGURE 3

However, the curve described by these equations does not trace the entire line. Since $x \geq 0$ for all values of t, only the portion of the line in Figure 3 is traced. Moreover, except for the point $\left(0, \frac{3}{2}\right)$, it is traced twice, once for t in $(-\infty, 0)$ and again for t in $(0, \infty)$.

EXAMPLE 3 Describe the curve given by the parametric equations.

a. $x = \cos t$ and $y = \sin t$, for $0 \le t \le 2\pi$

b. $x = \sin 2t$ and $y = \cos 2t$, for $0 \le t \le 2\pi$

Solution **a.** We can eliminate the parameter t by using the fact that

$$x^2 + y^2 = (\cos t)^2 + (\sin t)^2 = 1.$$

This implies that all the points (x, y) lie on the unit circle. As t increases from 0 to 2π, the points $(x, y) = (\cos t, \sin t)$ make one complete revolution counterclockwise around the circle, starting at $(1, 0)$ and returning to the starting point when $t = 2\pi$. (See Figure 4(a).)

If a graphing device is used when sketching a parametric curve, observe how the curve is traced.

b. As in part (a), eliminating the parameter gives

$$x^2 + y^2 = (\sin 2t)^2 + (\cos 2t)^2 = 1,$$

which again traces the unit circle. But this time, as t increases from 0 to 2π the points $(x, y) = (\sin 2t, \cos 2t)$ start at $(0, 1)$ and make *two* complete revolutions *clockwise* around the circle, as shown in Figure 4(b). One revolution is traced when t is in $[0, \pi]$, and the other is traced when t is in $[\pi, 2\pi]$. ■

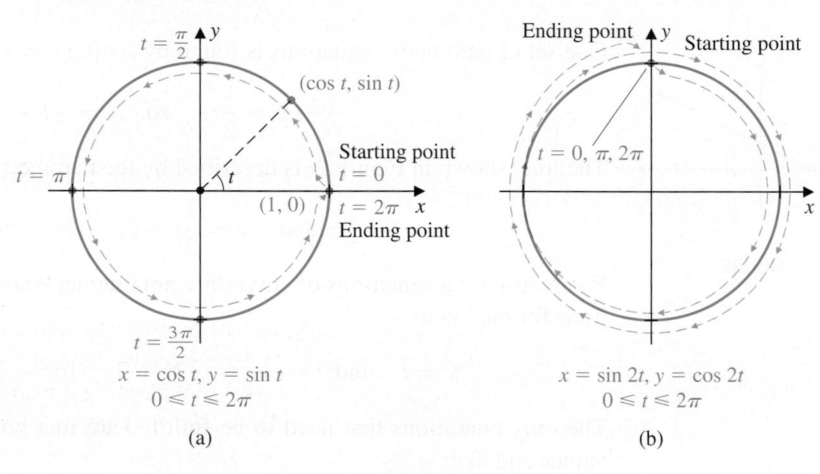

FIGURE 4

EXAMPLE 4 Describe the curve given by the parametric equations

$$x = t^2 - 4 \quad \text{and} \quad y = t^3 - 4t.$$

Solution Solving for t in terms of x gives

$$t = \pm\sqrt{x + 4},$$

and substituting this into the equation for y to eliminate the parameter t gives

$$y = (x + 4)^{3/2} - 4\sqrt{x + 4} \quad \text{or} \quad y = -(x + 4)^{3/2} + 4\sqrt{x + 4}.$$

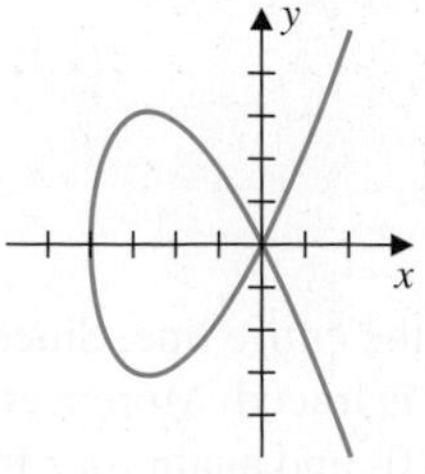

These curves are difficult to plot, but using a graphing device to plot the parametric equations shows that the curve has a loop, as shown in Figure 5.

FIGURE 5

The sign chart in Figure 6 indicates how the points on the curve are given by the parameter t. The graph in the fourth quadrant is given by $-\infty < t < -2$. Then it moves to the second quadrant for $-2 < t < 0$, to the third quadrant for $0 < t < 2$, and finally to the first quadrant for $2 < t$. ■

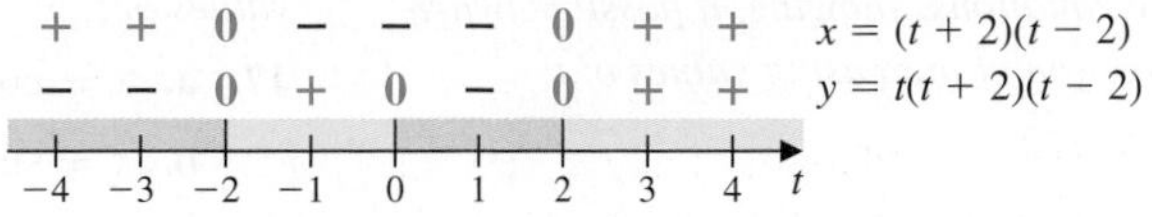

+	+	0	−	−	−	0	+	+		$x = (t + 2)(t - 2)$
−	−	0	+	0	−	0	+	+		$y = t(t + 2)(t - 2)$
−4	−3	−2	−1	0	1	2	3	4	t	

FIGURE 6

Polar coordinates were introduced in Section 6.5 and were used to describe conics in Section 6.6. The conversion formulas from polar to rectangular coordinates,

$$x = r\cos\theta \quad \text{and} \quad y = r\sin\theta,$$

can also be used to convert a polar equation of the form

$$r = f(\theta), \quad \text{for } \alpha \le \theta \le \beta,$$

into parametric equations. Simply replace r with $f(\theta)$ to get

$$x = f(\theta)\cos\theta \quad \text{and} \quad y = f(\theta)\sin\theta, \quad \text{for } \alpha \le \theta \le \beta.$$

For example, parametric equations of the cardioid

$$r = 1 + \sin\theta, \quad \text{for } 0 \le \theta \le 2\pi,$$

are

$$x = (1 + \sin\theta)\cos\theta \quad \text{and} \quad y = (1 + \sin\theta)\sin\theta, \quad \text{for } 0 \le \theta \le 2\pi.$$

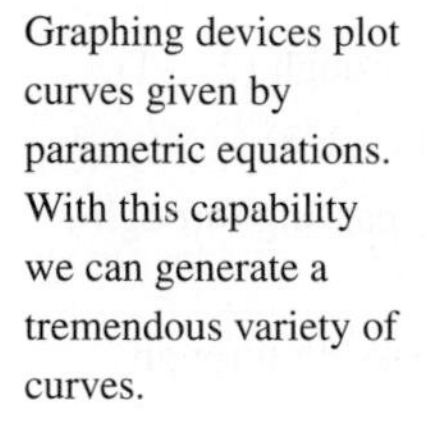
Graphing devices plot curves given by parametric equations. With this capability we can generate a tremendous variety of curves.

Exploiting this flexibility can provide important tools for studying curves.

Computer algebra systems and graphing calculators can graph curves given parametrically since they employ simple point-plotting methods. As a consequence, complicated curves, such as those in Figure 7, are easily generated by computer, but would be very difficult to generate otherwise.

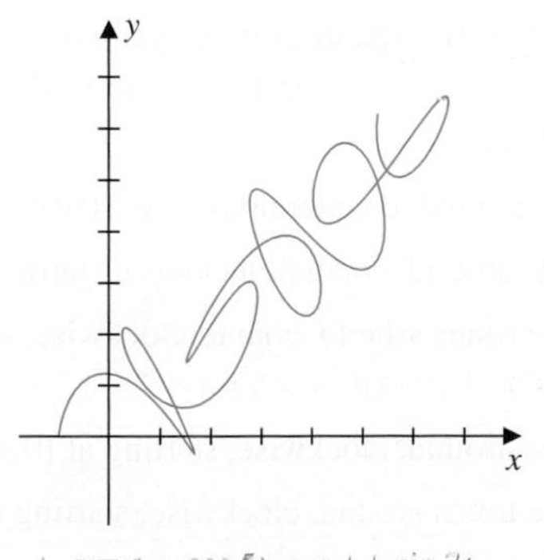

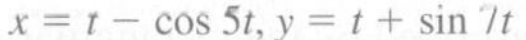
$x = t - \cos 5t,\ y = t + \sin 7t$

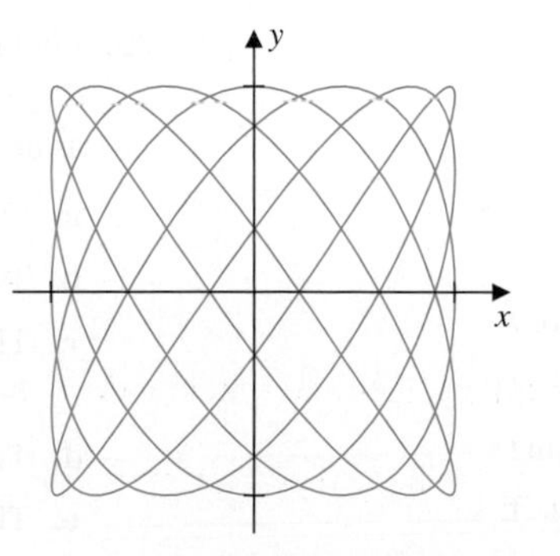

$x = \cos 5t,\ y = \sin 7t$

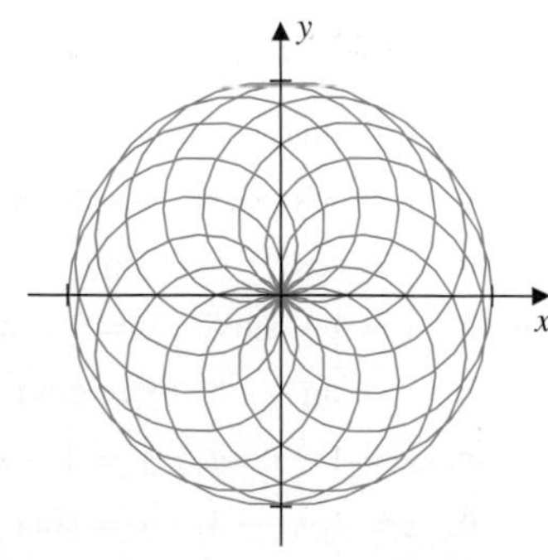

$x = \sin 7t\cos 10t,\ y = \sin 7t\sin 10t$

FIGURE 7

EXERCISE SET 6.7

*In Exercises 1–12, **(a)** find a corresponding rectangular equation, and **(b)** sketch the graph of the curve described by the parametric equations, showing, if possible, when $t = 0$ and the direction of increasing values of t.*

1. $x = 3t, \quad y = t/2$
2. $x = 2t + 1, \quad y = 3t - 2$
3. $x = \sqrt{t}, \quad y = t + 1$
4. $x = t + 2, \quad y = -3\sqrt{t}$
5. $x = \sin t, \quad y = (\cos t)^2$
6. $x = (\cos t)^2, \quad y = \sin t$
7. $x = \sec t, \quad y = \tan t$
8. $x = 3 \sin t, \quad y = 4 \cos t$
9. $x = 3 \cos t, \quad y = 2 \sin t$
10. $x = \ln t, \quad y = \ln \sqrt{t}$
11. $x = e^t, \quad y = e^{-t}$
12. $x = e^t - 2, \quad y = e^{2t} + 3$

In Exercises 13–16, sketch the graph of the conic described by the parametric equations, showing, if possible, when $t = 0$ and the direction of increasing values of t.

13. **a.** $x = t, \quad y = t^2$
    b. $x = t^2, \quad y = t$
    c. $x = t^2, \quad y = t^4$
    d. $x = t^4, \quad y = t^2$
14. **a.** $x = 2t + 1, \quad y = 3t$
    b. $x = 3t + 1, \quad y = 2t$
    c. $x = 2t, \quad y = 3t + 1$
    d. $x = 3t, \quad y = 2t + 1$
15. **a.** $x = \sin t, \quad y = \cos t$
    b. $x = \sin t, \quad y = \cos t + 1$
    c. $x = \cos t + 1, \quad y = \sin t$
    d. $x = \cos t, \quad y = \sin t + 1$
16. **a.** $x = 1 - \sin t, \quad y = 1 - \cos t$
    b. $x = \sin t - 1, \quad y = \cos t - 1$
    c. $x = 1 - \cos t, \quad y = 1 - \sin t$
    d. $x = \cos t - 1, \quad y = \sin t - 1$

In Exercises 17–20, the graphs of the parametric equations in each part represent a portion of the same curve. Sketch the graphs of these equations, showing, if possible, when $t = 0$ and the direction of increasing values of t.

17. **a.** $x = \cos t, \quad y = \sin t$
    b. $x = \sin t, \quad y = \cos t$
    c. $x = t, \quad y = \sqrt{1 - t^2}$
    d. $x = -t, \quad y = \sqrt{1 - t^2}$
18. **a.** $x = t + 1, \quad y = 2t + 3$
    b. $x = t^2 - \frac{1}{2}, \quad y = 2t^2$
    c. $x = \ln t, \quad y = 1 + \ln t^2$
    d. $x = \sin t, \quad y = 1 + 2 \sin t$
19. **a.** $x = t, \quad y = \ln t$
    b. $x = e^t, \quad y = t$
    c. $x = t^2, \quad y = 2 \ln t$
    d. $x = 1/t, \quad y = -\ln t$
20. **a.** $x = t, \quad y = 1/t$
    b. $x = e^t, \quad y = e^{-t}$
    c. $x = \sin t, \quad y = \csc t$
    d. $x = \tan t, \quad y = \cot t$

In Exercises 21–24, find parametric equations for curves satisfying the given conditions.

21. A line with slope $\frac{1}{3}$, and passing through $(2, -1)$
22. A line passing through $(5, 3)$ and $(-2, 7)$
23. A parabola with vertex at $(1, -2)$ passing through $(0, 0)$ and $(2, 0)$
24. A parabola with vertex at $(2, 1)$ passing through $(0, -2)$ and $(0, 4)$
25. Find parametric equations in the parameter t, where $0 \leq t \leq 2\pi$, for the circle $x^2 + y^2 = r^2$ that is traced as described.
    a. Once around, counterclockwise, starting at $(r, 0)$
    b. Twice around, counterclockwise, starting at $(r, 0)$
    c. Three times around, counterclockwise, starting at $(-r, 0)$
    d. Twice around, clockwise, starting at $(0, r)$
    e. Three times around, clockwise, starting at $(0, r)$
26. The curve described by the rectangular equation $x^3 + y^3 = xy$ is called the *folium of Descartes*.

a. Show that this curve is described by the parametric equations

$$x = \frac{t^2}{1+t^3}, \quad y = \frac{t}{1+t^3}, \quad \text{for } t \neq -1.$$

b. Sketch the graph of the curve, and use arrows to indicate the direction of increasing t.

27. Use a graphing device to investigate the family of curves given by the parametric equations

$$x = (a - b \sin t) \cos t \quad \text{and} \quad y = (a - b \sin t) \sin t,$$

where a and b are real numbers and $0 \le t \le 2\pi$.

28. Use a graphing device to investigate the family of curves given by the parametric equations

$$x = a(\cos t)^3 \quad \text{and} \quad y = a(\sin t)^3,$$

where a is a real number. What effect does the parameter a have on the curve? These curves are called *four-cusp hypocycloids*.

29. A *cycloid* can be described as the path traced out by a point on a circle as the circle rolls along a line. If the radius of the circle is a, the parametric equations of the cycloid are given by

$$x = a(t - \sin t) \quad \text{and} \quad y = a(1 - \cos t).$$

Use a graphing device to sketch the cycloid for different values of a. What effect does the parameter a have on the curve? For what values of t does the curve touch the x-axis?

REVIEW EXERCISES FOR CHAPTER 6

In Exercises 1–4, sketch the graph showing the vertex, focus, and directrix of the parabola.

1. $4y - x^2 = 0$

2. $y^2 + 12x = 0$

3. $4x - y^2 + 6y - 17 = 0$

4. $x^2 + 4x + 8y - 4 = 0$

In Exercises 5–8, sketch the graph showing the vertices and focal points of the ellipse.

5. $x^2 + 4y^2 = 4$

6. $4x^2 + y^2 = 16$

7. $4(x - 1)^2 + 9(y + 2)^2 = 36$

8. $2(x + 1)^2 + (y - 1)^2 = 2$

In Exercises 9–12, sketch the graph showing the vertices, focal points, and asymptotes of the hyperbola.

9. $x^2 - 2y^2 = 4$

10. $4y^2 - x^2 = 16$

11. $2x^2 - 4x - 4y^2 + 1 = 0$

12. $9x^2 - 4y^2 - 18x + 16y - 43 = 0$

In Exercises 13–22, identify the type of curve and sketch the graph.

13. $x^2 = -2(y - 5)$

14. $9(x + 2)^2 - 4(y - 5)^2 = 36$

15. $16x^2 + 25y^2 = 400$

16. $y^2 = -16(x + 1)$

17. $x^2 - 2x - 4y - 11 = 0$

18. $y^2 - 2x + 2y + 7 = 0$

19. $9x^2 - 16y^2 = 144$

20. $9x^2 + 4y^2 - 90x - 16y + 205 = 0$

21. $16x^2 - 64x - 25y^2 + 150y = 561$

22. $4x^2 - 9y^2 - 16x - 90y - 173 = 0$

In Exercises 23–30, find an equation of the conic.

23. A parabola with focus at $(0, 0)$ and directrix $y = 2$.

24. An ellipse with foci at $(0, \pm 1)$ and vertices at $(0, \pm 3)$.

25. A hyperbola with foci at $(\pm 3, 0)$ and vertex at $(1, 0)$.

26. A parabola with focus at $(0, 1)$ and vertex at $(0, -1)$.

27. An ellipse with foci at $(0, \pm 5)$ and passing through the point $(4, 0)$.

28. A conic with focus at the origin, eccentricity $\frac{3}{4}$, and directrix $x = 2$. Identify the type of conic.

29. A conic with focus at the origin, eccentricity 3, and directrix $y = -2$. Identify the type of conic.

30. A conic with focus at the origin, eccentricity 1, and directrix $x = -3$. Identify the type of conic.

In Exercises 31–40, sketch the curve whose polar equation is given.

31. $r = 4 + 4 \cos \theta$

32. $r = 3 + 2 \sin \theta$

33. $r = 1 + 3\sin\theta$ **34.** $r = 3 + 3\cos\theta$

35. $r = 2\sin\theta$ **36.** $r = 2$

37. $r = 2\cos 2\theta$ **38.** $r = 2\sin 3\theta$

39. $\theta = \frac{1}{2}$ **40.** $\theta = -\pi/4$

In Exercises 41–44, sketch the conic section whose polar equation is given.

41. $r = \dfrac{3}{1 + \cos\theta}$ **42.** $r = \dfrac{3}{2 + 4\cos\theta}$

43. $r = \dfrac{4}{2 - \cos\theta}$ **44.** $r = \dfrac{2}{1 - \sin\theta}$

In Exercises 45–50, ***(a)*** *sketch the parametric curve, showing, if possible, when $t = 0$ and the direction of increasing t, and* ***(b)*** *eliminate the parameter to find a rectangular equation for the curve.*

45. $x = t^2 - 1, \quad y = t + 1$

46. $x = 2\cos t, \quad y = 3\sin t$

47. $x = e^t, \quad y = 1 + e^{-t}$

48. $x = t + 2, \quad y = (t - 1)^2 + 1$

49. $x = (\sin t)^2 + 1, \quad y = (\cos t)^2$

50. $x = \ln t^2, \quad y = \ln t^3 + 1$

51. Use a graphing device to sketch several curves from the family of curves

$$r = 1 + \sin 2n\theta + (\cos n\theta)^2$$

for positive integers n. How does the parameter n affect the curve?

52. Use a graphing device to sketch several curves from the family of curves

$$r = \sin m\theta + (\cos n\theta)^2$$

for m and n positive integers. Discuss the effect of the parameters m and n on the curve.

53. Use a graphing device to investigate the family of curves given by the parametric equations

$$x = (1 - a\sin t)\cos t, \quad y = (1 - a\sin t)\sin t,$$

where a is a real number and $0 \le t \le 2\pi$. What effect does the parameter a have on the curves?

54. Use a graphing device to investigate the family of curves given by the parametric equations

$$x = (a - b)\cos t + b\cos\frac{a - b}{b}t,$$

$$y = (a - b)\sin t - b\sin\frac{a - b}{b}t$$

for $0 \le t \le 2\pi$ and a and b real numbers with $a > b$. These curves are called *hypocycloids*.

55. Use a graphing device to investigate the family of curves given by the parametric equations

$$x = (a + b)\cos t - b\cos\frac{a + b}{b}t,$$

$$y = (a + b)\sin t - b\sin\frac{a + b}{b}t$$

for $0 \le t \le 2\pi$ and a and b real numbers with $a > b$. These curves are called *epicycloids*.

CHAPTER 6: EXERCISES FOR CALCULUS

1. a. Find an equation of the parabola with axis the y-axis and vertex $(0, 0)$ that passes through the point (x_1, y_1), where $x_1 \ne 0$ and $y_1 \ne 0$.

b. Use the result in part (a) to find an equation of the parabola with axis parallel to the y-axis and vertex (h, k) that passes through (x_1, y_1), where $x_1 \ne h$ and $y_1 \ne k$.

2. a. Find an equation of the parabola with axis the x-axis and vertex $(0, 0)$ that passes through the point (x_1, y_1), where $x_1 \ne 0$ and $y_1 \ne 0$.

b. Use the result in part (a) to find an equation of the parabola with axis parallel to the x-axis and vertex (h, k) that passes through (x_1, y_1), where $x_1 \ne h$ and $y_1 \ne k$.

3. Find an equation of the parabola with vertex at (h, k) and passing through $(h + 1, k + 1)$ with **(a)** a vertical axis and **(b)** a horizontal axis.

4. Find an equation of the parabola passing through the three points $(1, 0)$, $(0, 1)$, and $(2, 2)$ **(a)** with axis parallel to the y-axis and **(b)** with axis parallel to the x-axis.

In Exercises 5 and 6, use the figure, which shows an ellipse of the form $x^2/a^2 + y^2/b^2 = 1$ with a tangent line at the point (x_0, y_0) having equation

$$\frac{x_0 x}{a^2} + \frac{y_0 y}{b^2} = 1.$$

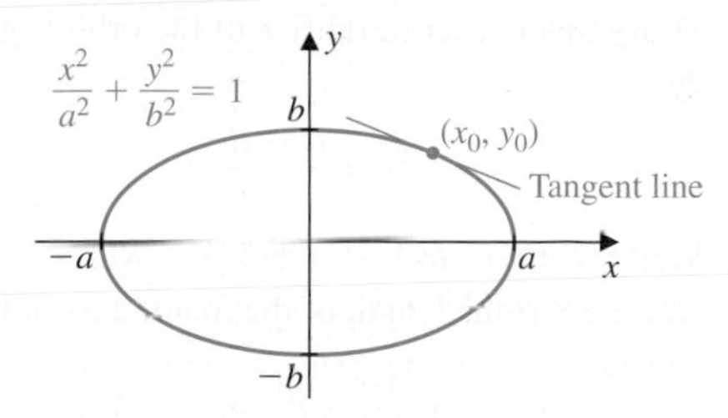

5. Find the equation of the tangent line to the ellipse $9x^2 + 4y^2 = 36$ at $(1, 3\sqrt{3}/2)$.
6. Find equations of the tangent lines to the ellipse $4x^2 + y^2 = 4$ that pass through the point $(3, 0)$.

In Exercises 7 and 8, use the figure, which shows a hyperbola of the form $x^2/a^2 - y^2/b^2 = 1$, with a tangent line at the point (x_0, y_0) having equation

$$\frac{x_0 x}{a^2} - \frac{y_0 y}{b^2} = 1.$$

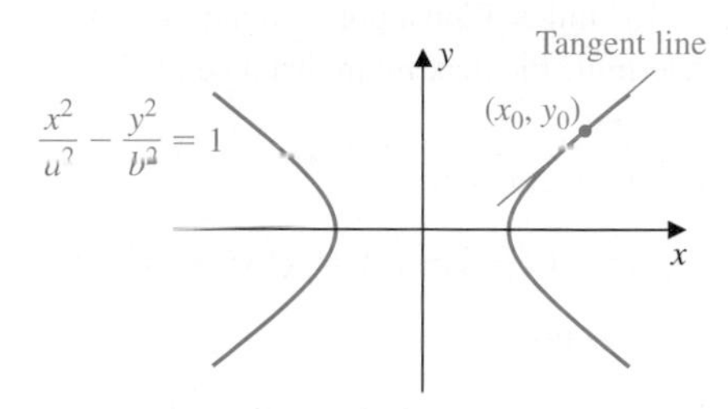

7. Find an equation of the tangent line to the hyperbola with equation $3x^2 - 4y^2 = 12$ at $(2\sqrt{2}, \sqrt{3})$.
8. Find equations of the tangent lines to the hyperbola $9x^2 - 4y^2 = 36$ that pass through $(1, 0)$. Sketch the hyperbola and show the tangent lines.
9. Describe the curve traced by a point 2 feet from the top of an 8-foot ladder as the bottom of the ladder moves away from a vertical wall.
10. A satellite dish receiver has its amplifier in line with the edge of the dish, as shown in the figure. The diameter of the dish at the edge is 18 inches. How deep is the dish?

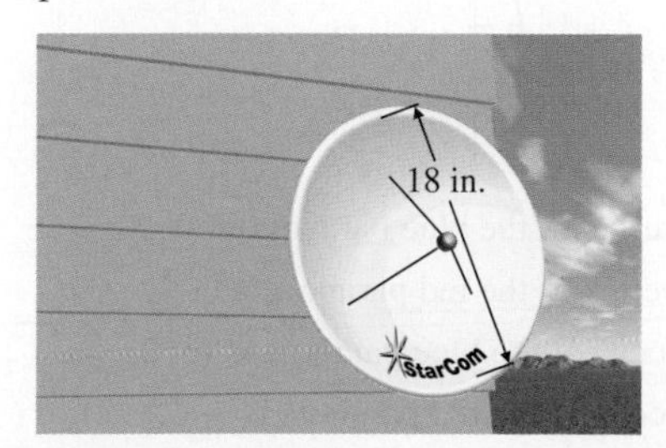

11. A reflecting dish is made by revolving the point of the parabola $4y = x^2$ below $y = 2$ about its axis of symmetry. Where should a light source be placed to produce a focused beam of light with parallel rays?
12. The shape of a flexible and inelastic cable supporting a load that is uniformly distributed horizontally, like the cables of a suspension bridge, is a parabola.
    a. Place the origin of an xy-coordinate system at the lowest point of a parabolic cable of a suspension bridge whose span is L and whose sag is h, as shown in the figure. Determine the equation for the cable.
    b. The George Washington Bridge across the Hudson River in New York has a span of 3500 feet and a sag of 316 feet. What is the equation of the parabolic cable?

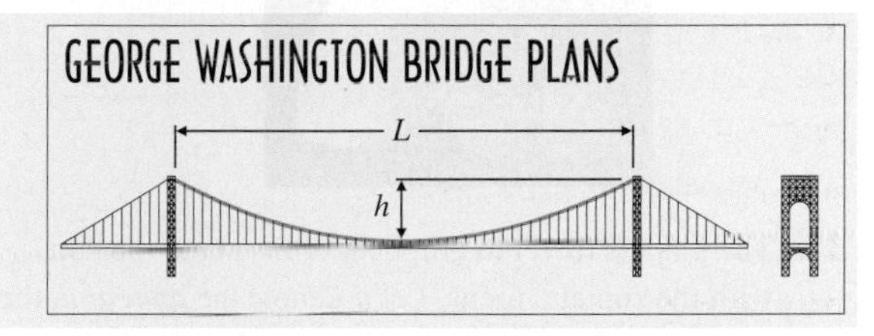

13. The Hale telescope, named to honor the American astronomer George Ellery Hale (1868–1939), is one of the world's largest compound reflecting telescopes. It is located at the Mount Palomar Observatory 45 miles northeast of San Diego, California. The main parabolic mirror of this telescope is 200 inches in diameter, with a depth of 3.75 inches from rim to vertex. A small cylindrical platform, for an observer to view and record the reflection from the telescope, is located within the tube of the telescope along the axis of the parabola. How far from the center of the mirror is the observer's viewing area located?
14. A satellite is placed in a position to make a parabolic flight past the moon, with the center of the moon at the focal point of the parabola. When the satellite is 5783 kilometers from the surface of the moon, it makes an angle of 60° with the axis of the parabola. The closest the satellite gets to the surface is 143 kilometers. What is the diameter of the moon? (Assume that the gravitational center of the moon is at its center, which is also the focus of the parabola. This assumption is not quite correct since the gravitational center is offset approximately 2 kilometers from the center toward the earth.)

15. Show that for constants a and b, the polar curve with equation $r = a\cos\theta + b\sin\theta$ is the circle through the origin with rectangular equation

$$\left(x - \frac{a}{2}\right)^2 + \left(y - \frac{b}{2}\right)^2 = \frac{a^2 + b^2}{4}.$$

16. The chambered nautilus (*Nautilus pompilus*) is a mollusk found in the Pacific and Indian Oceans. The outside of the shell grows in the form of an exponential spiral. A typical equation of such a spiral is $r = 2e^{-0.2\theta}$. Use a graphing device to sketch the graph of this spiral, and compare the resulting curve to the curve in the figure.

17. The planets travel in elliptical orbits about the sun, with the sun at a focus. Let a denote the *aphelion*, the greatest distance from the planet to the sun, and p the *perihelion*, the minimum distance from the planet to the sun.

a. Verify that the eccentricity e of the orbit is given by

$$e = \frac{a - p}{a + p}.$$

b. Verify that $a = R(1 + e)$ and $p = R(1 - e)$, where $2R$ is the length of the major axis of the ellipse.

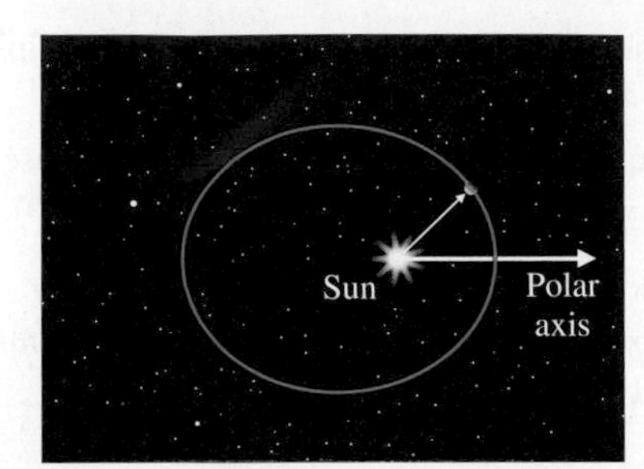

18. The comet Kohoutek has an elliptical orbit about the sun with eccentricity $e = 0.99993$ and a perihelion of 1.95×10^7 miles. Find a polar equation for the orbit, and determine the maximum distance of Kohoutek from the sun.

19. Show that the graph of

$$\sqrt{x^2 + (y - 1)^2} + 1 = \sqrt{x^2 + (y + 1)^2}$$

is a conic section.

CHAPTER 6: CHAPTER TEST

Determine whether the statement is true or false. If false, describe how the statement might be changed to make it true.

In Exercises 1–4, use the figure.

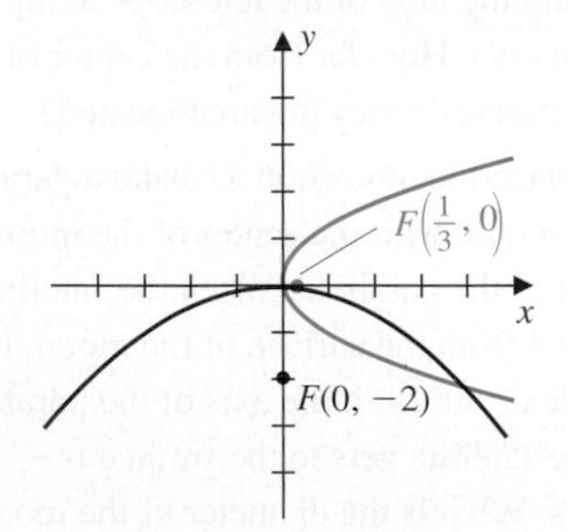

1. The directrix of the blue parabola is $y = -\frac{1}{3}$.

2. The equation of the blue parabola is $y = \frac{3}{4}x^2$.

3. The directrix of the red parabola is $y = 2$.

4. The equation of the red parabola is $y = -\frac{1}{8}x^2$.

In Exercises 5–10, use the figure.

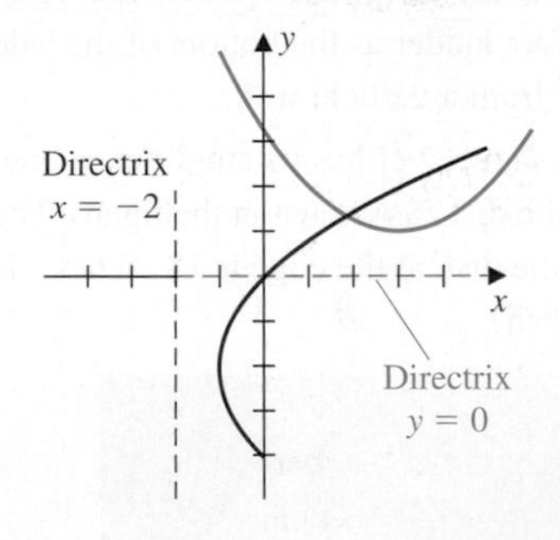

5. The vertex of the blue parabola is $(1, 3)$.

6. The vertex of the red parabola is $(-1, -2)$.

7. The focus of the blue parabola is $(3, 2)$.

8. The focus of the red parabola is $(-1, -2)$.

9. The equation of the blue parabola is $y - 1 = \frac{1}{4}(x + 3)^2$.

10. The equation of the red parabola is $x + 1 = \frac{1}{4}(y + 2)^2$.

In Exercises 11–14, use the figure.

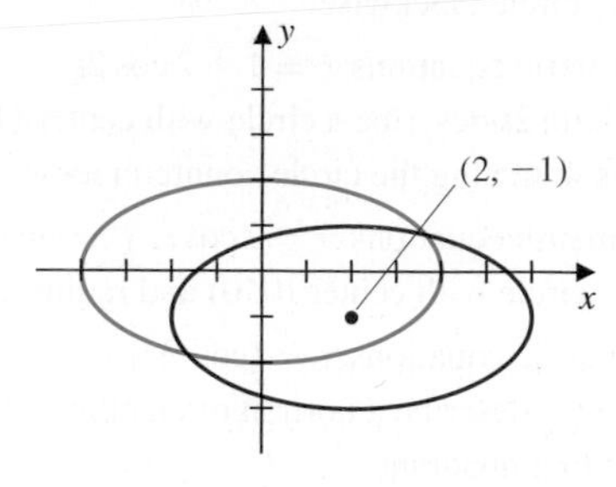

11. The focal points of the blue ellipse are (0, 3) and (0, −3).

12. The focal points of the red ellipse are (5, −2) and (−1, −2).

13. The equation of the blue ellipse is

$$\frac{x^2}{4} + \frac{y^2}{16} = 1.$$

14. The equation of the red ellipse is

$$\frac{(x-2)^2}{4} + \frac{(y-1)^2}{16} = 1.$$

In Exercises 15–18, use the figure.

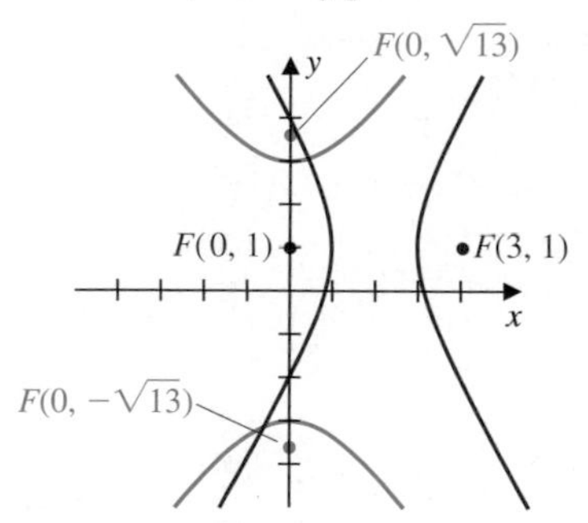

15. The equation of the blue hyperbola is

$$\frac{y^2}{4} - \frac{x^2}{9} = 1.$$

16. The equation of the red hyperbola is

$$\frac{(y-1)^2}{3} - (x-2)^2 = 1.$$

17. The asymptotes of the blue hyperbola are $y = \pm\frac{3}{2}x$.

18. The asymptotes of the red hyperbola are $y - 1 = \pm\frac{2}{3}(x - 2)$.

19. The polar coordinates $(3, \pi/3)$, $(3, -5\pi/3)$, and $(-3, 4\pi/3)$ all represent the same point.

20. The polar coordinates $(2, -3\pi/4)$, $(2, 5\pi/4)$, and $(2, -\pi/4)$ all represent the same point.

21. The point $(2, \pi/3)$ in polar coordinates is equivalent to the point $(1, \sqrt{3})$ in rectangular coordinates.

22. The point $(3, -3)$ in rectangular coordinates is equivalent to the point $(3\sqrt{2}, 7\pi/4)$ in polar coordinates.

23. The polar equation $\theta = 3\pi/4$ has the linear rectangular equation $y = -x$.

24. The polar equation

$$r = \frac{1}{2\sin\theta - \cos\theta}$$

has the linear rectangular equation $y = 2x - 1$.

25. The rectangular equation $x^2 + y^2 = 4$ has polar equation $r = 2$.

26. The polar equation $r = 3$ has the rectangular equation $x^2 + y^2 = 9$.

27. The polar equation $r = 2\sin\theta$ has the rectangular equation $(x - 1)^2 + y^2 = 1$.

28. The polar equation $r = 2\cos\theta$ has the rectangular equation $x^2 + y^2 - 2y = 0$.

In Exercises 29 and 30, use the figure.

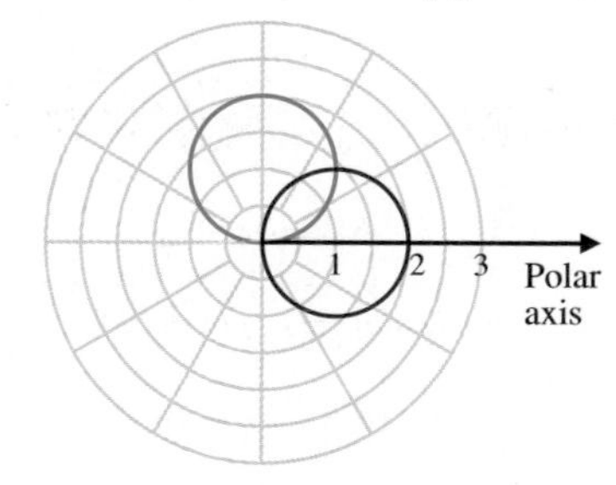

29. The red curve has polar equation $r = 2\sin\theta$.

30. The blue curve has polar equation $r = 2\cos\theta$.

In Exercises 31–34, use the figure.

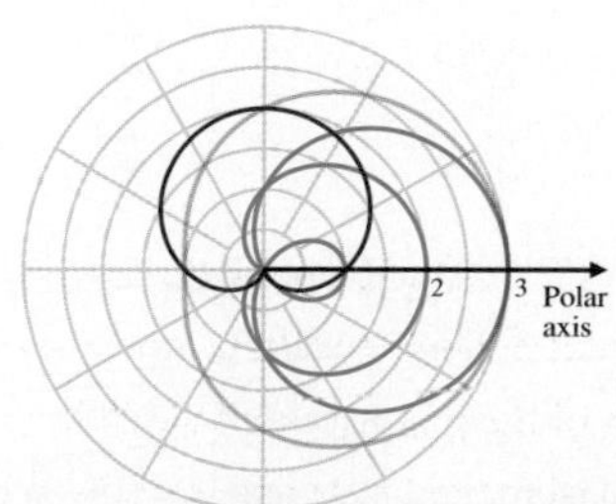

31. The blue curve has polar equation $r = 1 + \sin\theta$.

32. The red curve has polar equation $r = 1 + \cos\theta$.

33. The green curve has polar equation $r = 2 + \sin\theta$.

34. The yellow curve has polar equation $r = 1 + 2\sin\theta$.

35. The polar equation

$$r = \frac{1}{1 + \cos\theta}$$

describes a parabola opening to the right.

36. The polar equation

$$r = \frac{5}{5 - \cos\theta}$$

describes a hyperbola with horizontal directrix.

37. The polar curves $r = 1 + 2\sin\theta$ and $r = 1$ intersect at two points.

38. The polar curves $r = \cos\theta$ and $r = \sin\theta$ intersect at the two points $(\sqrt{2}/2, \pi/4)$ and $(0, 0)$.

39. A rectangular equation corresponding to the curve given by the parametric equations $x = 2t - 1$ and $y = t + 2$ is $y = \frac{1}{2}x + \frac{5}{2}$.

40. The parametric equations $x = 2t - 1$, $y = 3t + 2$ describe a line with slope $\frac{2}{3}$ and y-intercept $\frac{5}{2}$.

41. The parametric equations $x = t + 1$, $y = t^2 - 3$ describe a parabola with vertex $(1, -3)$ and opening upward.

42. The parametric equations $x = e^t + 1$, $y = e^{2t} - 1$ describe a parabola with vertex $(1, 1)$ and opening upward.

43. The parametric equations $x = 2\sin t$, $y = 2\cos t$ describe a circle with center $(0, 0)$ and radius 4, tracing the circle clockwise.

44. The parametric equations $x = 1 + 2\cos 2t$, $y = 2 + 2\sin 2t$ describe a circle with center $(1, 2)$ and radius 4, tracing the circle counterclockwise.

45. The parametric equations $x = 2\cos t$, $y = \sin t$ describe a circle with center $(0, 0)$ and radius $\sqrt{5}$.

46. The parametric equations $x = (\cos t)^2$, $y = (2\sin t)^2$ describe a portion of an ellipse that lies in the first quadrant.

47. One set of parametric equations for the line passing through $(3, 1)$ and $(1, 2)$ is $x = t + 2$, $y = -2t + 1$.

48. One set of parametric equations of the parabola with vertex $(1, -2)$ passing through $(2, 0)$ with axis parallel to the x-axis is $x = \frac{1}{4}t^2 + 1$, $y = t - 2$.

49. The parametric equations $x = e^{-t}$, $y = e^t$ and the rectangular equation $y = \frac{1}{x}$ describe the same curve.

50. The parametric equations $x = t^{1/3}$, $y = t + 1$ and the rectangular equation $y = x^3 + 1$ describe the same curve.

Answers to Odd Exercises

Chapter 1

Exercise Set 1.2 (page 12)

1. $-1 \le x \le 5$

3. $-\sqrt{3} < x \le \sqrt{2}$

5. $x < 3$

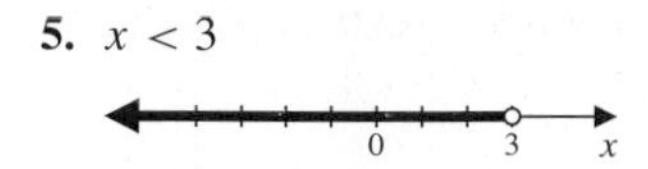

7. $x \ge \sqrt{2}$

9. $[-2, 3]$

11. $[2, 5)$

13. $(-\infty, 3)$

15. $[3, \infty)$

17. a. $|3 - 7| = |-4| = 4$

b. $\frac{1}{2}(3 + 7) = 5$

19. a. $|-3 - 5| = |-8| = 8$

b. $\frac{1}{2}(-3 + 5) = 1$

21. $(x + 1)(x + 2)$

23. $(x + 2)(x + 3)$

25. $(x - 2)(x + 6)$

27. a. $(0, 3]$ **b.** $[-1, 4)$

29. a. $(-2, 0)$ **b.** $(-\infty, 3]$

31. $(-\infty, 2)$

33. $[5, \infty)$

35. $(-\frac{1}{3}, \infty)$

37. $(-\infty, -4]$

39. $\left(\frac{2}{3}, 3\right)$

41. $(-\infty, -1] \cup [2, \infty)$

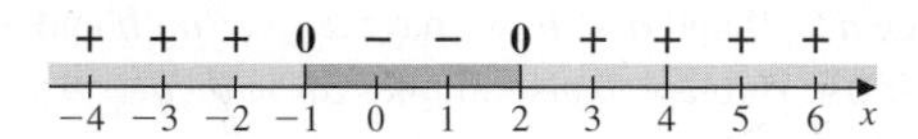

43. $[1, 3]$

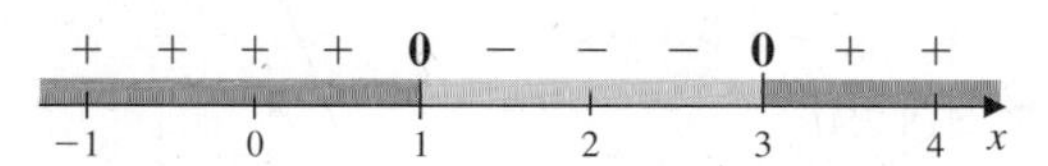

45. $(-\infty, -1] \cup [1, 2]$

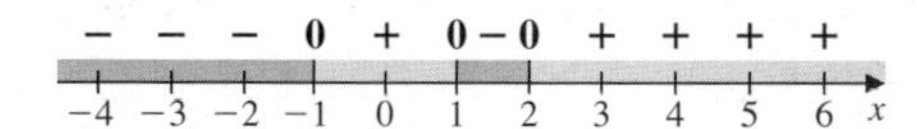

47. $[0, 1] \cup [2, \infty)$

49. $(-\infty, 0) \cup (0, 2)$

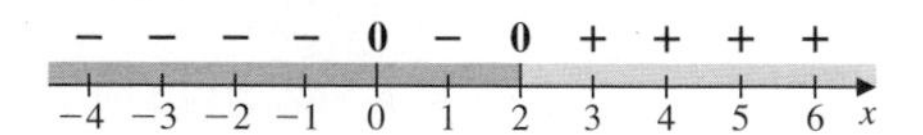

51. $(-\infty, -3] \cup (1, \infty)$

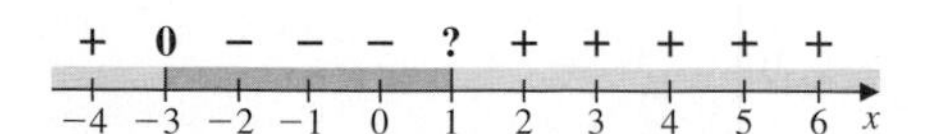

53. $(-\infty, -2] \cup [0, 2)$

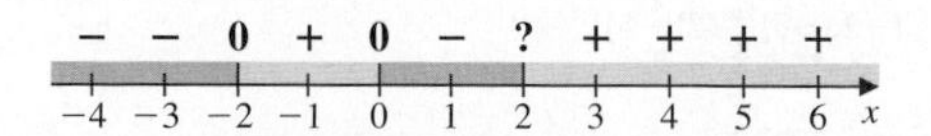

55. $(-2, -1) \cup (0, 1)$

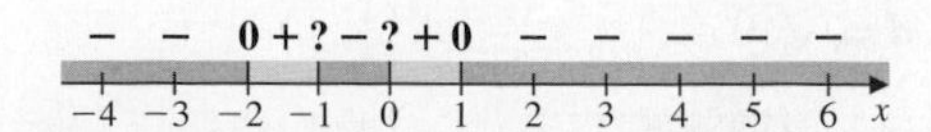

57. $(-\infty, 0) \cup [\frac{1}{5}, \infty)$

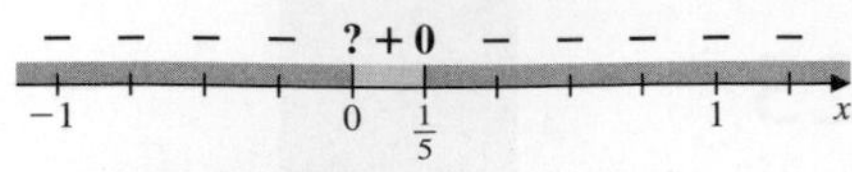

59. $(-\infty, -2) \cup (1, 7]$

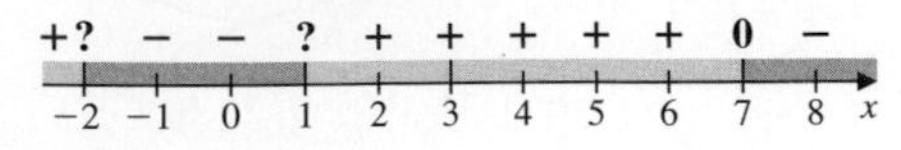

61. $x = 1, \quad x = \frac{1}{5}$

63. $x = -\frac{7}{3}, \quad x = -1$

65. $[3, 5]$

67. $(-\infty, 1] \cup [5, \infty)$

69. $(-\frac{11}{2}, -5) \cup (-5, -\frac{9}{2})$

71. $(-\infty, -2) \cup (-2, 2) \cup (2, \infty)$

73. Since $a > 0$ and $a < b$, we have $a \cdot a < a \cdot b$ and $a^2 < ab$. But $a < b$ also implies $ab < b \cdot b = b^2$. So $a^2 < b^2$.

75. **a.** $-\frac{20}{3} \leq C \leq 10$

b. $68 \leq F \leq 122$

77. $18,750

Exercise Set 1.3 (page 20)

1.

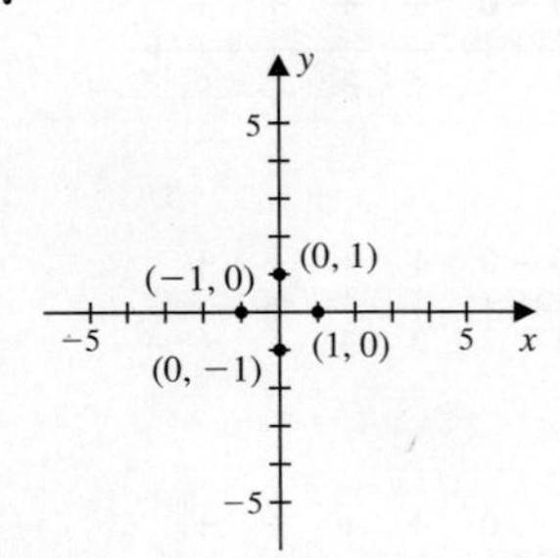

3.

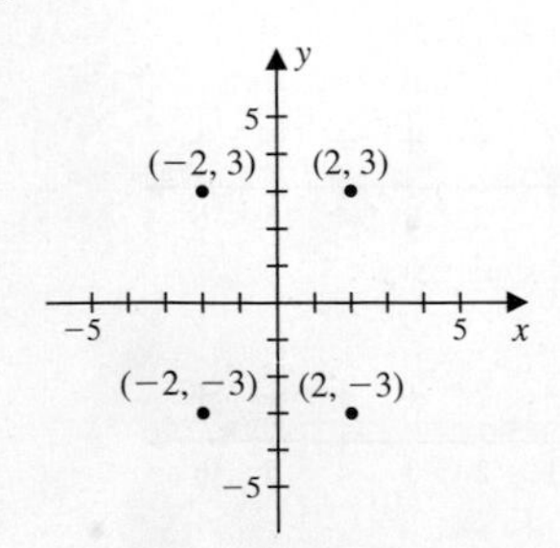

5. **a.** $d = \sqrt{10}$ **b.** $\left(\frac{1}{2}, \frac{7}{2}\right)$

7. **a.** $d = \sqrt{\pi^2 + 2\pi + 5}$ **b.** $\left(\frac{\pi-1}{2}, 1\right)$

9.

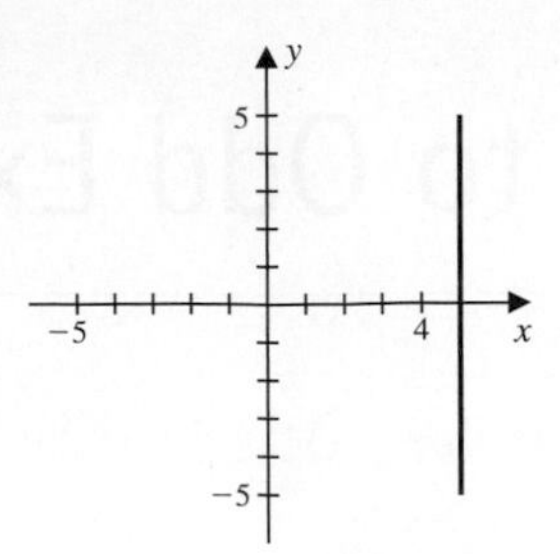

11.

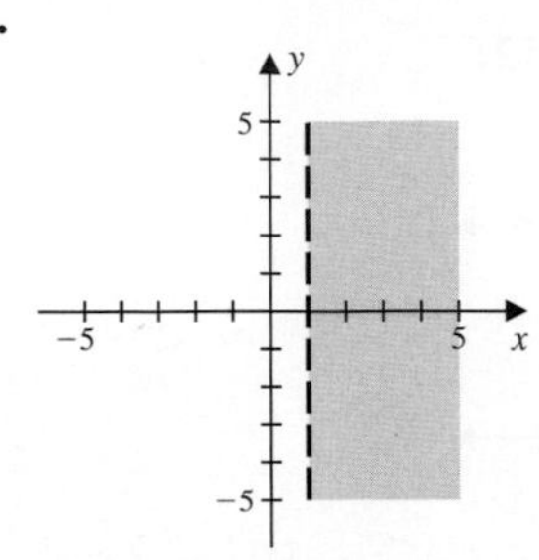

13.

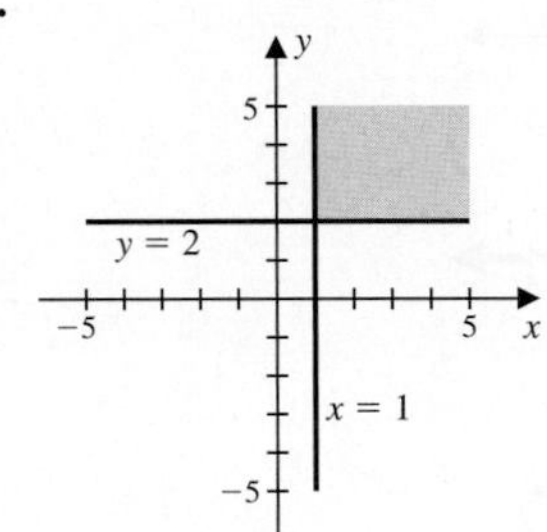

15.

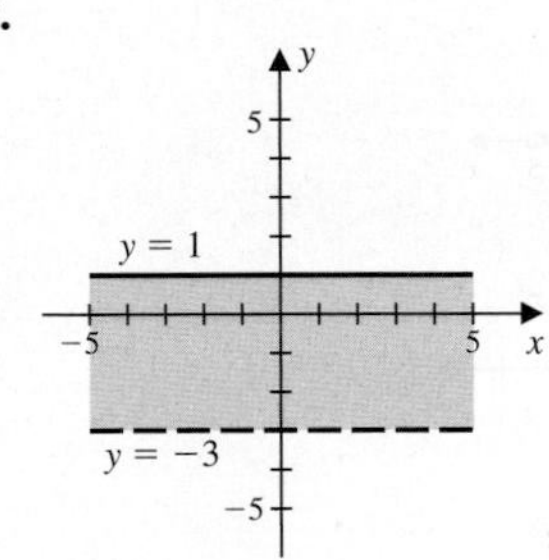

17.

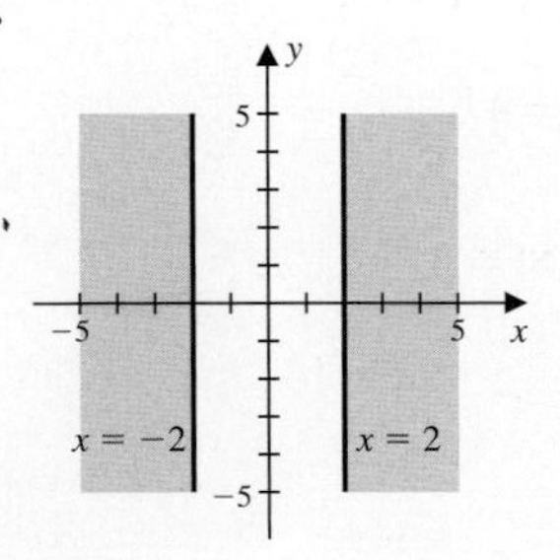

19.

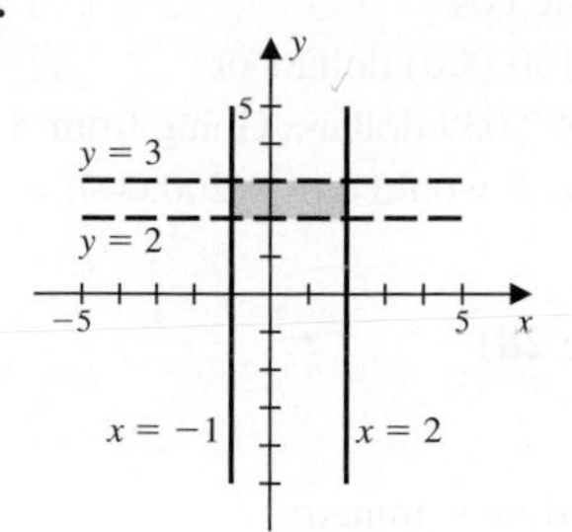

21.

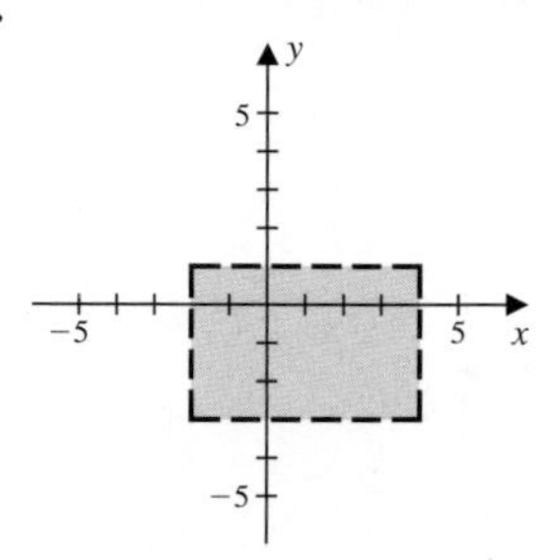

23.

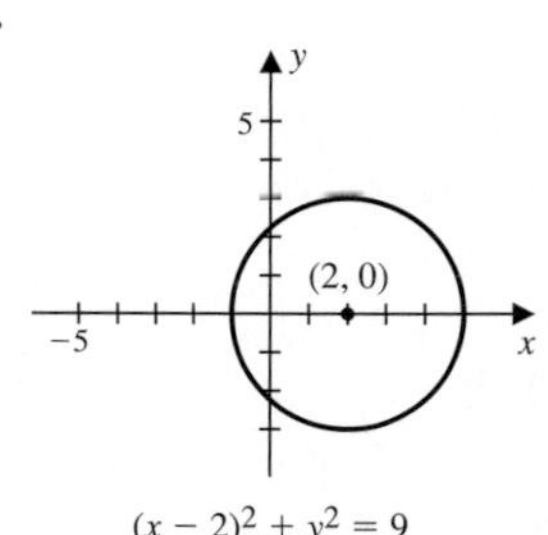

25.

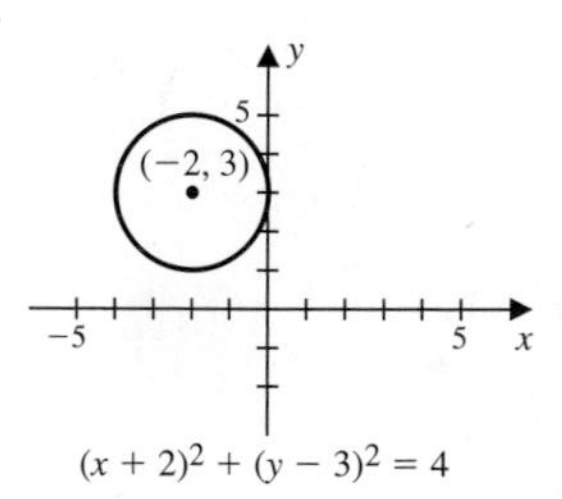

27.

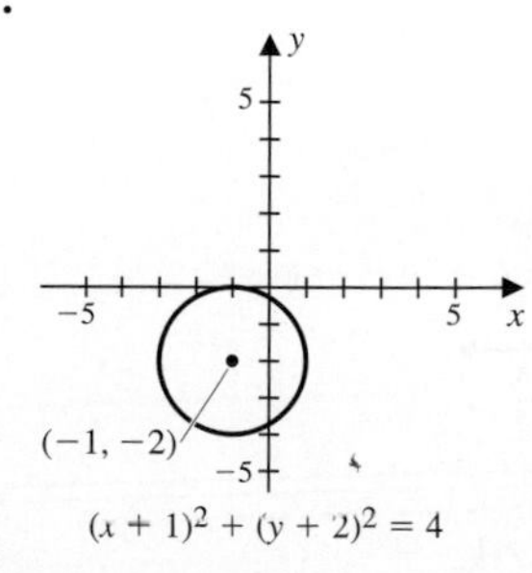

29.

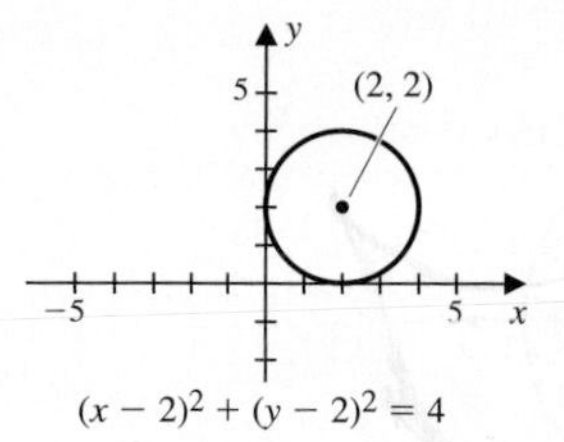

31. **a.** The center is (0, 0) and the radius is 3.

b.

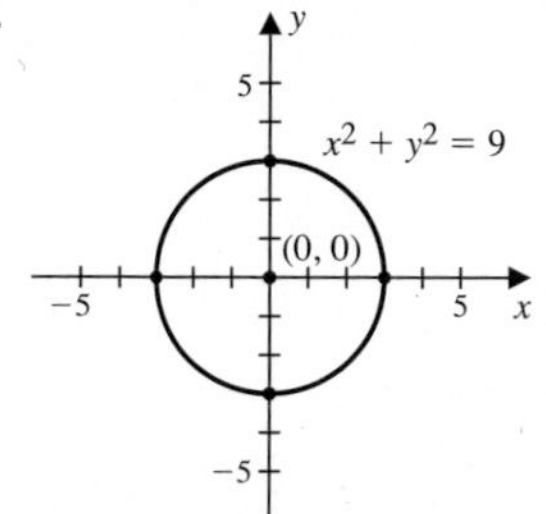

33. **a.** The center is (0, 1) and the radius is 1.

b.

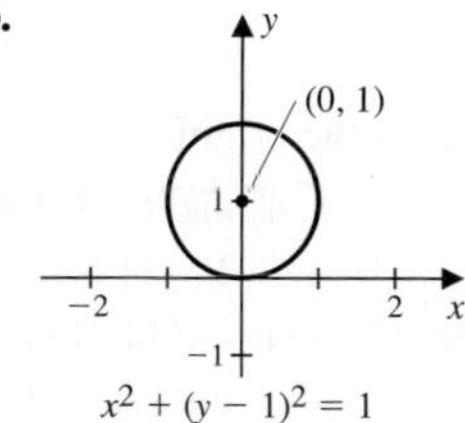

35. **a.** The center is (2, −1) and the radius is 3.

b.

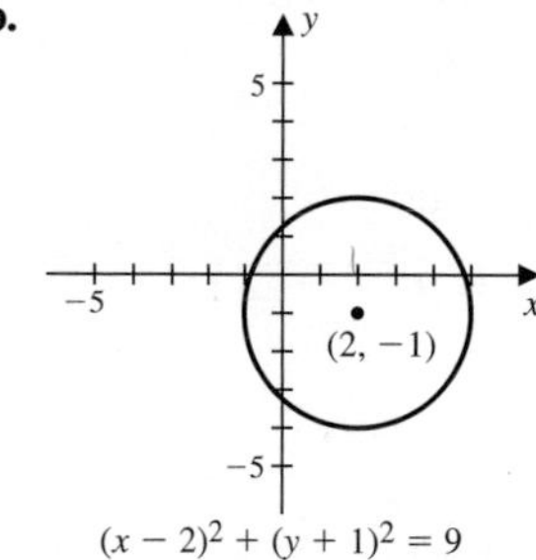

37. The center is (1, 0) and the radius is 2.

39. The center is (−1, 2) and the radius is 1.

41. The center is (2, 1) and the radius is 3.

43.

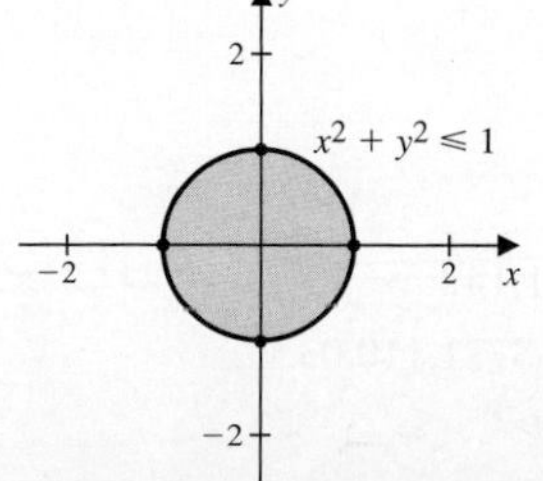

45.

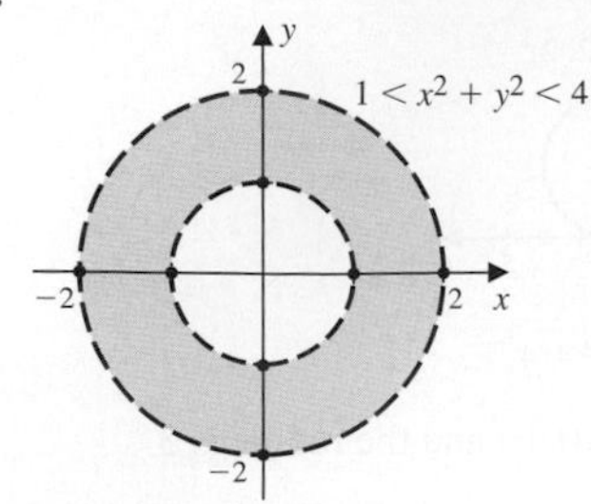

47.

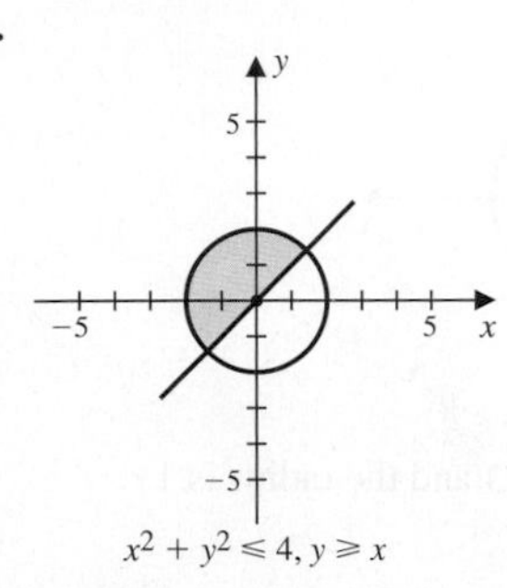

49. The point (6, 3) is closer to the origin.

51. **a.** The distance between $(-1, 4)$ and $(-3, -4)$ is $\sqrt{68}$, between $(-3, -4)$ and $(2, -1)$ is $\sqrt{34}$, and between $(-1, 4)$ and $(2, -1)$ is $\sqrt{34}$. Since $(\sqrt{68})^2 = (\sqrt{34})^2 + (\sqrt{34})^2$, the triangle is a right triangle.

b. The right angle is at $(2, -1)$.

53. The unique point is $(-6, 1)$.

55. $x^2 + y^2 = 13$

57. $(x - 3)^2 + (y - 7)^2 = 9$

59. $(x + 3)^2 + (y - 3)^2 = 9$

61. 8π

63.

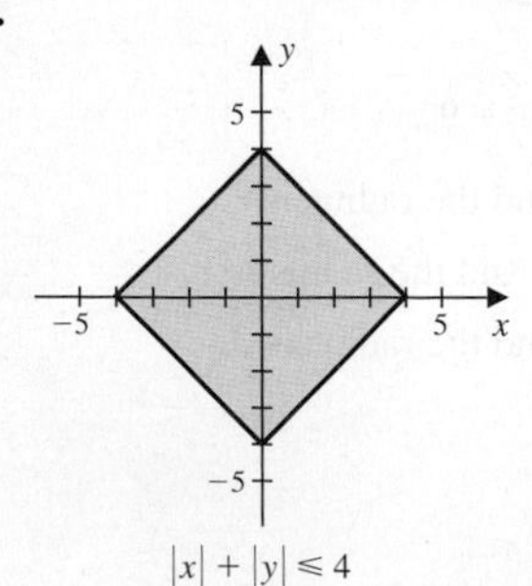

65. The area is $\pi - 2 \approx 1.142$.

67. The estimated cost is $771,110.03.

69. The direct route would cost $3(200{,}000) + \sqrt{20}(150{,}000)$ dollars or approximately 127,082,039 dollars. Going from A to C and then from C to B would cost $9(200{,}000) =$ 1,800,000 dollars.

Exercise Set 1.4 (page 28)

1. x-axis symmetry

3. y-axis, x-axis, and origin symmetry

5. origin symmetry **7.** x-axis symmetry

9. no symmetry

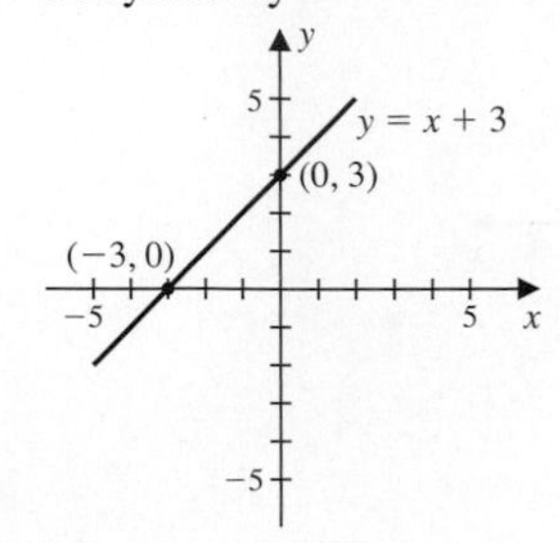

11. no symmetry

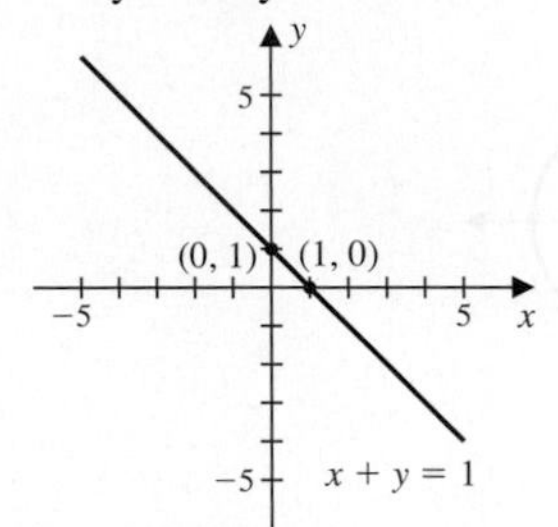

13. y-axis symmetry

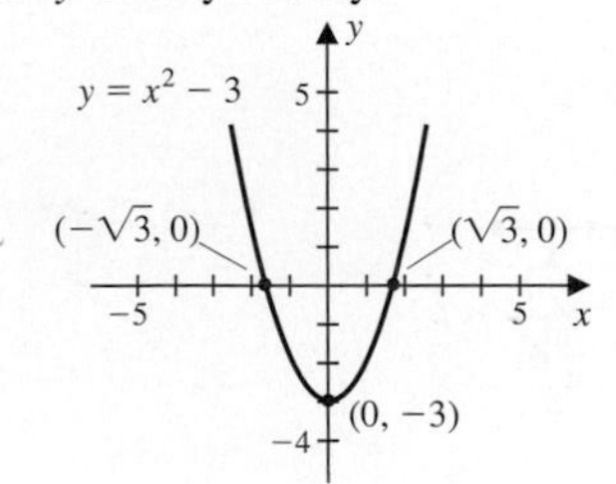

15. y-axis symmetry

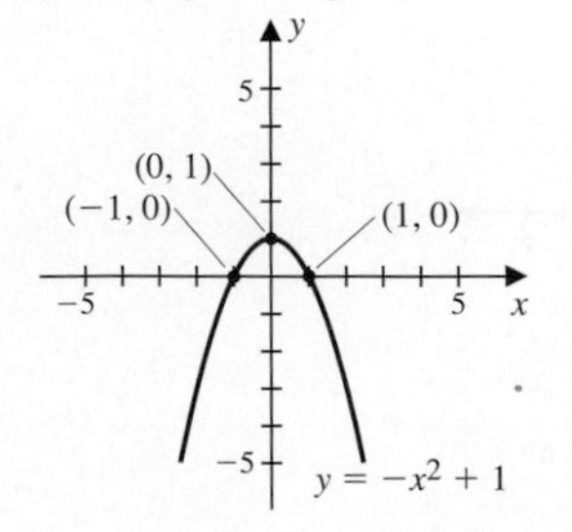

17. y-axis symmetry

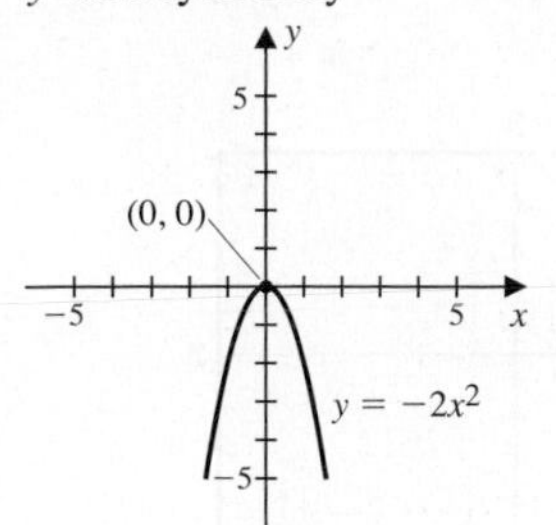

19. x-axis symmetry

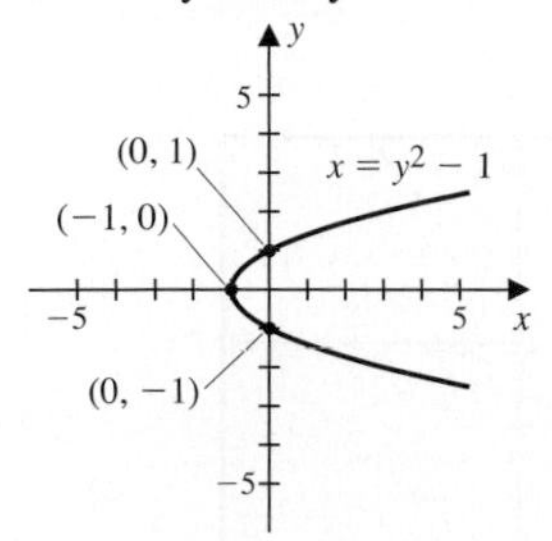

21. no symmetry

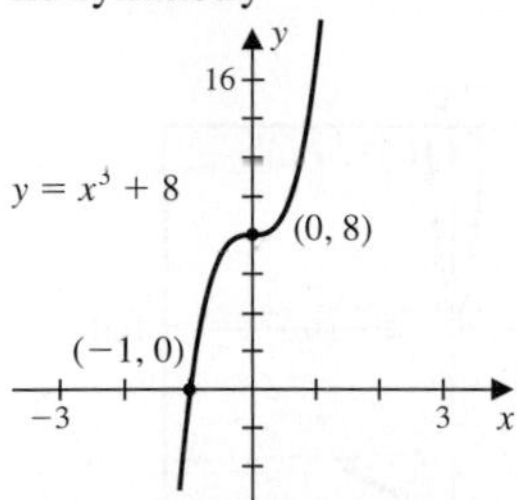

23. origin symmetry

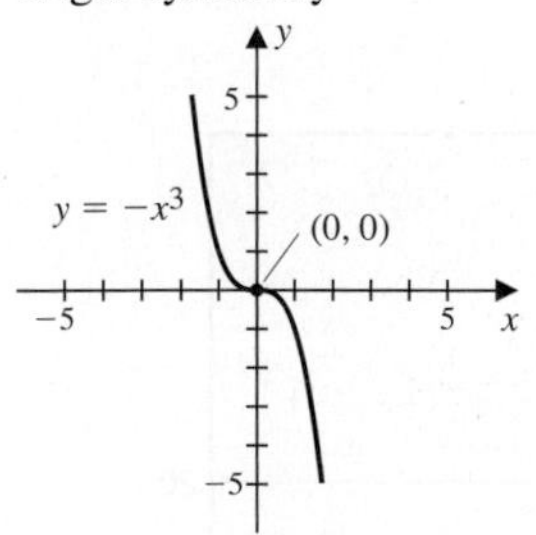

25. no symmetry

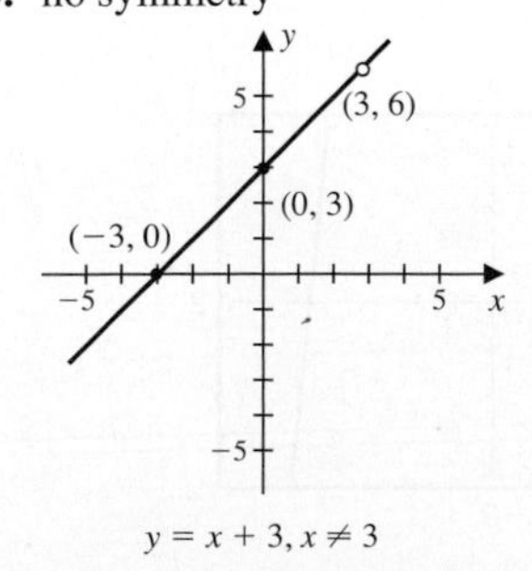

$y = x + 3, x \neq 3$

27. no symmetry

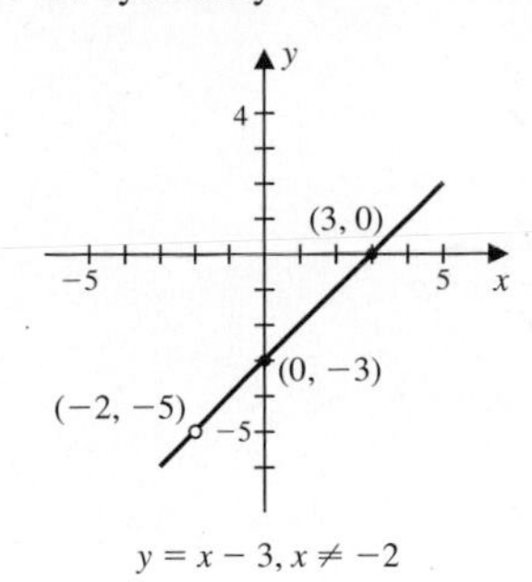

$y = x - 3, x \neq -2$

29. no symmetry

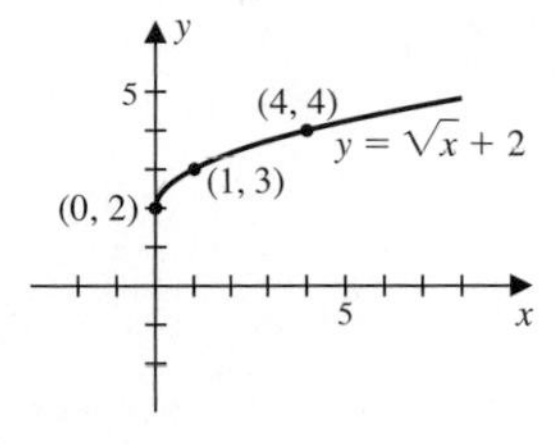

31. x-axis, y-axis, and origin symmetry

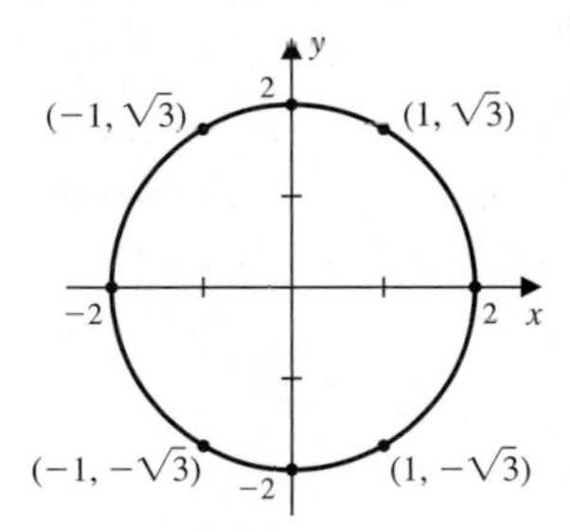

33. y-axis symmetry

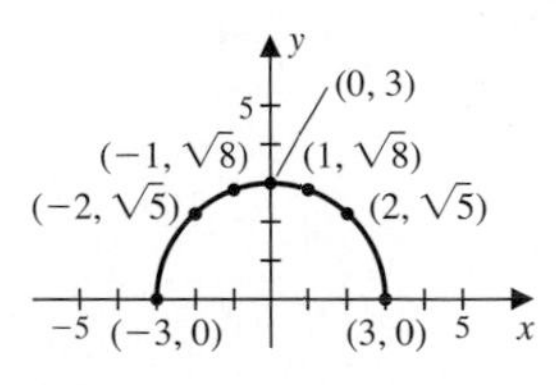

35. y-axis symmetry

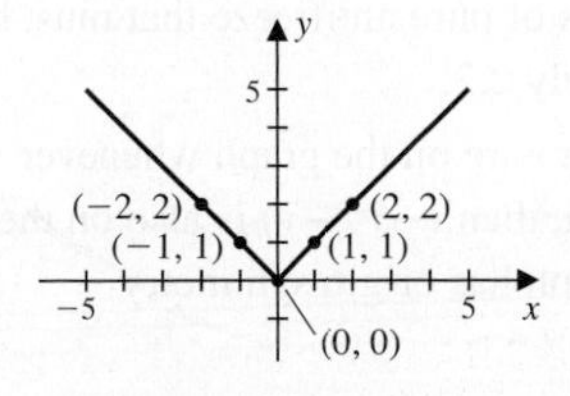

37. y-axis symmetry

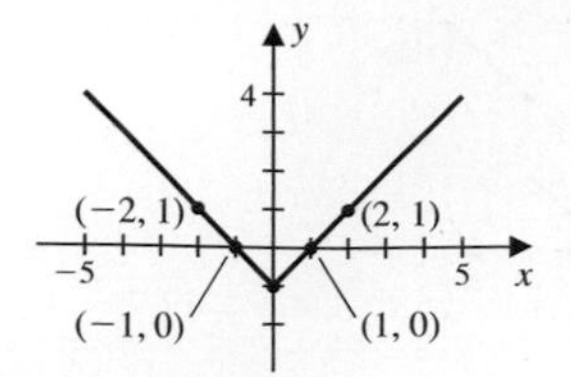

39.

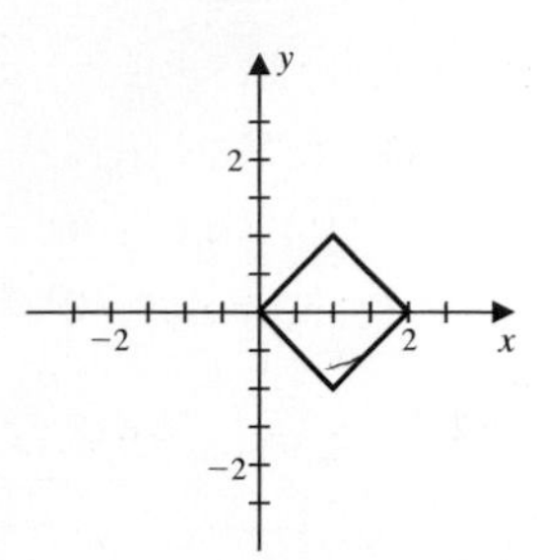

41.

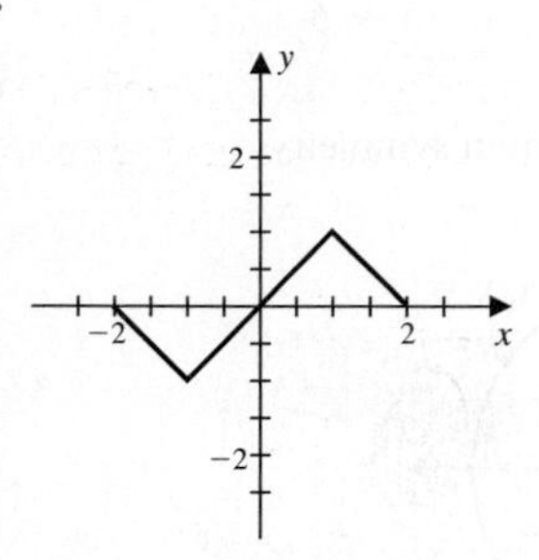

43. $d((1, 2), (-1, 2)) = 2$

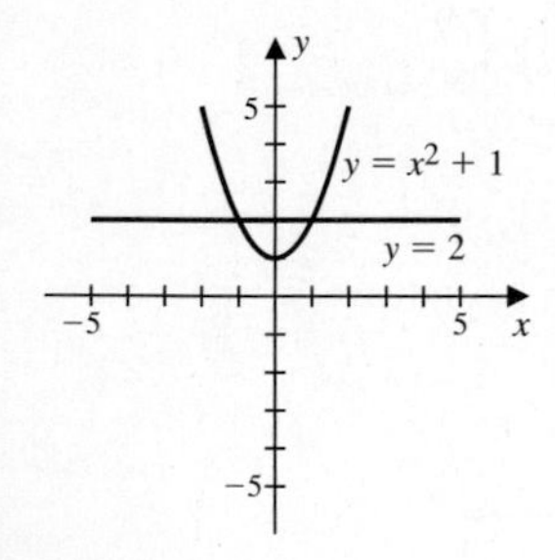

45. 51, 52, 53

47. 3 by 4

49. The number of quarts of pure antifreeze that must be added is approximately 2.2.

51. If $(x, -y)$ and $(-x, y)$ are on the graph whenever (x, y) is on the graph, then $(-x, -y)$ is also on the graph. Hence the graph has origin symmetry as well.

Exercise Set 1.5 (page 35)

1. a.

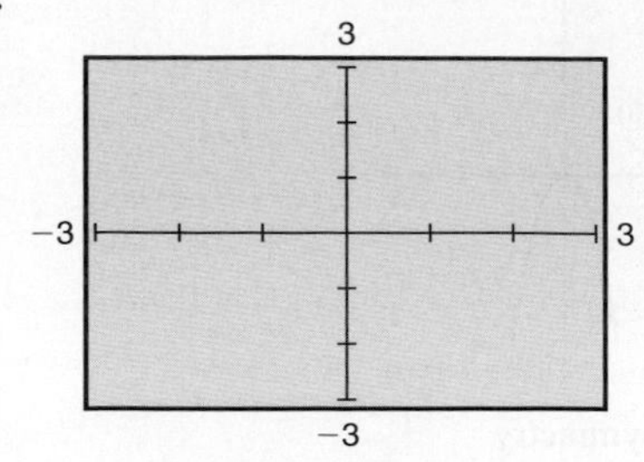

b.

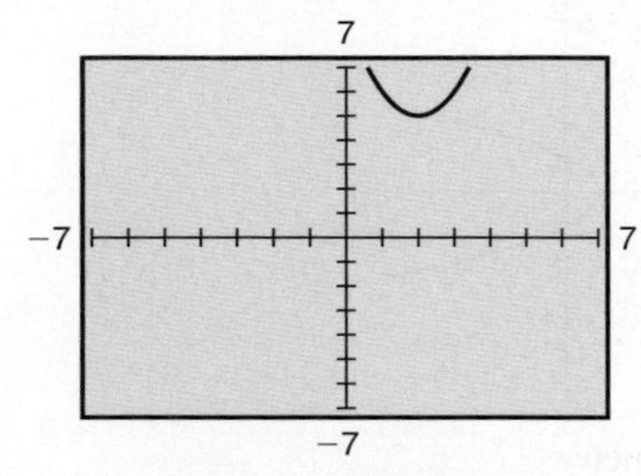

c.

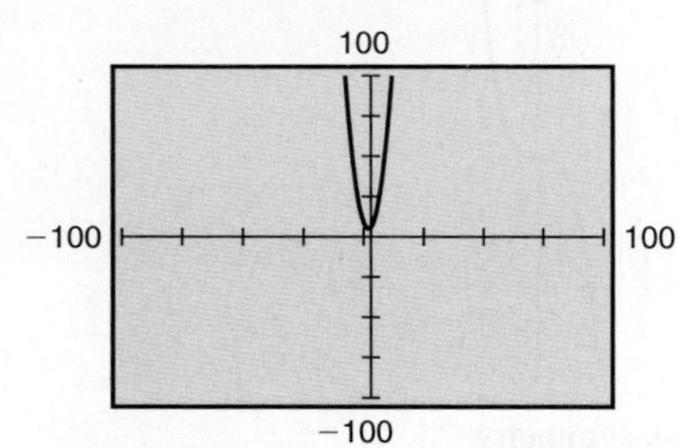

d.

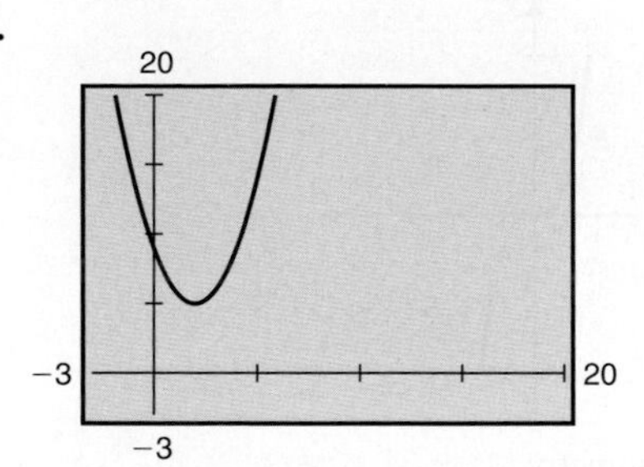

3. a.

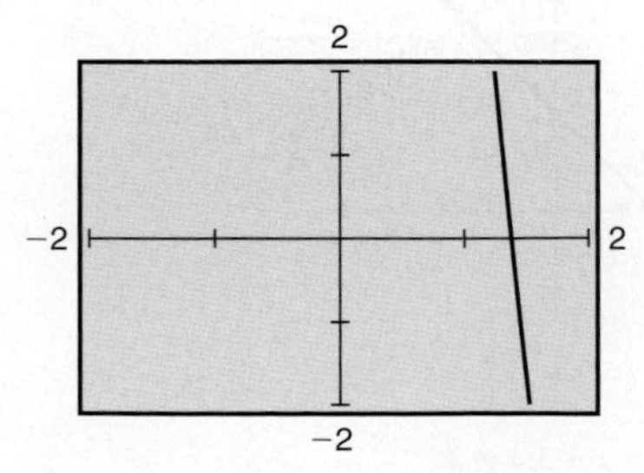

b.

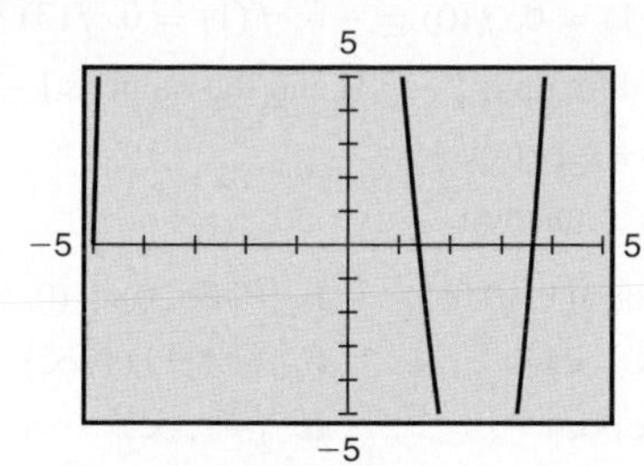

c.

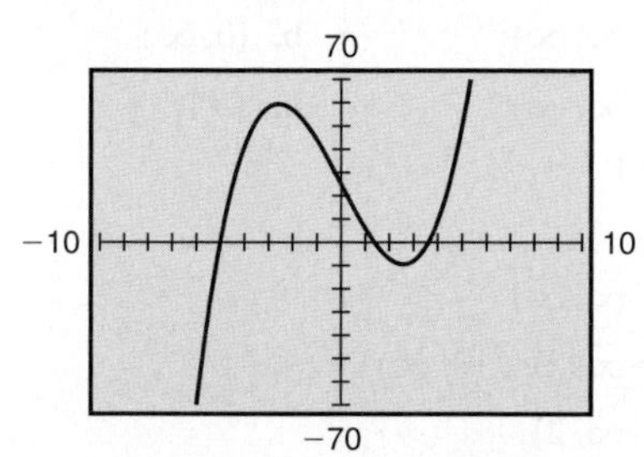

d.

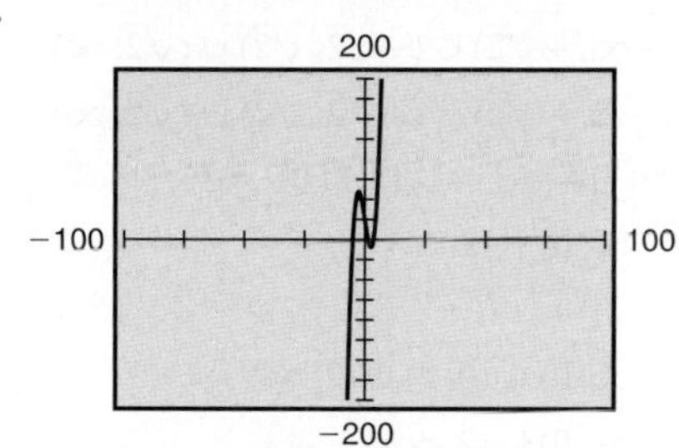

5. One choice is $[-10, 10] \times [-10, 10]$.

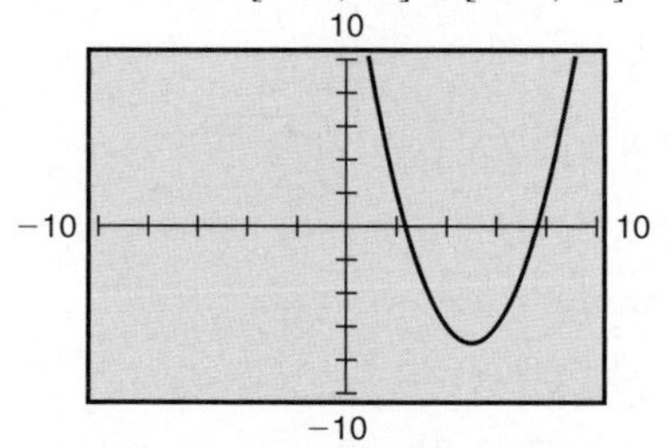

7. One choice is $[0, 20] \times [0, 10]$.

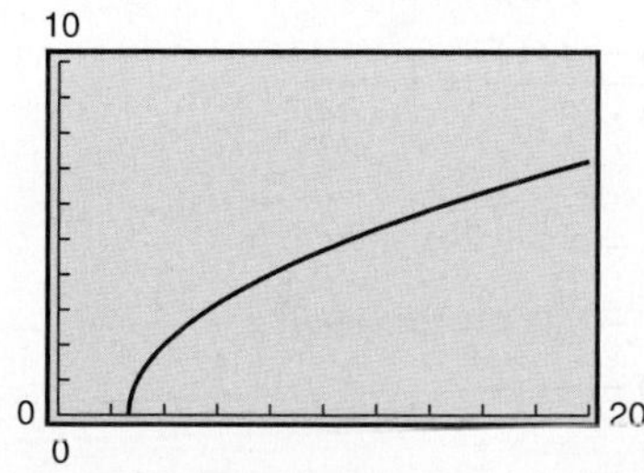

9. One choice is $[-10, 10] \times [-10, 10]$.

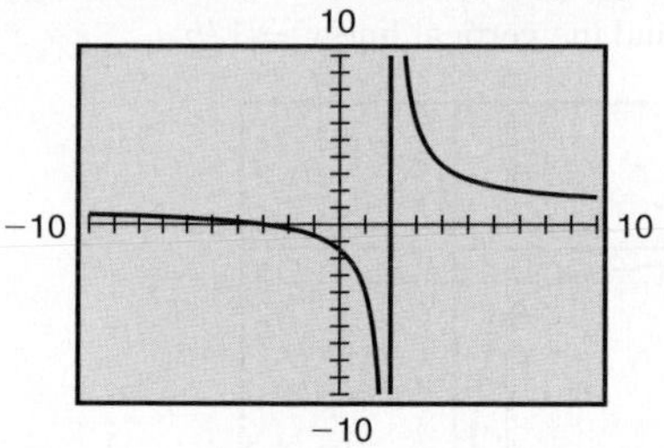

11. x is approximately 0.87.

13. x is approximately -0.93 and 2.82.

15. The points of intersection are $(1, 6)$ and $(-1, 0)$.

17. The points of intersection are approximately $(0, -1)$ and $(0.45, -0.79)$.

19. **a.** The inequality $x^2 + 3x - 2 \geq 0$ is satisfied for $x \leq -3.56$ and for $0.56 \leq x$.

b. The inequality $x^3 - 2x^2 - 6x + 9 < 0$ is satisfied for $x < -2.3$ and $1.3 < x < 3$.

21. For x very large, the graphs are almost identical. That is, as x grows without bound, the values of the two expressions get closer to each other.

a.

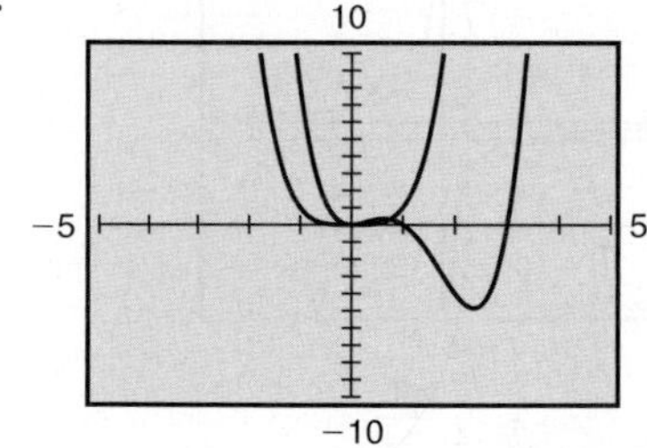

b.

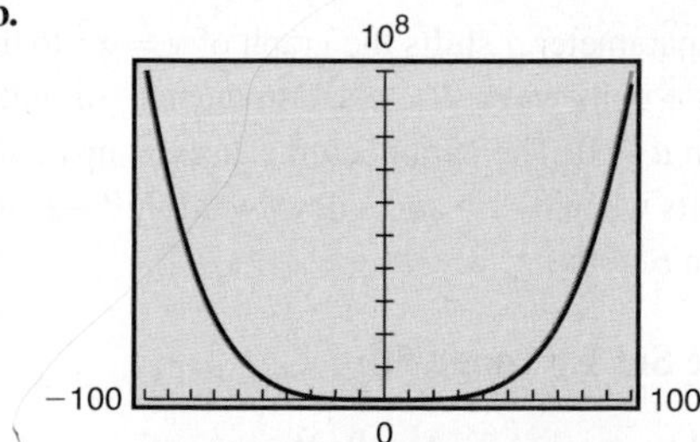

c.

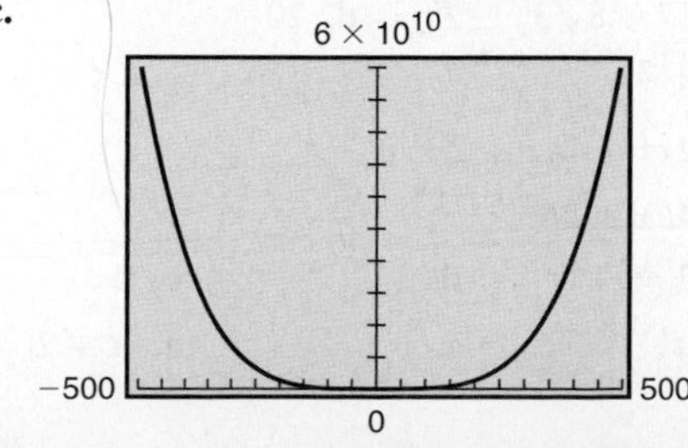

23. The graph appears to approach the horizontal line $y = a/b$ and the vertical line $x = 1/b$.

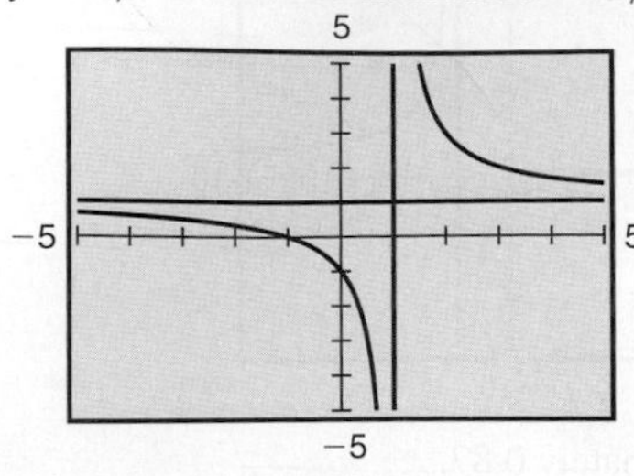

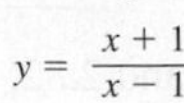
$y = \frac{x+1}{x-1}$

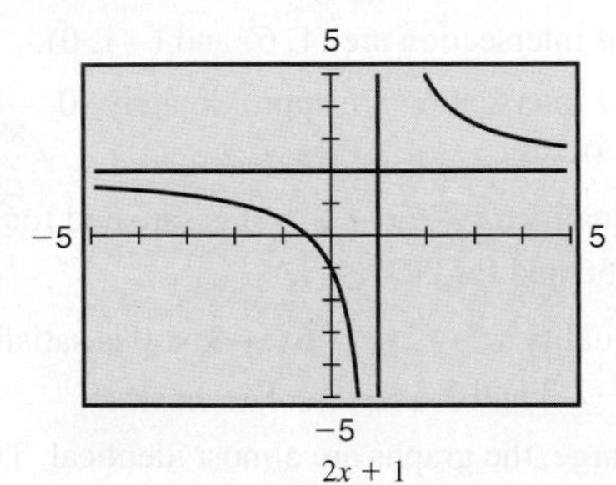

$y = \frac{2x+1}{x-1}$

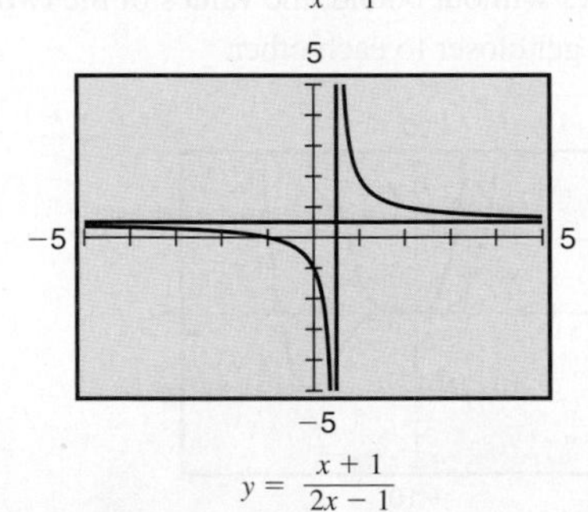

$y = \frac{x+1}{2x-1}$

25. The parameter a shifts the graph of $y = x^2$ to the right a units when $0 < a$ and to the left $-a$ units when $a < 0$. The parameter b causes an upward shift b units when $0 < b$ and a downward shift $-b$ units when $b < 0$.

Exercise Set 1.6 (page 50)

1. **a.** 11 **b.** 9 **c.** $17 + 8\sqrt{3}$ **d.** 20 **e.** $8x^2 + 3$ **f.** $2x^2 - 4x + 5$ **g.** $2x^2 + 4hx + 2h^2 + 3$ **h.** $4hx + 2h^2$

3. **a.** 2 **b.** 1 **c.** 2 **d.** $|t|$ **e.** t^2 **f.** $|t + 2|$

5. **a.** $f(-1) = 0, f(0) = -\frac{1}{2}, f(1) = 0, f(3) = -2$ **b.** The domain is $[-2, 3]$ and the range is $[-2, 3]$.

7. Yes, it is a function.

9. Yes, it is a function.

11. **a.** $(-\infty, 0) \cup (0, \infty)$ **b.** $(-\infty, 0) \cup (0, \infty)$

13. **a.** $(-\infty, \infty)$ **b.** $\{-1\} \cup (1, \infty)$

15. **a.** $(-\infty, \infty)$ **b.** $[-1, \infty)$

17. **a.** $[0, \infty)$ **b.** $[4, \infty)$

19. **a.** $(-\infty, \infty)$ **b.** $[0, \infty)$

21. **a.** $(-\infty, \infty)$ **b.** $\{-1, 1\}$

23. $a = \pm 1$

25. $a = \frac{5}{4}$

27. **a.** $(-\infty, \infty)$ **b.** $(-\infty, 2) \cup (2, \infty)$ **c.** $(-\infty, 2]$ **d.** $(-\infty, 2)$

29. **a.** $(-\infty, -\sqrt{2}) \cup (-\sqrt{2}, \sqrt{2}) \cup (\sqrt{2}, \infty)$ **b.** $(-\infty, -\sqrt{2}) \cup (-\sqrt{2}, \sqrt{2}) \cup (\sqrt{2}, \infty)$ **c.** $\{0\} \cup (-\infty, -\sqrt{2}) \cup (\sqrt{2}, \infty)$

31. **a.** $(-\infty, 0] \cup [2, \infty)$ **b.** $[0, 2]$ **c.** $(-\infty, 0) \cup (0, 2) \cup (2, \infty)$ **d.** $(-\infty, 0) \cup (2, \infty)$

33. $f(-x) = x^2 + 2; -f(x) = -x^2 - 2;$ $f\left(\frac{1}{x}\right) = \frac{1}{x^2} + 2; \frac{1}{f(x)} = \frac{1}{x^2+2};$ $f(\sqrt{x}) = x + 2; \sqrt{f(x)} = \sqrt{x^2 + 2}$

35. $f(-x) = -\frac{1}{x}; -f(x) = -\frac{1}{x};$ $f\left(\frac{1}{x}\right) = x; \frac{1}{f(x)} = x;$ $f(\sqrt{x}) = \frac{1}{\sqrt{x}}; \sqrt{f(x)} = \frac{1}{\sqrt{x}}$

37. **a.** $f(x + h) = 3x + 3h - 2$ **b.** $f(x + h) - f(x) = 3h$ **c.** $\frac{f(x+h) - f(x)}{h} = 3$ **d.** 3

39. **a.** $f(x + h) = x^2 + 2hx + h^2$ **b.** $f(x + h) - f(x) = 2hx + h^2$ **c.** $\frac{f(x+h) - f(x)}{h} = 2x + h$ **d.** $2x$

41. **a.** $f(x+h) = 2 - x - h - x^2 - 2hx - h^2$

b. $f(x+h) - f(x) = -h - 2hx - h^2$

c. $\dfrac{f(x+h) - f(x)}{h} = -1 - 2x - h$

d. $-1 - 2x$

43. **a.** $f(x+h) = \dfrac{1}{x+h}$

b. $f(x+h) - f(x) = \dfrac{1}{x+h} - \dfrac{1}{x} = -\dfrac{h}{x(x+h)}$

c. $\dfrac{f(x+h) - f(x)}{h} = -\dfrac{1}{x(x+h)}$

d. $-\dfrac{1}{x^2}$

45. **a.** $f(x+h) = \dfrac{x+h}{x+h-3}$

b. $f(x+h) - f(x) = -\dfrac{3h}{(x-3)(x+h-3)}$

c. $\dfrac{f(x+h) - f(x)}{h} = -\dfrac{3}{(x+h-3)(x-3)}$

d. $-\dfrac{3}{(x-3)^2}$

47. **a.** $f(x+h) = x^3 + 3x^2h + 3h^2x + h^3$

b. $f(x+h) - f(x) = 3x^2h + 3xh^2 + h^3$

c. $\dfrac{f(x+h) - f(x)}{h} = 3x^2 + 3hx + h^2$

d. $3x^2$

49. **a.** $\dfrac{f(x+h) - f(x)}{h} = \dfrac{1}{\sqrt{x+h} + \sqrt{x}}$

b. $\dfrac{1}{2\sqrt{x}}$

51. even **53.** neither

55. **a.**

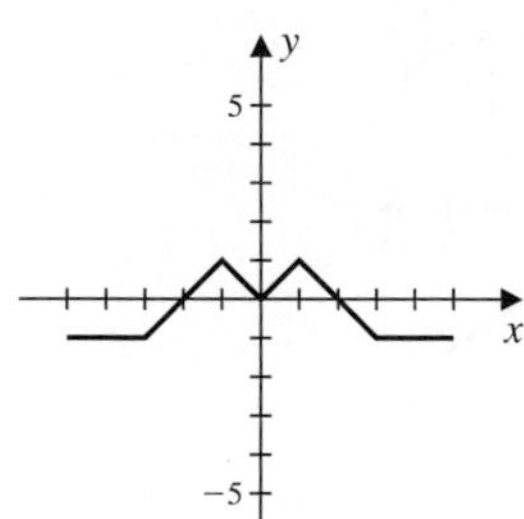

b.

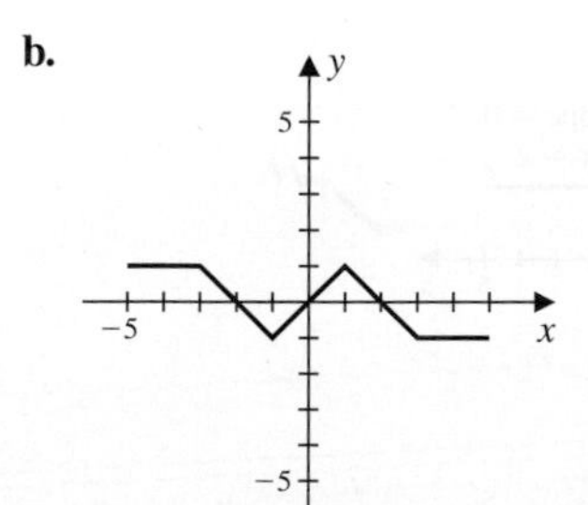

57. **a.**

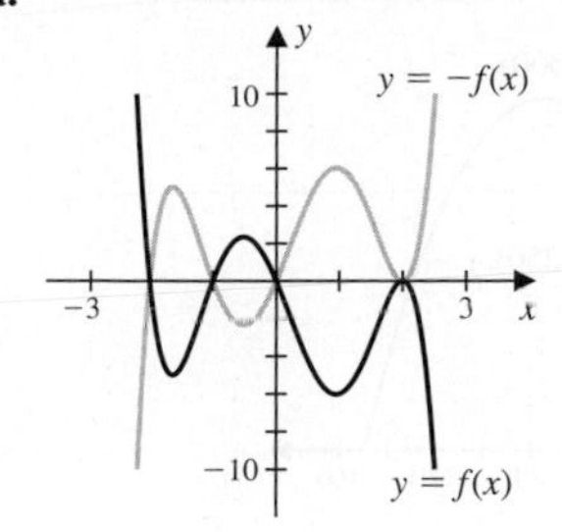

b.

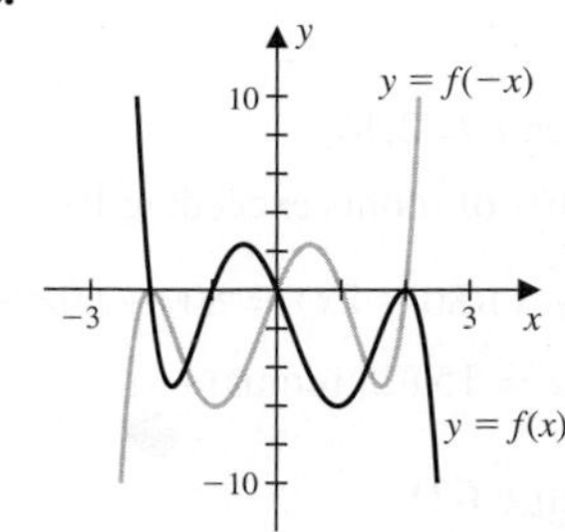

c.

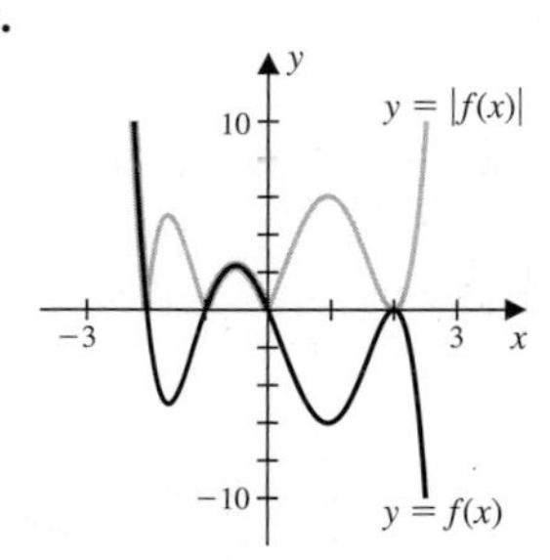

59. $d(t) = \sqrt{325t^2 + 600t + 900}$

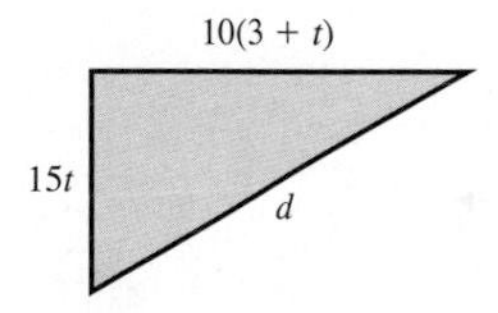

61. If one side is s, $P = 2s + \frac{128}{s}$.

63. $6V^{2/3}$

65. $A = 2x\sqrt{r^2 - \frac{x^2}{4}} = x\sqrt{4r^2 - x^2}$

67. **a.** $\dfrac{P(x+h) - P(x)}{h} = 300 - 4x - 2h$

b. \$150

c. \$200

d.

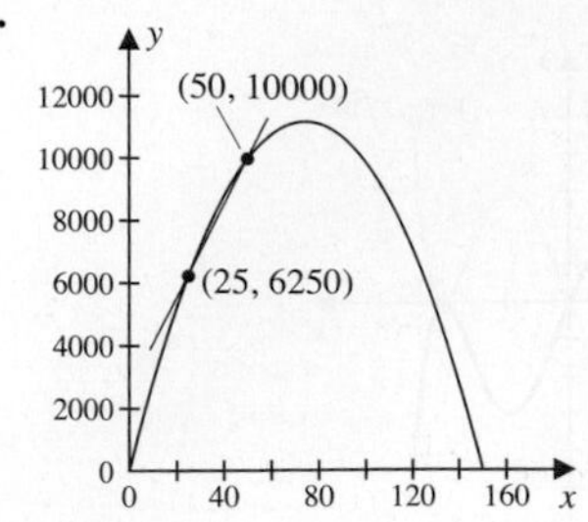

69. **a.** $S = 2\pi r^2 + \frac{1800}{r}$ **b.** $r \approx 5.2, h \approx 10.6$

71. **a.** $A = r - \pi r^2$

b. $A \approx 0.08$ when $r \approx 0.16$

73. Let x be the number of rooms exceeding 10.

a. $R(x) = (10 + x)(80 - 2x) = 800 + 60x - 2x^2$

b. \$1250 when $x = 15$ (25 rooms)

Exercise Set 1.7 (page 62)

1. $y = 3x - 2$

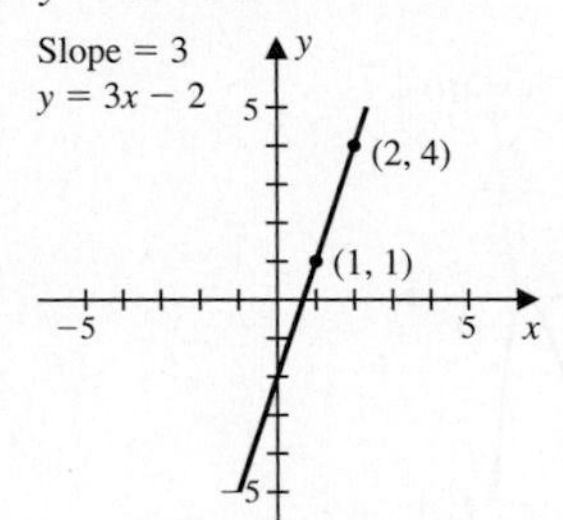

3. $y = -\frac{5}{4}x - \frac{3}{2}$

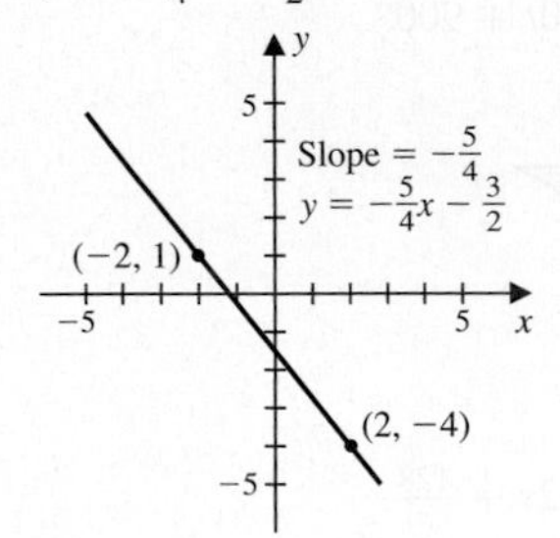

5. $y = 3$

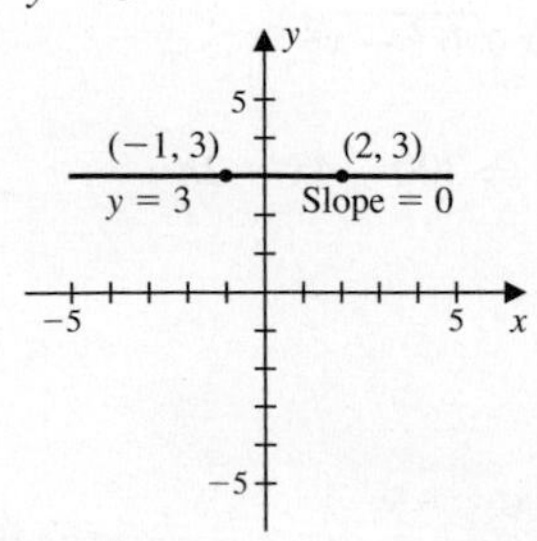

7. $x = -1$

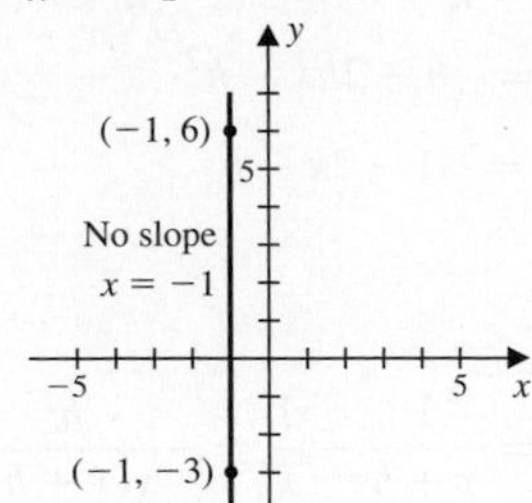

9. $m = 1$

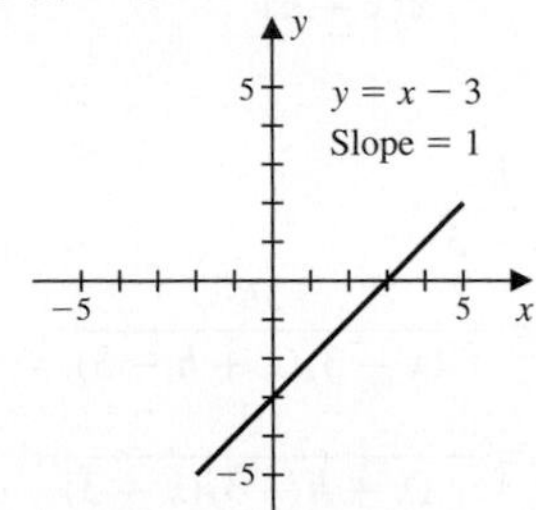

11. $m = 3$

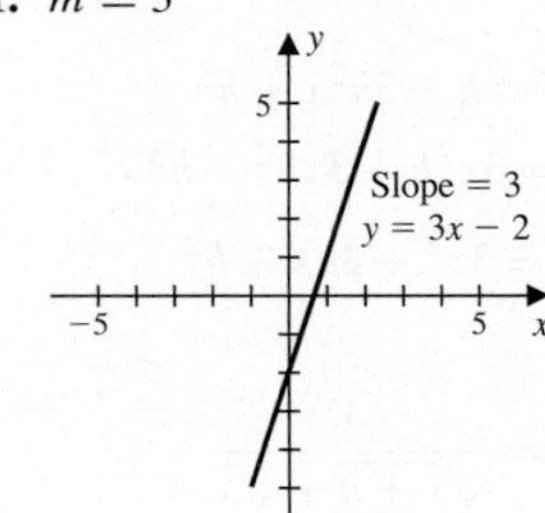

13. $m = -\frac{3}{4}$

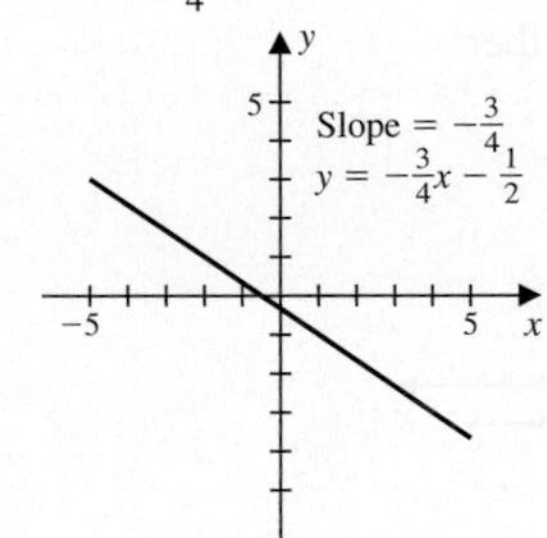

15. $m = 0$

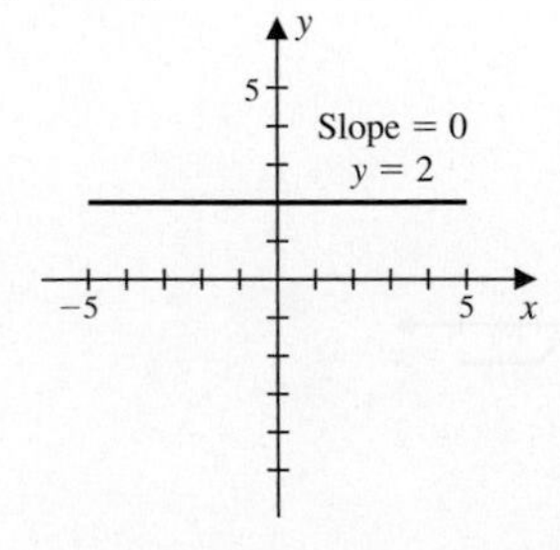

17. a. $y = x + 3$

b. $y = -x + 5$

c. $y = 3x + 1$

d. $y = \frac{1}{3}x + \frac{11}{3}$

19. Parallel lines: (a) and (f); (c) and (g); (b), (d), (i), and (j); and (e) and (h).

21. a. $y = 2x$ **b.** $y = -\frac{1}{2}x$

23. a. $y = -2x$

b. $y = \frac{1}{2}x + \frac{5}{2}$

25. $y = -x + 2$

27. $y = 3x - 5$

29. $y = -3x + 3$

31. $y = -1$

33. $y = \frac{2}{3}x + \frac{1}{3}$

35. $y = x$

37. a. $y = -\frac{\sqrt{2}}{2}x + \frac{3\sqrt{2}}{2}$

b. $(-1, -\sqrt{2})$

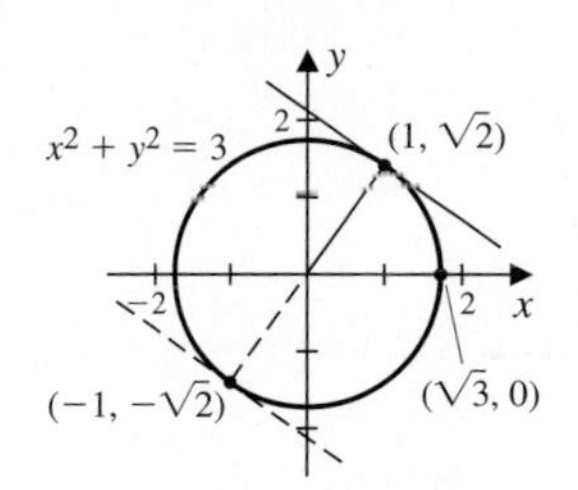

39. The velocity is -224 ft/s.

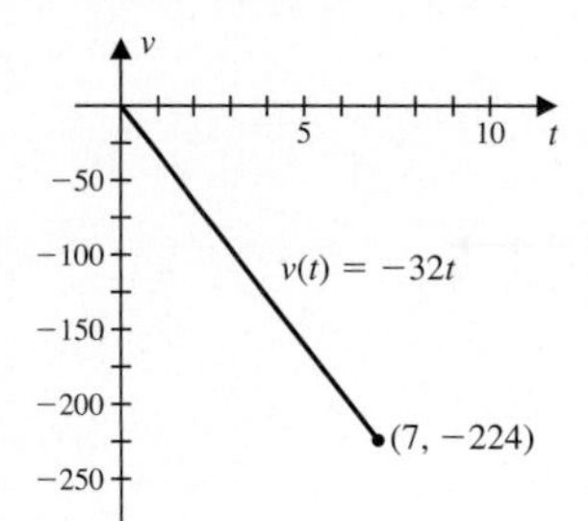

41. a.

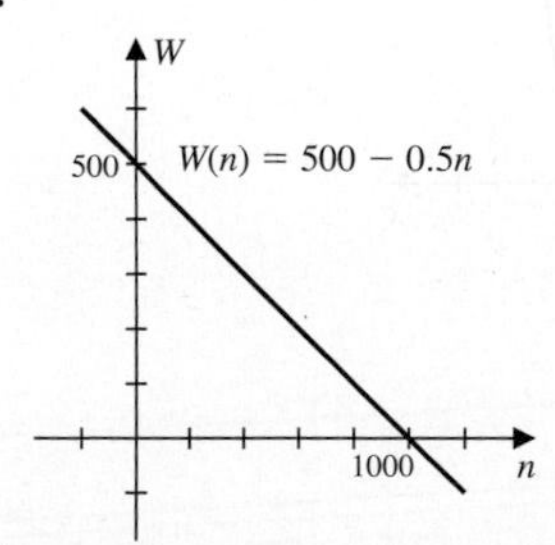

b. The total weight is $500n - 0.5n^2$.

c. There are no fish when $n \geq 1000$.

43. a. $V(t) = 28{,}000 - 4000t$ dollars

b. $V(5) \approx \$8000$.

c. In 7 years.

Exercise Set 1.8 (page 73)

1.

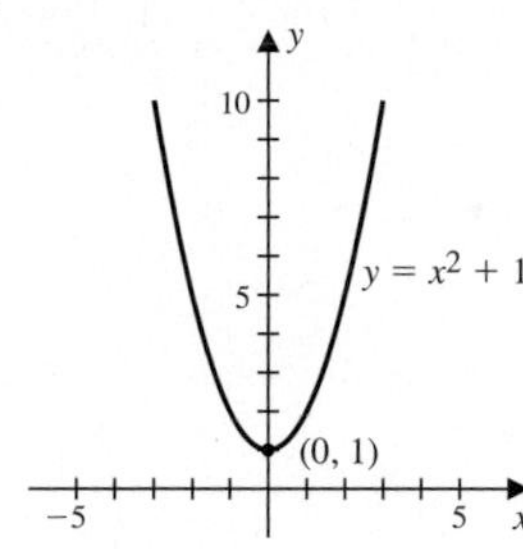

3.

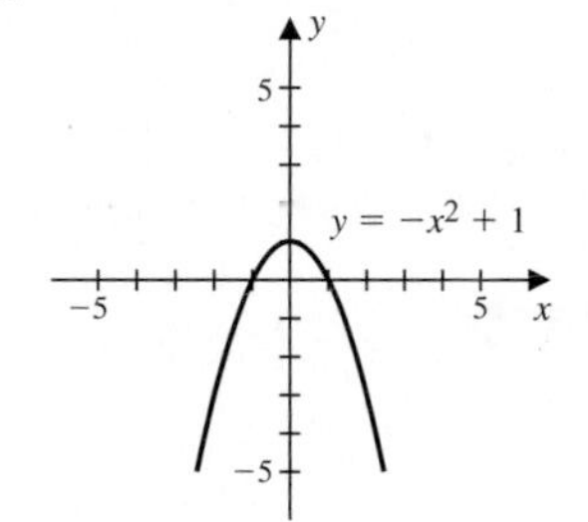

5.

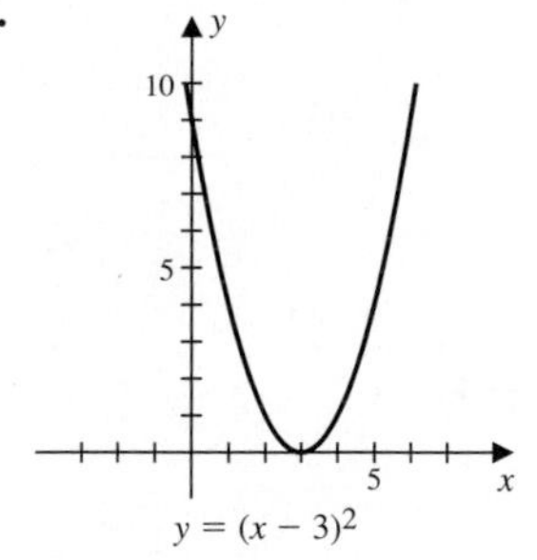

7.

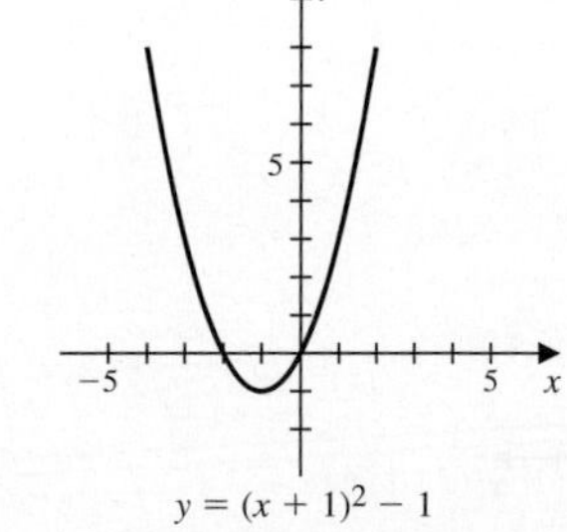

9.

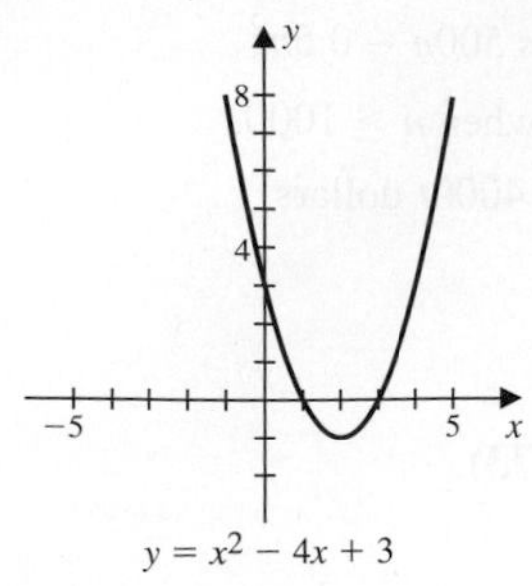

11.

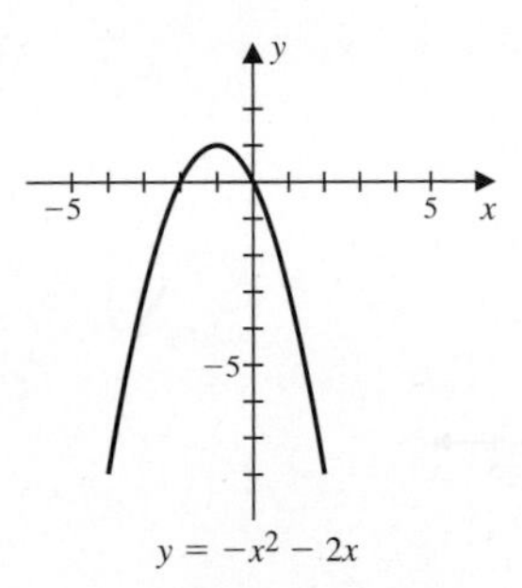

13.

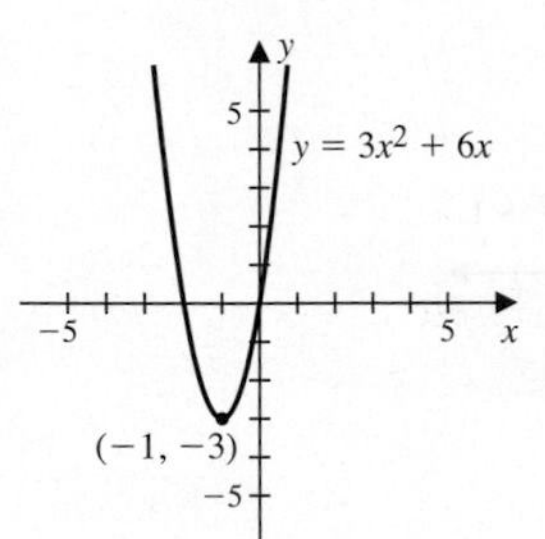

15.

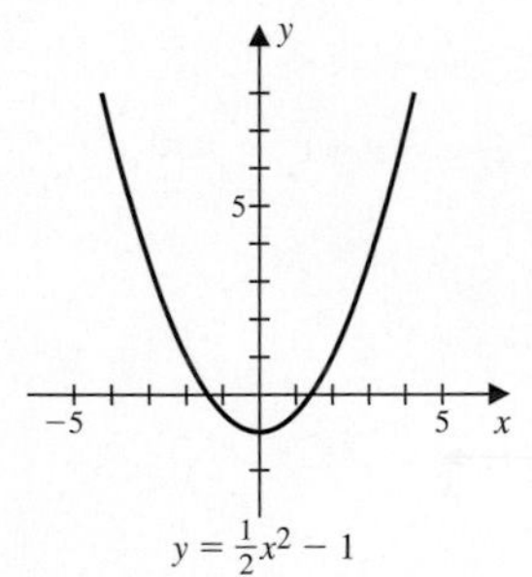

17.

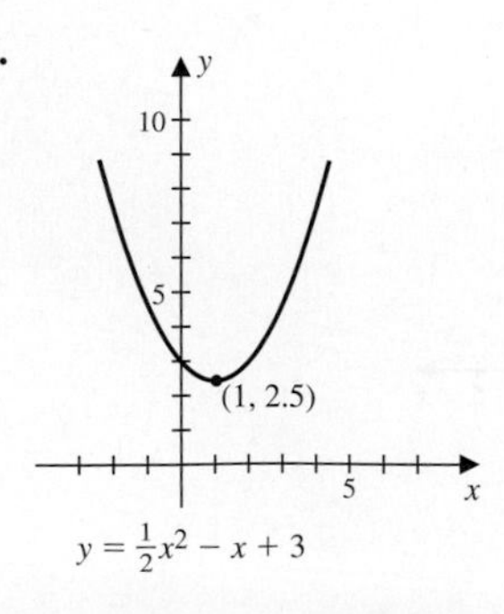

19. **a.** $(x-3)^2-2$

b. $x = 3 \pm \sqrt{2}$, $y = 7$

c. Minimum: -2 at $x = 3$

21. **a.** $-(x-2)^2+10$

b. $x = \sqrt{10}+2$, $x = -\sqrt{10}+2$, $y = 6$

c. Maximum: 10 at $x = 2$

23. $x = \frac{1}{2}$ and $x = \frac{1}{3}$

25. $x = 1 + \frac{\sqrt{2}}{2}$ and $x = 1 - \frac{\sqrt{2}}{2}$

27. no x-intercepts

29. **a.**

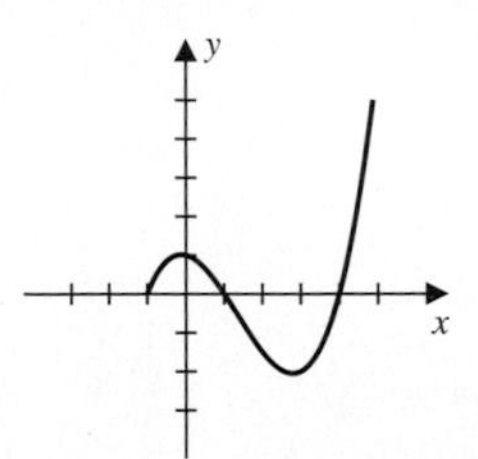

b.

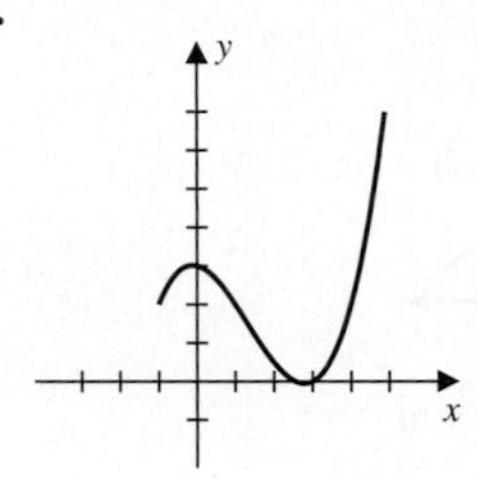

c.

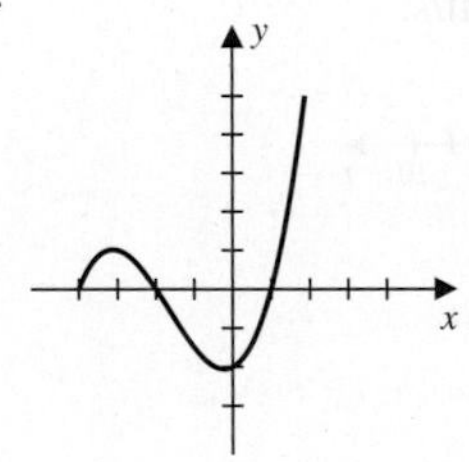

d.

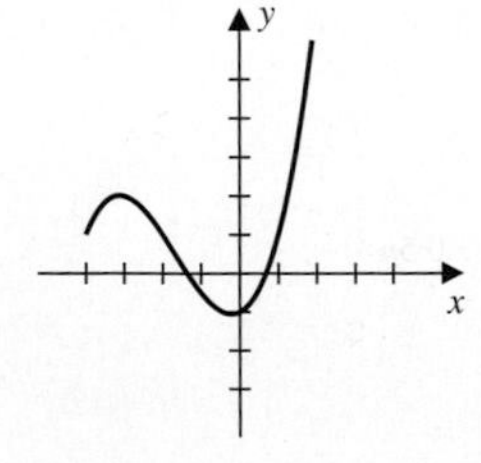

e.

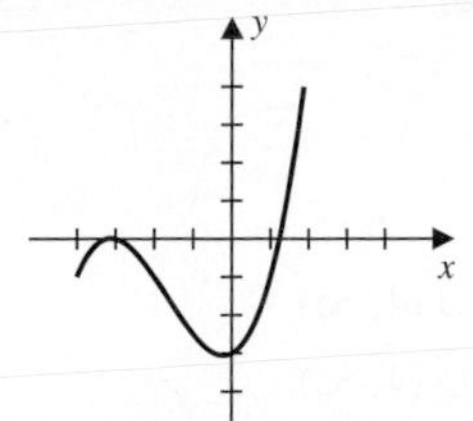

f.

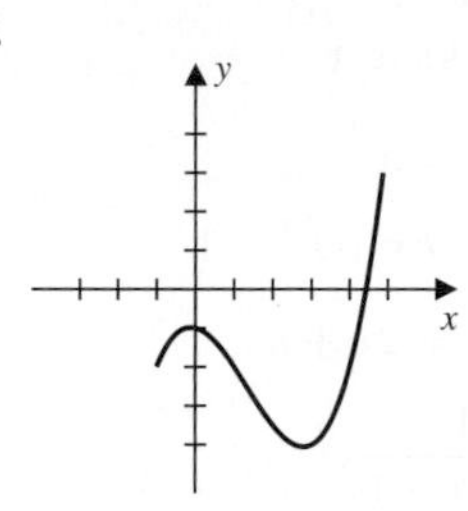

31. $y = \frac{2}{9}(x-1)^2 + 3$

33. **a.** The domain of $f(x) = \sqrt{x^2 - 3}$ is $(-\infty, -\sqrt{3}] \cup [\sqrt{3}, \infty)$.

b. The domain of $f(x) = \sqrt{x^2 - 1/2x}$ is $(-\infty, 0] \cup [\frac{1}{2}, \infty)$.

35. **a.** $a + b + c = 1$

b. $c = 6$

c. $b = -2a$, and $c = a + 1$

d. $a = 5$, $b = -10$, $c = 6$, and $f(x) = 5x^2 - 10x + 6$

37. **a.**

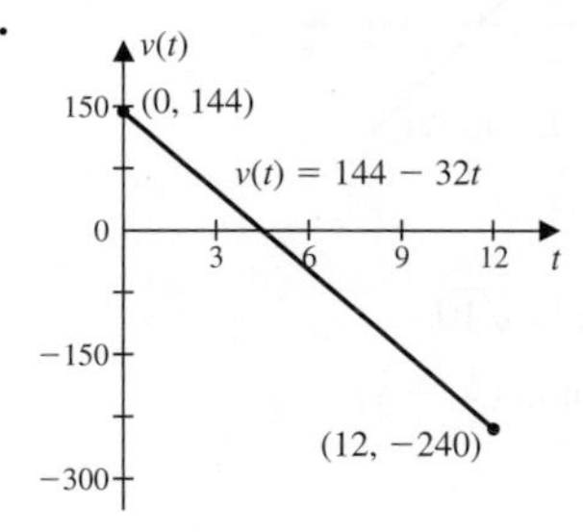

b. $v(t) = -240$ ft/s

c. The domain is $[0, 12]$, and the range is $[-240, 144]$.

39. **a.** The company should produce 800 terminals.

b. The maximum profit is $44,000.

41. **a.** $2x(4 - x^2) = 8x - 2x^3$

b. width ≈ 2.3, height ≈ 2.7

43. $R = kP(M - P)$ where k is the constant of proportionality.

45. **a.**

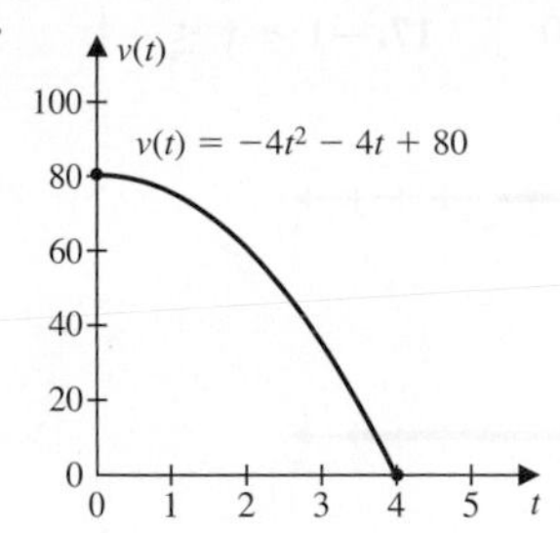

b. The car comes to rest after $t = 4$ seconds.

c.

t	0	$\frac{1}{2}$	1	$\frac{3}{2}$	2	$\frac{5}{2}$	3	$\frac{7}{2}$	4
$v(t)$	80	77	72	65	56	45	32	17	0

d. slowest: 77 feet/second; fastest: 80 feet/second

e.

	slowest	fastest
second 1/2 sec	72	77
third 1/2 sec	65	72

f.

	minimum distance	maximum distance
first 1/2 sec	$77 \times 1/2 = 38.5$ ft	$80 \times 1/2 = 40$ ft
second 1/2 sec	$72 \times 1/2 = 36$ ft	$77 \times 1/2 = 38.5$ ft

g. Lower bound:182; upper bound: 222

Review Exercises for Chapter 1 (page 76)

1. **a.** $-1 \le x \le 7$

b. −1 0 7 x

3. **a.** $-\infty < x < 7$

b. 0 7 x

5. **a.** $(-4, \infty)$

b. −4 0 x

7. **a.** $[2, 10)$

b. 0 2 10 x

9. $x \ge \frac{1}{2}$

11. $x \le -2$ or $x \ge 0$

13. $-2 \le x \le 1$ or $x \ge 2$

15. $x < -3$ or $x > 0$

17. $-1 < x \leq -\frac{1}{4}$

19. **a.** $(-1, 4)$

b. −1 0 4 x

21. **a.** $[-1, 7]$

b. −1 0 7 x

23.

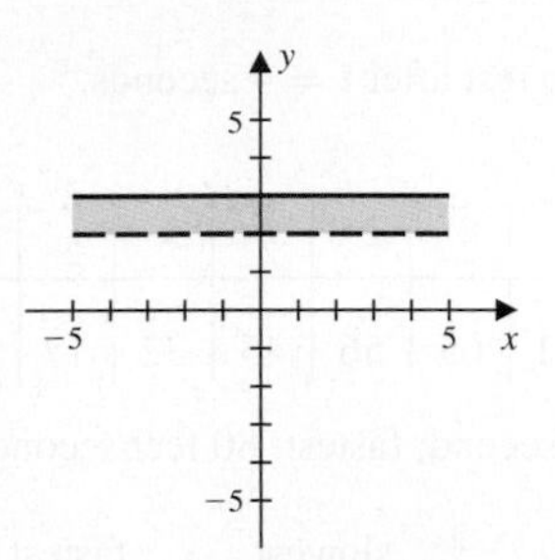

25.

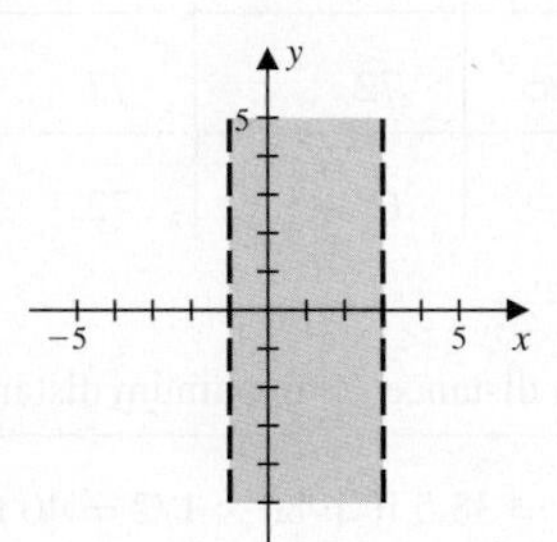

27.

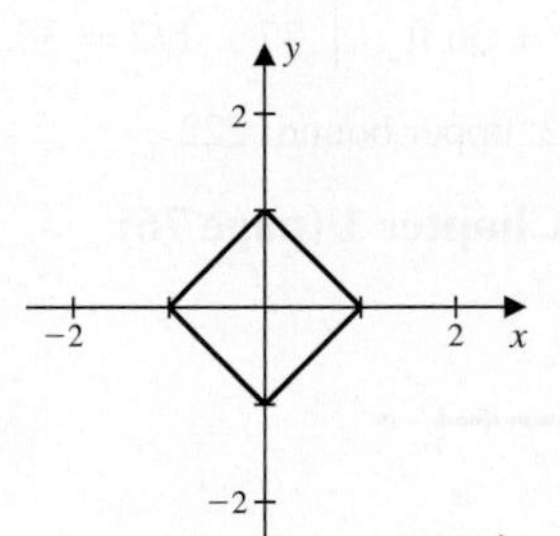

29.

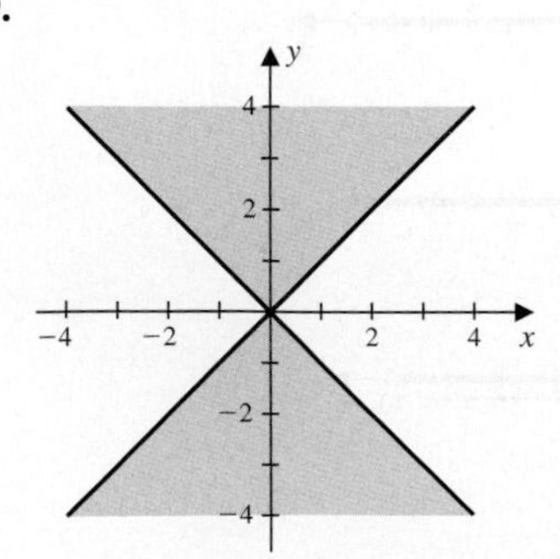

31. **a.** Domain: $(-\infty, \infty)$

b. Range: $[-3, \infty)$

33. **a.** Domain: $[2, \infty)$

b. Range: $[2, \infty)$

35. **a.** $(-\infty, 2) \cup (2, 4) \cup (4, \infty)$

b. $(-\infty, 2) \cup (2, 4) \cup (4, \infty)$

c. $(-\infty, 2) \cup (4, \infty)$

37. **a.** $f(x+h) = 5x + 5h + 3$

b. $\dfrac{f(x+h) - f(x)}{h} = 5$

39. **a.** $f(x+h) = x^2 + 2hx + h^2 - 1$

b. $\dfrac{f(x+h) - f(x)}{h} = 2x + h$

41. **a.** $f(x+h) = \dfrac{1}{x+h-1}$

b. $\dfrac{f(x+h) - f(x)}{h} = \dfrac{-1}{(x+h-1)(x-1)}$

43. **a.** ii **b.** v **c.** vi

d. iv **e.** i **f.** iii

45. **b.** The distance is $3\sqrt{2}$.

c. The midpoint is $(\frac{5}{2}, -\frac{1}{2})$.

e. $y = -x + 2$

a, d.

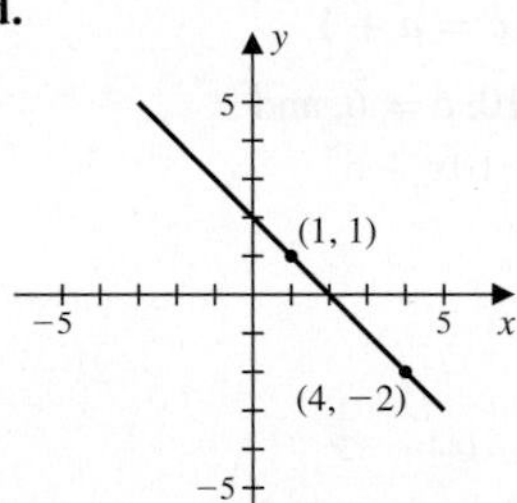

47. **b.** The distance is $\sqrt{10}$.

c. The midpoint is $(\frac{1}{2}, -\frac{5}{2})$.

e. $y = -\frac{1}{3}x - \frac{7}{3}$

a, d.

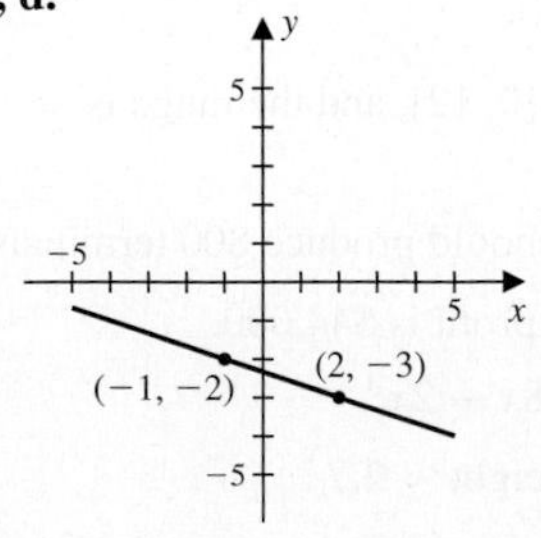

49. **a.** $y = 4x$

b. $y = -\frac{1}{4}x$

51. **a.** $y = -\frac{7}{5}x - \frac{22}{5}$

b. $y = \frac{5}{7}x - \frac{16}{7}$

53. **a.**

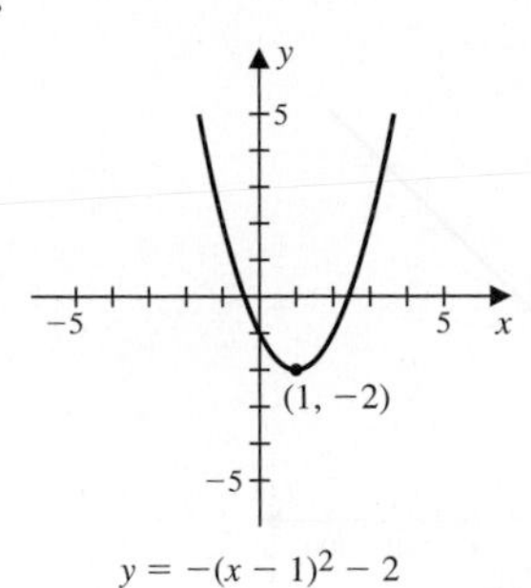

b. $[-2, \infty)$

c. The x-intercepts are at $(1 - \sqrt{2}, 0)$ and $(1 + \sqrt{2}, 0)$, and the y-intercept is at $(0, -1)$.

d. The minimum is -2 at $x = 1$.

55. **a.**

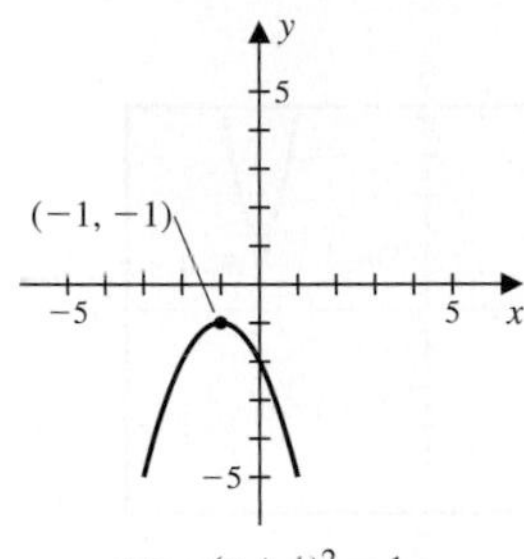

b. $(-\infty, -1]$

c. There are no x-intercepts, and the y-intercept is at $(0, -2)$.

d. The maximum is -1 at $x = -1$.

57. **a.**

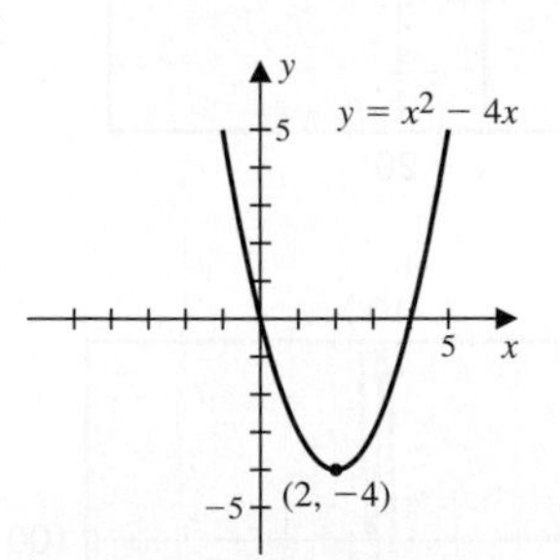

b. Range: $[-4, \infty)$

c. The x-intercepts are at $(0, 0)$ and $(4, 0)$, and the y-intercept is at $(0, 0)$.

d. The minimum is -4 at $x = 2$.

59. **a.**

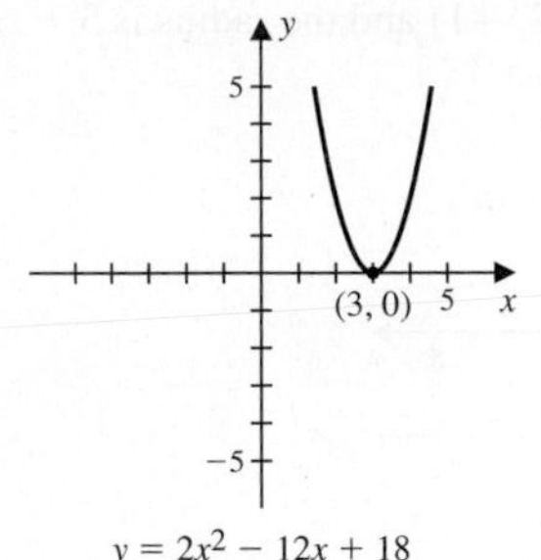

b. Range: $[0, \infty)$

c. The x-intercept is at $(3, 0)$, and the y-intercept is at $(0, 18)$.

d. The minimum is 0 at $x = 3$.

61. **a.**

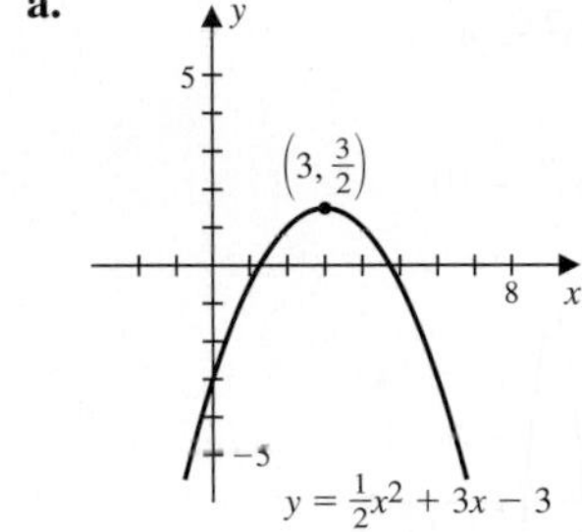

b. Range: $(-\infty, \frac{3}{2}]$

c. The x-intercepts are at $(3 \pm \sqrt{3}, 0)$, and the y-intercept is at $(0, -3)$.

d. The maximum is $\frac{3}{2}$ at $x = 3$.

63. The center is $(0, 0)$ and the radius is 4.

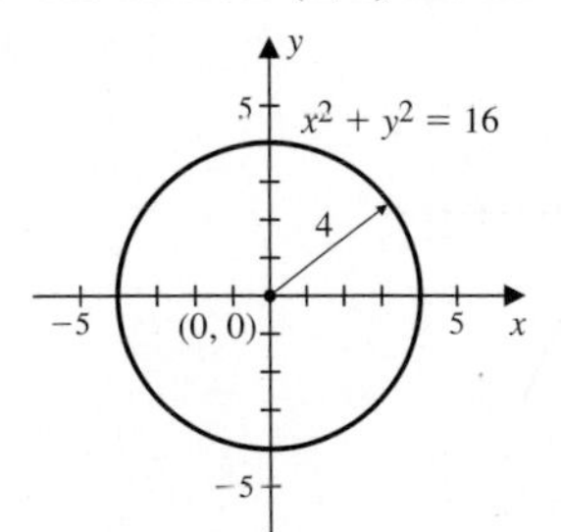

65. The center is $(-2, 1)$ and the radius is 3.

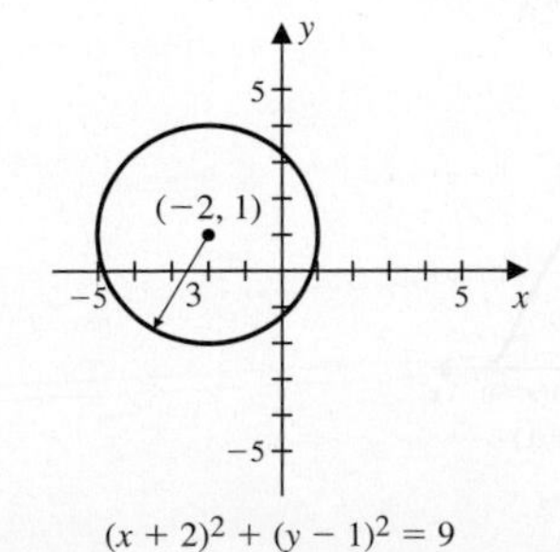

67. The center is $(-2, -1)$ and the radius is 3.

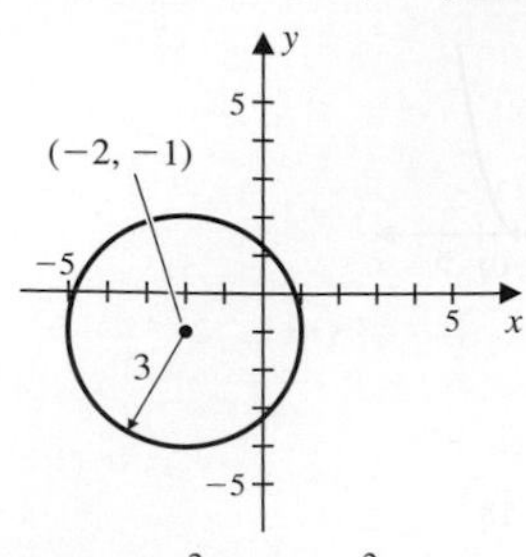

69. The fourth vertex is $(1, 6)$.

71. The line through $(-2, -1)$ and perpendicular to $y - 2x = 2$ is $y = -\frac{1}{2}x - 2$. The lines intersect at $(0, -2)$ and the distance from $(-2, -1)$ to $(0, -2)$ is $\sqrt{5}$.

73. $(x - 4)^2 + (y - 2)^2 = 4$

75. $\left(\frac{21}{10}, 0\right)$

77. **a.** $y = -4(x - 3)^2 + 10$

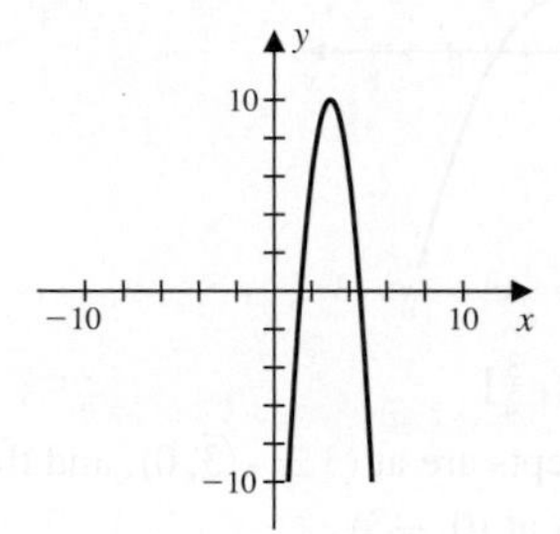

b. $y = -2x + 1$

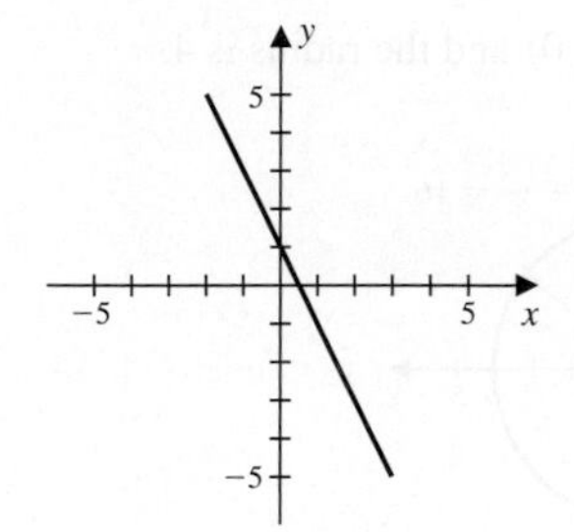

79. $43.00

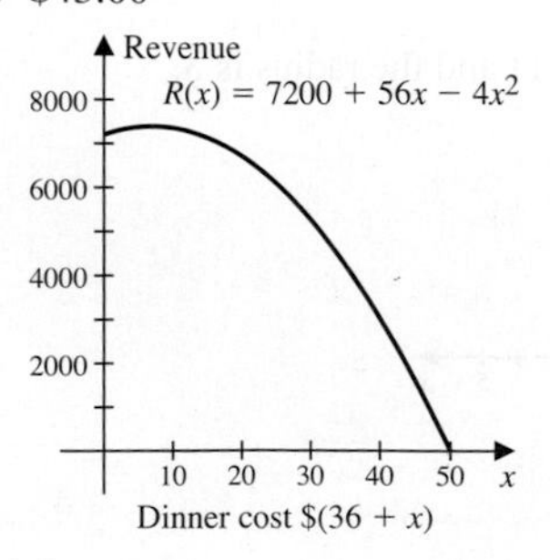

81. **a.** $F(l) = 3l + (864/l)$

b. The domain of F is $(0, \infty)$.

83. **a.** $y = 16x + 1600$

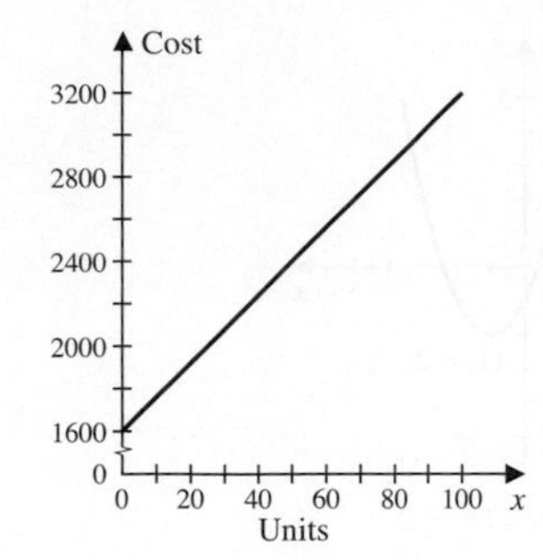

b. The slope tells how much it costs to produce one unit once production has begun.

c. The y-intercept is 1600, and it tells how much it costs to begin production.

85. Part (d) appears to give the best representation.

a.

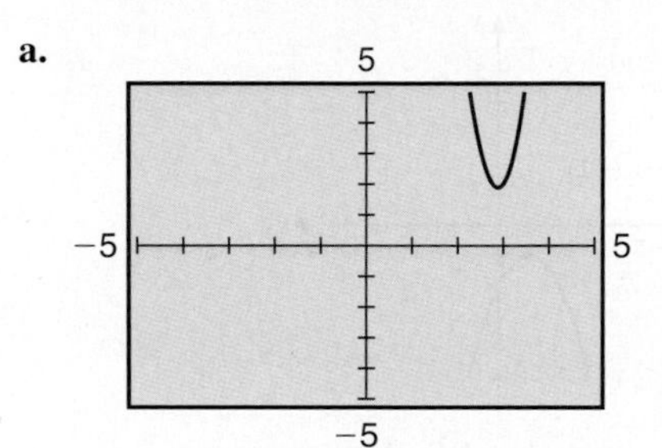

b.

20
−20
20
−20

c.

100
−100
100
−100

d.

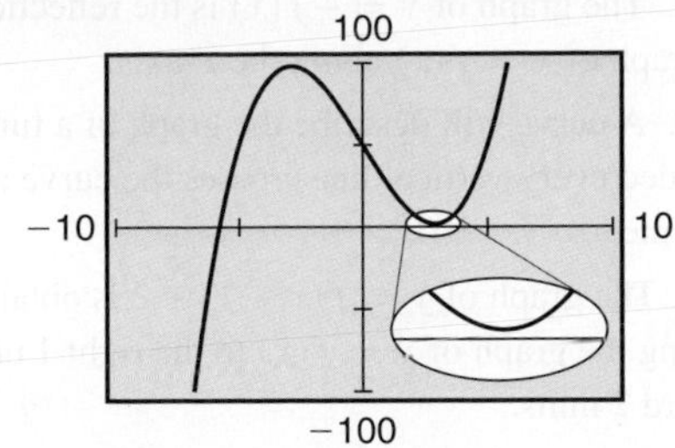

87. **a.** between A and B and to the right of E

b. between C and D

c. between B and C

d. between D and E

Chapter 1: Exercises for Calculus (page 79)

1. $f(x) = \frac{1}{2}x^2$; $g(x) = 3x + 8$

3. **a.** $14 \le n \le 16$

b. $8 \le n \le 9$

5. The graph of the temperature might be as follows:

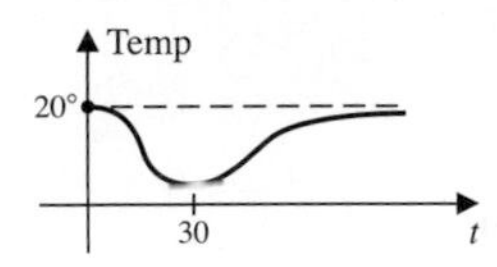

7. Graphs for the airplane are shown below.

a.

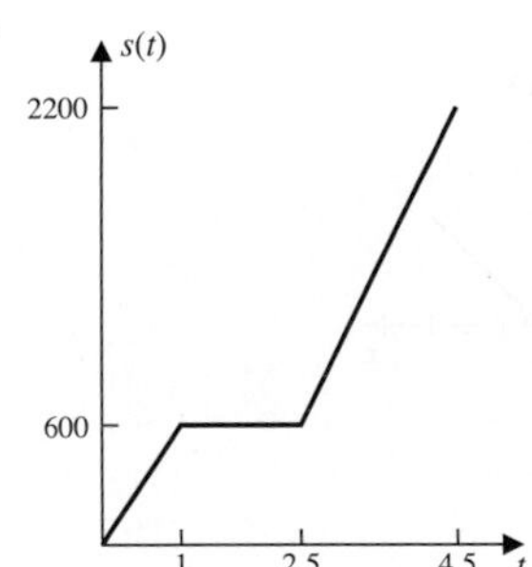

b.

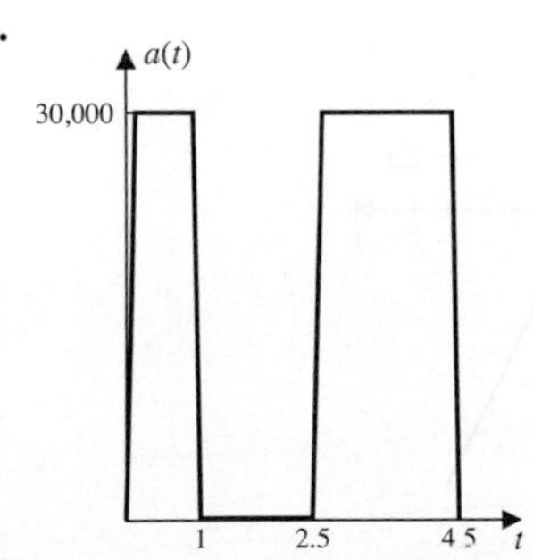

c.

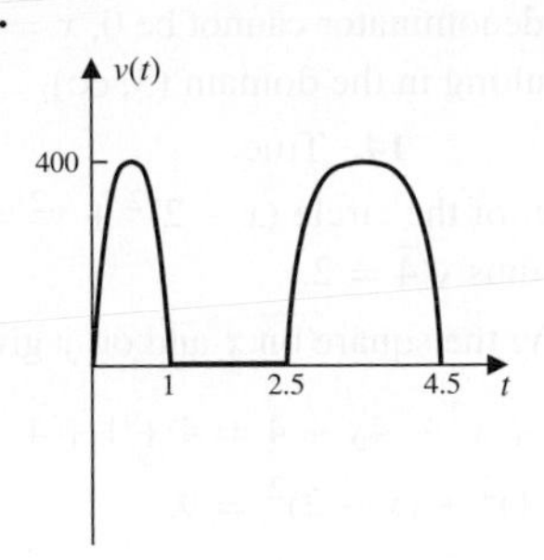

d.

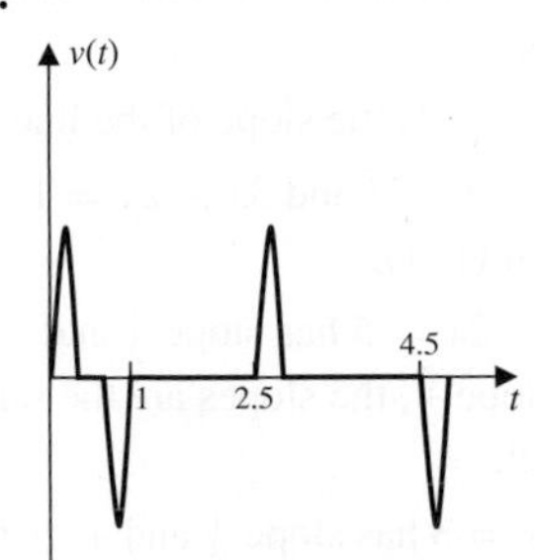

9. **a.** $P(x) = \begin{cases} 300 & \text{if } x \le 100 \\ 400 - x & \text{if } 100 < x \le 150 \\ 225 & \text{if } x > 150 \end{cases}$

b.

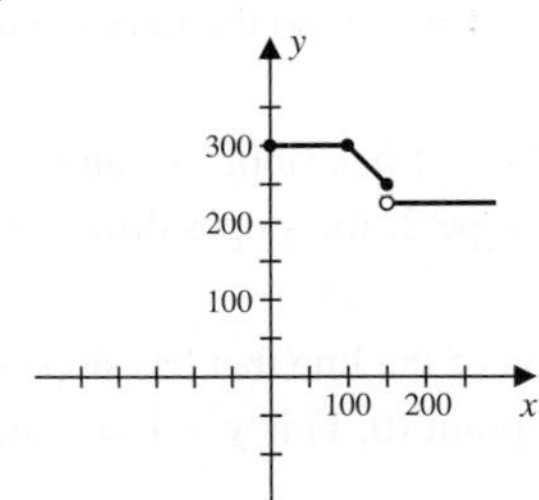

11. **a.** The length of the rod at time t is $L(t) = 2 + 2(11 \times 10^{-6})t = 2 + (22 \times 10^{-6})t$.

b. $L(1000) = 2 + 22 \times 10^{-3}$

Chapter 1: Chapter Test (page 81)

1. False. The solution to the equation $3x - 2 = 4$ is $x = 2$.

2. False. Since $x^2 - 3x + 2 = (x - 1)(x - 2)$, the solutions are $x = 2$ and $x = 1$.

3. True. **4.** True. **5.** True. **6.** True.

7. False. Another solution is $x = 2$.

8. True. **9.** True. **10.** True.

11. False. Since $x - 3 \ge 0$ implies $x \ge 3$, the domain is $[3, \infty)$.

12. False. Since the denominator cannot be 0, $x = 3$ must be excluded, resulting in the domain $(3, \infty)$.

13. True. **14.** True.

15. False. The center of the circle $(x - 2)^2 + y^2 = 4$ is $(2, 0)$, but the radius $\sqrt{4} = 2$.

16. False. Completing the square on x and on y gives

$$x^2 + 2x + 1 + y^2 - 4y + 4 = 4 + 1 + 4 \quad \text{or}$$
$$(x + 1)^2 + (y - 2)^2 = 9.$$

This is the equation of a circle with radius 3, but the center is at $(-1, 2)$.

17. False. Since $y = \frac{2}{3}x - \frac{4}{3}$, the slope of the line is $\frac{2}{3}$.

18. False. The lines $x + y = 2$ and $3x - 2y = 1$ intersect at the point $(1, 1)$.

19. False. Since $-3x + 2y = 5$ has slope $\frac{3}{2}$ and $4y = 6x + 7$ has slope $\frac{3}{2}$, the slopes are the same and the lines are parallel.

20. False. Since $x - 3y = 3$ has slope $\frac{1}{3}$ and $4x - 6y = 5$ has slope $\frac{2}{3}$, the slopes differ and the lines are not parallel.

21. False. Since $2x + y = 2$ has slope -2 and $2y + x = -1$ has slope $-\frac{1}{2}$, the products of the slopes $(-2)(-\frac{1}{2}) = 1 \neq -1$, so the lines are not perpendicular.

22. False. Since $x + 2y = 1$ has slope $-\frac{1}{2}$ and $-2x + y = 3$ has slope 2, the slopes differ and the lines are not parallel.

23. False. The equation of the line that has slope -3 and passes through the point $(0, 1)$ is $y - 1 = -3(x - 0)$, or $y = -3x + 1$.

24. False. The equation of the line that passes through the two points $(2, 1)$ and $(-3, 2)$ has slope $(2 - 1)/(-3 - 2) = -\frac{1}{5}$ and equation $5y + x = 7$.

25. False. Since $y = 1$ is not included, the shaded region is described by $y < 1$.

26. False. Since $y = 1$ is included, the region outside the shaded region is described by $y \geq 1$.

27. True. **28.** True. **29.** True.

30. True. **31.** True.

32. False. The parabola $y = -(x + 1)^2 - 2$ has a maximum point at $(-1, -2)$.

33. True. **34.** True. **35.** True. **36.** True.

37. False. The graph of an even function is symmetric with respect to the y-axis and the graph of an odd function is symmetric with respect to the origin.

38. False. The graph of $y = -f(x)$ is the reflection of the graph of $y = f(x)$ about the x-axis.

39. False. A curve will describe the graph of a function provided every vertical line crosses the curve at most one time.

40. False. The graph of $y = f(x - 1) + 2$ is obtained by shifting the graph of $y = f(x)$ to the right 1 unit and upward 2 units.

41. True.

42. True.

43. True.

44. True.

45. True.

Chapter 2

Exercise Set 2.2 (page 91)

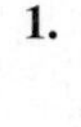
1.

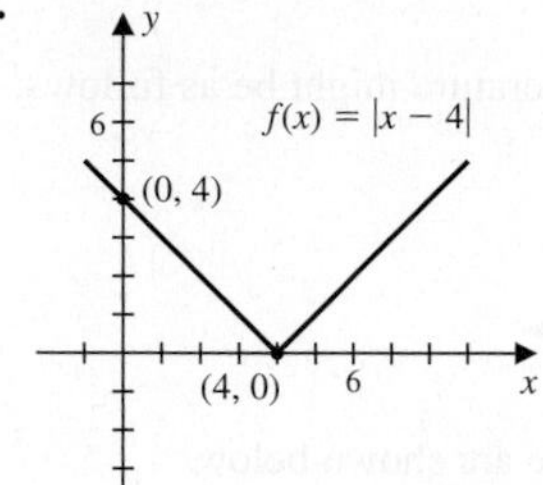

3.

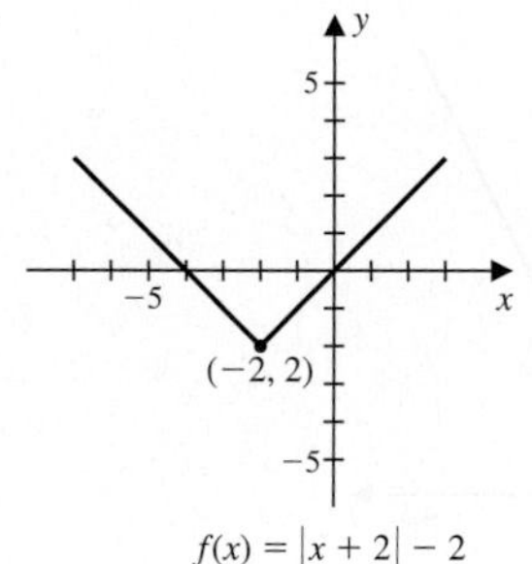

5.

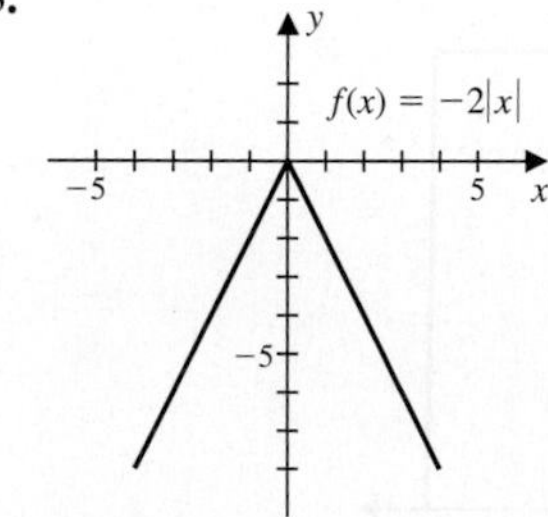

7.

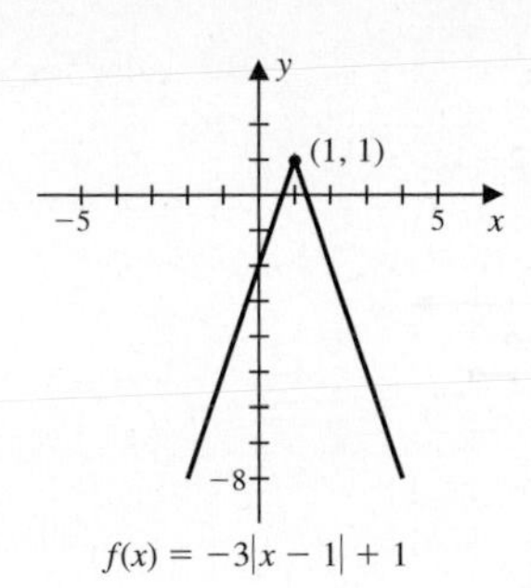

9. a.

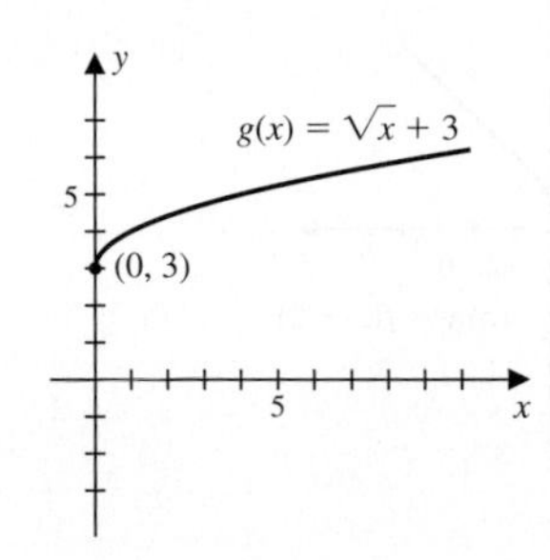

b. Domain: $[0, \infty)$; range: $[3, \infty)$

11. a.

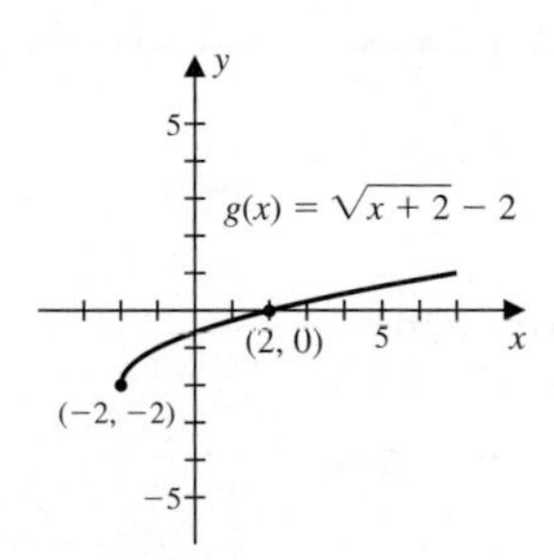

b. Domain: $[-2, \infty)$; range: $[-2, \infty)$

13. a.

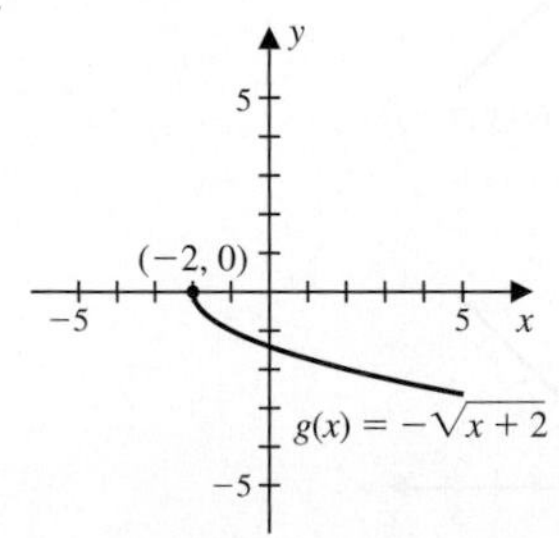

b. Domain: $[-2, \infty)$; range: $(-\infty, 0]$

15. a.

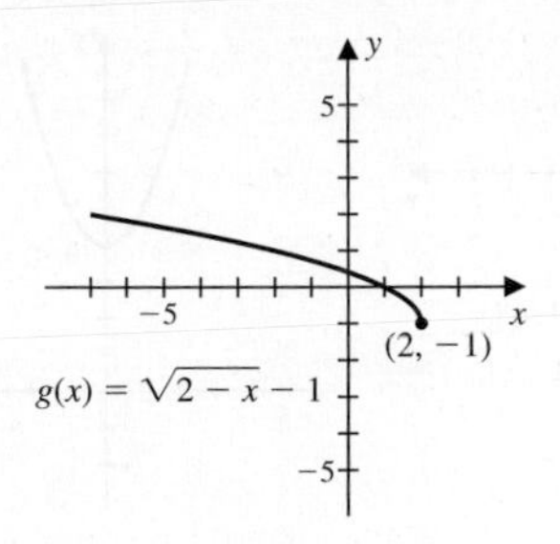

b. Domain: $(-\infty, 2]$; range: $[-1, \infty)$

17. a, b.

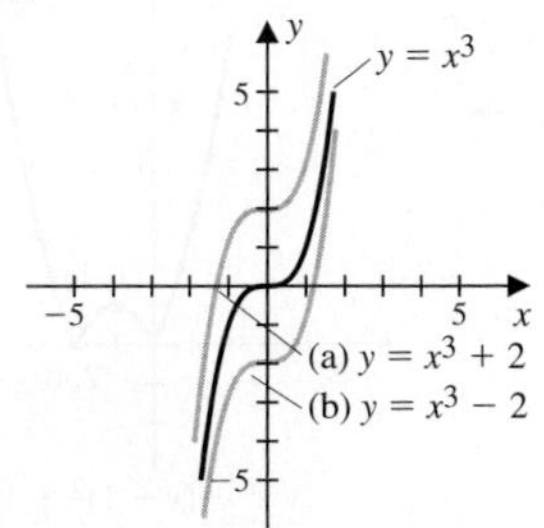

c, d.

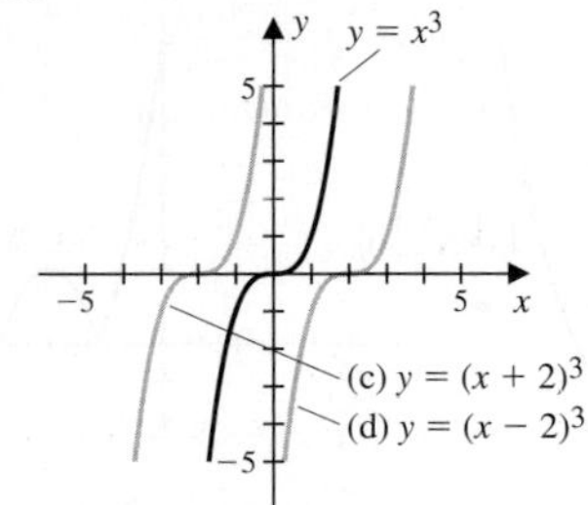

e, f.

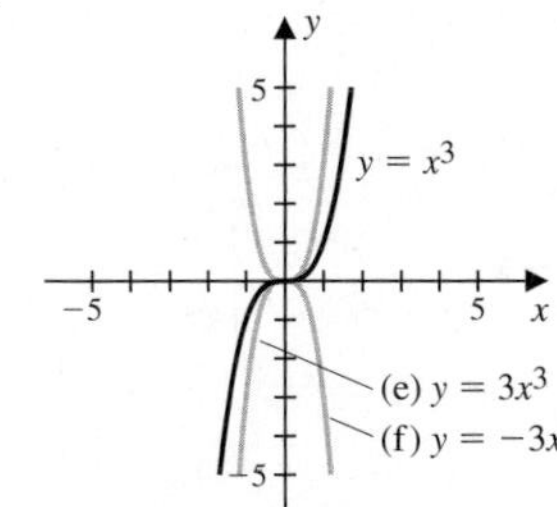

19.

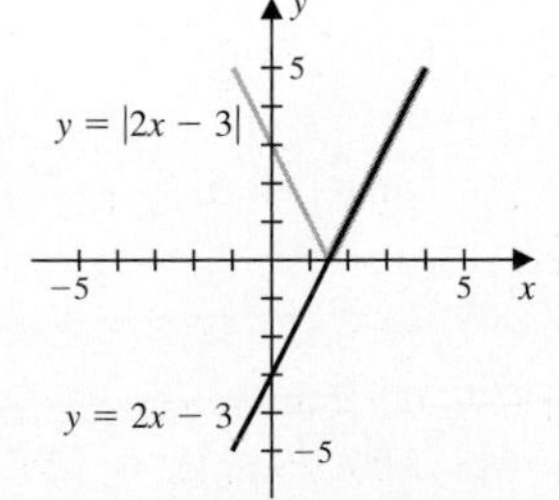

21.

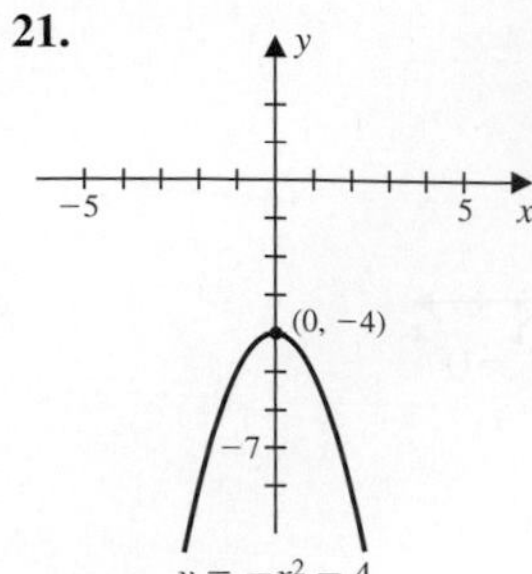

$y = -x^2 - 4$

(0, 4)

$y = |-x^2 - 4|$

23.

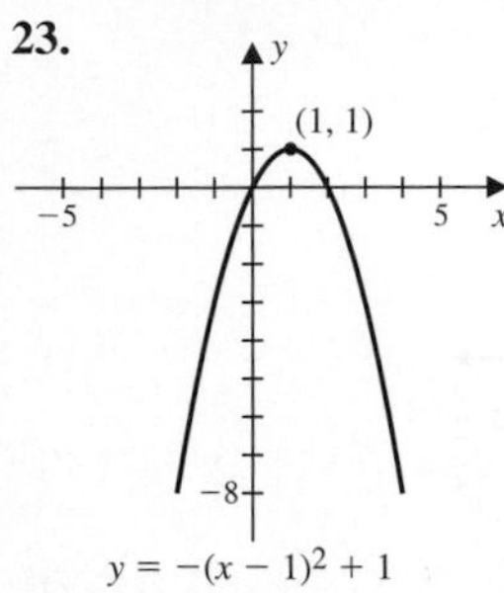

$y = -(x - 1)^2 + 1$

(1, 1)

(2, 0)

$y = |-(x - 1)^2 + 1|$

25.

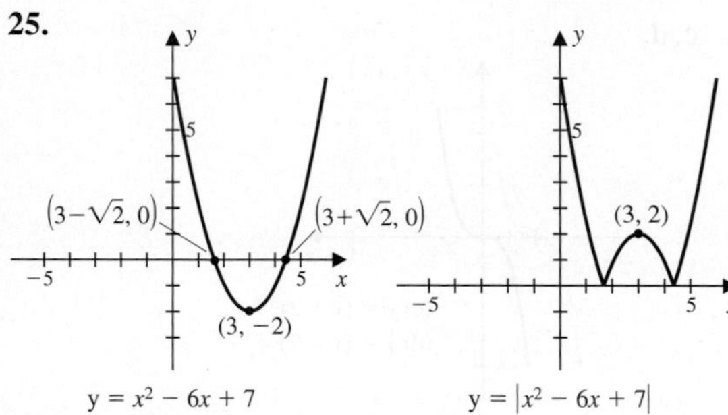

$y = x^2 - 6x + 7$

$y = |x^2 - 6x + 7|$

27.

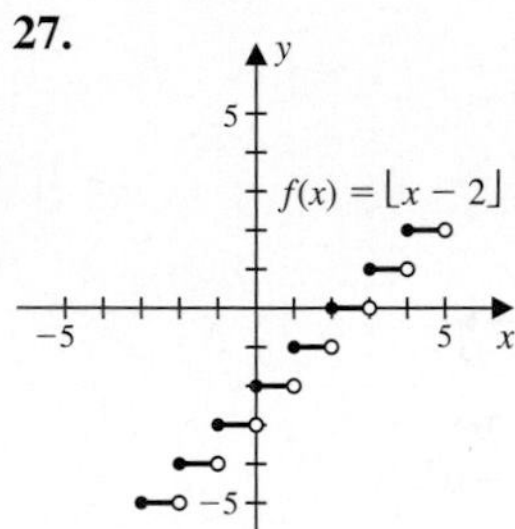

29.

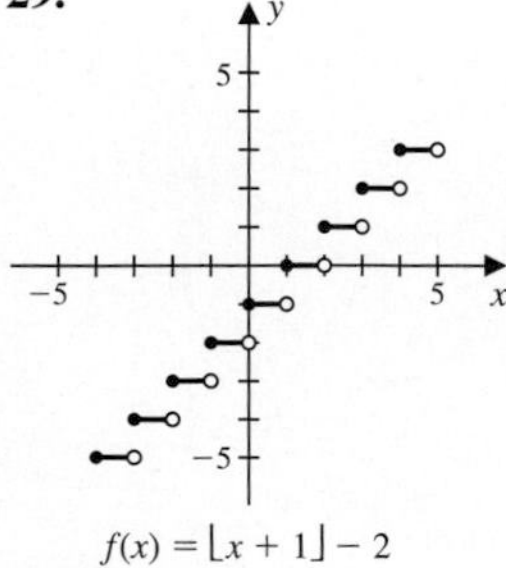

$f(x) = \lfloor x + 1 \rfloor - 2$

31.

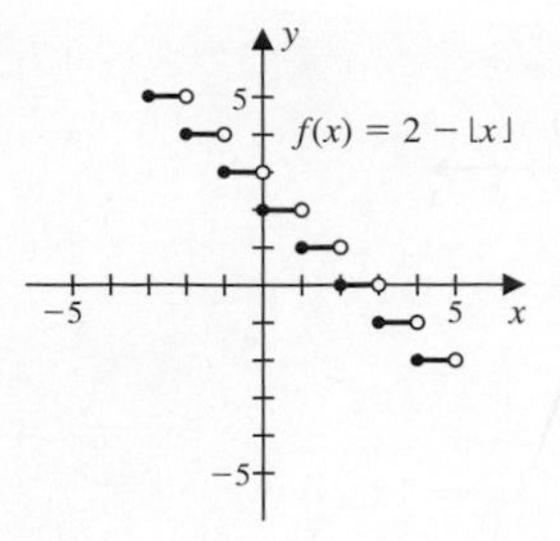

33.

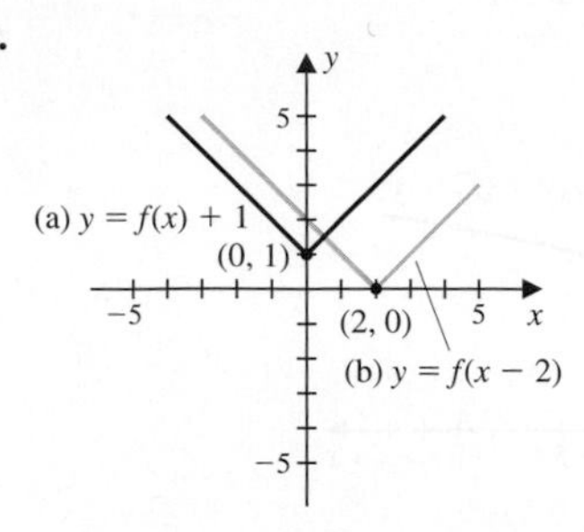

(c) and (d) $y = 2f(x) = f(2x)$

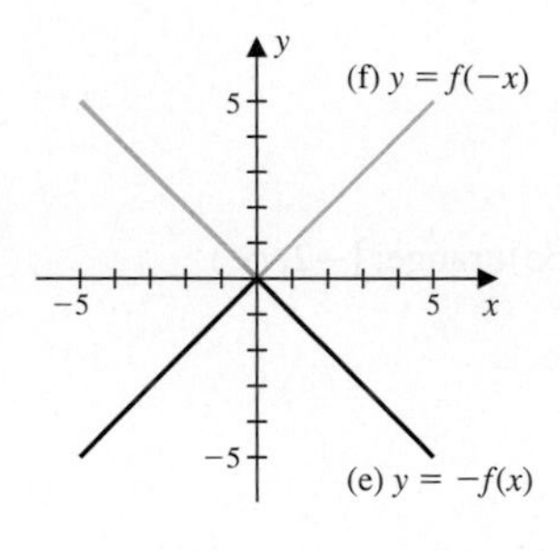

(g) and (h) $y = |f(x)| = f(|x|)$

35.

a.

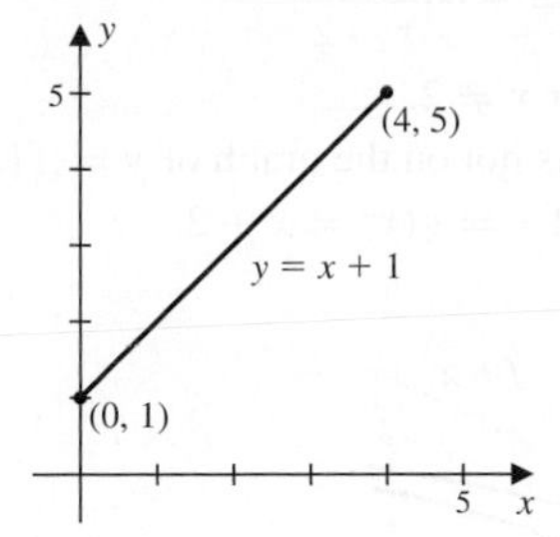

b. $A(t) = \frac{1}{2}t^2 + t$

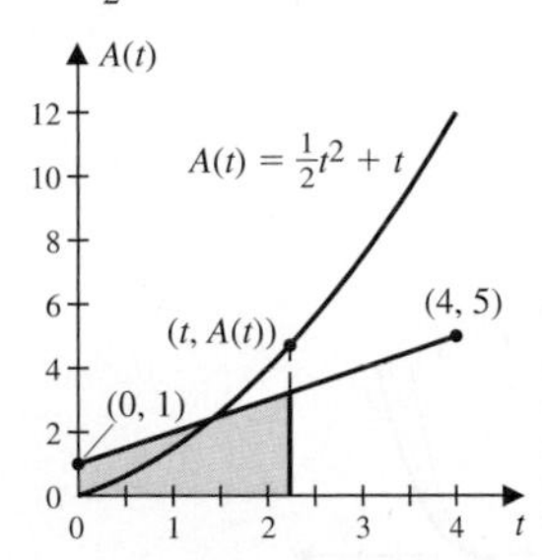

c. $d(x) = \sqrt{2x^2 + 2x + 1}$

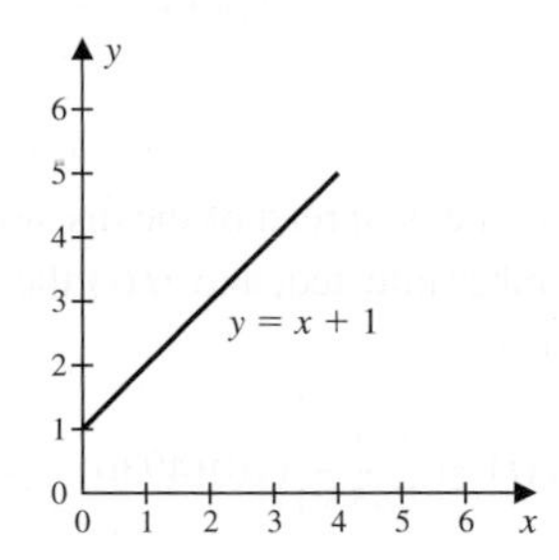

37.

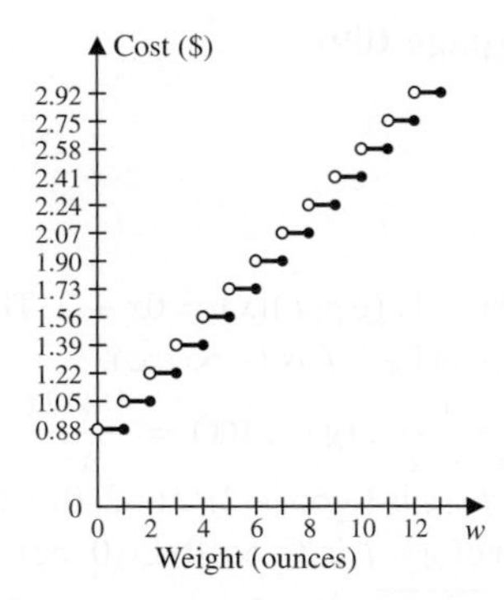

Domain: (0, 13]; range: $\{0.88 + 0.17w,$ $w = 0, 1, 2, \ldots 12\}$

39. **a.** 1.414 **b.** 3.606

43. **a.** \$7.97, rounded to the nearest cent.

b. \$9.11, rounded to the nearest cent.

c. \$12.26, rounded to the nearest cent.

d.

$$C(x) = \begin{cases} 6.58 + 0.0365x & \text{if } 0 \le x \le 50, \\ 7.0050 + 0.028x & \text{if } 50 < x \le 100 \\ 8.405 + 0.014x & \text{if } x > 100. \end{cases}$$

Exercise Set 2.3 (page 100)

1. $(f + g)(x) = x^2 + 2x + 1$; $(f - g)(x) = -x^2 + 2x - 1$; $(f \cdot g)(x) = 2x^3 + 2x$; $(f/g)(x) = 2x/x^2 + 1$: The domain of $f + g$, $f - g$, $f \cdot g$, and f/g is $(-\infty, \infty)$.

3. $(f + g)(x) = \dfrac{1}{x} + \dfrac{x}{x-2}$; $(f - g)(x) = \dfrac{1}{x} - \dfrac{x}{x-2}$; $(f \cdot g)(x) = \dfrac{1}{x-2}$; $(f/g)(x) = \dfrac{x-2}{x^2}$: The domain of $f + g$, $f - g$, $f \cdot g$ and f/g is $(-\infty, 0) \cup (0, 2) \cup (2, \infty)$.

5. $(f + g)(x) = \sqrt{x+1} + \sqrt{3-x}$; $(f - g)(x) = \sqrt{x+1} - \sqrt{3-x}$; $(f \cdot g)(x) = \sqrt{3 + 2x - x^2}$; $(f/g)(x) = \dfrac{\sqrt{x+1}}{\sqrt{3-x}}$: The domain of $f + g$, $f - g$, and $f \cdot g$ is $[-1, 3]$. The domain of f/g is $[-1, 3)$.

7.

$$(f + g)(x) = \begin{cases} 0, & \text{if } x < 0 \\ 1, & \text{if } x \ge 0 \end{cases}$$

$$(f - g)(x) = \begin{cases} -2, & \text{if } x < 0 \\ 1, & \text{if } x \ge 0; \end{cases}$$

$$(f \cdot g)(x) = \begin{cases} -1, & \text{if } x < 0 \\ 0, & \text{if } x \ge 0 \end{cases}$$

$$(f/g)(x) = -1, \quad \text{if } x < 0.$$

The domain of $f + g$, $f - g$, and $f \cdot g$ is $(-\infty, \infty)$. The domain of f/g is $(-\infty, 0)$.

9.

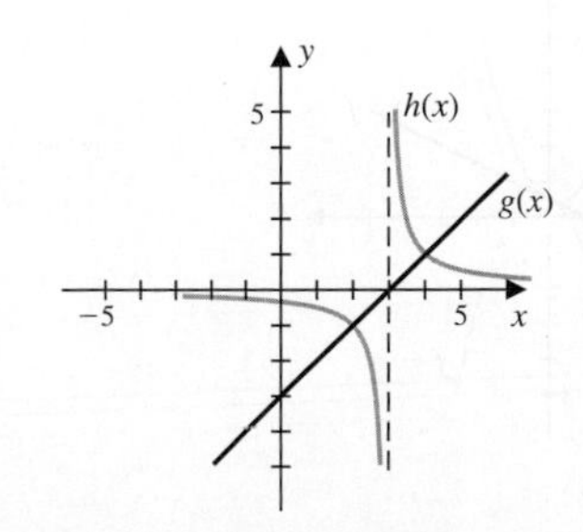

11.

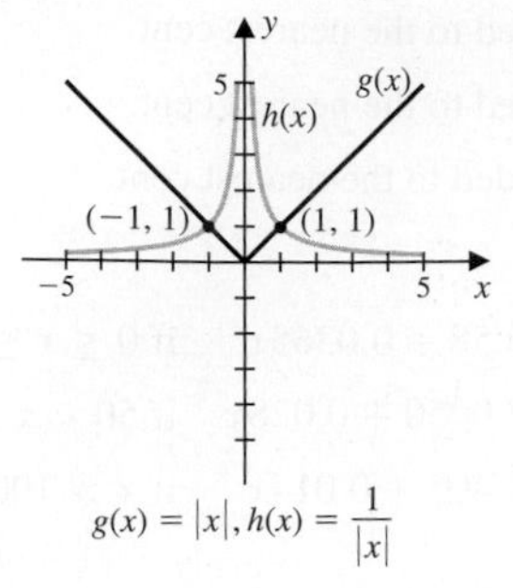

$g(x) = |x|, h(x) = \dfrac{1}{|x|}$

13.

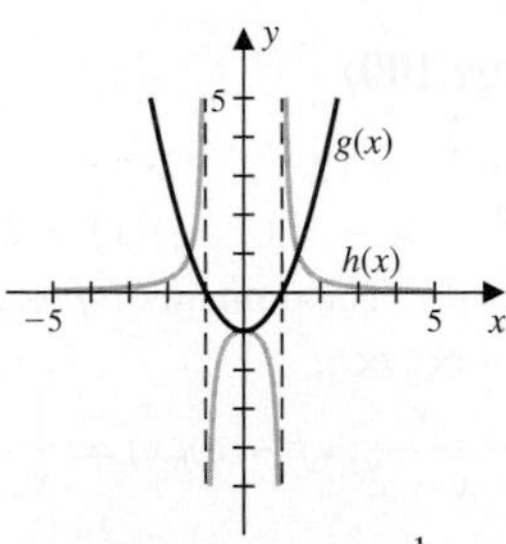

$g(x) = x^2 - 1, h(x) = \dfrac{1}{x^2 - 1}$

15.

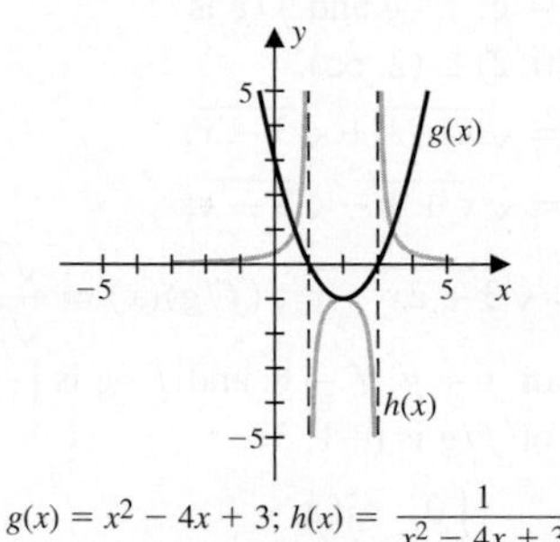

$g(x) = x^2 - 4x + 3;\ h(x) = \dfrac{1}{x^2 - 4x + 3}$

17.

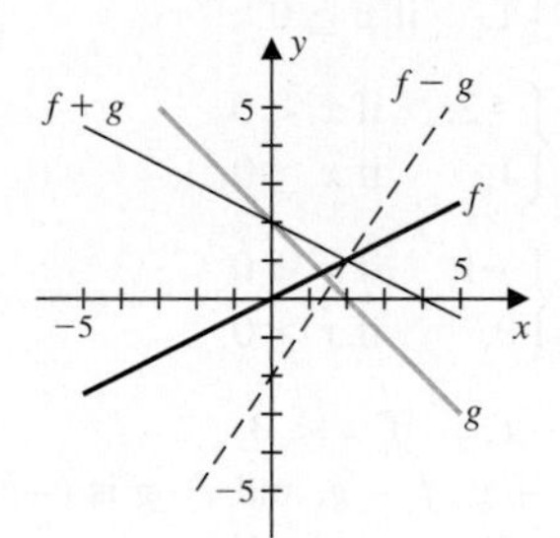

19.

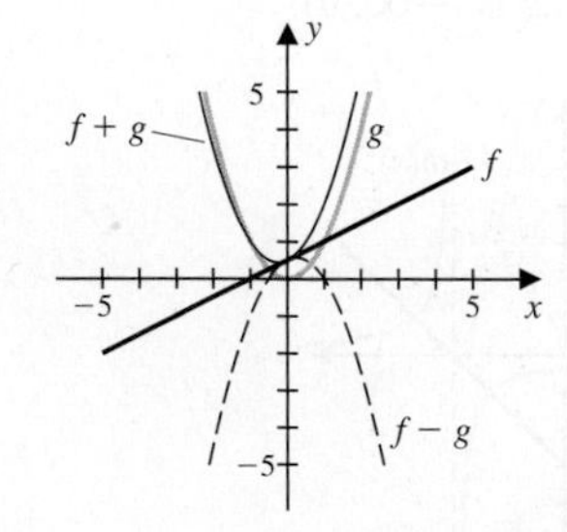

21. $f(x) = \dfrac{x^2 - 4}{x - 2} = \dfrac{(x - 2)(x + 2)}{x - 2}$
$= x + 2$ for $x \neq 2$.
The point $(2, 4)$ is not on the graph of $y = f(x)$, but is on the graph of $y = g(x) = x + 2$.

23.

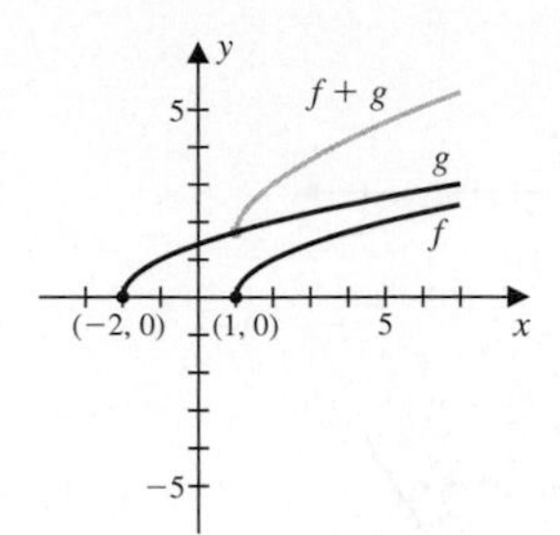

25.

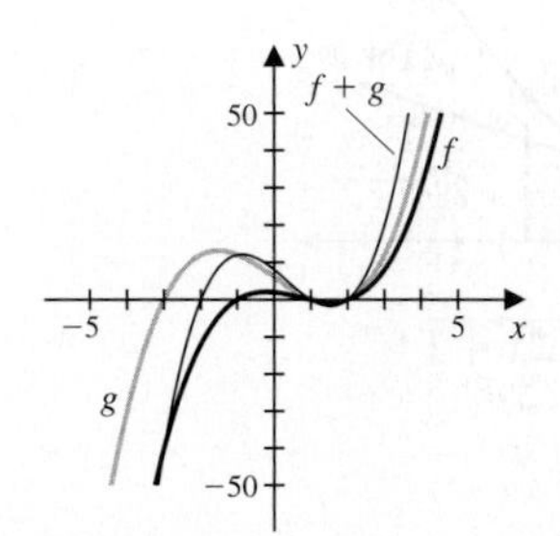

27. Let $R(t)$ denote the rate of spread of the disease at time t, $I(t)$ the number infected, and $H(t)$ the number healthy. Then

$$R(t) = kI(t)H(t) = \frac{1}{74550}(70)(4930) \approx 4.6.$$

Exercise Set 2.4 (page 109)

1. 19

3. 19

5. -17

7. $(f \circ g)(x) = 6x - 1$; $(g \circ f)(x) = 6x + 2$. The domain of $f \circ g$ and $g \circ f$ is $(-\infty, \infty)$.

9. $(f \circ g)(x) = \dfrac{1}{x^2 + 2x}$, $(g \circ f)(x) = \dfrac{1 + 2x}{x^2}$.
The domain of $f \circ g$ is $(-\infty, -2) \cup (-2, 0) \cup (0, \infty)$, and the domain of $g \circ f$ is $(-\infty, 0) \cup (0, \infty)$.

11. $(f \circ g)(x) = \sqrt{x^2 - 4}$, $(g \circ f)(x) = x - 4$. The domain of $f \circ g$ is $(-\infty, -2] \cup [2, \infty)$, and the domain of $g \circ f$ is $[1, \infty)$.

13. $(f \circ g)(x) = x + 1$, $(g \circ f)(x) = \frac{x}{x+1}$. The domain of $f \circ g$ is $(-\infty, -1) \cup (-1, \infty)$, and the domain of $g \circ f$ is $(-\infty, -1) \cup (-1, 0) \cup (0, \infty)$.

15. $f(x) = x^4$; $g(x) = 3x^2 - 2$

17. $f(x) = \sqrt[3]{x}$; $g(x) = x - 4$

19. $f(x) = 1/x$; $g(x) = x + 2$

21. a. **(i)** $(f \circ g)(-2) = 2$
(ii) $(g \circ f)(-2) = 0$
(iii) $(g \circ f)(2) = 0$
(iv) $(f \circ g)(2) = -2$

b. **(i)** $x = -4, 0, 4$
(ii) $x = -2, 0, 2$

23.

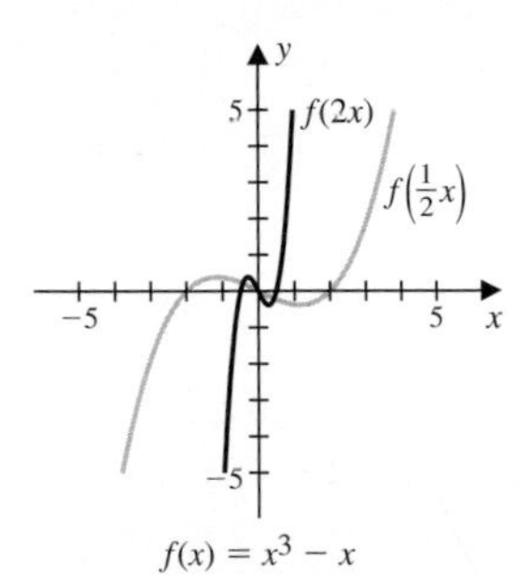

$f(x) = x^3 - x$

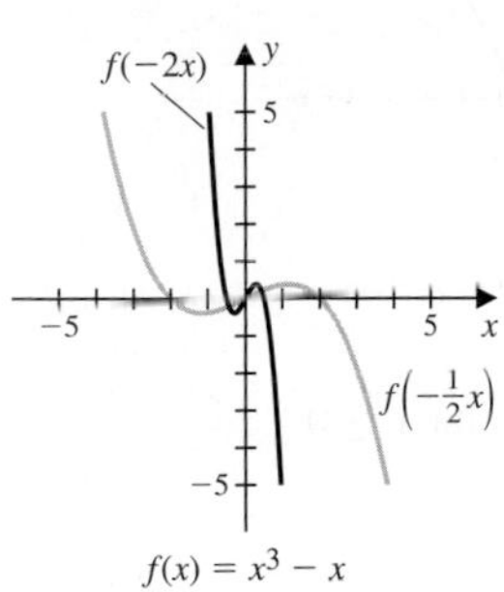

$f(x) = x^3 - x$

25. a, b.

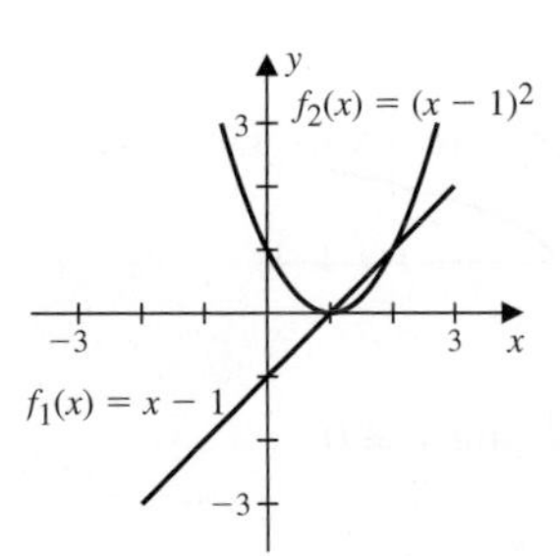

c, d.

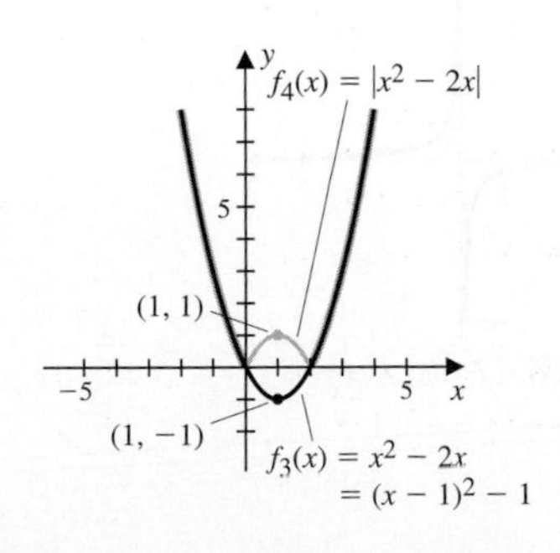

e, f.

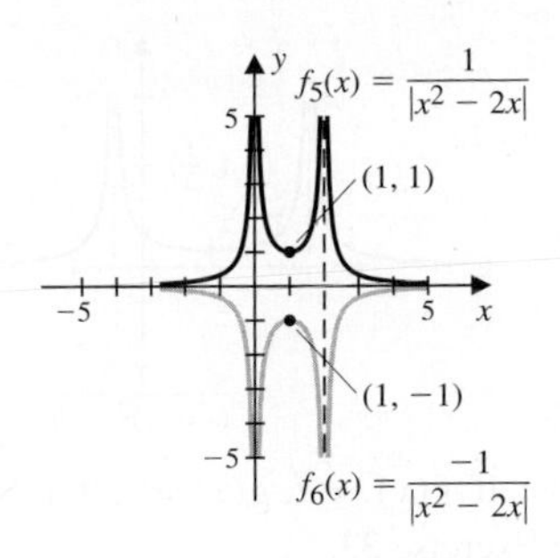

27. a. $g_2(g_1(x)) = (x-3)^2 + 1 = x^2 - 6x + 10$,
$g_3(g_2(g_1(x))) = |x^2 - 6x + 10|$, and

$$g_4(g_3(g_2(g_1(x)))) = \frac{1}{|x^2 - 6x + 10|} = f(x).$$

b. **(i)**

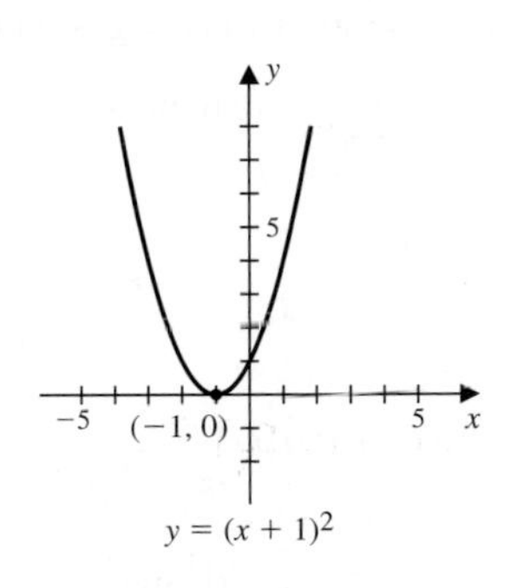

$y = (x + 1)^2$

(ii)

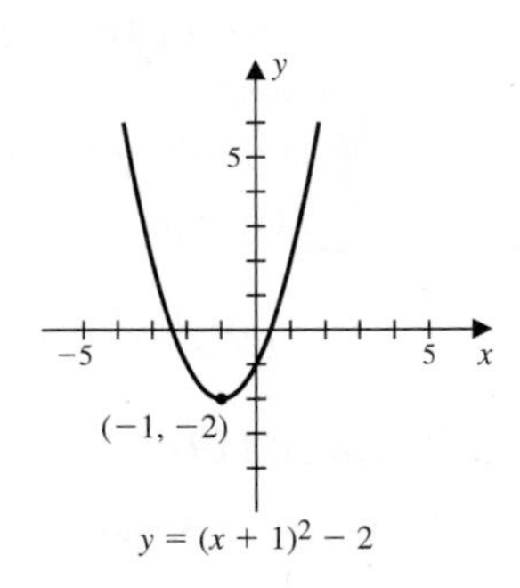

$y = (x + 1)^2 - 2$

(iii)

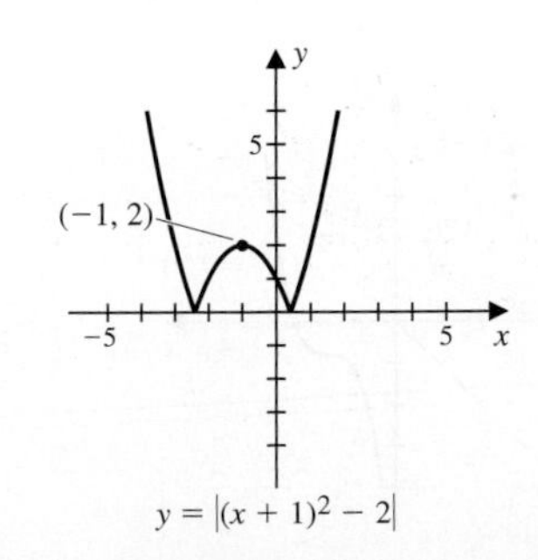

$y = |(x + 1)^2 - 2|$

(iv)

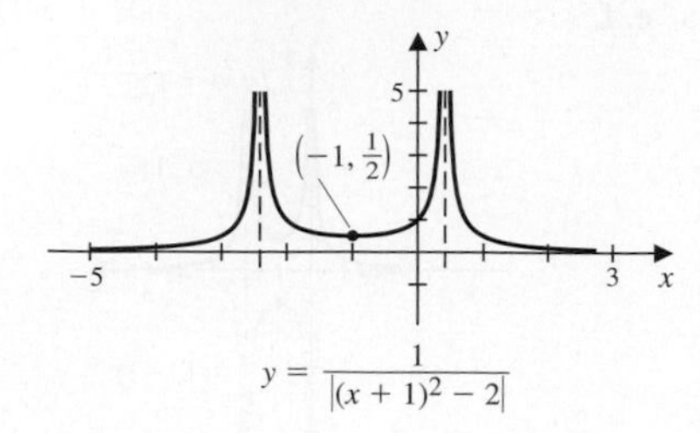

29. For example, if $f(x) = x$, $g(x) = x + 1$. See also the answers to Exercise 33.

31. Let f be an odd and g an even function, so that $f(-x) = -f(x)$ and $g(-x) = g(x)$. Then

$$(f \circ g)(-x) = f(g(-x)) = f(g(x)) = (f \circ g)(x)$$

and

$$(g \circ f)(-x) = g(f(-x)) = g(-f(x)) = (g \circ f)(x).$$

So $f \circ g$ and $g \circ f$ are both even.

33. **a.** $ad + b = bc + d$

b. $c = 1$ and $d = 0$

c. $a = 1$ and $b = 0$

35. $g(x) = 3f(x - 1) + 2$

37. $V(t) = \frac{4}{3}\pi(3 + 0.01t)^3$

39. **a.** $V(t) = \frac{4}{3}\pi(3\sqrt{t} + 5)^3$ cm^3

b. $s(t) = 4\pi(3\sqrt{t} + 5)^2$ cm^2

41. **a.** Value after 2 years: $(1.04)^2x + P(1 + 1.04)$;
Value after 3 years:
$(1.04)^3x + P\left[1 + 1.04 + (1.04)^2\right]$.

b. $\underbrace{V \circ V \circ \cdots \circ V}_{n \text{ times}}(x) = (1.04)^n x + P\left[1 + 1.04 + (1.04)^2 + \cdots + (1.04)^{n-1}\right]$.

Exercise Set 2.5 (page 119)

1. yes **3.** no **5.** no **7.** yes

9. no **11.** no **13.** yes

15.

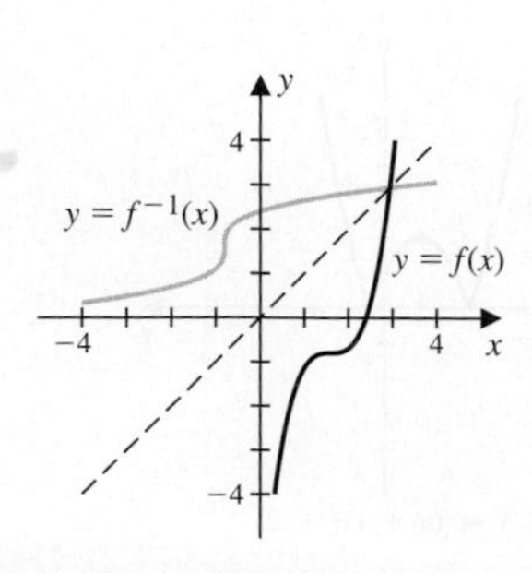

17.

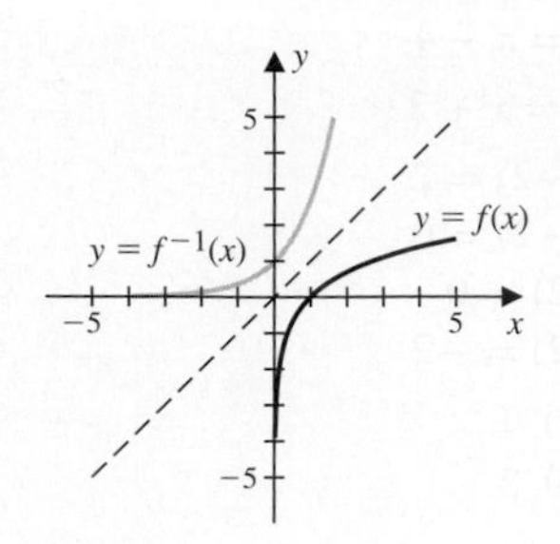

19. $f^{-1}(2) = 1$

21. $f^{-1}(1) = \frac{3}{2}$

23. $f^{-1}(2) = \frac{1}{2}$

25. $f^{-1}(x) = \frac{1}{2}x + \frac{1}{2}$

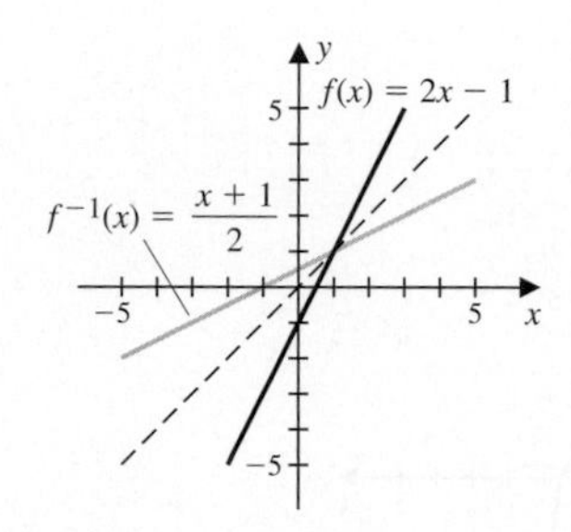

27. $f^{-1}(x) = x^2 + 3$, for $x \geq 0$

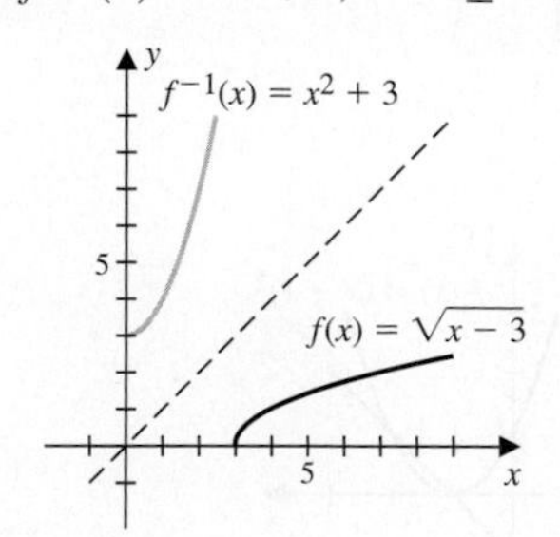

29. $f^{-1}(x) = \frac{1}{2x}$, for $x \neq 0$

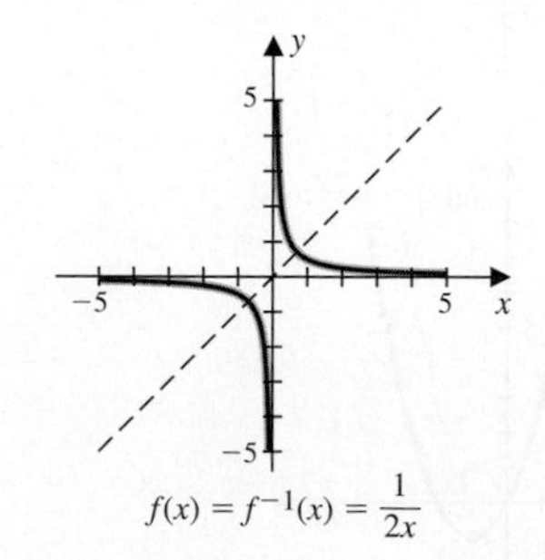

31. $f^{-1}(x) = 1/x^2$, for $x > 0$

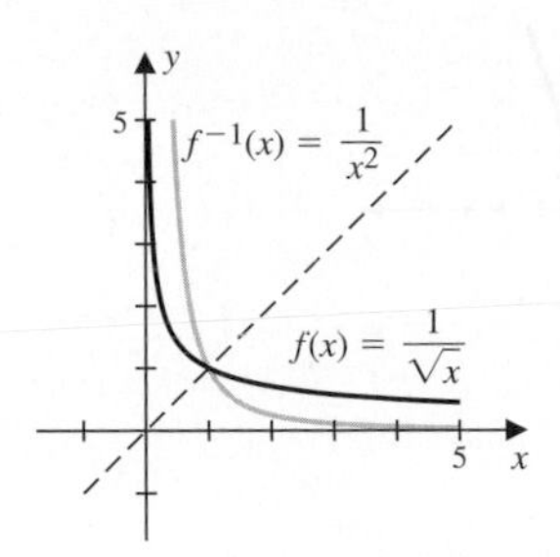

33. $f^{-1}(x) = (x - 1)^{1/3}$

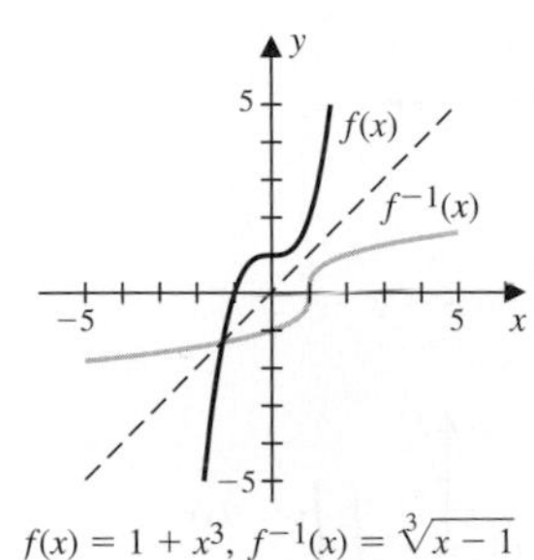

$f(x) = 1 + x^3$, $f^{-1}(x) = \sqrt[3]{x - 1}$

35. $f^{-1}(x) = \sqrt{x - 1}$, for $x \geq 1$

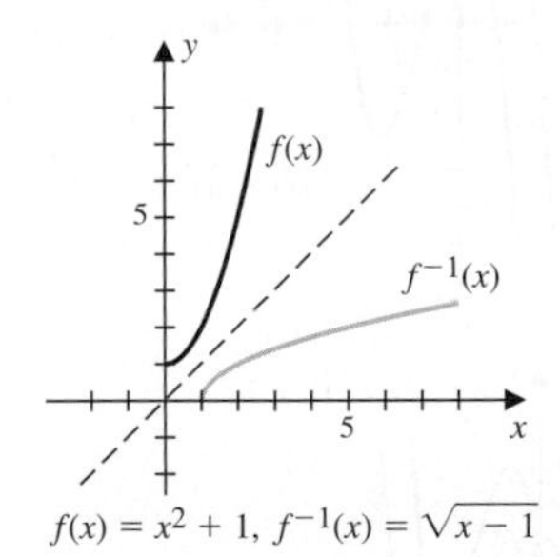

$f(x) = x^2 + 1$, $f^{-1}(x) = \sqrt{x - 1}$

37. **a.** For example, $f(0) = f(4)$

b. $[2, \infty)$; $f^{-1}(x) = x + 2$, for $x \geq 0$

39. **a.** For example, $f(0) = f(2)$

b. $[1, \infty)$; $f^{-1}(x) = 1 + \sqrt{x + 1}$, for $x \geq -1$

41. For all $m \neq 0$, $f(x) = mx + b$ is one-to-one, and $f^{-1}(x) = (x - b)/m$, for $m \neq 0$.

Review Exercises for Chapter 2 (page 120)

1. **a.** ii **b.** i **c.** iv **d.** iii

3. **a.** $(f + g)(x) = 3x + 1$; $(f - g)(x) = x - 3$; $(fg)(x) = 2x^2 + 3x - 2$; $(f/g)(x) = \dfrac{2x - 1}{x + 2}$; $(f \circ g)(x) = 2x + 3$; $(g \circ f)(x) = 2x + 1$; $(f \circ f)(x) = 4x - 3$; $(g \circ g)(x) = x + 4$

b. The domain of $f + g$, $f - g$, fg, $f \circ g$, $g \circ f$, $f \circ f$ and $g \circ g$ is $(-\infty, \infty)$. The domain of f/g is $(-\infty, -2) \cup (-2, \infty)$.

5. **a.** $(f + g)(x) = x + 1 + \sqrt{x - 1}$;
$(f - g)(x) = x + 1 - \sqrt{x - 1}$;
$(fg)(x) = (x + 1)\sqrt{x - 1}$;
$(f/g)(x) = \frac{x+1}{\sqrt{x-1}}$; $(f \circ g)(x) = \sqrt{x - 1} + 1$;
$(g \circ f)(x) = \sqrt{x}$; $(f \circ f)(x) = x + 2$;
$(g \circ g)(x) = \sqrt{\sqrt{x - 1} - 1}$;

b. The domain of $f + g$, $f - g$, fg and $f \circ g$ is $[1, \infty)$. The domain of f/g is $(1, \infty)$. The domain of $g \circ f$ is $[0, \infty)$. The domain of $f \circ g$ is $[1, \infty)$. The domain of $f \circ f$ is $(-\infty, \infty)$, and the domain of $g \circ g$ is $[2, \infty)$.

7. One choice is $f(x) = x^6$ and $g(x) = x^3 - 2x + 1$.

9. One choice is $f(x) = \sqrt{x}$ and $g(x) = 3x + 3$.

11. One choice is $f(x) = |x|$ and $g(x) = (x - 1)^2$

13.

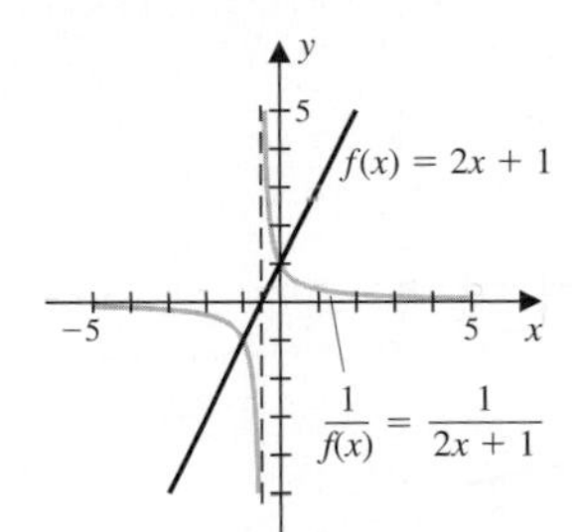

15.

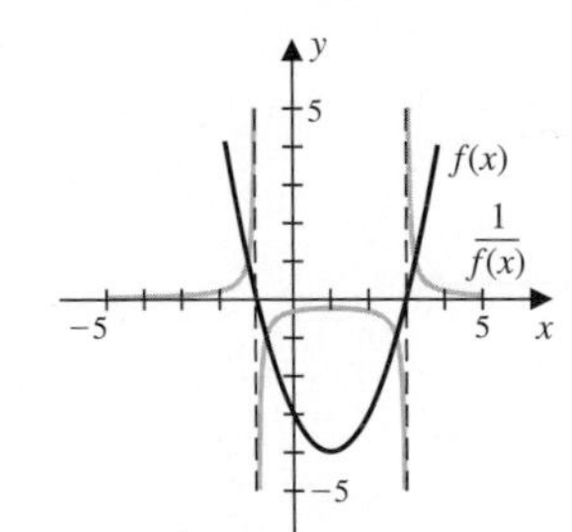

$f(x) = x^2 - 2x - 3$
$\frac{1}{f(x)} = \frac{1}{x^2 - 2x - 3}$

17.

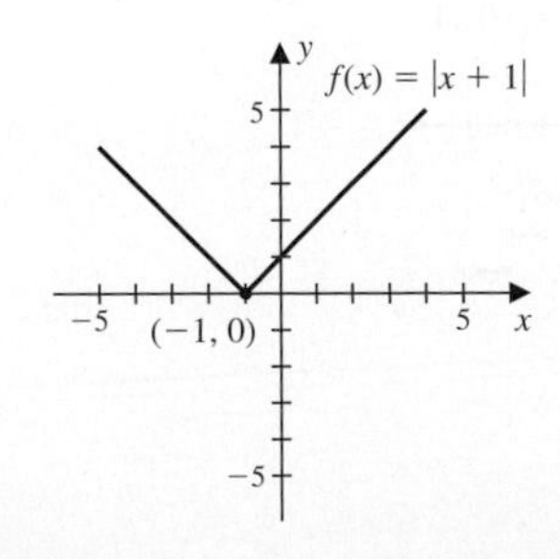

19.

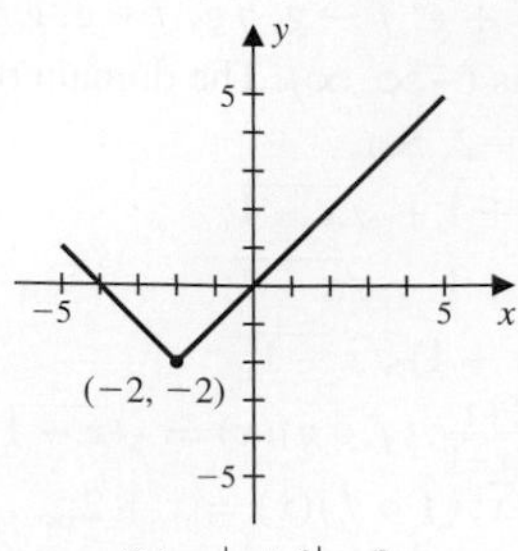

21.

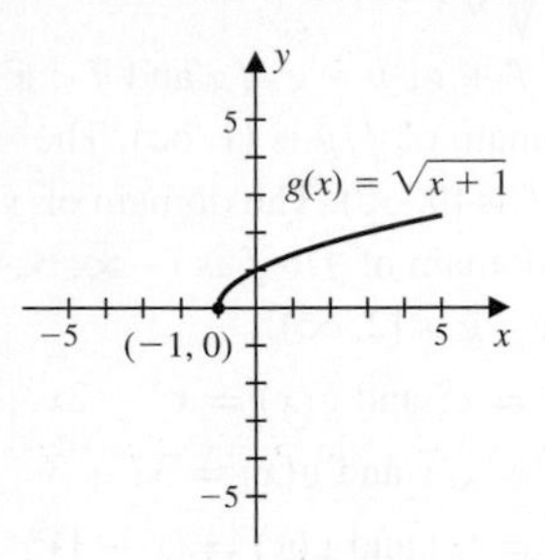

23.

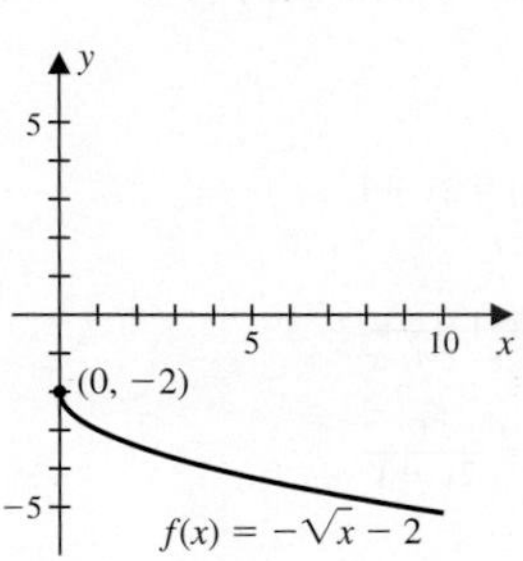

25.

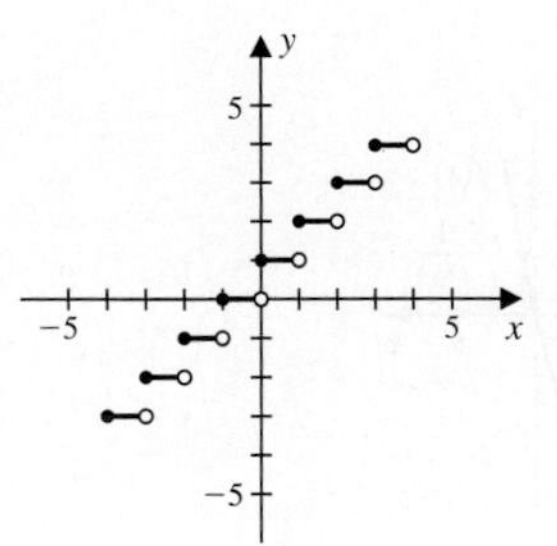

27.

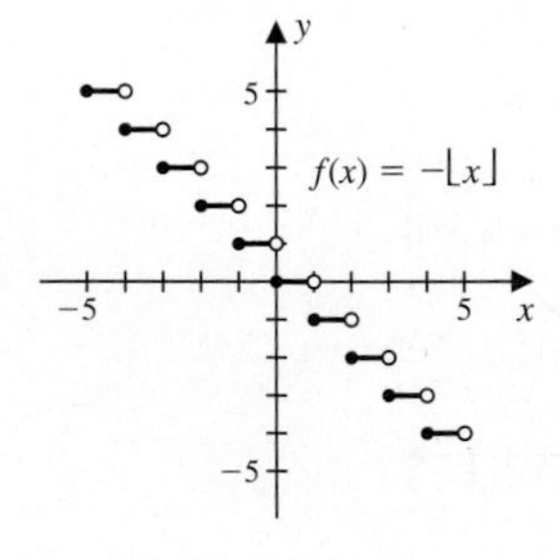

29.

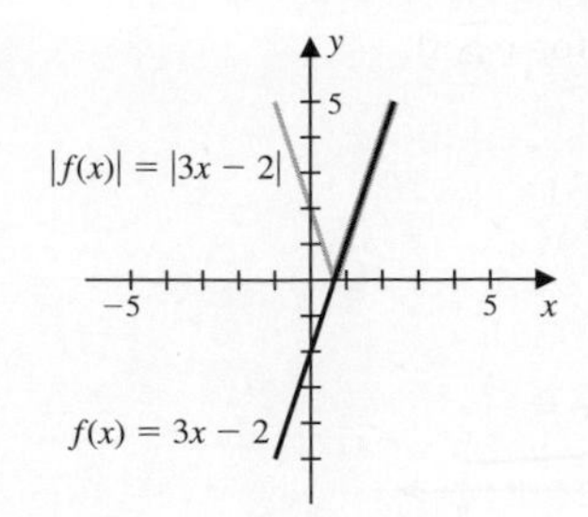

31.

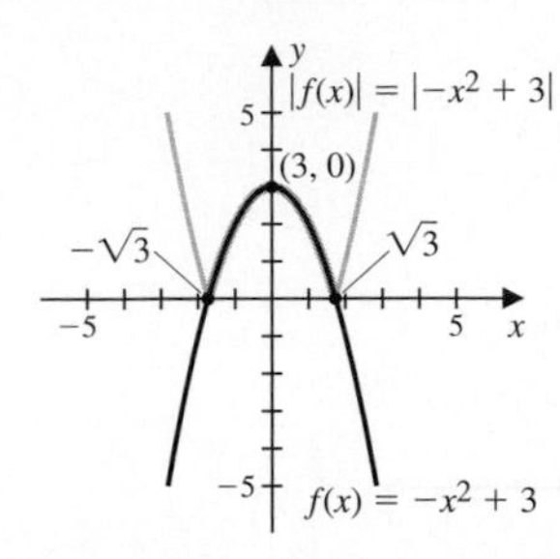

33. a, b

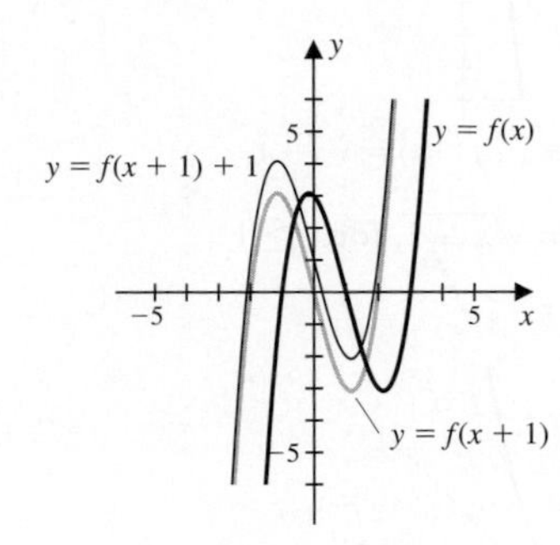

c, d

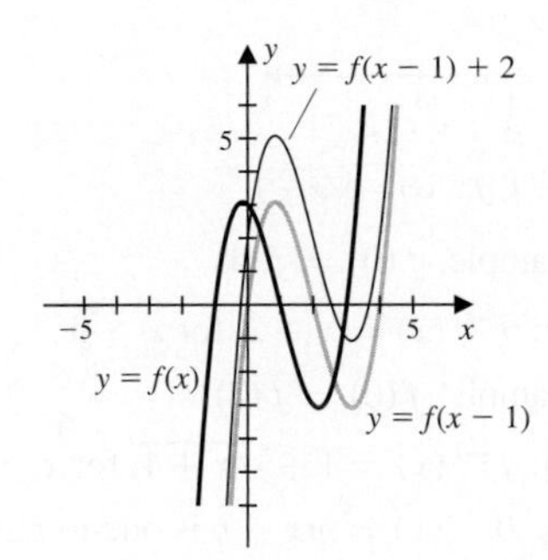

e, f

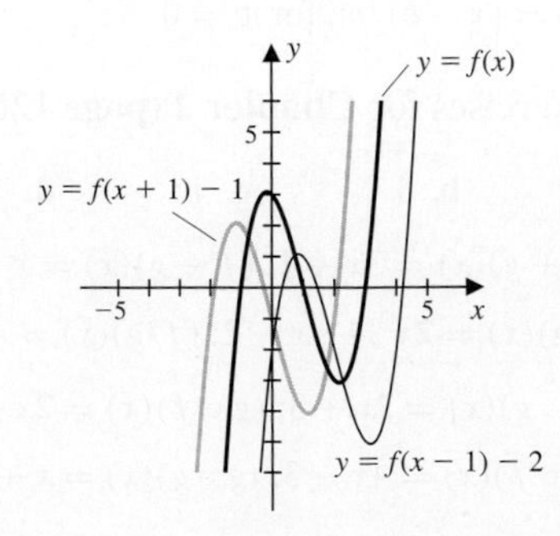

g, h

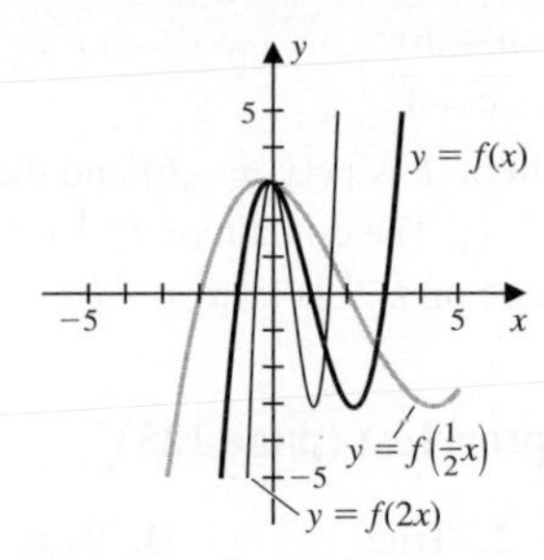

35. **a.** $f(-x) = 2x^2 - 3; -f(x) = -2x^2 + 3;$ $f(1/x) = 2/x^2 - 3; 1/f(x) = 1/(2x^2 - 3);$ $f(\sqrt{x}) = 2x - 3; \sqrt{f(x)} = \sqrt{2x^2 - 3}$

b. $f(-x) = 1/x^2; -f(x) = -1/x^2; f(1/x) = x^2;$ $1/f(x) = x^2; f(\sqrt{x}) = 1/x; \sqrt{f(x)} = 1/x$

37. $g(x) = \frac{1}{2}f(x + 2) - 3$

39. **a.** $f^{-1}(x) = \dfrac{x - 1}{2}$

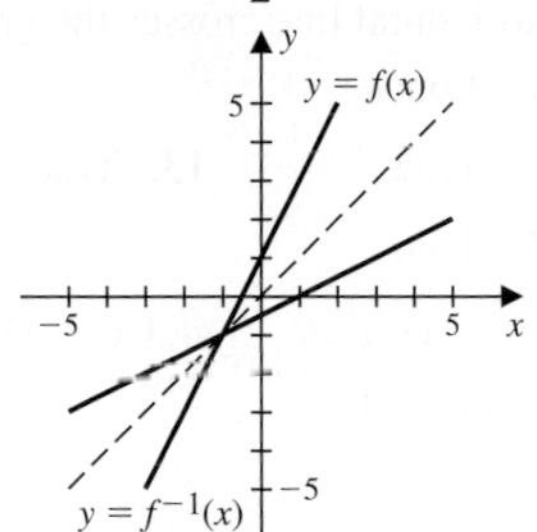

b. The function is not one-to-one.

c. $f^{-1}(x) = \sqrt[3]{x - 2}$

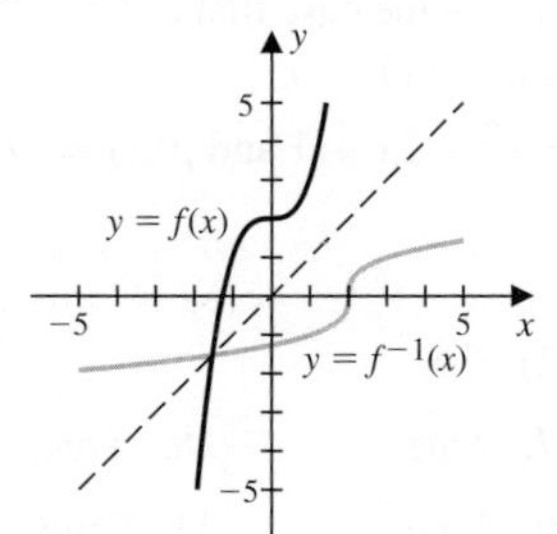

d. The function is not one-to-one.

e. The function is not one-to-one.

f. $f^{-1}(x) = (x - 2)^2 + 1$

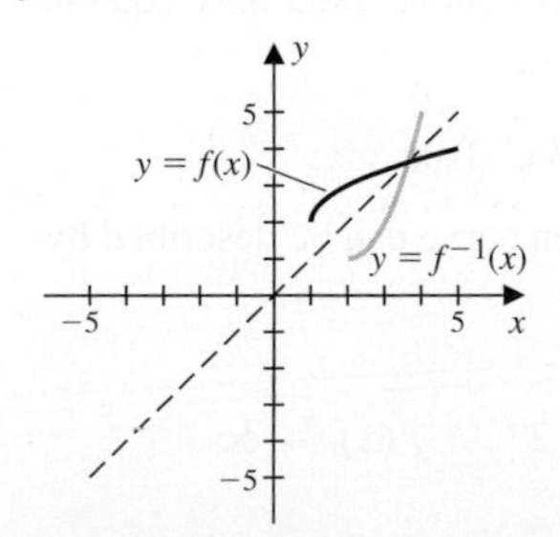

41. **a.** $[2, \infty)$

b.

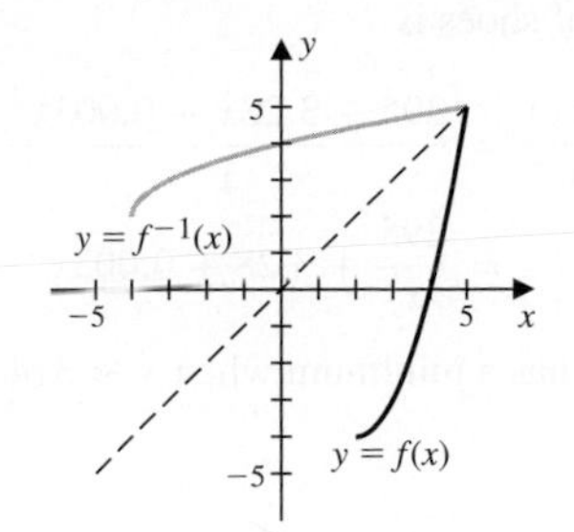

c. $f^{-1}(x) = 2 + \sqrt{x + 4}$. The domain of f^{-1} is $[-4, \infty)$.

43. **a.** $P(x) = -0.1x^2 + 90x - 1500$

b. 450

c. \$18,750

Chapter 2: Exercises for Calculus (page 124)

1. The sign of $f(x)$ and the zeros of $f(x)$ give the graph of $1/f(x)$.

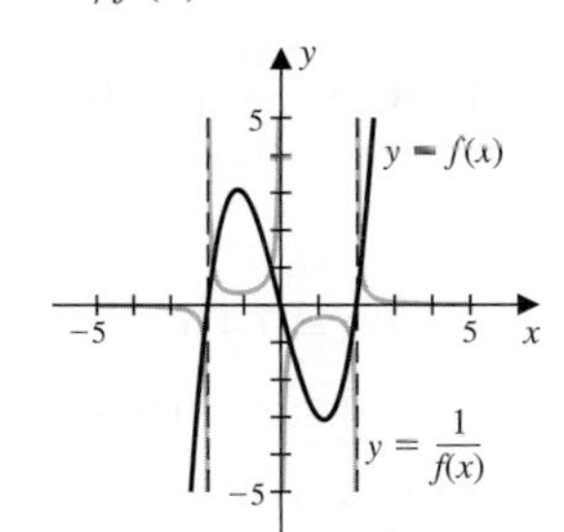

3. **a.**

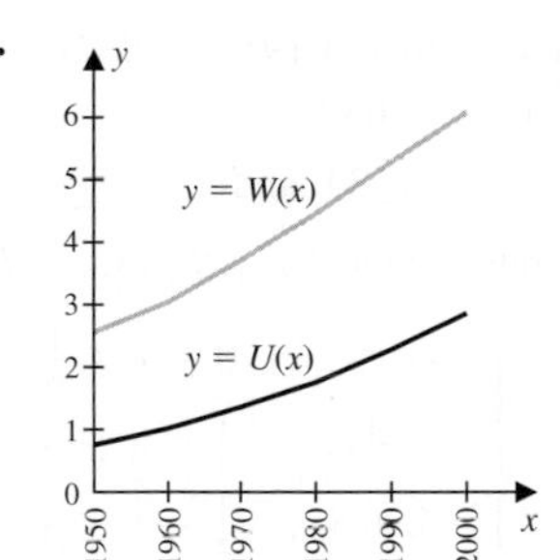

b.

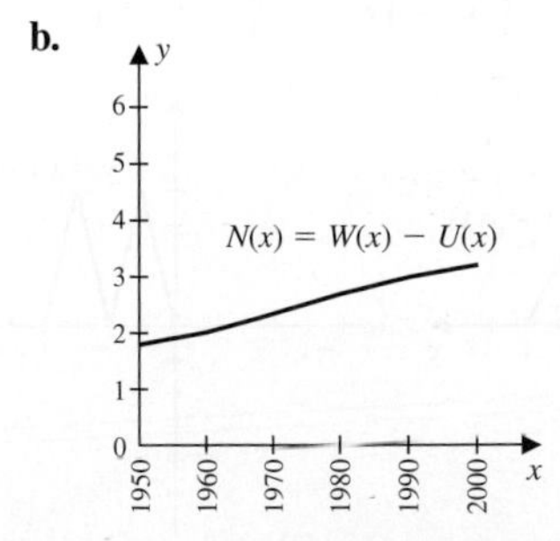

5. **a.** If x pairs of shoes are produced, the average cost for a pair of shoes is

$$\frac{C(x)}{x} = \frac{295 + 3.28x + 0.003x^2}{x} = \frac{295}{x} + 3.28 + 0.003x.$$

The graph has a minimum when $x \approx 314$.

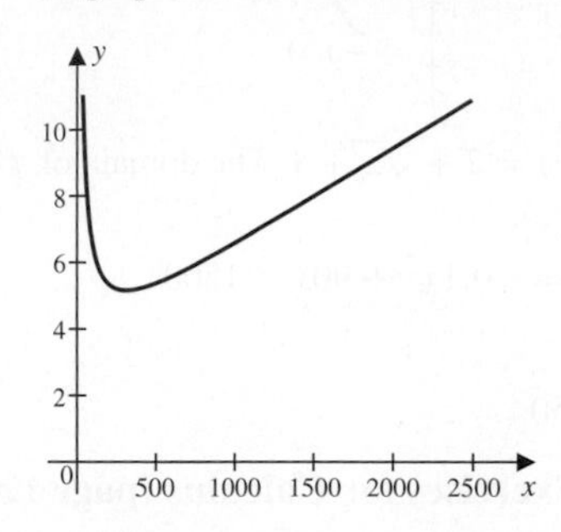

b. The revenue to the company if x pairs of shoes are sold is

$$R(x) = xp(x) = x\left(7.47 + \frac{321}{x}\right) = 7.47x + 321.$$

The profit is

$$\begin{aligned} P(x) &= R(x) - C(x) = 7.47x + 321 \\ &\quad -(295 + 3.28x + 0.003x^2) \\ &= 26 + 4.19x - 0.003x^2. \end{aligned}$$

The graph of $y = 26 + 4.19x - 0.003x^2$ has a maximum when $x \approx 698$.

7. **a.** 1, 1, 2, 3, 5, 8, 13, 21, 34, and 55

b. Show that d must divide $F_{n-1}, F_{n-2}, \ldots, F_1 = 1$.

9. **a.** $$(f \circ f)(x) = \begin{cases} 4x, & 0 \le x \le \frac{1}{4} \\ 2 - 4x, & \frac{1}{4} < x \le \frac{1}{2} \\ -2 + 4x, & \frac{1}{2} < x \le \frac{3}{4} \\ 4 - 4x, & \frac{3}{4} < x \le 1 \end{cases}$$

b.

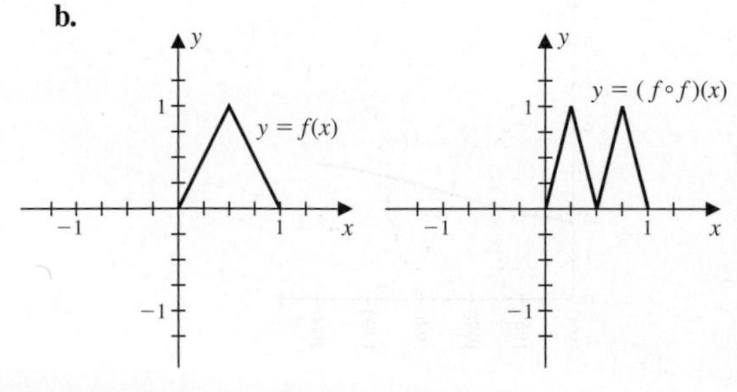

11. **a.** $f^{-1}(x) = \dfrac{a - bx}{x - 1}$

b. The domain of f is $\{x|x \neq -b\}$ and the range of f is $\{x|x \neq 1\}$. The domain of f^{-1} is $\{x|x \neq 1\}$ and the range of f^{-1} is $\{x|x \neq -b\}$.

Chapter 2: Chapter Test (page 125)

1. True. **2.** True. **3.** True.

4. False. The domain is $(-\infty, \infty)$ and the range is $(-\infty, 2]$.

5. True. **6.** True.

7. False. The value of $f(2)$ is 1.

8. False. The function is constantly 2.

9. False. The function is constantly -1.

10. False. A function will have an inverse function provided every horizontal line crosses the graph of the function at most once.

11. True. **12.** True. **13.** True.

14. True. **15.** True.

16. False. The domain is $\{x|x \neq -3 \text{ and } x \neq 1\}$.

17. False. $(f \circ g)(-1) = -3$

18. True.

19. False. The value of $(f \circ f)(2) = 4$.

20. True. **21.** True.

22. False. It is sometimes the case that $(f \circ g)(x) - (g \circ f)(x) = 0$.

23. False. If $f(x) = x^2 - 2x + 1$ and $g(x) = \sqrt{x}$, then $(f \circ g)(4) = 1$.

24. True.

25. False. $(f + g)(2) = 6$

26. True. **27.** True. **28.** True.

29. True. **30.** True. **31.** True.

32. False. First the curve is reflected about the x-axis.

33. True.

34. False. The yellow curve could have equation $y = 2|x - 1| + 2$.

35. True. **36.** True.

37. False. The green curve can be described by $y = f(x) - 2$.

38. True.

39. False. As $x \to 2^+$, $f(x) \to \infty$.

40. True.

41. False. As $x \to 3$, the function $f(x) = \dfrac{2x+1}{(x-3)^2} \to \infty$.

42. False. As $x \to \infty$, the function $f(x) = \frac{1}{x} + 2 \to 2$.

Chapter 3

Exercise Set 3.2 (page 142)

1. a. iii **b.** iv **c.** i **d.** ii

3. The lowest possible degree is 3.

5. The lowest possible degree is 4.

7.

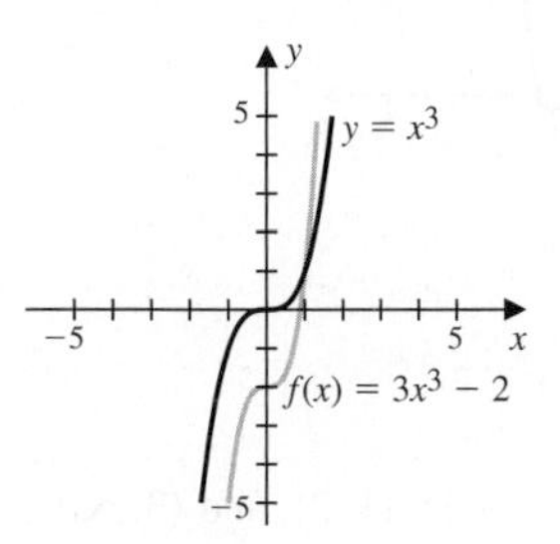

9.

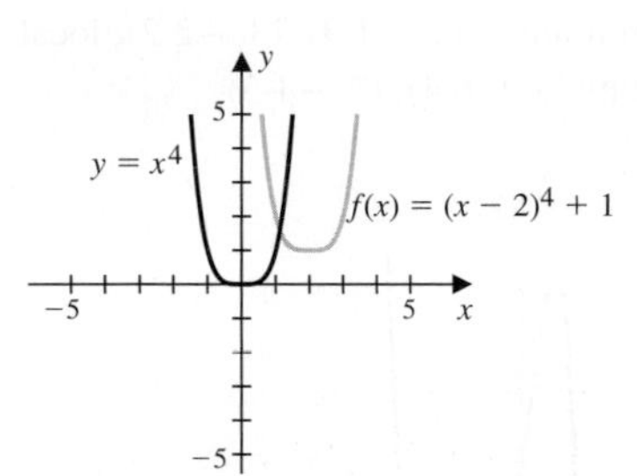

11.

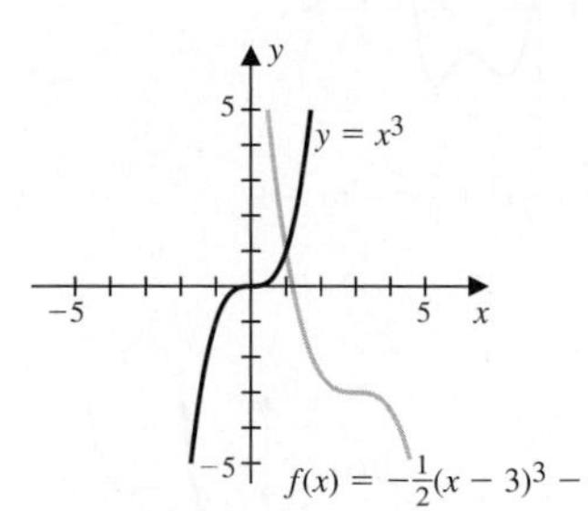

13.

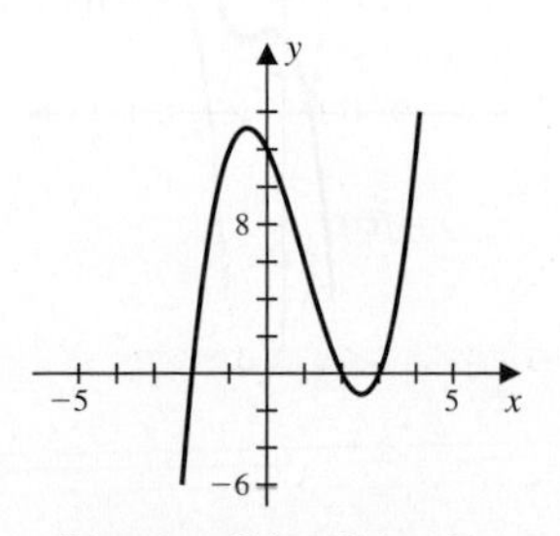

$f(x) = (x - 2)(x + 2)(x - 3)$

15.

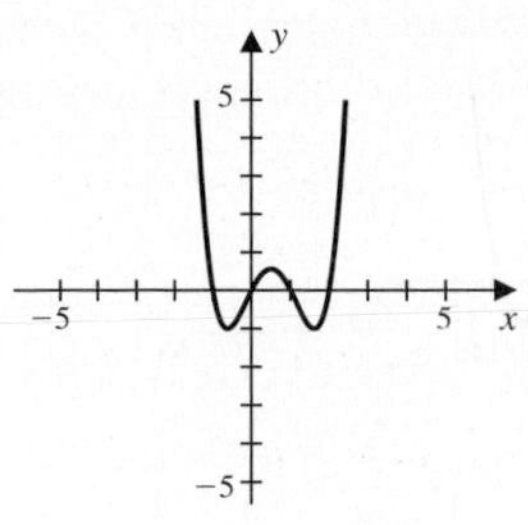

$f(x) = x(x + 1)(x - 1)(x - 2)$

17.

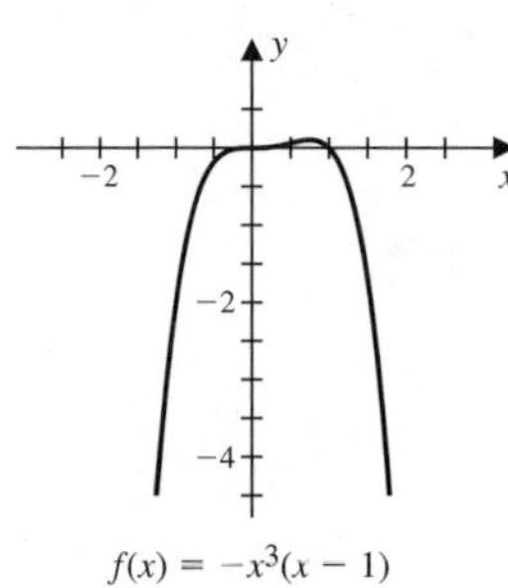

$f(x) = -x^3(x - 1)$

19.

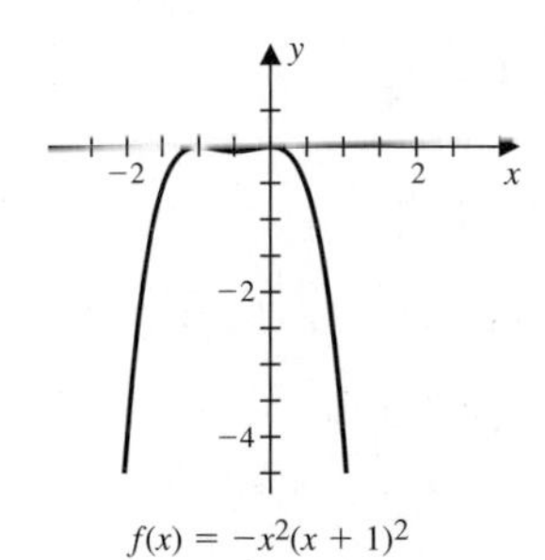

$f(x) = -x^2(x + 1)^2$

21.

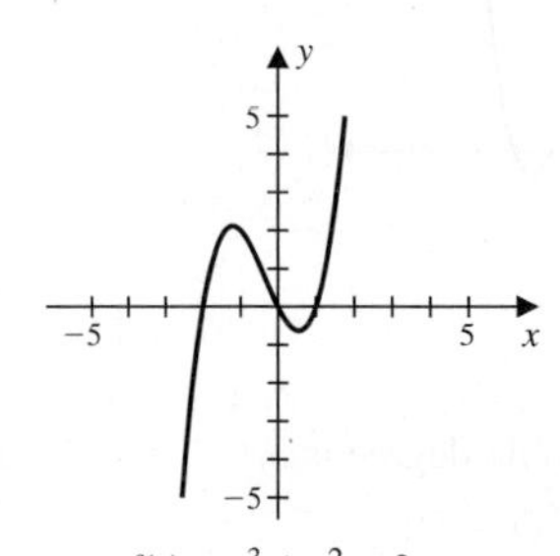

$f(x) = x^3 + x^2 - 2x$

23.

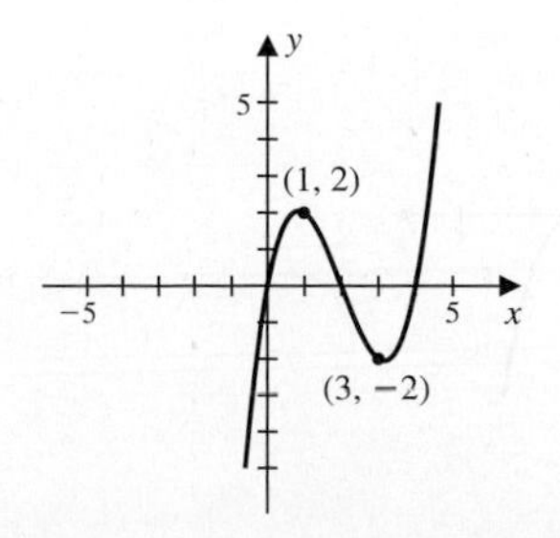

25.

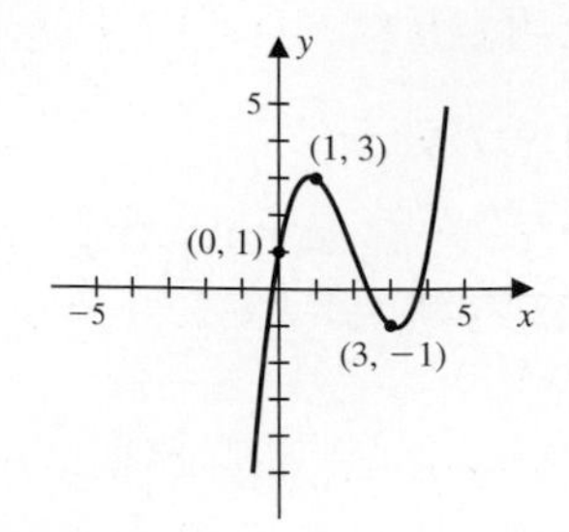

27.

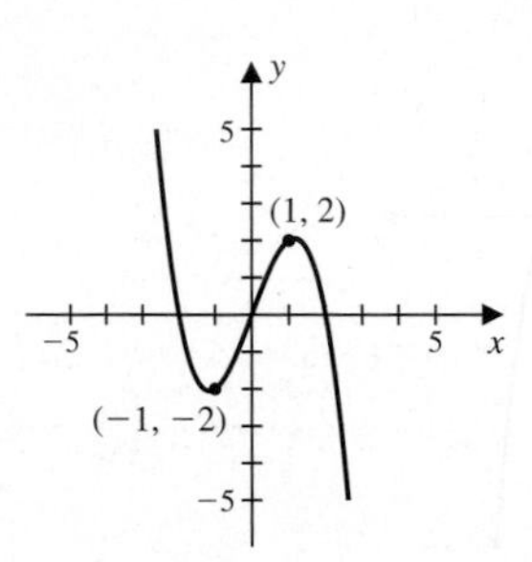

29.

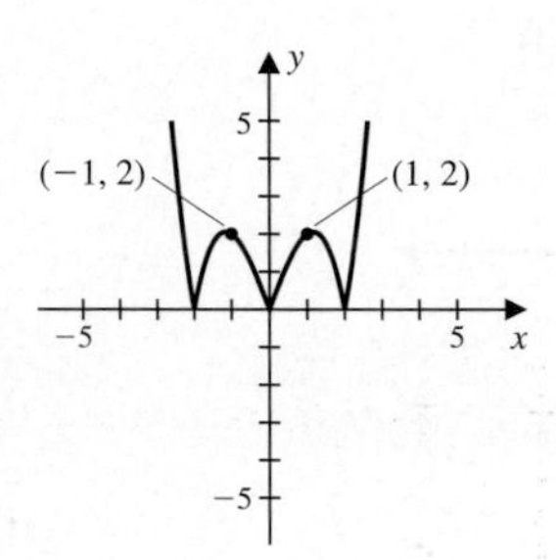

31. **a.**

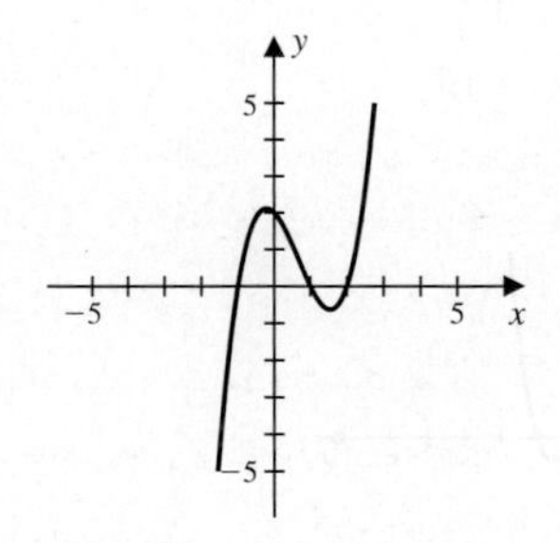

b. The lowest possible degree is 3.

33. $f(x) = -\frac{1}{3}(x-1)^2(x-2)$

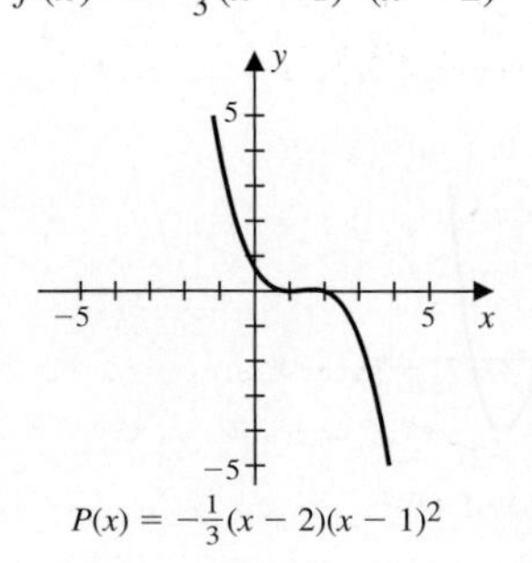

$P(x) = -\frac{1}{3}(x-2)(x-1)^2$

35. $f(x) = 2x(x-1)(x-2) + 2$

37. $P(x) = -\frac{2}{3}x^4 + \frac{8}{3}x^2$

39. On $(0, 1)$ we have $x^n > x^{n+1}$, and on $(1, \infty)$ we have $x^n < x^{n+1}$.

41. **a.** Increasing: $(-1.4, 0)$ and $(0.7, \infty)$

b. Local minimum: $(-1.4, -2.9)$, $(0.7, -0.4)$; local maximum: $(0, 0)$

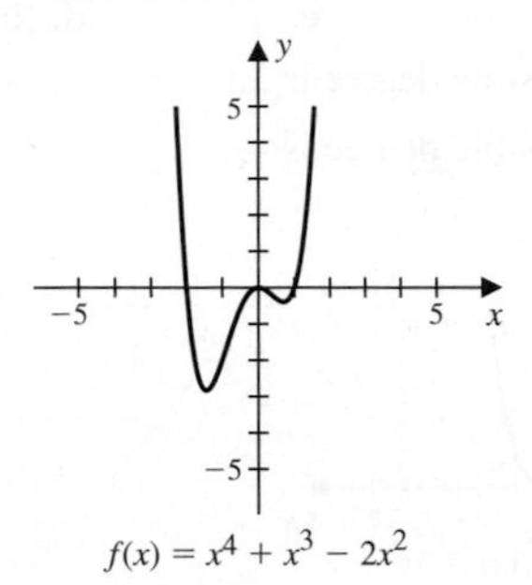

$f(x) = x^4 + x^3 - 2x^2$

43. **a.** Increasing: $(-\infty, -1)$, $(1, 2)$, and $(3, \infty)$

b. Local minimum: $(1, -1.9)$, $(3, -2.2)$; local maximum: $(-1, 6.4)$, $(2, -1.3)$

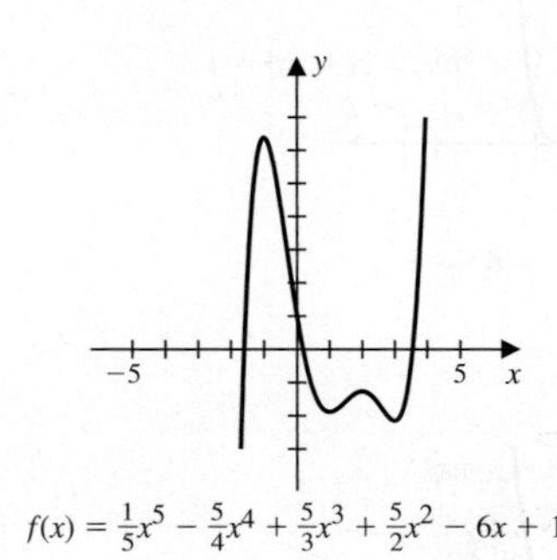

$f(x) = \frac{1}{5}x^5 - \frac{5}{4}x^4 + \frac{5}{3}x^3 + \frac{5}{2}x^2 - 6x + 1$

45.

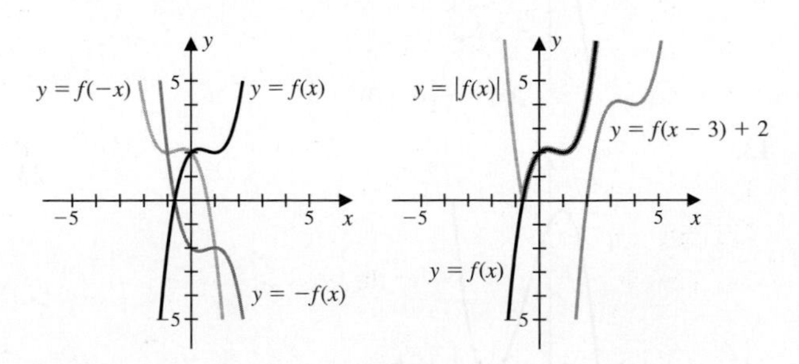

47.

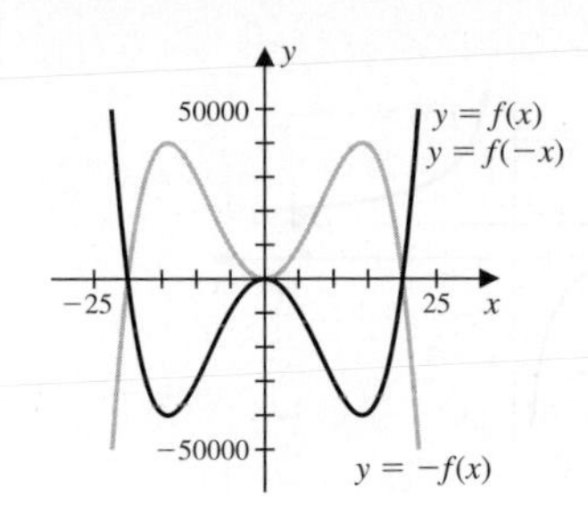

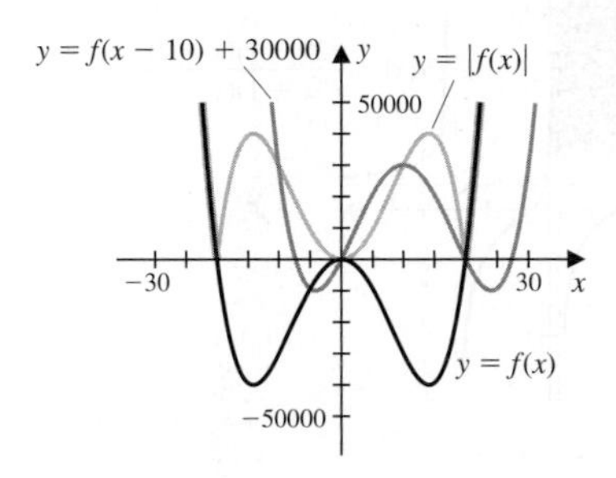

49. **a.** $V(x) = x(20 - 2x)^2$

b. $x \approx 1.9$ and $x = 5$

51.

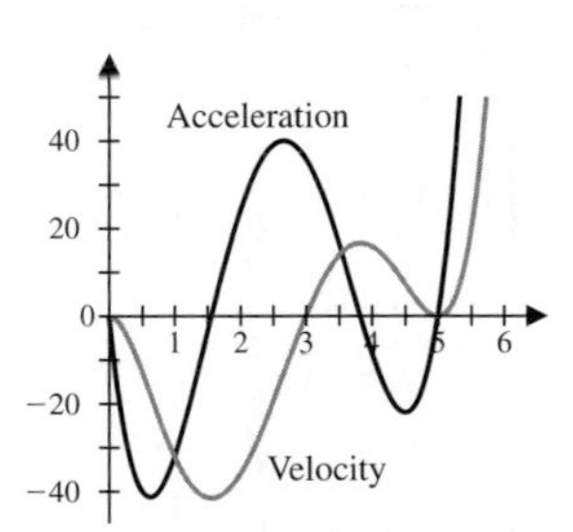

Exercise Set 3.3 (page 154)

1. $Q(x) = 2x + 7$; $R(x) = 11$
3. $Q(x) = x^2 + 2x + 2$; $R(x) = 0$
5. $Q(x) = 2x^2 + 6$; $R(x) = 7x + 4$
7. $P(x) = (x - 1)(x - 2)^2$
9. $P(x) = (x + 1)(x + 3)(2x - 1)(x - 2)$
11. $P(x) = (3x - 2)(x + 2)(x - 1)$
13. $\pm 1, \pm 2, \pm 4$
15. First divide by 2 to reduce the possibilities; $\pm 1, \pm \frac{1}{5}, \pm 2, \pm \frac{2}{5}$
17. $\pm 1, \pm 2, \pm 4, \pm 8, \pm \frac{1}{3}, \pm \frac{2}{3}, \pm \frac{4}{3}, \pm \frac{8}{3}$
19. Zeros: $-1, 2$; $P(x) = (x + 1)(x - 2)^2$
21. Zeros: $0, -1, \frac{1}{2}, 2$; $P(x) = x(x + 1)(2x - 1)(x - 2)$
23. Zeros: $1, -\frac{1}{2}, -3$; $P(x) = (x - 1)^3(2x + 1)(x + 3)$
25. Zeros: $\dfrac{-3 \pm \sqrt{13}}{2}, 1$; $P(x) = (x - 1)\left(x - \left(\frac{-3+\sqrt{13}}{2}\right)\right)\left(x - \left(\frac{-3-\sqrt{13}}{2}\right)\right)$
27. Zeros: $1, 1 + \frac{\sqrt{3}}{2}, 1 - \frac{\sqrt{3}}{2}$; $P(x) = 4(x - 1)(x - (1 + \frac{\sqrt{3}}{2}))(x - (1 - \frac{\sqrt{3}}{2}))$

29.

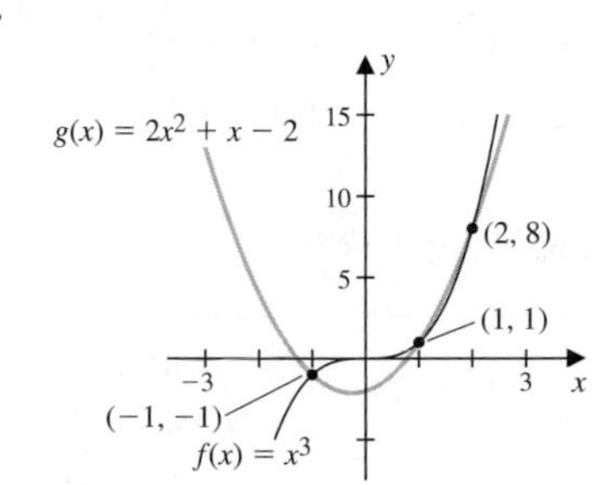

31.

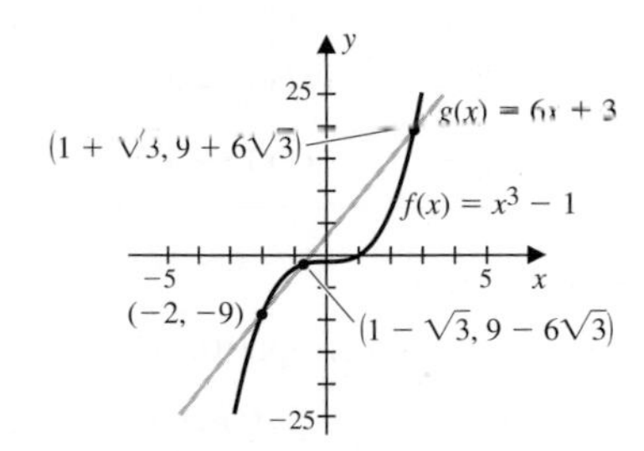

33. $P(x) = \frac{1}{10}(x + 3)(x + 1)(x - 1)(x - 4)$
35. $P(x) = -5x^4 - 10x^3 + 5x^2 + 10x$
37. $P(x) = (x - 2)^3 \cdot x^2(x - 1)$
39. **a.** $x = -1, x = 1$, and $x = 4$

    b. $[-1, 1]$ and $[4, \infty)$
41. **a.** $x = -1, x = 1$, and $x = 4$

    b. $[-1, 1]$ and $[4, \infty)$

Exercise Set 3.4 (page 168)

1. **a.** $(-\infty, -1) \cup (-1, \infty)$

   b. $(-\infty, 0) \cup (0, \infty)$

   c. $x = -1, y = 0$
3. **a.** $(-\infty, -1) \cup (-1, 1) \cup (1, \infty)$

   b. $(-\infty, 1) \cup [2, \infty)$

   c. $x = -1, x = 1, y = 1$

5. **a.** $(-\infty, 1)$ **b.** $(-\infty, \infty)$

b. $x = 1$; no horizontal asymptote

7. **a.** $x = -2, x = 2$ **b.** $y = 1$

$x \to -2^-$	$\frac{(-)(-)}{(-)(-)}$	$f(x) \to \infty$
$x \to -2^+$	$\frac{(-)(-)}{(-)(+)}$	$f(x) \to -\infty$
$x \to 2^-$	$\frac{(+)(+)}{(-)(+)}$	$f(x) \to -\infty$
$x \to 2^+$	$\frac{(+)(+)}{(+)(+)}$	$f(x) \to \infty$
$x \to -\infty$	$\frac{(-)(-)}{(-)(-)}$	$f(x) \to 1$
$x \to \infty$	$\frac{(+)(+)}{(+)(+)}$	$f(x) \to 1$

9. **a.** $x = 1$ **b.** $y = 1$

$x \to -1^-$	$\frac{(-)(-)}{(-)(-)}$	$f(x) \to 0$
$x \to -1^+$	$\frac{(+)(-)}{(-)(-)}$	$f(x) \to 0$
$x \to 1^-$	$\frac{(+)(-)}{(-)(-)}$	$f(x) \to -\infty$
$x \to 1^+$	$\frac{(+)(+)}{(+)(+)}$	$f(x) \to \infty$
$x \to -\infty$	$\frac{(-)(-)}{(-)(-)}$	$f(x) \to 1$
$x \to \infty$	$\frac{(+)(+)}{(+)(+)}$	$f(x) \to 1$

11. **a.** $(-\infty, -1) \cup (-1, \infty)$

b. x-intercept: $(3, 0)$; y-intercept: $(0, -3)$

c. $x = -1, y = 1$

13. **a.** $(-\infty, -1) \cup (-1, 1) \cup (1, \infty)$

b. x-intercepts: $(4, 0)$, $(-\frac{1}{3}, 0)$, y-intercept: $(0, 4)$

c. $x = -1, x = 1, y = 3$

15. **a.** $(-\infty, 0) \cup (0, \infty)$

b. x-intercepts: $(1, 0)$, $(2, 0)$, $(-1, 0)$; no y-intercept

c. $x = 0$; no horizontal asymptote

17.

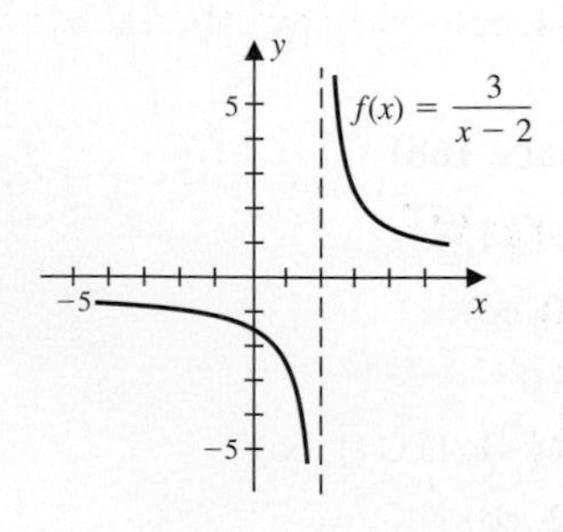

19.

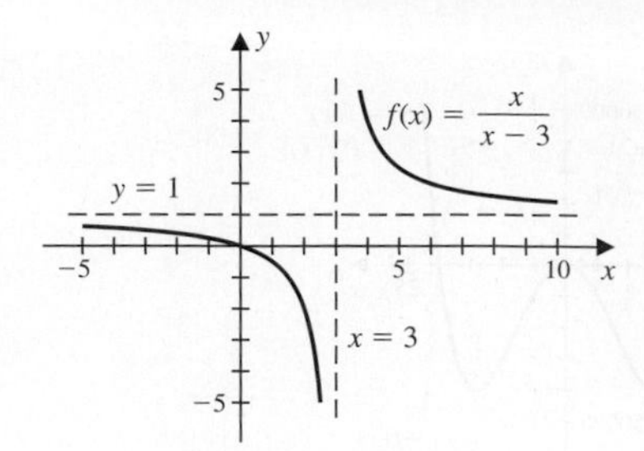

21.

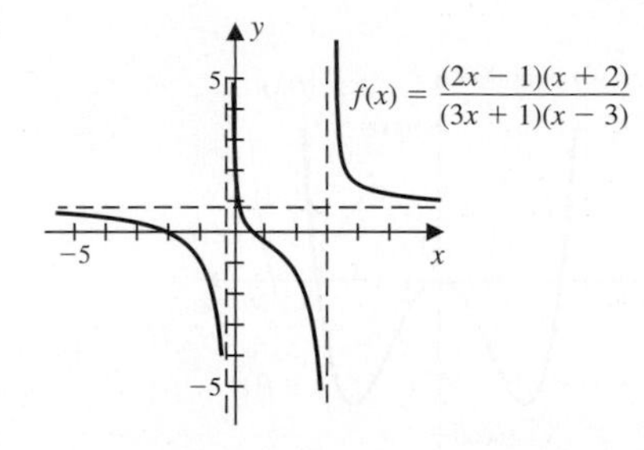

23.

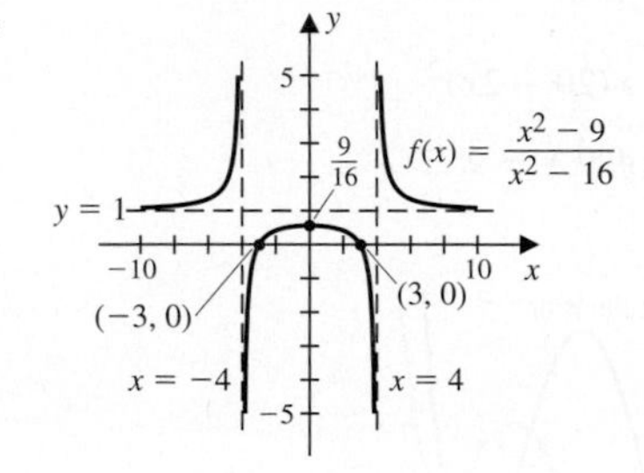

25.

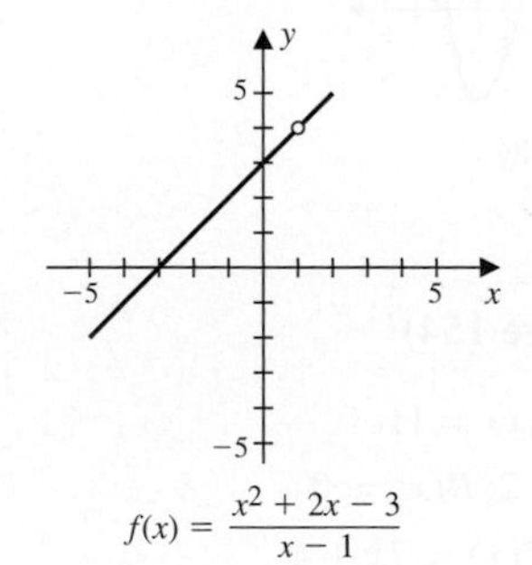

27.

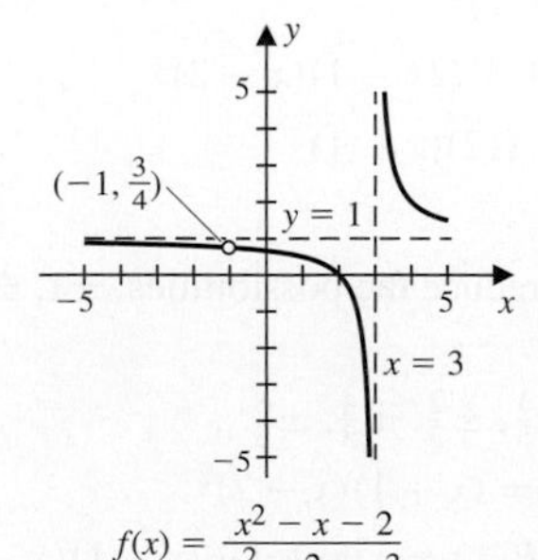

29.

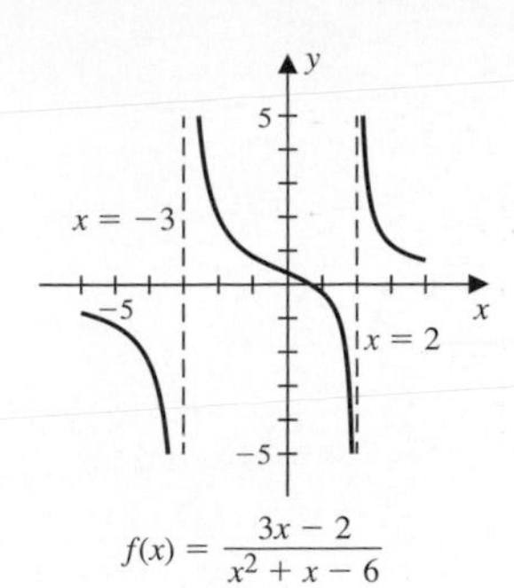

31. One example is $f(x) = \dfrac{1}{x-1}$.

33. One example is $f(x) = \frac{(x-3)(x+4)}{(x+3)(x-2)}$.

35. One example is $f(x) = \dfrac{3x-6}{x}$.

37.

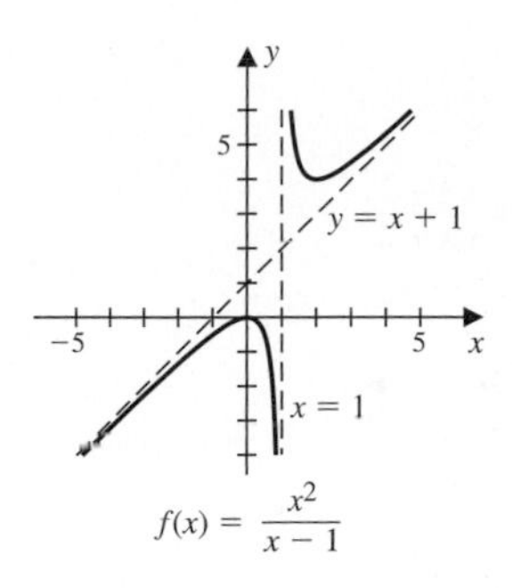

39.

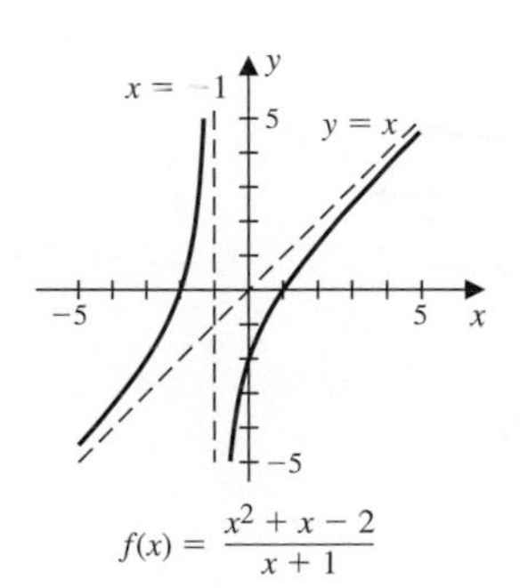

41.

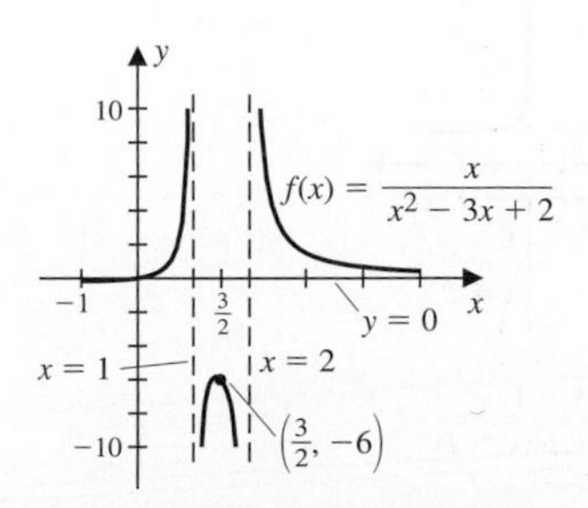

43.

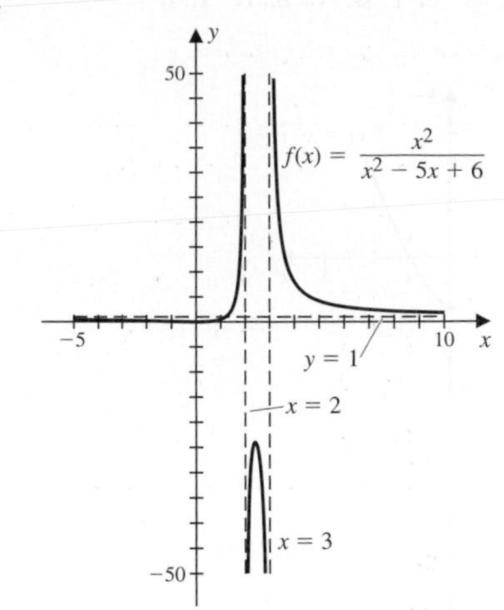

45. $r \approx 4.66, \quad h \approx 4.66$

47. **a.** The population does become stable.

b. The stabilizing level is 30,000.

c. $t > \frac{\sqrt{6}}{2}$

49. **a.** The parameter a determines the height of the curve. That is, the curve has a horizontal asymptote $y = a$. If $a > 0$, the curve lies above the x-axis and if $a < 0$, the curve lies below the x-axis.

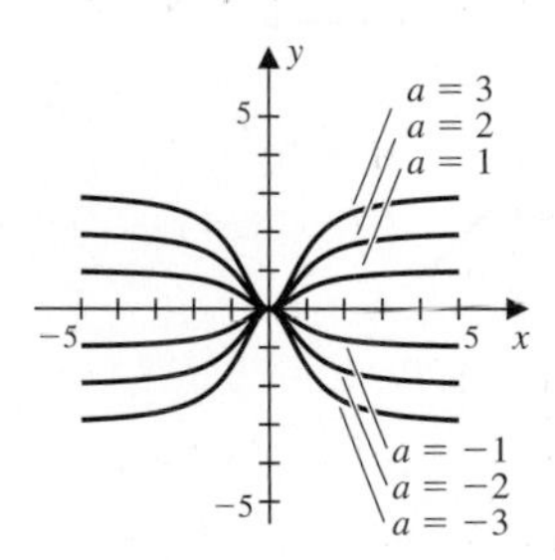

b. As t increases, the population stabilizes to the value a.

Exercise Set 3.5 (page 175)

1. $(-\infty, -5] \cup [3, \infty)$

3. $[-2, \infty)$

5. $[-4, -1] \cup [3, \infty)$

7. $(-3, 1]$

9. $(-3, -1] \cup (1, \infty)$

11. a. iii **b.** vi **c.** i **d.** v **e.** iv **f.** ii

13.

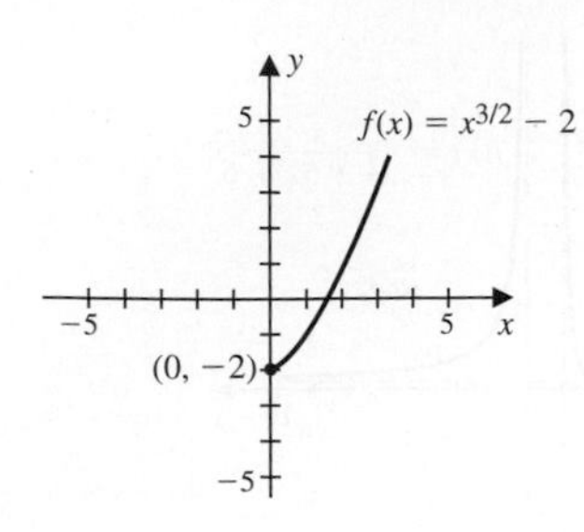

15.

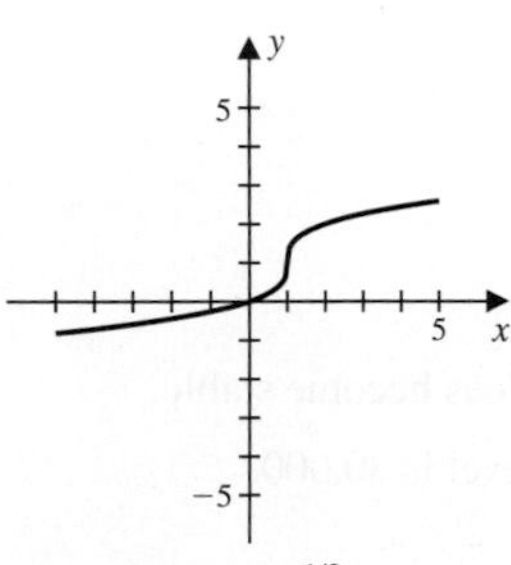

$f(x) = (x - 1)^{1/3} + 1$

17.

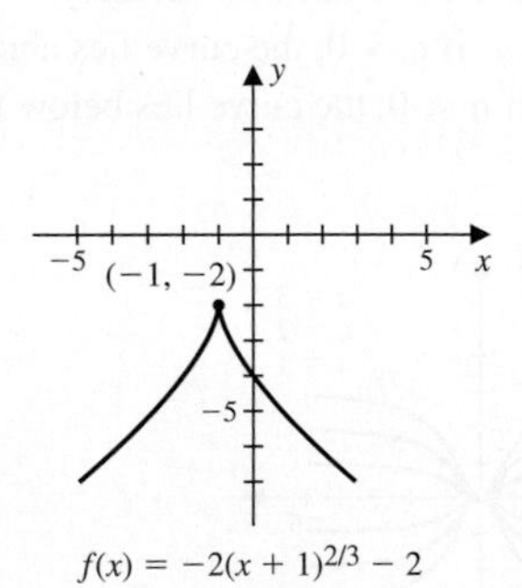

$f(x) = -2(x + 1)^{2/3} - 2$

19.

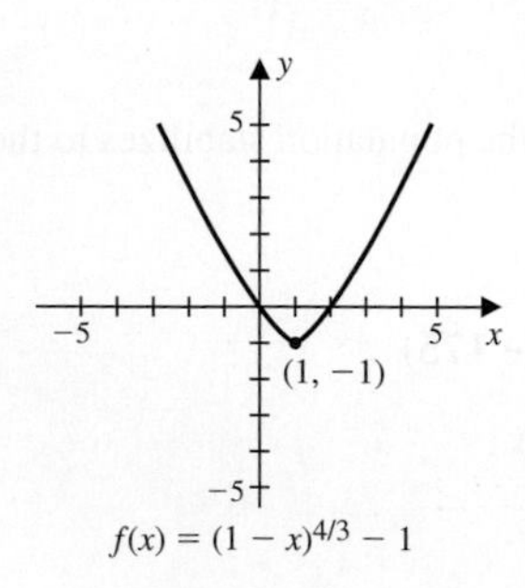

$f(x) = (1 - x)^{4/3} - 1$

21.

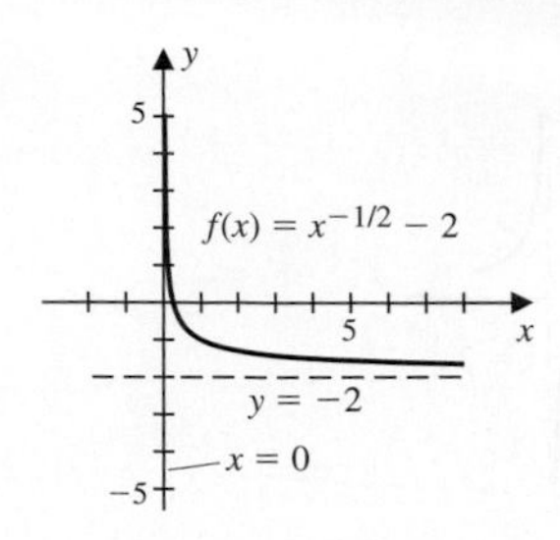

23.

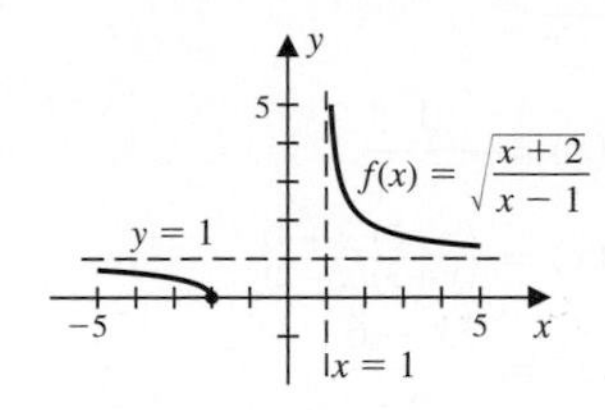

25.

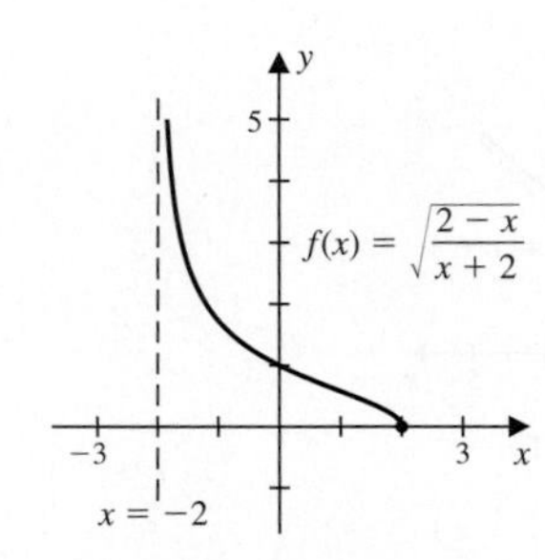

27.

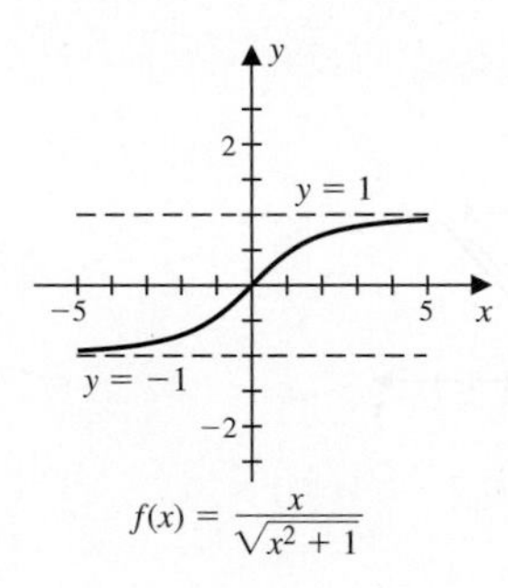

$f(x) = \dfrac{x}{\sqrt{x^2 + 1}}$

29.

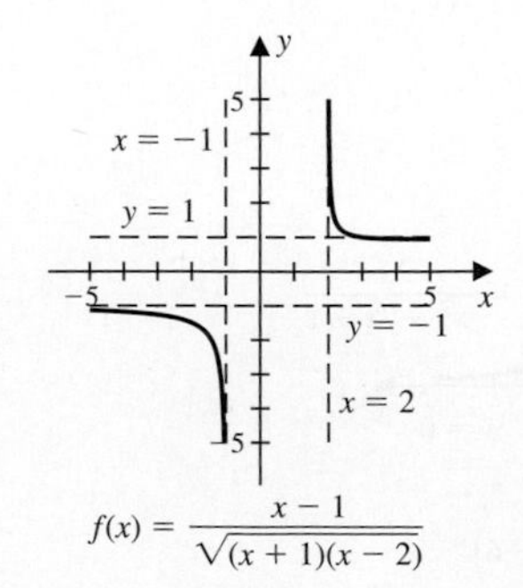

$f(x) = \dfrac{x - 1}{\sqrt{(x + 1)(x - 2)}}$

31.

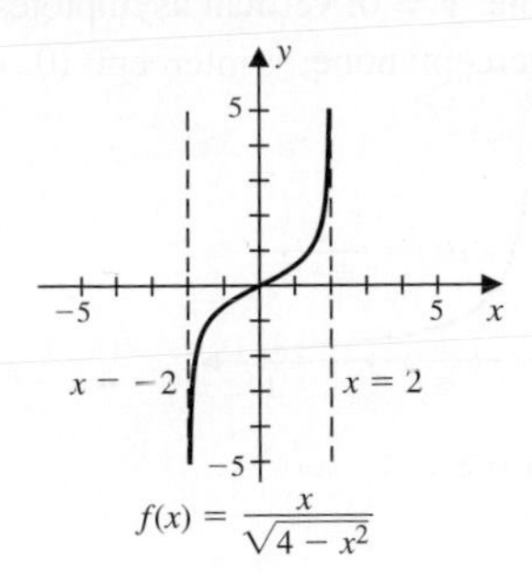

33. $w \approx 0.398$ kg

Exercise Set 3.6 (page 184)

1. $-1-2i$ **3.** $-5+8i$

5. $-3+6i$ **7.** $7-i$

9. $-8-27i$ **11.** 13

13. $-2-19i$ **15.** i

17. 1 **19.** $3i$

21. $(2-\sqrt{5})+(2+\sqrt{5})i$

23. $\frac{4}{3}i$ **25.** $\frac{1}{2}+\frac{1}{2}i$

27. $\frac{2}{13}-\frac{3}{13}i$

29. $-\frac{15}{17}-\frac{8}{17}i$

31. a. $x=\pm 2i$

b. $f(x)=(x-2i)(x+2i)$

33. a. $x=1\pm i$

b. $f(x)=(x-(1+i))(x-(1-i))$

35. a. $x=\frac{1\pm\sqrt{15}i}{4}$

b. $f(x)=\left(x-\left(\frac{1}{4}+\frac{\sqrt{15}}{4}i\right)\right)\left(x-\left(\frac{1}{4}-\frac{\sqrt{15}}{4}i\right)\right)$

37. $x=-\sqrt{2}, \sqrt{2}, -\sqrt{2}i, \sqrt{2}i$

39. $x=\sqrt{3}, -\sqrt{3}, \sqrt{2}i, -\sqrt{2}i$

41. $x=-2, 1+\sqrt{3}i, 1-\sqrt{3}i$

43. $x=3i, -3i, 2$

45. $x=i, -i, 3, -1$

47. $x=3, 3i, -3i$

49. $x=1$ (of multiplicity 2), $-1+i, -1-i$

51. x^3-2x^2+4x-8

53. x^4+12x^2+27

55. $x^5+2x^4+x^3+12x^2+20x$

57. To have a real number solution it must be true that $a \geq 2b$.

Review Exercises for Chapter 3 (page 185)

1. a. ii **b.** iii **c.** iv **d.** i

3.

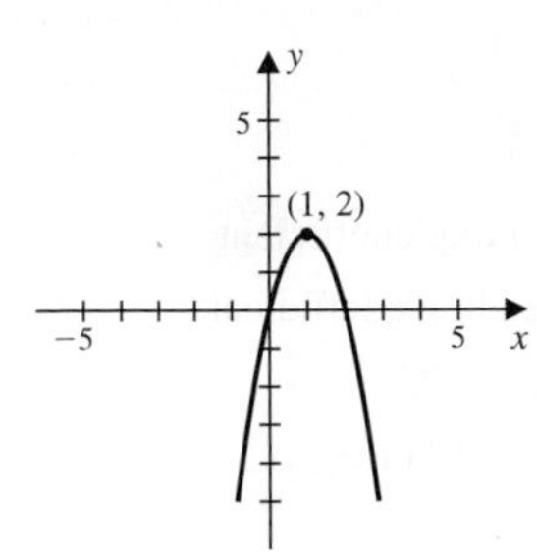

5.

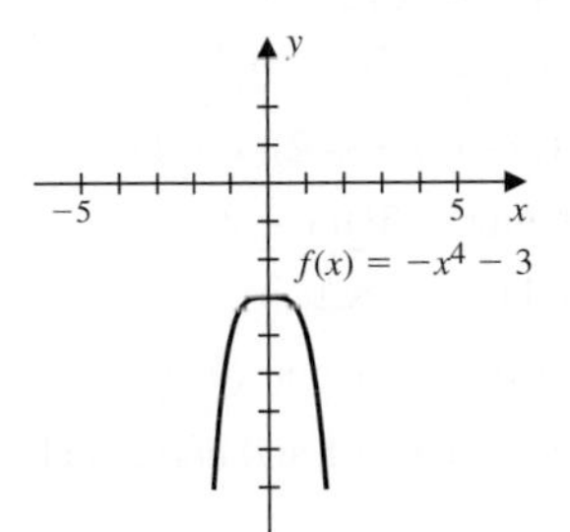

7.

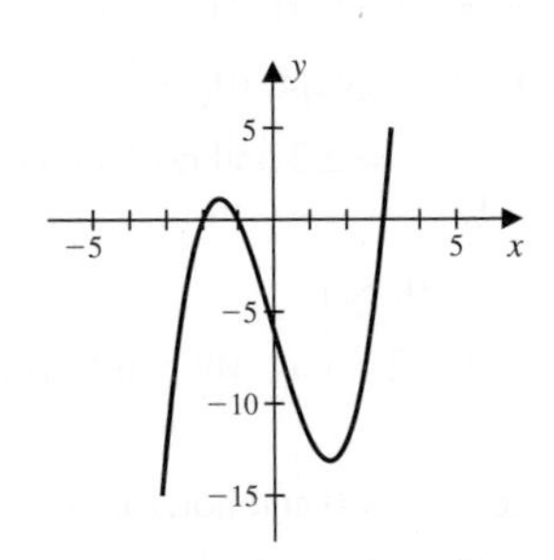

9.

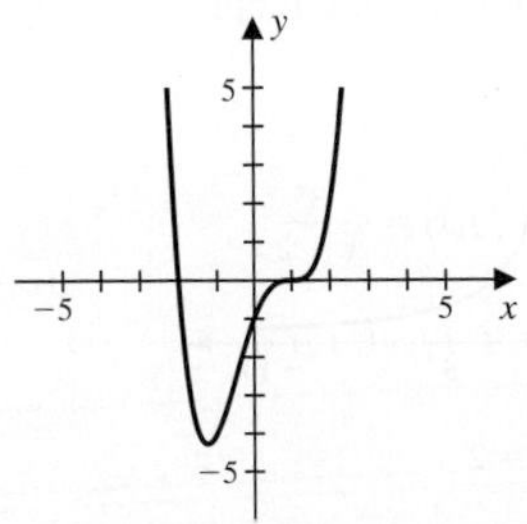

11.

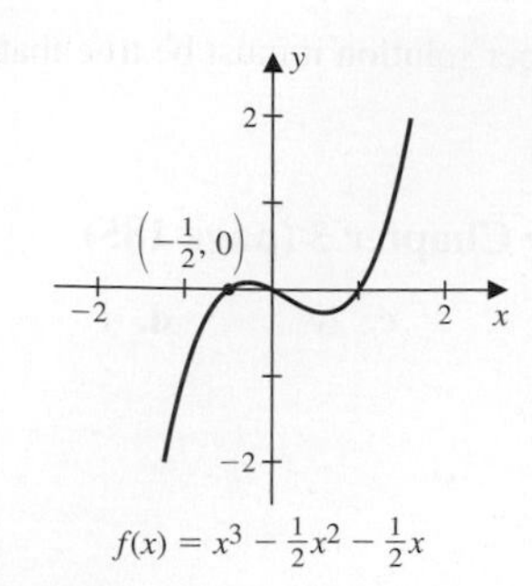

13. Degree 3; positive leading coefficient

15. Degree 4; negative leading coefficient

17. $Q(x) = 2x + 9$; $R(x) = 23$

19. $Q(x) = 3x^2 - 4x + 6$; $R(x) = -11$

21. **a.** $\pm 1, \pm 3, \pm\frac{1}{3}$

c. $P(x) = (x - 3)(x - 1)(3x^2 + 3x + 1)$

23. **a.** $\pm 1, \pm 2, \pm 3, \pm 4, \pm 6, \pm 12$

c. $P(x) = (x + 3)(x - 1)^2(x - 2)^2$

25. $(x - (3 + i))(x - (3 - i))(x + 2)(x - 1)$

27. $(x - 1)(x - i)(x + i)(x - 3i)(x + 3i)$

29. **a.** Domain: $(-\infty, 1) \cup (1, \infty)$

b. x-intercept: $(4, 0)$; y-intercept: $(0, 4)$

c. Vertical asymptote at $x = 1$ and horizontal asymptote at $y = 1$

31. **a.** Domain: $(-\infty, -3) \cup (-3, 3) \cup (3, \infty)$

b. x-intercept: $(1, 0)$; y-intercept: $(0, -\frac{1}{18})$

c. Vertical asymptote at $x = \pm 3$ and horizontal asymptote at $y = \frac{1}{2}$

33. **a.** Domain: $(-\infty, 0) \cup (0, \infty)$

b. x-intercepts: $(-2, 0)$, $(2, 0)$, $(1, 0)$; y-intercept: none

c. Vertical asymptote at $x = 0$ and horizontal asymptote at $y = 1$

35. Horizontal asymptote: $y = 0$; vertical asymptote: $x = 2$; x-intercept: none; y-intercept: $(0, -\frac{3}{2})$

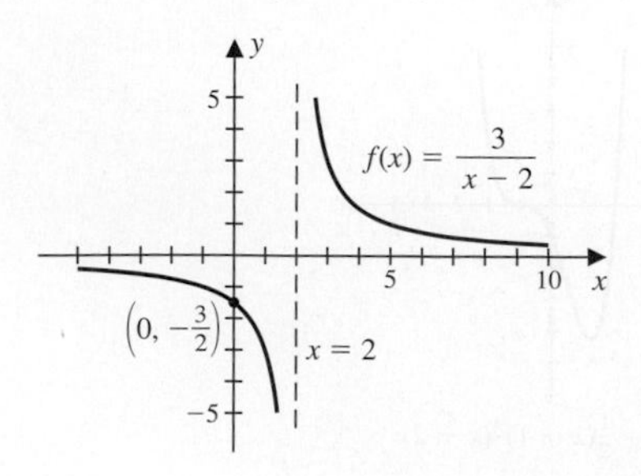

37. Horizontal asymptote: $y = 0$; vertical asymptotes: $x = 1$, $x = 2$; x-intercept: none; y-intercept: $(0, -1)$

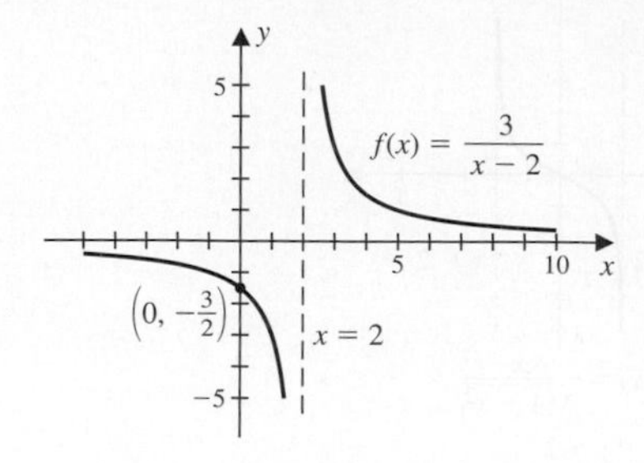

39. Horizontal asymptote: $y = 0$; vertical asymptotes: $x = -2$, $x = 2$; x-intercept: none; y-intercept: $(0, -1)$

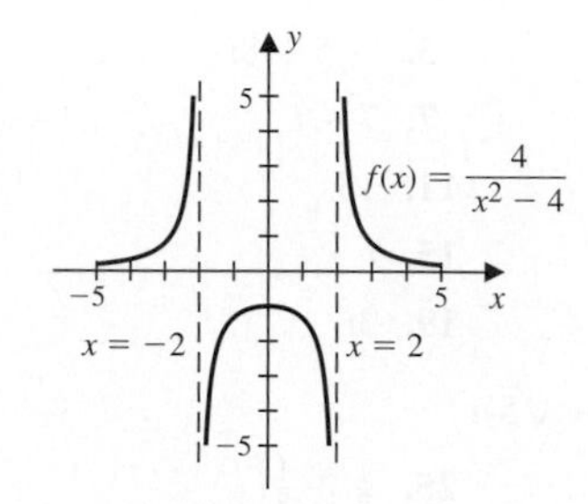

41. Horizontal asymptote: $y = 1$; vertical asymptote: $x = -5$, $x = 0$; x-intercepts: $(2, 0)$, $(-2, 0)$; y-intercept: none

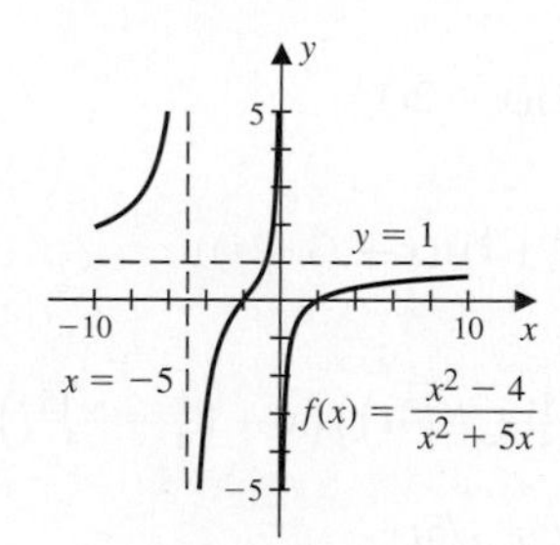

43. Vertical asymptote: $x = -1$; slant asymptote: $y = x - 3$; x-intercept: $(1, 0)$; y-intercept: $(0, 1)$

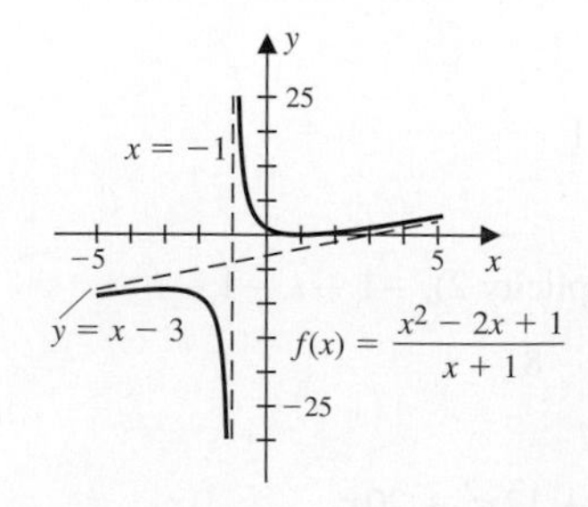

45. **a.** $(-\infty, -1) \cup [2, \infty)$

b.

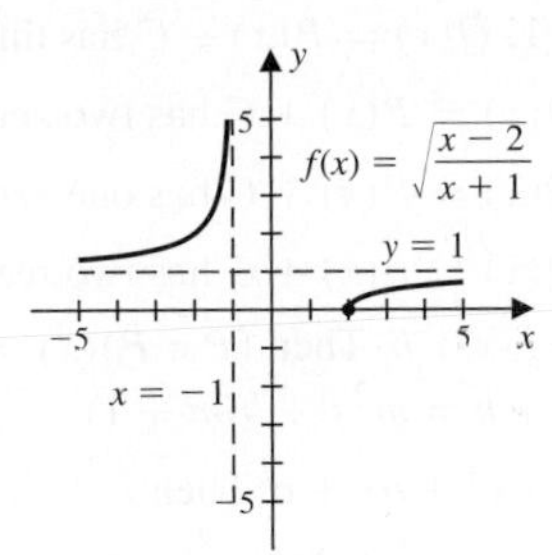

47. **a.** $(-4, 4]$

b.

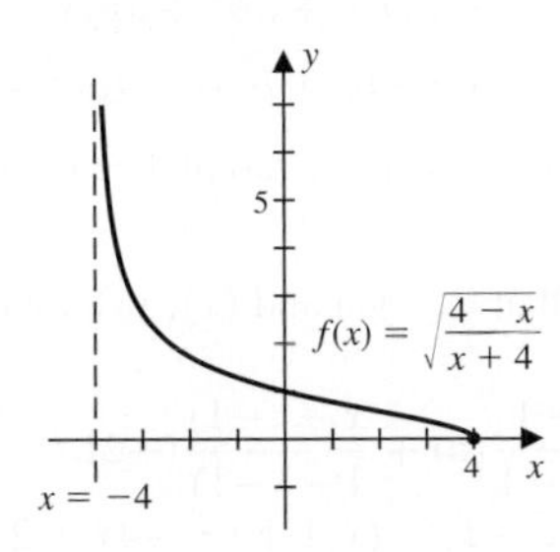

49. **a.** $(-3, 3)$

b.

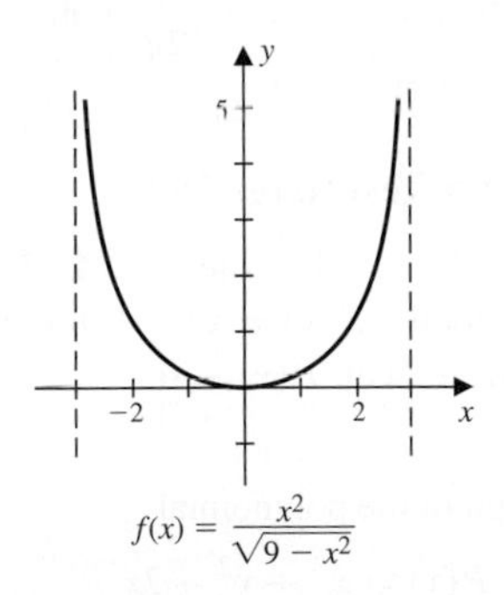

$f(x) = \frac{x^2}{\sqrt{9-x^2}}$

51. **a.** Increasing: $(-\infty, -0.2)$ and $(1.5, \infty)$; decreasing: $(-0.2, 1.5)$

b. Local maximum: $(-0.2, 2.1)$; local minimum: $(1.5, 0.6)$

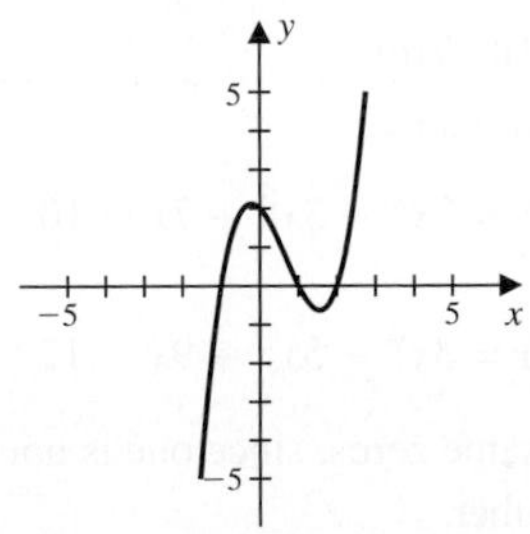

$f(x) = x^3 - 2x^2 - x + 2$

53. **a.** Increasing: $(-\infty, -4.7)$ and $(1.3, \infty)$; decreasing: $(-4.7, 1.3)$

b. Local maximum: $(-4.7, 64.5)$; local minimum: $(1.3, -8.2)$

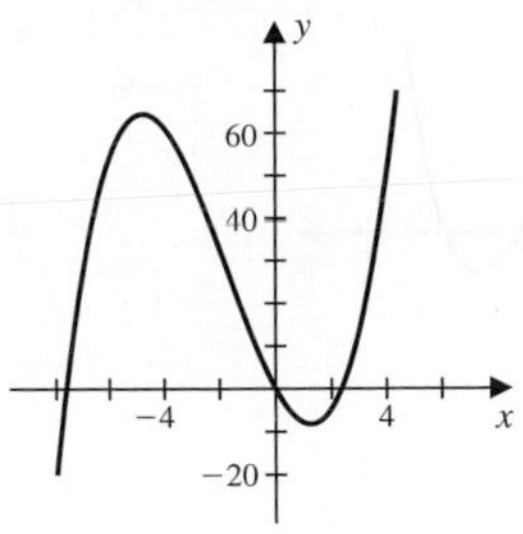

$f(x) = \frac{2}{3}x^3 + \frac{7}{2}x^2 - 12x$

55. $\frac{1}{2} + \frac{\sqrt{2}}{4}i$

57. $-1 + i$

59. $3 - 4i$

61. 1

63. $-\frac{1}{5} + \frac{2}{5}i$

65. **a.** $(2, -4)$ **b.** $(2, -5)$ **c.** $(0, 5)$ **d.** $(1, 4)$

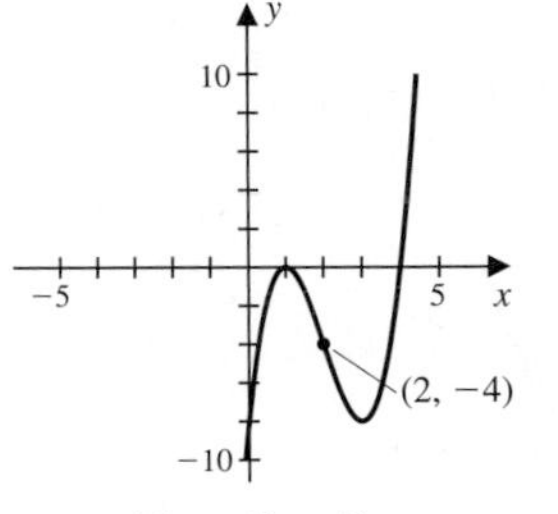

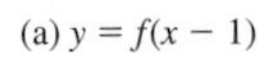

(a) $y = f(x - 1)$

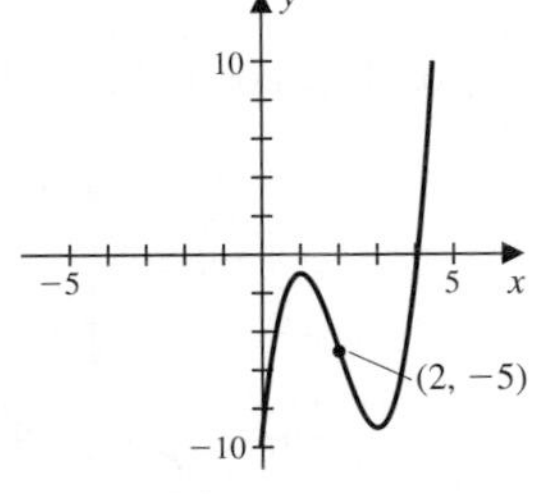

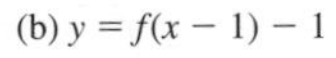

(b) $y = f(x - 1) - 1$

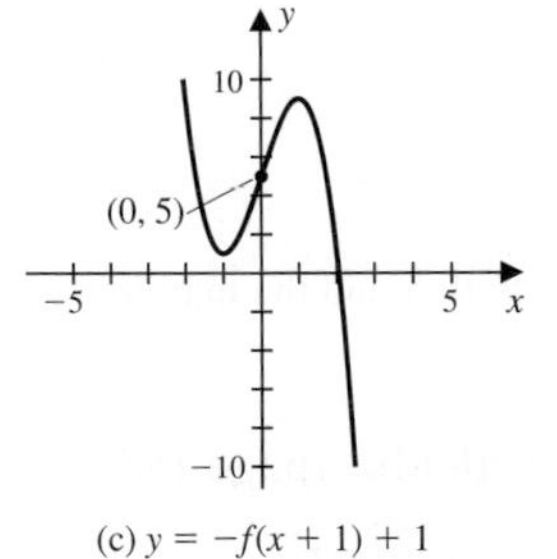

(c) $y = -f(x + 1) + 1$

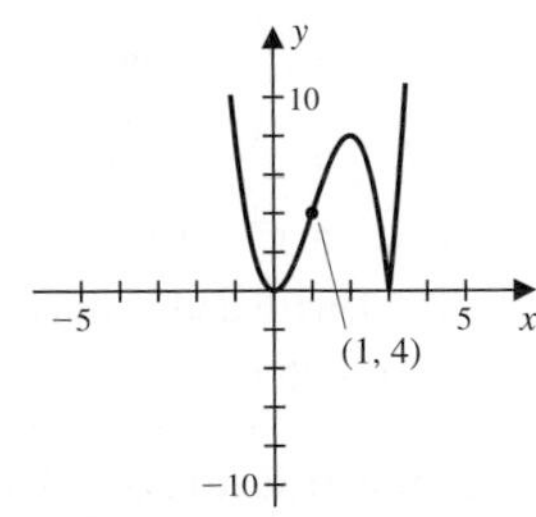

(d) $y = |f(x)|$

67.

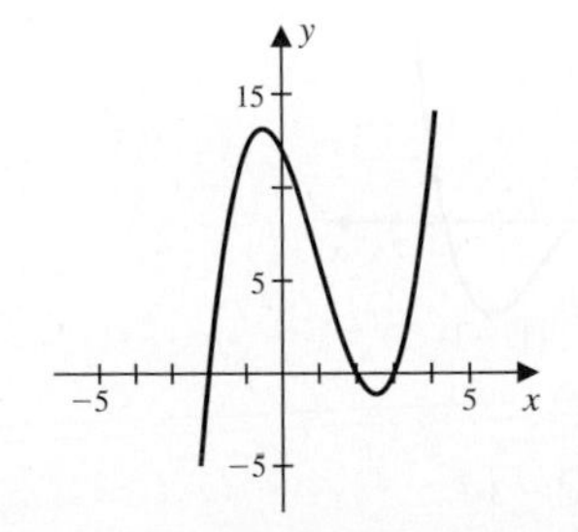

69. $P(x) = \frac{1}{3}x^3 - x^2 - \frac{1}{3}x + 1$

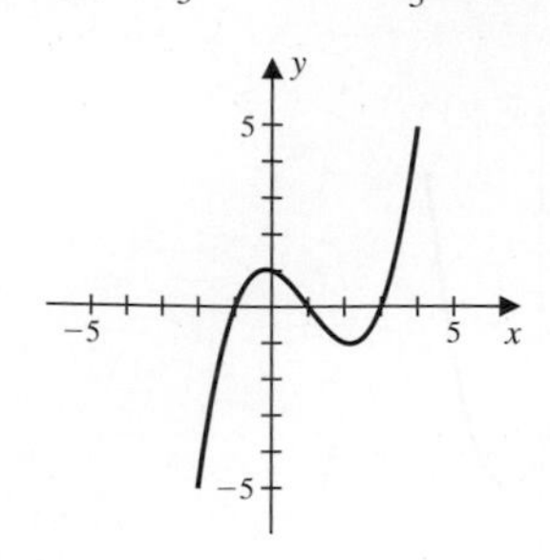

71. **(a)**

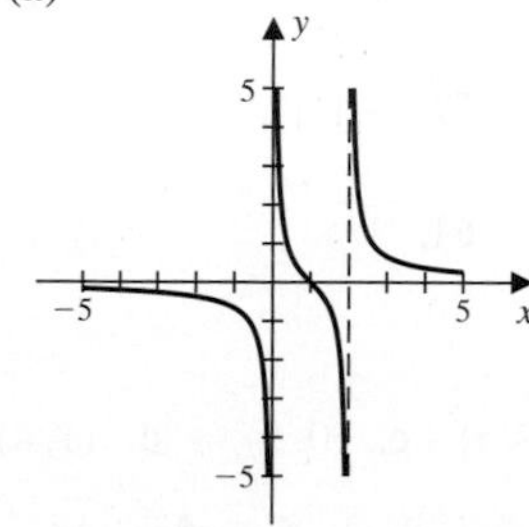

(b) One possibility is $f(x) = \frac{x-1}{x^2-2x}$.

73.

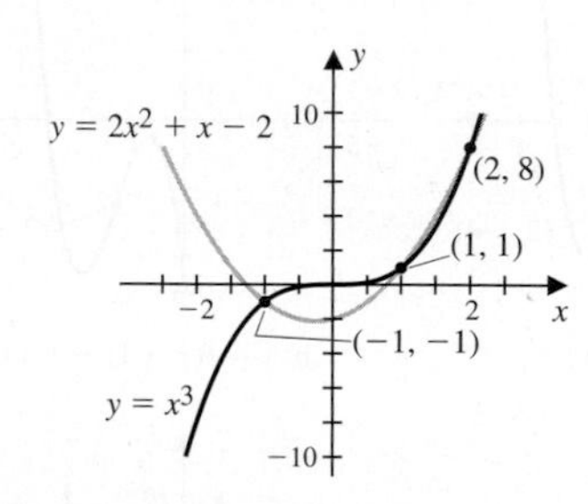

75. $P(x) = x^3 - x^2 + x - 1$

77. $P(x) = x^4 + 6x^2 + 8$

79. The length and width are both 5 and the maximum area is 25 ft^2.

Chapter 3: Exercises for Calculus (page 188)

1. **a.** $P(x) = 2x^3 - 3x^2 = x^2(2x - 3)$

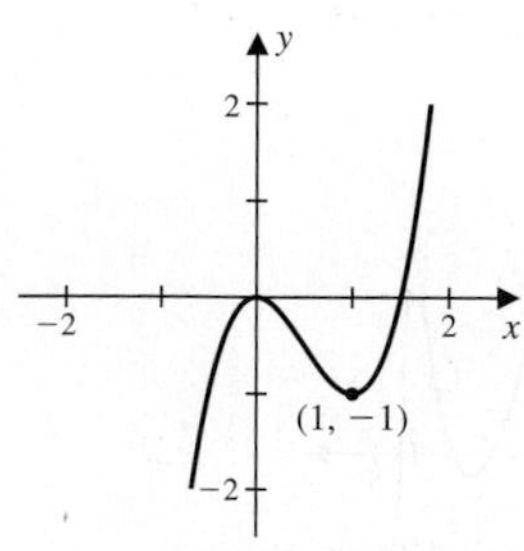

b. If $C < 0$, $Q(x) = P(x) + C$ has one zero.
If $0 < C < 1$, $Q(x) = P(x) + C$ has three zeros.
If $C = 1$, $Q(x) = P(x) + C$ has two zeros.
If $C > 1$, $Q(x) = P(x) + C$ has one zero.
If $C = 0$, $Q(x) = P(x) + C$ has two real zeros.

3. **a.** Let $P(x) = mx + b$. Then $(P \circ P)(x) = m(mx + b) + b = m^2x + b(m + 1)$.

b. Let $P(x) = ax^2 + bx + c$. Then

$$(P \circ P)(x) = a^3x^4 + 2a^2bx^3 + (2a^2c + ab^2 + ab)x^2 + (2abc + b^2)x + (ac^2 + bc + c).$$

5. The equation has at least one rational zero for $k = 1$, -3, -11, and -7.

7. **a.** Just verify that (x_1, y_1) and (x_2, y_2) are on the line.

b. $y = \frac{x-1}{-1-1}(6) + \frac{x-(-1)}{1-(-1)}(-2) = -3(x-1) - (x+1) = -4x + 2$

9. **a.** $s(0) = 0$ **b.** $t = \frac{2v_0}{g}$ **c.** $\frac{v_0^2}{2g}$ **d.** $t = \frac{v_0}{g}$

11. $V(x) = x(10 - x)(50 - 3x)$

Chapter 3: Chapter Test (page 190)

1. True. **2.** True. **3.** True. **4.** True.

5. False. The polynomial $P(x) = x^2 + 1$, for example, has no real numbers with $P(x) = 0$.

6. True.

7. False. The graph of the polynomial

$$P(x) = x^3 + x^2 - 2x$$

crosses the x-axis at $x = 1$, $x = -2$, and $x = 0$.

8. False. The graph of the polynomial $P(x) = x(x-1)^3(x+1)^2$ crosses the x-axis twice. It touches the axis at $x = -1$, but does not cross.

9. True. **10.** True.

11. False. The polynomials

$$P(x) = 2x^4 - 3x^3 + 7x - 10$$

and

$$Q(x) = 3x^4 - 5x^3 + 9x - 12$$

do not have the same zeros, since one is not a multiple of the other.

12. False. The polynomials $P(x) = x^4 - 2x^3 + 6x - 9$ and $Q(x) = x^4 - 2x^3 + 6x - 5$ do not have the same zeros,

13. False. The possible rational zeros of the polynomial $P(x) = 2x^5 - 2x^3 + x^2 - 3x + 5$ are

$$\pm 1, \pm\tfrac{1}{2}, \pm 5, \pm\tfrac{5}{2}.$$

14. False. The possible rational zeros of the polynomial $P(x) = 3x^4 - 2x^3 + x^2 - 2x + 4$ are

$$\pm 1, \pm 2, \pm 4, \pm\tfrac{1}{3}, \pm\tfrac{2}{3}, \pm\tfrac{4}{3}.$$

15. True.

16. False. The polynomial shown in the figure has a zero of odd multiplicity at $x = 2$.

17. True.

18. False. The degree of the polynomial shown in the figure is at least 6.

19. False. The polynomial shown in the figure has zeros at $x = -1$, $x = 1$, and $x = 2$.

20. False. The end-behavior of the polynomial shown in the figure is similar to the end-behavior of $-x^3$, but the given graph approaches the end-behavior more rapidly because its degree is higher.

21. True. **22.** True. **23.** True.

24. False. The graph has a horizontal asymptote $y = 1$.

25. False. It satisfies

$$f(x) \to 1 \text{ as } x \to \infty,$$
$$f(x) \to 1 \text{ as } x \to -\infty,$$
$$f(x) \to \infty \text{ as } x \to -2^+ \text{ and}$$
$$f(x) \to -\infty \text{ as } x \to -2^-.$$

26. False. The rational function

$$f(x) = \frac{x-1}{x^2-4}$$

has vertical asymptotes $x = 2$ and $x = -2$, as shown in the figure, but it has a horizontal asymptote $y = 0$ and y-intercept $(0, \frac{1}{4})$, which does not agree with the figure. Since the figure has a horizontal asymptote $y = 1$ and y-intercept $(0, \frac{5}{4})$, a possible equation is

$$f(x) = \frac{x-1}{x^2-4} + 1.$$

27. False. The graph of f has vertical asymptotes $x = 1$ and $x = -1$.

28. True.

29. False. The domain of the rational function is $(-\infty, -1) \cup (-1, 1) \cup (1, \infty)$.

30. False. The range of the rational function is $(-\infty, 1] \cup (2, \infty)$.

31. True. **32.** True. **33.** True.

34. False. One zero of the polynomial $P(x) = x^5 - 5x^4 + 6x^3 + 4x^2 - 8x$ is $x = 0$, and another is $x = -1$. Factoring $P(x)$ using these observations gives

$$x^5 - 5x^4 + 6x^3 + 4x^2 - 8x$$
$$= x(x+1)(x^3 - 6x^2 + 12x - 8)$$

and the right-most term has a zero at $x = 2$. Continuing we have

$$x(x+1)(x^3 - 6x^2 + 12x - 8)$$
$$= x(x+1)(x-2)(x^2 - 4x + 4)$$
$$= x(x+1)(x-2)^3.$$

So, $P(x)$ has a zero of multiplicity 3 at $x = 2$.

35. False. The rational function

$$f(x) = \frac{1-x^2}{x^2+1}$$

has no vertical asymptotes, but it has a horizontal asymptote at $y = -1$.

36. True.

37. False. The rational function

$$f(x) = \frac{x^2-1}{x^2-4}$$

has vertical asymptotes at $x = 2$ and $x = -2$.

38. False. The rational function

$$f(x) = \frac{x^2+5x+4}{x^2-2x-3} = \frac{x+4}{x-3}, \text{ if } x \neq -1,$$

has a vertical asymptote at $x = 3$.

39. False. The function

$$f(x) = \frac{x^2+1}{x^2-3x+2}$$

satisfies

$$f(x) \to -\infty \text{ as } x \to 1^+$$
$$f(x) \to \infty \text{ as } x \to 1^-.$$

40. True.

41. True.

42. False. The function

$$f(x) = \frac{2x^3 - 3x^2 + x - 1}{3x^3 - 4x^2 - 2}$$

has a horizontal asymptote $y = \frac{2}{3}$.

43. True.

44. False. For large values of x,

$$\frac{10x^2 - x + 1}{5x^2} \geq \frac{5x^3 - x^2 - 2x + 1}{3x^3 + 1},$$

since

$$\frac{10x^2 - x + 1}{5x^2} \to 2, \text{ as } x \to \infty$$

and

$$\frac{5x^3 - x - 2x + 1}{3x^3 + 1} \to \frac{5}{3}, \text{ as } x \to \infty.$$

Chapter 4

Exercise Set 4.2 (page 200)

1. $\pi/6$ **3.** $5\pi/6$ **5.** $5\pi/4$

7. $-2\pi/5$ **9.** $135°$ **11.** $-330°$

13. $810°$ **15.** yes **17.** yes **19.** no

21. a.

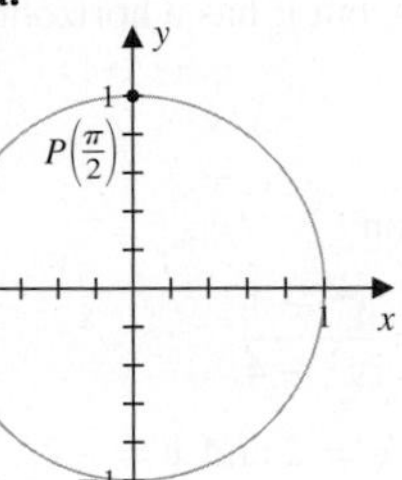

b.

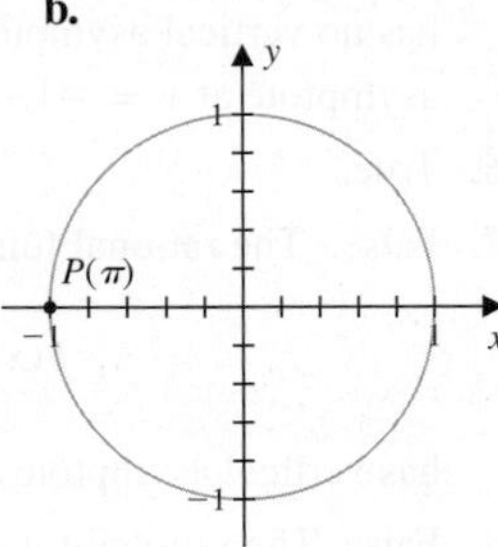

c.

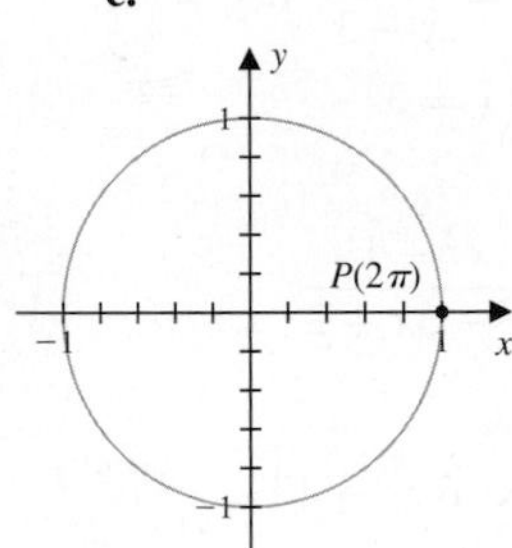

d.

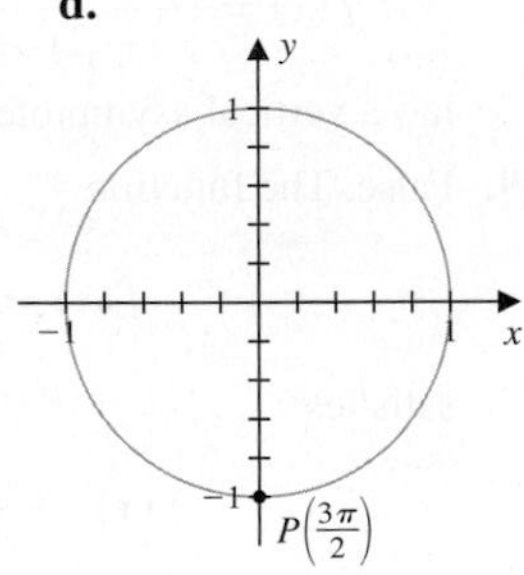

23. a.

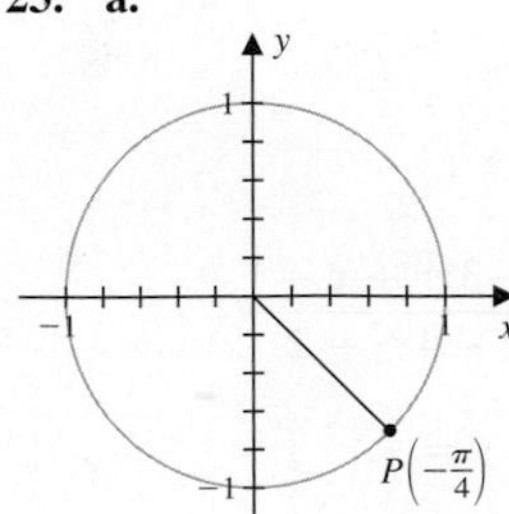

b.

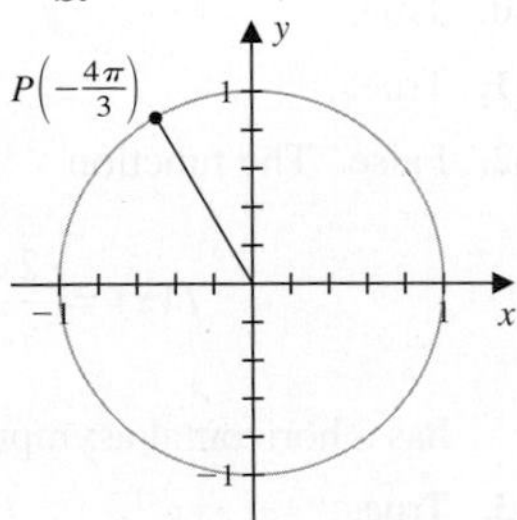

c. **d.**

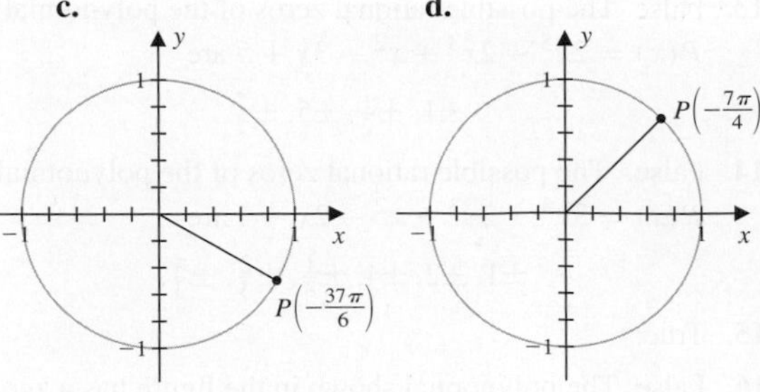

25. a. $P(t + \pi) = \left(-\frac{3}{5}, -\frac{4}{5}\right)$

b. $P(-t) = \left(\frac{3}{5}, -\frac{4}{5}\right)$

c. $P(t - \pi) = \left(-\frac{3}{5}, -\frac{4}{5}\right)$

d. $P(-t - \pi) = \left(-\frac{3}{5}, \frac{4}{5}\right)$

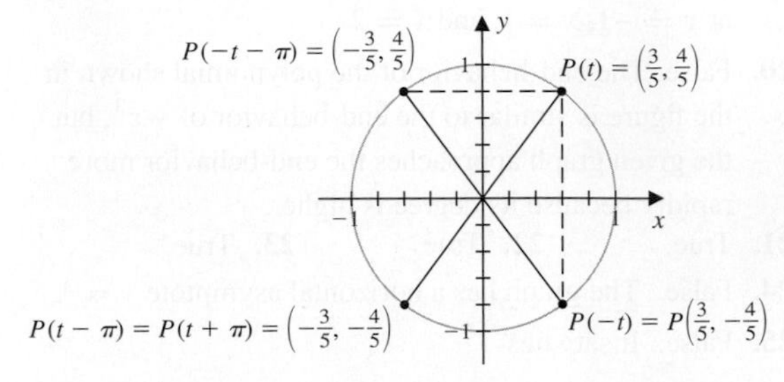

27. a. $P(t + \pi) = \left(\frac{\sqrt{5}}{3}, -\frac{2}{3}\right)$

b. $P(-t) = \left(-\frac{\sqrt{5}}{3}, -\frac{2}{3}\right)$

c. $P(t - \pi) = \left(\frac{\sqrt{5}}{3}, -\frac{2}{3}\right)$

d. $P(-t - \pi) = \left(\frac{\sqrt{5}}{3}, \frac{2}{3}\right)$

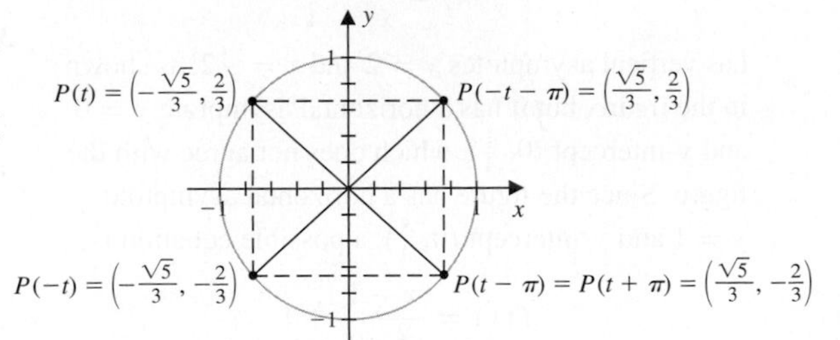

29. 2π

31. $\theta = 2 \text{ radians} = 2\left(\frac{180}{\pi}\right)^{\circ} = \frac{360}{\pi}^{\circ}$

33. $s = 4\left(110 \frac{\pi}{180}\right) = \frac{22\pi}{9}$

35. 25 **37.** $r = \frac{4}{\sqrt{\pi}}$ **39.** $\frac{26\pi}{9}$

41. $3960(43 - 36.5)\frac{\pi}{180} = 143\pi$

43. Approximately 330π miles per hour.

45. a. $4\rho = \theta$ **b.** $5/4$ revolutions per second

Exercise Set 4.3 (page 206)

1. $\sin\theta = \frac{2}{5}$; $\cos\theta = \frac{\sqrt{21}}{5}$; $\tan\theta = \frac{2\sqrt{21}}{21}$; $\cot\theta = \frac{\sqrt{21}}{2}$; $\sec\theta = \frac{5\sqrt{21}}{21}$; $\csc\theta = \frac{5}{2}$

3. $\sin\theta = \frac{1}{2}$; $\cos\theta = \frac{\sqrt{3}}{2}$; $\tan\theta = \frac{\sqrt{3}}{3}$; $\cot\theta = \sqrt{3}$; $\sec\theta = \frac{2\sqrt{3}}{3}$; $\csc\theta = 2$

5. $\sin\theta = \frac{5}{13}$; $\cos\theta = \frac{12}{13}$; $\tan\theta = \frac{5}{12}$; $\cot\theta = \frac{12}{5}$; $\sec\theta = \frac{13}{12}$; $\csc\theta = \frac{13}{5}$

7. $x = 8\sqrt{3}$

9. $x = 7\sqrt{2}$

11. $x = 2\sqrt{3}$

13. $\beta = 60°$; $\overline{AB} = 8\sqrt{3}$; $\overline{AC} = 16$

15. $\alpha = 45°$; $\overline{AB} = \overline{BC} = \frac{5\sqrt{2}}{2}$

17. a. $\cos\theta = \frac{\sqrt{5}}{3}$; $\tan\theta = \frac{2\sqrt{5}}{5}$; $\cot\theta = \frac{\sqrt{5}}{2}$; $\sec\theta = \frac{3\sqrt{5}}{5}$; $\csc\theta = \frac{3}{2}$

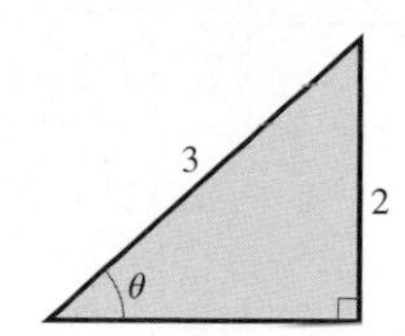

b. $\sin\theta = \frac{2\sqrt{6}}{5}$; $\tan\theta = 2\sqrt{6}$; $\cot\theta = \frac{\sqrt{6}}{12}$; $\sec\theta = 5$; $\csc\theta = \frac{5\sqrt{6}}{12}$

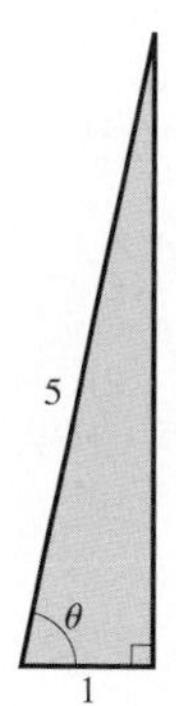

c. $\sin\theta = \frac{4}{5}$; $\cos\theta = \frac{3}{5}$; $\cot\theta = \frac{3}{4}$; $\sec\theta = \frac{5}{3}$; $\csc\theta = \frac{5}{4}$

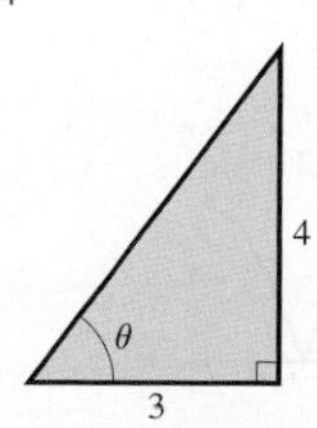

d. $\sin\theta = \frac{2}{3}$; $\cos\theta = \frac{\sqrt{5}}{3}$; $\tan\theta = \frac{2\sqrt{5}}{5}$; $\cot\theta = \frac{\sqrt{5}}{2}$; $\sec\theta = \frac{3\sqrt{5}}{5}$

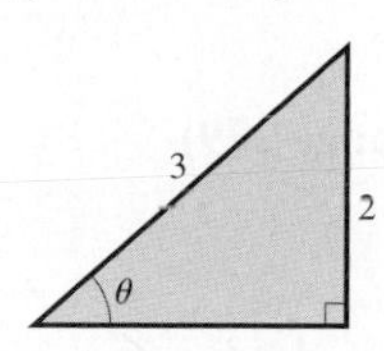

19. About $3200\sqrt{2} \approx 4530$ km.

21. $h = \tan 30° \approx 0.577$ mi ≈ 3048 ft

23. $h = 40\tan 60° = 40\sqrt{3} \approx 69.3$ ft

25. $x = 80\cot 10.5° \approx 431.64$ ft

Exercise Set 4.4 (page 219)

1. a. $r = \pi/4$ **b.** $r = \pi/3$ **c.** $r = \pi/3$ **d.** $r = \pi/4$

3. a. $r = \pi/3$ **b.** $r = \pi/6$ **c.** $r = \pi/3$ **d.** $r = \pi/4$

5. a. $r = 45°$ **b.** $r = 30°$ **c.** $r = 60°$ **d.** $r = 30°$

7. $\sin\frac{\pi}{6} = \frac{1}{2}$; $\cos\frac{\pi}{6} = \frac{\sqrt{3}}{2}$

9. $\sin\frac{5\pi}{6} = \sin\frac{\pi}{6} = \frac{1}{2}$; $\cos\frac{5\pi}{6} = -\cos\frac{\pi}{6} = -\frac{\sqrt{3}}{2}$

11. $\sin\frac{4\pi}{3} = -\frac{\sqrt{3}}{2}$; $\cos\frac{4\pi}{3} = -\frac{1}{2}$

13. $\sin\frac{13\pi}{6} = \sin\frac{\pi}{6} = \frac{1}{2}$; $\cos\frac{13\pi}{6} = \cos\frac{\pi}{6} = \frac{\sqrt{3}}{2}$

15. $\sin\left(-\frac{\pi}{3}\right) = -\frac{\sqrt{3}}{2}$; $\cos\left(-\frac{\pi}{3}\right) = \frac{1}{2}$

17. $\sin\left(-\frac{5\pi}{6}\right) = -\frac{1}{2}$; $\cos\left(-\frac{5\pi}{6}\right) = -\frac{\sqrt{3}}{2}$

19. $\sin\left(-\frac{5\pi}{4}\right) = \frac{\sqrt{2}}{2}$; $\cos\left(-\frac{5\pi}{4}\right) = -\frac{\sqrt{2}}{2}$

21. $\sin\left(-\frac{3\pi}{2}\right) = 1$; $\cos\left(-\frac{3\pi}{2}\right) = 0$

23. $\sin(-7\pi) = 0$; $\cos(-7\pi) = -1$

25. $\sin(45°) = \frac{\sqrt{2}}{2}$; $\cos(45°) = \frac{\sqrt{2}}{2}$

27. $\sin(150°) = \frac{1}{2}$; $\cos(150°) = -\frac{\sqrt{3}}{2}$

29. $\sin(240°) = -\frac{\sqrt{3}}{2}$; $\cos(240°) = -\frac{1}{2}$

31. $\sin(330°) = -\frac{1}{2}$; $\cos(330°) = \frac{\sqrt{3}}{2}$

33. $\sin(-60°) = -\frac{\sqrt{3}}{2}$; $\cos(-60°) = \frac{1}{2}$

35. $\sin(-240°) = \frac{\sqrt{3}}{2}$; $\cos(-240°) = -\frac{1}{2}$

37. $\sin 540 = 0$; $\cos 540 = -1$

39. $t = \frac{\pi}{4}, \frac{7\pi}{4}$

41. $t = \frac{7\pi}{6}, \frac{11\pi}{6}$

43. $t = 0, 2\pi$

45. $t = \frac{2\pi}{3}$

47. Even

49. Odd

51. a. $\sin(t+\pi) = \frac{2\sqrt{2}}{3}$; $\cos(t+\pi) = -\frac{1}{3}$

b. $\sin(-t) = \frac{2\sqrt{2}}{3}$; $\cos(-t) = \frac{1}{3}$

c. $\sin(t+\frac{\pi}{2}) = \frac{1}{3}$; $\cos(t+\frac{\pi}{2}) = \frac{2\sqrt{2}}{3}$

d. $\sin(-t+\pi/2) = \frac{1}{3}$; $\cos(-t+\frac{\pi}{2}) = -\frac{2\sqrt{2}}{3}$

53. $t = \frac{\pi}{6}, \frac{5\pi}{6}, \frac{3\pi}{2}$ **55.** $0, \frac{\pi}{2}, 2\pi$

57. $4000\sqrt{3}$ ft **59.** 30,000 ft

Exercise Set 4.5 (page 229)

1. a.

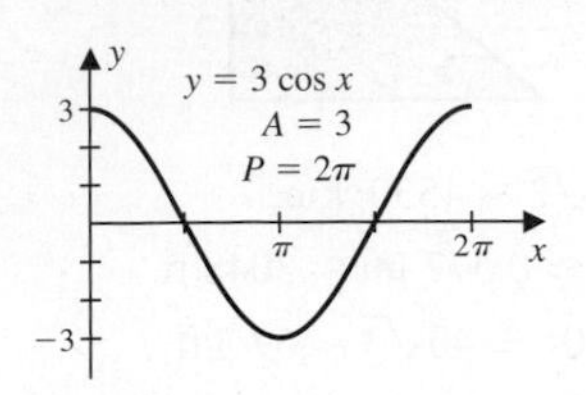

b.

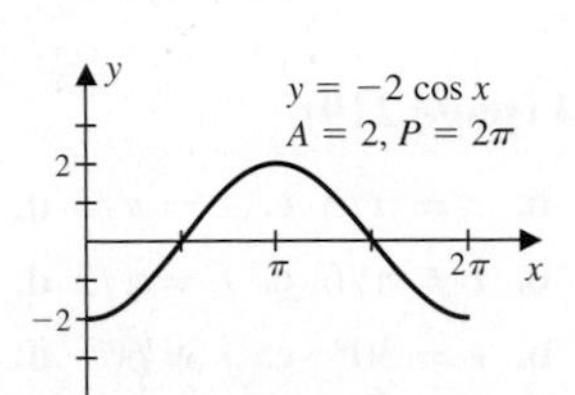

c.

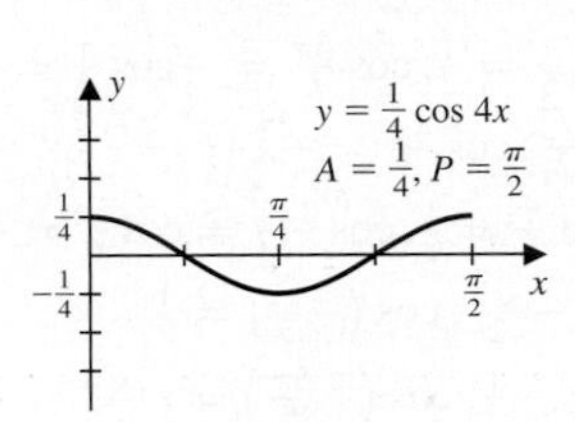

3. a.

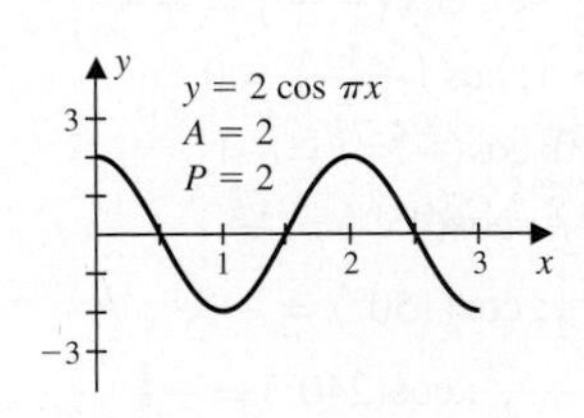

b.

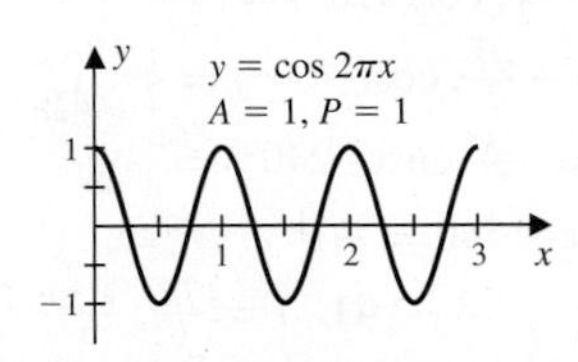

c.

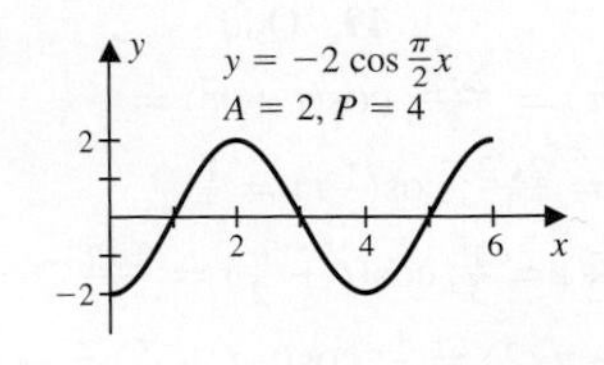

5.

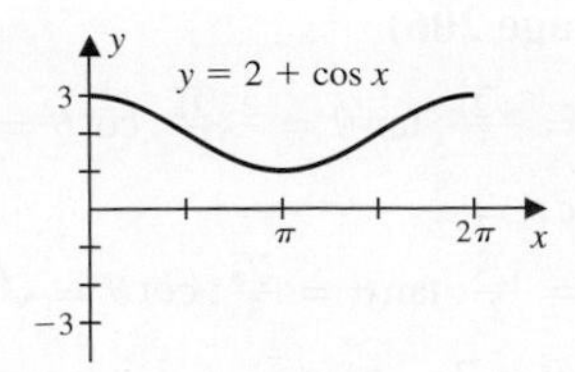

7.

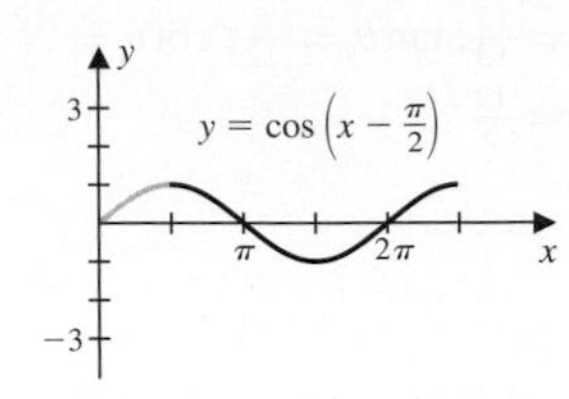

9.

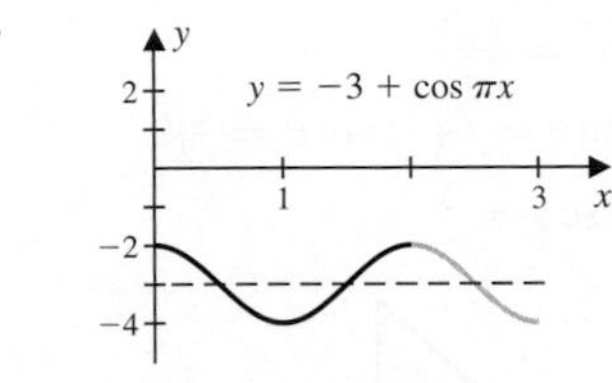

11.

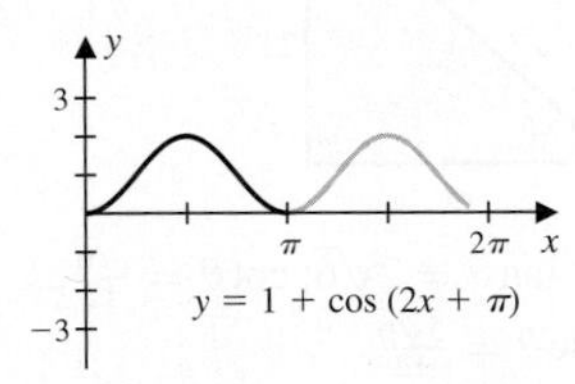

13.

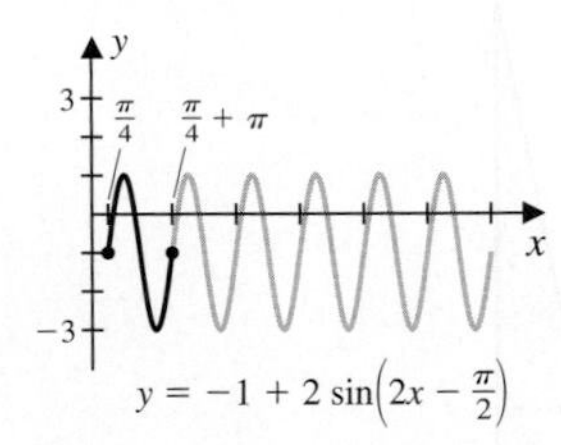

15.

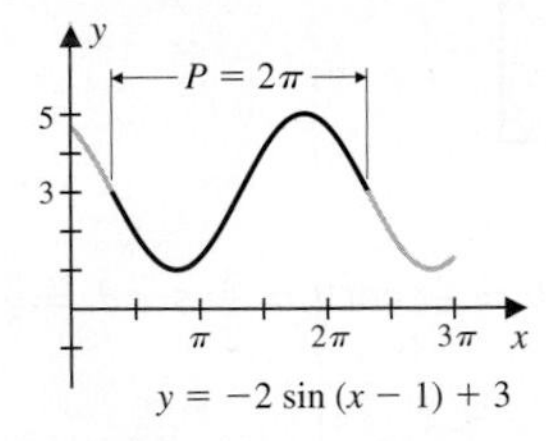

17.

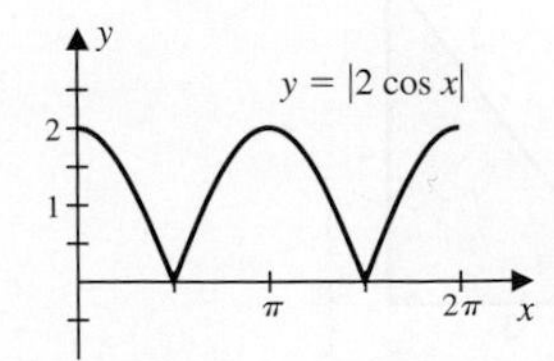

19. **a.** $y = 3\cos 2x$

b. $y = 3\sin 2(x + \frac{\pi}{4})$

21. **a.** $y = 2\cos\frac{\pi}{5}\left(x - \frac{9}{2}\right) + 1$

b. $y = 2\sin\frac{\pi}{5}(x - 2) + 1$

23. **a.** iv **b.** i **c.** iii **d.** ii

25. $y = -1 + \frac{1}{2}\sin(x - 2)$

27. $\left[-\frac{\pi}{50}, \frac{\pi}{50}\right] \times [-1, 1]$

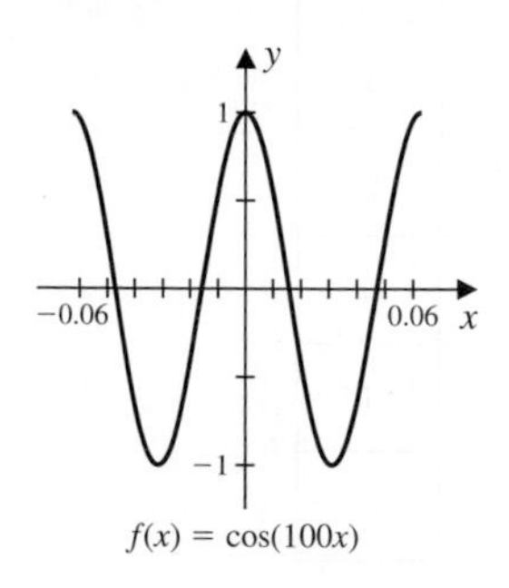

$f(x) = \cos(100x)$

29. $[-100\pi, 100\pi] \times [-1, 1]$

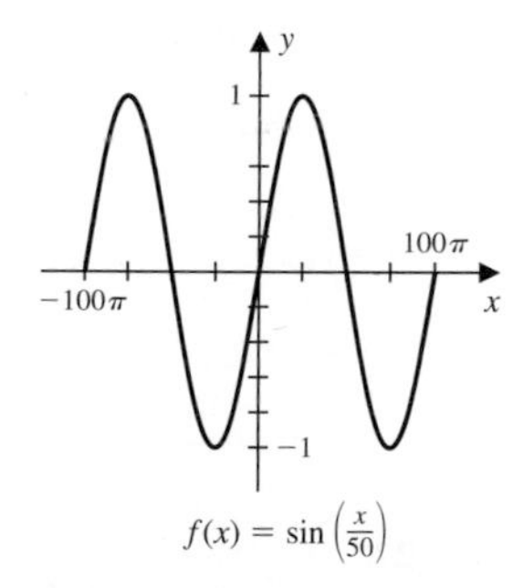

$f(x) = \sin\left(\frac{x}{50}\right)$

31. The curves intersect when $x \approx 0.7$.

33. The curves intersect when $x \approx 1.3$.

35. **a.** $t = \frac{\pi}{12}$ **b.** $t = \frac{\pi}{4}$ **c.** $t = \frac{7\pi}{12}$

37. $y = 1 + 2\sin\frac{\pi}{2}x$

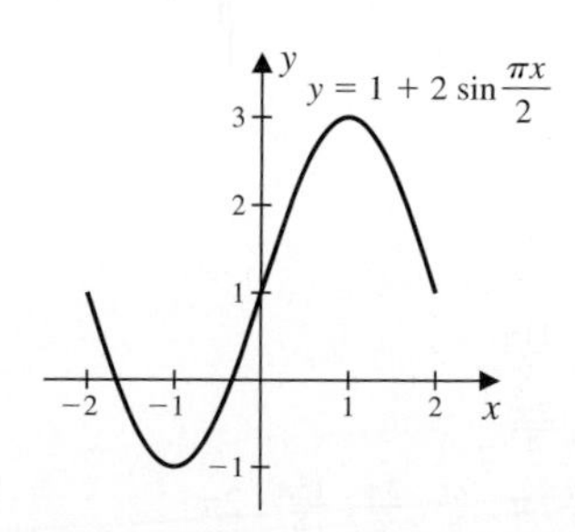

39. **a.**

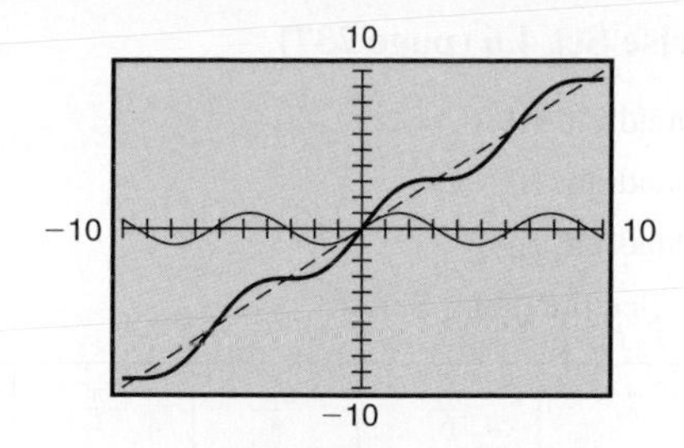

b.

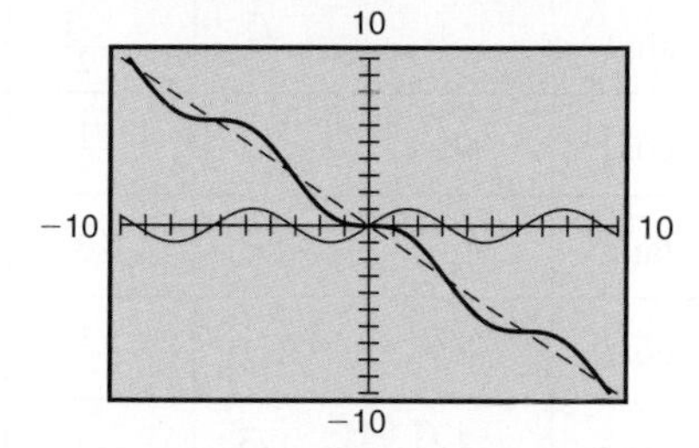

c.

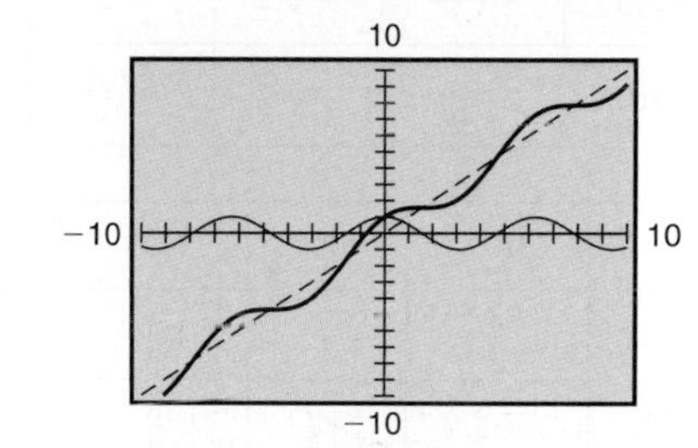

d.

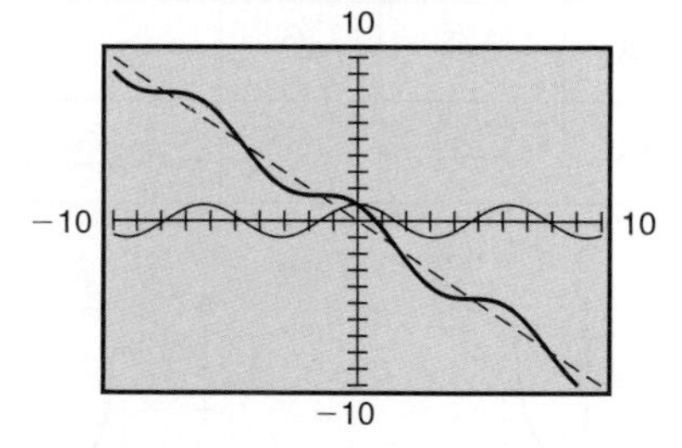

41. **a.** It vertically stretches the curve by a factor of 2. That is, it doubles the amplitude.

b. It horizontally compresses the curve by a factor of 2. That is, the period is halved.

c. The horizontal shift is doubled.

43. **a.** $f(x) = 22\sin\left(\frac{\pi}{6}(x - 4.5)\right) + 48$

b.

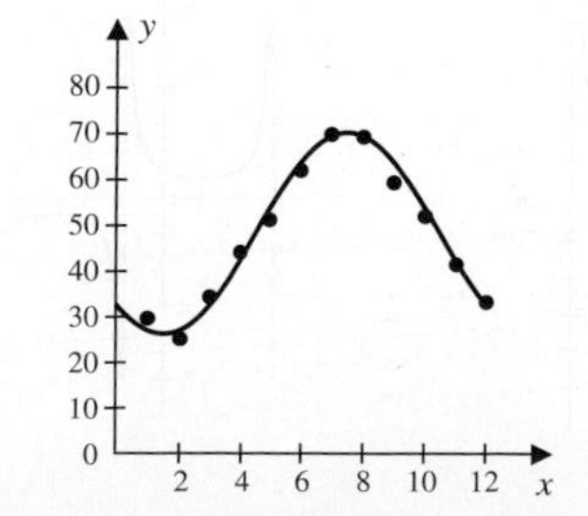

Exercise Set 4.6 (page 237)

1. Quadrant III
3. Quadrant III
5. Quadrant I

7–15. See the tables below.

7.	$\frac{7\pi}{6}$	$\frac{5\pi}{4}$	$\frac{4\pi}{3}$	$\frac{3\pi}{2}$	$\frac{5\pi}{3}$	$\frac{7\pi}{4}$	$\frac{11\pi}{6}$	2π
$\sin t$	$-\frac{1}{2}$	$-\frac{\sqrt{2}}{2}$	$-\frac{\sqrt{3}}{2}$	-1	$-\frac{\sqrt{3}}{2}$	$-\frac{\sqrt{2}}{2}$	$-\frac{1}{2}$	0
$\cos t$	$-\frac{\sqrt{3}}{2}$	$-\frac{\sqrt{2}}{2}$	$-\frac{1}{2}$	0	$\frac{1}{2}$	$\frac{\sqrt{2}}{2}$	$\frac{\sqrt{3}}{2}$	1
$\tan t$	$\frac{\sqrt{3}}{3}$	1	$\sqrt{3}$	—	$-\sqrt{3}$	-1	$-\frac{\sqrt{3}}{3}$	0
$\cot t$	$\sqrt{3}$	1	$\frac{\sqrt{3}}{3}$	0	$-\frac{\sqrt{3}}{3}$	-1	$-\sqrt{3}$	—
$\sec t$	$-\frac{2\sqrt{3}}{3}$	$-\sqrt{2}$	-2	—	2	$\sqrt{2}$	$\frac{2\sqrt{3}}{3}$	1
$\csc t$	-2	$-\sqrt{2}$	$-\frac{2\sqrt{3}}{3}$	-1	$-\frac{2\sqrt{3}}{3}$	$-\sqrt{2}$	-2	—

	$\sin t$	$\cos t$	$\tan t$	$\cot t$	$\sec t$	$\csc t$
9.	$\frac{4}{5}$	$\frac{3}{5}$	$\frac{4}{3}$	$\frac{3}{4}$	$\frac{5}{3}$	$\frac{5}{4}$
11.	$-\frac{3}{5}$	$-\frac{4}{5}$	$\frac{3}{4}$	$\frac{4}{3}$	$-\frac{5}{4}$	$-\frac{5}{3}$
13.	$\frac{2\sqrt{5}}{5}$	$\frac{\sqrt{5}}{5}$	2	$\frac{1}{2}$	$\sqrt{5}$	$\frac{\sqrt{5}}{2}$
15.	$-\frac{2\sqrt{2}}{3}$	$\frac{1}{3}$	$-2\sqrt{2}$	$-\frac{\sqrt{2}}{4}$	3	$-\frac{3\sqrt{2}}{4}$

17.
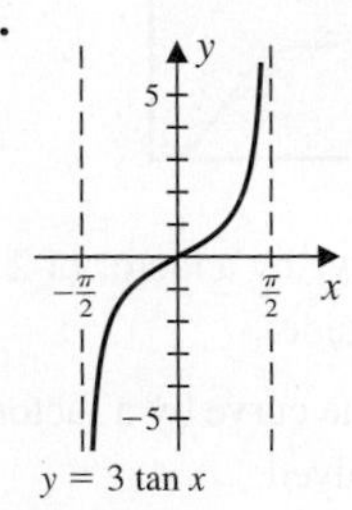

19.
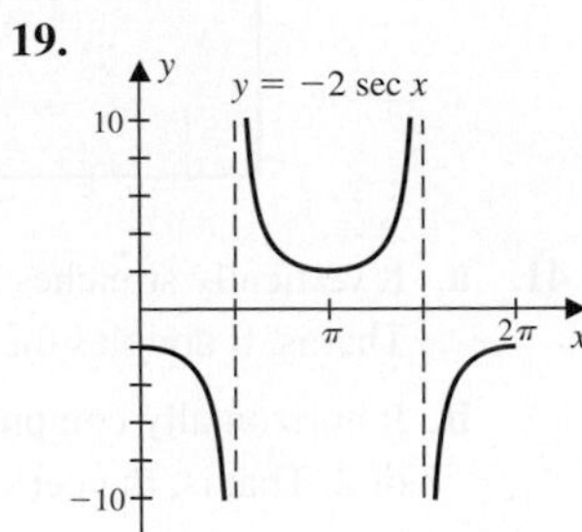

25.
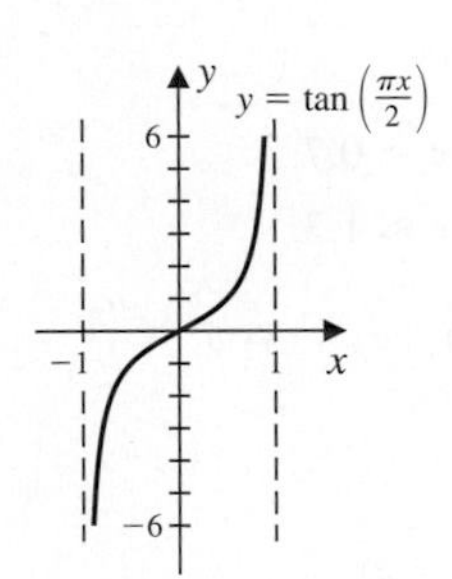

27.
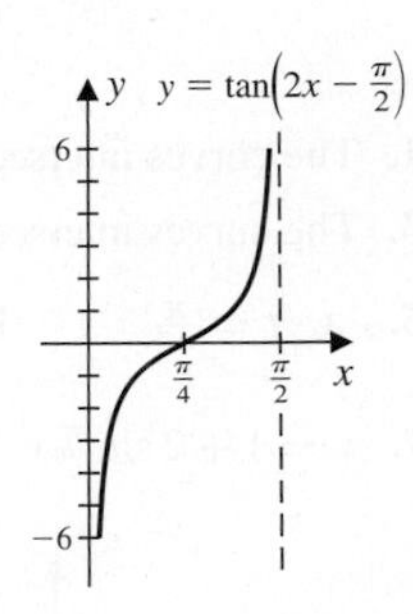

21.
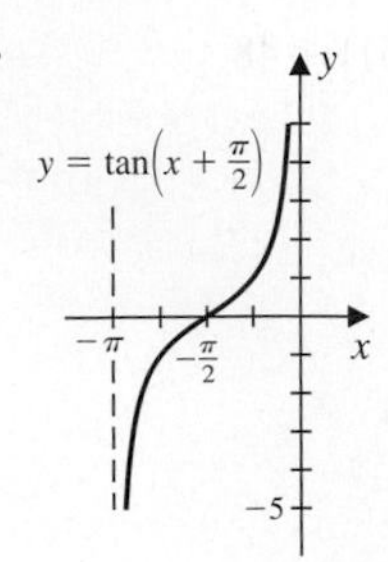

23.
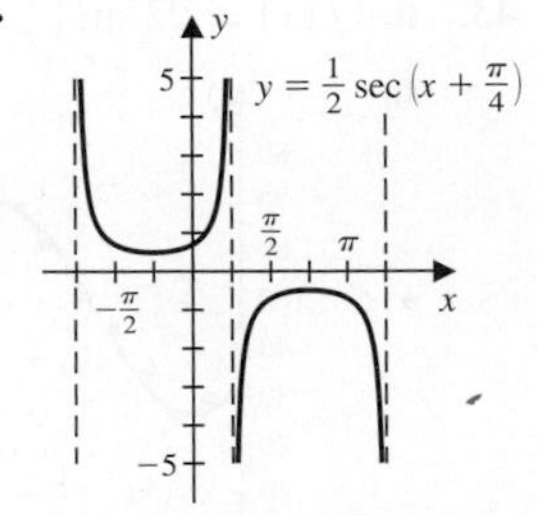

29. $t = \frac{3\pi}{4}, \frac{7\pi}{4}$

31. $t = \frac{\pi}{6}, \frac{5\pi}{6}, \frac{7\pi}{6}, \frac{11\pi}{6}$

33. $t = \frac{\pi}{4}, \frac{3\pi}{4}, \frac{5\pi}{4}, \frac{7\pi}{4}$

35. $t = 0, \frac{\pi}{8}, \frac{\pi}{2}, \frac{7\pi}{8}, \pi, \frac{9\pi}{8}, \frac{3\pi}{2}, \frac{15\pi}{8}, 2\pi$

37. One choice is $\left[-\frac{\pi}{10}, \frac{\pi}{10}\right] \times [-5, 5]$.

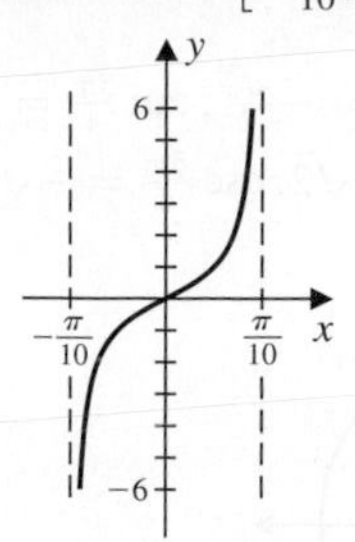

39. One choice is $\left[0, \frac{\pi}{50}\right] \times [-5, 5]$.

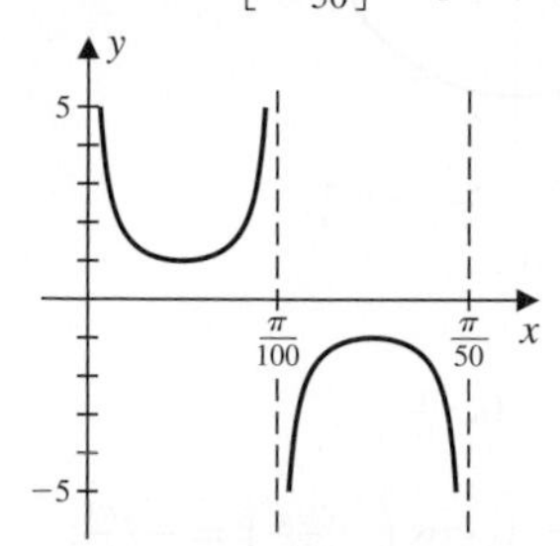

41. One choice is $[-50\pi, 50\pi] \times [-5, 5]$.

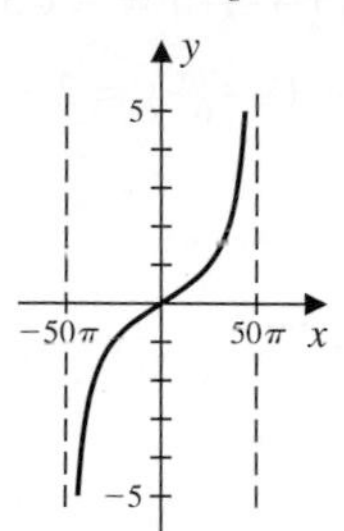

43. $\sin t = -\frac{2\sqrt{5}}{5}$; $\cos t = \frac{\sqrt{5}}{5}$; $\tan t = -2$; $\cot t = -\frac{1}{2}$; $\sec t = \sqrt{5}$; $\csc t = -\frac{\sqrt{5}}{2}$

Exercise Set 4.7 (page 250)

1. $\frac{1}{2}$

3. $\frac{\sqrt{2}}{4}(1 - \sqrt{3})$

5. $\frac{\sqrt{3}-1}{\sqrt{3}+1}$

7. $\frac{\sqrt{2+\sqrt{3}}}{2}$

9. $-\frac{\sqrt{2-\sqrt{2}}}{2}$

11. $-\frac{\sqrt{2-\sqrt{3}}}{2}$

13. **a.** $\cos 2t = \frac{7}{25}$ **b.** $\sin 2t = \frac{24}{25}$

c. $\cos \frac{t}{2} = \frac{3\sqrt{10}}{10}$ **d.** $\sin \frac{t}{2} = \frac{\sqrt{10}}{10}$

15. **a.** $\cos 2t = \frac{119}{169}$ **b.** $\sin 2t = \frac{120}{169}$

c. $\cos \frac{t}{2} = -\frac{\sqrt{26}}{26}$ **d.** $\sin \frac{t}{2} = \frac{5\sqrt{26}}{26}$

23. $(\cos 2x)^2 = \frac{1}{2} + \frac{1}{2}\cos 4x$

25. $(\sin x)^4 = \frac{3}{8} - \frac{1}{2}\cos 2x + \frac{1}{8}\cos 4x$

27. $\frac{1}{2}\sin t + \frac{1}{2}\sin 11t$

29. $\frac{1}{2}\cos t + \frac{1}{2}\cos 5t$

39. $x = 0, \frac{\pi}{3}, \pi, \frac{5\pi}{3}, 2\pi$

41. $x = 0, \frac{2\pi}{3}, \frac{4\pi}{3}, 2\pi$

43. The equation is an identity, so it holds for all values of x except at integer multiples of $\frac{\pi}{2}$, when it is undefined.

49. It is an identity because the graphs coincide.

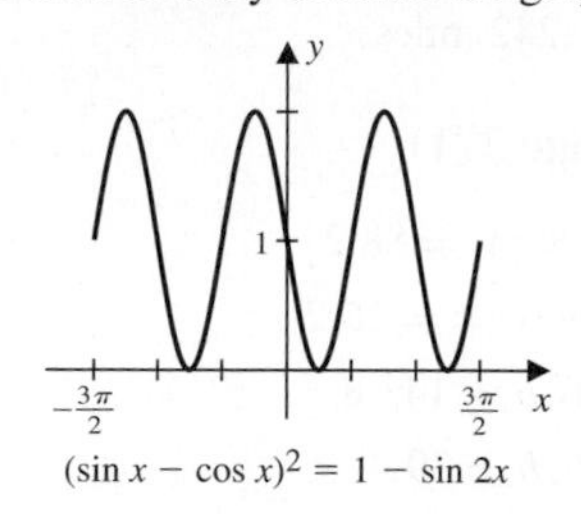

$(\sin x - \cos x)^2 = 1 - \sin 2x$

51. It is an identity because the graphs coincide.

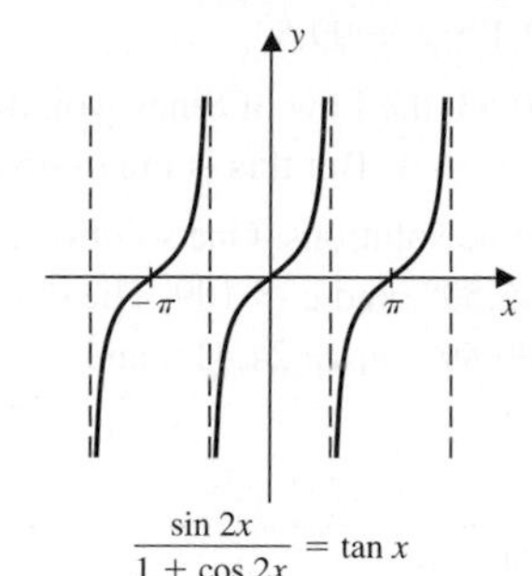

$\frac{\sin 2x}{1 + \cos 2x} = \tan x$

53. It is not an identity because the graphs do not coincide.

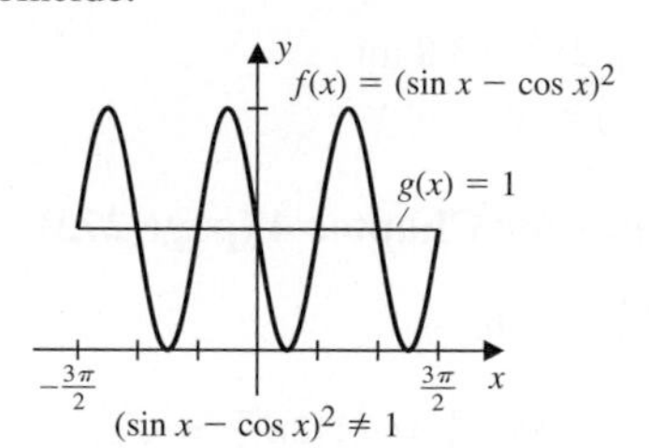

$(\sin x - \cos x)^2 \neq 1$

55. Approximately 102.8 ft/sec

Exercise Set 4.8 (page 259)

1. $\frac{\pi}{6}$ **3.** $\frac{\pi}{2}$ **5.** $\frac{3\pi}{4}$

7. $\frac{\pi}{3}$ **9.** $-\frac{\pi}{4}$ **11.** $\frac{\pi}{4}$

13. Undefined, since $\cos x$ is always between -1 and 1.

15. $\frac{1}{2}$ **17.** $\frac{\sqrt{2}}{2}$ **19.** $\sqrt{3}$

21. $-\frac{\pi}{4}$ **23.** $\frac{\pi}{4}$ **25.** $\frac{\pi}{3}$

27. $\frac{\pi}{6}$ **29.** $\frac{3}{5}$ **31.** $\frac{3}{4}$

33. $\frac{\sqrt{17}}{17}$ **35.** $\frac{7}{25}$

43. a. $x = \arctan(-1)$, and $x = \arctan 2$

b. $x = \arctan(-1) \approx -0.785$ and $x = \arctan 2 \approx 1.107$

45. $\theta = \arctan(x/4)$

47. Approximately 11,242 miles.

Exercise Set 4.9 (page 271)

1. $a = 16.4; \beta = 36.8°; \gamma = 88.2°$

3. $\alpha = 50.4°; \gamma = 99.6°; b = 16.2$

5. $\gamma = 53°; a = 92.6; b = 143.8$

7. $\alpha = 50°; a = 30.6; b = 39.4$

9. $\beta = 20.4°; \gamma = 29.6; a = 11$

11. $\alpha = 37.4°; \beta = 43.1°; \gamma = 99.5°$

13. Since $\sin 25.4° \approx 0.43$, the Law of Sines would imply $\sin\beta = \frac{10}{3}0.43 > 1$. But this is impossible.

15. There are two possible solutions. One solution is $\beta = 79.41°$, $\gamma = 45.59°$, and $c \approx 109$. The other solution is $\beta_1 = 100.59°$, $\gamma_1 = 24.41°$, and $c \approx 63.06$.

17. $A \approx 81.3$

19. a. Avoid the swamp.

b. Approximately \$115,880.

21. $d(A, B) \approx 55.3$ ft

23. $d(\text{ship 1, ship 2}) \approx 83.8$ mi

25. $d(B, \text{fire}) \approx 9.7$ mi

Review Exercises for Chapter 4 (page 273)

1. a. $\left(\frac{1}{2}, \frac{\sqrt{3}}{2}\right)$ **b.** $\frac{\pi}{3}$

c. $\sin\frac{\pi}{3} = \frac{\sqrt{3}}{2}$; $\cos\frac{\pi}{3} = \frac{1}{2}$; $\tan\frac{\pi}{3} = \sqrt{3}$; $\cot\frac{\pi}{3} = \frac{\sqrt{3}}{3}$; $\sec\frac{\pi}{3} = 2$; $\csc\frac{\pi}{3} = \frac{2\sqrt{3}}{3}$

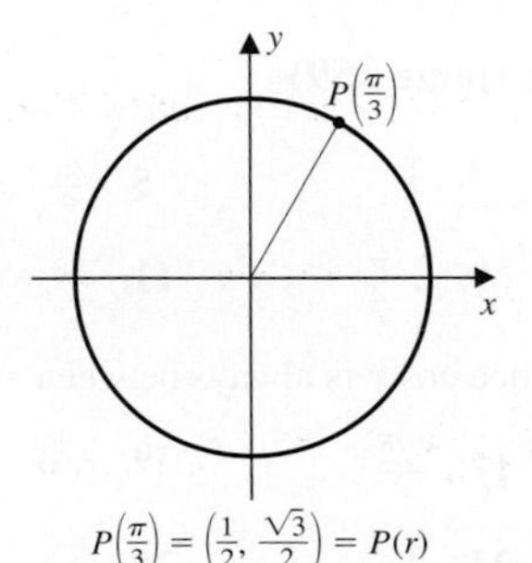

$P\left(\frac{\pi}{3}\right) = \left(\frac{1}{2}, \frac{\sqrt{3}}{2}\right) = P(r)$

3. a. $\left(-\frac{\sqrt{2}}{2}, -\frac{\sqrt{2}}{2}\right)$ **b.** $\frac{\pi}{4}$

c. $\sin\frac{5\pi}{4} = -\frac{\sqrt{2}}{2}$; $\cos\frac{5\pi}{4} = -\frac{\sqrt{2}}{2}$; $\tan\frac{5\pi}{4} = 1$; $\cot\frac{5\pi}{4} = 1$; $\sec\frac{5\pi}{4} = -\sqrt{2}$; $\csc\frac{5\pi}{4} = -\sqrt{2}$

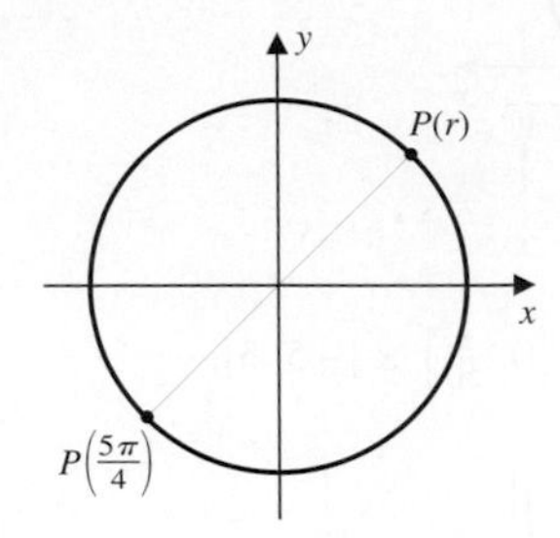

$P(r) = \left(\frac{\sqrt{2}}{2}, \frac{\sqrt{2}}{2}\right) = P\left(\frac{\pi}{4}\right)$

$P\left(\frac{5\pi}{4}\right) = \left(-\frac{\sqrt{2}}{2}, -\frac{\sqrt{2}}{2}\right)$

5. a. $\left(-\frac{\sqrt{3}}{2}, \frac{1}{2}\right)$ **b.** $\frac{\pi}{6}$

c. $\sin\left(-\frac{19\pi}{6}\right) = \frac{1}{2}$; $\cos\left(-\frac{19\pi}{6}\right) = -\frac{\sqrt{3}}{2}$; $\tan\left(-\frac{19\pi}{6}\right) = -\frac{\sqrt{3}}{3}$; $\cot\left(-\frac{19\pi}{6}\right) = -\sqrt{3}$; $\sec\left(-\frac{19\pi}{6}\right) = -\frac{2\sqrt{3}}{3}$; $\csc\left(-\frac{19\pi}{6}\right) = 2$

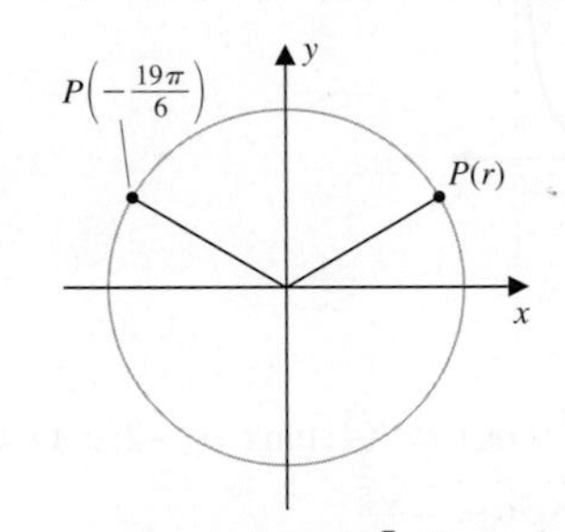

$P(r) = P\left(\frac{\pi}{6}\right) = \left(\frac{\sqrt{3}}{2}, \frac{1}{2}\right)$

$P\left(-\frac{19\pi}{6}\right) = \left(-\frac{\sqrt{3}}{2}, \frac{1}{2}\right)$

7. $\sin t = -\frac{4}{5}$; $\cos t = \frac{3}{5}$; $\tan t = -\frac{4}{3}$; $\cot t = -\frac{3}{4}$; $\sec t = \frac{5}{3}$; $\csc t = -\frac{5}{4}$

9. $\sin t = \frac{\sqrt{17}}{17}$; $\cos t = \frac{4\sqrt{17}}{17}$; $\tan t = \frac{1}{4}$; $\cot t = 4$; $\sec t = \frac{\sqrt{17}}{4}$; $\csc t = \sqrt{17}$

11. $x = \pi$

13. $x = \frac{\pi}{6}$, $x = \frac{\pi}{2}$ and $x = \frac{5\pi}{6}$

15. $x = 0$, $x = \pi$, $x = \arctan 2 \approx 1.11$, and $x = \pi - \arctan 2 \approx 2.03$

17. There are no solutions.

19. Odd.

21. Even.

23. **a.** ii **b.** iv **c.** i **d.** iii

25.

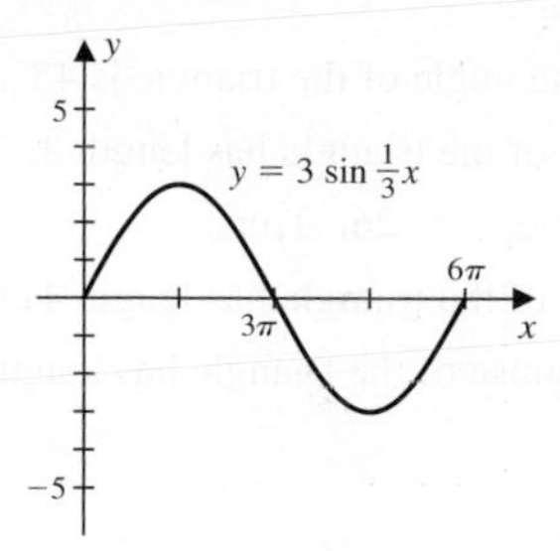

27.

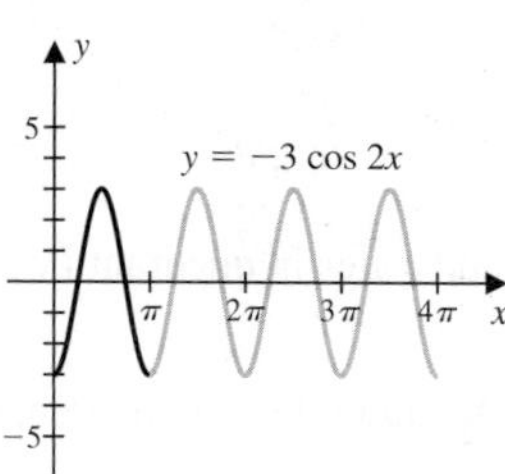

29.

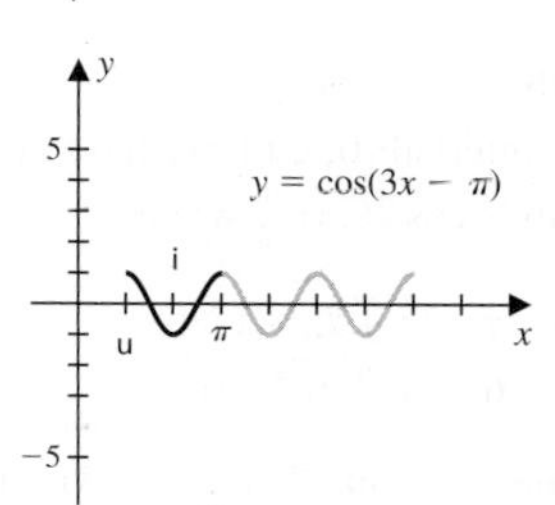

31.

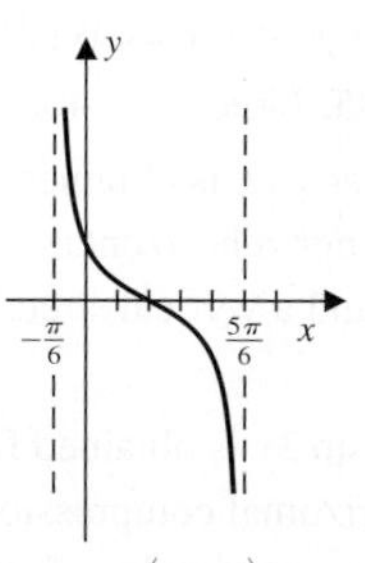

33.

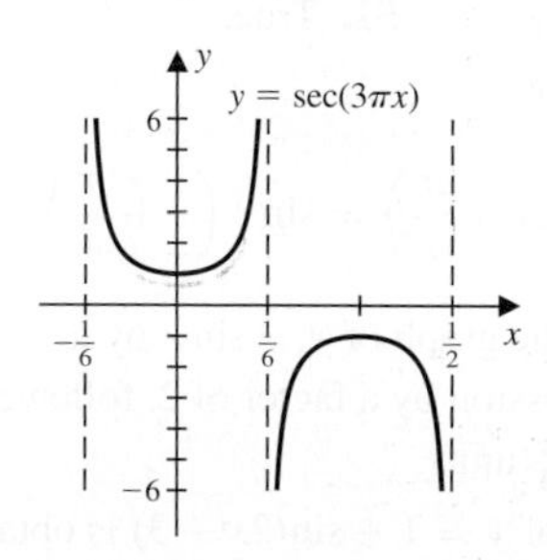

35. **a.** $y = \sin 4x$ **b.** $y = \cos 4\left(x - \frac{\pi}{8}\right)$

37. **a.** $y = 3\sin 2\pi\left(x + \frac{1}{4}\right)$ **b.** $y = 3\cos 2\pi x$

39. **a.** $AB = \frac{13}{2}, BC = \frac{13}{2}\sqrt{3}, \gamma = 30^\circ$

b. $\sin\alpha = \frac{\sqrt{3}}{2}, \cos\alpha = \frac{1}{2}, \tan\alpha = \sqrt{3},$
$\csc\alpha = \frac{2\sqrt{3}}{3}, \sec\alpha = 2, \cot\alpha = \frac{\sqrt{3}}{3}$

41. $-\frac{\sqrt{2}}{4}\left(1 + \sqrt{3}\right)$

43. $-\frac{\sqrt{2}}{4}(\sqrt{3} + 1)$

45. $\frac{\sqrt{2+\sqrt{2}}}{2}$

47. $-2 - \sqrt{3}$

53. $\frac{1}{2} + \frac{1}{2}\cos 10x$

55. $\frac{3}{8} - \frac{1}{2}\cos 2x + \frac{1}{8}\cos 4x$

57. $\frac{1}{2}[\sin 7t - \sin t]$

59. $\frac{1}{2}\cos 2t + \frac{1}{2}\cos 6t$

61. $2\sin 4t\cos 2t$

63. $2\cos 3t\cos t$

69. $\frac{\sqrt{3}}{2}$

71. $\frac{16}{65}$

73. $\beta = 29.1^\circ; \gamma = 125.9^\circ; a = 10.4$

75. $\alpha = 36.9^\circ; \beta = 53.1^\circ; \gamma = 90^\circ$

77. $\alpha = 54^\circ; a = 8.8; c = 8.3$

79. $x = 0$, and $x \approx 0.88$

81. One choice is $\left[-\frac{2\pi}{125}, \frac{2\pi}{125}\right] \times [-4, 4]$.

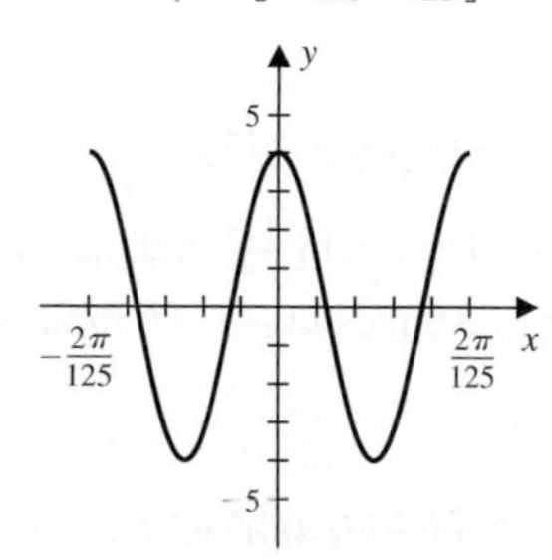

83. $A = b^2(1 + \cos\theta)\sin\theta$

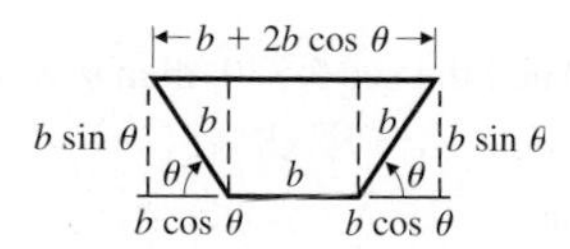

85. Approximately 475 miles east and 822.7 miles south

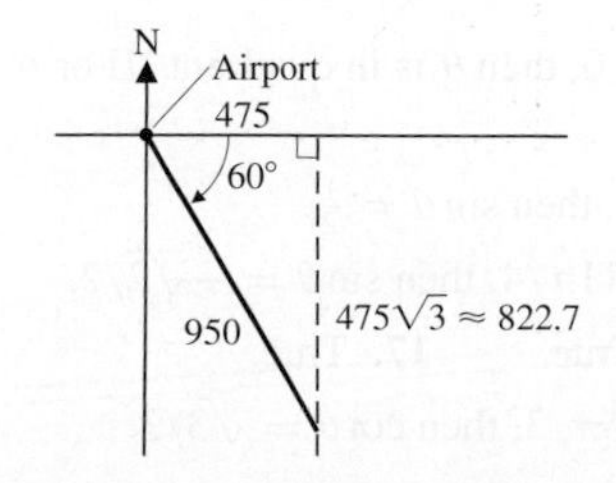

Chapter 4: Exercises for Calculus (page 275)

1. The minimum occurs at $x = -2$ with a minimum value of approximately -1.09. The maximum occurs at $x = 3$ with a maximum value of approximately 2.9.

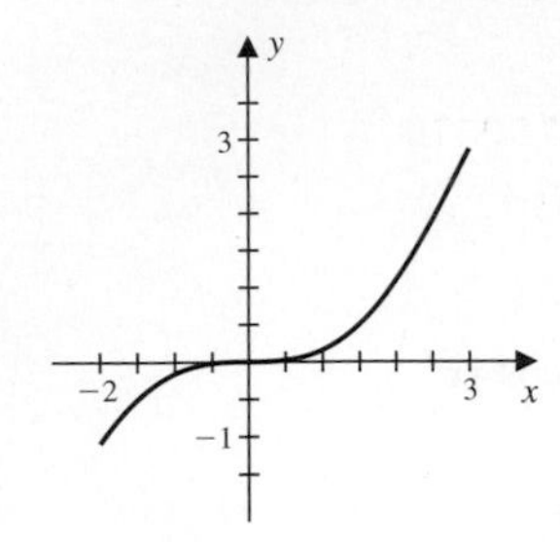

3. **a.** $\sqrt{a^2 - u^2} = a \cos t$

b. $\sqrt{u^2 + a^2} = a \sec t$

c. $\sqrt{u^2 - a^2} = a \tan t$

d. $\frac{\sqrt{u^2-a^2}}{u} = \sin t$

5. The Law of Cosines and Heron's Formula give $A \approx 34.87 + 51.59 \approx 86.5 \text{ ft}^2$.

7. The two ships are approximately 22,000 feet apart.

9. The distance for A to C is $\dfrac{300 \sin 59^\circ}{\sin 41^\circ} \approx 392$ feet.

11. **b.** π

Chapter 4: Chapter Test (page 277)

1. False. The number of degrees in $\frac{11\pi}{4}$ radians is 495°.

2. False. The number of degrees in $-5\pi/6$ radians is -150°.

3. True. **4.** True. **5.** True.

6. False. The reference angle for 480° is 60°.

7. False. If $\cos\theta > 0$ and the $\tan\theta < 0$, then θ is in quadrant IV.

8. False. If $\sin\theta > 0$ and the $\sec\theta > 0$, then θ is in quadrant I.

9. True. **10.** True.

11. False. If $\tan\theta > 0$, then θ is in quadrant I or in quadrant III.

12. False. If $\csc\theta < 0$, then θ is in quadrant III or in quadrant IV.

13. False. If $\theta = \frac{5\pi}{6}$, then $\sin\theta = \frac{1}{2}$.

14. False. If $\theta = -11\pi/4$, then $\sin\theta = -\sqrt{2}/2$.

15. True. **16.** True. **17.** True.

18. False. If $\theta = -5\pi/3$, then $\cot\theta = \sqrt{3}/3$.

19. False. If $\theta = \pi$, then $\sec\theta = -1$.

20. True. **21.** True.

22. False. The missing angle of the triangle is 45°.

23. False. The side a of the triangle has length 3.

24. True. **25.** True. **26.** True.

27. False. The side a of the triangle has length $4\tan 60^\circ$.

28. False. The hypotenuse of the triangle has length $x = 4\csc 30^\circ$.

29. True.

30. False. $\cos\theta = \frac{\sqrt{5}}{3}$

31. False. $\tan\theta = \frac{2\sqrt{5}}{5}$

32. True.

33. False. If $\cos\theta = \frac{4}{5}$ and θ lies in quadrant IV, then $\sin\theta = -\frac{3}{5}$.

34. False. If $\tan\theta = -\frac{\sqrt{3}}{2}$ and θ lies in quadrant II, then $\cos\theta = -\frac{2\sqrt{7}}{7}$.

35. True. **36.** True.

37. False. For θ in the interval $[0, 2\pi]$, we have 2θ in $[0, 4\pi]$. So θ satisfies $\cos 2\theta = \frac{1}{2}$ when

$$\theta = \frac{\pi}{6}, \frac{5\pi}{6}, \frac{7\pi}{6}, \frac{11\pi}{6}.$$

38. True. **39.** True. **40.** True. **41.** True.

42. False. The curve is given by $y = 1 + 2\cos\left(2x - \frac{\pi}{2}\right)$.

43. True. **44.** True. **45.** True. **46.** True.

47. False. The graph of $y = 2\cos\frac{1}{2}x$ is obtained from the graph of $y = \cos x$ through a horizontal stretching by a factor of 2 and a vertical stretching by a factor of 2.

48. False. The graph of $y = \frac{1}{2}\sin 2x$ is obtained from the graph of $y = \sin x$ by a horizontal compression by a factor of 2 and a vertical compression by a factor of 2.

49. True. **50.** True. **51.** True.

52. False. The graph of

$$y = \sin\left(2x + \frac{\pi}{2}\right) = \sin 2\left(x + \frac{\pi}{4}\right)$$

is obtained from the graph of $y = \sin x$ by a horizontal compression by a factor of 2, followed by a shift to the left $\frac{\pi}{4}$ units.

53. False. The graph of $y = 1 + \sin(2x - 3)$ is obtained from the graph of $y = \sin x$ from a horizontal compression by a factor of 2, followed by a shift to the right $\frac{3}{2}$ units and upward 1 unit.

54. False. The graph of $y = -1 + 2\cos\left(2x - \frac{\pi}{3}\right)$ is obtained from the graph of $y = \cos x$ from a horizontal compression by a factor of 2 and a vertical stretching by a factor of 2, followed by a shift to the right $\frac{\pi}{6}$ units and downward 1 unit.

55. True.

56. False. The only values of θ in the interval $[0, 2\pi]$ that satisfy $2\sin 2\theta = 4\sin\theta$ are $\theta = 0, \pi, 2\pi$.

57. True.

58. False. $\cos\left(\frac{\pi}{12}\right) = \frac{\sqrt{2}}{4}\left(\sqrt{3} + 1\right)$

59. True.

60. False. For all θ, $\sin\left(\theta - \frac{\pi}{2}\right) = -\cos\theta$.

61. True. **62.** True.

63. False. $\arccos\left(\frac{\sqrt{3}}{2}\right) = \frac{\pi}{6}$.

64. False. Since the values of $\arcsin x$ are in the interval $[-\frac{\pi}{2}, \frac{\pi}{2}]$, we have $\arcsin(-\frac{1}{2}) = -\frac{\pi}{6}$.

65. True.

66. False. If $\theta = \arctan\left(\frac{1}{3}\right)$, then $\sin\theta = \frac{\sqrt{10}}{10}$.

67. True. **68.** True. **69.** True. **70.** True.

Chapter 5

Exercise Set 5.2 (page 292)

1.

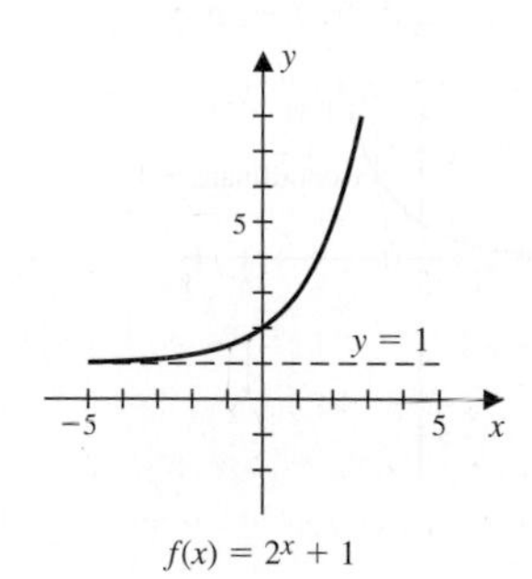

3.

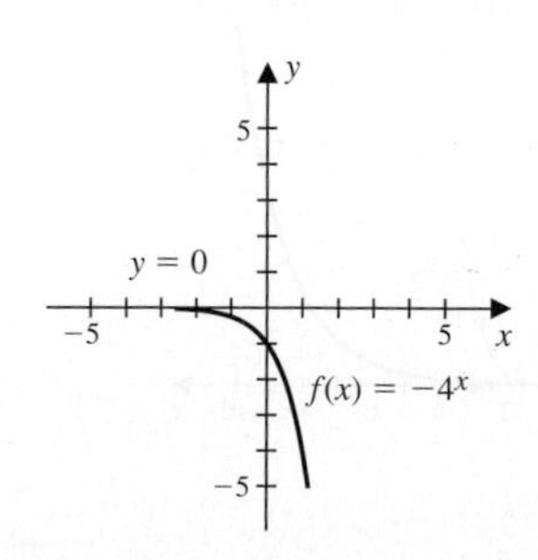

5.

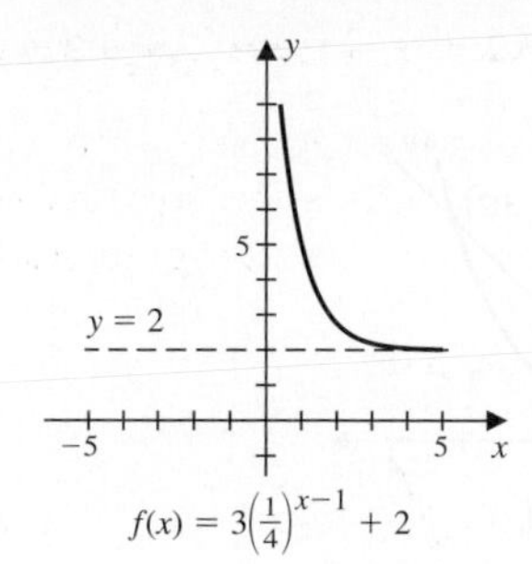

7.

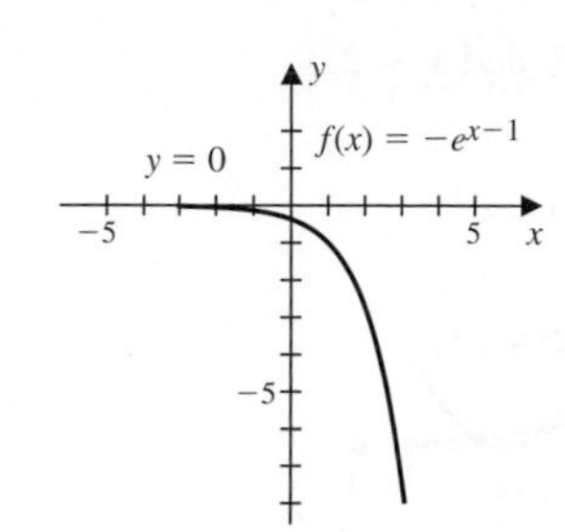

9.

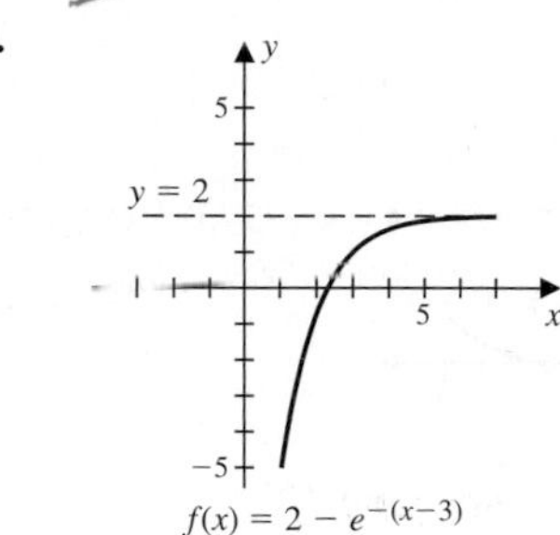

11.

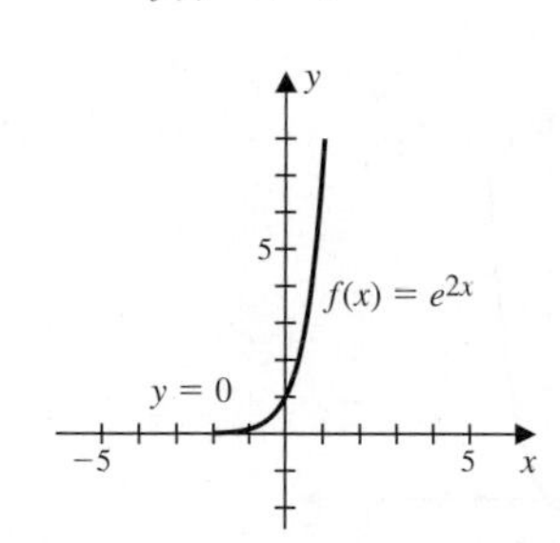

13.

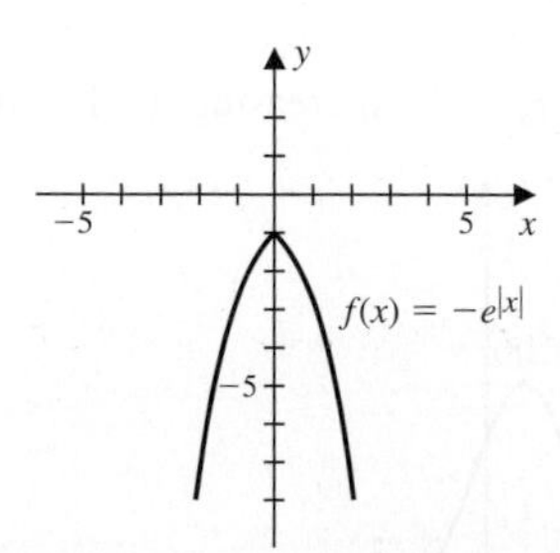

15. $x > 0$ **17.** $x \leq -\frac{1}{2}$

19. a. ii **b.** iii **c.** iv **d.** i

21. $x \approx 0.2$ and $x \approx 3.2$

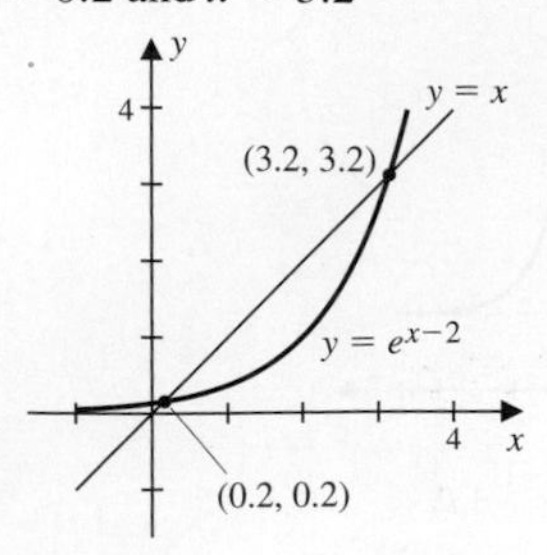

23. $x \approx -3.3$, $x \approx 1.5$, and $x \approx 2.3$

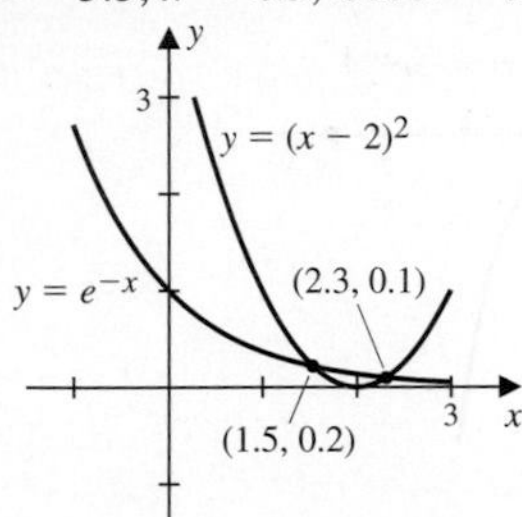

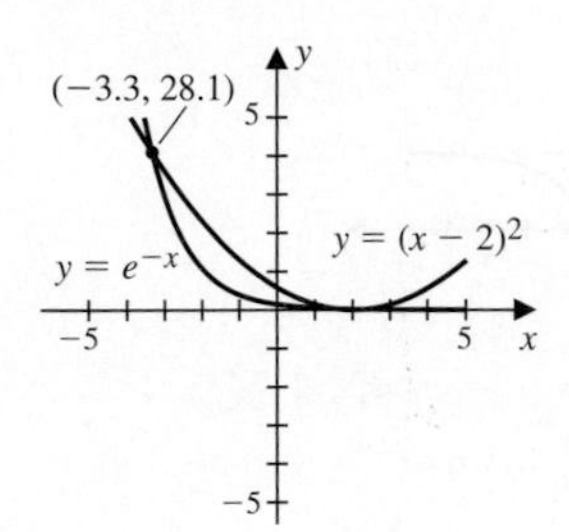

25. a.

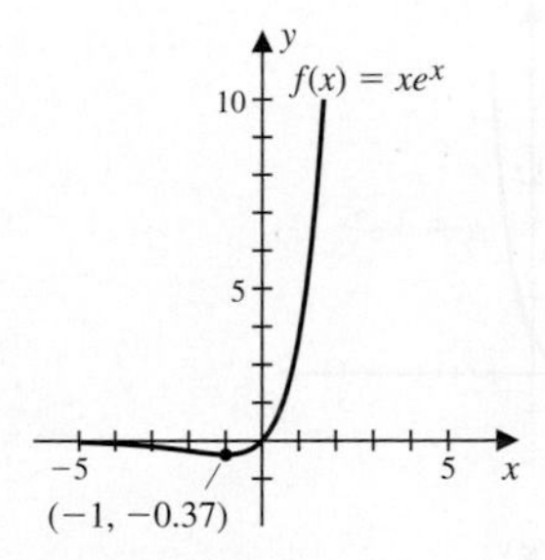

b. Decreasing: $(-\infty, -1)$; increasing: $(-1, \infty)$

27. a.

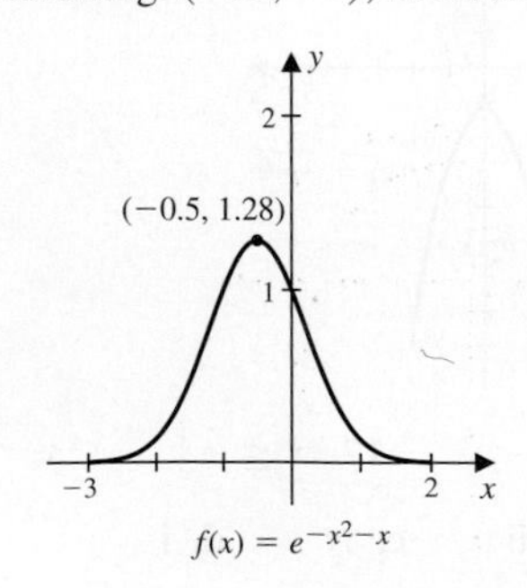

b. Decreasing: $(-0.5, \infty)$; increasing: $(-\infty, -0.5)$

29. a.

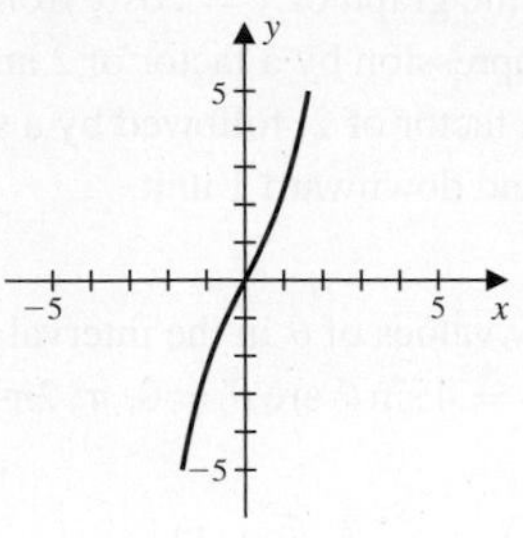

b. increasing on $(-\infty, \infty)$

31. $k \approx 1.1$

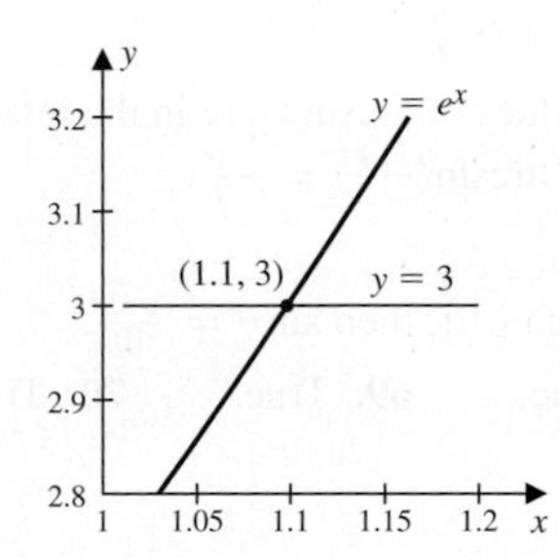

33. a. $f(x)$ grows faster than $g(x)$ when $x > 22.4$.

i.

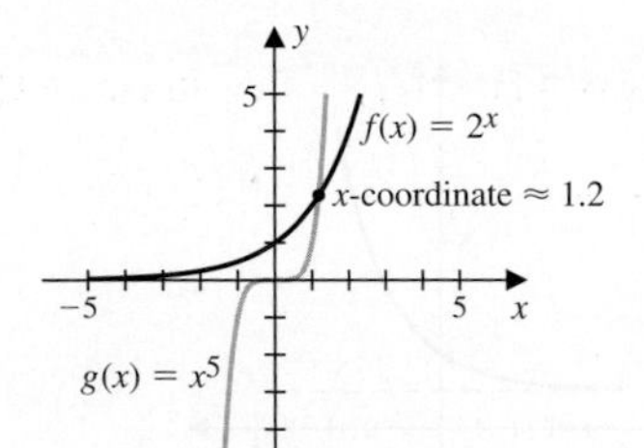

ii.

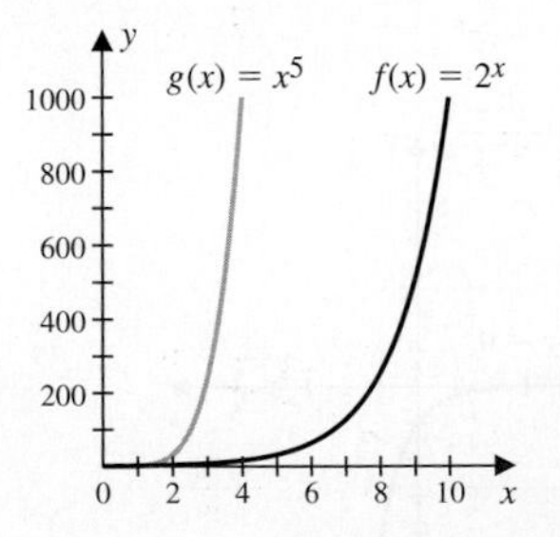

iii.

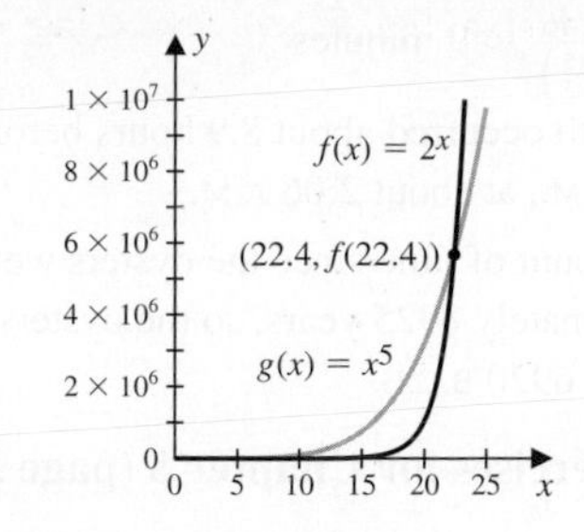

b. The curves intersect at $x \approx 1.2$ and $x \approx 22.4$.

35. **a.** \$6850.43 **b.** \$6914.09
c. \$6919.95 **d.** \$6920.15

37. **a.** \$14,859.47 **b.** \$13,804.20
c. \$13,468.55 **d.** \$13,140.67

39. \$6756

41. $A(t) = 1000(1.02)^t$, so $A(7) = 11{,}487$.

Exercise Set 5.3 (page 303)

1. 3 **3.** 3 **5.** $\frac{1}{2}$ **7.** -3

9. -3 **11.** 5 **13.** $\frac{1}{3}$ **15.** π^2

17. $\ln x + \ln(x+1)$

19. $4\log_3 x - \log_3(x+1)$

21. $\ln 2 + 3\ln x - 2\ln(x+4)$

23. $\frac{3}{2}\log_3(3x+2) + 3\log_3(x-1) - \log_3 x - \frac{1}{2}\log_3(x+1)$

25. $\frac{1}{2}\ln x + \frac{1}{4}\ln(x+1)$ **27.** $\ln x(x+1)^2$

29. $\ln\left(\frac{\sqrt{x}}{(x-1)^2}\right)$ **31.** $\ln\left(\frac{x-1}{x^{3/2}}\right)$

33. $x = 81$ **35.** $x = 4$

37. $x = 2$ **39.** $x = 2 - e^4$

41. $x = \frac{9}{2}$ **43.** $x = 3$

45. $x = \frac{4}{3}$ **47.** $x = \frac{1}{2}$ and $x = -9$

49. $x = \log_4 3 = \frac{\ln 3}{\ln 4}$ **51.** $x = \frac{4\ln 3}{\ln 3 - 2}$

53. $x = \frac{\ln 2}{3\ln 2 + \ln 3}$

55.

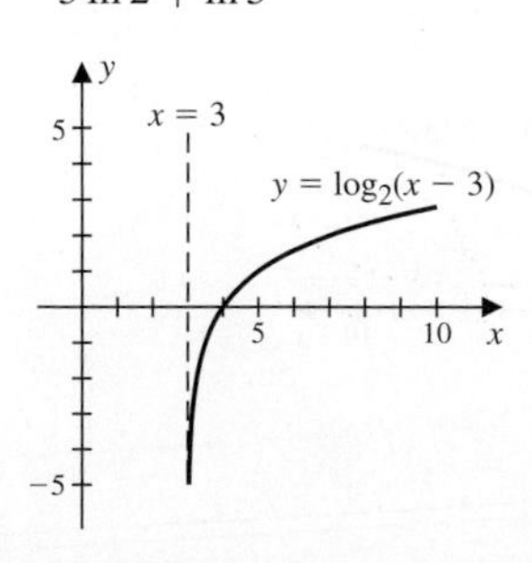

57.

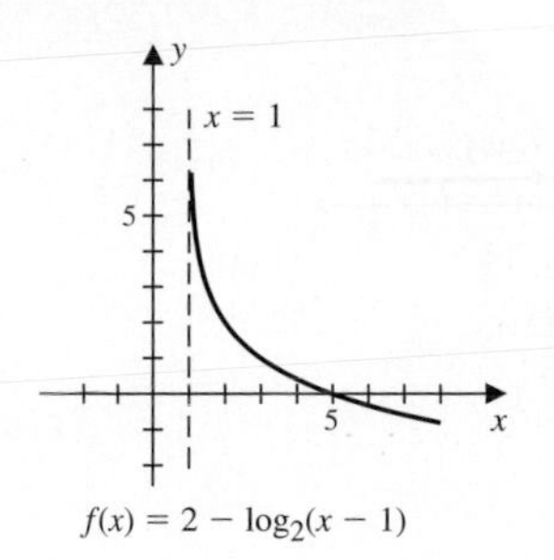

59.

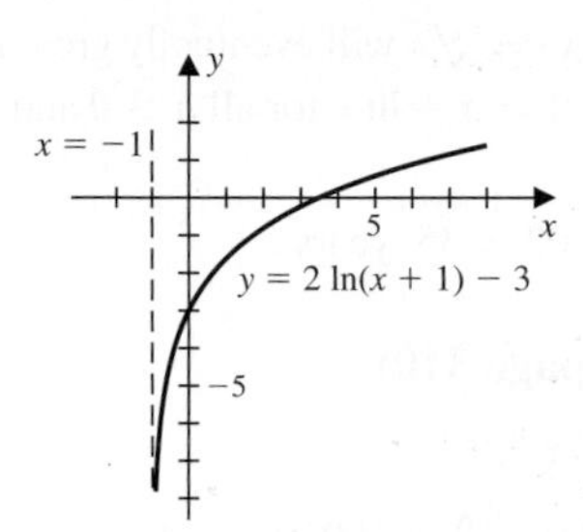

61.

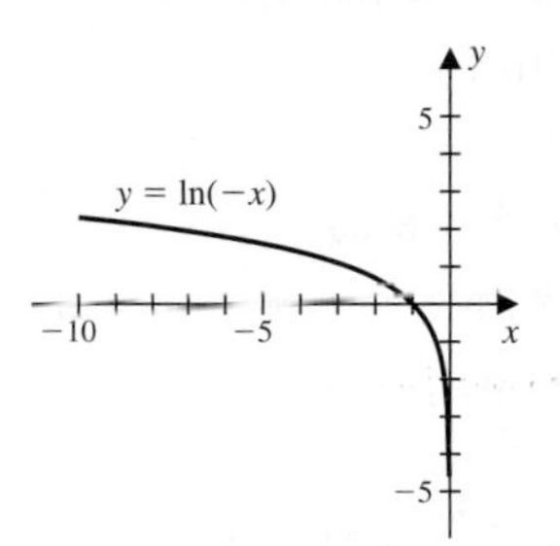

63.

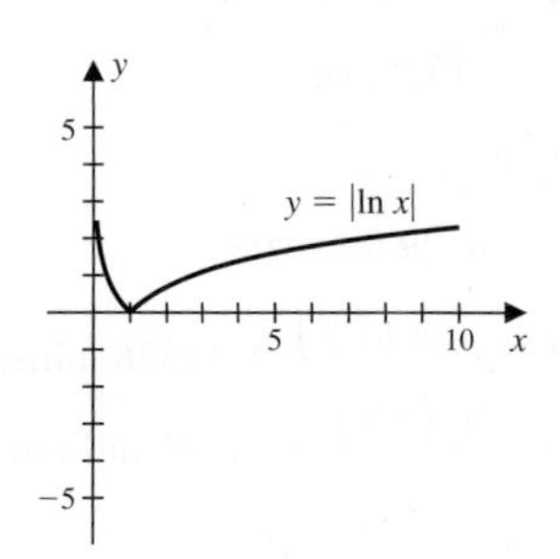

65. **a.** iii **b.** iv **c.** i **d.** ii

67.

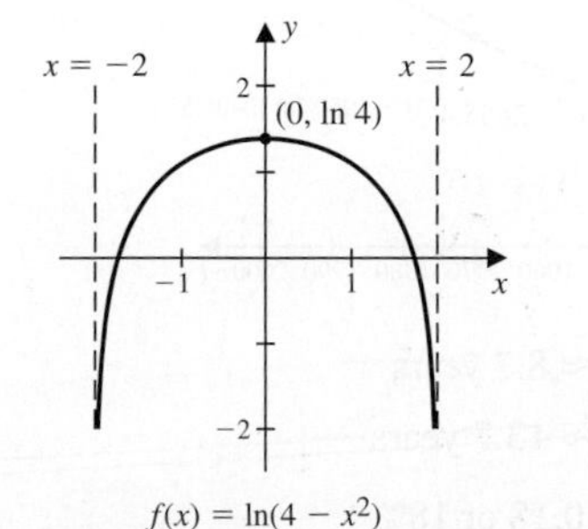

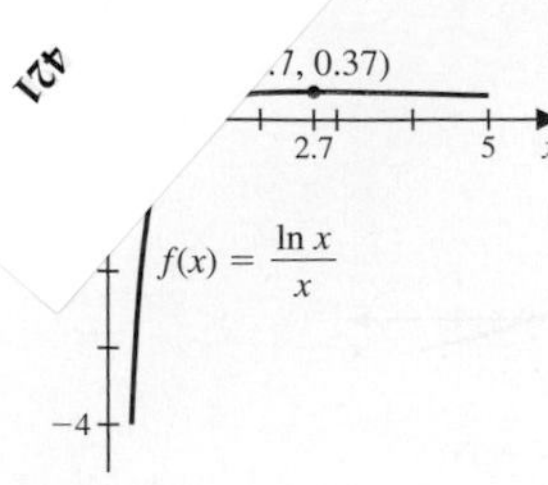

71. The function $g(x) = \sqrt[n]{x}$ will eventually grow more rapidly than $f(x) = a + \ln x$ for all $n > 0$ and real numbers a.

73. About $\ln 2/\ln 1.02 = 35$ years.

Exercise Set 5.4 (page 310)

1. **a.** $Q(t) = 2000e^{\frac{\ln 2}{3}t}$

b. $Q(7) = 2000e^{\frac{\ln 2}{3}6} \approx 10079.4$

c. $t = \frac{3\ln 11}{\ln 2} \approx 10.4\,\text{h}$

3. **a.** $Q(t) = 500e^{\frac{\ln 8}{5}t}$

b. $Q(6) = 500e^{\frac{6\ln 8}{5}} \approx 6063$

c. $t = \frac{5\ln 30}{\ln 8} \approx 8.2\,\text{h}$

d. $t = \frac{5\ln 2}{\ln 8} \approx 1.7\,\text{h}$

5. **a.** $Q(t) = 64e^{-\frac{\ln 2}{578}t}$

b. $Q(75) = 64e^{-\frac{\ln 2}{578}75} \approx 58.5\,\text{mg}$

c. $t = -\frac{578\ln\frac{3}{16}}{\ln 2} \approx 1395.9\,\text{h}$

7. $t = \frac{2\ln 3}{\ln 2} \approx 3.2$ hours

9. $t = -\dfrac{4\ln 2}{\ln 10 - \ln 11} \approx 29.09$ years

11. **a.** $Q(100) = 2555e^{10\ln\left(\frac{3040}{2555}\right)} \approx 14528$ million

b. $Q(60) = 5275e^{6\ln\left(\frac{6079}{5275}\right)} \approx 12356$ million

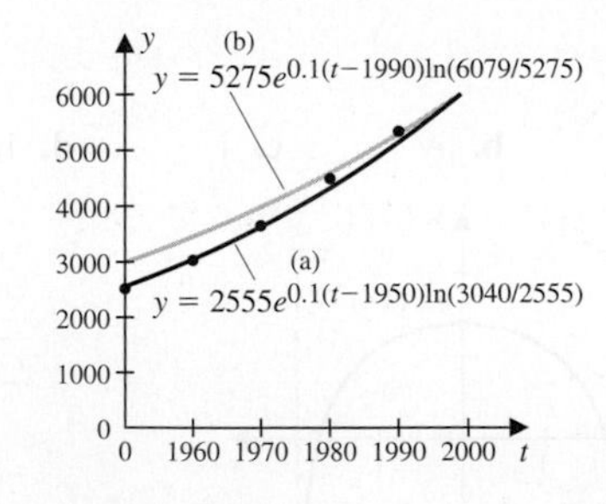

13. **a.** $t = \frac{\ln 2}{0.08} \approx 8.7$ years

b. $t = \frac{\ln 3}{0.08} \approx 13.7$ years

15. $i = \frac{1}{5}\ln\frac{5}{2} \approx 0.18$ or 18%

17. $t = \dfrac{-\ln 46}{\ln\left(\frac{15}{23}\right)} \approx 9$ minutes

19. The death occurred about 8.9 hours before 11:00 A.M., at about 2:06 A.M.

21. The amount of time since the oysters were alive is approximately 8925 years, so the oysters were eaten in about 6920 B.C.E.

Review Exercises for Chapter 5 (page 311)

1. **a.** iii **b.** i **c.** iv **d.** ii

3.

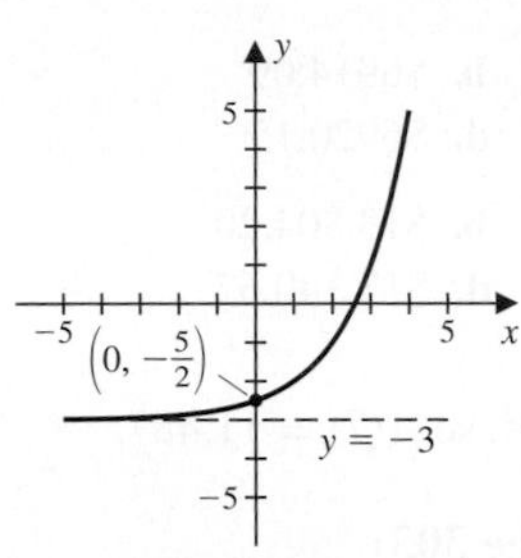

5.

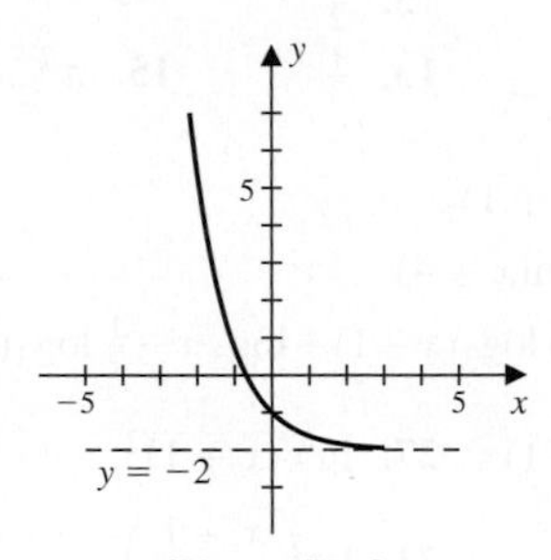

7.

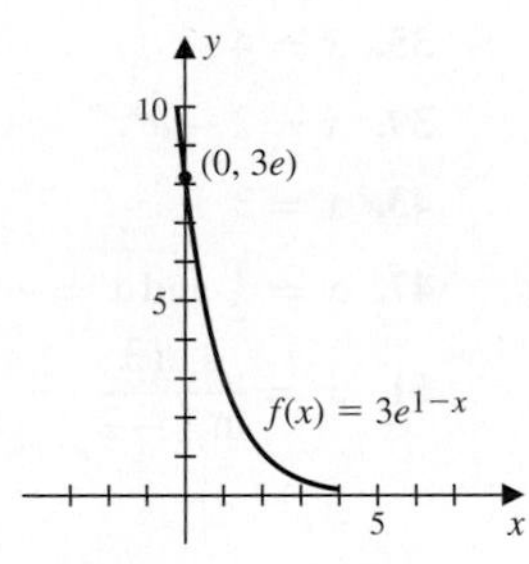

9.

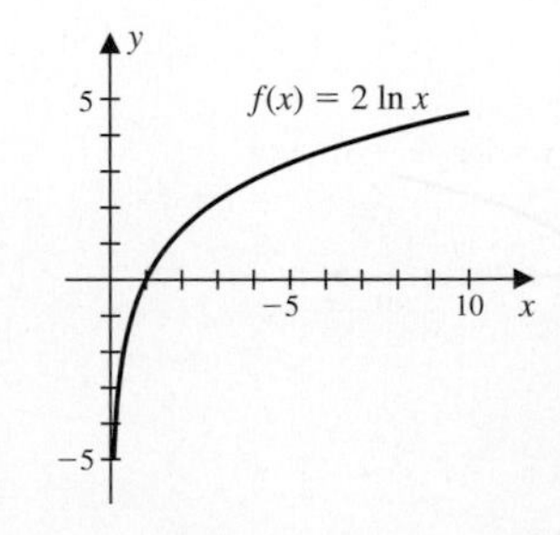

11.

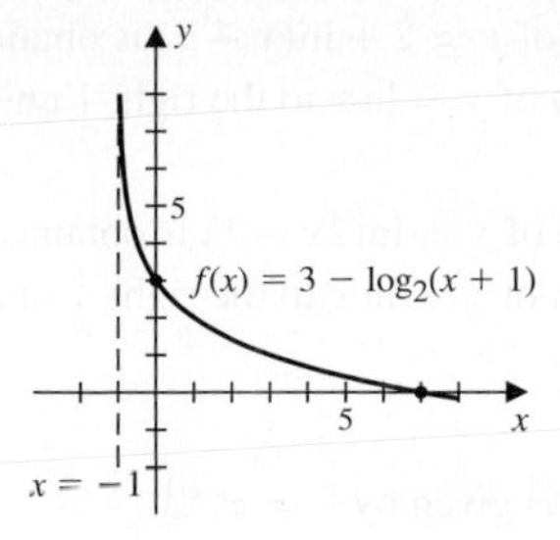

13.

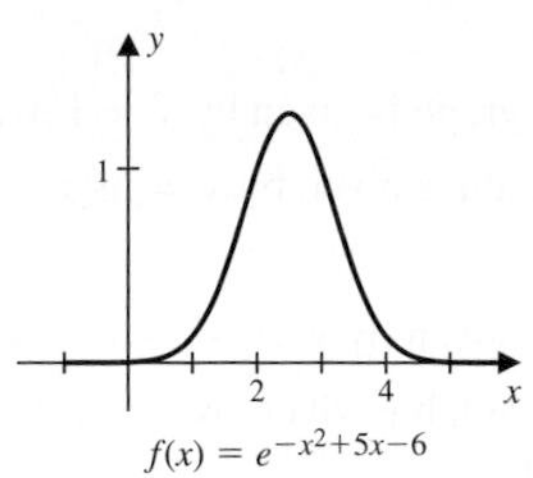

15. 0 **17.** 15 **19.** $\frac{1}{2}$ **21.** 64

23. $\ln 3 + 2\ln x - \frac{1}{2}\ln(x-1)$

25. $\frac{1}{2}\log_{10}(x+1) + \frac{1}{3}\log_{10}(x-1) - \log_{10} x - \frac{5}{2}\log_{10}(x+3)$

27. $\ln\left(x^{4/3}(x+1)^{1/3}(x-1)^2\right)$

29. $\ln\left(\dfrac{5(x^3+2)^3}{\sqrt{x^5-1}}\right)$ **31.** $x = \dfrac{3+e^4}{2}$

33. $x = \frac{5}{3}$ **35.** $x = \dfrac{2\ln 5}{\ln 5 - 3\ln 3}$

37. $x = -\frac{1}{2}$ and $x = 1$

39. No intersections

41. $-0.81 < x < 1.43$ and $x > 8.6$

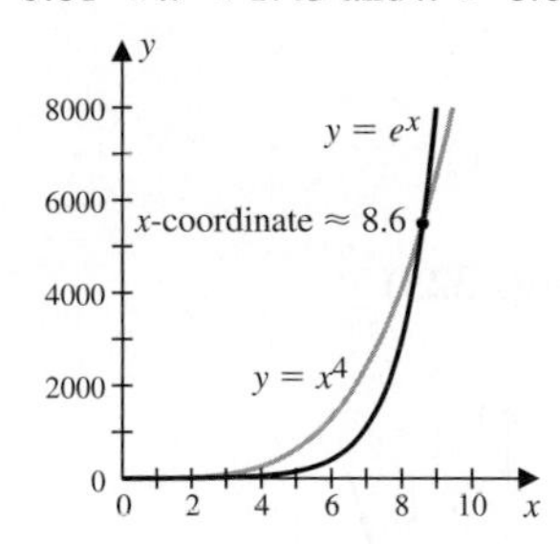

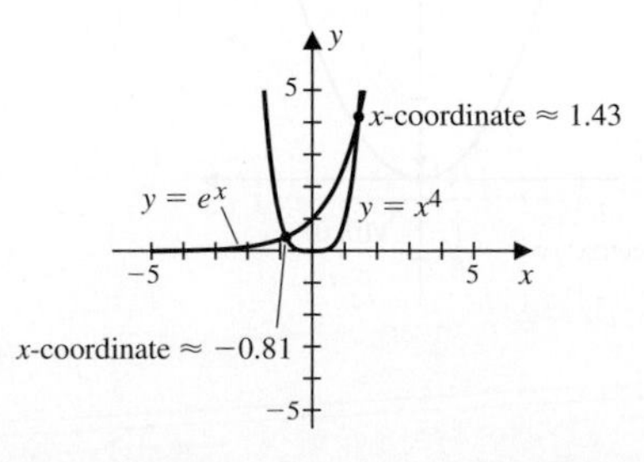

43. $-1.2 < x < 14.3$

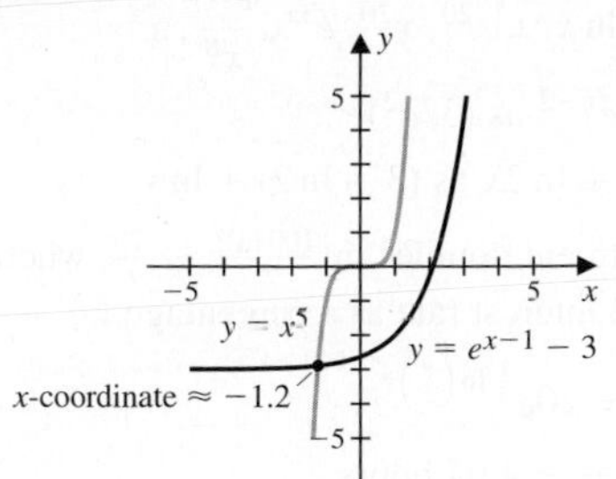

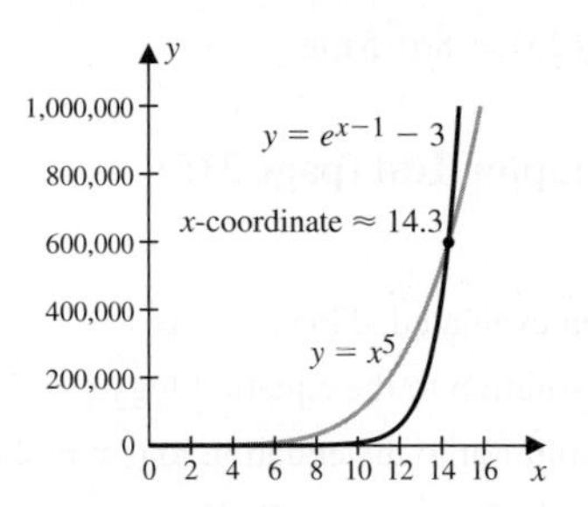

45. Increasing: $(-\infty, -1)$ and $(0, 1)$; decreasing: $(-1, 0)$ and $(1, \infty)$; local maxima: $(1, 1)$ and $(-1, 1)$; local minimum: $(0, 0)$

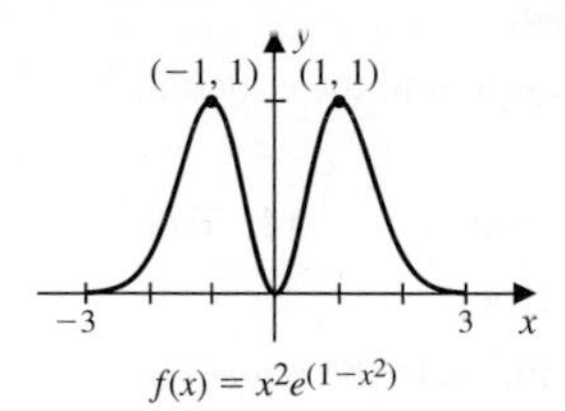

47. $\frac{\ln 2}{0.06} \approx 11.55$ years

49. **a.** $Q(t) = \frac{1000}{\sqrt[3]{3}} e^{\frac{t}{3}\ln 3}$

b. $Q(5) \approx 4327$

c. $Q(t) = 20000$ when $t = 1 + 3\frac{\ln 20}{\ln 3} \approx 9.2$ hours

d. $t = 3$ hours

51. The investment doubles in approximately 7.02 years.

53. The investment doubles in approximately 7.7 years.

Chapter 5: Exercises for Calculus (page 314)

1. **a.** $f = g^{-1}$

b. $f \neq g^{-1}$

c. Neither has an inverse.

d. $f = g^{-1}$

e. $f = g^{-1}$

f. $f \neq g^{-1}$

3. $\frac{x^{10}}{e^x}, \frac{1}{x^4}, \ln x, x^{1/20}, x^{20}, e^{3x}, \frac{e^{6x}}{x^8}, x^x$

5. Rewrite $3e^{x-2}$ as $(3/e^2)e^x$.

7. Rewrite $3 + \ln 2x$ as $(3 + \ln 2) + \ln x$.

9. The investment doubles in $\frac{100 \ln 2}{r} \approx \frac{70}{r}$, where r is the annual interest rate as a percentage.

11. **a.** $C(t) = 20e^{\frac{1}{3}\ln\left(\frac{3}{5}\right)t}$

b. $-\frac{3 \ln 2}{\ln\left(\frac{3}{5}\right)} \approx 4.07$ hours

c. $(34.46)(25) = 861.5$ mg

Chapter 5: Chapter Test (page 315)

1. True.
2. False. When evaluated, $2 \log_a a^{1/2}$ is 1.
3. False. The solution to the equation $\log_2 x = 5$ is 32.
4. False. The solution to the equation $\log_3 x = 2$ is 9.
5. True. **6.** True. **7.** True.
8. False. The range of the function $f(x) = 1 + e^{x-2}$ is $(1, \infty)$.
9. True. **10.** True.
11. False. The only solution to the equation $3^{2x+5} = 27$ is $x = -1$.
12. True. **13.** True. **14.** True.
15. True. **16.** True.
17. False. The only solution to the equation $\ln x + \ln(x - 1) = \ln(4x + 6)$ is $x = 6$, since $x = -1$ is not in the domain.
18. False. The only solution to the equation $\ln(x + 6) - \ln(x + 1) = \ln(x - 2)$ is $x = 4$.
19. False. The graph of $y = e^x$ has a horizontal asymptote $y = 0$.
20. True.
21. False. The graph of $y = -1 + e^{x-2}$ is obtained by shifting the graph of $y = e^x$ to the right 2 units and downward 1 unit.
22. False. The graph of $y = e^{x+3}$ can be obtained by shifting the graph of $y = e^x$ to the left 1 unit and vertically stretching the result by a factor of e^2.
23. False. The graph of $y = \ln x$ has a vertical asymptote $x = 0$.
24. False. The graph of $y = -3 + \ln(x - 1)$ has a vertical asymptote $x = 1$.
25. False. The graph of $y = 2 + \ln(x - 1)$ is obtained by shifting the graph of $y = \ln x$ to the right 1 unit and upward 2 units.
26. False. The graph of $y = \ln(2x - 1)$ is obtained by shifting the graph of $y = \ln x$ to the right $\frac{1}{2}$ units and upward $\ln 2$ units.
27. True.
28. False. The graph is given by $y = e^{x+1}$.
29. True.
30. False. The yellow graph is given by $y = 1 + 2e^{x-1}$.
31. False. The blue graph is given by $y = \ln(x - 1)$.
32. True.
33. False. The graph is given by $y = -1 - \ln(x + 1)$.
34. False. The yellow graph is given by $y = 1 + 2\ln(x - 1)$.
35. True.
36. True.
37. False. If a bacteria culture starts with 1000 bacteria and after 3 hours there are 2500 bacteria, the number of bacteria at any time t is given by $Q(t) = 1000e^{(t \ln 2.5)/3}$.
38. False. If a radioactive substance decays 5% in 10 years, half of the initial mass will decay in $t = -\frac{10 \ln 2}{\ln(0.95)}$ years.
39. False. An initial investment deposited in an account returning 7% interest compounded continuously will double in approximately 10 years (rounded to the nearest year).
40. True.

Chapter 6

Exercise Set 6.2 (page 325)

1.

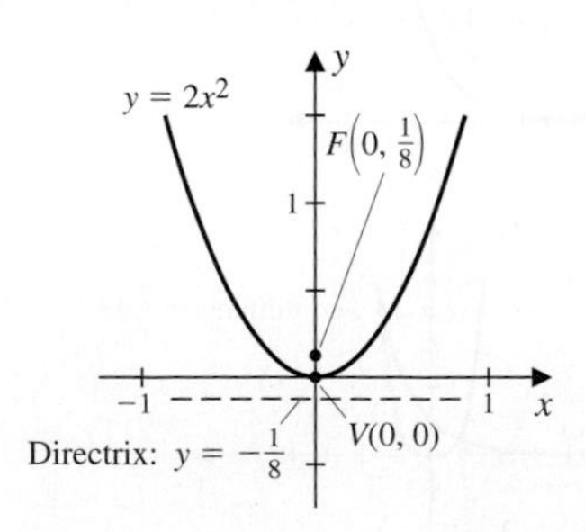

3.

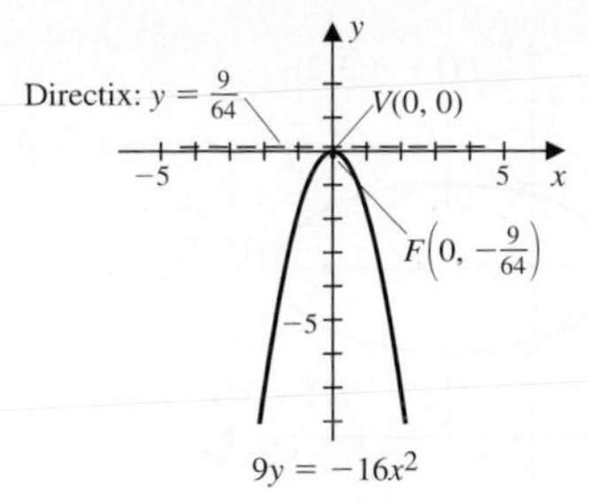

5.

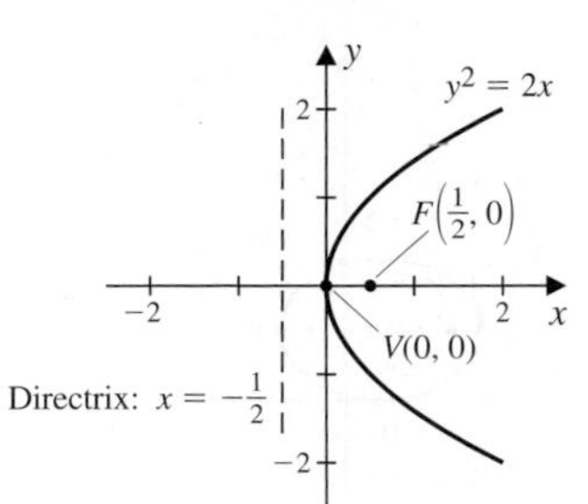

7.

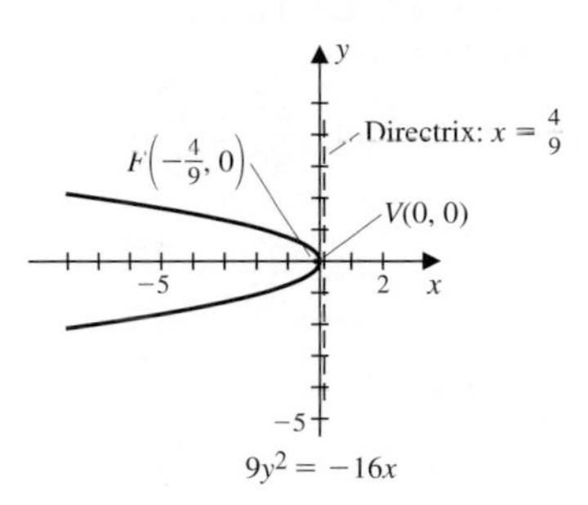

9.

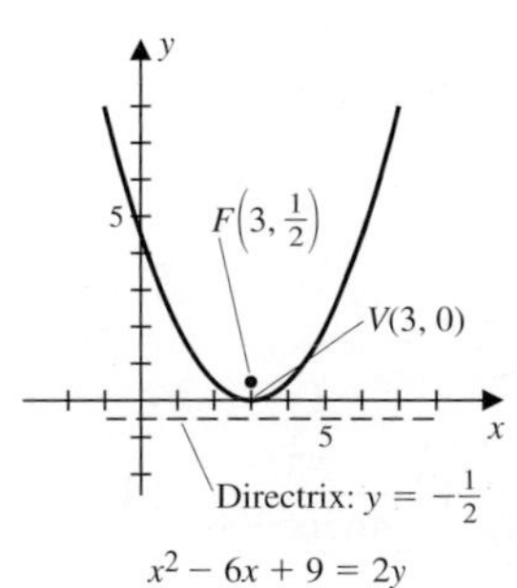

11.

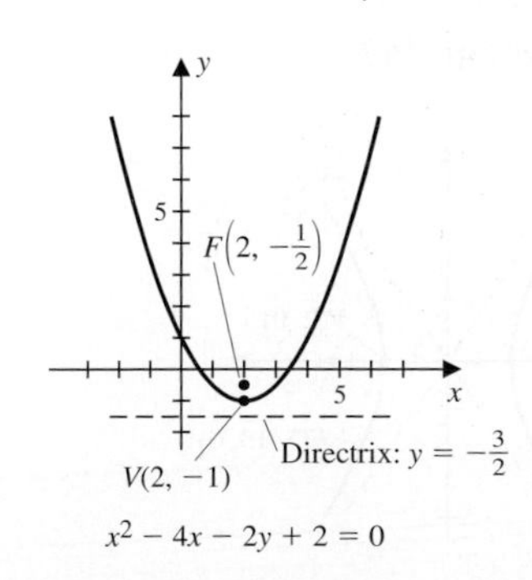

13.

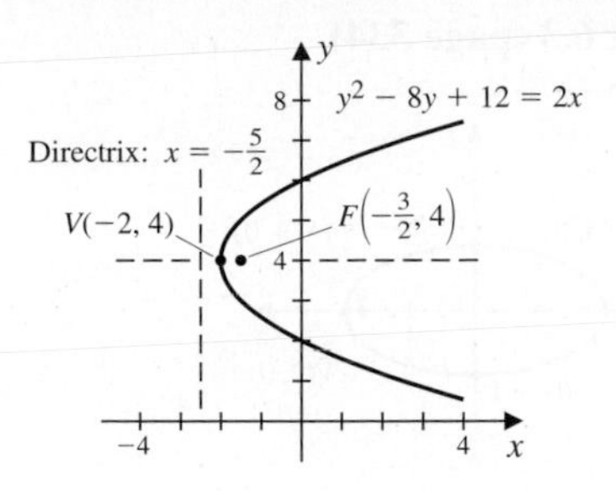

15.

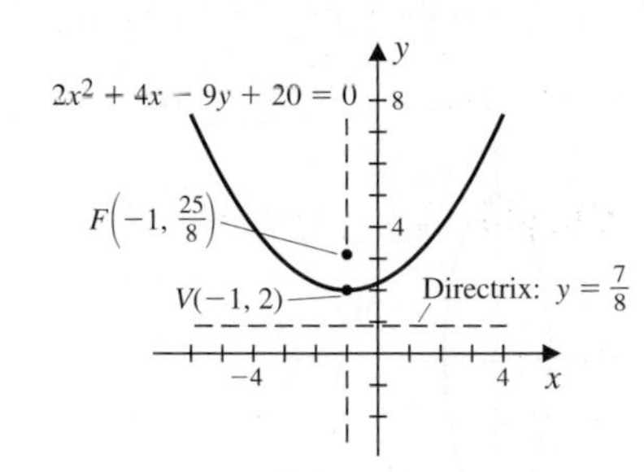

17.

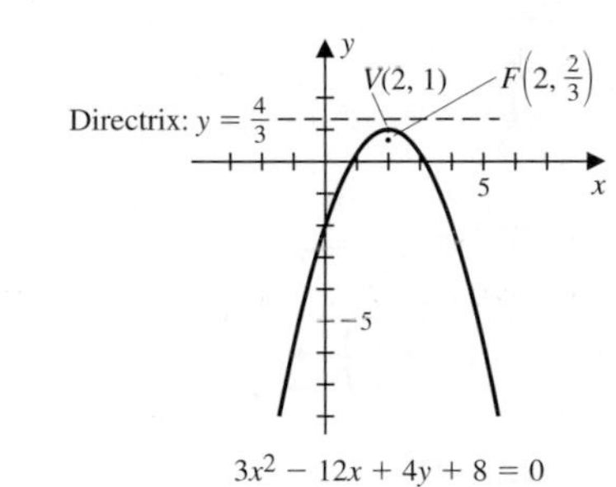

19. $y = \frac{1}{4}x^2$

21. $x = \frac{1}{4}y^2$

23. $y = \frac{1}{8}(x + 2)^2$

25. $x + 2 = -\frac{1}{16}(y - 2)^2$

27. $y - 2 = -\frac{1}{8}(x + 2)^2$

29. $x + 2 = -\frac{1}{8}(y - 2)^2$

31. **a.** $y = \frac{3}{8}x^2$

b. $y = \frac{3}{8}(x - 1)^2$

c. $y - 2 = \frac{3}{8}(x - 1)^2$

d. $y - 2 = \frac{3}{8}x^2$

33. $y - b = (2 - b)x^2$

35. Place the light $\frac{1}{2}$ inch from the vertex of the light.

37. **a.** The ball hits the ground 200 feet from the building.

b. $\sqrt{(1450)(625)} \approx 952$ ft

Exercise Set 6.3 (page 334)

1.

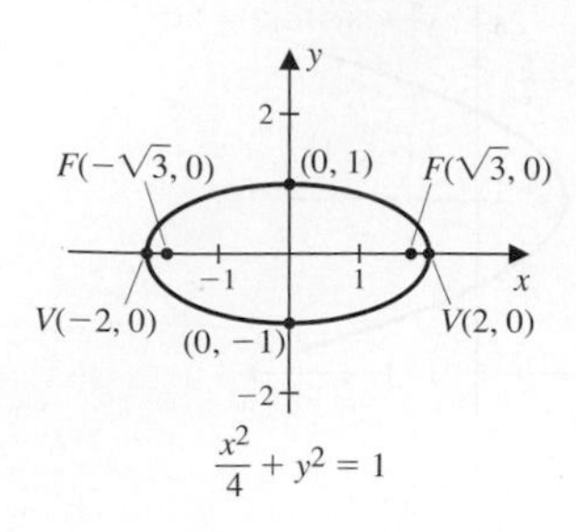

3.

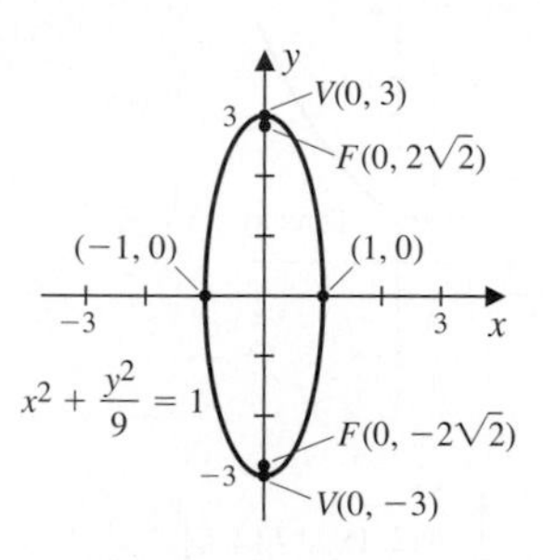

5.

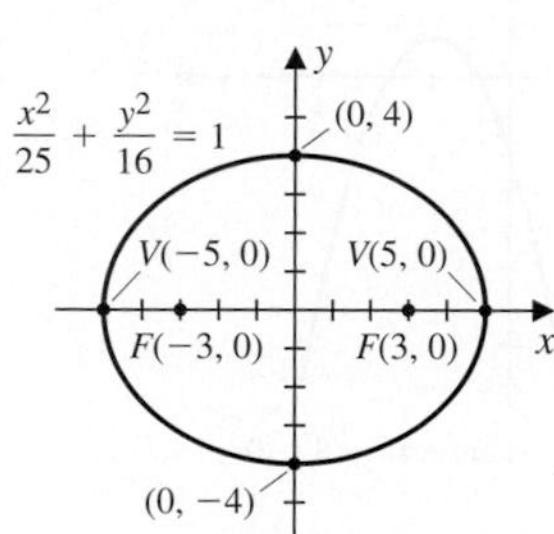

7.

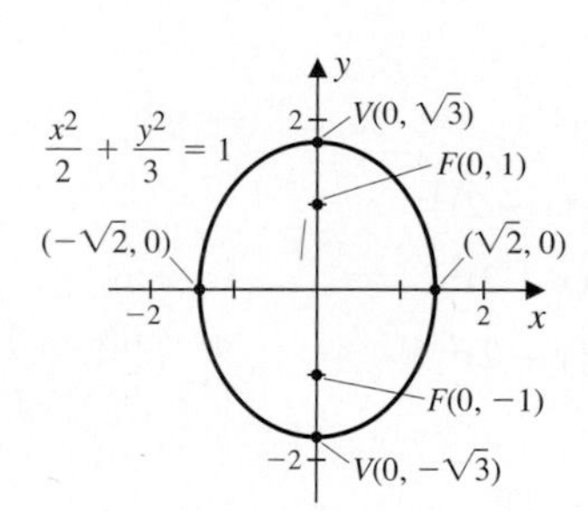

9.

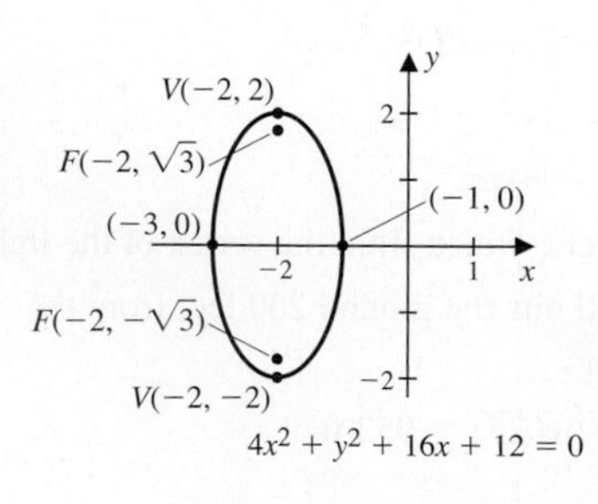

11.

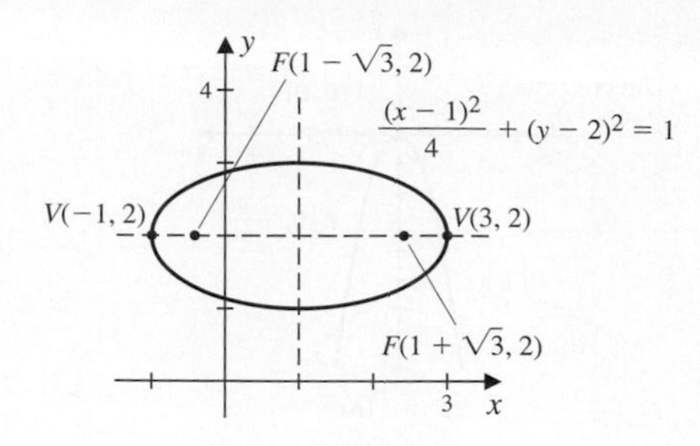

13.

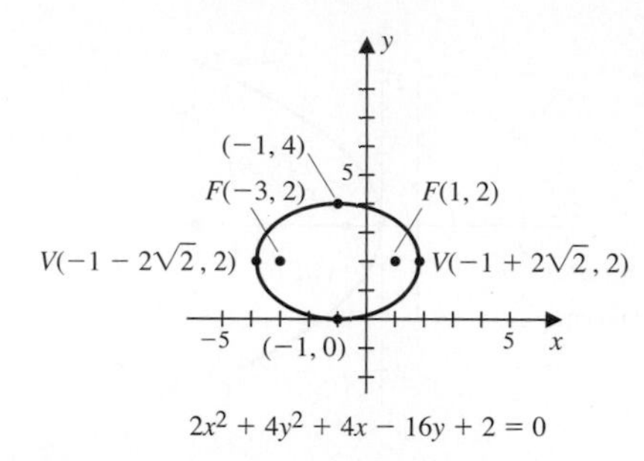

15. $\dfrac{x^2}{16} + \dfrac{y^2}{9} = 1$

17. $\dfrac{x^2}{9} + \dfrac{y^2}{5} = 1$

19. $\dfrac{x^2}{4} + \dfrac{y^2}{5} = 1$

21. $\dfrac{4x^2}{25} + \dfrac{4y^2}{9} = 1$

23. $\dfrac{(x-2)^2}{4} + \dfrac{y^2}{3} = 1$

25. $\dfrac{(x+4)^2}{16} + \dfrac{(y-3)^2}{25} = 1$

27. $(x-3)^2 + \dfrac{(y-1)^2}{4} = 1$

29. $\dfrac{2b^2}{a}$

31. $\dfrac{x^2}{57600} + \dfrac{y^2}{19600} = 1$

33. $e = 0.6$; $\dfrac{x^2}{209091600} + \dfrac{y^2}{146365200} = 1$

Exercise Set 6.4 (page 342)

1.

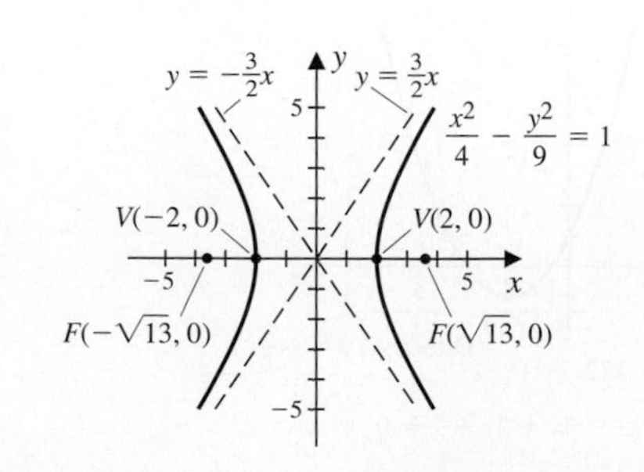

3.

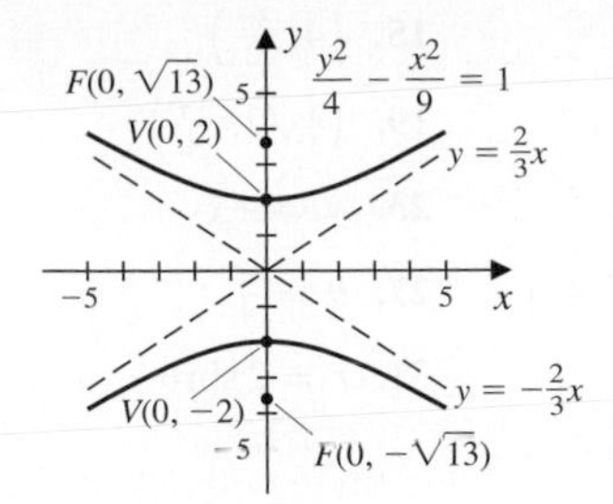

5.

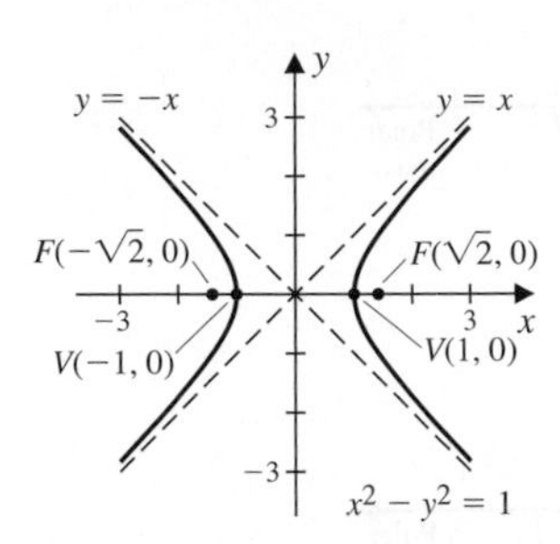

7.

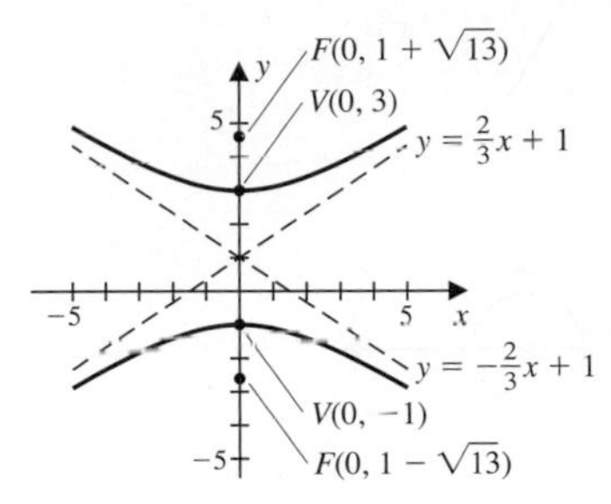

9.

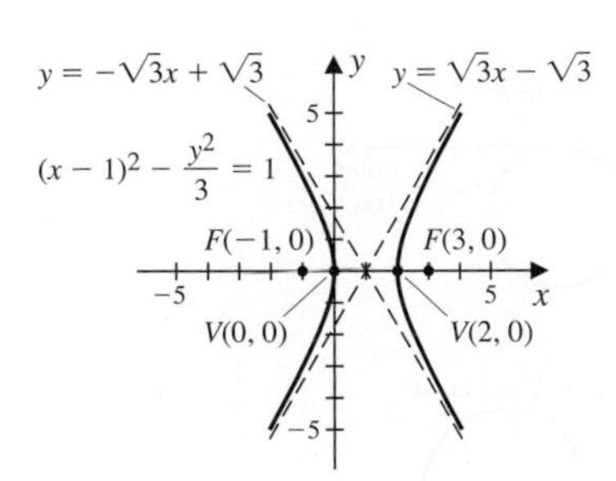

11.

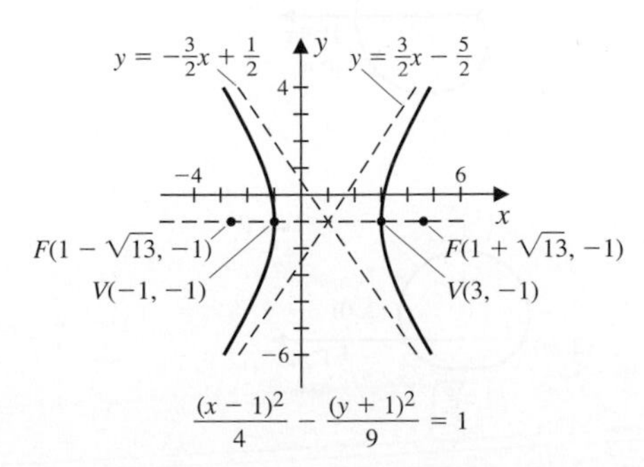

13.

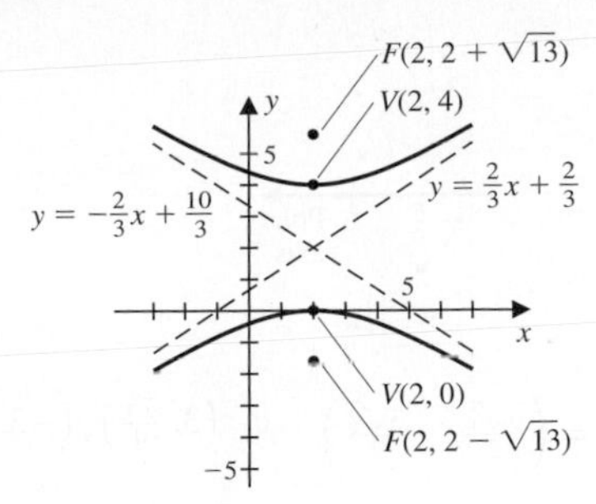

$9y^2 - 4x^2 - 36y + 16x - 16 = 0$

15. $\dfrac{x^2}{9} - \dfrac{y^2}{16} = 1$ **17.** $\dfrac{y^2}{16} - \dfrac{x^2}{9} = 1$

19. $\dfrac{(y+1)^2}{4} - \dfrac{(x-2)^2}{5} = 1$

21. $\dfrac{(x-2)^2}{4} - \dfrac{(y-4)^2}{5} = 1$

23. $\dfrac{y^2}{4} - \dfrac{x^2}{3} = 1$ **25.** $\dfrac{x^2}{9} - \dfrac{y^2}{16} = 1$

27. $\dfrac{2b^2}{a}$

29. The axes of the hyperbolas are perpendicular to each other and they share asymptotes. An example is shown.

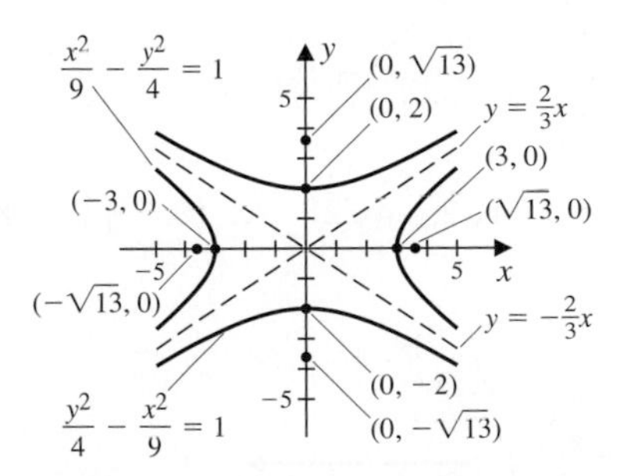

31. If $d(B, C) = d(A, C) + 130$, then ship from either plant. If $d(B, C) < d(A, C) + 130$, then ship from plant B. If $d(B, C) > d(A, C) + 130$, then ship from plant A.

Exercise Set 6.5 (page 353)

1. a.

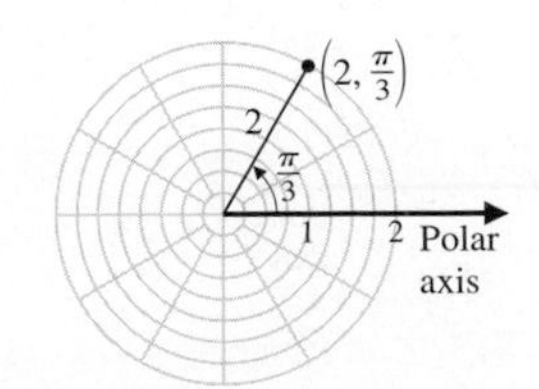

b. $(x, y) = (1, \sqrt{3})$ **c.** $\left(2, \frac{7\pi}{3}\right), \left(-2, \frac{4\pi}{3}\right)$

3. a.

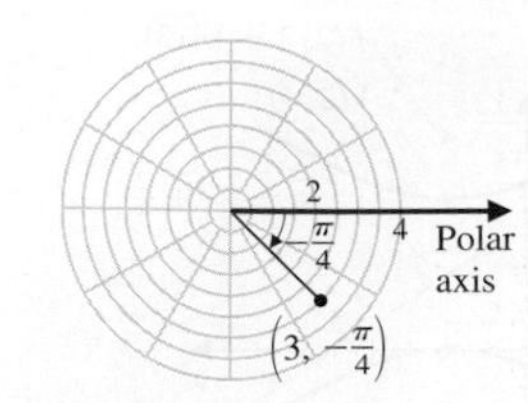

b. $(x, y) = \left(\frac{3\sqrt{2}}{2}, -\frac{3\sqrt{2}}{2}\right)$ **c.** $\left(3, \frac{7\pi}{4}\right), \left(-3, \frac{3\pi}{4}\right)$

5. a.

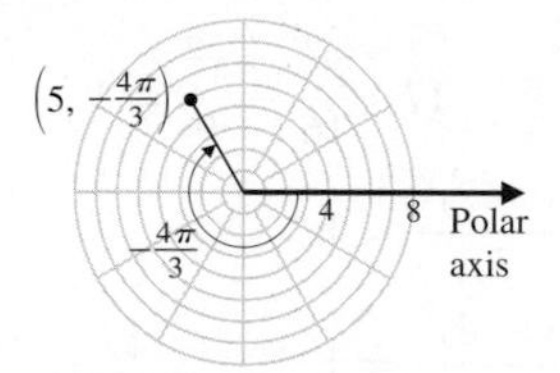

b. $(x, y) = \left(-\frac{5}{2}, \frac{5\sqrt{3}}{2}\right)$ **c.** $\left(5, \frac{2\pi}{3}\right), \left(-5, \frac{5\pi}{3}\right)$

7. a.

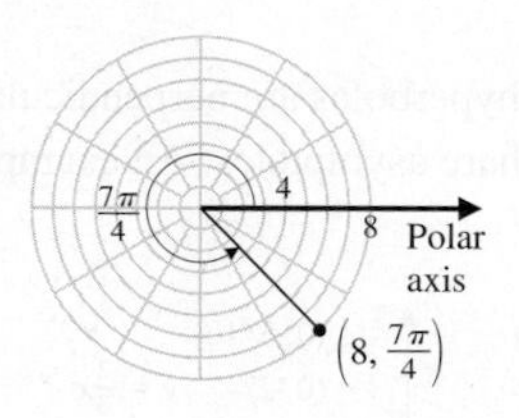

b. $(x, y) = (4\sqrt{2}, -4\sqrt{2})$ **c.** $\left(8, -\frac{\pi}{4}\right), \left(-8, \frac{3\pi}{4}\right)$

9. a.

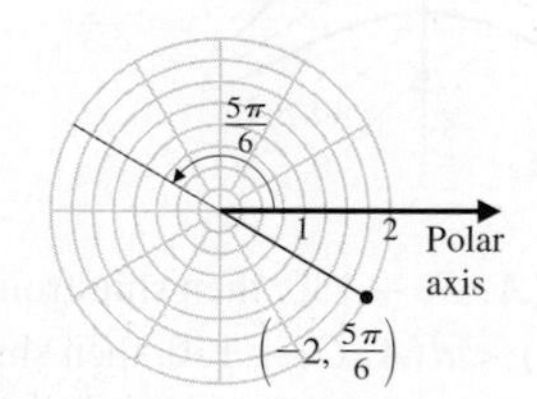

b. $(x, y) = (\sqrt{3}, -1)$ **c.** $\left(2, \frac{11\pi}{6}\right), \left(-2, -\frac{7\pi}{6}\right)$

11. a.

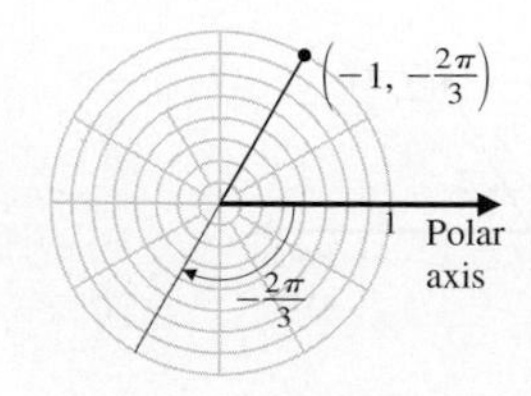

b. $(x, y) = \left(\frac{1}{2}, \frac{\sqrt{3}}{2}\right)$ **c.** $\left(1, \frac{\pi}{3}\right), \left(-1, \frac{4\pi}{3}\right)$

13. $(2, 0)$

15. $\left(4, \frac{3\pi}{2}\right)$

17. $\left(2, \frac{5\pi}{3}\right)$

19. $\left(4\sqrt{2}, \frac{3\pi}{4}\right)$

21. $x^2 + y^2 = 16$

23. $y = -x$

25. $(x - 1)^2 + y^2 = 1$

27. $\theta = \frac{\pi}{4}$

29. $r = 3$

31. $r = 2 \sin \theta$

33.

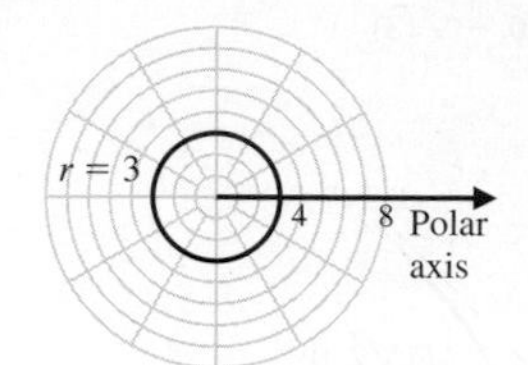

35.

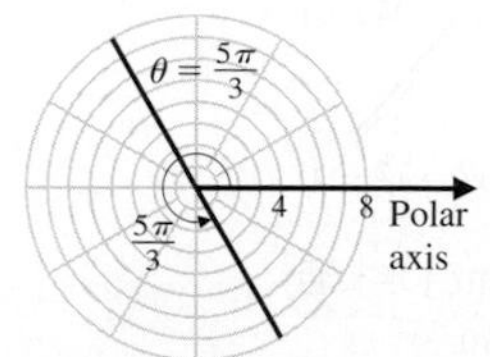

37.

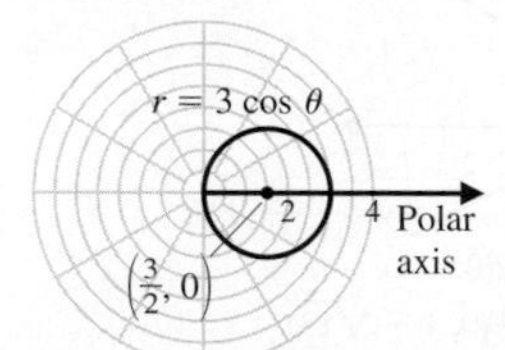

39.

$r = -4 \sin \theta$
2
4
Polar axis

41.

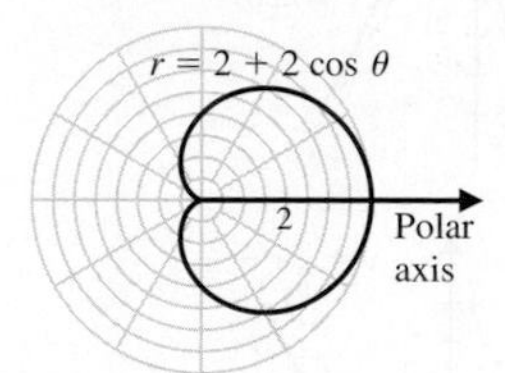

43.

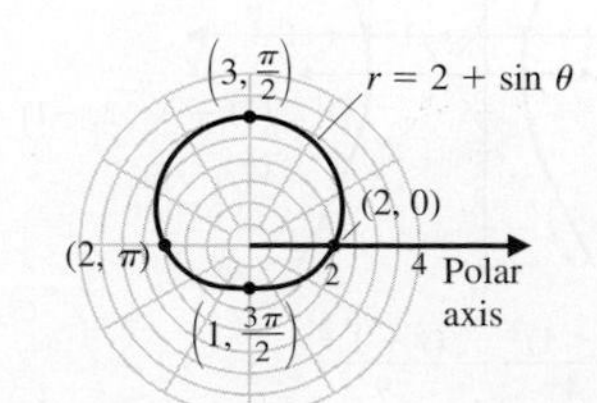

45.

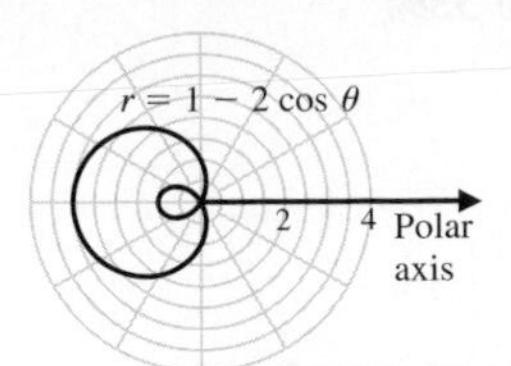

47.

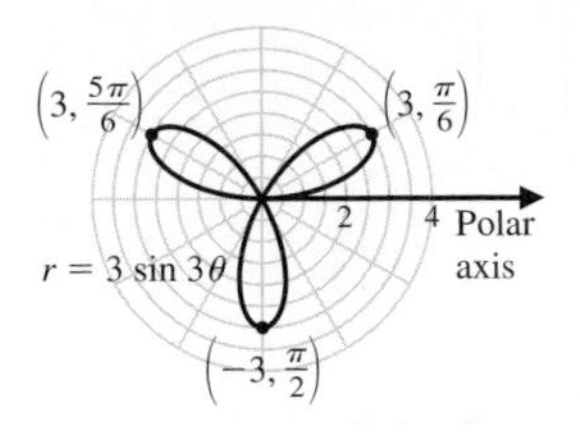

49.

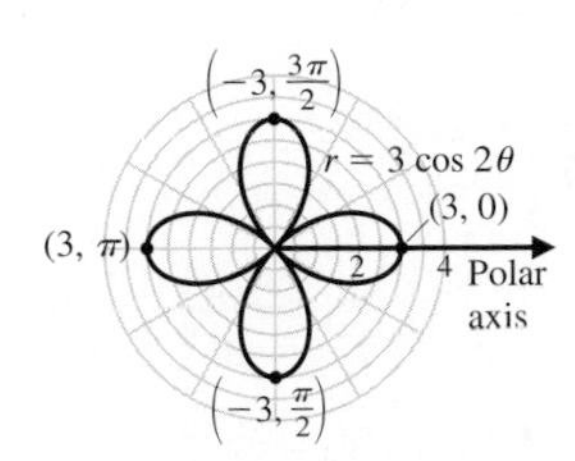

51.

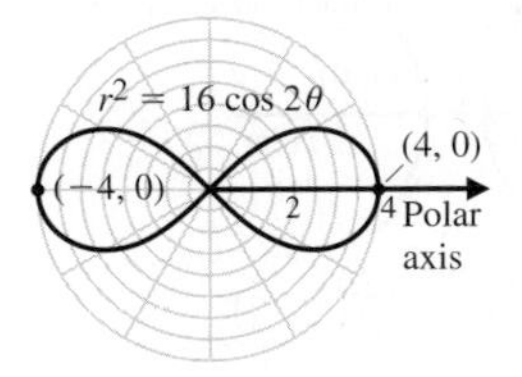

53.

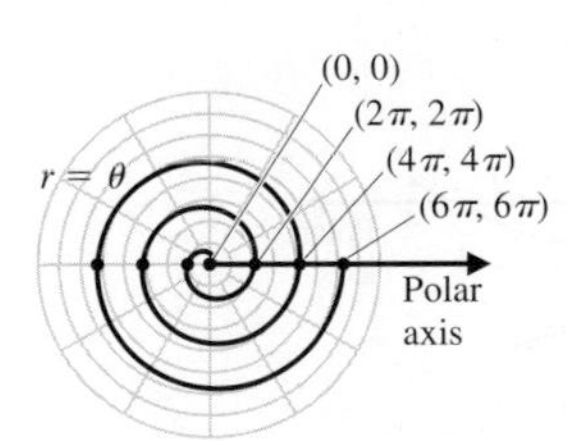

55.

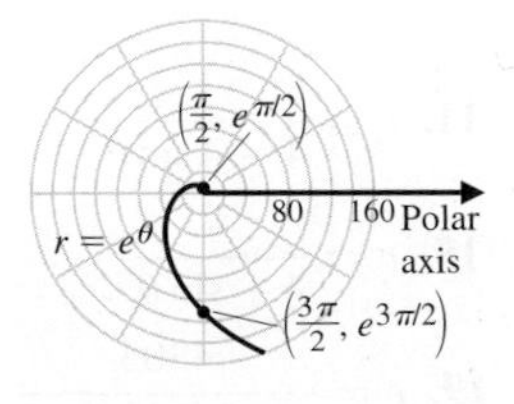

57.

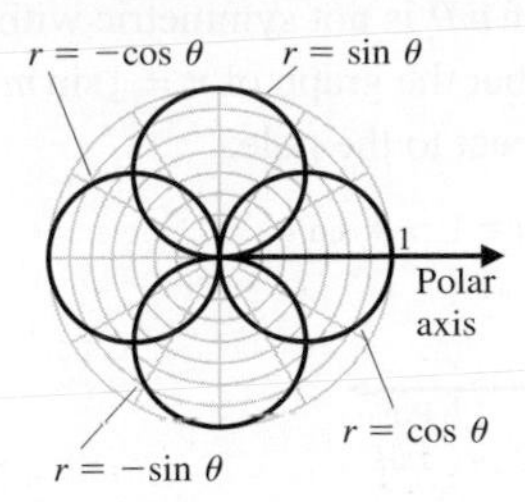

59.

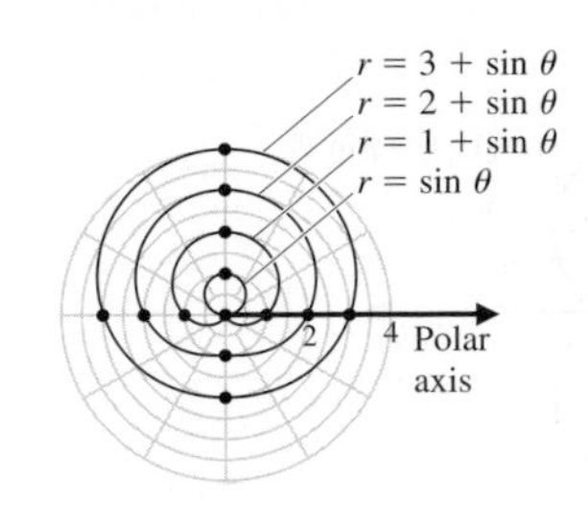

61. a.

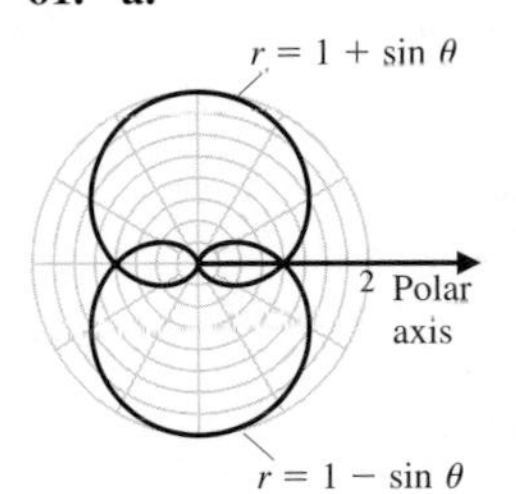

b.

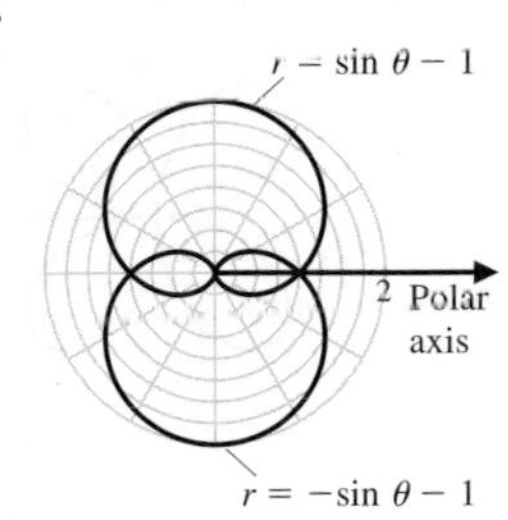

63. a. $(0, 0)$, $(90, \frac{\pi}{4})$, $(90\sqrt{2}, \frac{\pi}{2})$, $(90, \frac{3\pi}{4})$

b. $(0, 0)$, $(90, 0)$, $(90\sqrt{2}, \frac{\pi}{4})$, $(90, \frac{\pi}{2})$

65. $(1, 0)$, $(0, 1)$, $(0, -1)$

67. $(0, 0)$, $(\frac{\sqrt{3}}{2}, \frac{1}{2})$, $(-\frac{\sqrt{3}}{2}, \frac{1}{2})$

69.

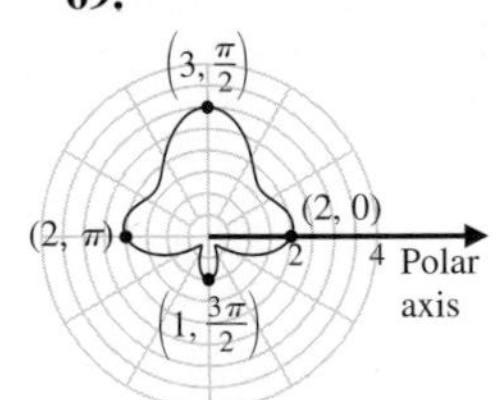

$n = 1$: $r = 1 + \sin\theta + \cos^2 2\theta$

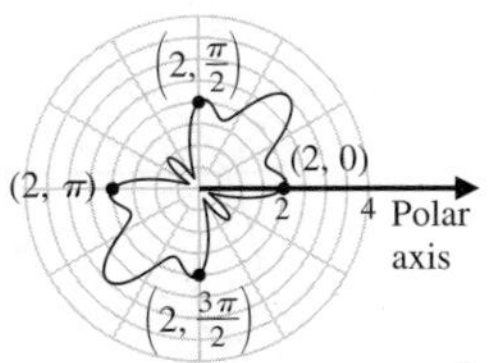

$n = 2$: $r = 1 + \sin 2\theta + \cos^2 4\theta$

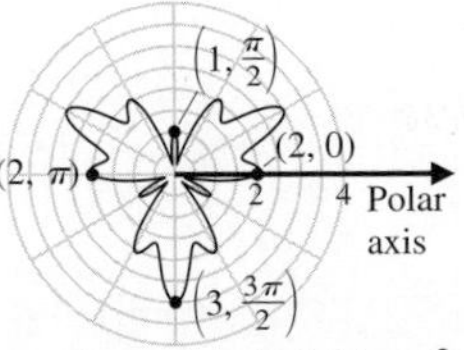

$n = 3$: $r = 1 + \sin 3\theta + \cos^2 6\theta$

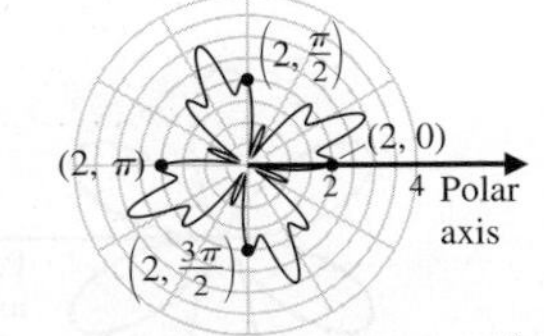

$n = 4$: $r = 1 + \sin 4\theta + \cos^2 8\theta$

71. The graph of $r = \sin m\theta$ is not symmetric with respect to the pole, but the graph of $r = |\sin m\theta|$ is symmetric with respect to the pole.

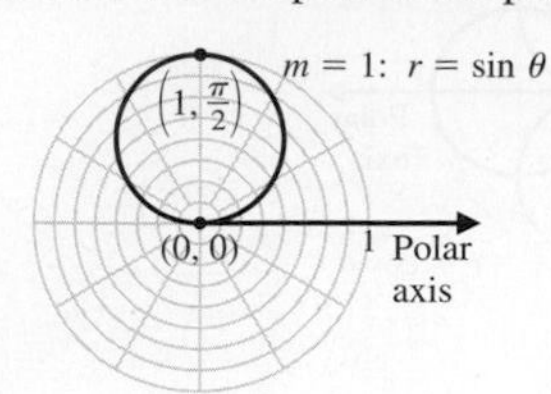

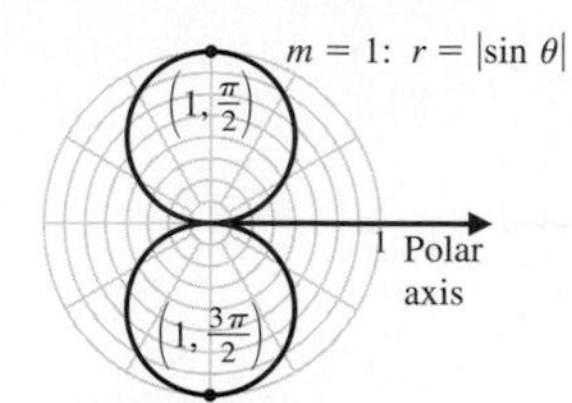

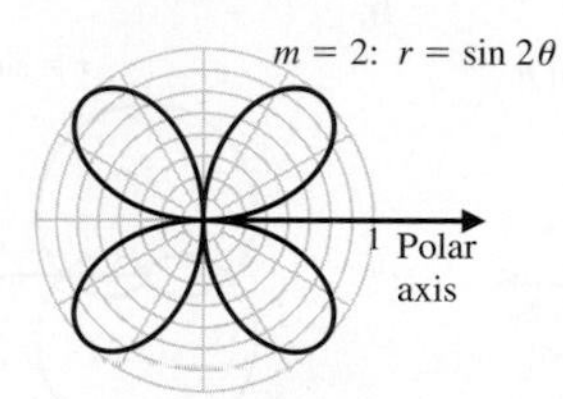

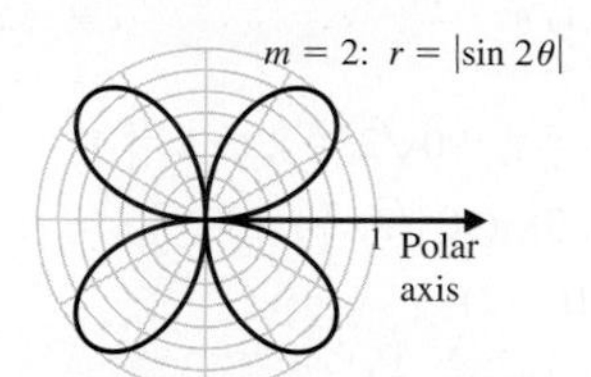

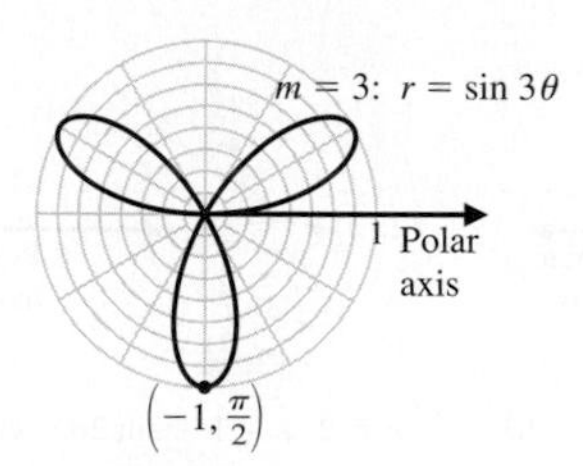

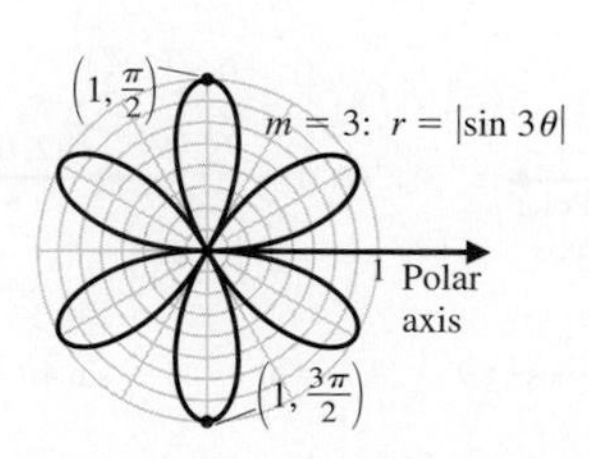

Exercise Set 6.6 (page 359)

1.

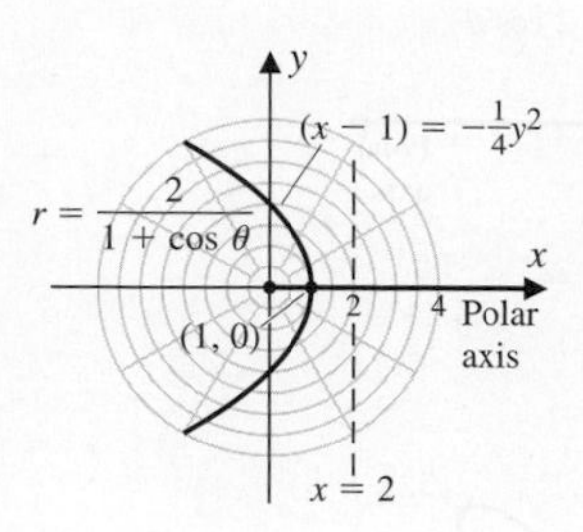

3.

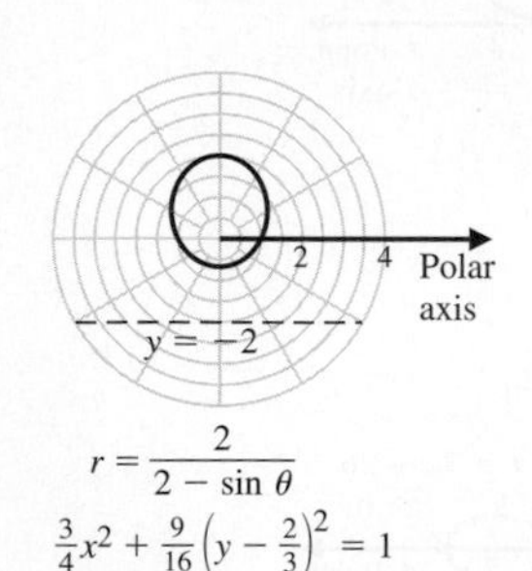

$r = \dfrac{2}{2 - \sin\theta}$

$\frac{3}{4}x^2 + \frac{9}{16}\left(y - \frac{2}{3}\right)^2 = 1$

5.

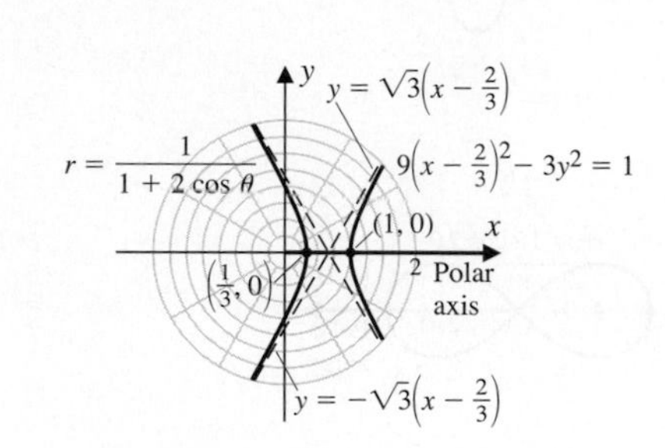

7.

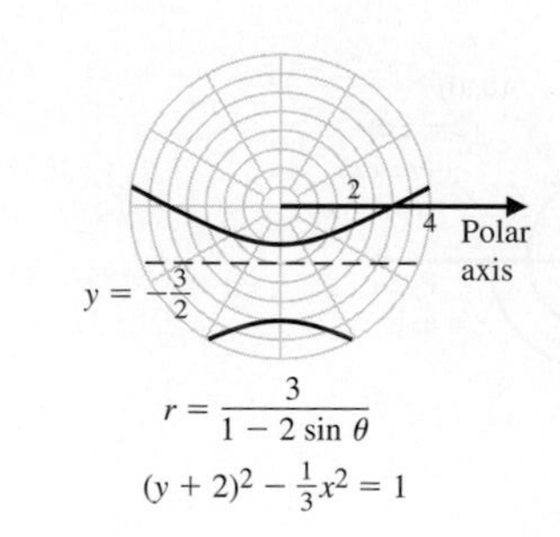

$r = \dfrac{3}{1 - 2\sin\theta}$

$(y + 2)^2 - \frac{1}{3}x^2 = 1$

9. $r = \dfrac{8}{1 - 2\cos\theta}$

11. $r = \dfrac{1}{4 - 4\sin\theta}$

13. $r = \dfrac{6}{1 + 3\cos\theta}$

15. $r = \dfrac{1}{4 + \sin\theta}$

17. $r = \dfrac{3}{2 + \cos\theta}$

19. $r = \dfrac{4303}{1 - 0.048\cos\theta}$

Exercise Set 6.7 (page 364)

1.

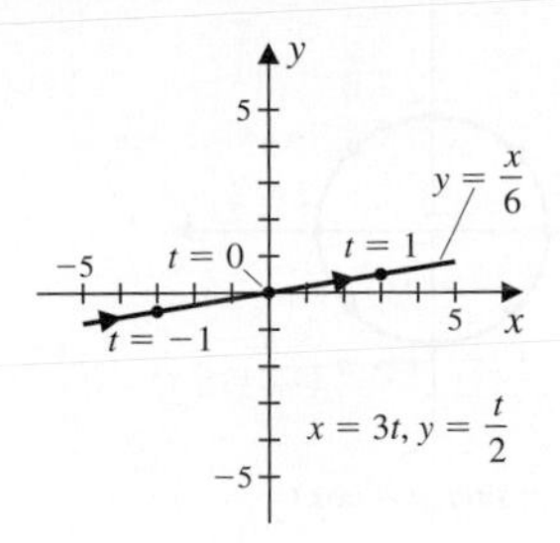

3.

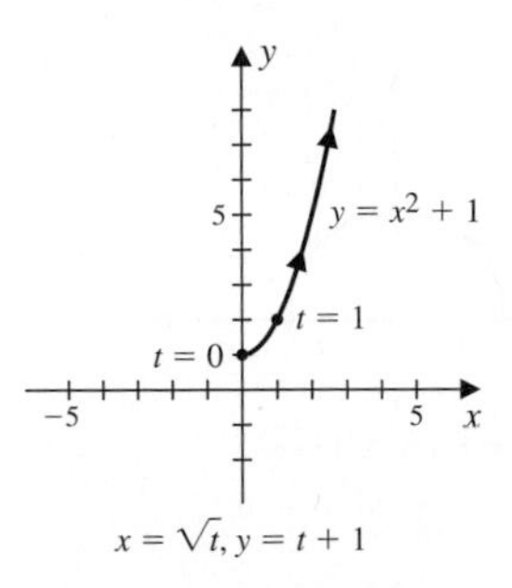

5.

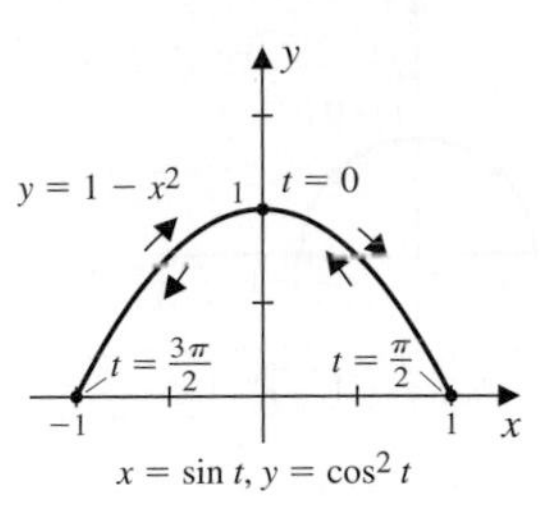

7.

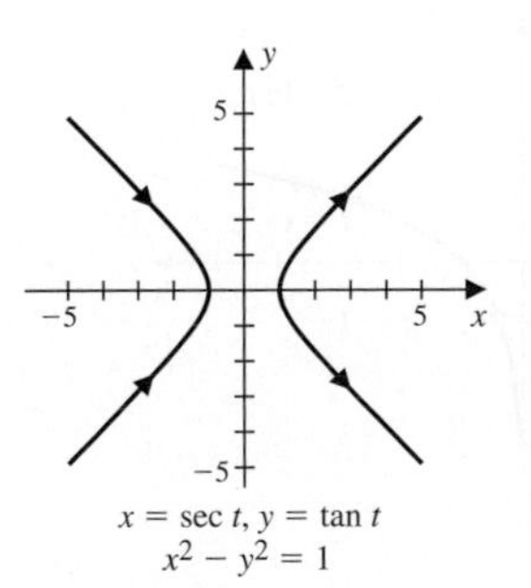

9.

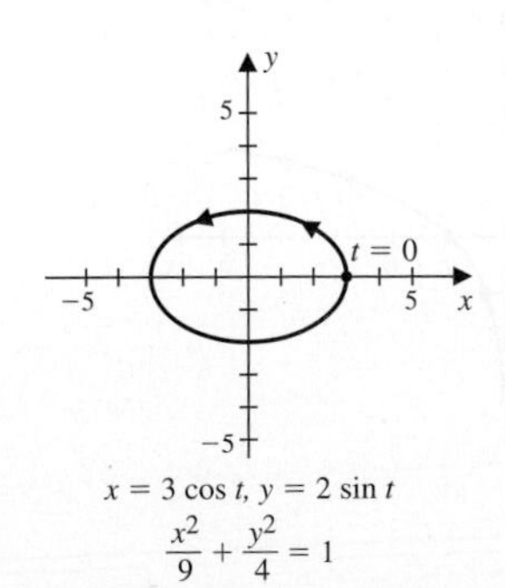

11.

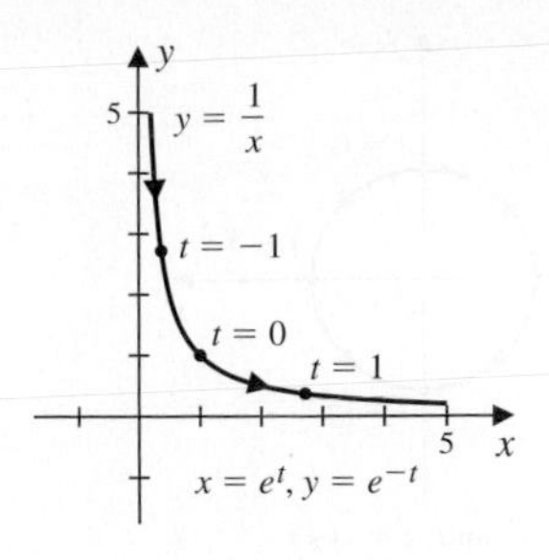

13. a.

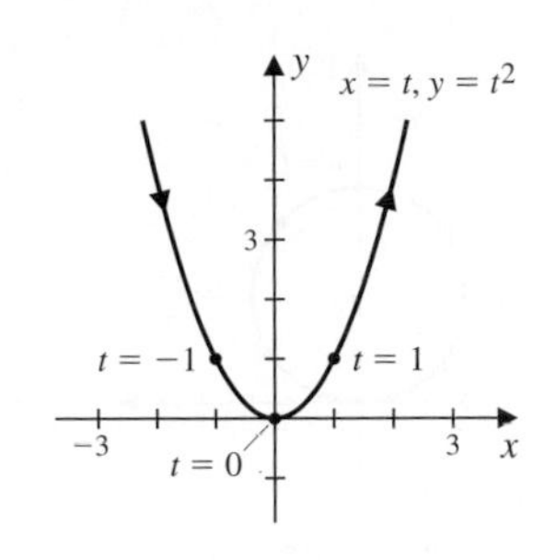

b.

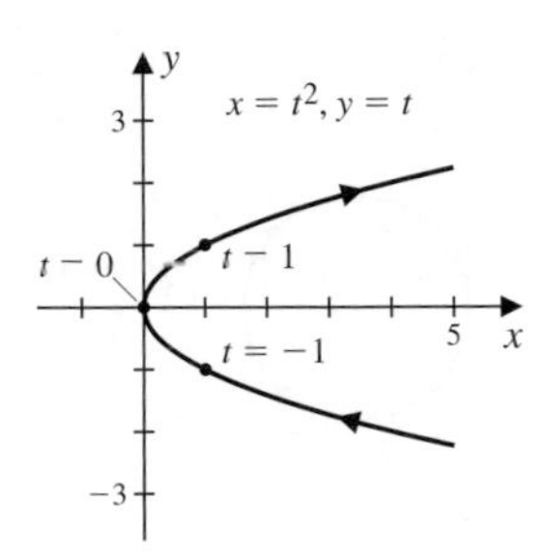

c.

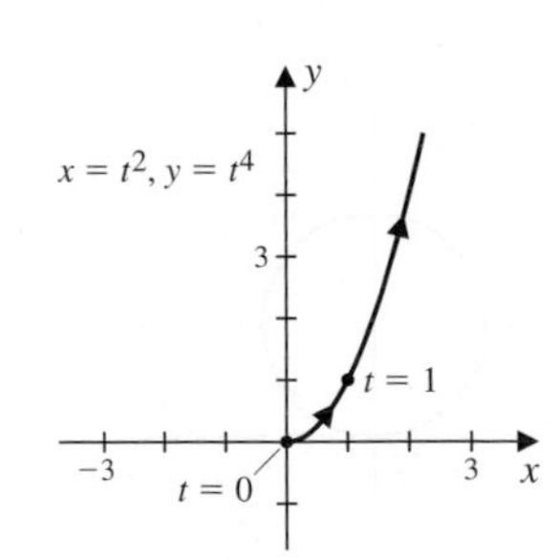

d.

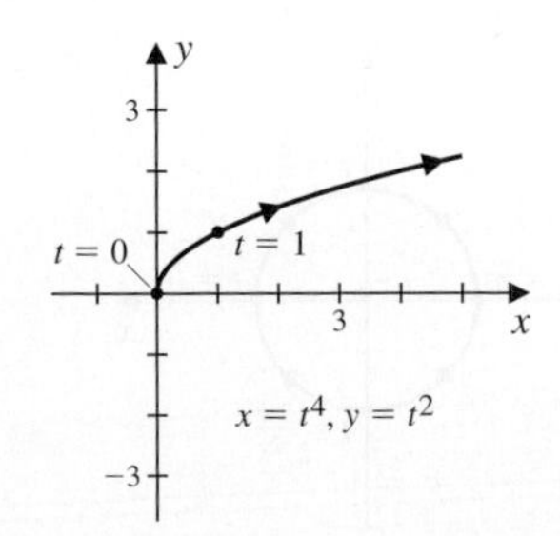

15. a.

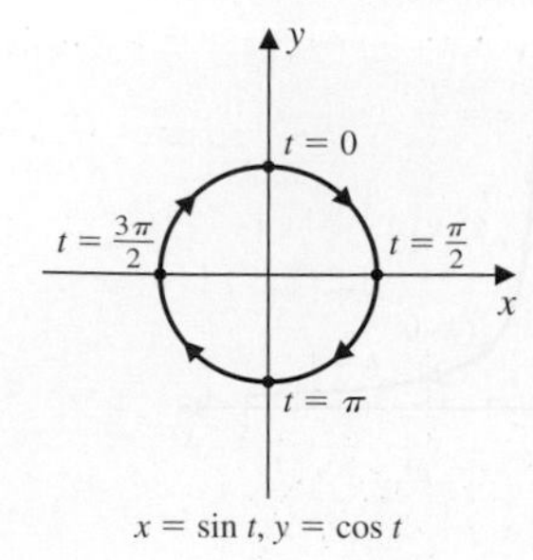

$x = \sin t, y = \cos t$

b.

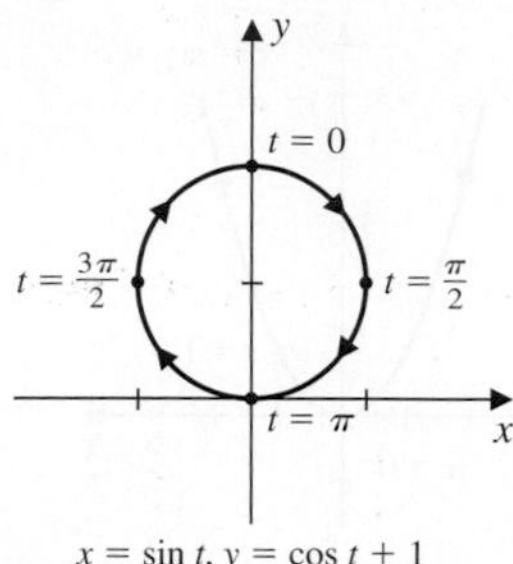

$x = \sin t, y = \cos t + 1$

c.

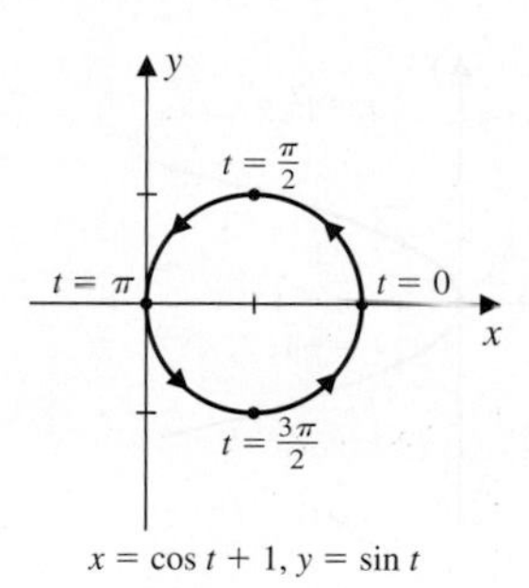

$x = \cos t + 1, y = \sin t$

d.

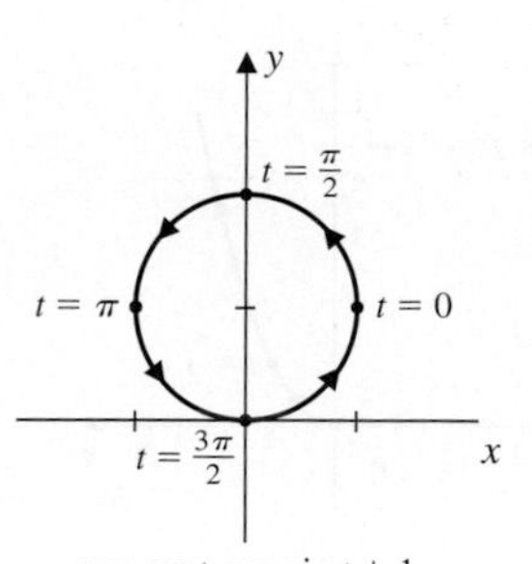

$x = \cos t, y = \sin t + 1$

17. a.

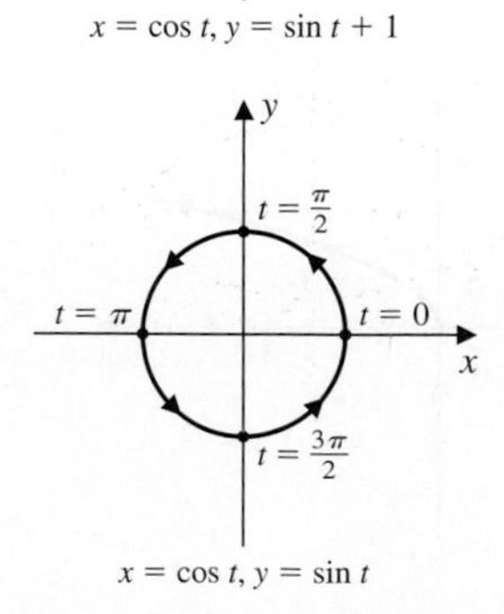

$x = \cos t, y = \sin t$

b.

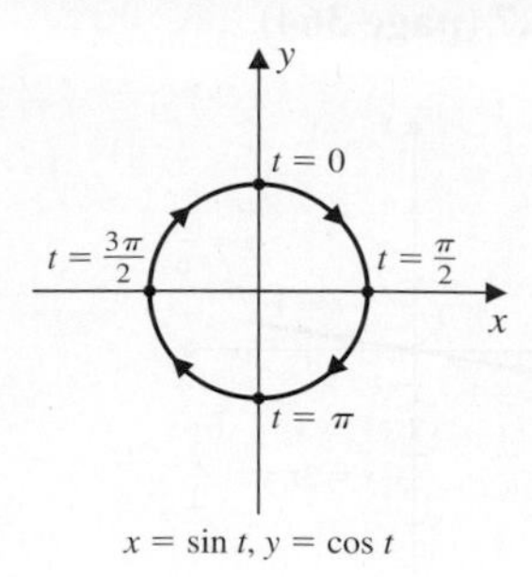

$x = \sin t, y = \cos t$

c.

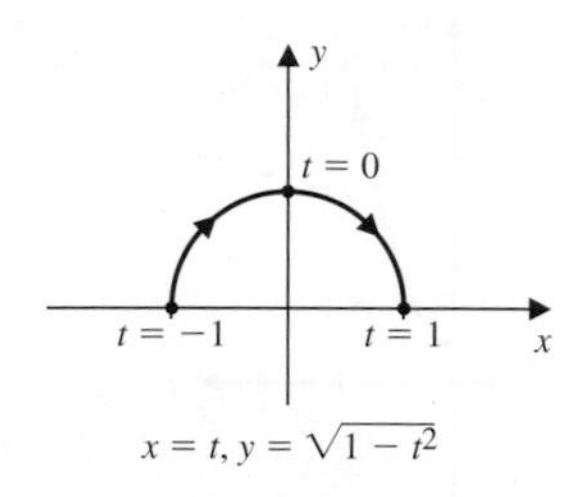

$x = t, y = \sqrt{1 - t^2}$

d.

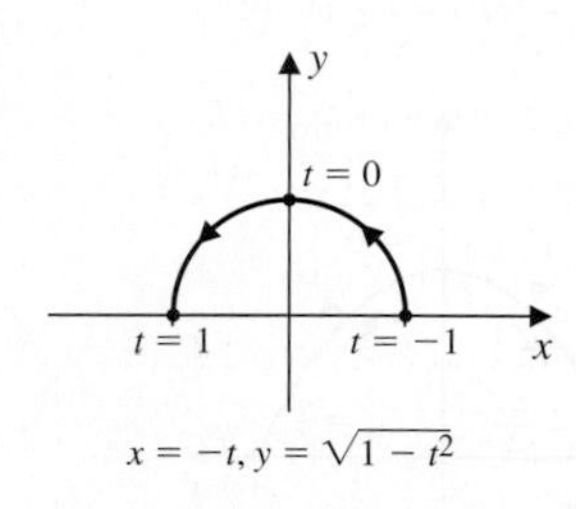

$x = -t, y = \sqrt{1 - t^2}$

19. a.

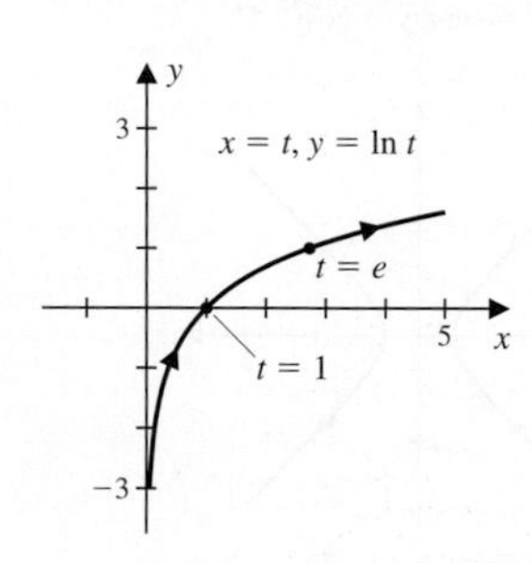

b.

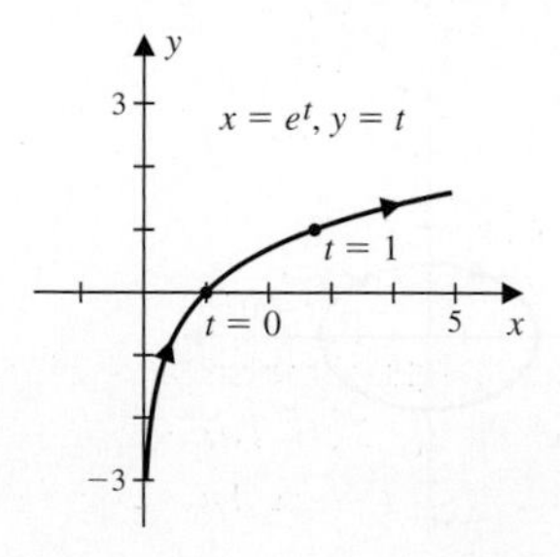

c.

d.

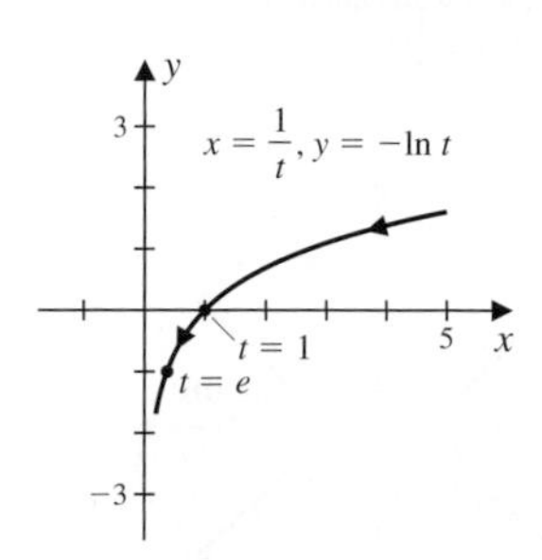

21. One choice is $x = t + 2$ and $y = \frac{1}{3}t - 1$.

23. One choice is $x = t + 1$ and $y = 2t^2 - 2$.

25. **a.** $x = r\cos t,\ y = r\sin t$

b. $x = r\cos 2t,\ y = r\sin 2t$

c. $x = -r\cos 3t,\ y = -r\sin 3t$

d. $x = r\sin 2t,\ y = r\cos 2t$

e. $x = r\sin 3t,\ y = r\cos 3t$

27. $a = 1; b = -1$

$a = 1; b = 1$

$a = 1; b = 2$

$a = 1; b = 3$

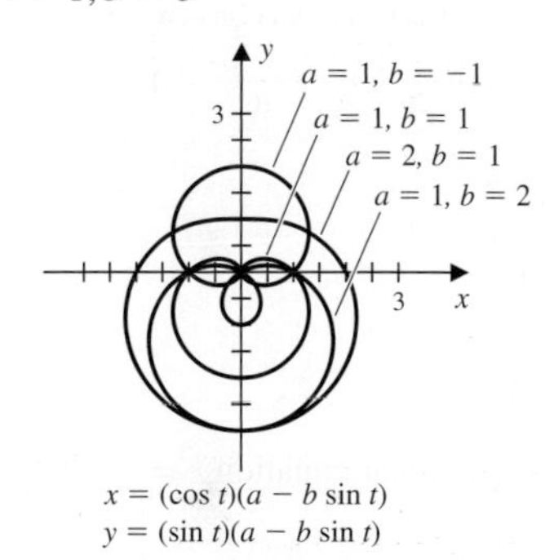

29. The maximum height of the curve is $2a$, and the period of the curve is $2\pi a$. The curve touches the x-axis for those t that make the y-coordinate 0. So,

$$a(1 - \cos t) = 0 \text{ implies } \cos t = 1, \text{ and}$$
$$t = +2k\pi, \quad \text{for} \quad k = 0, 1, 2 \ldots.$$

The curve touches the x-axis at the x-values:

$$x = a(t - \sin t) = a(\pm 2k\pi - \sin(\pm 2k\pi))$$
$$= \pm 2ak\pi, \quad \text{for} \quad k = 0, 1, 2 \ldots.$$

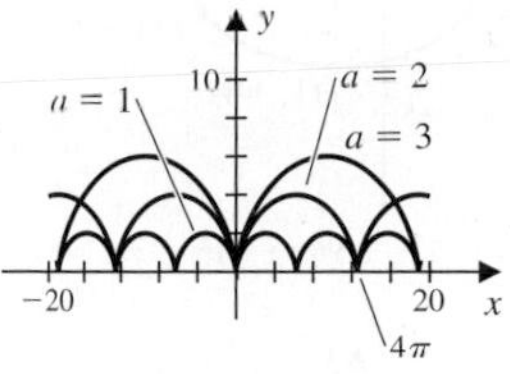

Review Exercises for Chapter 6 (page 365)

1.

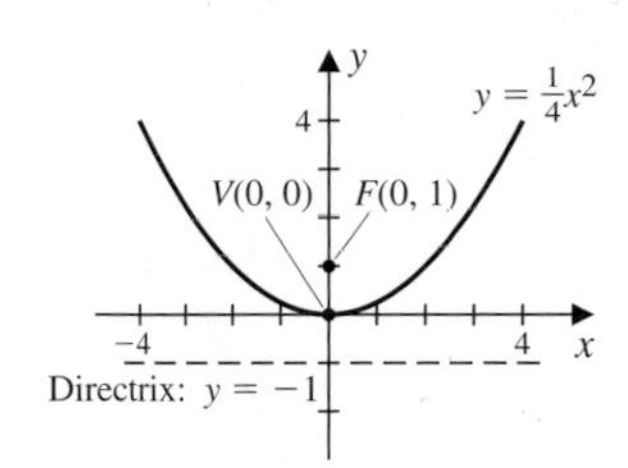

3.

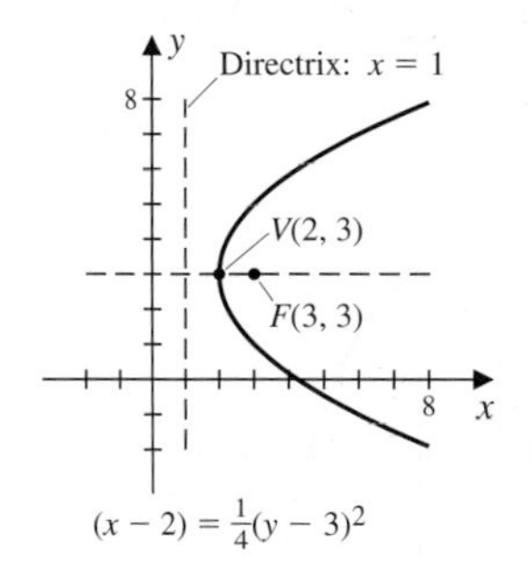

5.

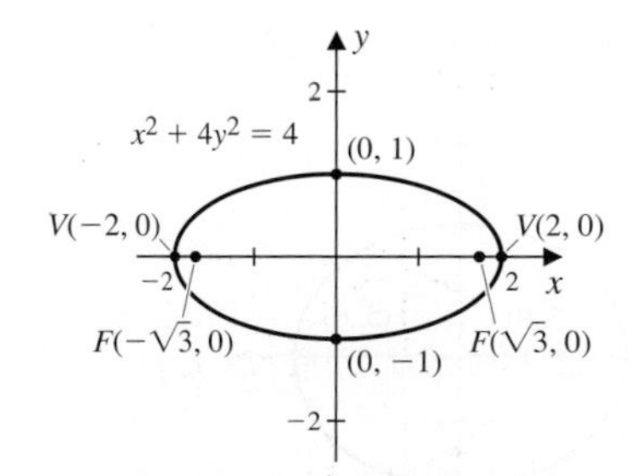

7.

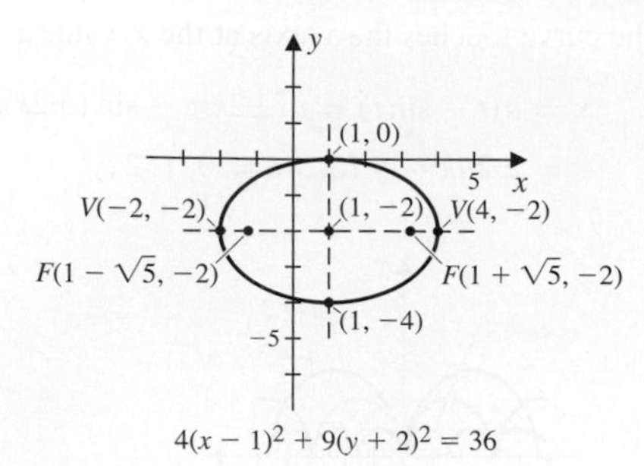

9.

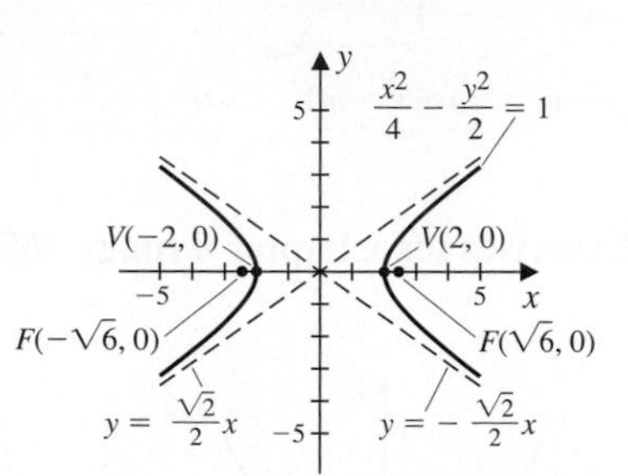

11.

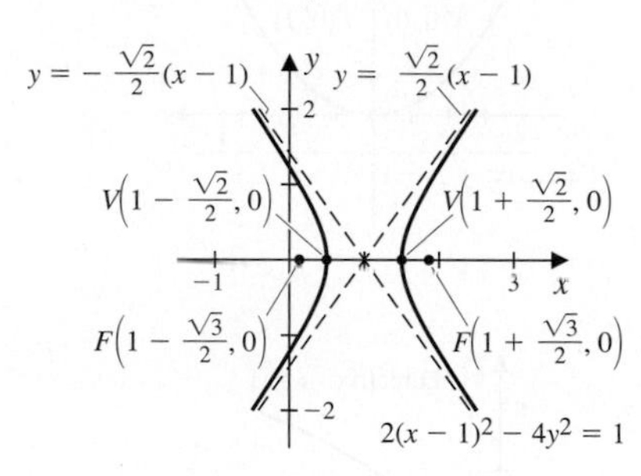

13.

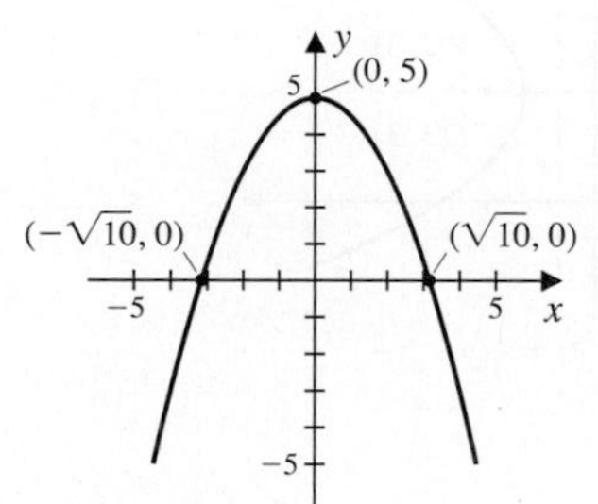

Parabola with equation $y - 5 = -\frac{1}{2}x^2$

15.

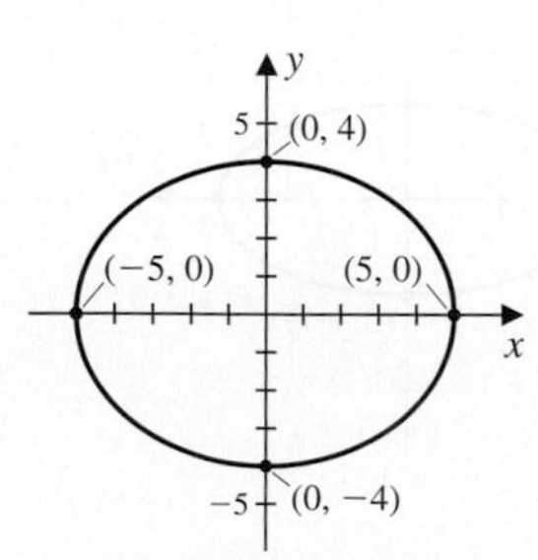

Ellipse with equation
$16x^2 + 25y^2 = 400$

17.

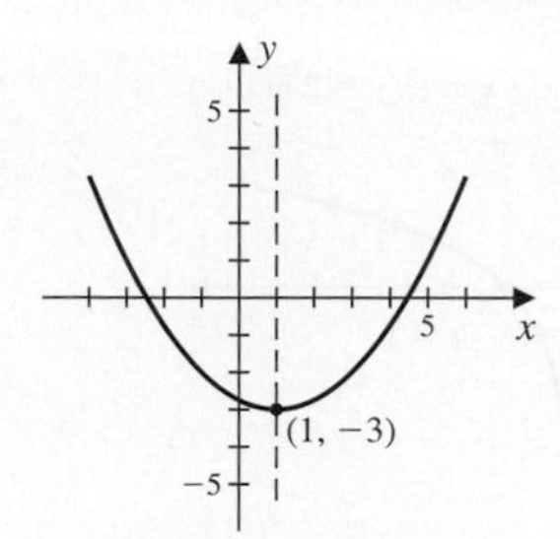

Parabola with equation
$y + 3 = \frac{1}{4}(x - 1)^2$

19.

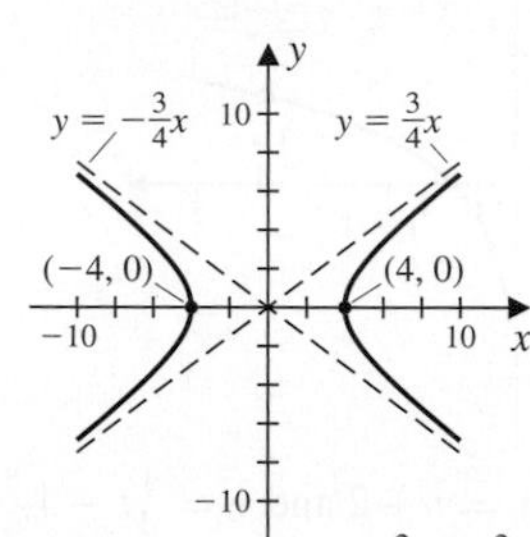

Hyperbola with equation $\frac{x^2}{16} - \frac{y^2}{9} = 1$

21.

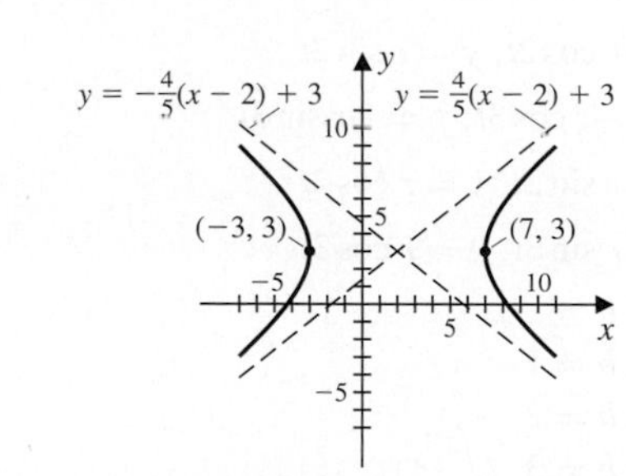

Hyperbola with equation
$\frac{(x-2)^2}{25} - \frac{(y-3)^2}{16} = 1$

23. $y - 1 = -\frac{1}{4}x^2$

25. $x^2 - \frac{y^2}{8} = 1$

27. $\frac{y^2}{41} + \frac{x^2}{16} = 1$

29. A hyperbola with polar equation $r = \frac{6}{1 - 3\sin\theta}$

31.

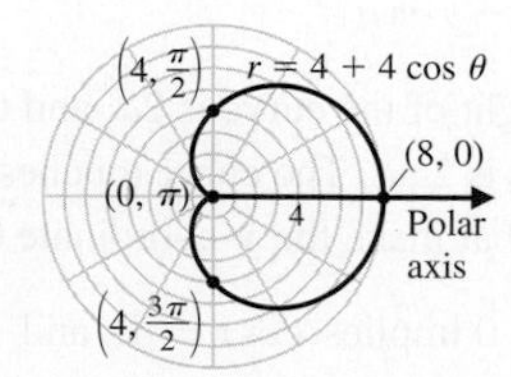

33.

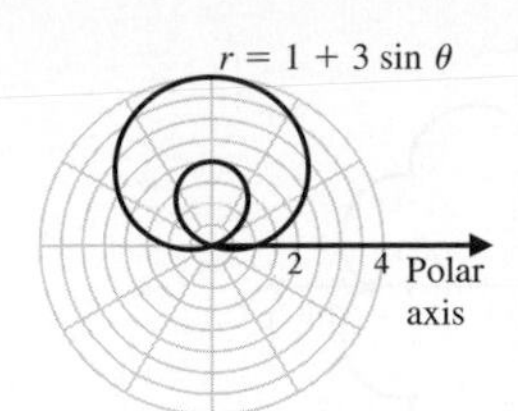

35.

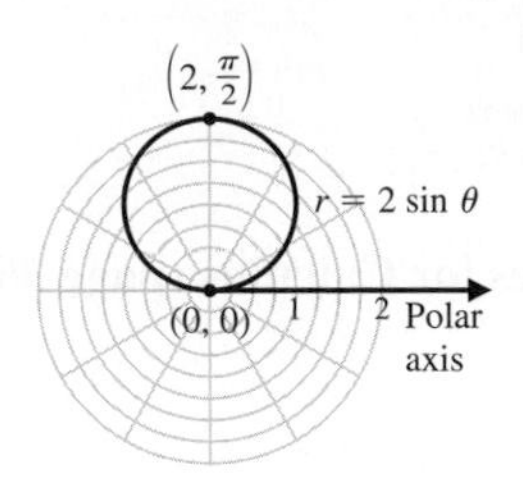

37.

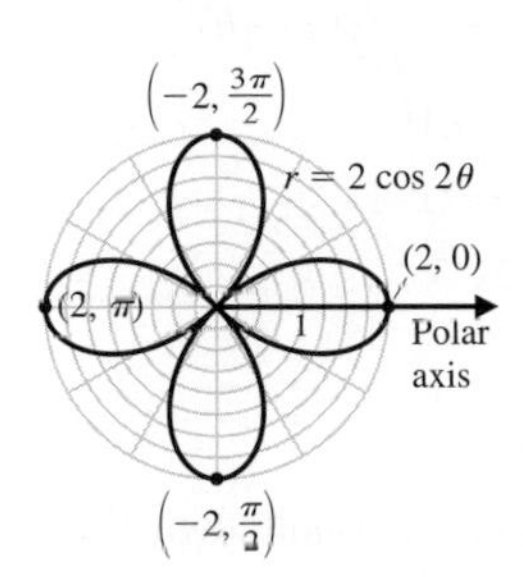

39.

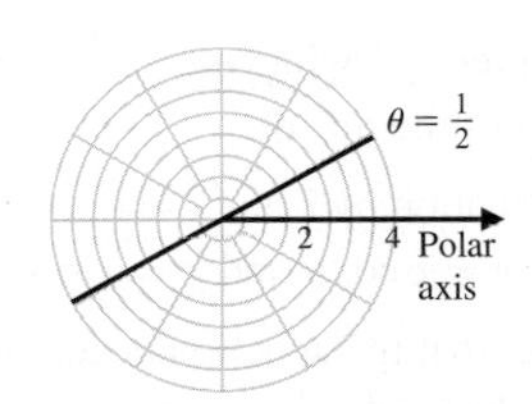

41.

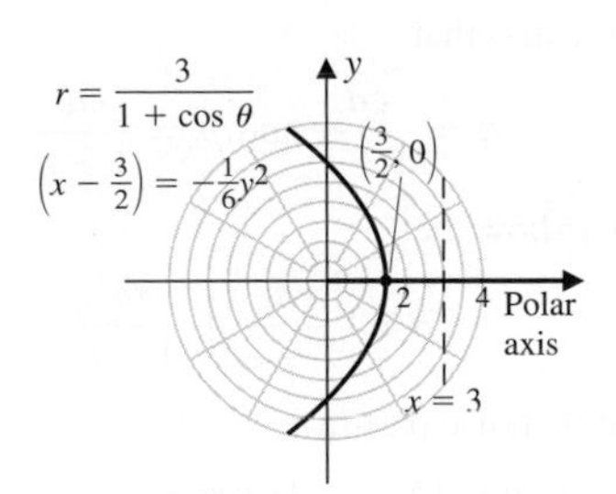

43.

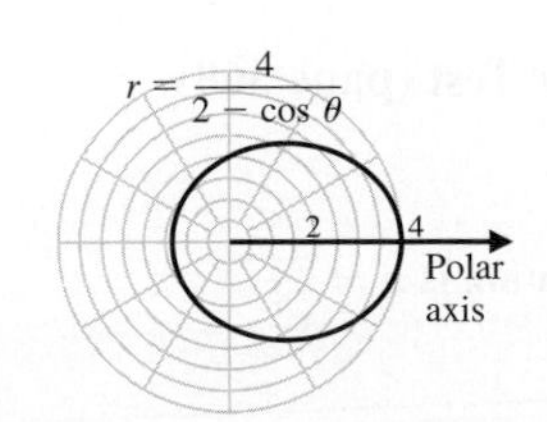

45. a.

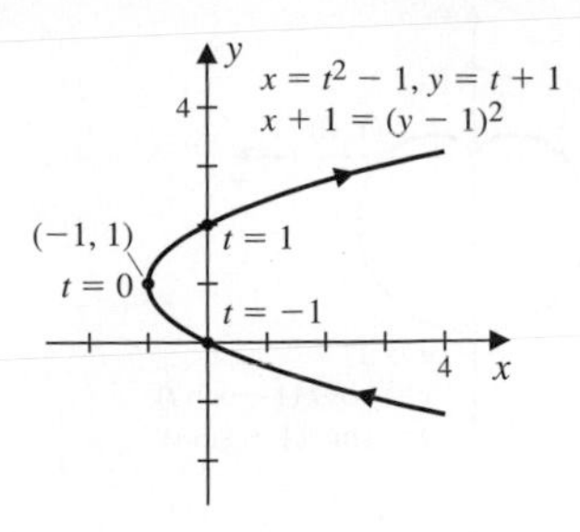

b. $x + 1 = (y - 1)^2$

47. a.

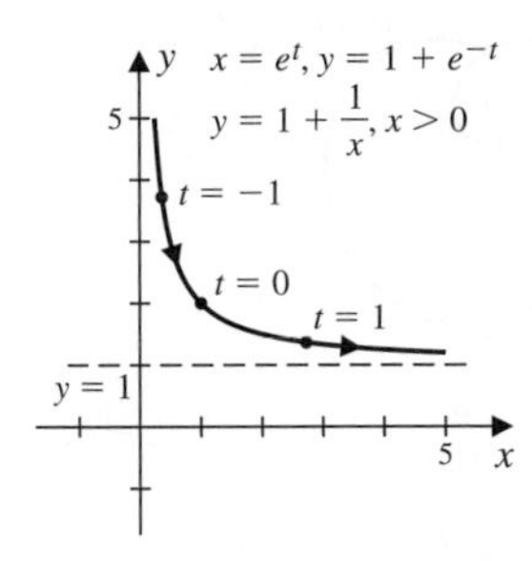

b. $y = 1 + \frac{1}{x}$, when $x > 0$

49. a.

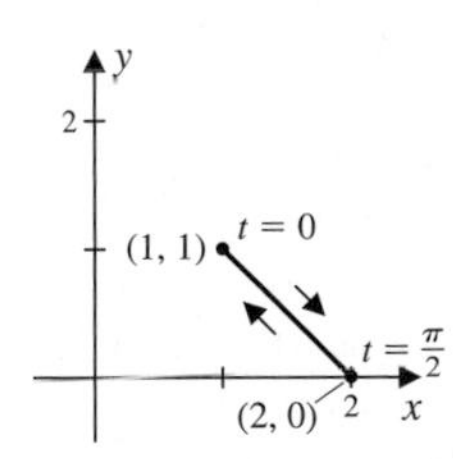

$x = (\sin t)^2 + 1,\ y = (\cos t)^2$
$y = 2 - x,\ 1 \leq x \leq 2,\ 0 \leq y \leq 1$

b. $y = 2 - x$, when $1 \leq x \leq 2$, and $0 \leq y \leq 1$

51.

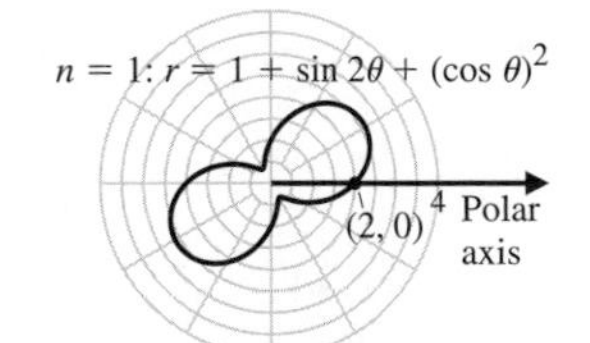

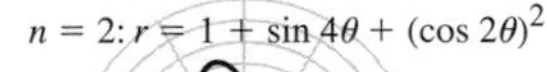

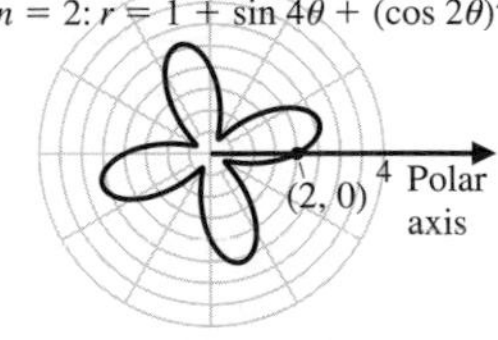

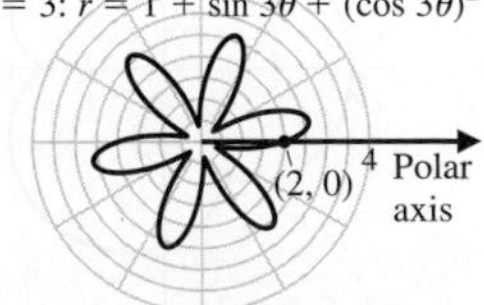

53.

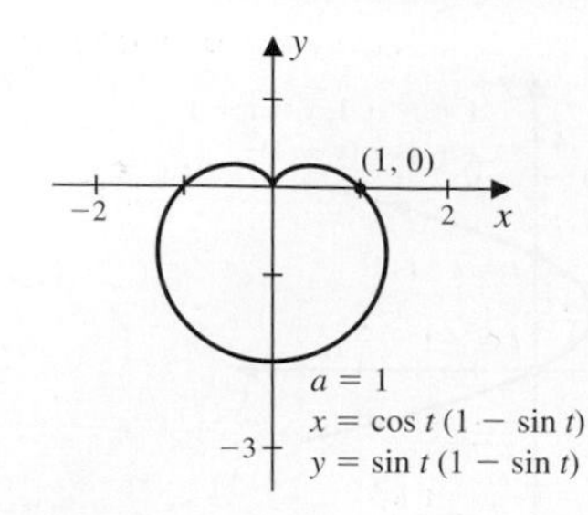

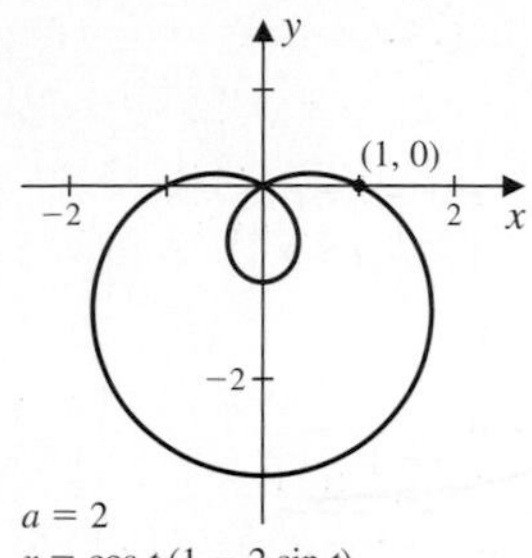

$a = 2$
$x = \cos t\,(1 - 2\sin t)$
$y = \sin t\,(1 - 2\sin t)$

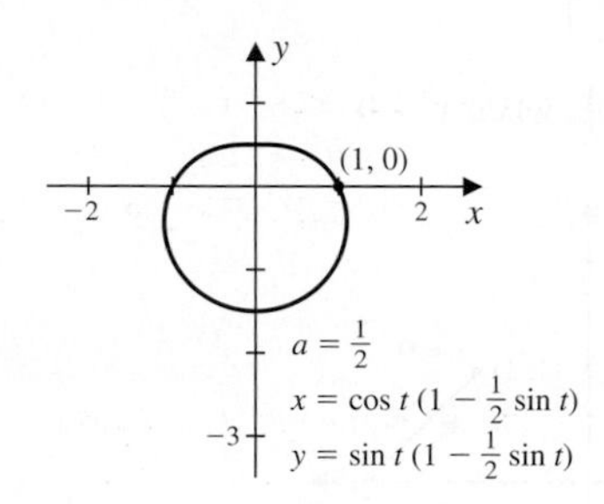

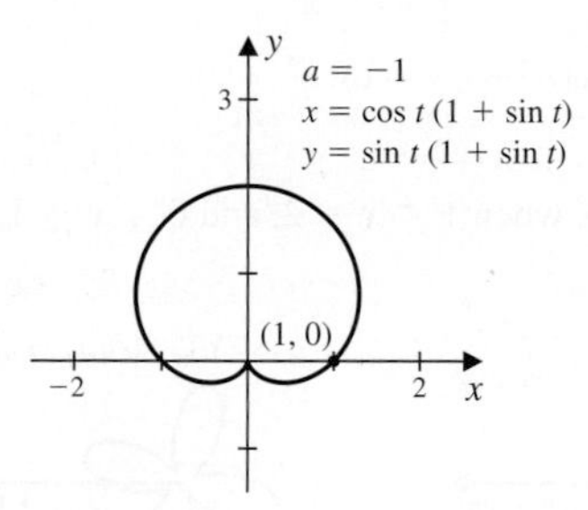

55.

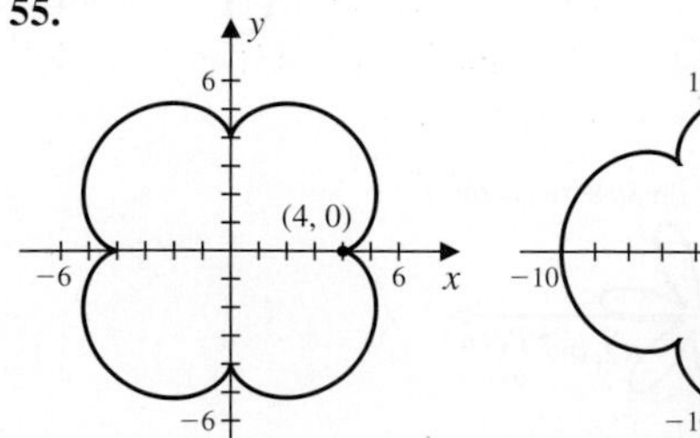

$a = 4, b = 1$
$x = 5\cos t - \cos 5t$
$y = 5\sin t - \sin 5t$

$a = 6, b = 2$
$x = 8\cos t - 2\cos 4t$
$y = 8\sin t - 2\sin 4t$

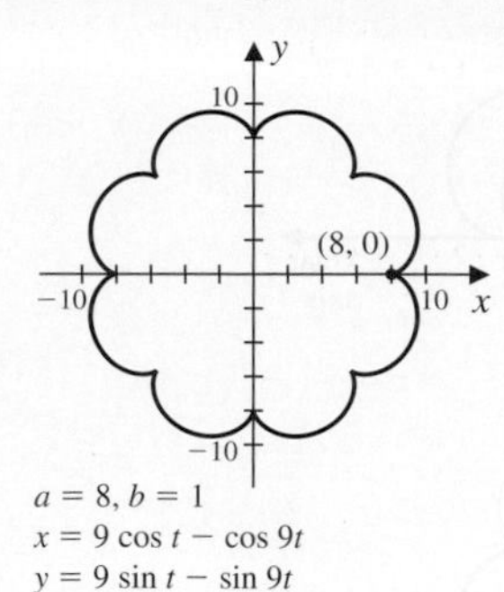

$a = 8, b = 1$
$x = 9\cos t - \cos 9t$
$y = 9\sin t - \sin 9t$

Chapter 6: Exercises for Calculus (page 366)

1. a. $y = \dfrac{y_1}{x_1^2}x^2$

b. $y - k = \dfrac{y_1 - k}{(x_1 - h)^2}(x - h)^2$

3. a. $y - k = (x - h)^2$

b. $x - h = (y - k)^2$

5. $y = -\frac{\sqrt{3}}{2}x + 2\sqrt{3}$

7. $y = \frac{\sqrt{6}}{2}x - \sqrt{3}$

9. $9x^2 + y^2 = 36$

11. Place the light source 1 unit along the axis from the vertex.

13. Place the observer's viewing area $\frac{2000}{3}$ inches from the center of the mirror.

15. Multiply the equation by r and use the fact that $r^2 = x^2 + y^2$, $x = r\cos\theta$ and $y = r\sin\theta$.

17. a. Use the fact that the aphelion occurs when $\theta = 0$ and the perihelion when $\theta = \pi$ to show first that

$$a = \frac{ed}{1-e} \text{ and } p = \frac{ed}{1+e}.$$

b. First show that

$$2R = \frac{ed}{1-e} + \frac{ed}{1+e},$$

and then use part (a).

19. The equation is $12y^2 - 4x^2 = 3$.

Chapter 6: Chapter Test (page 368)

1. True.

2. False. The equation is $x = \frac{3y^2}{4}$.

3. True.

4. True.

5. False. The vertex of the parabola is $(3, 1)$.

6. True.

7. True.

8. False. The focus is $(0, -2)$.

9. False. The equation of the parabola is $y - 1 = \frac{1}{4}(x - 3)^2$.

10. True.

11. False. The focal points are $(2\sqrt{3}, 0)$ and $(-2\sqrt{3}, 0)$.

12. True.

13. False. The equation is $\frac{x^2}{16} + \frac{y^2}{4} = 1$.

14. False. The major axis is on the x-axis.

15. False. The equation is $\frac{y^2}{9} - \frac{x^2}{4} = 1$.

16. False. The equation is $(x - 1)^2 - \frac{(y-1)^2}{3} = 1$.

17. False The asymptotes are $y = \pm\frac{2x}{3}$.

18. False. The asymptotes of the hyperbola are $y - 1 = \pm\frac{3}{2}(x - 3)$.

19. True.

20. False. The polar coordinates $(2, -\frac{3\pi}{4})$ and $(2, \frac{5\pi}{4})$ represent the same point, but $(2, -\frac{\pi}{4})$ does not. To be correct, it could be replaced by $(-2, \frac{\pi}{4})$.

21. True.

22. True.

23. True.

24. False. The linear equation is $x = 2y - 1$.

25. True.

26. True.

27. False. The rectangular equation is $x^2 + (y - 1)^2 = 1$.

28. True.

29. False. The curve could have polar equation $r = 2\cos\theta$.

30. False. The curve could have polar equation $r = 2\sin\theta$.

31. False. The curve could have polar equation $r = 1 + \cos\theta$.

32. False. The curve could have polar equation $r = 1 + \sin\theta$.

33. False. The curve could have polar equation $r = 1 + 2\cos\theta$.

34. False. The curve could have polar equation $r = 2 + \cos\theta$.

35. False. It is a parabola that opens to the left.

36. False. The polar equation $r = \frac{5}{5 - \cos\theta}$ describes an ellipse with vertical directrix.

37. False. The polar curves $r = 1 + 2\sin\theta$ and $r = 1$ intersect at three points. These points have rectangular coordinates $(1, 0)$, $(0, 1)$, and $(-1, 0)$.

38. True.

39. True.

40. False. The parametric equations $x = 2t - 1$, $y = 3t + 2$ describe a line with slope $\frac{3}{2}$ and y-intercept $\frac{7}{2}$.

41. True.

42. False. The parametric equations $x = e^t + 1$, $y = e^{2t} - 1$ describe a parabola with vertex $(1, -1)$ and opening upward, but only when $x > 1$.

43. True.

44. True.

45. False. The parametric equations $x = 2\cos t$, $y = \sin t$ describe an ellipse with rectangular equation $x^2 + 4y^2 = 4$.

46. False. The parametric equations $x = (\cos t)^2$, $y = (2\sin t)^2$ describe the portion of the line with equation $4x + y = 4$ when $x \geq 0$.

47. False. One set of parametric equations for the line passing through $(3, 1)$ and $(1, 2)$ is $x = 2t + 1$, $y = -t + 2$.

48. True.

49. False. The parametric equations $x = e^t$, $y = e^{-t}$ and the rectangular equation $y = 1/x$ describe the same curve only for $x > 0$.

50. True.

Index

TRIGONOMETRY

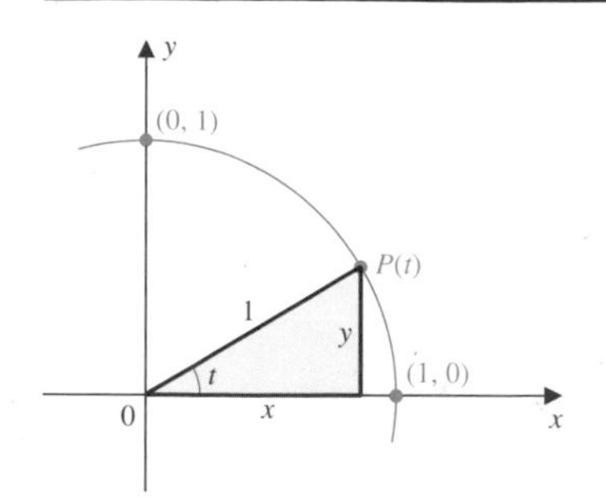

$$\sin t = y \qquad \cos t = x \qquad \tan t = \frac{\sin t}{\cos t} = \frac{y}{x}$$

$$\cot t = \frac{\cos t}{\sin t} = \frac{x}{y} \qquad \sec t = \frac{1}{\cos t} = \frac{1}{x} \qquad \csc t = \frac{1}{\sin t} = \frac{1}{y}$$

$$(\sin t)^2 + (\cos t)^2 = 1$$
$$1 + (\tan t)^2 = (\sec t)^2$$
$$1 + (\cot t)^2 = (\csc t)^2$$

$$\sin(-t) = -\sin t$$
$$\cos(-t) = \cos t$$
$$\tan(-t) = -\tan t$$

$$\cot(-t) = -\cot t$$
$$\sec(-t) = \sec t$$
$$\csc(-t) = -\csc t$$

$$\sin(t_1 \pm t_2) = \sin t_1 \cos t_2 \pm \cos t_1 \sin t_2$$
$$\cos(t_1 \pm t_2) = \cos t_1 \cos t_2 \mp \sin t_1 \sin t_2$$
$$\tan(t_1 \pm t_2) = \frac{\tan t_1 \pm \tan t_2}{1 \mp \tan t_1 \tan t_2}$$

$$\sin 2t = 2 \sin t \cos t$$
$$\cos 2t = (\cos t)^2 - (\sin t)^2 = 1 - 2(\sin t)^2 = 2(\cos t)^2 - 1$$
$$(\cos t)^2 = \frac{1 + \cos 2t}{2} \qquad (\sin t)^2 = \frac{1 - \cos 2t}{2}$$

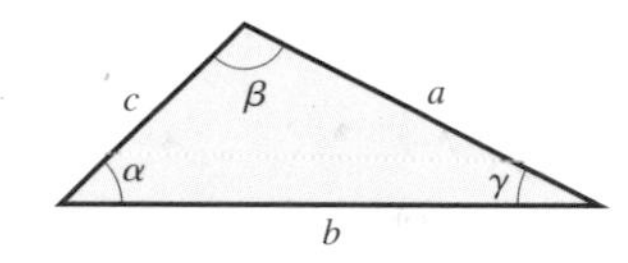

LAW OF COSINES

$$a^2 = b^2 + c^2 - 2bc \cos \alpha$$
$$b^2 = a^2 + c^2 - 2ac \cos \beta$$
$$c^2 = a^2 + b^2 - 2ab \cos \gamma$$

LAW OF SINES

$$\frac{\sin \alpha}{a} = \frac{\sin \beta}{b} = \frac{\sin \gamma}{c}$$

SPECIAL RIGHT TRIANGLES

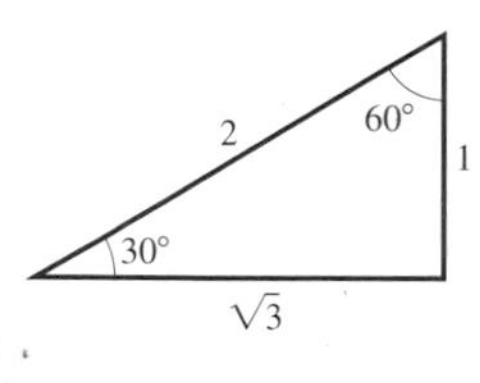

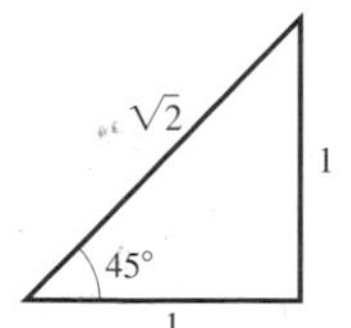

SPECIAL VALUES OF TRIGONOMETRIC FUNCTIONS

θ (radians)	θ (degrees)	$\sin\theta$	$\cos\theta$	$\tan\theta$	$\cot\theta$	$\sec\theta$	$\csc\theta$
0	0°	0	1	0	—	1	—
$\frac{\pi}{6}$	30°	$\frac{1}{2}$	$\frac{\sqrt{3}}{2}$	$\frac{\sqrt{3}}{3}$	$\sqrt{3}$	$\frac{2\sqrt{3}}{3}$	2
$\frac{\pi}{4}$	45°	$\frac{\sqrt{2}}{2}$	$\frac{\sqrt{2}}{2}$	1	1	$\sqrt{2}$	$\sqrt{2}$
$\frac{\pi}{3}$	60°	$\frac{\sqrt{3}}{2}$	$\frac{1}{2}$	$\sqrt{3}$	$\frac{\sqrt{3}}{3}$	2	$\frac{2\sqrt{3}}{3}$
$\frac{\pi}{2}$	90°	1	0	—	0	—	1